**In many cases, developments in chemical analysis have played a central role in work for which a Nobel Prize has been awarded. Some notable examples are listed below.**

| ***Year, award*** [a] | ***Recipients*** [b] | ***Analytical development*** [c] |
|---|---|---|
| 1902, C | Emil Fischer | Phenyl h |
| 1906, P | J. J. Thomson | Fundame |
| 1910, C | Otto Wallach | Analyses |
| 1914, C | T. W. Richards | Precise te determin |
| 1914, P | Max von Laue | Diffraction of X-rays |
| 1915, C | Richard Willstätter | New techniques of adsorption chromatography |
| 1915, P | W. H. Bragg, W. L. Bragg | X-ray crystallography |
| 1922, C | F. W. Aston | Practical mass spectrometers |
| 1923, C | Fritz Pregl | Organic microanalysis |
| 1924, P | K. M. G. Siegbahn | X-ray spectroscopy |
| 1925, P | James Franck, Gustav Hertz | Atomic emission spectroscopy |
| 1926, C | Theodor Svedberg | Ultracentrifugation |
| 1930, P | C. V. Raman | Raman effect |
| 1931, C | Carl Bosch,* F. Bergius | Process-control analysis and instrumentation |
| 1931, MP | Otto Warburg | Manometric measurements in metabolic studies |
| 1943, C | Georg von Hevesy | Introduction of isotopic tracers |
| 1946, C | J. B. Summer,* J. H. Northrop, W. M. Stanley | Purification and crystallization of enzymes |
| 1948, C | Arne Tiselius | Electrophoresis |
| 1952, C | A. J. P. Martin, R. L. M. Synge | Partition chromatography |
| 1952, P | F. Block, E. M. Purcell | Nuclear magnetic resonance |
| 1953, C | Hermann Staudinger | Characterization of polymers |
| 1958, C | Frederick Sanger | Protein end-group determination |
| 1959, C | Jaroslav Heyrovsky | Polarography |
| 1960, C | W. F. Libby | $^{14}C$ analysis |
| 1968, MP | R. W. Holley,* H. G. Khorana, M. W. Nirenberg | Characterization of polynucleotides |
| 1972, C | Stanford Moore,* W. H. Stein,* C. B. Anfinsen | Ion-exchange chromatography of amino acids |



---

[a] C indicates chemistry; P, physics; MP, medicine and physiology.
[b] In cases where there has been more than one recipient but not all have been involved in the analytical developments, an asterisk is used to indicate those most responsible for the analytical innovations.
[c] The "analytical development" is, in many cases, not the specific accomplishment for which the prize was awarded. The relationship between the award and the development cited is very strong in all cases, however.

DENNIS G. PETERS
JOHN M. HAYES
GARY M. HIEFTJE

*Department of Chemistry*
*Indiana University*
*Bloomington, Indiana*

# A Brief Introduction to MODERN CHEMICAL ANALYSIS

*SAUNDERS GOLDEN SUNBURST SERIES*

**1976**
W. B. SAUNDERS COMPANY
Philadelphia • London • Toronto

W. B. Saunders Company: West Washington Square
Philadelphia, PA 19105

833 Oxford Street
Toronto, Ontario M8Z 5T9, Canada

1 St. Anne's Road
Eastbourne, East Sussex BN21 3UN, England



About the Cover:

The superimposed images of a high-resolution gas chromatogram and the glass capillary column tubing essential in its production. The chromatogram happens to be that of marijuana smoke. The full recording, not shown here, reveals well over 200 components. Only 2.5 ml of unenriched smoke was taken as the sample. The column temperature was programmed from –70° to +130°C. The average width at half height of the peaks is 9 seconds. In the chromatogram as shown, time is increasing linearly from right to left, with retention times varying from about 20 to 60 min. Chromatogram and column provided by M. V. Novotny and M. L. Lee, Indiana University. (See M. V. Novotny and M. L. Lee: Detection of Marijuana Smoke in the Atmosphere of a Room. Experientia *29*:1038, 1973.)

**Library of Congress Cataloging in Publication Data**

Peters, Dennis G.

A Brief Introduction to Modern Chemical Analysis

(Saunders Golden Sunburst Series)

1. Chemistry, Analytic. I. Hayes, John Marion, 1909 – joint author. II. Hieftje, Gary M., joint author. III. Title.

QD75.2.P47 543 75-25272

ISBN 0-7216-7202-7

A Brief Introduction to
Modern Chemical Analysis ISBN 0-7216-7202-7



Last digit is the print number: 9 8 7 6 5 4 3 2 1

# PREFACE

This book is a shortened version of our previous text, *Chemical Separations and Measurements*, published in 1974, and is intended for one-semester courses in analytical chemistry. In achieving a reduction in size, we have tried to avoid anything that could be termed a serious omission in the context of modern chemical analysis. At the same time, we have worked to retain important details and have attempted to avoid compressing our discussions to the point of mystery. Some reallocation of emphasis has been inevitable, and a useful coverage of separations and spectrochemical analysis has been preserved at the expense of some electroanalytical techniques.

We are convinced that analytical chemistry merits serious and systematic study, and would like to describe our point of view. As a basis for commentary, we have compiled the names of thirty-seven Nobel laureates inside the front cover of this book. The selections are our own, and are, no doubt, imperfect, but we have tried to choose those who have contributed most directly to progress on the various frontiers of analytical chemistry. Together with each name, we have listed the particular accomplishment which we regard as having analytical significance. This has not always been the work for which the Nobel prize was awarded, but it has been very strongly related in almost every case.

Even a casual examination of this list makes our first point, which is entirely scientific and quite perfectly obvious: analytical chemistry is useful. That is, there is a classical link between improvements in analytical technology and advances in chemical knowledge. To understand nature better, we must be able to measure or *see* more precisely how it behaves. On a more routine level, the acquisition of new chemical knowledge often requires the application of analytical techniques. Obviously, not every chemist has to become so wrapped-up in analysis that the development of new methodology becomes an end in itself; but it is not easy to do much organic chemistry, for example, without gas-liquid chromatography.

The list of Nobel laureates also reinforces our second point, which is basically pedagogic: a large portion of what we can regard as "chemistry" is *analytical chemistry*. It seems to us that, to an increasing extent, treatments of analytical technology are being squeezed out of chemistry courses by the pressure of an ever-expanding body of fundamental material. Given the practical value of analytical chemistry, we are sure that its topics cannot be cut from the curriculum without doing serious damage. Chemistry is, after all, an *experimental* science, and the tools, methods, and procedures of experimental chemistry are inextricably related to analytical operations. Historically, the applications of experimental techniques have, sooner or later, crossed most of the boundaries between the traditional areas of chemistry. It seems, therefore, that there is a strong justification (quite apart from real or presumed synergism) for bringing together discussions of analytical techniques ranging (in their *present* applications) from inorganic to physiological chemistry.

These considerations do little to restrict the dimensions of analytical chemistry. To discuss a useful fraction of the field will take more than half an academic year, and it is not clear that there is widespread agreement on how this time could best be used. However, we hope that this text will provide a good introduction which can be used either in the first part of a multi-course sequence or, if necessary, as the backbone of a single course in modern analytical chemistry.

In addition to the many friends and colleagues who helped with the preparation of *Chemical Separations and Measurements* and who are acknowledged in the preface to that book, we would like to thank Professors Dennis Evans, University of Wisconsin; Gordon Parker, University of Toledo; and Robert Osteryoung, Colorado State University; who reviewed the longer text and made numerous constructive comments on how it might be reduced in size.

We hope that this textbook will be worthy of revision, and we invite letters of comment and criticism from all readers, especially students.

Dennis G. Peters

John M. Hayes

Gary M. Hieftje

# CONTENTS

# 1 NATURE OF ANALYTICAL CHEMISTRY

In this book we hope to provide an introductory coverage of the field of analytical chemistry. Yet even an introduction will require us to deal in some depth with a broad range of topics. This is because analytical chemistry, like most scientific disciplines, has become so intertwined with other sciences that it is difficult to define the field.

## WHAT IS ANALYTICAL CHEMISTRY?

Analytical chemists working in industrial control laboratories are often responsible for maintaining, by means of appropriate chemical measurements, the quality of an outgoing product or an incoming raw material. In addition, there are analytical chemists employed in clinical laboratories, where important tests are performed to establish the condition of health of a patient or to serve as an indispensable aid to medical diagnosis. Most of these chemists concentrate on applying the knowledge and findings of chemistry to the analysis of real samples. Generally, the tools and techniques utilized are well known and established, with little room for innovation or error.

Another kind of activity is the development of new analytical techniques and the improvement of existing ones. New discoveries and instrumentation in chemistry, physics, or engineering may frequently lead to new analytical techniques. For example, the mass spectrometer, originally developed as a tool of nuclear physics, has been adapted to a wide variety of chemical uses ranging from structure determinations in organic chemistry to quantitative analysis of isotopes in tracer experiments. Similarly, both x-ray spectroscopy and electron spectroscopy are presently being utilized primarily in chemical applications.

Development of a new analytical method often occurs in response to a specific need. In many instances, the new technique takes the form of an extension of some existing principle of measurement. Thus, for the detection, determination, and monitoring of environmental pollutants, chromatographic and electrochemical techniques have been reshaped and made more sensitive as dictated by the particular analytical problem. Improvements in existing methods can be especially worthwhile if they improve the accuracy or lower the unit cost of an analysis. Very frequently, both of the preceding objectives can be met through automation of an already established technique. Control of an experiment as well as the processing of data can be taken over by a small computer which costs less and performs more reliably than any alternative system.

There is more to technique development than just engineering. To be sure, once the concept and design underlying a method have been formulated, engineering can play a major role in the successful execution of experiments. However, the design of a system for a chemical measurement requires a familiarity both with the basic principles and capabilities of the technique being developed and with the goals and practical aspects of the chemical experiments. Thus, an analytical chemist involved in technique development provides, for example, an interface between physicists who have perfected a mass spectrometer and organic chemists interested in determining the structure of organic compounds. Such a role is stimulating and important, for the development of new investigative techniques is often a crucial step in the discovery of new chemical knowledge.

Today, analytical chemists are confronted with such a bewildering array of established, modified, and new techniques that it is practically impossible to keep up with the entire field. For this reason, most analytical chemists at the advanced level choose one subdiscipline such as analytical separations, spectrochemical analysis, or electrochemical analysis to which they devote their major attention, while still maintaining a working knowledge of other subdisciplines. Within each of the numerous but still broad subdisciplines, active areas of research are usually even more narrowly defined. For example, an electroanalytical chemist may opt to study the mechanisms of organic electrode reactions, to apply ion-selective electrodes to pollution monitoring or biochemical analysis, or to perfect new coulometric techniques for nonaqueous titrations of organic compounds. Someone working in the field of analytical separations might concentrate his attention on designing novel detectors for use in liquid chromatography, on the development of ion-exchange resins to separate optically active compounds, or on the chromatographic examination of body fluids containing drug metabolites. Spectrochemical analysis might center on fundamental investigations of how processes occurring in flames affect the precision and accuracy of atomic absorption spectrometry, on spectroscopic methods for the rapid and sensitive measurement of substances of biological and pharmaceutical interest, or on the construction and evaluation of new instrumentation for the spectroscopic determination of molecular structure.

## ORGANIZATION OF THE TEXTBOOK

A glance at the table of contents of this textbook shows that we have placed emphasis on practical chemical measurements and on separation techniques. This selection of topics has broad utility for chemists who go on to concentrate in any of the non-theoretical areas of chemistry, and forms an introductory coverage of the field for chemists who plan to specialize in chemical instrumentation or other aspects of problem-solving methodology.

Because the treatment and interpretation of data are so vital to all kinds of chemical experiments, Chapter 2 describes in detail how to express the accuracy and precision of an analytical result and how to assess errors in measurements through the application of rigorous statistical and mathematical concepts; in addition, this material provides a solid base for the discussion of chromatographic separations in later chapters. In Chapter 3, the behavior of solutes in aqueous media is examined, and some of the principles of chemical equilibrium upon which material in subsequent chapters will depend are discussed. Chapters 4 and 5 cover the topics of acid-base reactions in aqueous and nonaqueous systems; this treatment is essential to the quantitative descriptions of the solubility of precipitates in various solvents and of the chemical interactions involved in analytical methods based on complexation and solvent-extraction phenomena. In Chapter 6, the theory and analytical uses of complex-formation reactions are considered, and a foundation is laid for applying these concepts to such analytical techniques as direct potentiometry, coulometric titrations, polarography, and chromatography. Analytical methods based on the formation of precipitates are discussed in Chapter 7.

A thorough description of oxidation-reduction processes is presented in Chapters 8 through 11; starting with a discussion of galvanic cells, the material progresses to titrimetric redox methods involving both inorganic and organic species and then to surveys of electrometric procedures, including potentiometry, coulometry, and polarography. Chapters 12 through 14 offer an excellent introduction to all aspects of analytical separations; there is information about the nature of phase equilibria and extraction, and the theory and applications of various chromatographic techniques for the resolution of mixtures of inorganic, organic, and biological species. In the final portion of the textbook, Chapters 15 through 17 present some of the most prominent techniques employed in modern spectrochemical analysis—interactions of ultraviolet and visible radiation with atomic and molecular species leading to absorption, emission, and fluorescence.

Some of the topics in this textbook seem far less glamorous than many of the activities of analytical chemists listed earlier. However, in order to engage in productive and meaningful work, an analytical chemist must be well versed in the fundamental principles upon which so many modern techniques rely. This then is the rationale for the existence of the present textbook.

## LITERATURE OF ANALYTICAL CHEMISTRY

If there is any doubt that analytical chemistry is a diverse and active area of endeavor, one need only examine the large number of general and specialized scientific journals that publish both review articles and original research papers in the field to realize the tremendous importance and impact of modern analytical chemistry. All the specialized areas of analytical chemistry—including clinical analysis, spectrochemical analysis, electrochemical analysis, chromatography, instrumentation, computer applications, kinetics, mass spectrometry, and the collection, storage, retrieval, and processing of experimental data—are represented by at least several specialized journals that publish theoretical and practical aspects of a subdiscipline. In addition, there are a number of journals that bring together papers covering innovations in all these specialized areas as well as new developments in gravimetric and volumetric methods of analysis. It is undoubtedly conservative to assert that approximately 15,000 articles pertaining to analytical chemistry are published annually.

A professional analytical chemist intending to keep abreast of the rapid advances in his field must develop the habit of regularly reading the literature. We can only urge the student beginning a first serious excursion into analytical chemistry to adopt a similar attitude as well as to recognize the virtue of consulting several books and articles on any one subject in order to reach a balanced point of view.

Among the prominent journals that cater to all branches of analytical chemistry are *The Analyst, Analytica Chimica Acta, Analytical Chemistry, Analusis* (French), *Zhurnal Analiticheskoi Khimii* (Russian, but available in English translation), *Talanta,* and *Zeitschrift für analytische Chemie* (German). Biennial reviews published in the April issue of *Analytical Chemistry* in each even-numbered year offer comprehensive summaries and bibliographies of fundamental research written by experts in almost every subdiscipline of analytical chemistry; an April issue in each odd-numbered year is devoted to a series of reviews of analytical applications in such fields as air pollution, clinical chemistry, food, petroleum, pharmaceuticals and drugs, and water analysis. *Analytical Letters* is an international journal for the rapid dissemination of new results in all branches of analytical chemistry, and *Analytical Abstracts* is a monthly publication that presents short summaries of papers appearing in journals around the world in all subdisciplines of analytical chemistry.

It is almost impossible to cite all the specialty journals in the field of analytical chemistry, but some idea of the scope of subdisciplinary research—and of the important way that analytical chemistry impinges upon and contributes to other sciences—is provided by the following list of publications, the titles of which are self-descriptive: *Acta Crystallographica, Analytical Biochemistry, Applied Spectroscopy, Chemical Instrumentation, Chromatographic Reviews, Clinica Chimica Acta, Clinical Chemistry, Journal of Chromatographic Science, Journal of Electroanalytical Chemistry and Interfacial Electrochemistry, Journal of the Electrochemical Society, Organic Mass Spectrometry,* and *Spectrochimica Acta.*

Several monthly magazines regularly present articles as well as informative advertisements on the latest developments and trends in spectroscopy, electroanalytical chemistry, instrumentation, and computers. Of particular interest are *Industrial Research* (published by Industrial Research, Inc., Chicago, Illinois), *American Laboratory* (published by American Laboratory, Inc., Greens Farms, Connecticut), and *Research/Development* (published by Technical Publishing Company, Barrington, Illinois).

Subjects discussed in the monthly feature sections as well as in the regularly contributed articles of the *Journal of Chemical Education,* published by the Division of Chemical Education of the American Chemical Society, frequently include topics which are of interest to analytical chemists.

Many textbooks and reference works deal with one or more specific areas of analytical chemistry. Some of these are listed as suggestions for additional reading at the ends of the chapters in this book. Among the individual books and series of volumes of general interest throughout many areas of analytical chemistry are the following:

1. *Standard Methods of Chemical Analysis,* sixth edition, D. Van Nostrand Company, Princeton, New Jersey. Volume I (The Elements) was edited by N. H. Furman and appeared in 1962. Volume II (Industrial and Natural Products and Noninstrumental Methods) and Volume III (Instrumental Methods) were both edited by F. J. Welcher and were published in 1963 and 1966, respectively. These books serve as a reliable source of analytical information for practical use in the laboratory.

2. *Handbook of Analytical Chemistry,* edited by L. Meites, McGraw-Hill Book Company, New York, 1963. This one-volume compendium provides summaries of fundamental data and practical laboratory procedures.

3. *Treatise on Analytical Chemistry,* edited by I. M. Kolthoff and P. J. Elving, Wiley-Interscience Publishers, New York (Volume 1 of Part I in 1959 and others subsequently). This three-part, multivolume series offers a thorough coverage of virtually every phase of analytical chemistry, including theory and practice of analytical techniques, analytical chemistry of the elements, and analysis of industrial products.

4. *Comprehensive Analytical Chemistry,* edited by C. L. Wilson and D. W. Wilson, Elsevier Publishing Company, Amsterdam (Volume I-A in 1959 and others subsequently). This is another multivolume series dealing with most aspects of the theory and practice of analytical chemistry.

5. *Advances in Analytical Chemistry and Instrumentation,* edited by C. N. Reilley and others, Wiley-Interscience Publishers, New York (Volume 1 in 1960 and others subsequently). Early volumes in this series contain a selection of reviews by outstanding authorities in various fields of analytical chemistry. Since 1968 each volume has been devoted to a different single field of analytical chemistry, with several experts contributing chapters in their own fields of expertise.

# 2 TREATMENT OF ANALYTICAL DATA

Imagine that you are determining by analysis some variable quantity $x$, and that $x$ is *supposed to* take some value $A$. In a set of quadruplicate measurements, you obtain the data shown in Figure 2-1. The question is, do the results show that $x$ has deviated from the specified value? An accurate, but useless, answer is "Maybe." *It is possible to assess quantitatively the confidence with which you might instead answer "Yes" or "No."* It is pathetic to guess at the answer to such questions–incredible wastes of time and money can result, or needless risks can be taken. For example, the material analyzed might represent a large batch of a manufactured product. If the value of $x$ truly differs from $A$, the batch must be discarded at great expense. Alternatively, $x$ might represent the bilirubin content in the blood of a newborn child. If the concentration truly exceeds the limit $A$, an extensive and somewhat dangerous treatment must be started. Anyone who undertakes chemical measurements either can understand the statistical methods which allow useful answers to practical questions, or can accept the role of an analytical instrument, a mindless performer of laboratory procedures.

Consider a second example: the quantity $x$ has been determined by measurement of $a$, $b$, $c$, and $d$.

$$x = ab + \frac{c}{d}$$

How do uncertainties in $a$, $b$, $c$, and $d$ affect the result? How can the uncertainty in $x$ be computed from the known uncertainties in the measured quantities?

Questions posed above are very practical. Procedures which provide useful answers are easily mastered, and this chapter offers a brief treatment of some basic statistical techniques.

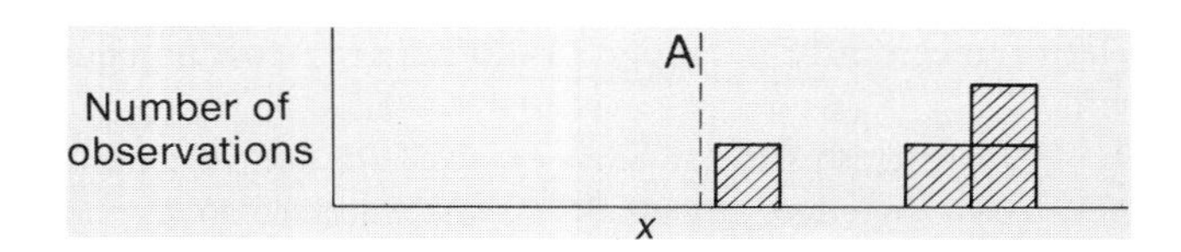


*Figure 2-1.* Histogram showing results for the determination of the variable $x$.

## NATURE OF QUANTITATIVE MEASUREMENTS

### Significant Figures

In the examples below, the number of significant figures in each quantity is given in parentheses.

| | | | | | |
|---|---|---|---|---|---|
| **1** | 12.270 g | (5) | **7** | 43,100 ml | (3) |
| **2** | 12.3 g | (3) | **8** | 40,000 ml | (1) |
| **3** | 10 g | (1) | **9** | 100.00 | (5) |
| **4** | 0.00524 *M* | (3) | **10** | 1.00 | (3) |
| **5** | 0.005 *M* | (1) | **11** | 0.010 | (2) |
| **6** | 43,062 ml | (5) | | | |

The number of significant figures is always equal to or greater than the number of nonzero digits. When zeroes are used only to locate the decimal point, they do not count as significant figures. Thus, in **1**, **9**, **10**, and **11**, the trailing zeroes are not required to locate the decimal point but instead indicate that the measurement has been made with the indicated accuracy. Similarly, in **6** the interior zero is significant.

In **3**, **7**, and **8**, the trailing zeroes may be significant, or may be used only to locate the decimal point. Scientific notation furnishes the most convenient way of overcoming this ambiguity, although an alternative used by some workers is the marking of significant trailing zeroes; for example, $4\bar{0}\bar{0}$ indicates that all three figures are significant. In scientific notation, the same number would be written $4.00 \times 10^2$. If there were only one significant figure, the number would be written $4 \times 10^2$.

To overcome uncertainties associated with various round-off conventions, some workers include the first insignificant digit in numerical data. The insignificant figure is frequently written as a subscript. For example, in the number $4.05_2$, the 2 is insignificant. (Caution: number bases are sometimes noted in the same way—thus, $624_8$ might represent an octal rather than a decimal number.)

**Absolute Uncertainty and Relative Uncertainty.** Uncertainty in measured values may be considered from two distinct viewpoints. *Absolute uncertainty* is expressed directly in units of the measurement. A weight expressed as 10.2 g is presumably valid within a tenth of a gram, so the absolute uncertainty is one tenth of a gram. Similarly, a volume measurement written as 46.26 ml indicates an absolute uncertainty of one hundredth of a milliliter.

*Relative uncertainty* is expressed in terms of relative magnitude—for example, as a percentage. The weight 10.2 g is valid to within one tenth of a gram and the entire quantity represents 102 tenths of a gram, so the relative uncertainty is about one part in 100 parts, or one per cent. The volume written as 46.26 ml is correct to within one hundredth of a milliliter in 4626 hundredths of a milliliter, so the relative uncertainty is one part in 4626 parts, or about 0.2 part in a thousand. It is customary, but by no means necessary, to express relative uncertainties as parts per hundred (per cent), as parts per thousand, or as parts per million. Relative uncertainties do not have dimensions because a relative uncertainty is simply a ratio between two numbers having the same dimensional units.

To distinguish further between absolute and relative uncertainty, consider the weighing of two different objects on an analytical balance, one object weighing 0.0021 g and the other 0.5432 g. As written, the absolute uncertainty of each number is one ten-thousandth of a gram; yet, the relative uncertainties differ widely—one part in 20 for the first weight and one part in approximately 5000 for the other value.

**Significant Figures in Mathematical Operations.** *In addition and subtraction the absolute uncertainty in the result must be equal to the largest absolute uncertainty among the components.* Consider three examples:

| | | |
|---:|---:|---:|
| 10.0051 | | 0.5362 |
| 1.9724 | 42598 | 0.0014 |
| +0.0003 | −42595 | +0.25 |
| 11.9778 | 3 | 0.79 |

In the first two cases, all the component numbers have the same absolute uncertainty, and determination of the correct number of significant figures in the result is a simple matter, although it has the somewhat surprising consequence that in the first case a component with only one significant figure contributes to a result with six, and, in the second case, two components with five significant figures yield a result with only one. In the third example, the absolute uncertainties are not equal, and the number of significant figures in the result is determined by the absolute uncertainty in the third number added.

*In multiplication and division, the relative uncertainty in the result must be equal to the largest relative uncertainty among the components.* For example, in the operation $0.12 \times 9.678234$, the correct product is 1.2. Expressing the result as 1.1614 would be unjustified because the relative uncertainty in the first factor is one part in twelve.

## Precision and Accuracy

The terms, *precision* and *accuracy,* are not synonymous, and are defined here by contrasting examples. A balance which on repeated trials gives the weight of an object as 1.307, 1.308, 1.305, and 1.307 g is more *precise* than one which gives the weights 1.302, 1.316, 1.305, and 1.310 g. *Precision* relates to the degree of scatter in a set of data, the scatter in these examples clearly being greater in the second set. A balance which gives the weight of a known 10.000-g standard as 10.001 g is more *accurate* than one which gives the weight of the same standard as 10.008 g. *Accuracy* relates to the difference between a measured quantity and its true value.

## Errors

**Systematic Errors.** Systematic errors are frequently related to improper design or adjustment of experimental apparatus; such errors reduce accuracy by systematically skewing or offsetting the observed data. A careful study of measurements of some standard can reveal the nature of the error, and accuracy can be regained by adjusting the apparatus or applying some correction. For example, if a balance showed the weight of a ten-gram standard to be 10.080 g and of a one-gram standard to be 1.008 g, we could conclude that the balance was misadjusted so that it weighed 0.80 per cent high. It would be possible to apply a correction to all the data, but it would be far better to repair the balance. Alternatively, a balance might show the weight of a one-gram standard as 1.050 g and the weight of a ten-gram standard as 10.050 g. In this case, it could be concluded that the error was not proportional to the weight, but was, instead, a constant 0.050-g offset. Again, a correction could be applied or, preferably, the balance could be repaired. (These examples also illustrate the value of systematic observations during instrument trouble-shooting, because the nature of the errors observed would naturally lead to certain types of adjustments. The constant 0.050-g offset, for example, must be due to a misplaced zero point on the mass scale; the 0.80 per cent relative error is probably due to improper positioning of the balance arm.)

**Random Errors.** Random errors result from insufficiently controlled variations in measurement conditions. Many different effects acting together lead to small variations in the observed value. Repeated observations will scatter randomly around the true value, and it is thus clear that the size and frequency of these random errors will determine the precision of a given measurement. For example, in the case of the analytical balance, sources of random error include fluctuations in room temperature and humidity, small variations in the placement of weights, variations in the positioning of the knife edge on its bearing, and the subjectivity of the operator who reads the final weight from some uniformly graduated scale.

Careful consideration of the conditions of measurement can substantially reduce random errors, but, unlike systematic errors, they cannot be eliminated, nor can some formula be derived to correct an observation for their influence. As we shall learn in the following sections, random errors are amenable to statistical treatment, and repeated measurements of the same variable can have the effect of reducing their importance.

## BASIC STATISTICAL CONCEPTS

### Frequency Distributions

**Random Errors and the Normal Distribution.** Consider the quantitative determination of, for example, iron in ferric oxide. Figure 2–2 shows a sequence of three graphs illustrating the effect of random errors. In the first case, only four quite widely scattered observations were obtained. More observations were obtained in the second and third cases, and it is clear that the distribution of results is tending toward the smooth curve drawn at the bottom of the figure. The smooth curve represents the **universe** of all possible determinations of iron in iron oxide, and each of the smaller sets of observations is a **sample** drawn from that universe. The height of the universe distribution at any given value of per cent iron is a measure of the *frequency* with which observations of that value will be obtained during sampling. The relationship is known generally as a *frequency distribution.* The shape of the frequency distribution is given by the normal law of error, and the curve is known as the **normal distribution.** Many other names are also applied, but indicate exactly the same curve. Some of the most commonly used synonyms are "normal curve of error," "Gaussian distribution," and "probability distribution."

### Average and Measures of Dispersion

For any set of data, our aim is to summarize it quickly and usefully. Everyone is familiar with summarizing a set of measurements by computing an average, but this is strictly a measurement of *central tendency,* and tells nothing about the *distribution* of the measurements. In order to summarize a set of data, we require in addition some measurement of spread or dispersion of the individual data points.

**Central Tendency.** The most common measurement of central tendency is the **mean** or average. Usually, we denote the mean value of some variable by placing a bar over its symbol; thus,

$$\bar{x} = \frac{\sum x}{n}$$

where $\bar{x}$ is the mean of $n$ observations of $x$. We must distinguish between $\bar{x}$, the mean of some finite sample, and $\mu$, the true value or universe mean. The value of $\bar{x}$ is an *estimate*

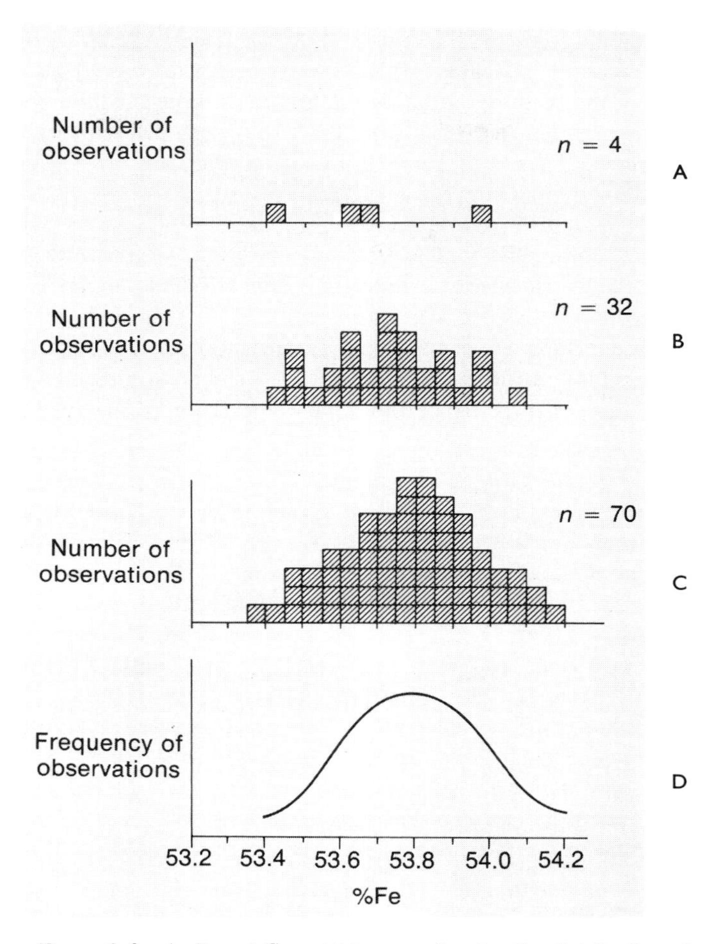


*Figure 2-2.* A, B, and C are histograms showing the distribution of results for a series of iron analyses. D shows the normal distribution which would be obtained if an infinite number of observations could be made.

of the value of $\mu$. The quality, or reliability, of this estimate increases with $n$, the number of observations used to find $x$.

**Measures of Dispersion.** The simplest measure of dispersion is the **range** of values found in some sample. It is, however, a poor indicator of the shape of the distribution because it depends only on the highest and lowest values. By far the best and most useful measure of dispersion is given by the **variance**, which is the mean square deviation of all observations, where the "deviation" is the difference between an observation and the mean:*

$$\sigma_x^{\,2} = \frac{\sum \text{(deviation from true mean)}^2}{n} = \frac{\sum (x - \mu)^2}{n} \tag{2-1}$$

*A better representation of the numerator would be given by $\sum_{i=1}^{n} (x_i - \mu)^2$, but it is conventional in unambiguous cases to omit the subscripts and interval notations for $\Sigma$.

The variance has the dimensions of $x^2$ (*e.g.*, $cm^2$ if $x$ is a length) and its value does not convey a direct feeling for the amount of scatter in $x$. Thus, the **standard deviation**, or root-mean-square deviation, is often used. It is assigned the symbol $\sigma_x$ and is also defined by equation (2–1). Notice that this formula can be used only when the universe mean, $\mu$, is known. Thus, equation (2–1) relates only to the universe variance and standard deviation, $\sigma_x{}^2$ and $\sigma_x$.

**Degrees of Freedom.** Just as the sample mean, $\bar{x}$, provides an estimate of $\mu$, the universe mean, the sample standard deviation, $s_x$, provides an *estimate* of $\sigma_x$, the universe standard deviation. Notice that Greek letters indicate universe parameters, whereas the corresponding Roman letters are used to represent their estimates. In practice, the universe parameters remain unknown because their exact evaluation would require the (impossible) performance of an infinite number of measurements. An expression for the calculation of the variance and standard deviation of the *sample* is

$$V_x = s_x{}^2 = \frac{\sum (x - \bar{x})^2}{n - 1} \qquad (2\text{–}2a)$$

where $V_x$ is the variance of $x$. This expression differs from that given for $\sigma_x{}^2$ in two ways. First, $\bar{x}$ has replaced $\mu$ and, second, the denominator is $(n - 1)$ rather than $n$. The effect of the second change, which decreases the denominator, is to increase the value for $s_x$. This is appropriate, because in equation (2–2a) we have "stacked the deck" by substituting $\bar{x}$ for $\mu$. This substitution minimizes the numerator because we have calculated $\bar{x}$ from the individual $x$ values; thus, $\bar{x}$ represents the center of our sample, but not necessarily that of the universe. More fundamentally, $(n - 1)$ represents the number of **degrees of freedom**, or independent deviation calculations, which are possible within the sample after $\bar{x}$ has been calculated. Consider a sample with $n$ data points. First, we calculate $\bar{x}$ and then we begin to calculate deviations from the mean, $(x - \bar{x})$. When we reach the $n$th data point, the comparison is no longer an independent one, because the $x$ value for that $n$th point could be calculated before we even saw it, given $\bar{x}$ and the $(n - 1)$ preceding data points. In general,

$$\varphi = \text{degrees of freedom} = [n - (\text{number of constants calculated from data})]$$

$$s^2 = \frac{\text{sum of (deviations)}^2}{\varphi} \qquad (2\text{–}3)$$

**Alternative Expressions for the Standard Deviation.** The expressions given above are inconvenient for the rapid computation of the sample variance. The following relations are entirely equivalent and permit much more rapid calculation:

$$s_x{}^2 = \frac{(\sum x^2) - n\bar{x}^2}{n - 1} = \frac{\sum x^2 - \dfrac{(\sum x)^2}{n}}{n - 1} \qquad (2\text{–}2b, 2\text{–}2c)$$

*Example 2-1.* Compute the sample standard deviation for the following results obtained for the analysis of carbon in lunar soil (Apollo 11, fine-grained material): 130, 162, 160, 122 ppm. It is convenient to arrange the data in a small table, as shown below:

| $x$ | $x^2$ | |
|---|---|---|
| 130 | 16,900 | |
| 162 | 26,244 | |
| 160 | 25,600 | |
| 122 | 14,884 | |
| $\sum x = 574$ | $\sum x^2 = 83,628$ | $n = 4$ |

Substituting these results in equation (2-2c), we obtain

$$s_x^2 = \frac{\sum x^2 - \frac{(\sum x)^2}{n}}{n - 1} = \frac{83,628 - \frac{(574)^2}{4}}{3}$$

$$= \frac{83,628 - 82,369}{3} = 419.7\ (\text{ppm})^2$$

$$s_x = \sqrt{419.7} = 20.5\ \text{ppm}$$

Notice that the numerator is the difference between two large numbers. Thus, many significant figures must be carried in the calculation. Slide-rule accuracy is usually inadequate.

### Probabilities Derived from the Normal Distribution

In Figure 2-2 we have indicated that random errors cause experimental results to be scattered so that they fall within the normal distribution. Imagine that we perform some analytical procedure which has been applied hundreds of times, and that we can now regard $s_x$ as essentially equivalent to $\sigma_x$. Can we not then state the probability that an individual measurement is within some given distance of $\mu$? This is easily possible, but the method requires some introduction.

**Mathematical Expression for the Normal Distribution.** The expression for the normal distribution curve has been found theoretically, and we shall learn how it can be used to evaluate the probability of various deviations as a function of their magnitude. As a preliminary step, we must understand the "reduced variable," $u$, which is defined by equation (2-4):

$$u = \frac{(x - \mu)}{\sigma_x} \tag{2-4}$$

By evaluating $u$ for any set of $x$, $\mu$, and $\sigma_x$ values, we determine the magnitude of any deviation $(x - \mu)$ in terms of the standard deviation. The following examples should make this point clear.

*Example 2-2.* In the series of iron analyses noted above (Figure 2-2), $\mu = 53.78\%$ Fe and $\sigma_x = 0.20\%$. Calculate $u$ for $x = 53.58\%$ Fe.

Substitution in equation (2–4) yields

$$u = \frac{(x - \mu)}{\sigma_x} = \frac{(53.58 - 53.78)}{0.20} = \frac{-0.20}{0.20} = -1.0$$

The value $x = 53.58\%$ Fe would be termed "one standard deviation out," that is, one standard deviation, one $u$ unit, away from the mean.

*Example 2–3.* For the final examination in a certain chemistry course, $\mu = 75$ points and $\sigma_x = 10$ points. Calculate $u$ for $x = 100$ points.

Substituting the appropriate values in equation (2–4), we obtain

$$u = \frac{(x - \mu)}{\sigma_x} = \frac{(100 - 75)}{10} = \frac{25}{10} = 2.5$$

On this examination a perfect score is "2.5 standard deviations out." We shall soon see what the probability of this occurrence is.

**Relationship between Probability and Area.** Every observation in any universe lies somewhere under the normal distribution curve for that universe. Expressing this certainty mathematically, we assign a value of 1.0 to the total area beneath the normal distribution curve.

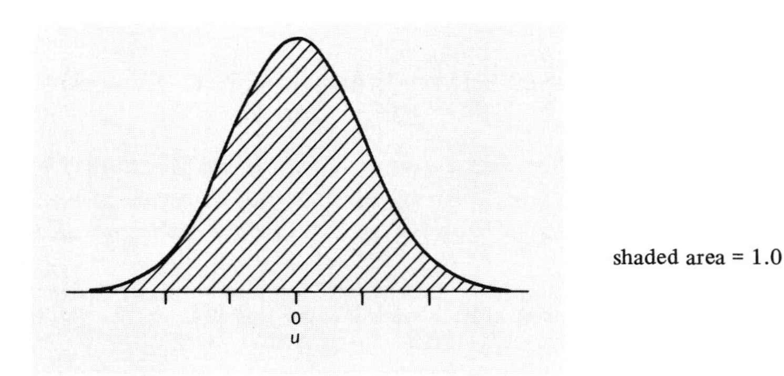


shaded area = 1.0

Similarly, we could observe that *half* of the observations will lie on each side of the mean. Expressing this probability in terms of areas beneath each half of the normal distribution curve, we can note that each of the shaded areas below has a value of 0.5.

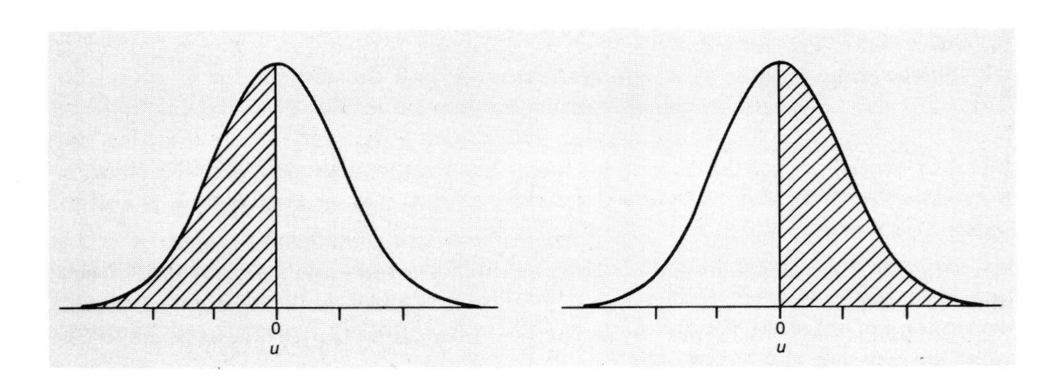

To further extend the correspondence between probabilities and areas, we can consider the probability that $x$ will take some value such that $u \geqslant 2.0$. That is, what is the chance that $x$ will be two or more standard deviations out on the positive side? This probability is represented by the area under the normal distribution curve in the range $2.0 \leqslant u \leqslant \infty$, and is shown graphically as the shaded area in Figure 2-3A. The value of this area can be determined by consulting a table giving areas under the normal distribution curve. Such tables can appear in either of two forms, here designated arbitrarily as "Form A" and "Form B."

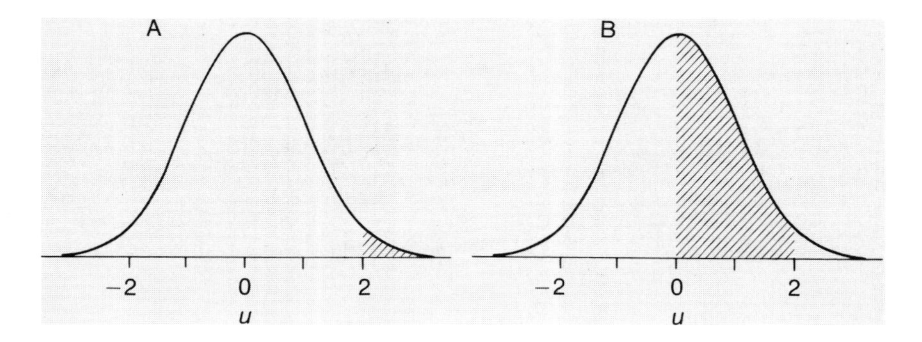


*Figure 2–3.* Normal error curves showing the areas which are given in the two alternative tabulation forms for the normal distribution. (See text for discussion.)

Although most mathematical handbooks list only one or the other, we present both forms in Table 2–1. For a tabulation of form A, the probability that $x$ will lie two or more standard deviations above the mean is given directly by the entry for $u = 2.0$, representing the area under the curve from $u = 2.0$ to $u = \infty$. This area is given in Table 2–1 as 0.0227, providing directly the information that there is a 2.27 per cent chance that an observation will fall two or more standard deviations away from the mean. When a table of form B is consulted, the area tabulated at $u = 2.0$ represents the probability that $x$ will lie between $\mu$ and $(\mu + 2\sigma_x)$. This area is given in Table 2–1 as 0.4773, and the probability that $x$ lies *outside* this range is determined by subtraction: $0.5000 - 0.4773 = 0.0227$.

Notice that, because it is symmetrical, only half the distribution is given. In Table 2–1 this is stressed by tabulating the absolute value of $u$. Practical calculations based on equation (2–4) give $u$ some sign which must be ignored when consulting the table. Conversely, when a certain $|u|$ is found to correspond to some specified area, it is necessary to think about whether the desired $x$ is less than or greater than $\mu$ and to assign a plus or minus sign to $u$ accordingly. An example follows.

*Example 2–4.* Your company makes steel-belted radial ply tires. Extensive company tests find an average tire life of $\mu = 50{,}000$ miles and a standard deviation on the distribution of observed tire lives of $\sigma_x = 4300$ miles. For only 1 per cent redemptions, what mileage can you guarantee?

We require $X$ to be chosen so that the shaded area in the sketch on the next page is 0.01, or 1 per cent of the total. We begin by consulting a table of the normal distribution. If it is of form A, we look for area = 0.0100. If it is of form B, we look for area = 0.4900. In either case, we find $|u| = 2.33 = (X - \mu)/\sigma_x$. Then, for $X < \mu$, $X = \mu - 2.33\sigma_x = 40{,}000$ miles.

**Table 2-1. The Normal Distribution**

$$\text{Form A: area} = \frac{1}{\sqrt{2\pi}} \int_u^{\infty} \exp\left(-\frac{u^2}{2}\right) du$$

| $\lvert u \rvert$ | area | $\lvert u \rvert$ | area | $\lvert u \rvert$ | area | $\lvert u \rvert$ | area |
|---|---|---|---|---|---|---|---|
| 0.0 | 0.5000 | 1.0 | 0.1587 | 2.0 | 0.0227 | 3.0 | $1.3 \times 10^{-3}$ |
| 0.1 | 0.4602 | 1.1 | 0.1357 | 2.1 | 0.0179 | 3.2 | $6.9 \times 10^{-4}$ |
| 0.2 | 0.4207 | 1.2 | 0.1151 | 2.2 | 0.0139 | 3.5 | $2.3 \times 10^{-4}$ |
| 0.3 | 0.3821 | 1.3 | 0.0968 | 2.3 | 0.0107 | 4.0 | $3.2 \times 10^{-5}$ |
| 0.4 | 0.3446 | 1.4 | 0.0808 | 2.4 | 0.0082 | 4.5 | $3.4 \times 10^{-6}$ |
| 0.5 | 0.3085 | 1.5 | 0.0668 | 2.5 | 0.0062 | 5.0 | $2.9 \times 10^{-7}$ |
| 0.6 | 0.2743 | 1.6 | 0.0548 | 2.6 | 0.0047 | 5.5 | $1.9 \times 10^{-8}$ |
| 0.7 | 0.2420 | 1.7 | 0.0446 | 2.7 | 0.0035 | 6.0 | $9.9 \times 10^{-10}$ |
| 0.8 | 0.2119 | 1.8 | 0.0359 | 2.8 | 0.0026 | 8.0 | $6.2 \times 10^{-16}$ |
| 0.9 | 0.1841 | 1.9 | 0.0287 | 2.9 | 0.0019 | 10.0 | $7.6 \times 10^{-24}$ |

$$\text{Form B: area} = \frac{1}{\sqrt{2\pi}} \int_0^{u} \exp\left(-\frac{u^2}{2}\right) du$$

| $\lvert u \rvert$ | area | $\lvert u \rvert$ | area | $\lvert u \rvert$ | area |
|---|---|---|---|---|---|
| 0.0 | 0.0000 | 1.0 | 0.3413 | 2.0 | 0.4773 |
| 0.1 | 0.0398 | 1.1 | 0.3643 | 2.1 | 0.4821 |
| 0.2 | 0.0793 | 1.2 | 0.3849 | 2.2 | 0.4861 |
| 0.3 | 0.1179 | 1.3 | 0.4032 | 2.3 | 0.4893 |
| 0.4 | 0.1554 | 1.4 | 0.4192 | 2.4 | 0.4918 |
| 0.5 | 0.1915 | 1.5 | 0.4332 | 2.5 | 0.4938 |
| 0.6 | 0.2258 | 1.6 | 0.4452 | 2.6 | 0.4953 |
| 0.7 | 0.2580 | 1.7 | 0.4554 | 2.7 | 0.4965 |
| 0.8 | 0.2881 | 1.8 | 0.4641 | 2.8 | 0.4974 |
| 0.9 | 0.3159 | 1.9 | 0.4713 | 3.0 | 0.4987 |

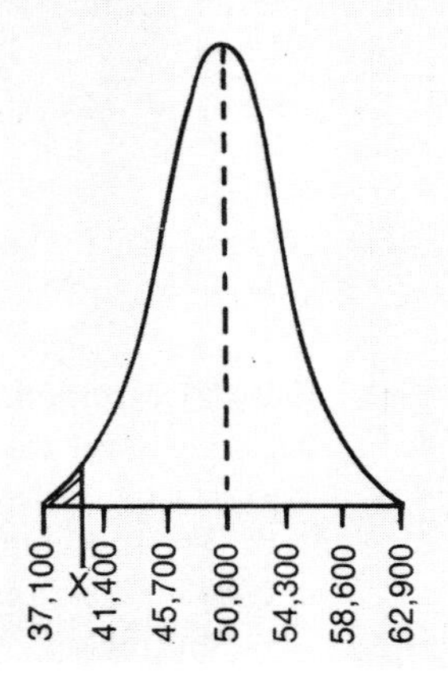


*Example 2–5.* For the examination mentioned in Example 2–3, what fraction of the students can be expected to obtain a perfect score?

We have already found $u$ to be 2.5. Consulting a table of form A, we find the area to be 0.0062, indicating that 0.62 per cent of the students will get a perfect score. The area found in a table of form B is 0.4938.

*Example 2–6.* What is the probability that any single data point lies within two standard deviations of the mean?

This probability is given by the area between $u = 2.0$ and $u = -2.0$. Half this area, namely that between $u = 0$ and $u = 2.0$, is listed in a table of form B, from which we find a value of 0.4773. Thus, the probability that a single observation lies within two standard deviations of the mean is 2(0.4773), or 95.46 per cent.

Notice that we can restate this result by saying "we are 95 per cent confident that $\mu$ is within $2\sigma_X$ of any individual $x$." Thus, if we have, for example, $x = 25.0$ and $\sigma_X = 0.1$, we can be 95 per cent confident that $(x - 2\sigma_X) \leqslant \mu \leqslant (x + 2\sigma_X)$ or $24.8 \leqslant \mu \leqslant 25.2$. These *95 per cent confidence limits* are often expressed for an experimental result in this way: $25.0 \pm 0.2$. Confidence limits are invaluable in the interpretation of experimental data, but we will not stress the concept at this point because the relations we have thus far derived require knowledge of the true $\sigma_X$. This is unlikely. In practical situations we usually deal with $s_X$; and a useful technique for the computation of confidence limits based on $s_X$ will be given in a later section of this chapter.

## PROPAGATION OF ERRORS

### The General Case

Consider some result $w$, obtained as a function of three experimentally determined independent variables, $x$, $y$, and $z$. The variances of $x$, $y$, and $z$ are known. How can they be combined to determine the variance of $w$? Here, it is an advantage to be familiar with differential calculus, because, if you are, you need only remember that an approximate expression is given by

$$V_w = \left(\frac{\partial w}{\partial x}\right)^2 V_x + \left(\frac{\partial w}{\partial y}\right)^2 V_y + \left(\frac{\partial w}{\partial z}\right)^2 V_z \qquad (2\text{–}5)$$

Equation (2–5) is applicable to calculations involving all types of arithmetic manipulations (sequential addition, multiplication, and division, for example). For cases involving only addition and subtraction, however, it is easy to bypass the calculus, and we discuss this below.

### Specific Cases

**Addition and Subtraction.** Consider the case $w = x \pm y \pm z$. Then the general relationship given in equation (2–5) takes the form

$$V_w = (1)^2 V_x + (1)^2 V_y + (1)^2 V_z = V_x + V_y + V_z \qquad (2\text{–}6a)$$

Rewriting equation (2–6a) in terms of standard deviations, we obtain

$$s_w^2 = s_x^2 + s_y^2 + s_z^2 \qquad (2\text{–}6b)$$

$$s_w = \sqrt{s_x^2 + s_y^2 + s_z^2} \qquad (2\text{–}6c)$$

Notice that $s_w$ *is not* $(s_x + s_y + s_z)$, but is instead considerably smaller. This occurs because it is unlikely that $x$, $y$, and $z$ will simultaneously take values far above their means. For example, at the same time $x$ happens to be two standard deviations high, $y$ and $z$ are likely to be much nearer their mean values, and possibly even negative with respect to their means. This opportunity for random errors in one variable to offset random errors in another variable is responsible for the form of the expression for $V_w$.

When coefficients are involved, the calculation is only slightly more complex. If $w = ax \pm by \pm cz$, then

$$V_w = a^2V_x + b^2V_y + c^2V_z \tag{2–7}$$

Note that, if the signs of any of the coefficients happened to be negative (*i.e.*, if we were dealing with subtraction), there would be no difference in the result.

*Example 2–7.* Suppose that the weight of a sample ($w_s$) is determined by noting first the weight of a sample bottle plus sample ($w_{s+b}$), and then the weight of the sample bottle alone ($w_b$):

$$w_s = w_{s+b} - w_b$$

Assume that the standard deviation of single-weight observations on the balance used is known to be $s_w = 1$ mg. Calculate $s_{w_s}$.

In accord with equation (2–6a), we can write

$$V_{w_s} = V_{w_{s+b}} + V_{w_b}$$

Since $w_{s+b}$ and $w_b$ are both single-weight observations, $V_{w_{s+b}} = V_{w_b} = s_w{}^2 = 1\ \text{mg}^2$. Thus, $V_{w_s} = 1 + 1 = 2\ \text{mg}^2$, and $s_{w_s} = \sqrt{2} = 1.4$ mg.

*Example 2–8.* An analysis is made by means of a single measurement, $X$, from which $3B$ (where $B$ is a background correction factor) must be subtracted. We can write

$$P = X - 3B$$

where $P$ is the analytical result. Suppose that the variable $B$ has been quite precisely measured ($s_B = 0.01$), whereas the quantity $X$ is known more poorly ($s_X = 0.1$). Calculate $s_P$.

Using equation (2–7), we obtain

$$V_P = (1)^2V_X + (3)^2V_B$$

$$V_P = 10^{-2} + 9(10^{-4}) = 1.09 \times 10^{-2}$$

$$s_P \sim 1 \times 10^{-1}$$

**Standard Deviation of the Mean.** An interesting application of equation (2–7) allows calculation of the standard deviation of some mean, $\bar{x}$, derived from $n$ observations:

$$\bar{x} = \frac{1}{n}(x_1 + x_2 + x_3 + \cdots x_n)$$

Then, from equation (2–7), we obtain

$$V_{\bar{x}} = n\left(\frac{1}{n}\right)^2 V_x = \frac{V_x}{n} \quad \text{or} \quad s_{\bar{x}} = \frac{s_x}{\sqrt{n}} \tag{2–8}$$

As the number of observations in a sample increases, the standard deviation of the mean decreases. In this sense, four measurements are twice as good as one, but 16 measurements are required if the precision is to be doubled again. The inverse square-root dependence of $s_{\bar{x}}$ on $n$ establishes this sequence of diminishing returns.

*Example 2–9.* In Example 2–1 above, the standard deviation ($s_x$) of a sample of lunar carbon measurements was calculated to be 20.5 ppm. Calculate the standard deviation of the mean, $\bar{x} = 574/4 = 144$ ppm. The number of observations, $n$, is four.

Applying equation (2–8), we obtain

$$s_{\bar{x}} = \frac{s_x}{\sqrt{n}} = \frac{20.5}{2} = 10.2 \text{ ppm}$$

## THE $t$-DISTRIBUTION AND CONFIDENCE LIMITS

Although we have been careful to define the difference between $s_x$ and $\sigma_x$, we have performed a number of calculations in which it was assumed that $\sigma_x$ was known. If 1000 (or even 100) observations have been made, this is effectively true. The $s_x$ calculated from such a large sample is bound to be a very good estimate of $\sigma_x$. However, in the typical situation where a sample contains only two, three, or four observations, the calculated $s_x$ is a very uncertain estimate of $\sigma_x$. When one tries to calculate confidence limits, this uncertainty in $s_x$ must be taken into account, and the $t$-distribution provides the only way of doing so.

**Confidence Limits Using $\sigma_x$.** In Example 2–6 above, we calculated 95 per cent confidence limits based on $\sigma_x$. In order to stress this important concept, let us repeat the procedure here in a slightly different and more formal way.

1. We begin by asking, "What limits must be placed on $u$ in order to include 95 per cent of the area under the normal distribution?"

2. See Figure 2–4. The shaded area totals $2\alpha$. The unshaded area is, therefore, $1.0 - 2\alpha$.

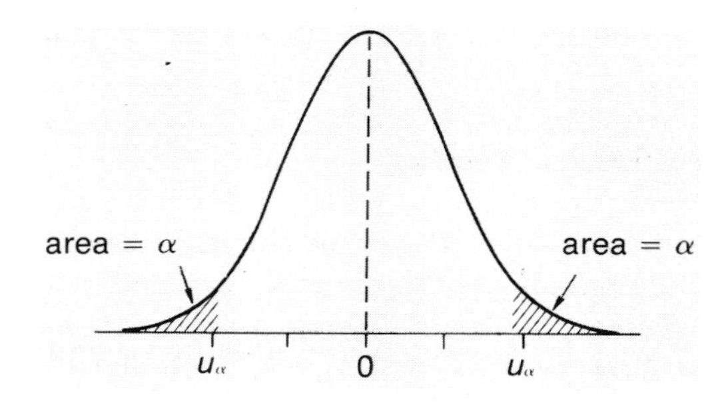


*Figure 2–4.* A view of the normal distribution used to clarify the concept of confidence limits. (See discussion in text.)

3. In order to include 95 per cent of the universe, we want to choose $u_\alpha$ such that $1.0 - 2\alpha = 0.95$; therefore $\alpha = 0.025$.

4. Searching a table of form A for area = 0.025, or one of form B for area = 0.4750, we would find that $u_\alpha = 1.96$; this limit is more accurate than the approximate value of $u = 2.0$ given in Example 2–6.

5. Thus, for any single observation drawn from the universe, we can be 95 per cent confident that $\mu$ is within $1.96\sigma_x$ of the observed $x$ and we can write the 95 per cent confidence limits on $x$ as

$$x \pm u_{0.025}\sigma_x = x \pm 1.96\sigma_x$$

There are only 5 chances in 100, or 1 chance in 20, that $\mu$ will be outside the 95 per cent confidence limits.

6. In general, if we wish to consider confidence levels other than 95 per cent, we can write the "100 (1 - 2$\alpha$) per cent confidence limits" as

$$x \pm u_\alpha \sigma_x \tag{2-9}$$

Note that considering other confidence levels requires only that we adjust $u_\alpha$. In terms of Figure 2–4, increasing $u_\alpha$ will move the shaded zones out farther toward the tails of the normal distribution, decreasing their area, and decreasing the chance that $\mu$ will fall outside the confidence limits.

7. An entirely equivalent expression holds for confidence limits on the mean. With reference to equation (2–8), this relation can be formulated as

$$\bar{x} \pm u_\alpha \sigma_{\bar{x}} = \bar{x} \pm \frac{u_\alpha \sigma_x}{\sqrt{n}} \tag{2-10}$$

**Confidence Limits Using $s_x$.** When $s_x$ replaces $\sigma_x$, a true value is replaced by an estimate, which may be in error. In order to take the uncertainty in $s_x$ into account, this substitution of $s_x$ for $\sigma_x$ requires an increase in the coefficient $u_\alpha$ in equations (2–9) and (2–10) in order to widen the confidence limits. If $n$ is large, the new coefficient can be about the same value as $u_\alpha$, because a large $n$ provides a relatively accurate estimate of $\sigma_x$. When $n$ is small, and $s_x$ is a relatively poor estimate of $\sigma_x$, the coefficient must be considerably larger than $u_\alpha$. The new coefficient is $t_{\alpha,\varphi}$, a function not only of $\alpha$, but also of $\varphi$, the number of degrees of freedom. For a sample from which one mean value has been calculated, $\varphi = n - 1$. The general expressions written above for the 100 (1 - 2$\alpha$) per cent confidence limits can be rewritten as

$$x \pm t_{\alpha,\varphi} s_x \tag{2-11}$$

and

$$\bar{x} \pm t_{\alpha,\varphi} s_{\bar{x}} = \bar{x} \pm \frac{t_{\alpha,\varphi} s_x}{\sqrt{n}} \tag{2-12}$$

A tabulation of the $t$-distribution is given in Table 2–2. It is instructive to examine how $t_{0.025,\varphi}$ changes as $\varphi$ varies. Note first that, for $\varphi = \infty$, $t_{0.025,\infty} = 1.96$, the same value that is obtained from the normal distribution. This occurs, of course, because $s_x = \sigma_x$ when $\varphi = \infty$. In general, $t_{\alpha,\infty} = u_\alpha$. As $\varphi$ decreases, $t$ increases, although $t$ is still within 10 per cent of $u$ for $\varphi = 13$. Further decreases in $\varphi$ result in sharp increases in $t$. In the extreme case where $\varphi = 1$ (that is, where $s_x$ is determined from a pair of observations), the confidence limits correctly calculated using $(t_{0.025,1})(s_x)$ exceed those which are sometimes *incorrectly calculated* using $(u_{0.025})(s_x)$ by a factor of (12.706/1.960) or 6.5.

**Table 2-2. The $t$-Distribution**

| $\varphi$ | $\alpha = 0.05$ | 0.025 | 0.005 | 0.0005 |
|---|---|---|---|---|
| 1 | 6.314 | 12.706 | 63.657 | 636.62 |
| 2 | 2.920 | 4.303 | 9.925 | 31.598 |
| 3 | 2.353 | 3.182 | 5.841 | 12.924 |
| 4 | 2.132 | 2.776 | 4.604 | 8.610 |
| 5 | 2.015 | 2.571 | 4.032 | 6.869 |
| 6 | 1.943 | 2.447 | 3.707 | 5.959 |
| 7 | 1.895 | 2.365 | 3.499 | 5.408 |
| 8 | 1.860 | 2.306 | 3.355 | 5.041 |
| 9 | 1.833 | 2.262 | 3.250 | 4.781 |
| 10 | 1.812 | 2.228 | 3.169 | 4.587 |
| 11 | 1.796 | 2.201 | 3.106 | 4.437 |
| 12 | 1.782 | 2.179 | 3.055 | 4.318 |
| 13 | 1.771 | 2.160 | 3.012 | 4.221 |
| 14 | 1.761 | 2.145 | 2.977 | 4.140 |
| 15 | 1.753 | 2.131 | 2.947 | 4.073 |
| 20 | 1.725 | 2.086 | 2.845 | 3.850 |
| 30 | 1.697 | 2.042 | 2.750 | 3.646 |
| 60 | 1.671 | 2.000 | 2.660 | 3.460 |
| $\infty$ | 1.645 | 1.960 | 2.576 | 3.291 |

*Example 2–10.* Calculate the 90, 95, and 99 per cent confidence limits on the mean carbon content in Apollo 11 lunar fines. The required data are in Examples 2–1 and 2–9 ($\bar{x}$ = 144 ppm, $s_{\bar{x}}$ = 10.2 ppm, $n$ = 4).

An expression for the confidence limits is given by equation (2–12). In this case, $\varphi = n - 1 = 3$. For the 90 per cent confidence limits, $\alpha = 0.05$; for the 95 per cent confidence limits, $\alpha = 0.025$; and, for the 99 per cent confidence limits, $\alpha = 0.005$. The desired results are as follows:

| Confidence Level | $t_{\alpha,3}$ | Confidence Limits |
|---|---|---|
| 90% | 2.353 | $\pm(2.353)(10.2) = \pm 24$ ppm |
| 95% | 3.182 | $\pm(3.182)(10.2) = \pm 32$ ppm |
| 99% | 5.841 | $\pm(5.841)(10.2) = \pm 60$ ppm |

Unless otherwise noted, the 95 per cent confidence limits are used in practical work.

## REGRESSION

It often happens that an analytical instrument is calibrated with a series of standard samples. A graph of the instrument output versus known sample input is constructed by drawing a line through the calibration points. A typical example is shown in Figure 2–5.

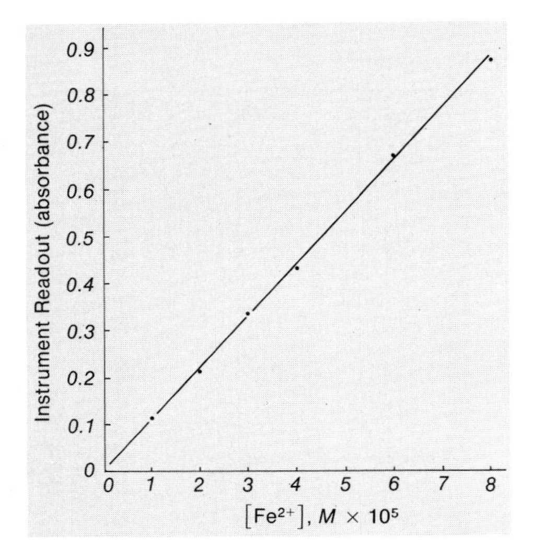


*Figure 2–5.* A typical instrument calibration graph; in this case, it is a plot of iron(II) concentration in an aqueous solution containing a complexing agent which gives an intense color with iron(II) versus the instrument readout (absorbance) at a certain wavelength. The statistical derivation of the calibration line is discussed in the text.

The simple graphic solution of such problems in instrument calibration is unsatisfactory in several respects. First, the graph must be very large if more than two significant figures are to be determined. Second, the placement of the line is an emotional process. The experimenter is very likely not to draw the line which best fits the points, and an improper fit can cause large errors. It is, in all cases, better to derive a mathematical relationship which expresses the instrument response function. Comparison of the derived coefficients on a day-to-day basis allows the experimenter to monitor instrument performance, and the unbiased mathematical procedure offers the only way to achieve line placement without prejudice.

The problem of deriving such relationships comes under the general heading of *regression analysis* in statistics. Here we shall be concerned with only the simplest case – the fitting of straight-line relationships by the **method of least squares.**

### The Distribution of Calibration Data

**The Known Chemical Inputs.** The chemical input (for example, the iron(II) solutions of known concentration used in establishing the graph of Figure 2–5) is regarded as a variable which is free of error; $\sigma = 0$. Everything which follows concerning the least-squares determination of calibration lines assumes that the standard chemical inputs are known with perfect accuracy, or, at least, that errors on this axis are insignificant when compared with errors on the instrument output axis. If this is not true, a different approach to data analysis is required.

**The Observed Instrument Response Points.** Each instrument output reading is a single observation drawn from the universe of all possible instrument outputs for a given chemical input. This situation is depicted in Figure 2–6. For the input standard with value 0.3, for example, the small normal distribution curve which is sketched on the graph represents the universe of all possible instrument outputs for a chemical input of 0.300. Point A, observed during calibration, is near the center of this distribution, well within the $\pm 2\sigma$ limits marked by the shading. In fact, we were quite lucky to obtain a calibration point so close to the universe mean. Even so, *we do not want the calibration line to go through that point,* but instead through the exact *center of the distribution,* which is marked by the arrow. Similar comments apply to the distributions indicated for standard inputs of 0.600 and 0.900.

## Principle of Operation

We might obtain estimates of the center of each distribution by recording several output observations for each input, but this is not necessary to obtain fullest possible accuracy from the method of least squares, which determines the locus of universe means by minimizing $\Sigma(\text{deviations})^2$ between the calibration data points and the line expressed by the function

$$y = a + bx \qquad (2\text{–}13)$$

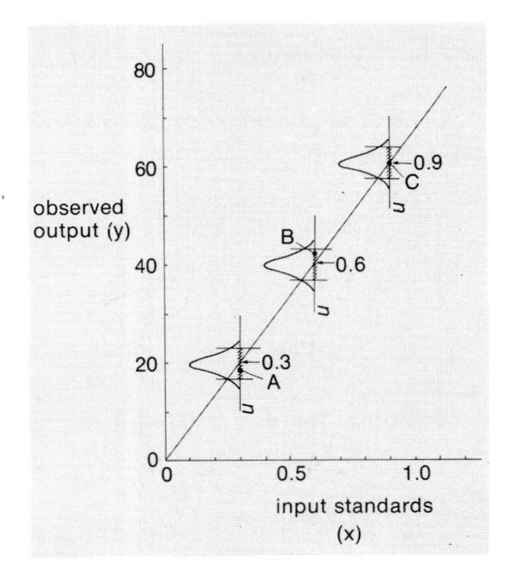


*Figure 2–6.* Another view of an instrument calibration graph. The small plots represent the normal distribution of outputs for each of three particular inputs. (See text for further discussion.)

The sum of squarcs of the deviations is minimized by adjusting $a$ and $b$; and, *if there is a linear functional relationship* between $x$ and $y$, this has the effect of putting the regression line through the best estimate of the true mean values. Thus (referring to Figure 2–6), points B and C as well as point A aid in the determination of $\mu_{0.3}$, and so on.

Mathematically, we derive solutions for $a$ and $b$ by expressing $\Sigma(\text{deviations})^2$ as a function of $a$ and $b$:

$$\sum (\text{deviations of experimental points from line})^2 = \sum [y - (a + bx)]^2 \quad (2\text{–}14)$$

Note that a "deviation" can be expressed mathematically simply as the difference between $y$ (an observed point) and $a + bx$ (the predicted value of $y$, determined by the application of the coefficients $a$ and $b$ to the knowr $x$ value). A derivation which cannot be repeated here shows that $\Sigma(\text{deviations})^2$ will be minimized when the values of $a$ and $b$ are given by

$$a = \frac{1}{n}\left(\sum y - b \sum x\right) \quad (2\text{–}15)$$

$$b = \frac{\sum xy - n\bar{x}\bar{y}}{\sum x^2 - n\bar{x}^2} \quad (2\text{–}16a)$$

where $\bar{x} = \Sigma\, x/n$ and $\bar{y} = \Sigma\, y/n$, just as in one-dimensional analysis.

The form of equation (2–14) shows clearly that it is only deviations parallel to the $y$ axis which are minimized. The $x$ values are regarded as fixed. This is the origin of the requirement mentioned above, that the input standards be very carefully prepared. When it happens that significant errors can occur on either axis, the reader is referred to the excellent (and simple) explanation given by York in the paper listed at the end of this chapter.

### Method of Calculation

**General Expressions for the Sums of Squares.** The data are presented as $n$ pairs of $x$ and $y$ values. For convenience in calculation, we define the following relationships:

$$\sum U^2 \equiv \sum (x - \bar{x})^2 = \sum x^2 - n\bar{x}^2 = \sum x^2 - \left(\sum x\right)^2/n \quad (2\text{–}17)$$

$$\sum V^2 \equiv \sum (y - \bar{y})^2 = \sum y^2 - n\bar{y}^2 = \sum y^2 - \left(\sum y\right)^2/n \quad (2\text{–}18)$$

$$\sum UV \equiv \sum (x - \bar{x})(y - \bar{y}) = \sum xy - n\bar{x}\bar{y} = \sum xy - \sum x \sum y/n \quad (2\text{–}19)$$

**Solutions for a and b.** Substitution of equations (2–17) and (2–19) in equation (2–16a) gives a convenient expression for the calculation of $b$:

$$\text{slope} = b = \frac{\sum UV}{\sum U^2} \quad (2\text{–}16b)$$

After $b$ has been calculated, it is substituted in equation (2–15), from which $a$ can be found.

### Uncertainties

**Confidence Limits for b.** It often happens that the slope of a regression line is predicted by theoretical considerations to have some given value. Occasionally, a set of observations which should define a flat, straight line ($b = 0$), indicating that $y$ is independent of $x$, is examined to see if some finite value of $b$ is found. In either case, it is vital to have a method for the assignment of confidence limits for $b$.

The first quantity required is termed the **variance about the regression**, defined as follows:

$$s_{y\cdot x}^2 = \frac{\sum V^2 - b^2 \sum U^2}{n - 2} \qquad (2\text{–}20)$$

Then, the **variance of the regression coefficient**, or **standard deviation of b**, is given by

$$s_{b_{y\cdot x}}^2 = \frac{s_{y\cdot x}^2}{\sum U^2} \qquad (2\text{–}21)$$

Confidence limits for $b$ are assigned in the usual way

$$\text{C.L.} = \pm t_{\alpha,\varphi} s_{b_{y\cdot x}} \qquad (2\text{–}22)$$

where $\varphi = n - 2$.

**Confidence Limits for a Regression Estimate.** In practice, the regression line is used to determine some estimate, $x_k$, of the chemical input which is responsible for an observed instrument output, $y_k$. The variance of an $x_k$ value which has been determined by the observation of $m$ values of $y_k$ is given by the expression

$$s_{x_k}^2 = \frac{s_{y\cdot x}^2}{b^2}\left[\left(\frac{1}{m} + \frac{1}{n}\right) + \frac{(\bar{y}_k - \bar{y})^2}{b^2 \sum U^2}\right] \qquad (2\text{–}23)$$

where, for clarity, the following points should be stressed:

(1) $x_k$ is the (unknown) chemical input corresponding to the observed instrument output. For example, in Figure 2–5, $x_k$ is the iron(II) concentration corresponding to a particular observed solution absorbance.

(2) $\bar{y}_k$ is $\Sigma\, y_k/m$, the average instrument output obtained from $m$ observations of the unknown chemical input. For example, in the spectrophotometric iron determination, if three samples of the same material have been processed and lead eventually to the observation of three absorbance values, then $m = 3$. Frequently, only a single sample is used and therefore $m = 1$.

(3) The quantities $s_{y\cdot x}^2$, $b$, $n$, $\bar{y}$, and $\Sigma\, U^2$ all relate to the instrument calibration data and have the same values as in equations (2–14) through (2–20).

Notice that $s_{x_k}^2$ becomes large as $\bar{y}_k$ gets farther away from $\bar{y}$. This is nothing but a mathematical expression of the sensible fact that the uncertainty in any regression estimate is lowest near the center of the calibration data, and that extrapolations are a particularly chancy business. The confidence limits for the desired $x_k$ are given by the relation

$$\text{C.L.} = \pm t_{\alpha,\varphi} s_{x_k} \qquad (2\text{–}24)$$

where $\varphi = (n - 2) \neq f(m)$.

*Example 2–11.* Use the method of least squares to construct a calibration line for the spectrophotometric determination of iron(II) described earlier.

(1) Actual data (which are plotted in Figure 2–5) appear in the table below:

| Standard Solution Concentration, $c$ moles/liter | Observed Absorbance, $A$ |
|---|---|
| $1.00 \times 10^{-5}$ | 0.114 |
| $2.00 \times 10^{-5}$ | 0.212 |
| $3.00 \times 10^{-5}$ | 0.335 |
| $4.00 \times 10^{-5}$ | 0.434 |
| $6.00 \times 10^{-5}$ | 0.670 |
| $8.00 \times 10^{-5}$ | 0.868 |

(2) The calibration line must fit an equation of the form $y = a + bx$, where $y$ is proportional to the absorbance and $x$ is proportional to the concentration. In order to simplify the arithmetic, we will take $y = 10A$ and $x = 10^5 c$. With $n = 6$, we obtain from equations (2–17), (2–18), and (2–19):

$$\sum x = 24.00 \qquad \bar{x} = 4.00 \qquad \sum x^2 = 130.00 \qquad \sum U^2 = 34$$

$$\sum y = 26.33 \qquad \bar{y} = 4.3883 \qquad \sum y^2 = 156.0845 \qquad \sum V^2 = 40.5397$$

$$\sum xy = 142.43 \qquad \sum UV = 37.11$$

In these calculations we have carried as many significant figures as the calculator will allow. This is done to minimize round-off errors and is a required procedure! It is not "synthetic precision" unless it is used in reporting some result.

(3) The coefficients $a$ and $b$ are calculated using equations (2–16b) and (2–15):

$$b = \frac{\sum UV}{\sum U^2} = 1.0914_7$$

$$a = \frac{1}{n}\left(\sum y - b \sum x\right) = 0.0224_5$$

The equation for the regression line thus takes the form

$$y = 0.0224_5 + 1.09_1 x$$

Substituting $x = 10^5 c$ and $y = 10A$ in the latter relation, we obtain the desired equation for the regression line:

$$10A = 0.0224_5 + (1.09_1 \times 10^5 c)$$

$$c = (-2.05_8 \times 10^{-7}) + (9.16_6 \times 10^{-5} A)$$

(4) To determine the confidence limits, we first obtain the variance about the regression according to equation (2–20). Because the numerator of equation (2–20) is the

difference between two large numbers, many significant figures must be carried in the computation:

$$s_{y \cdot x}^2 = \frac{\sum V^2 - b^2 \sum U^2}{n - 2} = 8.825 \times 10^{-3}$$

The standard deviation of $b$ can be calculated from equation (2–21):

$$s_{b_{y \cdot x}}^2 = \frac{s_{y \cdot x}^2}{\sum U^2} = 2.596 \times 10^{-4}; \qquad s_{b_{y \cdot x}} = 0.016$$

Next we determine confidence limits for $b$ by using equation (2–22) with $\alpha = 0.025$ and $\varphi = 4$:

$$\text{C.L.} = \pm t_{\alpha,\varphi} s_{b_{y \cdot x}} = \pm 0.044$$

Therefore, $b = 1.09_1 \pm 0.04_4$.

As an example of the calculation of confidence limits on a regression estimate, imagine

(a) That a single unknown sample gave an absorbance of 0.527.

(b) That five replicate samples gave an average absorbance of 0.527.

For either of the two situations:

$$c = (-2.05_8 \times 10^{-7}) + (9.16_6 \times 10^{-5} A) = 4.81_0 \times 10^{-5} M$$

The standard deviation of the regression estimate is obtained from equation (2–23), where, in case (a), $m = 1$; and in case (b), $m = 5$. In both situations, $\bar{y}_k = 10A = 5.27$. Inserting appropriate values in equation (2–23) for each of the two cases, we get

for (a): $$s_{x_k}^2 = 8.804 \times 10^{-3}; \qquad s_{x_k} = 0.093_8$$

and

for (b): $$s_{x_k}^2 = 2.878 \times 10^{-3}; \qquad s_{x_k} = 0.053_6$$

The confidence limits for the regression estimates are both computed from equation (2–24) with $\varphi = 4$:

(a) $$\text{C.L.} = \pm t_{0.025,4} s_{x_k} = \pm 0.26; \qquad c = (4.81 \pm 0.26) \times 10^{-5} M$$

(b) $$\text{C.L.} = \pm t_{0.025,4} s_{x_k} = \pm 0.15; \qquad c = (4.81 \pm 0.15) \times 10^{-5} M$$

Note that, because we have substituted $x = 10^5 c$, the confidence limits for $c$ are $10^{-5}$ times those for the regression estimates.

If nothing else, these results, in which the correctly calculated uncertainties amount to about 4 per cent, demonstrate that the method of least squares is no substitute for quality in the calibration data themselves. Many chemists learn and use the method of least squares only up to the point of calculating $b$ and $a$, and completely omit step (4) above. In this way, they misuse the statistical method and frequently deceive themselves and others by reporting too many significant figures in their results.

## REJECTION OF AN OBSERVATION

Consider this set of analytical results: 15.25, 15.28, 15.30, 16.43 per cent. It is logical to suspect that something is wrong with the fourth observation, and the temptation to reject it before calculating the mean or the standard deviation is bound to be very strong. Such points are called **outliers**, and, though we have provided a flagrant example for the sake of illustration, they can be the source of considerable agony, particularly when the case is not so clear-cut and when dropping one or two data points would make the result coincide with some expected value. In such cases, any human being is tempted to look around for some statistical oil with which to anoint his whim, but this should be attempted only as a last resort, since the statistical tests for the exclusion of outliers are reasonably unsatisfactory.

The first thing to do is go back and thoroughly recheck all the calculations which led to the offending result. Second, check the data. Did the same sample bottle which weighed 15.145 g for samples 1, 2, and 3 weigh 14.145 g for sample 4? If so, the latter weight must be incorrectly recorded. Is there a note that the fourth sample included a peculiar green lump? These questions are aimed at uncovering some clear-cut *assignable cause* for the large deviation. If such a cause can be found, it can either be corrected, in the case of some arithmetic blunder, or used as a perfectly good reason for excluding the outlier.

Sometimes no clear-cut assignable cause is found, and the experimenter asks himself, "Don't I remember spilling a bit of that solution?" or, "I seem to remember that the analytical balance was acting up." This is only more evidence of our humanity, but in any real situation it has to be met with cold-hearted honesty. A test of the sincerity of such suggestions can be made if the experimenter will ask himself whether he is willing to resolve at that moment to discard any future result which might have been similarly affected, *regardless of its value.* Another useful policy to consider is the promise to discard any future sample which might have been similarly affected before the analysis is even completed.

When there are only three or four observations in a sample, it is nearly impossible to provide a useful statistical test for the exclusion of outliers. When there are five or more, it is sometimes helpful to practice the technique of **interior averaging**, in which *both* the highest and lowest data points in the sample are discarded.

## SUMMARY

In this chapter, we have attempted to provide a solid foundation for the understanding and use of statistical techniques in chemical experimentation. The discussions are not mathematically rigorous, but neither are they useless trivializations which force the reader into memorizing statistical formulas if anything is to be gained. We hope the reader will avoid memorization and instead work to understand a few cardinal points:

1. The differences between a sample and a universe, between $s$ and $\sigma$, between $\bar{x}$ and $\mu$
2. The generalization $s^2 = \dfrac{\text{sum of (deviations)}^2}{\varphi}$
3. The use of the normal distribution tables
4. Why the $t$-distribution is needed, and the generalization "confidence limits on *anything*" $= \pm\, t_{\alpha,\varphi} s_{anything}$
5. The principle of regression analysis, and the concept that poor data cannot be improved by its use

6. The fact that outliers *must* have some assignable cause; the idea is to find it, not to throw away data

## QUESTIONS AND PROBLEMS

1. Define and illustrate each of the following terms: significant figures, precision, accuracy, systematic error, random error, average, deviation, confidence limits, absolute uncertainty, relative uncertainty.
2. Assuming that the following quantities are all determined to within ±1 digit in the least significant figure, express the relative uncertainty in each in parts per thousand, or, where appropriate, in parts per million: (a) 0.104 g, (b) 1204.3 ml, (c) 1.007825 atomic mass units (the mass of a $^{1}H$ atom), (d) 56.9354 atomic mass units (the mass of a $^{57}Fe$ atom), (e) $4.56 \times 10^{9}$ yr (the age of the earth), (f) 5730 yr (the half-life of $^{14}C$).
3. Are equations (2–1) and (2–3) entirely equivalent? If not, why not? What about equations (2–2a) and (2–3)?
4. The following results were obtained for replicate determinations of the percentage of chloride in a solid chloride sample: 59.83, 60.04, 60.45, 59.88, 60.33, 60.24, 60.28, 59.77. Calculate (a) the arithmetic mean, (b) the standard deviation, and (c) the relative standard deviation (in per cent).
5. Assuming the sample of problem 4 was pure sodium chloride, calculate the absolute and relative errors of the arithmetic mean.
6. (Drill problems on the use of the normal distribution tables.) Given a normally distributed universe of quantitative observations, what is the probability that (a) some observation will fall more than $0.8\sigma$ above the mean, (b) some observation will fall more than $5\sigma$ below the mean, (c) some observed $x$ will be in the range $(\mu - 0.4\sigma) \leqslant x \leqslant (\mu + 1.3\sigma)$, (d) an observation will fall in the range $-1.60 \leqslant u \leqslant 0.24$?
7. A 10.0000-g object is weighed repeatedly using a procedure with a known $\sigma = 1.0$ mg. What fraction of the observations will be greater than or equal to 10.0016 g?
8. Given a population with a birth rate of $10^{7}$/yr, and an average height at full growth of 175 cm with a standard deviation of 15 cm, what will be the average time interval between births of individuals destined to grow to 225 cm?
9. The mass of a particular lunar rock sample is determined by adding the masses of seven fragments. The standard deviation of a single weighing is 3 mg. What is the standard deviation of the resulting sum?
10. For a single weighing by a particular procedure, $\sigma_x = 1.0$ mg. (a) Calculate $\sigma_{\bar{x}}$ for the mean of four weighings. (b) How many weighings are required for $\sigma_{\bar{x}} = 0.1$ mg? (c) What is the probability that the mean of five weighings is within 0.3 mg of the true weight?
11. A manufacturer supplying glass electrodes for an instrument which continuously monitors pH finds and advertises that his electrodes have an average operating life of 8000 hours with a standard deviation of 200 hours. (a) If you were monitoring the pH of an important production process and wished to prevent breakdowns, yet avoid the expense of

replacing electrodes which were still perfectly good, how long could you use an electrode and expect no greater than (i) 5 per cent probability of failure, (ii) 1 per cent probability of failure? (b) If you get a shipment of four electrodes and find that their average life is 7700 hours, do you have cause for complaint? Why? (c) What is the probability that a given electrode will last at least 8400 hours?

12. Five observations of the chloride content of a potable water sample give an average of 29 ppm $Cl^-$ with a standard deviation of 3.4 ppm. What are the 95 per cent confidence limits of the mean?
13. Six measurements of the percentage of $TiO_2$ in a carload of titanium ore average 58.6% $TiO_2$ with an estimated standard deviation, $s$, of 0.7% $TiO_2$. (a) Find the 90 per cent confidence limits for the average result. (b) Compare with the 95 per cent confidence limits. (c) What would the 90 per cent confidence limits have been had the average been based on only four samples from the carload?
14. A continuous process is operated for the production of dichlorobutadiene from chlorobutadiene. A small amount of (undesirable) trichlorobutene is always formed. Long operating experience has shown that the product averages 1.60 per cent trichlorobutene. An experimental change in operating conditions is made and the analysis of six samples, taken at five-hour intervals, gives the following results: 1.46, 1.62, 1.37, 1.71, 1.52, and 1.40%. Does the change in operating conditions lead to any real change in the trichlorobutene content of the product?
15. By the method of least squares, calculate the equation of the best-fit straight line representing the following spectrophotometric analysis:

| Concentration (ppm) | 0.20 | 0.40 | 0.60 | 0.80 | 1.00 |
|---|---|---|---|---|---|
| Absorbance | 0.077 | 0.126 | 0.176 | 0.230 | 0.280 |

16. (a) Determine the equation of the best-fit straight line for the flame emission photometric data given below:

| Na concentration (ppm) | 0.02 | 0.20 | 0.40 | 0.70 | 1.00 | 1.50 |
|---|---|---|---|---|---|---|
| output (relative units) | 1.20 | 2.60 | 4.40 | 7.60 | 10.80 | 15.60 |

(b) An unknown sample gives an instrument output of 5.35 units. What is the apparent concentration of Na in the sample? (c) Calculate (i) the standard deviation of the result calculated in (b) above, (ii) the 95 per cent confidence limits for the result calculated in (b) above.

## SUGGESTIONS FOR ADDITIONAL READING

1. P. R. Bevington: *Data Reduction and Error Analysis for the Physical Sciences.* McGraw-Hill Book Company, New York, 1969.
2. O. L. Davies, ed.: *Statistical Methods in Research and Production.* Third edition, Hafner, New York, 1967.
3. W. J. Dixon and F. J. Massey, Jr.: *Introduction to Statistical Analysis.* Third edition, McGraw-Hill Book Company, New York, 1969.
4. R. A. Fisher: *Statistical Methods for Research Workers.* Fourteenth edition, Hafner, New York, 1969.
5. S. L. Meyer: *Data Analysis for Scientists and Engineers.* John Wiley and Sons, New York, 1975.
6. E. B. Wilson, Jr.: *An Introduction to Scientific Research.* McGraw-Hill Book Company, New York, 1952.

*Paper*

1. D. York: Least squares fitting of a straight line. *Canad. Jour. Phys., 44*:1079, 1966.

# 3 WATER, SOLUTES, AND CHEMICAL EQUILIBRIUM

Many systems encountered in analytical chemistry are aqueous solutions, so it is appropriate to examine the behavior of solutes in an aqueous medium. Accordingly, we shall begin this chapter with a discussion of interactions between ionic solutes and water. Afterward, we will consider the subject of chemical equilibrium as well as factors which influence the equilibrium state of a system.

## INTERACTIONS BETWEEN IONIC SOLUTES AND WATER

Dissolution of an ionic substance in water is a complex process, resulting in the evolution or absorption of heat and leading to profound changes in the environment of the solute species and in the structure and properties of the solvent. Every ionic species in solution is surrounded by a layer of water molecules which, under the influence of short-range ion-dipole forces, form a primary hydration shell. Within the primary hydration layer, water molecules are drawn by the electrostatic field around the ion into a new, more densely packed structure than the hydrogen-bonded arrangement found in pure liquid water. Beyond the primary solvent sheath is a secondary zone, in which the normal water structure has been disrupted by the still-effective electrostatic field of the central ion. However, as the distance from the central ion increases, the structure of the secondary region becomes less and less perturbed by the electrostatic field, and more and more like normal liquid water.

### Methods of Expressing Concentrations of Solutions

There are at least three ways to specify the concentration of a particular substance in solution – *molarity, formality,* and *normality.* We define the **molarity** of a solution, or the **molar concentration** of a solute, as the number of gram molecular weights (moles) of solute per liter of *solution,* the symbol $M$ being used to designate this concentration unit. For example, the molecular weight of sodium hydroxide is 40.00, so a 0.5000 $M$ solution contains 20.00 g of sodium hydroxide per liter of solution. Alternatively, the **formality** of a solution, or the **formal concentration** of a solute, is the number of gram formula weights of solute dissolved in one liter of *solution,* and is indicated by the symbol $F$. Since the formula weight of acetic acid, represented by the chemical formula $CH_3COOH$, is 60.05, a solution prepared by dissolution of 60.05 g, or 1.000 gram formula weight, of acetic acid in enough water to provide a final volume of exactly one liter is referred to as a 1.000 $F$ solution of acetic acid.

There is a distinction between molar and formal concentration units. If we specify, for example, that the concentration of acetic acid in a particular aqueous solution is 0.001000 $M$, it is explicitly meant that the concentration of the *molecular species*

$CH_3COOH$ is 0.001000 mole per liter of solution. However, acetic acid undergoes dissociation into hydrogen ions and acetate ions. Therefore, if we start with 0.001000 mole or formula weight of acetic acid per liter of solution, the true concentration of molecular acetic acid at equilibrium is 0.000876 $M$ whereas, as a consequence of dissociation, the concentrations of $H^+$ and $CH_3COO^-$ are each 0.000124 $M$. However, the solution still contains the original 0.001000 formula weight of acetic acid, although both molecular and dissociated acetic acid are present, and can correctly be specified as a 0.001000 $F$ solution of acetic acid. Specifying a *formal* concentration provides definite information about how a solution is originally prepared, but it does not necessarily imply what happens to the particular solute after it dissolves. *Molarity* refers to concentrations of actual molecules and ions present in solution at equilibrium, and might or might not be synonymous with formality.

Finally, the **normality** of a solution, or the **normal concentration** of a solute, symbolized by $N$, is the number of gram equivalent weights of solute per liter of *solution*. Normality must always be specified with respect to a definite reaction. For example, if a 0.02000 $F$ solution of potassium permanganate is utilized in a strongly alkaline medium in which permanganate ($MnO_4^-$) undergoes a one-electron reduction to manganate ($MnO_4^{2-}$), so that the equivalent weight of potassium permanganate is the *same* as its formula weight, the solution is said to be 0.02000 $N$ in potassium permanganate. However, if the 0.02000 $F$ $KMnO_4$ solution is employed in a strongly acidic medium in which permanganate is reduced to manganous ion ($Mn^{2+}$) in a five-electron process, so that the equivalent weight of potassium permanganate is *one-fifth* of its formula weight, the solution is 0.1000 $N$ in potassium permanganate. When 0.1000 $F$ sulfuric acid is titrated with sodium hydroxide solution, both protons of $H_2SO_4$ are neutralized, the equivalent weight of $H_2SO_4$ is *one-half* of its formula weight, and the sulfuric acid solution is 0.2000 $N$.

### Activities and Activity Coefficients

If a solution containing 0.1 $M$ hydrogen ion and 0.1 $M$ chloride ion behaved in an ideal manner – that is, if each hydrogen ion and chloride ion was totally independent of its environment – each species would act as if its concentration were 0.1 $M$. In fact, however, there is electrostatic attraction between a certain fraction of the hydrogen ions and chloride ions, so that the solution exhibits the properties of one in which the effective or apparent concentrations of the ions are less than 0.1 $M$. Depending on the relative importance of factors such as ion-ion attractions and ion-solvent interactions, the effective concentration of a solute dissolved in water may appear to be less than, equal to, or greater than the molar or formal concentration.

To distinguish between the molar or formal concentration of a substance and the effective concentration of a substance, which accounts for its nonideal behavior, the latter is called the **activity** of the species. It is customary to relate the activity of a species to its concentration through the expression

$$a_i = f_i C_i$$

in which $a_i$ is the activity of substance $i$, $f_i$ is the **activity coefficient** of substance $i$, and $C_i$ is the molar or formal concentration of substance $i$. For example, we may write the activity of hydrogen ion as

$$a_{H^+} = f_{H^+}[H^+]$$

and that for chloride ion as

$$a_{Cl^-} = f_{Cl^-}[Cl^-]$$

where $[H^+]$ and $[Cl^-]$ represent the molar concentrations of hydrogen ion and chloride ion, respectively. For aqueous solutions, both activity and concentration are expressed in identical units, such as moles per liter, so that the activity coefficient is dimensionless.

If any doubt exists that interionic attractions and ion-solvent interactions can cause marked differences between activity and concentration, the data in Table 3–1 are especially impressive. Serious errors may arise if one attempts to make predictions based on concentrations instead of activities. Figure 3–1 shows a plot of the quantity $f_\pm$ versus the concentration of hydrochloric acid, the data for the graph having been taken from Table 3–1. Instead of $f_i$ we have plotted on the ordinate of this graph a new parameter, $f_\pm$, called the **mean activity coefficient**. Although the significance of the mean activity coefficient will become evident later, $f_\pm$ is employed because the individual behavior of hydrogen ion and chloride ion in a hydrochloric acid medium cannot be sorted out, since each ion influences the activity of the other species.

**Table 3-1. Concentrations and Activities of Hydrochloric Acid Solutions**

| Concentration, *F* | Activity, *F* |
|---|---|
| 0.00500 | 0.00465 |
| 0.0100 | 0.00906 |
| 0.0200 | 0.0176 |
| 0.0500 | 0.0417 |
| 0.100 | 0.0799 |
| 0.200 | 0.154 |
| 0.500 | 0.382 |
| 1.00 | 0.828 |
| 2.00 | 2.15 |
| 3.00 | 4.47 |
| 5.00 | 15.6 |
| 6.00 | 28.2 |
| 8.00 | 89.6 |
| 10.0 | 254. |
| 12.0 | 695. |

Figure 3–1 reveals that the mean activity coefficient for hydrochloric acid tends toward unity as the concentration approaches zero. However, as the hydrochloric acid concentration is increased, the activity coefficient decreases, reaching a minimum value of approximately 0.76 at a concentration near 0.5 *F*. If the concentration of hydrochloric acid is increased above 0.5 *F*, the mean activity coefficient climbs until it attains a value of unity at a concentration slightly less than 2 *F*. For concentrations greater than 2 *F*, the activity coefficient rises very abruptly.

For an approximately 2 *F* hydrochloric acid solution, the activity coefficient is unity because two opposing phenomena offset each other perfectly. First, ion-ion attractions between hydrogen ion and chloride ion cause the activity to be *less* than the concentration of hydrochloric acid and the activity coefficient to be *less* than unity. Second, the solvation of hydrogen ions and chloride ions by water molecules decreases

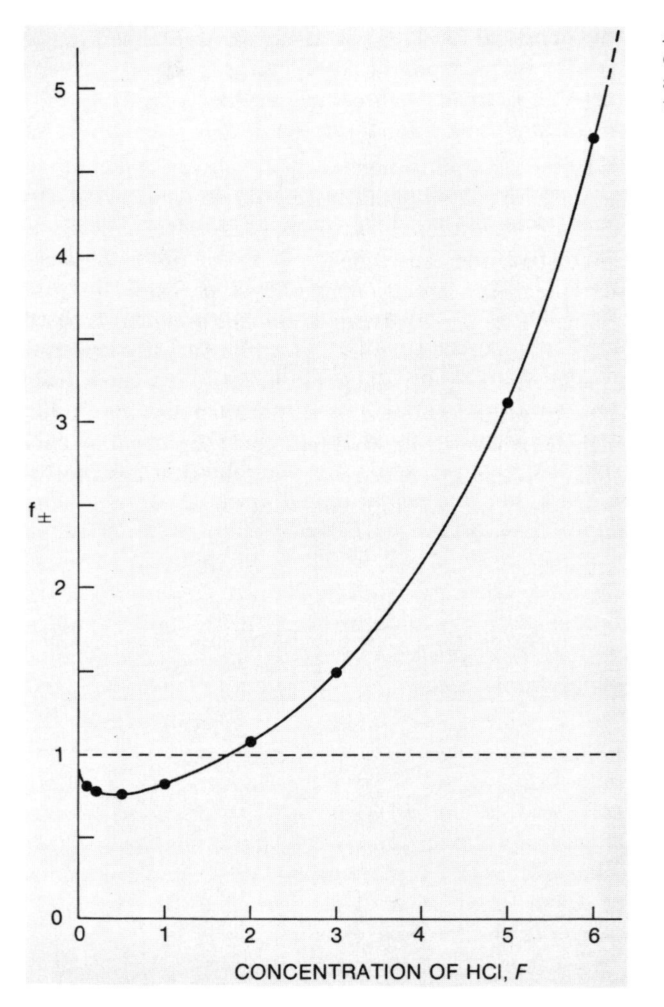


*Figure 3–1.* Mean activity coefficient $f_{\pm}$ for hydrochloric acid as a function of the acid concentration. (See text for discussion.)

the available free or unbound solvent, thereby tending to make the activity *greater* than the concentration and the activity coefficient *greater* than unity. For hydrochloric acid solutions less concentrated than 2 *F*, the first factor is of greater importance, whereas the second factor predominates in hydrochloric acid media more concentrated than 2 *F*.

To understand better why the mean activity coefficient becomes so much larger than unity in very concentrated hydrochloric acid solutions, consider the following crude picture. As shown in Table 3–1, 12 *F* hydrochloric acid behaves as if the acid concentration is 695 *F*, indicating that the mean activity coefficient is approximately 58. If we have one liter of 12 *F* hydrochloric acid, the solution contains 12 moles of the acid and, to a rough approximation, 56 moles of water. If the hydrochloric acid is completely dissociated, 12 moles each of hydrogen ion and chloride ion will be present, and these ions will be solvated by water. If we assume arbitrarily that each ion is solvated on the average by 2.3 water molecules, the total quantity of *bound* water will be (24)(2.3) or 55.2 moles, leaving only 0.8 mole of water not bound to the ions. Water coordinated to these ions belongs to the solute species and is definitely not part of the solvent. Now, if 56 moles of water occupies one liter, 0.8 mole would occupy a volume of about 14 ml. Imagine that 12 moles of hydrochloric acid is dissolved in only 14 ml of water – the acid concentration would be 857 *F*, which is not so far from 695 *F*. Despite the fact that this model is oversimplified, very high activity coefficients in concentrated solutions do arise because most of the solvent is actually bound to the solute species.

For most electrolytes, a plot of activity coefficient versus concentration is qualitatively similar to that for hydrochloric acid shown in Figure 3–1. Naturally, the absolute values of the activity coefficients, as well as the solute concentration at which the activity coefficient is minimal and at which the activity coefficient rises to unity again, depend on the electrolyte.

### Debye-Hückel Limiting Law

One of the most significant contributions to our understanding of the behavior of electrolyte solutions is the Debye-Hückel limiting law, the derivation of which was published in 1923. A key assumption involved in the formulation of the limiting law is that interactions between charged solute species are electrostatic in character. Electrostatic forces are strictly long-range interactions. All short-range forces, including those due to van der Waals attractions and ion-dipole interactions, are ignored.

Second among the assumptions underlying the limiting law is that, although long-range electrostatic forces are operative, ions in solution are subject to random, thermal motion which disrupts the orientation of oppositely charged species caused by interionic attractions. Debye and Hückel simplified their model of electrolyte solutions by proposing that ions can be regarded as point charges. Furthermore, the limiting law presumes that the dielectric constant of an electrolyte solution is uniform and independent of the actual concentration of the dissolved solute and that the dielectric constant of pure water can be used in all calculations.

No attempt will be made to derive the Debye-Hückel limiting law because, in spite of simplifications discussed above, the final result is obtained only after considerable effort. For a single ionic species such as hydrogen ion or chloride ion, the Debye-Hückel limiting law is usually written in the form

$$-\log f_i = A z_i^2 \sqrt{I}$$

in which $f_i$ is the activity coefficient, $z_i$ is the charge of the ion of interest, $A$ is a collection of constants (including Avogadro's number, Boltzmann's constant, the charge of an electron, the dielectric constant of the solvent, and the absolute temperature), and $I$ is the **ionic strength** of the solution, defined as one-half the sum of the concentration $C_i$ multiplied by the square of the charge $z_i$ for each ionic species in the solution, that is

$$I = \tfrac{1}{2} \sum_i C_i z_i^2$$

For water at 25°C, the value of $A$ is 0.512, so the Debye-Hückel limiting law for aqueous solutions at this temperature becomes

$$-\log f_i = 0.512 z_i^2 \sqrt{I}$$

Although the latter relation may be employed to evaluate the activity coefficient for any single ionic species such as $Na^+$, this expression cannot be tested experimentally because it is impossible to prepare a solution containing just one kind of ion. By defining a new term, called the **mean activity coefficient** ($f_\pm$), we obtain a parameter which can be both experimentally and theoretically evaluated. For a *binary* salt, $M_m N_n$, whose cation M has a charge $z_M$ and whose anion N has a charge $z_N$, the mean activity coefficient is given by the relation

$$-\log f_{\pm} = 0.512 z_M z_N \sqrt{I}$$

in which only the *absolute* magnitudes of $z_M$ and $z_N$ are used.

At this point it is instructive to consider some actual calculations involving the use of the Debye-Hückel limiting law. For example, let us determine the mean activity coefficient for 0.10 $F$ hydrochloric acid. First, we must evaluate the ionic strength of the solution. Hydrogen ion and chloride ion are the only species present, so it follows that

$$I = \tfrac{1}{2} \sum_i C_i z_i^2 = \tfrac{1}{2}[C_{H^+} z_{H^+}^2 + C_{Cl^-} z_{Cl^-}^2]$$
$$I = \tfrac{1}{2}[(0.10)(1)^2 + (0.10)(-1)^2] = 0.10$$

Although ionic strength should be expressed in concentration units, such as moles per liter, it is commonly reported as a dimensionless number. Next, since the absolute values of the charges on hydrogen ion and chloride ion are unity, the mean activity coefficient can be obtained from the expression

$$-\log f_{\pm} = 0.512 z_M z_N \sqrt{I} = 0.512(1)(1)\sqrt{0.10}$$
$$-\log f_{\pm} = 0.512(1)(1)(0.316) = 0.162$$
$$\log f_{\pm} = -0.162 = 0.838 - 1.000$$
$$f_{\pm} = 0.689$$

For purposes of comparison, the mean activity coefficient for 0.10 $F$ hydrochloric acid has been experimentally measured and found to be 0.799.

As a second example, we shall calculate the mean activity coefficient for 0.10 $F$ aluminum chloride solution. Notice that the ionic strength

$$I = \tfrac{1}{2}[C_{Al^{3+}} z_{Al^{3+}}^2 + C_{Cl^-} z_{Cl^-}^2]$$
$$I = \tfrac{1}{2}[(0.10)(3)^2 + (0.30)(-1)^2] = 0.60$$

is six times greater than the concentration of the aluminum chloride solution. Substitution of the values for ionic strength and for the charges of the ions into the Debye-Hückel limiting law yields

$$-\log f_{\pm} = 0.512(3)(1)\sqrt{0.60} = 1.189$$
$$\log f_{\pm} = -1.189 = 0.811 - 2.000$$
$$f_{\pm} = 0.0647$$

This predicted mean activity coefficient is more than five times smaller than the experimentally determined value of 0.337.

Numerous similar calculations were performed for other concentrations of hydrochloric acid and aluminum chloride, and the results are shown in Figure 3–2. Included in the plot are the actual experimental values of the mean activity coefficients. Perhaps the most important conclusion to be drawn from Figure 3–2 is that the Debye-Hückel limiting law is quantitatively useful only for very dilute solutions. A second feature of the plot is that the mean activity coefficients calculated from the limiting law become

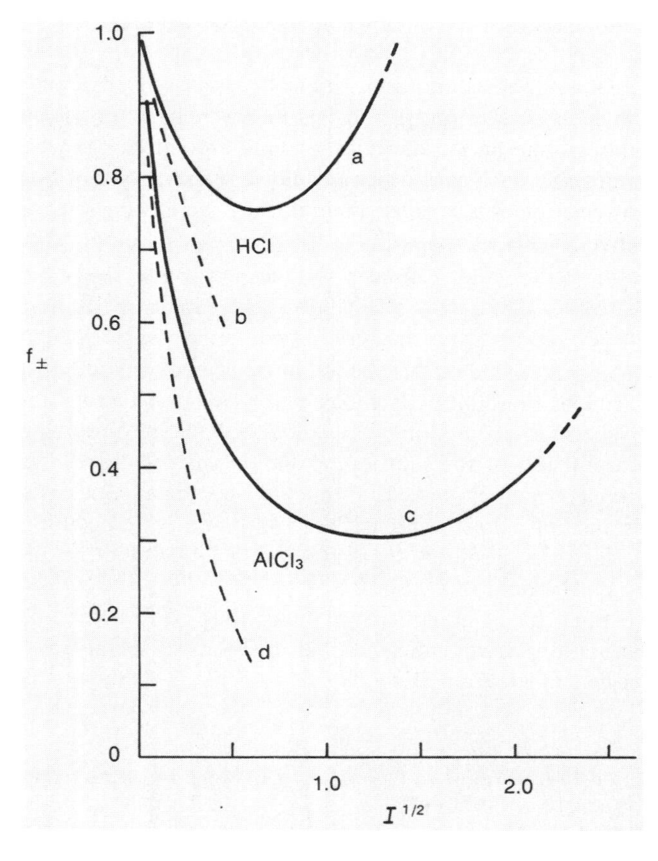


*Figure 3–2.* Comparison of the observed mean activity coefficients for hydrochloric acid (HCl) and aluminum chloride ($AlCl_3$) with the theoretical mean activity coefficients calculated from the Debye-Hückel limiting law, plotted as a function of the square root of ionic strength. Curve *a*: observed mean activity coefficients for HCl. Curve *b*: calculated mean activity coefficients for HCl. Curve *c*: observed mean activity coefficients for $AlCl_3$. Curve *d*: calculated mean activity coefficients for $AlCl_3$.

significantly smaller than the true or observed values as ionic strength increases. Furthermore, the Debye-Hückel limiting law cannot account at all for the upward trend in activity coefficients at relatively high ionic strengths. Nevertheless, the theoretical and experimental curves do converge at low ionic strengths and extrapolate to a mean activity coefficient of unity at zero ionic strength, so the assumptions made by Debye and Hückel are valid for small electrolyte concentrations – particularly the ideas that ions act as point charges in a solution of uniform dielectric constant and that ions undergo simple electrostatic interactions with their neighbors.

In general, activity coefficients for salts consisting of singly charged species can be evaluated by means of the Debye-Hückel limiting law with an uncertainty not exceeding about 5 per cent for ionic strengths up to 0.05. For salts which consist of doubly charged ions, the limiting law gives results reliable to within several per cent up to ionic strengths of perhaps 0.01, whereas, for triply charged species like $Al^{3+}$, the limiting law fails above ionic strengths of approximately 0.005. However, these limits concerning the applicability of the limiting law are only rough guidelines, because the specific nature of the ions in question has a great influence on the behavior of an electrolyte.

## Modifications of the Simple Debye-Hückel Theory

One of the assumptions originally used by Debye and Hückel was that ions are point charges. However, every ion has its own characteristic radius, so that the electrostatic fields around an ion of finite size and a point charge will differ. Furthermore, two ions cannot approach each other any closer than the sum of their radii, so the coulombic force between them is smaller than if they were point charges. Accordingly, the **extended Debye-Hückel equation**

$$-\log f_{\pm} = 0.512\, z_M\, z_N \left[\frac{\sqrt{I}}{1 + Ba\sqrt{I}}\right]$$

incorporates the effect of finite ion size on activity coefficients. This expression resembles the limiting law except for two new factors in the denominator of the term in brackets: $B$, a function ($50.3/\sqrt{DT}$) of the absolute temperature ($T$) and the dielectric constant ($D$) of the solution, and $a$, an *empirically adjustable* number, called the **ion-size parameter** and expressed in units of angstroms, which is said to correspond to the mean distance of closest approach between the solvated cation and anion of an electrolyte. If we take the absolute temperature to be 298°K and the dielectric constant to be the value for pure water (78.5), $B$ is 0.328.

For ionic strengths not exceeding 0.1, experimental studies and theoretical calculations have indicated that the ion-size parameters for many hydrated inorganic species range from 3 to 11 Å, as Table 3–2 shows. To compute the mean activity

**Table 3–2. Single-Ion Activity Coefficients Based on the Extended Debye-Hückel Equation***

| Ion | Ion Size | Ionic Strength | | | | |
|---|---|---|---|---|---|---|
| | $a$, Å | 0.001 | 0.005 | 0.01 | 0.05 | 0.10 |
| $H^+$ | 9 | 0.967 | 0.933 | 0.914 | 0.86 | 0.83 |
| $Li^+$ | 6 | 0.965 | 0.930 | 0.909 | 0.845 | 0.81 |
| $Na^+$, $IO_3^-$, $HCO_3^-$, $HSO_3^-$, $H_2PO_4^-$, $H_2AsO_4^-$ | 4 | 0.964 | 0.927 | 0.901 | 0.815 | 0.77 |
| $K^+$, $Rb^+$, $Cs^+$, $Tl^+$, $Ag^+$, $NH_4^+$, $OH^-$, $F^-$, $SCN^-$, $HS^-$, $ClO_3^-$, $ClO_4^-$, $BrO_3^-$, $IO_4^-$, $MnO_4^-$, $Cl^-$, $Br^-$, $I^-$, $CN^-$, $NO_3^-$ | 3 | 0.964 | 0.925 | 0.899 | 0.805 | 0.755 |
| $Mg^{2+}$, $Be^{2+}$ | 8 | 0.872 | 0.755 | 0.69 | 0.52 | 0.45 |
| $Ca^{2+}$, $Cu^{2+}$, $Zn^{2+}$, $Sn^{2+}$, $Mn^{2+}$, $Fe^{2+}$, $Ni^{2+}$, $Co^{2+}$ | 6 | 0.870 | 0.749 | 0.675 | 0.485 | 0.405 |
| $Sr^{2+}$, $Ba^{2+}$, $Ra^{2+}$, $Cd^{2+}$, $Pb^{2+}$, $Hg^{2+}$, $S^{2-}$, $CO_3^{2-}$, $SO_3^{2-}$ | 5 | 0.868 | 0.744 | 0.67 | 0.465 | 0.38 |
| $Hg_2^{2+}$, $SO_4^{2-}$, $S_2O_3^{2-}$, $CrO_4^{2-}$, $HPO_4^{2-}$ | 4 | 0.867 | 0.740 | 0.660 | 0.445 | 0.355 |
| $Al^{3+}$, $Fe^{3+}$, $Cr^{3+}$, $Ce^{3+}$, $La^{3+}$ | 9 | 0.738 | 0.54 | 0.445 | 0.245 | 0.18 |
| $PO_4^{3-}$, $Fe(CN)_6^{3-}$ | 4 | 0.725 | 0.505 | 0.395 | 0.16 | 0.095 |
| $Th^{4+}$, $Zr^{4+}$, $Ce^{4+}$, $Sn^{4+}$ | 11 | 0.588 | 0.35 | 0.255 | 0.10 | 0.065 |
| $Fe(CN)_6^{4-}$ | 5 | 0.57 | 0.31 | 0.20 | 0.048 | 0.021 |

* Taken from the paper by J. Kielland: J. Amer. Chem. Soc., *59*: 1675, 1937.

coefficient for a binary electrolyte by means of the extended Debye-Hückel equation, the ion sizes of the cation and anion which comprise the solute may be averaged. Thus, for chloride ion and hydrogen ion, the respective ion sizes are 3 and 9 Å, so that 6 Å can be employed as the ion-size parameter of hydrochloric acid. As long as the ionic strength is less than 0.1, use of the ion-size parameters compiled in Table 3–2 provides good agreement between experimentally measured mean activity coefficients and those calculated from the extended Debye-Hückel equation.

To determine how well the predictions of the extended Debye-Hückel equation follow the true behavior of hydrochloric acid, we constructed the graph shown in Figure 3–3, which repeats some of the information given in Figure 3–2. As anticipated, the extended Debye-Hückel equation (EDHE) provides better correlation between theory and experiment than does the limiting law (DHLL).

Other modifications of the basic Debye-Hückel relationships have been proposed. However, most changes take the form of empirical correction terms added to or subtracted from the extended Debye-Hückel equation in order to improve the agreement

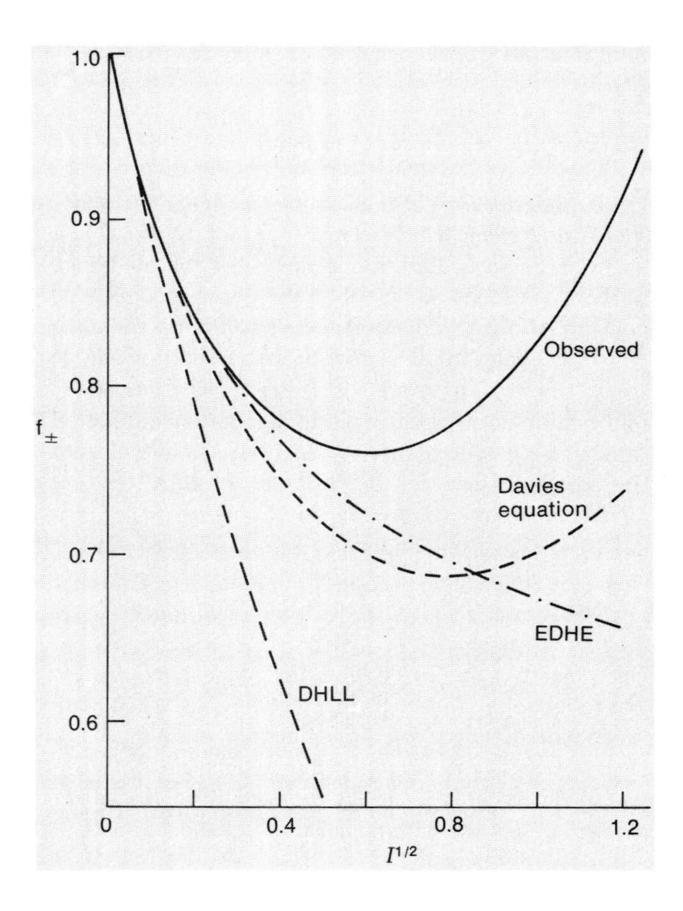


*Figure 3–3.* Comparison of mean activity coefficients for hydrochloric acid computed from the Debye-Hückel limiting law (DHLL), the extended Debye-Hückel equation (EDHE), and the Davies equation with observed mean activity coefficients. Results based on the extended Debye-Hückel equation correspond to $a = 6$ Å.

between observed and calculated activity coefficients. One such modified expression is the **Davies equation**

$$-\log f_{\pm} = 0.512\, z_M\, z_N \left[ \frac{\sqrt{I}}{1 + \sqrt{I}} - 0.2I \right]$$

in which the second term inside the brackets seeks to correct for the effects of ion-pair formation, nonuniform dielectric constant, ion polarizability, and other peculiarities of an individual solute on the mean activity coefficient. Mean activity coefficients for hydrochloric acid solutions calculated from the Davies equation are plotted in Figure 3–3.

## CHEMICAL EQUILIBRIUM

One of the fundamental laws of nature is that a physical or chemical system always tends to undergo a spontaneous and irreversible change from some initial, nonequilibrium state to a final, equilibrium state. Once the system reaches a state of equilibrium, no additional change will occur unless some new stress is placed upon the system.

### Thermodynamic Concept of Equilibrium

To describe equilibrium in thermodynamic terms, one must define a property of a chemical system which can be related to the concentrations of the various species involved in an equilibrium state – the thermodynamic parameter most useful in this respect is $G$, the **Gibbs free energy**. A chemical system which is initially in some nonequilibrium state has a tendency to change spontaneously until the free energy for the system becomes minimal and the system itself reaches equilibrium. It is strictly correct to speak only of a *tendency* for change to occur, since thermodynamics offers no information at all about the rate of a particular process. We can measure the driving force which causes a system to pass from an initial state to a certain final state by the free-energy change for that process. In turn, the free-energy change is defined as the difference between the free energies of the final and initial states, and is symbolized by $\Delta G$ when the system undergoes a finite change of state; that is,

$$\Delta G = G_{\text{final}} - G_{\text{initial}}$$

Our objectives are to evaluate the free-energy change for a particular chemical reaction and to relate the free-energy change to the state of equilibrium for that process.

**Standard States and Free Energies of Formation.** In general, the free energy of a chemical species is dependent upon temperature and pressure, as well as upon the nature and quantity of the substance itself. Therefore, standard-state conditions have been established for the various kinds of chemical species, so that values for the absolute free energy of an element or the free energy of formation of a compound can be precisely specified and tabulated.

Our choice of **standard state** for a substance depends on whether it is a gaseous, liquid, solid, or solute species; however, all standard states pertain to a temperature of 298°K or 25°C. A gaseous substance, if ideal, exists in its standard state when present at a pressure of exactly one atmosphere. For a liquid or solid, the standard state is taken to be the *pure* liquid or *pure* solid, respectively, at a pressure of one atmosphere. Dissolved or solute species are in their standard states if their activities are unity, again at a pressure of one atmosphere.

By convention, the most stable form of any pure element in its standard state is assigned a free energy of zero. For a compound or ion, the free-energy *change* for the reaction by which that compound or ion is formed from its elements, when all reactants and products are in their standard states, is termed the **standard free energy of formation,** $\Delta G_f{}^0$. In Table 3–3 are listed standard free energies of formation for some representative species, expressed in kilocalories per mole of the substance formed. Note that the free-energy change attending the formation of $n$ moles of a substance under standard-state conditions corresponds to a free-energy change of $n(\Delta G_f{}^0)$.

**Table 3-3. Standard Free Energies of Formation ($\Delta G_f{}^0$) for Various Substances at 298°K (Kilocalories/Mole)**

| Gases | | Solids | |
|---|---|---|---|
| $H_2O$ | −54.64 | AgI | −15.81 |
| $H_2O_2$ | −24.7 | $BaSO_4$ | −323.4 |
| $O_3$ | 39.06 | CaO | −144.4 |
| HBr | −12.72 | $CaCO_3$ | −269.8 |
| HI | 0.31 | $Ca(OH)_2$ | −214.3 |
| $SO_2$ | −71.79 | | |
| $SO_3$ | −88.52 | **Aqueous Ions** | |
| NO | 20.72 | | |
| $NO_2$ | 12.39 | $H^+$ | 0.0 |
| $NH_3$ | −3.97 | $Ag^+$ | 18.43 |
| CO | −32.81 | $Ba^{2+}$ | −134.0 |
| $CO_2$ | −94.26 | $Ca^{2+}$ | −132.18 |
| $C_2H_4$ | 16.28 | $OH^-$ | −37.59 |
| $C_2H_6$ | −7.86 | $I^-$ | −12.35 |
| | | $SO_4{}^{2-}$ | −177.34 |

**Free-Energy Change for a Reaction.** From information such as that compiled in Table 3–3, it is possible to compute the **standard free-energy change,** $\Delta G^0$, for each of a large number of chemical reactions from the general expression

$$\Delta G^0 = \sum \Delta G_f^0 \text{ (products)} - \sum \Delta G_f^0 \text{ (reactants)}$$

where the summation signs permit us to consider as many reactants and products as necessary. According to this defining equation, the standard free-energy change for a reaction is the sum of the standard free energies of formation of the products minus the sum of the standard free energies of formation of the reactants. For the reaction

$$a\text{A} + b\text{B} \rightleftharpoons c\text{C} + d\text{D}$$

we may write

$$\Delta G^0 = c[\Delta G_f^0(\text{C})] + d[\Delta G_f^0(\text{D})] - a[\Delta G_f^0(\text{A})] - b[\Delta G_f^0(\text{B})] \qquad (3\text{–}1)$$

However, in most chemical systems, we encounter reactant and product species under conditions of concentration different from those specified for their standard states. For example, gaseous compounds may be present at pressures other than one atmosphere, whereas the activities of dissolved substances are very likely not equal to unity.

To formulate an equation for the free-energy change associated with the previously mentioned general reaction, we must employ $\Delta G$ and $\Delta G_f$ rather than $\Delta G^0$ and $\Delta G_f{}^0$, since the latter terms pertain only to processes occurring under standard-state conditions. Therefore,

$$\Delta G = c[\Delta G_f(\mathrm{C})] + d[\Delta G_f(\mathrm{D})] - a[\Delta G_f(\mathrm{A})] - b[\Delta G_f(\mathrm{B})] \qquad (3\text{–}2)$$

Subtraction of equation (3–1) from equation (3–2), followed by rearrangement of terms, yields

$$\begin{aligned}\Delta G - \Delta G^0 &= c[\Delta G_f(\mathrm{C}) - \Delta G_f^0(\mathrm{C})] + d[\Delta G_f(\mathrm{D}) - \Delta G_f^0(\mathrm{D})] \\ &\quad - a[\Delta G_f(\mathrm{A}) - \Delta G_f^0(\mathrm{A})] - b[\Delta G_f(\mathrm{B}) - \Delta G_f^0(\mathrm{B})] \qquad (3\text{–}3)\end{aligned}$$

From the second law of thermodynamics, we can obtain an expression to relate $\Delta G_f$, the free energy of formation of a substance at *any* activity, to the standard free energy of formation, $\Delta G_f{}^0$, which is

$$\Delta G_f - \Delta G_f^0 = RT \ln a$$

where $R$, the universal gas constant, is 0.001987 kilocalorie per mole-degree, $T$ is the absolute temperature (298°K), and ln $a$ is the natural logarithm of the activity of the species of interest in moles per liter. Applying the latter expression to each of the four substances A, B, C, and D, and substituting the results into equation (3–3), one obtains

$$\Delta G - \Delta G^0 = cRT \ln a_C + dRT \ln a_D - aRT \ln a_A - bRT \ln a_B$$

Rearrangement and combination of logarithmic terms gives

$$\Delta G = \Delta G^0 + RT \ln \frac{(a_C)^c (a_D)^d}{(a_A)^a (a_B)^b} \qquad (3\text{–}4)$$

Thus, the free-energy change for a chemical reaction is dependent upon two factors – first, the standard free-energy change for the reaction which is related to the standard free energies of formation of the substances involved in the reaction and, second, the activities of the reactants and products.

**Free Energy and the Equilibrium State.** When a system is at equilibrium, the free-energy change, $\Delta G$, is zero, and equation (3–4) becomes

$$\Delta G^0 = -RT \ln \frac{(a_C)^c (a_D)^d}{(a_A)^a (a_B)^b} \qquad (3\text{–}5)$$

Recognizing that the ratio of factors in the logarithmic term of equation (3–5) contains the *equilibrium* activities of all reactants and products involved in the preceding general chemical reaction, we identify this ratio as a **thermodynamic equilibrium constant,** $\mathcal{K}$. Thus,

$$\mathcal{K} = \frac{(a_C)^c (a_D)^d}{(a_A)^a (a_B)^b}$$

and

$$\Delta G^0 = -RT \ln \mathscr{K}$$

Since the activity of a chemical species is defined as its concentration multiplied by an activity coefficient ($a_i = f_i C_i$), the relation for the thermodynamic equilibrium constant can be written as

$$\mathscr{K} = \frac{(a_C)^c (a_D)^d}{(a_A)^a (a_B)^b} = \frac{(f_C [C])^c (f_D [D])^d}{(f_A [A])^a (f_B [B])^b} = \frac{f_C{}^c f_D{}^d}{f_A{}^a f_B{}^b} \cdot \frac{[C]^c [D]^d}{[A]^a [B]^b} = \frac{f_C{}^c f_D{}^d}{f_A{}^a f_B{}^b} \cdot K \quad (3\text{–}6)$$

where $K$ is the common **equilibrium constant** based on concentration units:

$$K = \frac{[C]^c [D]^d}{[A]^a [B]^b}$$

Because $\mathscr{K}$ is a thermodynamic constant, it does not depend on such factors as ion-ion attractions or ion-solvent interactions, so that the *product* of the terms on the right-hand side of equation (3–6) is likewise constant. However, the individual activity coefficients for reactants and products do change as the composition and ionic strength of a chemical system vary, which means that $K$ is not truly constant.

When rigorously correct equilibrium calculations are to be performed, use of the thermodynamic equilibrium constant ($\mathscr{K}$) and the activities of the reactants and products is essential. Unfortunately, activity coefficients (and, consequently, activities) are difficult to evaluate accurately for aqueous solutions of analytical interest, which usually have high ionic strengths and which often contain more than one electrolyte, so that solute-solute and solute-solvent interactions cannot be assessed quantitatively. Therefore, we shall follow throughout most of this textbook the usual practice of ignoring activity coefficients as well as the distinction between $\mathscr{K}$ and $K$. Instead, we will be satisfied to assume the constancy of $K$ and to utilize equilibrium-constant expressions involving concentrations rather than activities.

One can frequently tell by glancing at the numerical magnitude of an equilibrium constant whether a reaction has a tendency to proceed far toward the right, or only slightly so, before reaching equilibrium. For a reaction having a large $K$, equilibrium is attained only after the reaction has proceeded far to the right. A two-to-one mixture of hydrogen and oxygen gases has a strong tendency to form water vapor

$$2\, H_2(g) + O_2(g) \rightleftharpoons 2\, H_2O(g)$$

although the reaction does not occur readily without a catalyst. At 25°C the equilibrium expression for the process is

$$\frac{[H_2O(g)]^2}{[H_2(g)]^2[O_2(g)]} = 1.7 \times 10^{80}$$

so that the numerator of the relation is exceedingly large compared to the denominator, indicating that very little hydrogen and oxygen remain unreacted at equilibrium.

Some familiar chemical processes have $K$ values much smaller than unity, so that equilibrium is reached before the reactions have gone very far toward the right. For example, the dissolution of silver chloride in water may be represented by the reaction

$$AgCl(s) \rightleftharpoons Ag^+ + Cl^-$$

for which the equilibrium expression* at 25°C is

$$[Ag^+][Cl^-] = 1.78 \times 10^{-10}$$

Such an equilibrium constant indicates that the concentrations of silver ion and chloride ion (equal to each other in a saturated aqueous solution of silver chloride) are quite small, which simply confirms our knowledge of the low solubility of silver chloride.

Some words of caution should be interjected at this point. Although the magnitude of an equilibrium constant is a criterion by which to judge whether a reaction has a pronounced or only a slight tendency to proceed from left to right as written, the initial concentrations of reactants and products provide a second and equally important criterion for deciding the direction in which a reaction will actually go in order to attain a state of equilibrium. Thus, a chemical reaction may have a *large* equilibrium constant, yet it may proceed from *right* to *left* if the initial concentrations of species on the right-hand side of the reaction happen to be much greater than the concentrations of substances on the left-hand side of the reaction.

### Factors Which Influence Chemical Equilibrium

In this section are discussed the effects of several factors, which may be experimentally adjusted and controlled, upon systems in a state of chemical equilibrium. Qualitative predictions of the influence of specific variables can be based on the **principle of Le Châtelier**, which states that, *when a stress is applied to a system at equilibrium, the position of equilibrium tends to shift in such a direction as to diminish or relieve that stress.* Quantitative changes in the equilibrium properties of a system due, for example, to pressure or temperature variations can be evaluated from consideration of the equilibrium-constant expression or other thermodynamic relationships.

**Effect of Concentration.** According to Le Châtelier's principle, an increase or decrease in the concentration of a reactant or product in a system initially at equilibrium will cause the system to shift in one direction or the other to establish a new position of equilibrium which minimizes that change. Such a change results in different concentrations of all species when equilibrium is restored, but the numerical value of the equilibrium constant remains unaltered. Consider the system in which the diammine-silver(I) complex is formed in an aqueous medium:

$$Ag^+ + 2\,NH_3 \rightleftharpoons Ag(NH_3)_2^+$$

When an additional amount of one of the reactants is introduced to the system at equilibrium, the system shifts toward a new position of equilibrium by consuming part of the added material. Thus, introduction of some silver ion, in the form of silver nitrate, would promote the formation of more $Ag(NH_3)_2^+$, the reaction proceeding from left to right as written.

**Effect of Catalysts.** By influencing the rate of a reaction, a catalyst hastens the approach of a chemical system to a state of equilibrium. However, catalysis always affects the rates of forward and backward reactions to the same extent, and never changes the value of the equilibrium constant. In general, the catalyst enters into a chemical or physical process at some step in the reaction mechanism and is regenerated in a

*The concentration (activity) of solid silver chloride does not appear in this equilibrium expression because, as described in Chapter 7, the concentration of a pure solid phase is *defined* as unity.

subsequent step. Although the physical properties of a catalyst may be modified, it undergoes no permanent chemical change. Thus, the catalyst does not appear as a reactant or product in the overall reaction.

An example of catalysis is encountered for the isotope-exchange reaction occurring in the vapor phase between water and deuterium molecules:

$$H_2O(g) + D_2(g) \rightleftharpoons HD(g) + HDO(g)$$

A mixture of $H_2O$ and $D_2$ gases in a clean vessel reacts only slowly, typically requiring hours to attain a state of equilibrium. In the presence of finely divided platinum oxide, however, the same initial mixture reaches equilibrium in a few minutes. Nevertheless, both equilibrium mixtures contain exactly the same concentrations of each of the four species, whether the catalyst is present or not.

**Effect of Temperature.** Chemical reactions absorb or liberate heat as they proceed, so that increases or decreases in temperature – which, respectively, supply or withdraw thermal energy from a system – produce substantial changes in the values of equilibrium constants. Usually, the quantity of heat liberated or absorbed during occurrence of a chemical reaction is referred to as the **enthalpy change** and is given the symbol $\Delta H$. In a manner similar to the treatment of free energy, each element in its standard state is assigned an enthalpy of zero, and a **standard enthalpy of formation**, $\Delta H_f^0$, is associated with the production of one mole of a compound in its standard state from the appropriate elements in their standard states. Furthermore, the **standard enthalpy change**, $\Delta H^0$, for any process is the sum of the standard enthalpies of formation for the products minus the sum of the standard enthalpies of formation for the reactants. Processes for which $\Delta H$ (or $\Delta H^0$) is negative are heat-liberating or **exothermic**, and the products of such reactions tend to be stable relative to the reactants. Conversely, **endothermic** reactions, for which the products are apt to be less stable than the reactants, absorb thermal energy and have positive $\Delta H$ (or $\Delta H^0$) values.

Le Châtelier's principle predicts that, if the temperature of a system is increased, the position of equilibrium will shift in such a direction as to consume at least part of the extra thermal energy. Accordingly, when the temperature is raised, an endothermic reaction will be favored over an exothermic one. If the process of interest is endothermic, the equilibrium constant increases with a rise in temperature and decreases when the temperature drops. On the other hand, if the reaction is exothermic, the equilibrium constant becomes smaller as the temperature increases, but larger as the temperature falls.

Thermodynamics provides an explicit equation to describe quantitatively how variations in temperature affect the magnitude of the equilibrium constant for a reaction. If $K_1$ and $K_2$ are equilibrium constants at absolute temperatures $T_1$ and $T_2$, respectively, we discover that

$$\log \frac{K_2}{K_1} = -\frac{\Delta H^0}{2.303\,R}\left(\frac{1}{T_2} - \frac{1}{T_1}\right) \qquad (3\text{–}7)$$

where $\Delta H^0$ is the standard enthalpy change for the reaction of interest and $R$ is the universal gas constant. If the equilibrium constant for a reaction at one temperature is known, along with the value of $\Delta H^0$, the equilibrium constant at another temperature can be determined. It is necessary to assume that $\Delta H^0$ remains constant within the temperature range of concern; but for most purposes no serious error is incurred, because $\Delta H^0$ changes only slightly with temperature. Let us apply these considerations to a specific chemical system.

*Example 3–1.* Given that the standard enthalpy change ($\Delta H^0$) for the reaction

$$2\ HI(g) \rightleftharpoons H_2(g) + I_2(g)$$

is +3.03 kilocalories and that the equilibrium constant is 0.0162 at 667°K, what is the value of the equilibrium constant at 760°K?

If $T_1$ = 667°K, $T_2$ = 760°K, and $K_1$ = 0.0162, and if these values are substituted into equation (3–7), we have

$$\log \frac{K_2}{(0.0162)} = -\frac{(3.03)}{(2.303)(0.001987)}\left(\frac{1}{760} - \frac{1}{667}\right) = 0.120$$

$$\frac{K_2}{(0.0162)} = 1.32; \quad K_2 = 0.0214$$

Let us consider the effect of temperature on the free-energy change for a chemical process. From the second law of thermodynamics, we obtain the **Gibbs-Helmholtz equation**

$$\Delta G = \Delta H - T\Delta S$$

which relates the free-energy change ($\Delta G$) for a reaction to the enthalpy change ($\Delta H$) and to the product of the absolute temperature ($T$) and $\Delta S$, the change in the **entropy** of the system. According to the Gibbs-Helmholtz equation, the entropy change is a factor which contributes to the driving force of a chemical reaction. Entropy is a measure of the *degree of disorder* in a system, and there is a natural tendency for a system to become disorganized. Such a tendency to attain maximal entropy or disorder is actually the major driving force for some chemical processes. As mentioned earlier, values of $\Delta H$ for various reactions are essentially independent of temperature. Likewise, values of $\Delta S$ for numerous processes are largely insensitive to temperature. Therefore, the $T\Delta S$ term is directly proportional to temperature, and $\Delta G$ is very strongly temperature-dependent. At room temperature, $T\Delta S$ for most reactions is much smaller than $\Delta H$. This means that the dominant component of $\Delta G$ is the $\Delta H$ term, which agrees with the fact that most exothermic processes ($\Delta H$ negative) are spontaneous ($\Delta G$ negative). However, at room temperature for a few reactions, and at higher temperatures for many reactions, the $T\Delta S$ term is sufficiently large compared to $\Delta H$, and is of the appropriate sign, that an exothermic process may be nonspontaneous and an endothermic process may be spontaneous.

At a pressure of one atmosphere and a temperature of 298°K, the reaction

$$4\ Ag(s) + O_2(g) \rightleftharpoons 2\ Ag_2O(s)$$

is exothermic, having a $\Delta H^0$ value of −14.62 kilocalories, whereas the standard entropy change ($\Delta S^0$) is −0.0316 kilocalorie per degree. To determine the influence of temperature on the direction of this reaction, we can evaluate the free-energy change at various temperatures with the aid of the Gibbs-Helmholtz equation. Figure 3–4 presents the results of a series of such calculations, based on the assumption that the values of $\Delta H^0$ and $\Delta S^0$ at 298°K are applicable over a relatively wide temperature range. At temperatures below 463°K, the reaction is spontaneous as written, because the free-energy change is negative. However, for temperatures above 463°K, the value of $T\Delta S$

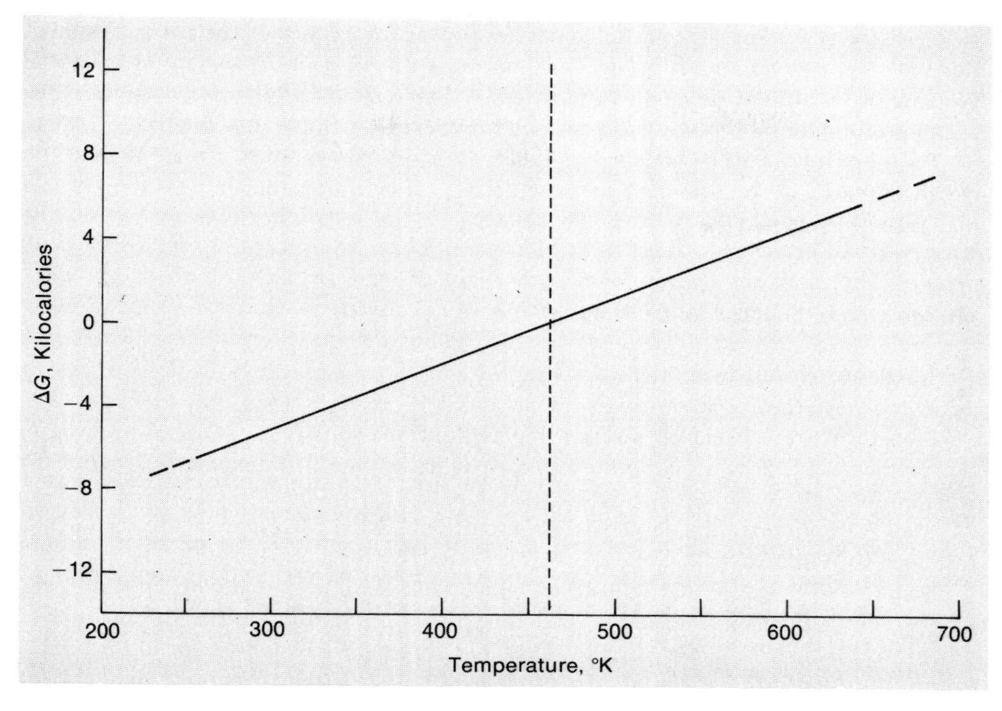


*Figure 3–4.* Free-energy change ($\Delta G$) for the reaction

$$4\,Ag(s) + O_2(g) \rightleftharpoons 2\,Ag_2O(s)$$

as a function of the absolute temperature. (See text for discussion.)

exceeds the magnitude of $\Delta H$, so that $\Delta G$ becomes positive and the *reverse* process occurs spontaneously at one atmosphere pressure.

**Effect of Pressure.** Changes in pressure can exert either a considerable influence upon the position of a chemical equilibrium or almost none at all. As should be expected, the effects are most pronounced for gas-phase systems. For example, a rise in the total pressure of a system in which the gas-phase equilibrium

$$2\,SO_2(g) + O_2(g) \rightleftharpoons 2\,SO_3(g)$$

prevails will cause a marked change in the position of equilibrium. Such a result is explicable on the basis of Le Châtelier's principle, that *any increase in pressure favors a shift in the equilibrium position in whichever direction tends to reduce the pressure.* From the properties of gases, we know that this will occur if the total number of gas molecules in the system is decreased. Notice that three molecules of reactants – two molecules of sulfur dioxide and one molecule of oxygen – combine to form two molecules of sulfur trioxide. Consequently, any rise in pressure will favor the formation of a higher percentage of sulfur trioxide, because the system can most readily relieve this stress through a decrease in the total number of molecules.

For a gas-phase equilibrium such as

$$2\,HI(g) \rightleftharpoons H_2(g) + I_2(g)$$

in which the *same* total number of molecules appears on both the reactant and product side of the reaction, no change in the number of molecules accompanies the decomposition of hydrogen iodide at a specified temperature and pressure. Therefore, the position of equilibrium is not influenced by variations in the total pressure of the system.

Equilibria occurring in condensed phases, such as aqueous solutions, are usually not greatly affected by variations in pressure because liquids are so much less compressible than gaseous systems.

### Introduction to Equilibrium Calculations

Numerical calculations involving chemical equilibria vary widely in their degree of difficulty. Nevertheless, all problems can be solved by means of a four-step plan of attack.

Step 1. Write a balanced reaction representing the state of equilibrium involved in the problem and, if two or more chemical equilibria prevail simultaneously, write the reaction for each.

Step 2. Formulate the equilibrium-constant expression corresponding to each of the reactions written in Step 1.

Step 3. Write additional equations from available information and data until the number of mathematical relations, including those of Step 2, equals the number of unknowns contained therein. Each of these equations must be independent of the others, in the sense that no two relationships can merely be rearrangements of one another. One source of information consists of numerical data given in the statement of the problem. A second source is the compilation of data in various reference tables and books, particularly listings of equilibrium constants.

A third source of information for additional equations is the **mass-balance relationship**. For example, if a 0.100 $F$ solution of silver nitrate is mixed with an equal volume of an aqueous ammonia solution, the total concentration of all dissolved silver-bearing species must be 0.0500 $M$. However, silver ion exists partly in the form of $Ag^+$ and partly as the two silver-ammine complexes, $Ag(NH_3)^+$ and $Ag(NH_3)_2^+$, so

$$0.0500 = [Ag^+] + [Ag(NH_3)^+] + [Ag(NH_3)_2^+]$$

Another example is provided by the dissolution of 0.0100 mole of potassium chromate ($K_2CrO_4$) in one liter of aqueous hydrochloric acid, in which some of the chromate remains as the $CrO_4^{2-}$ anion, a fraction of the chromate is protonated to yield $HCrO_4^-$,

$$CrO_4^{2-} + H^+ \rightleftharpoons HCrO_4^-$$

and part of the chromate undergoes dimerization to form dichromate ion:

$$2\ CrO_4^{2-} + 2\ H^+ \rightleftharpoons Cr_2O_7^{2-} + H_2O$$

Thus, the mass-balance equation is

$$0.0100 = [CrO_4^{2-}] + [HCrO_4^-] + 2[Cr_2O_7^{2-}]$$

the concentration of dichromate being multiplied by a factor of *two* because that species contains *two* chromium atoms.

A fourth source, used to treat ionic equilibrium, is the **charge-balance relationship**, which states that *the total concentration of positive charge must equal the total* *concentration of negative charge.* For example, in an aqueous potassium chloride medium, the cationic species are $K^+$ and $H^+$, whereas the anions are $Cl^-$ and $OH^-$. Accordingly, the charge-balance equation is

$$[K^+] + [H^+] = [Cl^-] + [OH^-]$$

Now, consider the hydrochloric acid solution of potassium chromate mentioned above, in which the positively charged species are $H^+$ and $K^+$ and the negatively charged species are $Cl^-$, $OH^-$, $CrO_4{}^{2-}$, $HCrO_4{}^-$, and $Cr_2O_7{}^{2-}$. In this situation, we write the charge-balance relation as

$$[K^+] + [H^+] = [Cl^-] + [OH^-] + 2[CrO_4{}^{2-}] + [HCrO_4{}^-] + 2[Cr_2O_7{}^{2-}]$$

where the concentrations of chromate and dichromate ions are multiplied by *two* because these species bear doubly negative charges. Finally, if we have an aqueous solution of lanthanum chloride ($LaCl_3$), the charge-balance expression is

$$3[La^{3+}] + [H^+] = [Cl^-] + [OH^-]$$

in which it is necessary to multiply the concentration of the triply charged lanthanum cation by *three.*

Step 4. Solve these equations for the desired unknown quantity or quantities. As long as Steps 2 and 3 result in the same number of equations as unknowns, it is possible to manipulate these relations algebraically to obtain a numerical value for any or all of the unknown quantities.

Mathematical operations encountered in Step 4 frequently become complex and time-consuming. To simplify the work, it is often possible to make assumptions so as to minimize the time and effort involved in the solution of a problem. It must be stressed, however, that the desire to simplify a calculation is *never a sufficient* reason for making such an assumption. Each assumption must be clearly justified or justifiable, either in advance or by subsequent checking, to ensure that the resultant error is fully within the limit of uncertainty which is acceptable in light of the precision of the data and the desired use of the end result of the calculations. Several methods of doing this will be encountered as we progress through this textbook.

We will now consider some problems which illustrate the use of these principles.

*Example 3–2.* Calculate the equilibrium constant for the reaction between iron(III) and iodide ion to yield iron(II) and triiodide ion, assuming that the *initial* iron(III) concentration is 0.200 *M,* the *initial* iodide concentration is 0.300 *M,* and the *equilibrium* triiodide concentration is 0.0866 *M.*

Let us write the *pertinent chemical reaction*

$$2\ Fe^{3+} + 3\ I^- \rightleftharpoons 2\ Fe^{2+} + I_3{}^-$$

and the corresponding *equilibrium expression*:

$$K = \frac{[Fe^{2+}]^2[I_3{}^-]}{[Fe^{3+}]^2[I^-]^3}$$

Since no iron(II) is originally present, the stoichiometry of the reaction tells us that the equilibrium iron(II) concentration must be *twice* that of the triiodide ion, two $Fe^{2+}$ ions being produced for every triiodide ion. Therefore,

$$[I_3^-] = 0.0866\ M$$
$$[Fe^{2+}] = 0.173\ M$$

To obtain the equilibrium concentration of iron(III), it is necessary to use the *mass-balance relation* for the total iron concentration:

$$[Fe^{2+}] + [Fe^{3+}] = 0.200\ M$$
$$[Fe^{3+}] = 0.200 - [Fe^{2+}].$$
$$[Fe^{3+}] = 0.200 - 0.173 = 0.027\ M$$

A *mass-balance expression* for the total iodine concentration can be utilized to compute the concentration of iodide ion at equilibrium. Thus,

$$[I^-] + 3[I_3^-] = 0.300\ M$$
$$[I^-] = 0.300 - 3[I_3^-]$$
$$[I^-] = 0.300 - 3(0.0866) = 0.040\ M$$

If we now substitute the values for the concentrations of all four species into the equilibrium expression, the result is

$$K = \frac{(0.173)^2(0.0866)}{(0.027)^2(0.040)^3} = 5.6 \times 10^4$$

*Example 3–3.* For the dissociation of hydrogen fluoride (HF) gas into the elemental gases, $H_2$ and $F_2$, the equilibrium constant is $1.00 \times 10^{-13}$ at 1000°C. If 1.00 mole of hydrogen fluoride is allowed to reach dissociation equilibrium in a 2.00-liter container at this temperature, what will be the final molar concentration of fluorine gas?

As a starting point for the solution of this problem, we should write the *pertinent reaction*

$$2\ HF(g) \rightleftharpoons H_2(g) + F_2(g)$$

along with the *equilibrium expression*:

$$K = \frac{[H_2(g)][F_2(g)]}{[HF(g)]^2} = 1.00 \times 10^{-13}$$

Based on the stoichiometry of the reaction which must occur before a state of equilibrium is reached, we can conclude that the final concentrations of hydrogen and fluorine gases are equal:

$$[H_2(g)] = [F_2(g)]$$

Next, we can formulate the *mass-balance relationship*

$$[HF(g)] + 2[F_2(g)] = \frac{1.00\ \text{mole}}{2.00\ \text{liters}} = 0.500\ M$$

which states that the *total* concentration of fluorine in both of its forms must be 0.500 $M$. From this equation it follows that the equilibrium concentration of hydrogen fluoride gas is

$$[HF(g)] = 0.500 - 2[F_2(g)]$$

If the expressions for the concentrations of hydrogen and hydrogen fluoride are inserted into the equilibrium-constant relation, we have

$$K = \frac{[F_2(g)]^2}{(0.500 - 2[F_2(g)])^2} = 1.00 \times 10^{-13}$$

This equation can be solved mathematically to obtain a numerical value for the concentration of $F_2$. However, we can simplify the expression by recognizing that, as a consequence of the very small equilibrium constant, only a tiny fraction of the HF dissociates. This means that the concentration of $F_2$ is far less than the concentration of HF, so the term $2[F_2(g)]$ may be neglected in the denominator, and the last preceding relation may be rewritten as

$$\frac{[F_2(g)]^2}{(0.500)^2} = 1.00 \times 10^{-13}$$

When solved for the fluorine concentration, this equation yields

$$[F_2(g)]^2 = (1.00 \times 10^{-13})(0.250) = 2.50 \times 10^{-14}$$

$$[F_2(g)] = 1.58 \times 10^{-7}\ M$$

It is necessary to verify the simplifying assumption which was made; twice $1.58 \times 10^{-7}$ $M$ is indeed much smaller than 0.500 $M$.

*Example 3–4.* If the equilibrium constant for the reaction

$$PCl_5(g) \rightleftharpoons PCl_3(g) + Cl_2(g)$$

is 33.3 at 760°K, calculate the equilibrium concentration of each species in the system, if the initial concentrations of phosphorus pentachloride and phosphorus trichloride are 0.050 and 5.00 $M$, respectively.

First, we can state that the *equilibrium* concentration of phosphorus pentachloride is equal to the *initial* concentration minus the concentration of $PCl_5$ which decomposes:

$$[PCl_5] = 0.050 - [Cl_2]$$

Notice that the equilibrium concentration of chlorine is a measure of how much phosphorus pentachloride decomposes because $PCl_5$ is the *only* source of $Cl_2$. Next, the following expression can be written for the equilibrium concentration of $PCl_3$:

$$[PCl_3] = 5.00 + [Cl_2]$$

This equation indicates that the final concentration of phosphorus trichloride is larger than its original value of 5.00 $M$ by a term equal to the concentration of phosphorus pentachloride which decomposes. However, we have already concluded that the chlorine concentration is equivalent to the concentration of decomposed $PCl_5$, so the use of $[Cl_2]$ in the last relationship is justified.

When we make appropriate substitutions into the equilibrium expression, the result is

$$K = \frac{[PCl_3][Cl_2]}{[PCl_5]} = \frac{(5.00 + [Cl_2])[Cl_2]}{(0.050 - [Cl_2])} = 33.3$$

If *all* the $PCl_5$ decomposed, the final concentrations of chlorine and phosphorus trichloride would be 0.050 and 5.05 *M*, respectively. Since the maximal concentration of $PCl_3$ is not much larger than the original value of 5.00 *M*, we will neglect $[Cl_2]$ which appears in the term $(5.00 + [Cl_2])$ in the preceding equation. This simplification, which corresponds to saying that the equilibrium $PCl_3$ concentration is essentially 5.00 *M*, leads to

$$\frac{(5.00)[Cl_2]}{(0.050 - [Cl_2])} = 33.3$$

which can be easily solved for the chlorine concentration:

$$5.00[Cl_2] = 1.67 - 33.3[Cl_2]$$

$$38.3[Cl_2] = 1.67$$

$$[Cl_2] = 0.0435\ M$$

Now, when the concentrations of the other two species are calculated, we have

$$[PCl_5] = 0.050 - [Cl_2] = 0.050 - 0.0435 = 0.0065\ M$$

$$[PCl_3] = 5.00 + [Cl_2] = 5.00 + 0.0435 = 5.04\ M$$

Our earlier assumption that the equilibrium concentration of $PCl_3$ was very close to 5.00 *M* is seen to be justified, so we can have confidence that the calculations are valid.

## QUESTIONS AND PROBLEMS

1. Calculate the formal concentration or formality of each of the following solutions:
   (a) 500.0 ml of solution containing 2.183 g of silver nitrate ($AgNO_3$)
   (b) 250.0 ml of solution containing 4.708 g of potassium permanganate ($KMnO_4$)
   (c) 20.00 ml of solution containing 8.879 mg of sodium chloride (NaCl)
   (d) 500.0 ml of solution containing 12.63 g of hydrogen bromide (HBr)
   (e) 25.00 ml of solution containing 372.1 mg of potassium sulfate ($K_2SO_4$)
2. What weight of solute in grams is needed to prepare each of the following solutions?
   (a) 500.0 ml of a 0.005124 *F* perchloric acid solution
   (b) 1.000 liter of a 0.1034 *F* barium hydroxide solution
   (c) 50.00 ml of a 0.06663 *F* potassium iodide solution

(d) 250.0 ml of a 0.2767 *F* copper(II) nitrate solution
(e) 100.0 ml of a 0.04839 *F* ammonium sulfate solution

3. For each of the following systems, a 0.1000 *F* solution of the *first* designated reactant is employed. State the normality of that solution in each situation.
(a) $Ba(OH)_2 + 2\ HCl \rightleftharpoons BaCl_2 + 2\ H_2O$
(b) $2\ KMnO_4 + 3\ MnSO_4 + 2\ H_2O \rightleftharpoons 5\ MnO_2 + K_2SO_4 + 2\ H_2SO_4$
(c) $6\ FeCl_2 + K_2Cr_2O_7 + 14\ HCl \rightleftharpoons 6\ FeCl_3 + 2\ CrCl_3 + 7\ H_2O + 2\ KCl$
(d) $H_3PO_4 + 3\ KOH \rightleftharpoons K_3PO_4 + 3\ H_2O$
(e) $2\ Na_2S_2O_3 + KI_3 \rightleftharpoons Na_2S_4O_6 + KI + 2\ NaI$

4. Write the equilibrium-constant expression for each of the following reactions:
(a) $H_3PO_4(aq) + PO_4^{3-}(aq) \rightleftharpoons H_2PO_4^-(aq) + HPO_4^{2-}(aq)$
(b) $N_2(g) + 3\ H_2(g) \rightleftharpoons 2\ NH_3(g)$
(c) $As_2S_3(s) + 6\ H_2O(l) \rightleftharpoons 2\ H_3AsO_3(aq) + 3\ H_2S(aq)$
(d) $2\ CO_2(g) \rightleftharpoons 2\ CO(g) + O_2(g)$
(e) $H^+(aq) + CO_3^{2-}(aq) \rightleftharpoons HCO_3^-(aq)$
(f) $Ag^+(aq) + 2\ NH_3(aq) \rightleftharpoons Ag(NH_3)_2^+(aq)$
(g) $I_2(aq) + 2\ S_2O_3^{2-}(aq) \rightleftharpoons 2\ I^-(aq) + S_4O_6^{2-}(aq)$
(h) $MgNH_4AsO_4(s) \rightleftharpoons Mg^{2+}(aq) + NH_4^+(aq) + AsO_4^{3-}(aq)$
Use terms involving molar concentrations and note that (aq), (g), (s), and (l) denote species present in an aqueous phase, in a gas phase, as a pure solid, and as a pure liquid, respectively.

5. For each of the following gas-phase reactions
(a) $2\ HI(g) \rightleftharpoons H_2(g) + I_2(g)$
(b) $PCl_5(g) \rightleftharpoons PCl_3(g) + Cl_2(g)$
(c) $N_2(g) + 3\ H_2(g) \rightleftharpoons 2\ NH_3(g)$
predict the direction in which the position of equilibrium will be shifted if the total pressure of the system is decreased.

6. If the system in which the reaction

$$Ag^+ + 2\ CN^- \rightleftharpoons Ag(CN)_2^-$$

occurs is at equilibrium, predict the effect of each of the following on the molar concentration of the $Ag(CN)_2^-$ complex: (a) solid silver nitrate is dissolved in the solution; (b) ammonia gas, $NH_3$, which can form a complex, $Ag(NH_3)_2^+$, with silver ion, is passed into the solution; (c) solid sodium iodide is dissolved in the solution, AgI being insoluble.

7. An equilibrium mixture at 377°C for the gas-phase system

$$2\ NH_3(g) \rightleftharpoons N_2(g) + 3\ H_2(g)$$

is found to contain, in one liter, 0.0100 mole of $NH_3$, 0.100 mole of $N_2$, and 0.162 mole of $H_2$. Evaluate the equilibrium constant, based on molar concentration units, for the reaction.

8. For the gas-phase reaction

$$2\ SO_2(g) + O_2(g) \rightleftharpoons 2\ SO_3(g)$$

the equilibrium constant, based on molar concentration units, is 5800 at 600°C. If 0.150 mole of sulfur dioxide and 6.00 moles of oxygen are admitted to a previously evacuated 6.00-liter chamber at 600°C, what will be the equilibrium concentration of $SO_3$?

9. If the reaction

$$N_2O_4(g) \rightleftharpoons 2\ NO_2(g)$$

has an equilibrium constant, based on molar concentration units, of 0.0510 at 320°K, how many grams of $NO_2$ are present in 5.00 liters of an equilibrium mixture which contains 10.0 grams of $N_2O_4$ gas?

10. If the equilibrium constant, based on molar concentration units, for the dissociation of hydrogen fluoride gas into elemental hydrogen and fluorine

$$2\ HF(g) \rightleftharpoons H_2(g) + F_2(g)$$

is $1.00 \times 10^{-13}$ at 1000°C, what will be the equilibrium molar concentration of fluorine if 1.00 mole of $H_2$ and 1.00 mole of $F_2$ are allowed to reach equilibrium in a 2.00-liter container at this temperature?

11. For the decomposition reaction

$$2\ NOCl(g) \rightleftharpoons 2\ NO(g) + Cl_2(g)$$

the equilibrium constant, based on molar concentration units, is $4.66 \times 10^{-4}$ at 500°K. If 0.200 mole of NOCl and 1.00 mole of $Cl_2$ are mixed in a 2.00-liter container at 500°K, how many grams of the original NOCl will have decomposed when equilibrium is attained?

12. From the data listed in Table 3–3 on page 39, compute the standard free-energy change ($\Delta G^0$) for each of the following reactions:
    (a) $3\ O_2(g) \rightleftharpoons 2\ O_3(g)$
    (b) $CaCO_3(s) \rightleftharpoons CaO(s) + CO_2(g)$
    (c) $SO_2(g) + ½\ O_2(g) \rightleftharpoons SO_3(g)$
    (d) $C_2H_4(g) + H_2(g) \rightleftharpoons C_2H_6(g)$
    (e) $H^+ + OH^- \rightleftharpoons H_2O(g)$
    (f) $2\ H_2O(g) + O_2(g) \rightleftharpoons 2\ H_2O_2(g)$
    (g) $BaSO_4(s) \rightleftharpoons Ba^{2+} + SO_4^{\ 2-}$

13. Evaluate the equilibrium constant at 25°C for each of the following reactions using information given in Table 3–3 on page 39:
    (a) $6\ NH_3(g) + 7\ O_3(g) \rightleftharpoons 6\ NO_2(g) + 9\ H_2O(g)$
    (b) $3\ H_2(g) + N_2(g) \rightleftharpoons 2\ NH_3(g)$
    (c) $CO_2(g) + NO(g) \rightleftharpoons CO(g) + NO_2(g)$
    (d) $H_2(g) + Br_2(l) \rightleftharpoons 2\ HBr(g)$
    (e) $AgI(s) \rightleftharpoons Ag^+ + I^-$
    (f) $C_2H_4(g) + 3\ O_2(g) \rightleftharpoons 2\ CO_2(g) + 2\ H_2O(g)$
    (g) $2\ NH_3(g) + 5\ SO_3(g) \rightleftharpoons 2\ NO(g) + 5\ SO_2(g) + 3\ H_2O(g)$

14. For the dissociation of $N_2O_4$ gas

$$N_2O_4(g) \rightleftharpoons 2\ NO_2(g)$$

the equilibrium constant, based on molar concentration units, is 0.0125 at 35°C and is 0.171 at 75°C. Is the dissociation of $N_2O_4$ an exothermic or an endothermic process? Explain.

15. If the equilibrium constant for the reaction

$$PCl_5(g) \rightleftharpoons PCl_3(g) + Cl_2(g)$$

is $1.8 \times 10^{-7}$ at 25°C and 1.8 at 250°C, determine the enthalpy change for the process.

16. At 1000°K the enthalpy changes for two different reactions are as follows:

$$H_2(g) + CO_2(g) \rightleftharpoons H_2O(g) + CO(g); \Delta H = +8.3 \text{ kcal}$$

$$2\ H_2S(g) \rightleftharpoons 2\ H_2(g) + S_2(g); \quad \Delta H = -43.1 \text{ kcal}$$

Suppose that each of these systems is at equilibrium at a temperature of 1000°K and a total pressure of 0.500 atmosphere. For each of the following experimental changes, state (i) whether no net reaction occurs, (ii) whether net reaction occurs to the right, or (iii) whether net reaction occurs to the left:

(a) Additional hydrogen gas is introduced into the system at 1000°K, and the total pressure is maintained constant by an appropriate increase in the volume.

(b) While the total pressure is kept at 0.500 atmosphere, the temperature of the equilibrium mixture is lowered to 500°K.

(c) Without any change in the temperature, the total pressure is increased to 5.00 atmospheres.

17. At a temperature of 760°K, the reaction for the gas-phase decomposition of phosphorus pentachloride into phosphorus trichloride and chlorine

$$PCl_5(g) \rightleftharpoons PCl_3(g) + Cl_2(g)$$

has an equilibrium constant (based on molar concentration units) of 33.3. If 2.00 moles of phosphorus pentachloride is admitted to a previously evacuated 1.00-liter vessel, what will be the equilibrium concentrations of all three species?

18. Calculate the ionic strength of each of the following systems:

(a) a solution containing 0.00300 $F$ lanthanum chloride

(b) a solution containing 0.00500 $F$ sodium sulfate and 0.00300 $F$ lanthanum nitrate

(c) a solution containing 0.0200 $F$ sodium nitrate, 0.0500 $F$ potassium sulfate, and 0.0300 $F$ cadmium chloride

(d) a solution containing 0.0600 $F$ aluminum sulfate and 0.1000 $F$ sodium ferricyanide

19. Calculate the mean activity coefficient for a 0.200 $F$ calcium chloride solution using (a) the Debye-Hückel limiting law, (b) the extended Debye-Hückel equation, and (c) the Davies equation. Compare the results with the experimentally measured value of 0.472.

20. Use the extended Debye-Hückel equation as well as information presented in Table 3–2 to compute the mean activity coefficient for the specified binary salt in each of the following solutions:

(a) 0.0250 $F$ calcium iodide ($CaI_2$) in water

(b) 0.00100 $F$ barium chloride ($BaCl_2$) in an aqueous 0.0300 $F$ potassium nitrate solution

(c) 0.00500 *F* aluminum chloride ($AlCl_3$) in an aqueous 0.0100 *F* sodium nitrate solution
(d) 0.0150 *F* sodium bicarbonate ($NaHCO_3$) in water
(e) 0.0300 *F* potassium chromate ($K_2CrO_4$) in an aqueous solution containing 0.100 *F* sodium hydroxide and 0.200 *F* lithium perchlorate

## SUGGESTIONS FOR ADDITIONAL READING

1. A. J. Bard: *Chemical Equilibrium.* Harper & Row, New York, 1966.
2. T. R. Blackburn: *Equilibrium: A Chemistry of Solutions.* Holt, Rinehart and Winston, New York, 1969.
3. J. N. Butler: *Ionic Equilibrium.* Addison-Wesley Publishing Company, Reading, Massachusetts, 1964.
4. G. M. Fleck: *Equilibria in Solution.* Holt, Rinehart and Winston, New York, 1966.
5. H. Freiser and Q. Fernando: *Ionic Equilibria in Analytical Chemistry.* John Wiley and Sons, New York, 1963, pp. 9–34.

# 4 AQUEOUS ACID-BASE EQUILIBRIA

Aqueous acid-base reactions are more easily characterizable than other kinds of chemical interactions in water, because they involve only the transfer of hydrogen ions from proton donors to proton acceptors. Thus, it is possible to acquire detailed information about the identities of the reactant and product species in a proton-transfer reaction and to know with some reliability the equilibria and equilibrium constants which govern the behavior of an acid-base system. In this chapter, we will consider the nature of acids and bases in water as well as the methods used to perform equilibrium calculations for aqueous acid-base systems.

## FUNDAMENTAL CONCEPTS OF ACIDITY AND BASICITY

**Definitions of Acids and Bases.** From the point of view of chemical analysis, the most useful definition of acids and bases is the one suggested independently in 1923 by Brønsted and by Lowry. A **Brønsted-Lowry acid** is a species having a tendency to lose or donate a proton; a **Brønsted-Lowry base** is a substance having a tendency to accept or gain a proton. Loss of a proton by a Brønsted-Lowry acid gives rise to the formation of a corresponding Brønsted-Lowry base, called the **conjugate base** of the parent acid. Addition of a proton to a Brønsted-Lowry base produces the **conjugate acid** of the original base. Accordingly, $H_2O$ is the conjugate base of the Brønsted-Lowry acid $H_3O^+$, but the conjugate acid of the Brønsted-Lowry base $OH^-$. Similarly, $HCO_3^-$ is the conjugate acid of $CO_3^{2-}$, whereas $NH_3$ is the conjugate base of $NH_4^+$. Two species such as $H_3O^+$ and $H_2O$ comprise a conjugate acid-base pair.

A proton does not exist in solution by itself, that is, independently of its surrounding environment. In aqueous media the proton is associated with one or more water molecules to give the **hydronium ion**, which is usually symbolized as $H_3O^+$. In addition, the simpler designation $H^+$ is used in chemical reactions and equilibrium expressions for aqueous systems where the hydronium ion is understood to be present; invariably when $H^+$ is employed, and often when $H_3O^+$ appears, the term **hydrogen ion** is introduced as a synonym for hydronium ion. In an aqueous medium, all other ions and molecules are hydrated (including hydroxide ion, $OH^-$, the conjugate base of the solvent) even though water is not written as part of the chemical formulas of these species.

**Autoprotolysis of Water.** Pure water is slightly ionized, because one water molecule is capable of accepting a proton from a second water molecule:

$$H_2O + H_2O \rightleftharpoons H_3O^+ + OH^-$$

Such a proton-transfer reaction involving only solvent molecules is termed **autoprotolysis**, and the equilibrium expression, written in terms of activities, is

$$K = \frac{(H_3O^+)(OH^-)}{(H_2O)^2}$$

This equation can be simplified if the thermodynamic convention, that a pure liquid is defined as having unit activity, is assumed to be valid for relatively dilute aqueous solutions. Furthermore, if values of unity are used for the activity coefficients of the hydronium and hydroxide ions, we can replace the activities of $H_3O^+$ and $OH^-$ by their respective concentrations to arrive at the familiar ion-product expression for water:

$$K_W = [H_3O^+][OH^-] = [H^+][OH^-]$$

Although the ion-product constant, $K_W$, also known as the autoprotolysis constant, is temperature dependent (ranging from $1.14 \times 10^{-15}$ at 0°C through $1.01 \times 10^{-14}$ at 25°C and $5.47 \times 10^{-14}$ at 50°C to approximately $5.4 \times 10^{-13}$ at 100°C), it will be assumed for all the calculations in this book that the temperature is 25°C and that the ion-product constant is precisely $1.00 \times 10^{-14}$.

**Definition of pH.** To specify the degree of acidity of an aqueous solution, pH is usually employed. Throughout most of this textbook, we will define the pH of an aqueous solution as the negative base-ten logarithm of the hydrogen ion concentration:

$$\text{pH} = -\log [H^+]$$

Actually, the pH of a solution is probably more accurately measured by the negative base-ten logarithm of the hydrogen ion *activity*, $a_{H^+}$.

If we return to the ion-product expression for water,

$$K_w = [H^+][OH^-] = 1.00 \times 10^{-14}$$

and take the negative logarithm of each member of this equation, the result is

$$-\log K_w = -\log [H^+] - \log [OH^-] = 14.00$$

or

$$\text{p}K_w = \text{pH} + \text{pOH} = 14.00$$

where $\text{p}K_W$ is the negative logarithm of the ion-product constant and pOH is the negative logarithm of the hydroxide ion concentration. This equation provides a simple way to calculate the pOH of a solution if the pH is determined and, conversely, the pH can be obtained if the pOH is known.

*Example 4–1.* Calculate the pH of two solutions, the hydrogen ion concentrations of which are 0.0023 $M$ and $8.4 \times 10^{-10} M$, respectively.

Because the pH of a solution is, by definition, the negative base-ten logarithm of the hydrogen ion concentration, it follows for the first solution that

$$\text{pH} = -\log(0.0023) = -\log(2.3 \times 10^{-3})$$

Recalling that the logarithm of a product is the *sum* of logarithms, we can write

$$\text{pH} = -[\log(2.3) + \log(10^{-3})] = -[(0.36) + (-3.00)]$$

$$\text{pH} = 2.64$$

For the second solution,

$$\text{pH} = -\log(8.4 \times 10^{-10}) = -[\log(8.4) + \log(10^{-10})]$$

$$\text{pH} = -[(0.92) + (-10.00)] = 9.08$$

*Example 4–2.* Calculate the hydrogen ion concentration in each of two solutions, the respective pH values of which are 4.76 and 7.28.

If we take the antilogarithm of the relation, pH = $-\log[H^+]$, the result is

$$[H^+] = 10^{-pH}$$

For the solution with a pH of 4.76, we have

$$[H^+] = 10^{-4.76} = 10^{0.24-5.00} = 10^{0.24} \times 10^{-5}$$

$$[H^+] = 1.74 \times 10^{-5}\ M$$

In the case of the second solution,

$$[H^+] = 10^{-7.28} = 10^{0.72-8.00} = 10^{0.72} \times 10^{-8}$$

$$[H^+] = 5.25 \times 10^{-8}\ M$$

*Example 4–3.* Calculate the hydrogen ion concentration in a solution whose pOH is 4.57.

Using the relationship previously mentioned,

$$pH + pOH = 14.00$$

we can obtain the pH as follows:

$$pH = 14.00 - pOH = 14.00 - 4.57 = 9.43$$

Then,

$$[H^+] = 10^{-9.43} = 10^{0.57} \times 10^{-10} = 3.7 \times 10^{-10} M$$

In pure water, or in an aqueous medium free from any acidic or basic solutes, autoprotolysis leads to the formation of equal concentrations of hydrogen ion and hydroxide ion; that is, $[H^+] = [OH^-]$. Substitution of this condition into the ion-product expression gives

$$[H^+]^2 = [OH^-]^2 = K_w = 1.00 \times 10^{-14}$$

$$[H^+] = [OH^-] = 1.00 \times 10^{-7}\ M$$

$$pH = pOH = -\log(1.00 \times 10^{-7}) = 7.00$$

At 25°C an aqueous solution in which the concentrations of hydrogen ion and hydroxide ion are both $1.00 \times 10^{-7}$ *M*, and for which the pH is 7.00, is stated to be *neutral.* However, if $[H^+]$ exceeds $[OH^-]$, the hydrogen ion concentration is larger than $1.00 \times 10^{-7}$ *M*, the pH is below 7.00, and the solution is *acidic.* On the other hand, the solution is *basic* or *alkaline,* the hydrogen ion concentration is smaller than $1.00 \times 10^{-7}$ *M*, and the pH is above 7.00, if $[H^+]$ is less than $[OH^-]$.

**Strengths of Acids and Bases in Water.** In aqueous solutions, the strength of an acid is determined by the extent to which it loses protons to water, whereas the strength of a base is governed by how completely it captures protons from water. Quantitatively, the strength of an acid or base is measured by the equilibrium constant for its proton-transfer reaction with the solvent – the larger the equilibrium constant, the greater the strength of the acid or base. In writing these equilibrium constants, one customarily uses $K_a$ for acids and $K_b$ for bases.

In water, the hydronium ion (the conjugate acid of the solvent) is the strongest acidic species that can exist and remain stable. Any acid stronger than the hydronium ion reacts completely with the solvent (acting as a Brønsted-Lowry base) to yield essentially stoichiometric quantities of $H_3O^+$ and the conjugate base of the original acid. Perchloric ($HClO_4$), sulfuric ($H_2SO_4$),* hydriodic (HI), nitric ($HNO_3$), hydrobromic (HBr), and hydrochloric (HCl) acids exhibit this kind of behavior. Hydrogen chloride dissolved in water

$$HCl + H_2O \rightleftharpoons Cl^- + H_3O^+; \qquad K_a \gg 1$$

is representative of these strong acids. Donation of a proton by HCl to $H_2O$ is far more successful than the donation of a proton by $H_3O^+$ to $Cl^-$. Thus, the position of equilibrium for the above proton-transfer reaction lies well toward the right ($K_a \gg 1$), and we say that HCl is a very strong acid in water.

Most acidic species are much weaker in an aqueous medium than the strong acids mentioned in the last paragraph. For example, the proton-donating ability of acetic acid is much less than that of the hydronium ion, as reflected by the small equilibrium constant for the reaction

$$CH_3COOH + H_2O \rightleftharpoons CH_3COO^- + H_3O^+; \qquad K_a = 1.75 \times 10^{-5}$$

In other words, acetate ion (the conjugate base of the parent acid) is a better proton acceptor than water; therefore, the proton stays with the acetate ion, and formation of $H_3O^+$ occurs only to a slight extent. Ammonium ion is even weaker than acetic acid,

$$NH_4^+ + H_2O \rightleftharpoons NH_3 + H_3O^+; \qquad K_a = 5.55 \times 10^{-10}$$

the equilibrium constant ($K_a$) for its reaction with water being more than 30,000 times smaller than that for the acetic acid-water system. Clearly, $NH_3$ holds onto a proton much better than either acetate ion or water. Finally, the hydrogen sulfide anion, $HS^-$ (which is the conjugate base of the weak acid $H_2S$), is an exceedingly feeble acid, showing almost no tendency at all to transfer a proton to water:

$$HS^- + H_2O \rightleftharpoons S^{2-} + H_3O^+; \qquad K_a = 1.1 \times 10^{-15}$$

Thus, sulfide ion is an excellent proton acceptor compared to water as well as ammonia and acetate ion.

Two conclusions emerge from the preceding discussion. First, the family of Brønsted-Lowry acids includes cationic ($NH_4^+$), anionic ($HS^-$), and uncharged ($CH_3COOH$) members. Second, the strength of a solute acid (HB) is fundamentally determined by competition for protons between the solvent and the conjugate base ($B^-$) of the solute acid, as shown by the generalized equilibrium

$$HB + H_2O \rightleftharpoons B^- + H_3O^+$$

---

*Although sulfuric acid is diprotic, only the loss of the first proton to water occurs completely to yield $HSO_4^-$ and $H_3O^+$; the second proton is transferred to water with much more difficulty, inasmuch as it is necessary to separate a positively charged proton ($H^+$) from sulfate anion ($SO_4^{2-}$), a stronger Brønsted-Lowry base than $HSO_4^-$.

If water is a much better proton acceptor than $B^-$, the proton of HB will be donated to $H_2O$, and HB will appear to be a strong or completely dissociated acid. As the proton-accepting ability of $B^-$ increases, and eventually surpasses that of $H_2O$, the acid strength (dissociation) of HB decreases because the proton is retained rather than donated to the weaker base $H_2O$.

A general consequence of the Brønsted-Lowry concept of acid-base behavior is that, as the proton-donating ability of an acid *increases,* the proton-accepting tendency of its conjugate base *decreases.* Thus, the conjugate base ($Cl^-$) of a strong acid (HCl) is an exceedingly weak proton acceptor. Conversely, the conjugate base ($S^{2-}$) of an extremely weak acid ($HS^-$) is a very strong proton acceptor. It follows that, if the relative proton-donating abilities of a series of Brønsted-Lowry acids are known, we can immediately list the relative proton-accepting abilities of the corresponding conjugate bases. Since the relative strengths of the four acids discussed in the second-preceding paragraph are $HCl > CH_3COOH > NH_4^+ > HS^-$, the relative strengths of the conjugate bases must be $Cl^- < CH_3COO^- < NH_3 < S^{2-}$. This predicted order of relative base strength can be confirmed by inspection of the $K_b$ value for the proton-transfer reaction between water and each of the four bases:

$$Cl^- + H_2O \rightleftharpoons HCl + OH^-; \qquad K_b \text{ is immeasurably small}$$

$$CH_3COO^- + H_2O \rightleftharpoons CH_3COOH + OH^-; \qquad K_b = 5.71 \times 10^{-10}$$

$$NH_3 + H_2O \rightleftharpoons NH_4^+ + OH^-; \qquad K_b = 1.80 \times 10^{-5}$$

$$S^{2-} + H_2O \rightleftharpoons HS^- + OH^-; \qquad K_b = 9.1$$

Sulfide ion is an especially strong base; for, if one attempts to prepare a solution containing $1.0\,M$ sulfide ion by dissolving one formula weight of sodium sulfide ($Na_2S$) in a liter of water, most of the sulfide reacts with the solvent to produce $HS^-$ and $OH^-$, only $0.09\,M$ sulfide ion remaining unprotonated at equilibrium. Similar efforts to obtain a $0.01\,M$ sulfide ion solution result in almost complete transfer of protons from water to sulfide ions, since the equilibrium sulfide concentration is but $1.1 \times 10^{-5}\,M$. Dissolution in water of sodium oxide ($Na_2O$), the salt of a much stronger base ($O^{2-}$) than sulfide or even hydroxide ion, is a violently exothermic process leading to the quantitative conversion of oxide ion to hydroxide ion:

$$O^{2-} + H_2O \rightleftharpoons OH^- + OH^-; \qquad K_b \gg 1$$

Thus, the strongest Brønsted-Lowry base capable of stable existence in an aqueous medium is the hydrated hydroxide ion – the conjugate base of the solvent.

**Relation Between $K_a$ and $K_b$ for Conjugate Acid-Base Pairs.** A fundamental quantitative relationship exists between the equilibrium constant ($K_a$) for the proton-transfer reaction between water and a Brønsted-Lowry acid, and the equilibrium constant ($K_b$) for the proton-transfer reaction between water and the conjugate base of that acid. For example, if we consider acetic acid and its conjugate base, acetate ion, the appropriate equilibria are

$$CH_3COOH + H_2O \rightleftharpoons CH_3COO^- + H_3O^+; \qquad K_a = \frac{[CH_3COO^-][H^+]}{[CH_3COOH]}$$

(where $[H^+]$ symbolizes the concentration of the hydronium ion) and

$$CH_3COO^- + H_2O \rightleftharpoons CH_3COOH + OH^-; \qquad K_b = \frac{[CH_3COOH][OH^-]}{[CH_3COO^-]}$$

If the equilibrium expressions for $K_a$ and $K_b$ are multiplied together, the result is

$$K_a \cdot K_b = \frac{[CH_3COO^-][H^+]}{[CH_3COOH]} \cdot \frac{[CH_3COOH][OH^-]}{[CH_3COO^-]} = [H^+][OH^-]$$

and, since the product of the hydrogen ion and hydroxide ion concentrations is the ion-product constant ($K_w$) for water, we reach the significant and useful conclusion that

$$K_a K_b = K_w$$

This expression is applicable to any Brønsted-Lowry conjugate acid-base pair; if $K_a$ for an acid is available, the value of $K_b$ for its conjugate base can be computed, and, if $K_b$ for a base is known, $K_a$ for its conjugate acid is readily obtainable.

**Polyprotic Acids and Their Conjugate Bases.** Thus far, we have considered acids which possess only one transferable proton – that is, **monoprotic acids** – and bases which can accept only one proton.* There is an important class of acids, called **polyprotic acids**, which have two or more ionizable protons. Included in the list of polyprotic acids are phosphoric acid ($H_3PO_4$), arsenic acid ($H_3AsO_4$), oxalic acid ($H_2C_2O_4$), carbonic acid ($H_2CO_3$), malonic acid ($HOOCCH_2COOH$), tartaric acid ($HOOC(CHOH)_2COOH$), and glycinium ion ($^+NH_3CH_2COOH$) – the latter being the conjugate acid of glycine, one of the amino acids.

Phosphoric acid typifies the behavior of these substances, undergoing three stepwise dissociations as indicated by the following equilibria:

$$H_3PO_4 + H_2O \rightleftharpoons H_2PO_4^- + H_3O^+; \qquad K_{a1} = 7.5 \times 10^{-3}$$

$$H_2PO_4^- + H_2O \rightleftharpoons HPO_4^{2-} + H_3O^+; \qquad K_{a2} = 6.2 \times 10^{-8}$$

$$HPO_4^{2-} + H_2O \rightleftharpoons PO_4^{3-} + H_3O^+; \qquad K_{a3} = 4.8 \times 10^{-13}$$

From the magnitude of the first equilibrium constant ($K_{a1}$), we see that phosphoric acid itself is a moderately strong acid, being much stronger than acetic acid but not so strong as hydrochloric acid. Dihydrogen phosphate ($H_2PO_4^-$) and hydrogen phosphate ($HPO_4^{2-}$) ions are increasingly weak acids, primarily because separation of a positively charged proton from a doubly charged anion ($HPO_4^{2-}$) and from a triply charged anion ($PO_4^{3-}$) in the second and third steps of dissociation, respectively, requires that a successively larger electrostatic force of attraction be overcome.

Conjugate bases of polyprotic acids (often in the form of solutions of alkali metal salts) are capable of accepting two or more protons. Thus, sulfide, carbonate, tartrate, malonate, hydrogen phosphate, hydrogen arsenate, and glycinate anions can each combine with two hydrogen ions, whereas phosphate and arsenate ions may react with up to three protons. Oxalate ion can interact with water (or some other Brønsted-Lowry acid) to form the hydrogen oxalate ion

$$C_2O_4^{2-} + H_2O \rightleftharpoons HC_2O_4^- + OH^-; \qquad K_{b1} = \frac{K_w}{K_{a2}} = 1.64 \times 10^{-10}$$

*Sulfide ion is the one exception, since two protons can be added to form $H_2S$.

which can accept another proton to become oxalic acid:

$$HC_2O_4^- + H_2O \rightleftharpoons H_2C_2O_4 + OH^-; \qquad K_{b2} = \frac{K_w}{K_{a1}} = 1.54 \times 10^{-13}$$

Equilibrium constants for these two processes indicate that neither $C_2O_4^{2-}$ nor $HC_2O_4^-$ is a particularly strong base, since the conjugate acids of these anions – $HC_2O_4^-$ and $H_2C_2O_4$, respectively – are both stronger than acetic acid.

**Amphiprotic Substances.** A species is **amphiprotic** if it can accept a proton from a Brønsted-Lowry acid or donate a proton to a Brønsted-Lowry base. Water is the classic example of such a substance, because it can act as either a proton donor (acid) or a proton acceptor (base). Moreover, many solutes display amphiprotic character in water.

Species derived from the ionization of a polyprotic acid are included among the list of amphiprotic substances – hydrogen carbonate or bicarbonate ($HCO_3^-$), hydrogen phosphate ($HPO_4^{2-}$), dihydrogen arsenate ($H_2AsO_4^-$), hydrogen sulfide ($HS^-$), hydrogen oxalate ($HC_2O_4^-$), hydrogen sulfite ($HSO_3^-$), dihydrogen phosphite ($H_2PO_3^-$), hydrogen tartrate ($HOOC(CHOH)_2COO^-$), and hydrogen malonate ($HOOCCH_2COO^-$). Thus, bicarbonate can behave as an acid

$$HCO_3^- + H_2O \rightleftharpoons CO_3^{2-} + H_3O^+; \qquad K_{a2} = 4.68 \times 10^{-11}$$

or as a base

$$HCO_3^- + H_2O \rightleftharpoons H_2CO_3 + OH^-; \qquad K_{b2} = \frac{K_w}{K_{a1}} = 2.24 \times 10^{-8}$$

but an aqueous solution of sodium bicarbonate ($NaHCO_3$) will be distinctly alkaline – its pH will be greater than 7 – because the equilibrium constant for the second reaction (which produces $OH^-$ ions) is larger than that for the first reaction (which yields $H_3O^+$ ions). A solution of sodium hydrogen oxalate ($NaHC_2O_4$) is acidic, because the reaction showing $HC_2O_4^-$ as a proton donor

$$HC_2O_4^- + H_2O \rightleftharpoons C_2O_4^{2-} + H_3O^+; \qquad K_{a2} = 6.1 \times 10^{-5}$$

has an equilibrium constant ($K_{a2}$) much greater than that ($K_{b2}$) for the equilibrium representing $HC_2O_4^-$ as a proton acceptor:

$$HC_2O_4^- + H_2O \rightleftharpoons H_2C_2O_4 + OH^-; \qquad K_{b2} = \frac{K_w}{K_{a1}} = 1.54 \times 10^{-13}$$

A substance such as ammonium acetate is amphiprotic in an aqueous medium, because ammonium ion is a Brønsted-Lowry acid

$$NH_4^+ + H_2O \rightleftharpoons NH_3 + H_3O^+; \qquad K_a = 5.55 \times 10^{-10}$$

and acetate ion is a Brønsted-Lowry base:

$$CH_3COO^- + H_2O \rightleftharpoons CH_3COOH + OH^-; \qquad K_b = 5.71 \times 10^{-10}$$

Solutions of ammonium acetate in water are just barely alkaline, because $K_b$ for acetate ion is slightly greater than $K_a$ for ammonium ion.

Another important group of amphiprotic compounds is the family of amino acids, which are notable because they exist predominantly as **zwitterions** or dipolar ions in aqueous media. Alanine is representative of these species:

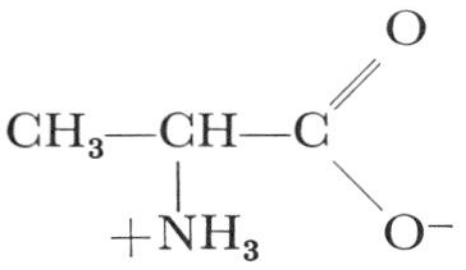

$$CH_3-CH(^{+}NH_3)-C(=O)O^-$$

zwitterion form of alanine

Alanine, as well as other amino acids, is amphiprotic because the zwitterion can be protonated or deprotonated according to the following equilibria:

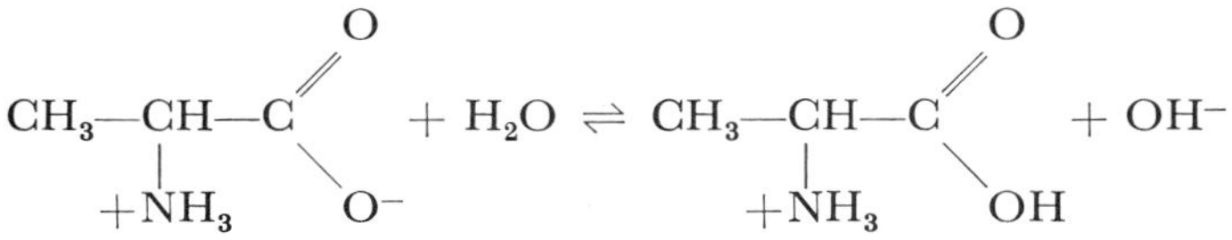

$$CH_3-CH(^{+}NH_3)-C(=O)O^- + H_2O \rightleftharpoons CH_3-CH(^{+}NH_3)-C(=O)OH + OH^-$$

conjugate acid of alanine

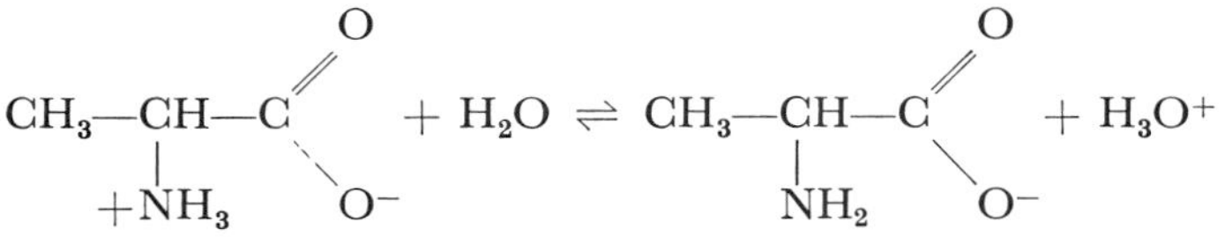

$$CH_3-CH(^{+}NH_3)-C(=O)O^- + H_2O \rightleftharpoons CH_3-CH(NH_2)-C(=O)O^- + H_3O^+$$

conjugate base of alanine

**Metal Cations as Brønsted-Lowry Acids.** All metal cations are hydrated in aqueous media, so it is realistic to speak about species such as $Ce(H_2O)_6^{4+}$, $Fe(H_2O)_6^{3+}$, $Cr(H_2O)_6^{3+}$, $Cu(H_2O)_4^{2+}$, $Mg(H_2O)_4^{2+}$, and $Na(H_2O)_4^{+}$ – although it is not always clear how many water molecules are coordinated to a particular metal ion. As a general rule, the acid strength of a hydrated metal ion *increases* with increasing charge and decreasing radius of the metal ion.

Singly positive metal cations such as $Ag(H_2O)_2^{+}$, $Na(H_2O)_4^{+}$, and $K(H_2O)_4^{+}$ are too weak as proton donors to display significant acidic properties in water, and so have no effect on the pH of an aqueous solution (aside from the secondary effect of ionic strength on the autoprotolysis of water). Cations of the alkaline earth elements – $Mg(H_2O)_4^{2+}$, $Ca(H_2O)_4^{2+}$, $Sr(H_2O)_4^{2+}$, and $Ba(H_2O)_4^{2+}$ – are essentially neutral and do not influence the pH of aqueous solutions. However, dipositive cations of the transition metals such as $Fe(H_2O)_6^{2+}$, $Cu(H_2O)_4^{2+}$, $Cd(H_2O)_4^{2+}$, $Zn(H_2O)_4^{2+}$, and $Ni(H_2O)_4^{2+}$ are roughly comparable in acid strength to the ammonium ion, so that proton-transfer reactions such as

$$Cu(H_2O)_4^{2+} + H_2O \rightleftharpoons Cu(H_2O)_3OH^+ + H_3O^+$$

occur to some extent.

Tripositive cations, especially those of the transition-metal series, show considerable acid strength in water. For example, $Fe(H_2O)_6{}^{3+}$ is a much stronger proton donor than acetic acid:

$$Fe(H_2O)_6{}^{3+} + H_2O \rightleftharpoons Fe(H_2O)_5OH^{2+} + H_3O^+; \qquad K_{a1} = 9.1 \times 10^{-4}$$

Other cations which show analogous behavior are $Cr(H_2O)_6{}^{3+}$, $Al(H_2O)_6{}^{3+}$, and $Bi(H_2O)_6{}^{3+}$. Quadripositive cations such as $Ti(H_2O)_6{}^{4+}$ and $Ce(H_2O)_6{}^{4+}$ are so acidic that it is practically impossible for these species to exist in aqueous solutions other than very concentrated sulfuric, nitric, and perchloric acid media. Therefore, proton-transfer processes such as

$$Ce(H_2O)_6{}^{4+} + H_2O \rightleftharpoons Ce(H_2O)_5OH^{3+} + H_3O^+$$

and

$$Ce(H_2O)_5OH^{3+} + H_2O \rightleftharpoons Ce(H_2O)_4(OH)_2{}^{2+} + H_3O^+$$

occur extensively, and precipitation of the hydrous oxides of these elements, $TiO_2 \cdot 2\,H_2O$ and $CeO_2 \cdot 2\,H_2O$, can easily take place in an acidic medium, as exemplified by the following reaction for titanium(IV):

$$Ti(H_2O)_6{}^{4+} + 4\,H_2O \rightleftharpoons TiO_2 \cdot 2\,H_2O + 4\,H_3O^+ + 2\,H_2O$$

**Neutral Species.** Dissolution of a number of inorganic salts in water does not influence the acidity or basicity of the resulting solution (other than an effect on the activity coefficients of hydrogen ion and hydroxide ion). Among these compounds are the chlorides, bromides, iodides, nitrates, sulfates, and perchlorates of lithium, sodium, and potassium – which consist of the anions of the very strong mineral acids and of the cations of very strong bases. An aqueous solution containing no solute other than one of these substances has a pH of essentially 7 at room temperature. Addition of such a salt to a solution already containing an acid or base does not alter the pH of that acid or base solution. Furthermore, although a cation or anion of one of these neutral substances often accompanies a weak acid ($NH_4Cl$) or a weak base ($NaCH_3COO$), the cation or anion has no effect on the pH of the solution.

## EQUILIBRIUM CALCULATIONS FOR SOLUTIONS OF ACIDS AND BASES IN WATER

This section of the chapter is intended to illustrate methods for calculating the pH of aqueous solutions containing monoprotic and polyprotic acids and their conjugate bases, amphiprotic substances, and buffers.

### Solutions of Strong Acids and Bases

A strong acid or base is one which is fully dissociated in an aqueous medium. Among the familiar strong acids are hydrochloric, hydrobromic, hydriodic, nitric, sulfuric, and perchloric acids, and the most common strong bases are the hydroxides of sodium, potassium, lithium, and barium.

**Strong Acid Solutions.** For a solution containing a strong monoprotic acid, the hydrogen ion concentration is equal to the original molar concentration of the acid. For example, a solution containing 0.1 mole of hydrogen chloride per liter is 0.1 $M$ in hydrogen (hydronium) ions and 0.1 $M$ in chloride ions.

*Example 4–4.* Calculate the pH of each of the following solutions: (a) 0.00150 $F$ HCl and (b) 1.000 liter of solution containing 0.1000 g of HCl.

(a) $[H^+] = 1.50 \times 10^{-3}\ M$

$$pH = -\log [H^+] = 2.82$$

(b) $$\text{Number of moles of HCl} = 0.1000 \text{ g} \times \frac{1 \text{ mole}}{36.46 \text{ g}}$$

$$= 0.00274 \text{ mole}$$

$$\text{Concentration of HCl} = 0.00274 \text{ mole/liter}$$

$$[H^+] = 2.74 \times 10^{-3}\ M$$

$$pH = -\log [H^+] = 2.56$$

**Strong Base Solutions.** In a strong base solution, the hydroxide ion concentration is directly related to the original concentration of the base in a manner analogous to the situation for a strong acid solution, and the pH can be calculated from the hydroxide ion concentration and the ion-product constant for water.

*Example 4–5.* Calculate the pH of a 0.0023 $F$ barium hydroxide solution. Since two moles of hydroxide ion are formed from the ionization of each gram formula weight of barium hydroxide originally present,

$$[OH^-] = 4.6 \times 10^{-3}\ M$$

$$pOH = -\log [OH^-] = 2.34$$

$$pH = 14.00 - pOH = 11.66$$

### Solutions of Weak Acids and Bases

Weak acids and bases are incompletely dissociated in aqueous solution. As a result, the hydrogen ion concentration or the hydroxide ion concentration is, respectively, always less than the original concentration of the weak acid or weak base. Calculation of the pH of a weak acid or weak base solution relies upon knowledge of the equilibrium constant ($K_a$ or $K_b$) for the proton-transfer reaction between the solute and water. Values of $K_a$ and $K_b$ for various species are tabulated in Appendix 2.

**Weak Acid Solutions.** Let us consider several procedures by which the pH of an aqueous solution of a monoprotic weak acid can be calculated. Acetic acid, formic acid, and the ammonium ion will serve as typical examples.

*Example 4–6.* Calculate the pH of a 0.100 $F$ acetic acid solution.

We must first identify the reactions which can produce hydrogen ions in this system, namely the dissociation of acetic acid

$$CH_3COOH \rightleftharpoons H^+ + CH_3COO^-; \qquad K_a = 1.75 \times 10^{-5}$$

and the autoprotolysis of water:

$$H_2O \rightleftharpoons H^+ + OH^-; \qquad K_w = 1.00 \times 10^{-14}$$

Each of these equilibria is an abbreviated form of the true acid-base reaction which occurs. Thus, for acetic acid we should show the transfer of a proton from acetic acid to water, whereas the autoprotolysis of water proceeds through the transfer of a proton from one water molecule to another. However, as long as we perform equilibrium calculations involving acid-base reactions in the *same* solvent – so that the same proton acceptor (water) is always involved – abbreviated versions of acid-base equilibria as well as the corresponding equilibrium expressions can be employed.

Although some hydrogen ions do come from the autoprotolysis of water, acetic acid is such a stronger acid that water is an insignificant source of hydrogen ions (except for extremely dilute acetic acid solutions). Furthermore, the hydrogen ions from acetic acid repress the autoprotolysis of water through the common ion effect. Accordingly, we will assume that the dissociation of acetic acid is the only important equilibrium

$$\frac{[H^+][CH_3COO^-]}{[CH_3COOH]} = K_a = 1.75 \times 10^{-5}$$

but it will be essential to justify the assumption after the calculations are completed. Each molecule of acetic acid which undergoes dissociation produces one hydrogen ion and one acetate ion. Therefore,

$$[H^+] = [CH_3COO^-]$$

At equilibrium the concentration of acetic acid is equal to the initial acid concentration (0.100 *F*) minus the amount dissociated (which is expressible by either the concentration of hydrogen ion or the concentration of acetate ion):

$$[CH_3COOH] = 0.100 - [H^+]$$

Substitution of the latter two relations into the equilibrium expression for acetic acid yields

$$\frac{[H^+]^2}{0.100 - [H^+]} = 1.75 \times 10^{-5}$$

There are at least two ways to solve this equation for the hydrogen ion concentration. Since this relation is a quadratic equation, the formula for the solution of a quadratic equation* can be employed, from which the answer is found to be $[H^+] = 1.32 \times$

---

*An equation of the type $aX^2 + bX + c = 0$ may be solved by means of the quadratic formula, which is

$$X = \frac{-b \pm \sqrt{b^2 - 4ac}}{2a}$$

In the present example, the equilibrium equation may be rearranged to $[H^+]^2 + 1.75 \times 10^{-5}[H^+] - 1.75 \times 10^{-6} = 0$. Applying the quadratic formula, we get

$$[H^+] = \frac{-(1.75 \times 10^{-5}) \pm \sqrt{(1.75 \times 10^{-5})^2 - 4(1)(-1.75 \times 10^{-6})}}{2} = 1.32 \times 10^{-3}\,M$$

(The other root is negative, an impossible situation.)

$10^{-3}$ $M$ or pH = 2.88. There is, however, an alternate, simpler procedure for solving the previous equation for $[H^+]$.

Let us assume that the equilibrium concentration of acetic acid is 0.100 $M$. In effect, we are saying that the value of $[H^+]$ is negligible in comparison to 0.100 $M$ or, in other words, that very nearly all the acetic acid remains undissociated. Since a weak acid is, by definition, one that is only slightly dissociated, this assumption might be expected to be reasonable. Ultimately, the validity of this assumption in any particular case must be demonstrated as shown below. On the basis of the approximation that

$$[CH_3COOH] = 0.100\ M$$

the equilibrium expression for acetic acid takes the form

$$\frac{[H^+]^2}{0.100} = 1.75 \times 10^{-5}$$

which can be solved as follows:

$$[H^+]^2 = 1.75 \times 10^{-6}$$

$$[H^+] = 1.32 \times 10^{-3}\ M; \qquad pH = 2.88$$

Results obtained by means of the quadratic formula and by the second, approximate approach are identical to the third significant figure. Notice that the hydrogen ion concentration ($1.32 \times 10^{-3}$ $M$) is only 1.3 per cent of the original concentration of acetic acid (0.100 $F$). Therefore, the actual equilibrium acetic acid concentration would be

$$[CH_3COOH] = 0.100 - 0.00132 = 0.099\ M$$

which is not significantly different from the assumed approximate value of 0.100 $M$. In general, if the original concentration of weak, monoprotic acid is at least 10,000 times greater than $K_a$, the approximate method will be accurate to within about 1 per cent. Furthermore, if the original acid concentration is only 1000 times larger than $K_a$, the approximate calculation will be in error by less than 2 per cent, which is an acceptable uncertainty for most purposes.

Another assumption made at the beginning of this problem was that the contribution of the autoprotolysis of water to the hydrogen ion concentration is negligible. If the hydrogen ion concentration is taken to be $1.32 \times 10^{-3}$ $M$, the value of $[OH^-]$ can be obtained from the relation

$$[OH^-] = \frac{K_w}{[H^+]} = \frac{1.00 \times 10^{-14}}{1.32 \times 10^{-3}} = 7.5 \times 10^{-12}\ M$$

However, since water is the only source of hydroxide ion in this system and since one hydrogen ion and one hydroxide ion are formed for each water molecule ionized, the $[H^+]$ furnished by water is only $7.5 \times 10^{-12}$ $M$.

One other technique used to solve equilibrium problems is called the **method of successive approximations.** We shall now use this procedure to consider the dissociation of formic acid.

*Example 4–7.* Calculate the pH of a 0.0250 $F$ formic acid solution.

As in the case of acetic acid, the important source of hydrogen ions is the dissociation of the weak acid:

$$HCOOH \rightleftharpoons H^+ + HCOO^-; \qquad K_a = 1.76 \times 10^{-4}$$

Accordingly, we can set the concentrations of hydrogen ion and formate ion equal to each other:

$$[H^+] = [HCOO^-]$$

At equilibrium, the concentration of undissociated formic acid is

$$[HCOOH] = 0.0250 - [H^+]$$

and the complete equilibrium expression for the dissociation of formic acid takes the form

$$\frac{[H^+][HCOO^-]}{[HCOOH]} = \frac{[H^+]^2}{0.0250 - [H^+]} = 1.76 \times 10^{-4}$$

To solve this equation by the method of successive approximations, we first ignore tentatively the $[H^+]$ term in the denominator of the equation (in spite of the fact that the initial formic acid concentration is only about 440 times larger than $K_a$). Therefore, we obtain

$$\frac{[H^+]^2}{0.0250} = 1.76 \times 10^{-4}$$

$$[H^+]^2 = 4.40 \times 10^{-6}$$

$$[H^+] = 2.10 \times 10^{-3}\ M$$

Obviously, this hydrogen ion concentration is hardly negligible in comparison to the initial formic acid concentration; but, instead of abandoning this approach and resorting to the quadratic formula, let us continue. A closer approximation to the formic acid concentration would be to subtract the $[H^+]$ from the original concentration of formic acid, *i.e.*,

$$[HCOOH] = 0.0250 - 0.0021 = 0.0229\ M$$

This new value for the concentration of formic acid is now substituted back into the equilibrium expression and the equation solved once again for $[H^+]$:

$$\frac{[H^+]^2}{0.0229} = 1.76 \times 10^{-4}$$

$$[H^+]^2 = 4.03 \times 10^{-6}$$

$$[H^+] = 2.01 \times 10^{-3}\ M$$

Through this method of successive approximations, the second approximation to the formic acid concentration (0.0229 $M$) has led to a self-consistent set of values for the concentrations of hydrogen ion and formic acid. In other words, if the formic acid concentration is assumed to be 0.0229 $M$ at equilibrium, the calculated $[H^+]$ is correct because, when the hydrogen ion concentration of 0.00201 $M$ is subtracted from the original formic acid concentration of 0.0250 $M$, the result (0.0230 $M$) agrees with the *assumed* value for [HCOOH]. Therefore, the value for $[H^+]$ of $2.01 \times 10^{-3}$ $M$ may be accepted as an accurate result, and the pH is 2.70.

*Example 4–8.* Calculate the pH of a 0.100 $F$ ammonium chloride solution.

Again, there are two sources of hydrogen ions. One is the autoprotolysis of water,

$$H_2O \rightleftharpoons H^+ + OH^-$$

whereas the second source is the dissociation of the ammonium ion,

$$NH_4^+ + H_2O \rightleftharpoons NH_3 + H_3O^+; \qquad K_a = 5.55 \times 10^{-10}$$

which may be represented in abbreviated form as follows:

$$NH_4^+ \rightleftharpoons NH_3 + H^+$$

However, the dissociation constant for ammonium ion ($K_a$) is more than 50,000 times greater than the ion-product constant for water ($K_w$), so the autoprotolysis of water should not contribute significantly to the hydrogen ion concentration. Dissociation of the ammonium ion produces stoichiometrically equal concentrations of $NH_3$ and $H^+$; therefore, neglect of the autoprotolysis of water leads to the relation

$$[NH_3] = [H^+]$$

In addition, ammonium ion is such a weak acid that we may safely approximate its equilibrium concentration as 0.100 $M$.

To calculate the concentration of hydrogen ion, and then the pH, it is only necessary to substitute the relations deduced from the previous arguments into the equilibrium expression:

$$\frac{[NH_3][H^+]}{[NH_4^+]} = \frac{[H^+]^2}{0.100} = 5.55 \times 10^{-10}$$

Solution of this equation gives the results

$$[H^+]^2 = 5.55 \times 10^{-11}$$
$$[H^+] = 7.45 \times 10^{-6}\ M$$
$$pH = 5.13$$

Note that our neglect of the autoprotolysis of water was justified; for, if we accept the hydrogen ion concentration as $7.45 \times 10^{-6}$ $M$, then

$$[OH^-] = \frac{K_w}{[H^+]} = \frac{1.00 \times 10^{-14}}{7.45 \times 10^{-6}} = 1.34 \times 10^{-9}\ M$$

which is equivalent to the quantity of hydrogen ion furnished by the autoprotolysis of water. Thus, the hydroxide ion concentration is indeed negligible compared to the concentration of hydrogen ion arising from dissociation of ammonium ions. Furthermore, the true equilibrium concentration of ammonium ion is not exactly 0.100 $M$, but is smaller by an amount essentially equal to the hydrogen ion concentration ($7.45 \times 10^{-6}$ $M$); however, the difference is negligible.

**Weak Base Solutions.** Equilibrium calculations involving solutions of weak bases differ little from those we have already considered. Pyridine and benzoate ion are representative of the behavior of Brønsted-Lowry bases in water.

*Example 4–9.* Calculate the pH of an aqueous solution of 0.0150 $F$ pyridine.

In problems concerned with solutions of weak bases in water, we must identify the sources of hydroxide ion. Most important is the proton-transfer reaction between water and pyridine to yield the pyridinium ion and hydroxide ion:

$$C_5H_5N + H_2O \rightleftharpoons C_5H_5NH^+ + OH^- ; \qquad K_b = 1.70 \times 10^{-9}$$

A second source of hydroxide ions is the autoprotolysis of water:

$$H_2O \rightleftharpoons H^+ + OH^-; \qquad K_w = 1.00 \times 10^{-14}$$

Although pyridine is a very weak base, it is still sufficiently stronger as a base than water that we can consider pyridine the only important source of hydroxide ions. Neglecting the autoprotolysis of water, we may equate the concentrations of the pyridinium and hydroxide ions formed from the pyridine-water reaction:

$$[C_5H_5NH^+] = [OH^-]$$

At equilibrium the concentration of pyridine is given by

$$[C_5H_5N] = 0.0150 - [OH^-]$$

Substitution of the latter two relations into the equilibrium expression gives

$$\frac{[C_5H_5NH^+][OH^-]}{[C_5H_5N]} = \frac{[OH^-]^2}{0.0150 - [OH^-]} = 1.70 \times 10^{-9}$$

This equation can be readily solved if we provisionally let the equilibrium concentration of pyridine be 0.0150 $M$. Such an approximation seems reasonable in view of the weakness of pyridine as a base and the relatively high concentration of pyridine. Therefore,

$$\frac{[OH^-]^2}{0.0150} = 1.70 \times 10^{-9}$$

$$[OH^-]^2 = 2.55 \times 10^{-11}; \qquad [OH^-] = 5.05 \times 10^{-6}\ M$$

$$pOH = -\log [OH^-] = 5.30$$

$$pH = 14.00 - pOH = 8.70$$

Let us check our approximations. Is the autoprotolysis of water a negligible source of hydroxide ions? *Yes*; the concentration of hydroxide ion from water is equal to the hydrogen ion concentration

$$[H^+] = \frac{K_w}{[OH^-]} = \frac{1.00 \times 10^{-14}}{5.05 \times 10^{-6}} = 1.98 \times 10^{-9}\ M$$

since water is the only proton source in this system. Is the equilibrium concentration of pyridine essentially 0.0150 *M*? *Yes*; the amount of pyridine which reacts is equivalent to the concentration of hydroxide ion, and $5.05 \times 10^{-6}$ *M* can be neglected in comparison to 0.0150 *M*.

We shall next consider the situation of a very weak base, present at a low concentration.

*Example 4–10.* Calculate the pH of a $2.00 \times 10^{-5}$ *F* sodium benzoate solution.

Although sodium ion is a neutral species, benzoate anion is a very weak base, and the pertinent proton-transfer process can be written as

$$C_6H_5COO^- + H_2O \rightleftharpoons C_6H_5COOH + OH^- \ ; \ K_b = 1.59 \times 10^{-10}$$

Once again, the autoprotolysis of water can provide hydroxide ions:

$$H_2O \rightleftharpoons H^+ + OH^-; \qquad K_w = 1.00 \times 10^{-14}$$

Let us attempt to calculate the pH of the sodium benzoate solution according to the method followed in the preceding problem (Example 4–9), by neglecting the autoprotolysis of water and by assuming that the equilibrium concentration of benzoate anion is $2.00 \times 10^{-5}$ *M*. Thus,

$$\frac{[C_6H_5COOH][OH^-]}{[C_6H_5COO^-]} = \frac{[OH^-]^2}{2.00 \times 10^{-5}} = 1.59 \times 10^{-10}$$

$$[OH^-]^2 = 3.18 \times 10^{-15}; [OH^-] = 5.64 \times 10^{-8}\ M; \text{pOH} = 7.25; \text{pH} = 6.75$$

These results make no sense at all. If benzoate anion were absent and we had pure water, we would expect $[OH^-] = 1.00 \times 10^{-7}$ *M* and pOH = 7.00. However, a small concentration of benzoate ion is present; so, although benzoate is a very weak base, the solution must be very slightly *alkaline,* not acidic as the approximate calculations above have suggested. Consequently, we cannot ignore the autoprotolysis of water.

In order to solve this problem correctly, we must note that one $H^+$ ion and one $OH^-$ ion are formed for each molecule of water which ionizes, and that each benzoate ion which reacts with water results in the production of one $C_6H_5COOH$ molecule and one $OH^-$ ion. Therefore, the true hydroxide ion concentration is the *sum* of the concentrations of benzoic acid and hydrogen ion:

$$[OH^-] = [C_6H_5COOH] + [H^+]$$

Upon rearrangement, this expression becomes

$$[C_6H_5COOH] = [OH^-] - [H^+]$$

Note that the benzoic acid concentration does *not* equal the hydroxide ion concentration as we had assumed previously when the autoprotolysis of water was neglected. Using the ion-product expression for water, we can rewrite the last equation as

$$[C_6H_5COOH] = [OH^-] - \frac{K_w}{[OH^-]}$$

If the latter relation for the benzoic acid concentration is substituted into the equilibrium expression for the benzoate-water reaction and if, because $K_b$ is so small, the approximation is retained that the equilibrium concentration of benzoate ion is $2.00 \times 10^{-5}$ *M,* we have

$$\frac{[C_6H_5COOH][OH^-]}{[C_6H_5COO^-]} = \frac{\left([OH^-] - \frac{K_w}{[OH^-]}\right)[OH^-]}{2.00 \times 10^{-5}} = 1.59 \times 10^{-10}$$

$$[OH^-]^2 - K_w = 3.18 \times 10^{-15}$$

$$[OH^-]^2 = 1.32 \times 10^{-14}$$

$$[OH^-] = 1.15 \times 10^{-7}\,M; \qquad pOH = 6.94$$

$$[H^+] = \frac{K_w}{[OH^-]} = \frac{1.00 \times 10^{-14}}{1.15 \times 10^{-7}} = 8.70 \times 10^{-8}\,M; \qquad pH = 7.06$$

To check the validity of the approximation regarding the benzoate concentration at equilibrium, we can compute the concentration of benzoic acid and compare the result to $2.00 \times 10^{-5}$ *M.* Earlier it was deduced that

$$[C_6H_5COOH] = [OH^-] - [H^+]$$

Substitution of the values for $[OH^-]$ and $[H^+]$ into this expression yields

$$[C_6H_5COOH] = (1.15 \times 10^{-7}) - (8.70 \times 10^{-8}) = 2.8 \times 10^{-8}\,M$$

Thus, only 0.14 per cent of the original quantity of benzoate ion reacted with water to form benzoic acid, so our calculations are reliable.

### Buffer Solutions

A solution consisting of a weak acid along with its conjugate base is called a **buffer solution** because it resists pH changes when diluted (or perhaps concentrated) or when various amounts of acid or base are added. Examples of buffer solutions include mixtures of acetic acid and sodium acetate, ammonium nitrate and ammonia, pyridinium chloride and pyridine, and the sodium salts of dihydrogen phosphate and monohydrogen phosphate, the latter being one of the principal buffers in human blood.

In the discussions which follow, we will calculate pH values for two representative buffer solutions, and we will determine what changes occur in the pH of a buffer upon dilution or upon addition of other acids and bases.

*Example 4–11.* Calculate the pH of a solution initially 0.100 $F$ in acetic acid and 0.100 $F$ in sodium acetate.

For the proton transfer between acetic acid and water, we can write abbreviated versions for both the reaction

$$CH_3COOH \rightleftharpoons CH_3COO^- + H^+ \;; \qquad K_a = 1.75 \times 10^{-5}$$

and the equilibrium expression:

$$\frac{[H^+][CH_3COO^-]}{[CH_3COOH]} = 1.75 \times 10^{-5}$$

To determine the hydrogen ion concentration from the latter equation, and thus the pH, we must obtain the acetate and acetic acid concentrations. Because acetic acid is a much stronger acid than water and is present in relatively large amount, we can ignore the autoprotolysis of water as a source of hydrogen ions. Therefore, the equilibrium acetic acid concentration is

$$[CH_3COOH] = 0.100 - [H^+]$$

At equilibrium, the concentration of acetate ion is the sum of the acetate added in the form of sodium acetate (0.100 *M*) and the acetate formed from the dissociation of acetic acid (given by the hydrogen ion concentration):

$$[CH_3COO^-] = 0.100 + [H^+]$$

Substitution of these relationships into the equilibrium expression gives the following equation:

$$\frac{[H^+](0.100 + [H^+])}{0.100 - [H^+]} = 1.75 \times 10^{-5}$$

It is reasonable to expect that the extent of dissociation of acetic acid (represented by $[H^+]$) is small compared to the original acid concentration because the concentration of acetic acid is much larger than $K_a$ and because the acetate ion furnished by the sodium acetate represses the dissociation of acetic acid. Accordingly, the above equation can be simplified to

$$\frac{[H^+](0.100)}{0.100} = 1.75 \times 10^{-5}$$

from which it is obvious that $[H^+] = 1.75 \times 10^{-5}$ *M*. Our assumption that $[H^+]$ is small in comparison to 0.100 *M* is wel' justified, and the pH of the solution is 4.76.

*Example 4–12.* Calculate the pH of 100 ml of a solution containing 0.0100 mole of ammonium nitrate and 0.0200 mole of ammonia.*

Ammonia, which can accept a proton from the solvent,

$$NH_3 + H_2O \rightleftharpoons NH_4^+ + OH^-$$

is a stronger base than water, so the autoprotolysis of water does not contribute appreciably tc the hydroxide ion concentration. We can solve the equilibrium expression

$$\frac{[NH_4^+][OH^-]}{[NH_3]} = K_b = 1.80 \times 10^{-5}$$

*In an aqu:ous medium, dissolved ammonia exists as a hydrated species, conceivably with four water molecules hydrogen bonded to it – the oxygen atoms of three $H_2O$ molecules bonded to the three hydrogens of $NH_3$, with a fourth $H_2O$ molecule sharing one of its hydrogen atoms with the lone electron pair on nitrogen.

for $[OH^-]$ after appropriate substitutions for $[NH_4^+]$ and $[NH_3]$ are made. From the information given, the initial concentrations of ammonium ion and ammonia are 0.100 *M* and 0.200 *M*, respectively. After equilibrium is established, the ammonium ion concentration is greater than 0.100 *M* by the amount of $NH_4^+$ formed from ammonia (which is measured in this case by $[OH^-]$)

$$[NH_4^+] = 0.100 + [OH^-]$$

and the concentration of ammonia is

$$[NH_3] = 0.200 - [OH^-]$$

However, as in the previous example of the acetic acid-sodium acetate buffer, these corrections to the ammonium ion and ammonia concentrations are insignificant, so the equilibrium expression may be written

$$\frac{(0.100)[OH^-]}{0.200} = 1.80 \times 10^{-5}$$

Therefore, the solution to this problem is as follows:

$$[OH^-] = 3.60 \times 10^{-5}\ M$$

$$pOH = 4.44$$

$$pH = 9.56$$

**Effect of Dilution.** Let us recall the problem discussed in Example 4–11, in which the pH of a solution 0.100 *F* in both acetic acid and sodium acetate was calculated. We represented the acetic acid concentration at equilibrium as

$$[CH_3COOH] = 0.100 - [H^+]$$

and the acetate ion concentration as

$$[CH_3COO^-] = 0.100 + [H^+]$$

but we assumed and later justified that the term $[H^+]$ in each of these relations was negligibly small, being $1.75 \times 10^{-5}$ *M*. What would happen to the pH if the solution were diluted one-hundred-fold, that is, if the original concentrations of acetic acid and sodium acetate each became 0.00100 *F*?

In this new problem, the rigorous expressions for the acetic acid and acetate ion concentrations would be as follows:

$$[CH_3COOH] = 0.00100 - [H^+]$$

$$[CH_3COO^-] = 0.00100 + [H^+]$$

If the $[H^+]$ term in each relation were negligible compared to 0.00100 *M*, we could assert that no pH change would be observed. This statement must be tested. If the hydrogen ion concentration is taken to be $1.75 \times 10^{-5}$ *M*, we discover that the latter rigorous relations yield the results

$$[CH_3COOH] = 0.00100 - 0.00002 = 0.00098\ M$$

$$[CH_3COO^-] = 0.00100 + 0.00002 = 0.00102\ M$$

provided that we round off the hydrogen ion concentration to $2 \times 10^{-5}$ *M* for these calculations. What we can now do is substitute these new values for $[CH_3COOH]$ and $[CH_3COO^-]$ into the equilibrium expression

$$\frac{[H^+][CH_3COO^-]}{[CH_3COOH]} = 1.75 \times 10^{-5}$$

and calculate a new value for the hydrogen ion concentration:

$$\frac{[H^+](0.00102)}{(0.00098)} = 1.75 \times 10^{-5}$$

$$[H^+] = 1.68 \times 10^{-5}\ M; \qquad pH = 4.77$$

This result must now be carefully interpreted.

First, the calculation is valid because the values of $[CH_3COOH]$ and $[CH_3COO^-]$ are essentially the same (0.00098 and 0.00102 *M*, respectively) regardless of which hydrogen ion concentration is employed to correct the original concentrations (0.00100 *M*) of the reagents. However, the difference in the hydrogen ion concentration is significant in two ways. Our calculations have shown that the pH of a buffer solution composed of 0.1 *M* reagents is 4.76, and that dilution by a factor of 100 causes the pH to increase only to 4.77. This result reveals how well a buffer solution does stabilize the pH even upon great dilution. Yet the change in pH cautions against indiscriminate dilution. One can foresee considerable difficulty if the buffer is so extensively diluted that the concentrations of acetic acid and acetate ion approach the value of $K_a$.

**Effect of Adding Acid or Base.** Let us assume that we have exactly 100 ml of a buffer solution which is 0.100 *F* in both acetic acid and sodium acetate. In Example 4–11, we found that the hydrogen ion concentration was $1.75 \times 10^{-5}$ *M*, corresponding to a pH of 4.76.

*Example 4–13.* Calculate the pH of the solution which results from the addition of 10.0 ml of 0.100 *F* HCl to 100 ml of the buffer solution described in Example 4–11.

First, we note that the position of equilibrium for the reaction

$$CH_3COOH \rightleftharpoons H^+ + CH_3COO^-$$

will be shifted toward the left by the addition of the hydrochloric acid. In this case the large excess of acetate ions can consume the hydrochloric acid through the formation of the much weaker acetic acid.

We can calculate the hydrogen ion concentration from the expression

$$\frac{[H^+][CH_3COO^-]}{[CH_3COOH]} = 1.75 \times 10^{-5}$$

In 100 ml of the buffer solution, there are initially (100)(0.100) or 10.0 millimoles of acetic acid and 10.0 millimoles of acetate ion. Addition of 10.0 ml of 0.100 *F* hydrochloric acid is equivalent to 1.00 millimole of strong acid, which we can assume reacts completely with acetate ion to form acetic acid. Therefore, after this reaction, 11.0 millimoles of acetic acid are present and 9.0 millimoles of acetate remain. Since we are dealing with the *ratio* of two concentrations, $[CH_3COOH]$ and $[CH_3COO^-]$, it is unnecessary to convert the number of millimoles of each species into molar concentration. Therefore, substitution of these values into the equilibrium expression yields

$$\frac{[H^+](9.0)}{11.0} = 1.75 \times 10^{-5}$$

from which we calculate that $[H^+] = 2.14 \times 10^{-5}$ *M* and pH = 4.67.

*Example 4–14.* Calculate the pH of the solution which results from the addition of 10.0 ml of 0.100 *F* NaOH to 100 ml of the buffer solution described in Example 4–11.

In this situation the principal equilibrium

$$CH_3COOH \rightleftharpoons H^+ + CH_3COO^-$$

will be shifted toward the right because the addition of strong base (NaOH) will consume or neutralize some of the acetic acid with the formation of an equivalent quantity of acetate ion.

Addition of 10.0 ml of a 0.100 *F* sodium hydroxide solution to the 100-ml sample of the buffer introduces 1.00 millimole of strong base which reacts with a stoichiometric quantity of acetic acid. Consequently, at equilibrium only 9.0 millimoles of acetic acid remain, whereas a total of 11.0 millimoles of acetate ion is present. From the equilibrium equation

$$\frac{[H^+][CH_3COO^-]}{[CH_3COOH]} = \frac{[H^+](11.0)}{9.0} = 1.75 \times 10^{-5}$$

the hydrogen ion concentration is found to be $1.43 \times 10^{-5}$ *M* and the pH is 4.84.

These calculations demonstrate how well a buffer solution can maintain constancy of pH – the pH of the solution, initially 4.76, changed only to 4.67 and to 4.84 by the addition, respectively, of 10 ml of 0.1 *F* hydrochloric acid and 10 ml of a 0.1 *F* sodium hydroxide solution.

There is a definite limit – called the **buffer capacity** – as to how much acid or base can be added to a given buffer solution before an appreciable change in pH results. Buffer capacity is set by the original amounts of the weak acid and its conjugate base, since one or the other is consumed by reaction with the added acid or base. In the preceding problem (Example 4–14), the original buffer solution contained 10 millimoles of acetic acid. Obviously, the addition of 10 millimoles of strong base would have neutralized all the acetic acid, and no buffering action would have remained. Calculations similar to those of Examples 4–13 and 4–14 reveal that the pH changes at a rate which increases more and more rapidly as the limiting buffer capacity is approached.

Another closely related point concerning buffer action is that a specific buffer is useful only in a narrow pH range around $pK_a$ for the conjugate acid-base pair which comprises the buffer. For the dissociation of acetic acid, $pK_a$ is 4.76, and the buffering action of an acetic acid-sodium acetate mixture is most effective in a narrow range near pH 4.76, where the concentrations of acetic acid and acetate ion are equal. A formic acid-sodium formate buffer works best in a pH range centered upon 3.75, the value of $pK_a$; since $pK_a$ for the ammonium ion is 9.26, an ammonium nitrate-ammonia buffer is most suitable for controlling the pH near this value.

**Amphiprotic Substances**

Earlier in this chapter, we identified several classes of amphiprotic compounds – species capable of behaving as either acids (proton donors) or bases (proton acceptors). We must now establish methods by which pH values for aqueous solutions of these substances can be determined.

*Example 4–15.* Calculate the pH of a 0.100 $F$ sodium bicarbonate ($NaHCO_3$) solution.

This system can be described by means of three competing equilibria – proton donation to water by $HCO_3^-$,

$$HCO_3^- + H_2O \rightleftharpoons CO_3^{2-} + H_3O^+; \qquad K_{a2} = \frac{[H^+][CO_3^{2-}]}{[HCO_3^-]} = 4.68 \times 10^{-11}$$

proton donation to $HCO_3^-$ by water,

$$HCO_3^- + H_2O \rightleftharpoons H_2CO_3 + OH^-; \quad K_{b2} = \frac{K_w}{K_{a1}} = \frac{[H_2CO_3][OH^-]}{[HCO_3^-]} = 2.24 \times 10^{-8}$$

and the autoprotolysis of water:

$$H_2O \rightleftharpoons H^+ + OH^-; \qquad K_w = [H^+][OH^-] = 1.00 \times 10^{-14}$$

In order to solve this problem in a rigorous fashion, we must utilize charge-balance and mass-balance relationships, as introduced in Chapter 3. Accordingly, because the concentration of positive charges in an aqueous solution of $NaHCO_3$ equals the concentration of negative charges, we can write

$$[Na^+] + [H^+] = [HCO_3^-] + [OH^-] + 2[CO_3^{2-}]$$

but, since the sodium ion concentration is 0.100 $M$,

$$0.100 + [H^+] = [HCO_3^-] + [OH^-] + 2[CO_3^{2-}]$$

In addition, the mass-balance expression requires that the *sum* of the concentrations of $H_2CO_3$, $HCO_3^-$, and $CO_3^{2-}$ be equal to the original concentration of $NaHCO_3$, which is the only source for these three species. Thus,

$$0.100 = [H_2CO_3] + [HCO_3^-] + [CO_3^{2-}]$$

Now let us subtract the mass-balance equation from the charge-balance relation:

$$[H^+] = [CO_3^{2-}] + [OH^-] - [H_2CO_3]$$

Next, we can re-express the preceding equation in terms of the hydrogen ion concentration, the bicarbonate concentration, and the various equilibrium constants:

$$[H^+] = \frac{K_{a2}[HCO_3^-]}{[H^+]} + \frac{K_w}{[H^+]} - \frac{[H^+][HCO_3^-]}{K_{a1}}$$

Rewriting,

$$[H^+]^2 = K_{a2}[HCO_3^-] + K_w - \frac{[H^+]^2[HCO_3^-]}{K_{a1}}$$

collecting terms,

$$[H^+]^2\left\{1 + \frac{[HCO_3^-]}{K_{a1}}\right\} = K_{a2}[HCO_3^-] + K_w$$

and solving for the hydrogen ion concentration, we obtain

$$[H^+] = \sqrt{\frac{K_{a2}[HCO_3^-] + K_w}{1 + \frac{[HCO_3^-]}{K_{a1}}}}$$

We must know the equilibrium concentration of $HCO_3^-$ to employ this relation for calculation of the hydrogen ion concentration. For the present situation, we can conclude that $[HCO_3^-]$ is essentially 0.100 *M*, because the equilibrium constants $K_{a2}$ and $K_{b2}$ for reactions leading to the consumption of bicarbonate are exceedingly small relative to the initial $HCO_3^-$ concentration.

Before we attempt to use this equation, let us examine two simplifications. First, if $[HCO_3^-]$ is significantly greater than $K_{a1}$ in the denominator, the "1" will be negligible in comparison to $[HCO_3^-]/K_{a1}$. Second, the numerator becomes simpler if $K_{a2}[HCO_3^-]$ is larger than $K_W$. If $[HCO_3^-] = 1.0\,M$, then $K_{a2}[HCO_3^-]$ is 4680 times greater than $K_W$. However, if $[HCO_3^-]$ is only 0.010 *M*, $K_{a2}[HCO_3^-]$ is still 46.8 times greater than $K_W$. For $[HCO_3^-] = 0.010\,M$, we can neglect the $K_W$ term in the numerator and incur an error of just 1 per cent, which is acceptable (especially when we intend to convert $[H^+]$ to pH). *In conclusion, under the specific condition that the bicarbonate ion concentration is no smaller than* 0.010 *M*, the complicated expression for $[H^+]$ becomes

$$[H^+] = \sqrt{K_{a1}K_{a2}}$$

and the pH of the solution is given by the expression

$$pH = \tfrac{1}{2}(pK_{a1} + pK_{a2}) = \tfrac{1}{2}(6.35 + 10.33) = 8.34$$

It is noteworthy that the pH is independent of the bicarbonate ion concentration, provided that this concentration is at least 0.010 *M*.

*Example 4–16.* Calculate the pH of a 0.0100 *F* ammonium formate solution.

This amphiprotic substance differs somewhat from $HCO_3^-$, in that the cation of the salt is a weak acid

$$NH_4^+ + H_2O \rightleftharpoons NH_3 + H_3O^+; \quad K_a = \frac{[NH_3][H^+]}{[NH_4^+]} = 5.55 \times 10^{-10}$$

and the anion is a weak base:

$$HCOO^- + H_2O \rightleftharpoons HCOOH + OH^-;$$

$$K_b = \frac{K_w}{K_a'} = \frac{[HCOOH][OH^-]}{[HCOO^-]} = 5.68 \times 10^{-11}$$

Note that $K_a'$ represents the acid dissociation constant of formic acid, the conjugate acid of formate anion. We can expect an aqueous solution of ammonium formate to have a pH less than 7, because the acid strength of ammonium ion exceeds the base strength of formate ion.

A single charge-balance relation

$$[NH_4^+] + [H^+] = [HCOO^-] + [OH^-]$$

and two mass-balance expressions

$$0.0100 = [NH_4^+] + [NH_3]$$

$$0.0100 = [HCOO^-] + [HCOOH]$$

are required for the present problem. If we equate the two mass-balance equations

$$[NH_4^+] + [NH_3] = [HCOO^-] + [HCOOH]$$

and subtract the result from the charge-balance relation, we obtain

$$[H^+] = [NH_3] + [OH^-] - [HCOOH]$$

We can rewrite the latter equation in terms of $[H^+]$, $[NH_4^+]$, and $[HCOO^-]$, as well as appropriate equilibrium constants, by employing the two equilibria mentioned above:

$$[H^+] = \frac{K_a[NH_4^+]}{[H^+]} + \frac{K_w}{[H^+]} - \frac{[HCOO^-][H^+]}{K_a'}$$

Solution of this expression for the hydrogen ion concentration leads to

$$[H^+] = \sqrt{\frac{K_a[NH_4^+] + K_w}{1 + \frac{[HCOO^-]}{K_a'}}}$$

Notice the similarity between this result and the equation presented earlier for the concentration of hydrogen ion in a sodium bicarbonate solution.

Now we should seek to simplify the rigorous equation for $[H^+]$. First, if $[HCOO^-]$ is larger than $K_a'$, the "1" in the denominator can be neglected. Second, if $K_a[NH_4^+]$ is greater than $K_w$ in the numerator, the latter term may be deleted. When $[HCOO^-] = [NH_4^+] = 0.0100\,M$, $[HCOO^-]/K_a' = (0.01)/(1.76 \times 10^{-4}) = 56.8$ (as compared to 1) and $K_a[NH_4^+] = (5.55 \times 10^{-10})(0.01) = 5.55 \times 10^{-12}$ (as compared to $1.00 \times 10^{-14}$). If the numerator and denominator are modified as suggested, we obtain

$$[H^+] = \sqrt{K_a K_a'}$$

where $K_a$ is the acid dissociation constant for the ammonium ion and $K_a'$ is the acid dissociation constant for formic acid. If this simple equation is compared to the rigorous expression for calculation of the hydrogen ion concentration in a 0.0100 $F$ ammonium formate solution, the results ($3.13 \times 10^{-7}$ and $3.10 \times 10^{-7}$ $M$, respectively) show that the approximate relation is only 1 per cent in error.

Since we are interested in calculating the pH of the ammonium formate solution, the last equation can be rewritten as

$$\mathrm{pH} = \tfrac{1}{2}(\mathrm{p}K_a + \mathrm{p}K_a')$$

Therefore,

$$\mathrm{pH} = \tfrac{1}{2}(9.26 + 3.75) = 6.50$$

As long as the approximations used to obtain the simplified relationships are valid – which requires in the present situation that the concentrations of $NH_4^+$ and $HCOO^-$ be not much less than 0.01 $M$ – the pH of an aqueous solution of ammonium formate (or any other similar compound) will be independent of concentration.

## Polyprotic Acids and Their Conjugate Bases

Solutions of polyprotic acids and their conjugate bases may contain several species, and the relative abundances of these species can vary tremendously as the pH changes. Therefore, it is useful to introduce a graphical method which can show at a glance how the composition of a polyprotic acid system is affected by pH.

As an example, let us consider the phosphoric acid system, which involves three proton-transfer equilibria:

$$H_3PO_4 + H_2O \rightleftharpoons H_2PO_4^- + H_3O^+; \quad K_{a1} = \frac{[H^+][H_2PO_4^-]}{[H_3PO_4]} = 7.5 \times 10^{-3}$$

$$H_2PO_4^- + H_2O \rightleftharpoons HPO_4^{2-} + H_3O^+; \quad K_{a2} = \frac{[H^+][HPO_4^{2-}]}{[H_2PO_4^-]} = 6.2 \times 10^{-8}$$

$$HPO_4^{2-} + H_2O \rightleftharpoons PO_4^{3-} + H_3O^+; \quad K_{a3} = \frac{[H^+][PO_4^{3-}]}{[HPO_4^{2-}]} = 4.8 \times 10^{-13}$$

For a given solution, we shall represent by $C$ the sum of the concentrations of all phosphate-containing species:

$$C = [H_3PO_4] + [H_2PO_4^-] + [HPO_4^{2-}] + [PO_4^{3-}]$$

Next, we will define the concentration of each individual phosphate species to be a certain fraction, $\alpha_n$, of the total concentration, $C$, the subscript $n$ designating the number of protons lost by the parent acid to form the species of interest. Thus,

$$[H_3PO_4] = \alpha_0 C, \quad [H_2PO_4^-] = \alpha_1 C, \quad [HPO_4^{2-}] = \alpha_2 C, \quad [PO_4^{3-}] = \alpha_3 C$$

where

$$\alpha_0 + \alpha_1 + \alpha_2 + \alpha_3 = 1$$

By proper combination of the equilibrium expressions for the stepwise dissociation of phosphoric . id, we can obtain relations for the concentrations of the three anionic phosphate species in terms of $[H^+]$, $[H_3PO_4]$, and the equilibrium constants:

$$[H_2PO_4^-] = \frac{K_{a1}[H_3PO_4]}{[H^+]}$$

$$[HPO_4^{2-}] = \frac{K_{a1}K_{a2}[H_3PO_4]}{[H^+]^2}$$

$$[PO_4^{3-}] = \frac{K_{a1}K_{a2}K_{a3}[H_3PO_4]}{[H^+]^3}$$

Substitution of these results term-by-term into the earlier mass-balance equation yields

$$C = [H_3PO_4] + \frac{K_{a1}[H_3PO_4]}{[H^+]} + \frac{K_{a1}K_{a2}[H_3PO_4]}{[H^+]^2} + \frac{K_{a1}K_{a2}K_{a3}[H_3PO_4]}{[H^+]^3}$$

which can be rearranged to give

$$[H_3PO_4] = \frac{C[H^+]^3}{[H^+]^3 + K_{a1}[H^+]^2 + K_{a1}K_{a2}[H^+] + K_{a1}K_{a2}K_{a3}}$$

Inserting the latter expression into the relation for the definition of $\alpha_0$, we get

$$\alpha_0 = \frac{[H^+]^3}{[H^+]^3 + K_{a1}[H^+]^2 + K_{a1}K_{a2}[H^+] + K_{a1}K_{a2}K_{a3}}$$

It can be similarly shown that

$$\alpha_1 = \frac{K_{a1}[H^+]^2}{[H^+]^3 + K_{a1}[H^+]^2 + K_{a1}K_{a2}[H^+] + K_{a1}K_{a2}K_{a3}}$$

$$\alpha_2 = \frac{K_{a1}K_{a2}[H^+]}{[H^+]^3 + K_{a1}[H^+]^2 + K_{a1}K_{a2}[H^+] + K_{a1}K_{a2}K_{a3}}$$

and

$$\alpha_3 = \frac{K_{a1}K_{a2}K_{a3}}{[H^+]^3 + K_{a1}[H^+]^2 + K_{a1}K_{a2}[H^+] + K_{a1}K_{a2}K_{a3}}$$

Notice that the denominator for each $\alpha_n$ value is the same function of the hydrogen ion concentration and the various dissociation constants; although the denominator will be different for every pH, it will be identical for each $\alpha_n$ value. Furthermore, the relations for the various $\alpha_n$ values emphasize the important fact that the fraction of each phosphate-containing species depends upon pH, but not on the total concentration, $C$.

We will now examine a typical situation which illustrates the use of these concepts.

*Example 4–17.* Calculate the individual concentrations of all phosphate species in 1.00 liter of solution containing a total of 0.100 mole of phosphate species at pH 8.00.

To begin, let us evaluate each of the four terms which appear in the various $\alpha_n$ expressions:

$$[H^+]^3 = (1.00 \times 10^{-8})^3 = 1.00 \times 10^{-24}$$

$$K_{a1}[H^+]^2 = (7.5 \times 10^{-3})(1.00 \times 10^{-8})^2 = 7.50 \times 10^{-19}$$

$$K_{a1}K_{a2}[H^+] = (7.5 \times 10^{-3})(6.2 \times 10^{-8})(1.00 \times 10^{-8}) = 4.65 \times 10^{-18}$$

$$K_{a1}K_{a2}K_{a3} = (7.5 \times 10^{-3})(6.2 \times 10^{-8})(4.8 \times 10^{-13}) = 2.23 \times 10^{-22}$$

Before proceeding, we should note that the first term is proportional to $[H_3PO_4]$, the second term to $[H_2PO_4^-]$, the third term to $[HPO_4^{2-}]$, and the fourth term to $[PO_4^{3-}]$. Since the magnitudes of these terms, from largest to smallest, are in the order, third, second, fourth, and first, we can conclude that the relative concentrations of the four phosphate species are $[HPO_4^{2-}] > [H_2PO_4^-] > [PO_4^{3-}] > [H_3PO_4]$.

We obtain the denominator of each $\alpha_n$ value by adding the four terms together; the result is

$$[H^+]^3 + K_{a1}[H^+]^2 + K_{a1}K_{a2}[H^+] + K_{a1}K_{a2}K_{a3} = 5.40 \times 10^{-18}$$

Each $\alpha_n$ value is, in turn, computed as follows:

$$\alpha_0 = \frac{1.00 \times 10^{-24}}{5.40 \times 10^{-18}} = 1.85 \times 10^{-7}$$

$$\alpha_1 = \frac{7.50 \times 10^{-19}}{5.40 \times 10^{-18}} = 0.139$$

$$\alpha_2 = \frac{4.65 \times 10^{-18}}{5.40 \times 10^{-18}} = 0.861$$

$$\alpha_3 = \frac{2.23 \times 10^{-22}}{5.40 \times 10^{-18}} = 4.13 \times 10^{-5}$$

Since a total of 0.100 mole of phosphate species is present in 1.00 liter of solution, $C = 0.100\,M$. Finally, by utilizing the fundamental definitions for the $\alpha_n$ values, we can calculate the desired concentrations of the individual phosphate species:

$$[H_3PO_4] = \alpha_0 C = (1.85 \times 10^{-7})(0.100) = 1.85 \times 10^{-8}\,M$$

$$[H_2PO_4^-] = \alpha_1 C = (0.139)(0.100) = 1.39 \times 10^{-2}\,M$$

$$[HPO_4^{2-}] = \alpha_2 C = (0.861)(0.100) = 8.61 \times 10^{-2}\,M$$

$$[PO_4^{3-}] = \alpha_3 C = (4.13 \times 10^{-5})(0.100) = 4.13 \times 10^{-6}\,M$$

Calculations of $\alpha_n$ values for the phosphoric acid system have been performed, according to the procedure described, for the entire pH range from 0 to 14. Results of these computations are seen in Figure 4–1 as four curves, each showing how the fraction of one of the phosphate species changes with pH. Once such a distribution diagram is constructed for a polyprotic acid system, it is possible to tell by inspection which species are predominant at any pH as well as their relative concentrations. It is clear that, between pH 0 and 4.6, the only important species are $H_3PO_4$ and $H_2PO_4^-$. In the region

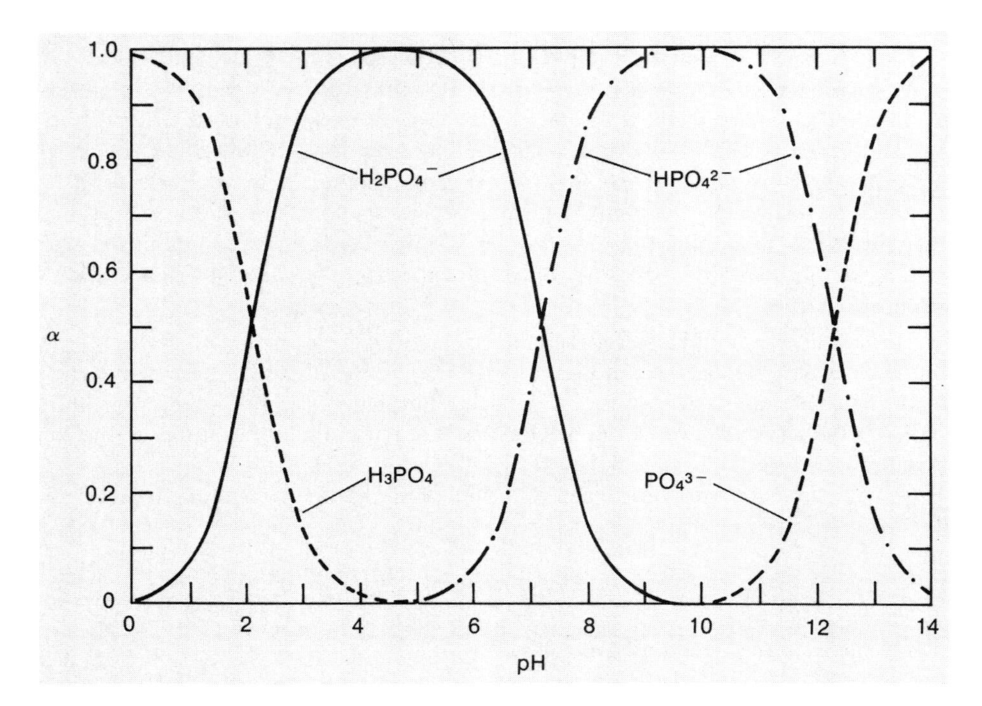


*Figure 4–1.* Distribution of various phosphate species as a function of pH.

from pH 4.6 to 9.8, only $H_2PO_4^-$ and $HPO_4^{2-}$ are present in significant amounts, whereas $HPO_4^{2-}$ and $PO_4^{3-}$ are the most abundant species between pH 9.8 and 14. In addition, notice that the curves for a conjugate acid-base pair intersect at $\alpha_n = 0.5$ and at a pH value numerically equal to $pK_a$ for the pertinent equilibrium. Thus, the curves for $H_2PO_4^-$ and $HPO_4^{2-}$ cross at pH 7.2, which is the numerical value of $pK_{a2}$ for the reaction

$$H_2PO_4^- + H_2O \rightleftharpoons HPO_4^{2-} + H_3O^+; \qquad K_{a2} = 6.2 \times 10^{-8}, \quad pK_{a2} = 7.2$$

As another illustration of the usefulness of Figure 4–1, let us consider the following problem.

*Example 4–18.* If 150 ml of an aqueous solution at pH 3.00 contains a total of 0.0250 mole of phosphate species, what are the concentrations of the *predominant* species?

Inspection of Figure 4–1 reveals that $H_3PO_4$ and $H_2PO_4^-$ are the only major phosphate species at pH 3.00, and that $\alpha_0$ and $\alpha_1$ are 0.12 and 0.88, respectively. In 150 ml of solution, there is 0.0250 mole of phosphate species, corresponding to a total concentration ($C$) of 0.167 $M$. Therefore,

$$[H_3PO_4] = \alpha_0 C = (0.12)(0.167) = 0.020\ M$$

and

$$[H_2PO_4^-] = \alpha_1 C = (0.88)(0.167) = 0.147\ M$$

In addition to the preceding examples, many equilibrium problems concerning polyprotic acid systems involve concepts already considered for weak monoprotic acids and their conjugate bases.

*Example 4–19.* Calculate the pH of a 0.1000 $F$ phosphoric acid solution.

As usual, we must consider the important sources of hydrogen ions in this system. In principle, phosphoric acid can furnish protons from each of its three dissociations, and water is also a potential source of hydrogen ions. However, in view of the comparatively large value of the first dissociation constant of phosphoric acid, it is reasonable to neglect at this point the other proton sources. Therefore, this problem is very similar to Examples 4–6, 4–7, and 4–8 which involve weak monoprotic acids. We can write the abbreviated equilibrium expression for the first dissociation of phosphoric acid as

$$\frac{[H^+][H_2PO_4^-]}{[H_3PO_4]} = K_{a1} = 7.5 \times 10^{-3}$$

At equilibrium, we shall write for this situation

$$[H^+] = [H_2PO_4^-]$$

and

$$[H_3PO_4] = 0.1000 - [H^+]$$

Since the first step of dissociation of phosphoric acid is too extensive for us to neglect the $[H^+]$ term in the preceding relation, the expression to be solved for $[H^+]$ is

$$\frac{[H^+]^2}{0.1000 - [H^+]} = 7.5 \times 10^{-3}$$

This last equation can be solved by means of the quadratic formula or the method of successive approximations, and the result is $[H^+] = 2.39 \times 10^{-2}\ M$; so pH = 1.62.

Dissociation of $H_2PO_4^-$ is the next most likely source of hydrogen ions. We can show that it and, hence, the other sources of protons can be neglected. If we provisionally take $[H^+] = 2.39 \times 10^{-2}\ M$ and $[H_2PO_4^-] = 2.39 \times 10^{-2}\ M$, the concentration of $HPO_4^{2-}$ can be calculated from a rearranged form of the second dissociation-constant expression for phosphoric acid. Thus,

$$[HPO_4^{2-}] = K_{a2}\frac{[H_2PO_4^-]}{[H^+]} = 6.2 \times 10^{-8}\frac{(0.0239)}{(0.0239)} = 6.2 \times 10^{-8}\ M$$

Since the equilibrium concentration of $HPO_4^{2-}$ is so small, it is apparent that an insignificant quantity of $H_2PO_4^-$ has dissociated under these conditions. Consequently, the pH is 1.62 as calculated.

Examples of other problems commonly encountered with polyprotic acid systems include (a) calculation of the pH of a mixture of $H_3PO_4$ and $H_2PO_4^-$ or a mixture of $HPO_4^{2-}$ and $PO_4^{3-}$, which is analogous to a calculation of the pH of an acetic acid-sodium acetate buffer (Example 4–11), and (b) calculation of the pH of a solution of $Na_2CO_3$ or $Na_3PO_4$, which is essentially identical to a calculation of the pH of a pyridine or sodium benzoate solution (Examples 4–9 and 4–10).

In the next chapter, we will apply many of the principles employed in the preceding discussions to acid-base titrations in aqueous and nonaqueous solvents.

## QUESTIONS AND PROBLEMS*

1. Convert the following pH values to $[H^+]$: 6.37; 4.83; 11.22; 8.54; 3.13.
2. Convert the following pOH values to $[H^+]$: 3.28; 9.86; 2.47; 12.19; 6.71.
3. Convert the following molar hydrogen ion concentrations to pH values: 0.0139; $2.07 \times 10^{-5}$; $7.34 \times 10^{-11}$; 2.41; $4.12 \times 10^{-7}$.
4. In a series of experiments, it was determined that the position of equilibrium for each of the proton-transfer reactions listed below lies well toward the right.

$$HCN + OH^- \rightleftharpoons H_2O + CN^-$$

$$HClO_4 + Cl^- \rightleftharpoons HCl + ClO_4^-$$

$$H_2S + CN^- \rightleftharpoons HCN + HS^-$$

$$(C_6H_5)_3C^- + C_4H_4NH \rightleftharpoons (C_6H_5)_3CH + C_4H_4N^-$$

$$CH_3COOH + HS^- \rightleftharpoons H_2S + CH_3COO^-$$

$$O^{2-} + (C_6H_5)_3CH \rightleftharpoons (C_6H_5)_3C^- + OH^-$$

$$C_4H_4N^- + H_2S \rightleftharpoons C_4H_4NH + HS^-$$

   Using the above information as well as some semiquantitative knowledge about the relative strengths of common acids (but without consulting tables of acid dissociation constants), arrange the following acids in order of *decreasing* acidity: (a) HCN (b) HCl (c) $HClO_4$ (d) $CH_3COOH$ (e) $(C_6H_5)_3CH$ (f) $C_4H_4NH$ (pyrrole) (g) $H_2S$ (h) $OH^-$ (i) $H_2O$.
5. Calculate the pH of each of the following systems, neglecting activity effects:
   (a) 40 ml of solution containing 1.00 g of nitric acid
   (b) 30 ml of solution containing 120 mg of barium hydroxide
   (c) 0.180 *F* benzoic acid solution
   (d) 0.250 *F* sodium dihydrogen phosphate solution
   (e) 0.700 *F* arsenic acid solution
   (f) 0.200 *F* ammonium cyanide solution
   (g) a solution prepared by dissolution of 0.267 mole of pyridine in 600 ml of water
   (h) a solution prepared by dissolution of 0.0235 mole of alanine (an amino acid) in 700 ml of water
   (i) $2.00 \times 10^{-5}$ *F* potassium acetate solution
   (j) 0.120 *F* chloroacetic acid solution
   (k) a solution prepared by dissolution of 0.500 g of *o*-phthalic acid in 125 ml of water
   (l) a solution prepared by dissolution of 0.200 mole of potassium nitrite and 0.00030 mole of potassium hydroxide in 800 ml of water.
6. Calculate the hydrogen ion concentration and the pH of a solution prepared by addition of 0.100 ml of 0.0100 *F* perchloric acid ($HClO_4$) to 10.0 liters of pure water.
7. In what ratio must acetic acid and sodium acetate be mixed to provide a solution of pH 6.20?

---

*A comprehensive table of dissociation constants for acids and bases is given in Appendix 2.

8. In what ratio must ammonia and ammonium chloride be mixed to provide a solution of pH 8.40?
9. Consider 100 ml of a 0.100 *F* acetic acid solution at 25°C.
   (a) Will the addition of 1 g of pure acetic acid cause the extent (percentage) of dissociation of acetic acid to increase, decrease, or remain the same?
   (b) Will the addition of 1 g of sodium acetate cause the pH of the solution to increase, decrease, or remain the same?
   (c) Will the addition of 1 g of sodium acetate cause the hydroxide ion concentration to increase, decrease, or remain the same?
10. Consider 50 ml of a 0.100 *F* aqueous ammonia solution at 25°C.
   (a) Will the solution at the end point of the titration of the ammonia solution with 0.100 *F* hydrochloric acid have a pH less than 7, equal to 7, or greater than 7?
   (b) Will the addition of 1 g of pure acetic acid cause the pH to increase, decrease, or remain the same?
   (c) Will the addition of 1 g of ammonium chloride cause the hydroxide ion concentration to increase, decrease, or remain the same?
11. Suppose you wish to prepare 1 liter of a buffer solution, containing sodium carbonate and sodium bicarbonate, having an initial pH of 9.70, and of such composition that, upon the formation of 60 millimoles of hydrogen ion during the course of a chemical reaction, the pH will not decrease below 9.30. What are the minimal original concentrations of sodium carbonate and sodium bicarbonate needed to prepare the desired buffer solution?
12. In what ratio must pyridine and pyridinium chloride be mixed to prepare a buffer solution with a pH of 5.00? If one liter of this buffer contains an initial pyridine concentration of 0.200 *M*, what will be the final pH of the solution after a chemical reaction occurring in the system consumes 50 millimoles of hydrogen ion?
13. Calculate the pH and the equilibrium concentration of $HPO_4^{2-}$ in a 0.500 *F* solution of phosphoric acid ($H_3PO_4$) in water.
14. Calculate the equilibrium concentrations of $NH_4^+$ and $NH_3$ and the pH of the solution obtained from dissolution of 0.085 mole of ammonium nitrate and 0.000060 mole of nitric acid in 300 ml of water.
15. Calculate the equilibrium concentrations of $H^+$, $C_2H_5COO^-$, and $C_2H_5COOH$ in a solution prepared by dissolution of 0.0400 mole of propanoic acid in 150 ml of water.
16. If 0.826 mole of pyridine and 0.492 mole of hydrochloric acid are mixed with water to give 500 ml of solution, what will be the pH of the resulting system?
17. Calculate the equilibrium concentrations of $H^+$, $NO_2^-$, and $HNO_2$ in a solution prepared by dissolution of 0.100 mole of nitrous acid in 200 ml of water.
18. Calculate the equilibrium concentrations of $F^-$ and HF and the pH of the solution prepared by dissolution of 0.100 mole of sodium fluoride in 150 ml of water.
19. Calculate the equilibrium concentrations of all ionic and molecular species except water, *i.e.*, $NH_3$, $NH_4^+$, $H^+$, and $OH^-$, in a solution which initially contains 0.00400 mole of ammonia in 150 ml of water.

20. Oxalic acid undergoes the following stepwise dissociation reactions:

$$H_2C_2O_4 \rightleftharpoons H^+ + HC_2O_4^-; \; K_{a1} = 6.5 \times 10^{-2}$$
$$HC_2O_4^- \rightleftharpoons H^+ + C_2O_4^{2-}; \; K_{a2} = 6.1 \times 10^{-5}$$

(a) Construct a distribution diagram showing how the fraction, $\alpha_n$, of each oxalate-containing species varies as a function of pH.

(b) For a solution in which the total concentration of all oxalate-containing species is 0.0100 *F*, what would the concentration of $HC_2O_4^-$ be at pH 3.50? For the same solution, what would be the concentration of $H_2C_2O_4$ at pH 2.35, and what would be the concentration of $C_2O_4^{2-}$ at pH 5.68?

(c) Suppose that 250 mg of anhydrous sodium oxalate ($Na_2C_2O_4$) is dissolved in 500 ml of an aqueous solution which is well buffered at a pH of 4.00. Calculate the equilibrium concentration of each oxalate-containing species.

## SUGGESTIONS FOR ADDITIONAL READING

1. C. Ayers: Acidimetry and alkalimetry. *In* C. L. Wilson and D. W. Wilson, eds.: *Comprehensive Analytical Chemistry.* Volume IB, Elsevier, New York, 1960, pp. 203–221.
2. R. G. Bates: Concept and determination of pH. *In* I. M. Kolthoff and P. J. Elving, eds.: *Treatise on Analytical Chemistry.* Part I, Volume 1, Wiley-Interscience, New York, 1959, pp. 361–404.
3. R. G. Bates: *Determination of pH.* John Wiley and Sons, New York, 1964.
4. R. P. Bell: *Acids and Bases.* John Wiley and Sons, New York, 1952.
5. S. Bruckenstein and I. M. Kolthoff: Acid-base strength and protolysis curves in water. *In* I. M. Kolthoff and P. J. Elving, eds.: *Treatise on Analytical Chemistry.* Part I, Volume 1, Wiley-Interscience, New York, 1959, pp. 421-474.
6. E. J. King: *Acid-Base Equilibria.* The Macmillan Company, New York, 1965.
7. I. M. Kolthoff: Concepts of acids and bases. *In* I. M. Kolthoff and P. J. Elving, eds.: *Treatise on Analytical Chemistry.* Part I, Volume 1, Wiley-Interscience, New York, 1959, pp. 405–420.
8. I. M. Kolthoff and V. A. Stenger: *Volumetric Analysis.* Second edition, Volume 2, Wiley-Interscience, New York, 1947, pp. 49–235.
9. C. A. VanderWerf: *Acids, Bases, and the Chemistry of the Covalent Bond.* Reinhold Publishing Corporation, New York, 1961.

# 5 ACID-BASE TITRATIONS IN AQUEOUS AND NONAQUEOUS SOLVENTS

In a **titration**, the amount of a substance is determined by measurement of the quantity of a reagent – called the **titrant** – required to react stoichiometrically with that substance. A titration involves the carefully measured and controlled addition of titrant to a solution of the substance to be determined until the reaction between the two species is judged to be complete. In classical volumetric analysis, the titrant is a solution of precisely known concentration – that is, a **standard solution** – and the amount of a substance is calculated from knowledge of the concentration and volume of the standard solution used in the titration.

That point during the course of a titration at which the amount of titrant added is stoichiometrically equivalent to the substance being determined is defined as the theoretical **equivalence point**. In practice, one looks for an abrupt change in some property of the solution being titrated to indicate that the equivalence point has been reached; the point at which this sharp change occurs is termed the experimental **end point**. Among the solution properties that can undergo change and thereby serve as a basis for end-point detection are color, refractive index, conductivity, and temperature; in addition, changes in the potential difference between indicator and reference electrodes immersed in a solution are often used to locate end points. Because the change in a particular solution property is not always adequately sharp, and because an abrupt change which does occur might not be detected properly, the end point seldom coincides perfectly with the theoretical equivalence point, although one strives to minimize the **titration error** – the difference between the end point and the equivalence point – as much as possible.

A chemical reaction must fulfill four requirements if it is to serve as the basis of a titrimetric method of analysis. First, the reaction must proceed rapidly, so that one can recognize without unnecessary waiting whether the end point has been reached or whether the titration should be continued. Second, the reaction must proceed essentially to completion – that is, it must have a large equilibrium constant – so that an adequately sharp end-point signal will be observed. Third, the reaction must proceed according to well-defined stoichiometry, so that the amount of substance being determined can be calculated from the titration data. Fourth, a convenient method of end-point detection must be available.

Acid-base systems meet these requirements exceptionally well. Proton transfer between an acid and a base is an exceedingly fast and stoichiometrically exact process. Moreover, equilibrium constants for many acid-base reactions are sufficiently large that sharp and reproducible end points are easily attainable. Finally, there are numerous acid-base indicators which undergo distinct color changes with a variation in pH and which can be employed for the accurate location of equivalence points.

## ACID-BASE TITRATIONS IN WATER

In this section of the chapter, we will deal with the preparation of standard solutions of acids and bases for proton-transfer titrations in water, the behavior of colored acid-base indicators, and a variety of practical titrations.

### Preparation and Standardization of Solutions of Strong Acids

Titrant solutions of sulfuric acid, nitric acid, perchloric acid, and hydrochloric acid in water are prepared by dilution of the concentrated, commercially available acids and can be standardized by a variety of methods. Sulfuric acid is not suitable, however, for titrations of solutions containing a cation that forms an insoluble sulfate precipitate because the end point may be obscured. Dilute nitric acid solutions near the 0.1 *F* concentration level and perchloric acid solutions as concentrated as 1.0 *F* are stable for long periods of time and make excellent standard acid solutions. However, hydrochloric acid is the most preferred titrant, because of the purity of the concentrated, commercially available reagent and because of the long shelf-life of dilute solutions. In addition, there are numerous highly precise methods for the accurate standardization of hydrochloric acid titrants.

Any one of several primary standard bases can be used for the standardization of a solution of hydrochloric acid or one of the other strong acids. Reagent-grade sodium carbonate is commercially available in very high purity for standardization purposes. An accurately weighed sample of sodium carbonate is dissolved in water, and the resulting solution is titrated with hydrochloric acid to form carbonic acid,

$$CO_3^{2-} + 2\,H^+ \rightleftharpoons H_2CO_3$$

the end point being easily determined with the aid of an acid-base indicator.

Tris(hydroxymethyl)aminomethane, sometimes abbreviated THAM and also known as "tris," is available in analytical-reagent-grade purity, and is an excellent primary standard. It reacts with hydrochloric acid, much the same as does ammonia, according to the equation

$$H^+ + (CH_2OH)_3CNH_2 \rightleftharpoons (CH_2OH)_3CNH_3^+$$

Other primary standards for the standardization of hydrochloric acid solutions include calcium carbonate and borax ($Na_2B_4O_7 \cdot 10\,H_2O$).

Another method to obtain a standard hydrochloric acid solution involves its preparation from constant-boiling hydrochloric acid. When concentrated (12 *F*) hydrochloric acid is boiled, hydrogen chloride and water molecules are distilled away until eventually the remaining solution approaches a definite, precisely known composition (close to 6.1 *F*) which is fixed by the prevailing barometric pressure. Thereafter, this constant-boiling mixture continues to distill off, and it can be collected and stored for future use. Although the concentration of the constant-boiling acid is rather high, quantitative dilution with distilled water suffices to bring it to a suitable concentration.

### Preparation and Standardization of Solutions of Strong Bases

Standard base solutions are usually prepared from sodium hydroxide. Two less important, but occasionally used, reagents for the preparation of standard base solutions are potassium hydroxide and barium hydroxide. None of these compounds is available in a pure form; solid sodium hydroxide, for example, is invariably contaminated by moisture and by small amounts (up to 2 per cent) of sodium carbonate as well as sodium chloride. Consequently, a sodium hydroxide solution of approximately the required concentration must be prepared and standardized.

Solutions of barium hydroxide absorb carbon dioxide from the atmosphere with the resultant precipitation of insoluble barium carbonate and the simultaneous decrease in the concentration of hydroxide ion:

$$CO_2 + 2\,OH^- \rightleftharpoons CO_3^{2-} + H_2O$$

$$Ba^{2+} + CO_3^{2-} \rightleftharpoons BaCO_3$$

Sodium and potassium hydroxide solutions absorb carbon dioxide as well, but the carbonate which is formed remains soluble. In the case of a sodium or potassium hydroxide titrant, the absorbed carbon dioxide does not necessarily cause a change in the effective concentration of the strong base solution, because a carbonate ion can react with two protons just as the two hydroxide ions which are neutralized when the carbon dioxide molecule is originally absorbed. However, sometimes an end-point indicator is employed which changes color at a pH corresponding only to the titration of carbonate to bicarbonate ion, so that the effective concentration of the base solution is smaller than expected. This difficulty can be remedied, however, if an end-point indicator is selected which changes color at a pH corresponding to the addition of *two* hydrogen ions to the carbonate ion.

It is highly desirable to remove carbonate ion from a sodium hydroxide solution prior to its standardization, and once standardized, the solution should be stored away from contact with the atmosphere. An excellent method for preparation of a carbonate-free solution takes advantage of the insolubility of sodium carbonate in very concentrated sodium hydroxide solutions. If a 50 per cent sodium hydroxide solution is prepared, the sodium carbonate impurity is virtually insoluble and will settle to the bottom of the container within a day or so. Then, the clear, carbonate-free sodium hydroxide syrup may be carefully decanted and diluted appropriately with freshly boiled and cooled distilled water. Since glass containers are attacked by sodium hydroxide solutions, they should not be employed as storage vessels; however, screw-cap polyethylene bottles are especially serviceable for the short-term storage and use of these solutions.

There are several primary standard acids, any one of which can be used to standardize a solution of sodium, potassium, or barium hydroxide. Potassium acid phthalate ($KHC_8H_4O_4$) has the advantages of high equivalent weight and purity, stability on drying, and ready availability. On the other hand, it has the disadvantage of being a weak acid so that it is generally suitable for standardization of carbonate-free base solutions only. Sulfamic acid ($NH_2SO_3H$) is a very good primary standard, being readily available in a pure form, rather inexpensive, and a strong acid. Oxalic acid dihydrate ($H_2C_2O_4 \cdot 2\,H_2O$) and benzoic acid ($C_6H_5COOH$) are occasionally used as primary standard acids.

### Acid-Base Indicators for End-Point Detection

Acid-base indicators are nothing more than highly colored organic dye molecules possessing acidic or basic functional groups. In a solution, variations in pH cause the acid-base indicator to lose or gain a proton and, because substantial structural rearrangement accompanies the proton-transfer process, a dramatic color change occurs.

We can understand the behavior of an acid-base indicator in terms of an equilibrium involving a weak monoprotic acid and its conjugate base

$$HIn + H_2O \rightleftharpoons In^- + H_3O^+$$

where HIn symbolizes the acid form of the indicator and $In^-$ denotes the conjugate-base form of the indicator. Needless to say, the colors of the two forms are different. Although we have represented the acid form as a neutral molecule, it often has a single negative charge, in which case the conjugate base is a doubly charged anion; the important point is that the conjugate-base form always bears one more negative charge than the acid form. An increase in the hydrogen ion concentration – or a decrease in pH – favors the formation of HIn and the appearance in solution of the color of the acid form. Conversely, an increase in pH shifts the proton-transfer equilibrium toward production of more $In^-$, which will impart its color to the solution.

Let us write the equilibrium expression for the proton-transfer reaction of the acid-base indicator

$$\frac{[In^-][H^+]}{[HIn]} = K_a$$

where $K_a$ is the acid dissociation constant for the indicator. Simple rearrangement of this relation yields

$$\frac{[In^-]}{[HIn]} = \frac{K_a}{[H^+]}$$

which shows that the ratio of the acid and base forms of an indicator varies with the hydrogen ion concentration. To the average human eye, a solution containing two colored species, HIn and $In^-$, will exhibit only the color of HIn if the concentration of HIn is at least 10 times greater than that of $In^-$. On the other hand, if the concentration of $In^-$ is 10 or more times larger than the HIn concentration, the solution will appear to have the color of $In^-$. Therefore, a solution of an acid-base indicator will show the color of the acid form if

$$\frac{[In^-]}{[HIn]} \leq 0.1$$

whereas the color of the base form will be observed if

$$\frac{[In^-]}{[HIn]} \geq 10$$

Only when the pH varies so that the ratio of $[In^-]$ to [HIn] shifts from at least 0.1 to 10 will the human eye detect a color change. For values of the ratio between these two limits, the solution will appear to have an intermediate color – a mixture of the colors of HIn and $In^-$.

Substitution of the preceding concentration ratios into the equilibrium expression for the acid-base indicator allows us to determine what variation in the hydrogen ion concentration is necessary to produce a color change. When a solution of the indicator exhibits the color of the *acid* form

$$\frac{[In^-]}{[HIn]} = \frac{K_a}{[H^+]} = 0.1; \quad [H^+] = 10K_a$$

but when the solution has the color of the *base* form of the indicator

$$\frac{[In^-]}{[HIn]} = \frac{K_a}{[H^+]} = 10; \quad [H^+] = 0.1K_a$$

These results may be conveniently expressed in terms of $pK_a$ and pH as follows:

If pH = $pK_a$ - 1, or less, the solution will have the color of the acid form (HIn).

If pH = $pK_a$ + 1, or more, the solution will have the color of the base form ($In^-$).

An acid-base indicator which has, for example, a dissociation constant ($K_a$) of $1.00 \times 10^{-8}$, or a $pK_a$ value of 8.00, will undergo a color change when the pH of a solution containing it varies from 7 to 9; the opposite color change occurs if the pH varies from 9 to 7.

A list of acid-base indicators is presented in Table 5–1, along with the colors of the acid and base forms of each indicator and the pH value at which one or the other form is sufficiently predominant to impart its color to the solution.

**Table 5–1. Selected Acid-Base Indicators**

| Indicator | Color Change | | pH Transition Interval | |
|---|---|---|---|---|
| | Acid Form | Base Form | Acid Form Predominant at pH | Base Form Predominant at pH |
| Picric acid | colorless | yellow | 0.1 | 0.8 |
| Paramethyl red | red | yellow | 1.0 | 3.0 |
| 2,6-Dinitrophenol | colorless | yellow | 2.0 | 4.0 |
| Bromphenol blue | yellow | blue | 3.0 | 4.6 |
| Congo red | blue | red | 3.0 | 5.0 |
| Methyl orange | red | yellow | 3.1 | 4.4 |
| Ethyl orange | red | yellow | 3.4 | 4.5 |
| Alizarin red S | yellow | purple | 3.7 | 5.0 |
| Bromcresol green | yellow | blue | 3.8 | 5.4 |
| Methyl red | red | yellow | 4.2 | 6.2 |
| Propyl red | red | yellow | 4.6 | 6.6 |
| Methyl purple | purple | green | 4.8 | 5.4 |
| Chlorophenol red | yellow | red | 4.8 | 6.4 |
| Paranitrophenol | colorless | yellow | 5.0 | 7.0 |
| Bromcresol purple | yellow | purple | 5.2 | 6.8 |
| Bromthymol blue | yellow | blue | 6.0 | 7.6 |
| Brilliant yellow | yellow | orange | 6.6 | 8.0 |
| Neutral red | red | amber | 6.7 | 8.0 |
| Phenol red | yellow | red | 6.7 | 8.4 |
| Metanitrophenol | colorless | yellow | 6.7 | 8.6 |
| Phenolphthalein | colorless | pink | 8.0 | 9.6 |
| Thymolphthalein | colorless | blue | 9.3 | 10.6 |
| 2,4,6-Trinitrotoluene | colorless | orange | 12.0 | 14.0 |

It is of interest to demonstrate the validity of the previous discussion by reference to a specific indicator. Methyl red, one of the azo-dye class of acid-base indicators, is an excellent example. According to Table 5–1, the red-colored acid form predominates at pH 4.2 and below, whereas the yellow conjugate base is the important form at and above pH 6.2. Structures of the two forms of methyl red, and the fundamental acid-base equilibrium, are given by

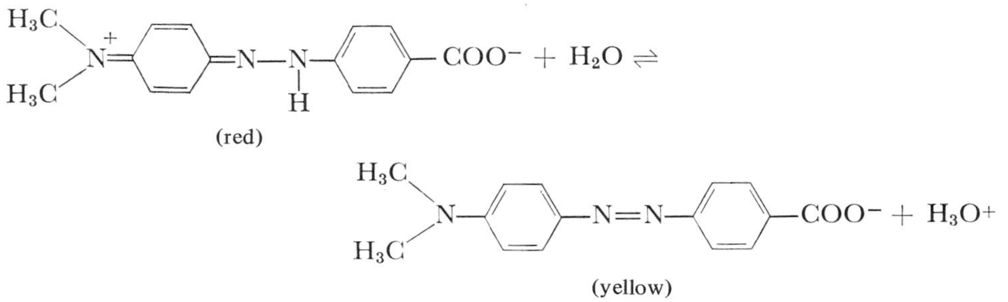

$$(H_3C)_2\overset{+}{N}{=}C_6H_4{=}N{-}\underset{H}{N}{-}C_6H_4{-}COO^- + H_2O \rightleftharpoons$$

(red)

$$(H_3C)_2N{-}C_6H_4{-}N{=}N{-}C_6H_4{-}COO^- + H_3O^+$$

(yellow)

and the equilibrium expression may be written as

$$\frac{[In^-][H^+]}{[HIn]} = K_a = 7.9 \times 10^{-6}; \quad pK_a = 5.1$$

where [HIn] is the concentration of the red acid form and $[In^-]$ is the concentration of the yellow base form. We can calculate the ratio $[HIn]/[In^-]$ at pH 4.2 and at pH 6.2 as follows:

At pH 4.2, $[H^+] = 6.3 \times 10^{-5}\ M$ and

$$\frac{[HIn]}{[In^-]} = \frac{6.3 \times 10^{-5}}{7.9 \times 10^{-6}} = 8.0$$

At pH 6.2, $[H^+] = 6.3 \times 10^{-7}\ M$ and

$$\frac{[HIn]}{[In^-]} = \frac{6.3 \times 10^{-7}}{7.9 \times 10^{-6}} = 0.080 = \frac{1}{12.5}$$

Thus, in the case of methyl red, we observe the red color of the indicator when there is 8.0 times more acid form than base form present. However, in order for us to see the indicator in its yellow form, there must be present 12.5 times more base form than acid form. In other words, at pH 4.2 (and below) the human eye recognizes that methyl red is present as the red acid form and that it exists as the yellow base form if the pH becomes 6.2 (and above). This range of pH values over which the acid-base indicator changes color is called the **pH transition interval.** Note that the pH transition interval for the indicators listed in Table 5–1 is often about two pH units, although it may be smaller than this for some indicators because the human eye responds more readily to some colors than to others and because one form of an indicator may be more intensely colored than the other, even at identical concentrations.

Conversion of an indicator from one form to another actually uses up some titrant. Clearly, the volume of titrant thus consumed must be negligible in comparison to the amount of acid or base involved in the main titration reaction; otherwise, a significant "indicator error" is introduced. Therefore, the indicator concentration should not be any larger than required to render the end-point color change visible. This "indicator error" is often of slight but measurable magnitude, and it is one reason why experimental end points and theoretical equivalence points may not coincide perfectly. A correction for the "indicator error" can be achieved by means of a **blank titration**. A solution of the same composition as the real sample, containing the indicator but not the acid or base being determined, is titrated to the same end point as in the real titration, and the small volume of titrant is subtracted from the total volume used in the real titration.

### Strong Acid-Strong Base Titrations

A titration curve for an acid-base reaction consists of a plot of pH versus the volume of titrant added. Such a curve can be constructed by consideration of the pertinent acid-base equilibria according to the methods developed in the preceding chapter. Theoretical titration curves are of importance because they provide information concerning the feasibility of a titration, and they are useful for purposes of choosing what end-point indicator should be employed.

**Titration Curve for a Strong Acid.** Let us consider the titration of 50.00 ml of 0.1000 $F$ hydrochloric acid with a 0.1000 $F$ sodium hydroxide solution. Prior to the equivalence point of the titration, there are two sources of hydrogen ions. These are the hydrochloric acid itself and the autoprotolysis of water. In general, the hydrogen ion concentration furnished by the hydrochloric acid is the number of millimoles of acid *remaining untitrated* divided by the solution volume in milliliters. Now, since the autoprotolysis of water yields equal concentrations of hydrogen ion and hydroxide ion, and since water is the *only* source of hydroxide ions, the hydroxide concentration is a direct measure of the hydrogen ion concentration from water. Therefore,

$$[H^+] = \frac{\text{millimoles of acid untitrated}}{\text{solution volume in milliliters}} + [OH^-]$$

or

$$[H^+] = \frac{\text{millimoles of acid untitrated}}{\text{solution volume in milliliters}} + \frac{K_w}{[H^+]}$$

Prior to the addition of any sodium hydroxide titrant, the hydrogen ion concentration contributed by hydrochloric acid is 0.1000 $M$. If we assume that the *total* hydrogen ion concentration is essentially 0.1000 $M$, the hydrogen ion contributed from water is exactly equal to the hydroxide ion concentration and is given by

$$[OH^-] = \frac{K_w}{[H^+]} = \frac{1.00 \times 10^{-14}}{0.1000} = 1.00 \times 10^{-13}\,M$$

Quite obviously, the concentration of hydrogen ions from the autoprotolysis of water is very unimportant. In fact, the autoprotolysis of water does not become significant until one has added *more* than 49.999 ml of the exactly 50 ml of titrant required, so we will neglect the contribution of water. Therefore, at the start of the titration, before any sodium hydroxide titrant has been added, the hydrogen ion concentration is 0.1000 $M$ and the pH is 1.00.

After the addition of 10.00 ml of 0.1000 $F$ sodium hydroxide (1.000 millimole of base), 4.000 millimoles of acid remain untitrated in 60.00 ml of solution, so

$$[H^+] = \frac{4.000 \text{ millimoles}}{60.00 \text{ milliliters}} = 6.67 \times 10^{-2}\, M; \qquad pH = 1.18$$

Similar calculations can be performed for the addition of various volumes of the sodium hydroxide titrant up to, and including, 49.999 ml. In Table 5–2 are presented the results of seven such calculations.

**Table 5–2. Variation of pH During Titration of 50.00 ml of 0.1000 $F$ Hydrochloric Acid with 0.1000 $F$ Sodium Hydroxide**

| Volume of NaOH, ml | pH |
|---|---|
| 0 | 1.00 |
| 10.00 | 1.18 |
| 20.00 | 1.37 |
| 30.00 | 1.60 |
| 40.00 | 1.95 |
| 49.00 | 3.00 |
| 49.90 | 4.00 |
| 49.99 | 5.00 |
| 49.999 | 6.00 |
| 50.000 | 7.00 |
| 50.001 | 8.00 |
| 50.01 | 9.00 |
| 50.10 | 10.00 |
| 51.00 | 11.00 |
| 55.00 | 11.68 |
| 60.00 | 11.96 |
| 65.00 | 12.12 |

After the addition of 50.00 ml of sodium hydroxide, which corresponds to the *equivalence point* of the titration, the hydrogen ion concentration is $1.00 \times 10^{-7}\, M$, water now being the only source of hydrogen ions, and the pH = 7.00.

Beyond the equivalence point of the titration, we are simply adding excess sodium hydroxide titrant to a solution of sodium chloride. There are two sources of hydroxide ions – the added sodium hydroxide solution and the autoprotolysis of water – and the *total* hydroxide ion concentration is given by the following equations:

$$[OH^-] = \frac{\text{millimoles of excess base}}{\text{solution volume in milliliters}} + [H^+]$$

$$[OH^-] = \frac{\text{millimoles of excess base}}{\text{solution volume in milliliters}} + \frac{K_w}{[OH^-]}$$

For volumes of sodium hydroxide titrant equal to or larger than 50.001 ml, the autoprotolysis of water is unimportant, so the second term on the right-hand side of each of the two preceding relations may be omitted. After the addition of 50.001 ml of 0.1000 $F$ sodium hydroxide, there is 0.0001000 millimole of excess base in essentially 100.0 ml of solution; thus,

$$[\mathrm{OH^-}] = \frac{0.0001000 \text{ millimole}}{100.0 \text{ milliliters}} = 1.00 \times 10^{-6}\ M$$

$$\mathrm{pOH} = 6.00; \qquad \mathrm{pH} = 8.00$$

Six more calculations, for different volumes of titrant up to 65.00 ml, provided the remaining pH values listed in Table 5–2.

In Figure 5–1 is shown the complete titration curve constructed from the data in Table 5–2. In addition, a titration curve for the neutralization of 50.00 ml of 0.001000 $F$ hydrochloric acid with 0.001000 $F$ sodium hydroxide is depicted. Notice that the change in pH near the equivalence point for the latter curve is approximately four pH units smaller than observed with reagents which are one hundred times more concentrated.

**Selection of Indicators.** In choosing end-point indicators, we should bear in mind several principles. First, in order for the indicator color change to be very sharply defined, the pH versus volume curve for the proposed titration must exhibit a steeply rising portion in the vicinity of the equivalence point. Second, this steep section of the titration curve must encompass an interval of pH values at least as large as the pH transition interval of an indicator. Third, the pH transition interval of the proposed end-point indicator must coincide with the steep portion of the titration curve. In addition, the

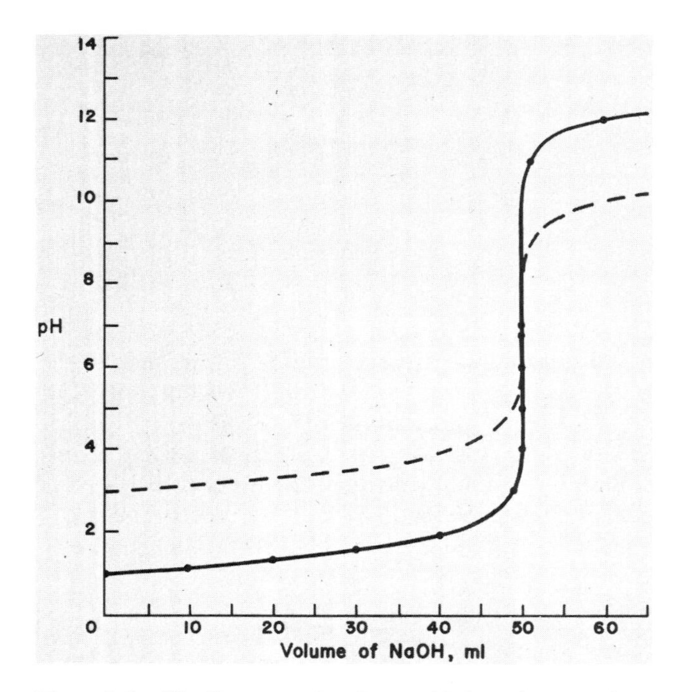


*Figure 5–1.* Titration curves for strong acid-strong base reactions. The solid line represents the titration of 50.00 ml of 0.1000 $F$ hydrochloric acid with a 0.1000 $F$ sodium hydroxide solution; titration data are listed in Table 5–2. The dashed line is the titration curve for the titration of 50.00 ml of 0.001000 $F$ hydrochloric acid with 0.001000 $F$ sodium hydroxide.

direction in which a titration is performed influences the choice of an end-point indicator. In the titration of a strong acid *with* a strong base, the indicator is initially present in its acid form, so the end point is signaled by the sudden appearance of the color of the *base* form of the indicator. In the reverse titration of a strong base *with* a strong acid, the indicator starts out in its base form, and the end point will be identified when the *acid* color of the indicator appears.

Which indicators listed in Table 5–1 would be suitable to detect the end point of the titration of 50.00 ml of 0.1000 $F$ hydrochloric acid with 0.1000 $F$ sodium hydroxide, if we do not want the titration error to exceed ±0.1 per cent? Accordingly, the permissible range of titrant volumes is from 49.95 to 50.05 ml. Using procedures discussed previously for the construction of titration curves, we can determine that the pH values after addition of 49.95 and 50.05 ml of the titrant are 4.30 and 9.70, respectively. Thus, to ensure the desired titration accuracy, the sudden appearance of the color of the *base* form of the indicator must occur between pH 4.30 and 9.70.

A number of indicators in Table 5–1 have *upper* extremes of their pH transition intervals between pH 4.30 and 9.70, but not all indicators in this category are equally useful. Let us examine Figure 5–2, which shows magnified portions of the equivalence-point regions for the titration curves in Figure 5–1. Superimposed upon the curves are the pH transition intervals for five common acid-base indicators – methyl orange, methyl red, bromthymol blue, phenol red, and phenolphthalein. Bromthymol blue and phenol red are excellent indicators for this titration because their pH transition intervals lie almost in the middle of the steep vertical part of the titration curve. Methyl red and phenolphthalein

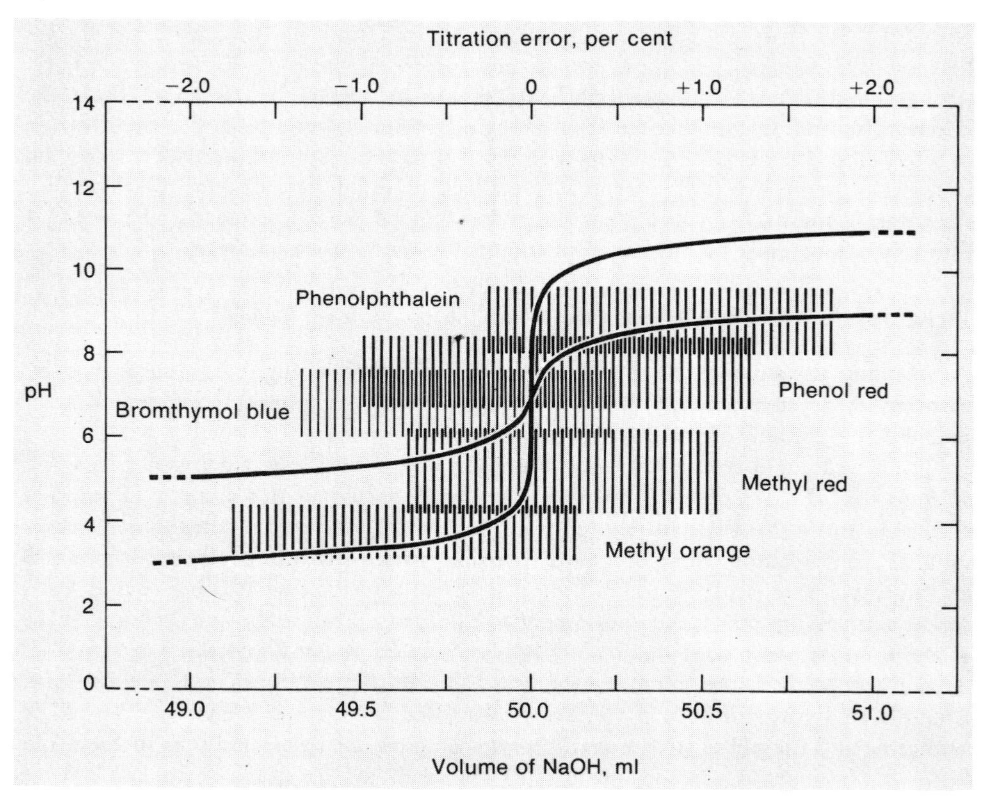


*Figure 5–2.* Magnified portions of equivalence-point regions for titration curves shown in Figure 5–1, along with pH transition intervals for methyl orange, methyl red, bromthymol blue, phenol red, and phenolphthalein.

are satisfactory indicators, since their pH transition intervals coincide reasonably well with the steeply rising portion of the titration curve. However, methyl orange begins to change color as early as one per cent before the equivalence point; although the *upper* limit of its pH transition interval is within the optimum pH range of 4.30 to 9.70, the color change from red to yellow would be too gradual to be acceptable. Choices of other acid-base indicators in Table 5–1 can be similarly evaluated.

For the titration of 50.00 ml of 0.001000 $F$ hydrochloric acid with 0.001000 $F$ sodium hydroxide solution, calculations reveal that, in order to restrict the titration error to ±0.1 per cent, an acid-base indicator must change to the color of its *base* form within the rather narrow range between pH 6.29 and 7.71. This information is conveyed graphically in Figure 5–2. Bromthymol blue, as well as several other indicators in Table 5–1, is a useful end-point indicator for this titration, whereas phenolphthalein, methyl orange, and methyl red are unsatisfactory. From Table 5–1 we see that phenol red would provide an end-point pH of 8.4. An inspection of Figure 5–2 reveals that a titration error of +0.5 per cent would result. However, if a blank titration were performed to correct for the "indicator error," phenol red could undoubtedly be employed as an end-point indicator.

**Titration Curves and Indicators for Strong Base-Strong Acid Titrations.** Calculation and construction of a titration curve for the titration of a strong base *with* a strong acid differ little from the preceding description for strong acid-strong base titrations, except that the roles of the strong acid and strong base are reversed. In fact, the data listed in Table 5–2 can immediately be employed to construct the pH-versus-volume curve for the titration of 50.00 ml of 0.1000 $F$ sodium hydroxide with 0.1000 $F$ hydrochloric acid, if it is recognized that the tabulated pH values become pOH values – which can be converted into pH readings – for the strong base-strong acid titration. End-point indicators may be chosen by means of the same criteria discussed for the titration of a strong acid with a strong base. For the titration of 0.1 $F$ sodium hydroxide solution with 0.1 $F$ hydrochloric acid, methyl orange and bromcresol green are both successfully utilized as indicators, despite the facts that the colors of their *acid* forms appear somewhat beyond the equivalence point and that blank titrations must be employed to correct for the "indicator errors."

### Titrations of Weak Monoprotic Acids with Strong Bases

Among the numerous weak monoprotic acids which might be determined by titration with a strong base are formic acid, acetic acid, propanoic acid, benzoic acid, salicylic acid, anilinium ion, and pyridinium ion.

In a weak acid-strong base titration, the sample is initially a solution of a weak acid, so that the pH can be calculated as shown in the preceding chapter (Examples 4–6, 4–7, and 4–8). Prior to the equivalence point, addition of titrant neutralizes some of the weak acid; therefore, we are dealing with a mixture of the weak acid and its conjugate base, and the pH is that of a buffer solution (Examples 4–11 and 4–12). At the equivalence point, the solution consists only of a weak base – the conjugate base of the original weak acid – and the pH is calculated as for Examples 4–9 and 4–10. Beyond the equivalence point, the solution contains the weak base and an excess of a strong base, hydroxide ion; accordingly, the effect of the weak base on the pH is negligible and the pH of such a solution may be evaluated essentially as in Example 4–5.

**Titration Curve for the Acetic Acid-Sodium Hydroxide Reaction.** Let us construct the theoretical titration curve for the titration of 25.00 ml of 0.1000 $F$ acetic acid with 0.1000 $F$ sodium hydroxide titrant.

Initially, before any titrant is added, the pH of a 0.1000 $F$ acetic acid solution is 2.88.

As each volume increment of sodium hydroxide solution is added, it reacts virtually completely with acetic acid to produce a stoichiometrically equivalent amount of acetate

$$CH_3COOH + OH^- \rightleftharpoons CH_3COO^- + H_2O$$

and the concentration of hydrogen ion is determined by the ratio of the acetic acid and acetate ion concentrations:

$$[H^+] = K_a \frac{[CH_3COOH]}{[CH_3COO^-]}$$

Since we have the *ratio* of two concentrations in this last equation, it suffices to substitute just the numbers of millimoles of acetic acid and acetate ion present at equilibrium after each portion of titrant is introduced. For example, after the addition of 5.00 ml of 0.1000 $F$ sodium hydroxide solution, 2.000 millimoles of acetic acid remain untitrated and 0.500 millimole of acetate ion is formed; hence,

$$[H^+] = 1.75 \times 10^{-5} \frac{(2.000)}{(0.500)} = 7.00 \times 10^{-5} M; \qquad pH = 4.15$$

In Table 5–3 are presented the results of other similar computations for titrant volumes up to 24.90 ml; in particular, note that the pH of the solution after addition of 12.50 ml of titrant is numerically equal to the $pK_a$ value of 4.76 for acetic acid, because at this point exactly one-half of the acetic acid has been titrated and the concentration ratio of acetic acid to acetate is unity. In performing these calculations, we neglected the autoprotolysis of water and we avoided any points on the titration curve exceedingly close to the equivalence point, where the concentration of acetic acid becomes so small that we would have to take account of the decrease in $[CH_3COOH]$ and the increase in $[CH_3COO^-]$ due to dissociation of acetic acid.

**Table 5–3. Variation of pH During Titration of 25.00 ml of 0.1000 *F* Acetic Acid with 0.1000 *F* Sodium Hydroxide**

| Volume of NaOH, ml | pH |
|---|---|
| 0 | 2.88 |
| 5.00 | 4.15 |
| 10.00 | 4.58 |
| 12.50 | 4.76 |
| 15.00 | 4.93 |
| 20.00 | 5.36 |
| 24.00 | 6.14 |
| 24.90 | 7.15 |
| 25.00 | 8.73 |
| 26.00 | 11.29 |
| 30.00 | 11.96 |
| 40.00 | 12.36 |

At the equivalence point, corresponding to the addition of 25.00 ml of sodium hydroxide, the solution is identical to one which could be prepared by dissolving 0.002500 mole of sodium acetate in 50.00 ml of water. Using the method discussed in Example 4–9, we calculate that the pH of a 0.05000 *F* sodium acetate solution is 8.73.

After the addition of 26.00 ml of 0.1000 *F* sodium hydroxide, there is an excess of 0.1000 millimole of strong base in a total volume of 51.00 ml. Therefore, $[OH^-] = 1.96 \times 10^{-3}$ *M*, pOH = 2.71, and pH = 11.29. Results of this and two more such calculations are listed in Table 5–3.

In Figure 5–3 is plotted the complete titration curve. On examining it closely, we note several characteristics which distinguish it from the strong acid-strong base titration curve of Figure 5–1. Prior to the equivalence point, the titration curve for a weak acid is displaced toward higher pH values (lower hydrogen ion concentrations) because only a fraction of the weak acid dissociates to furnish hydrogen ions, and the pH at the equivalence point is on the basic side of pH 7. In addition, the steep vertical portion of the curve encompasses a much smaller pH range than the analogous strong acid-strong base titration curve in Figure 5–1. There is one important similarity between the two titrations; beyond the equivalence point for a weak acid-strong base titration, where we are adding excess titrant, the system is essentially identical to a strong acid-strong base system.

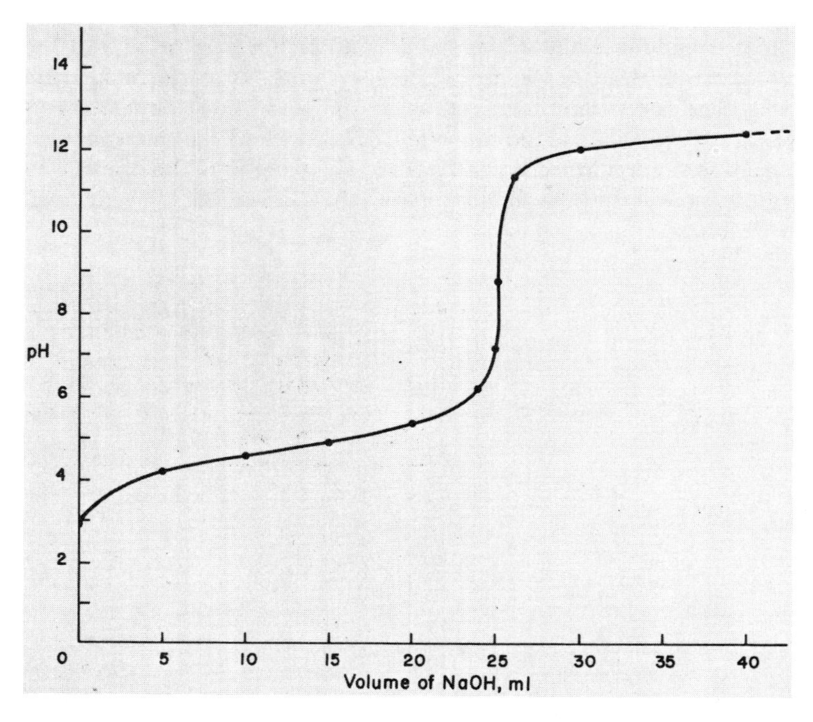


*Figure 5–3.* Titration curve for the titration of 25.00 ml of 0.1000 *F* acetic acid with a 0.1000 *F* sodium hydroxide solution. Titration data are listed in Table 5–3.

**Selection of Indicators.** In view of the characteristics of the titration curve for the acetic acid-sodium hydroxide system, which end-point indicators can be employed without causing a titration error greater than one part per thousand? We must calculate the pH of the solution 0.1 per cent before and after the equivalence point. One-tenth per cent *before* the equivalence point, 99.9 per cent of the original acetic acid has been converted to acetate ion, and 0.1 per cent remains as acetic acid molecules. Since the ratio of acetic acid to acetate ion is 1/999, or approximately 1/1000, we can write

$$[H^+] = K_a \frac{[CH_3COOH]}{[CH_3COO^-]} = \frac{1.75 \times 10^{-5}}{1000} = 1.75 \times 10^{-8} \, M; \qquad pH = 7.76$$

One-tenth per cent *after* the equivalence point, there is an excess of 0.025 ml of 0.1000 $F$ sodium hydroxide in a solution whose volume is essentially 50.00 ml, which corresponds to a pH of 9.70. Thus, to be suitable, an indicator must change abruptly from its acid form to its base form within the pH range from 7.76 to 9.70.

Figure 5–4 shows the magnified equivalence-point region of the titration curve for the acetic acid-sodium hydroxide system, along with the pH transition intervals for phenol red and phenolphthalein. It is obvious why phenolphthalein is the almost universal choice of indicator for the titration. Although the color of the base form of phenol red appears at pH 8.4, the gradual color change makes this indicator less satisfactory than phenolphthalein. Three other indicators listed in Table 5–1 – brilliant yellow, neutral red, and metanitrophenol – are comparable in behavior to phenol red.

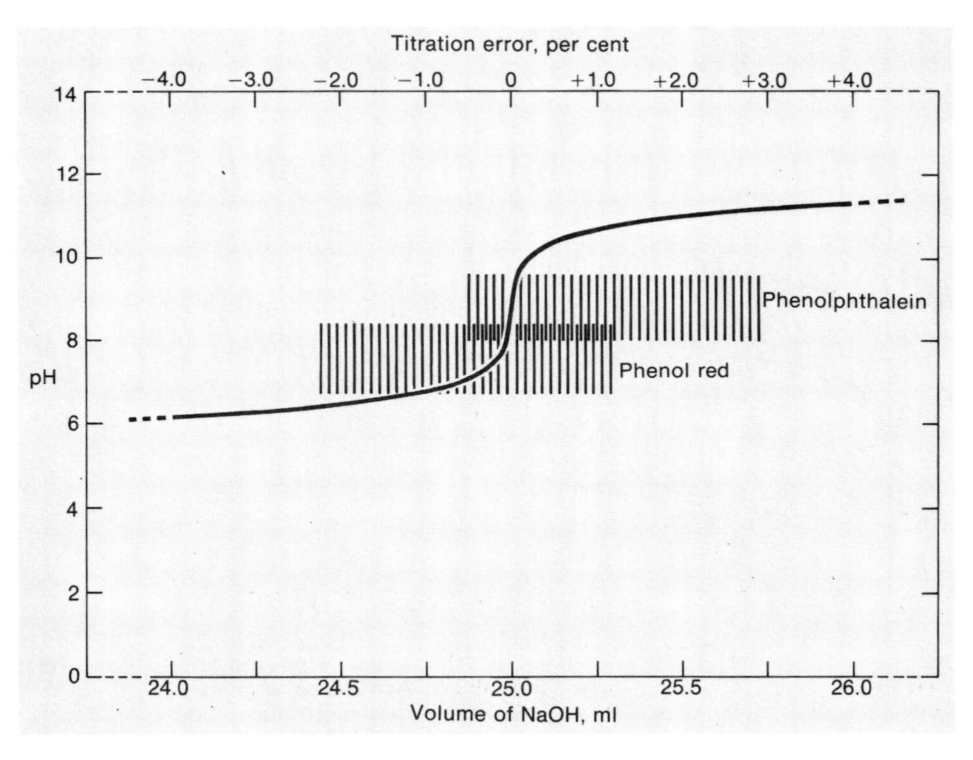


*Figure 5–4.* Magnified portion of equivalence-point region for titration of 25.00 ml of 0.1000 $F$ acetic acid with 0.1000 $F$ sodium hydroxide, showing pH transition intervals for phenolphthalein (the preferred end-point indicator) and phenol red.

**Feasibility of Titrations.** As Figure 5–5 reveals, the magnitude of the dissociation constant of a weak acid has a profound influence on the size and sharpness of the pH

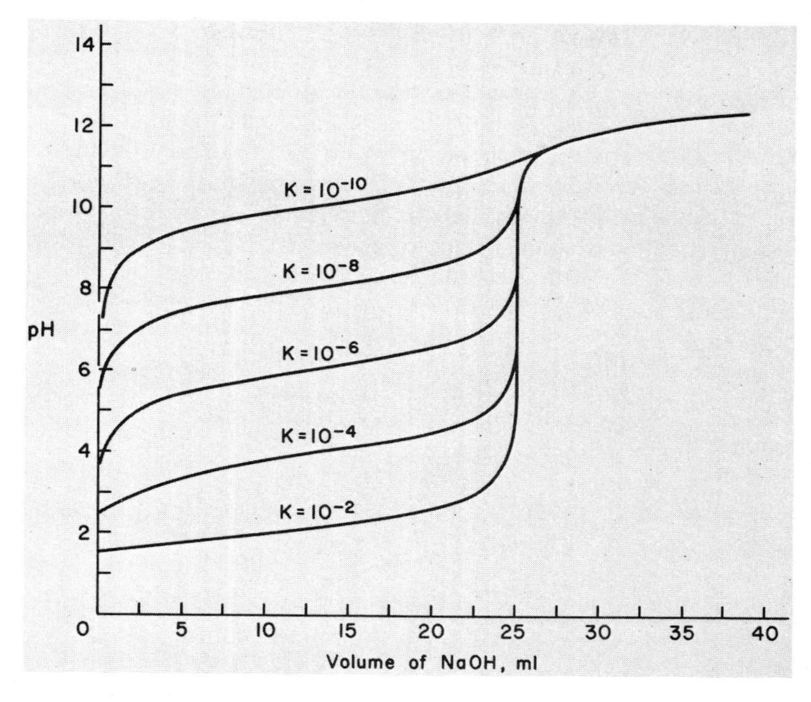


*Figure 5–5.* Family of titration curves for weak acid-strong base titrations, showing the effect of the value of the dissociation constant on the shape and position of the curve. For each titration curve, 25.00 ml of 0.1000 $F$ weak acid is titrated with 0.1000 $F$ sodium hydroxide.

change near the equivalence point of a titration. Furthermore, decreases in the concentrations of the weak acid and the titrant base diminish the pH change. Since most acid-base indicators change color over a two-pH-unit interval, the weakest acid which can be successfully titrated is one whose $K_a$ is about $10^{-7}$.

However, to obtain accurate analytical results, a special procedure must be followed. Suppose that it is desired to determine carbonic acid ($H_2CO_3$) – a very weak diprotic acid with a first dissociation constant ($K_{a1}$) of $4.47 \times 10^{-7}$ – by titration with a standard sodium hydroxide solution, according to the reaction

$$H_2CO_3 + OH^- \rightleftharpoons HCO_3^- + H_2O$$

Phenolphthalein is the usual choice for end-point indicator. At the end point of the titration, the solution is indistinguishable from one prepared by dissolution of sodium bicarbonate in water. From the results of the first titration, it is possible to calculate the concentration of sodium bicarbonate present at the end point. Now, in a separate titration flask, we reproduce artificially, but as accurately as possible, the conditions at the end point by adding the proper amounts of water, sodium bicarbonate, and phenolphthalein indicator. What we have, then, is a comparison solution in which the end-point pH is correct and in which a definite concentration of phenolphthalein is present in its pink base form. In subsequent titrations the titrant is carefully added until the end-point colors in the real and comparison titration flasks are matched as closely as possible.

## Titrations of Weak Bases with Strong Acids

By titration with a standard solution of a strong acid, many weak bases can be determined; ammonia, piperidine, ethanolamine, and triethylamine, as well as cyanide, borate, and phenoxide anions, are representative of this group of substances. In principle, the construction of theoretical titration curves for weak base-strong acid systems involves calculations similar to those already discussed for weak acid-strong base titrations. Figure 5–6 depicts the titration curve for the titration of 25.00 ml of 0.01000 $F$ sodium phenoxide solution with 0.01000 $F$ hydrochloric acid.

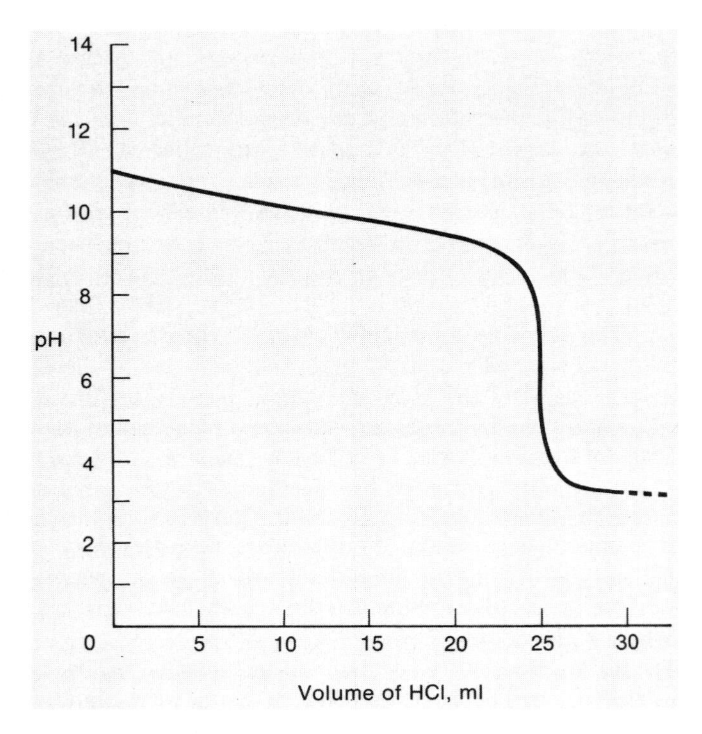


*Figure 5–6.* Titration curve for the titration of 25.00 ml of 0.01000 $F$ sodium phenoxide solution with 0.01000 $F$ hydrochloric acid.

## Titrations of Polyprotic Acids and Mixtures of Monoprotic Acids

It is not unusual for two or more acidic or basic species to be present in the same solution. Consider a mixture of two acids which is to be titrated with a strong base. If both acids are strong, such as a mixture of hydrochloric acid and perchloric acid, the titration curve would have a shape similar to that in Figure 5–1, and it would be impossible to determine the concentrations of the individual acids. A second possibility is that we have a mixture of two weak acids with nearly identical dissociation constants. Here, too, the two weak acids would behave quite similarly as hydrogen ion sources. Therefore, the resulting titration curve would exhibit the general characteristics of the curve shown in Figure 5–3, and it would not be feasible to analyze the mixture for its individual components.

If, however, the mixture contains one strong acid and one weak acid, or two weak acids of unequal strength, the extent to which titration of the stronger acid is completed before the second one begins to react with the titrant is determined by the relative strengths of the two acids. If this difference is sufficiently great, the titration curve will exhibit two distinct equivalence points, one for each of the two successive titrations.

**Titration of a Mixture of Hydrochloric and Acetic Acids.** Titration of a solution containing both hydrochloric acid and acetic acid with sodium hydroxide is an example of the kind of situation mentioned in the preceding paragraph. We may expect the hydrochloric acid to be titrated first because it is the stronger acid, followed by the reaction of sodium hydroxide with acetic acid.

A titration curve for the titration of 50.00 ml of a solution 0.1000 $F$ in both hydrochloric and acetic acids with 0.2000 $F$ sodium hydroxide is presented in Figure 5–7. We constructed the titration curve by considering the presence of only hydrochloric acid during the first step of the titration. This approximation is reasonable because hydrochloric acid represses the dissociation of acetic acid up to within a few per cent of the first equivalence point. At the first equivalence point, the pH may be calculated from the equilibrium expression for dissociation of acetic acid, for at this stage of the titration the solution consists of acetic acid and sodium chloride only. Beyond the first equivalence point, the titration curve is simply that of a weak acid-strong base system and is analogous to Figure 5–3.

An important characteristic of the titration curve is the absence of a sharply defined equivalence point at pH 3, because acetic acid starts to react before all the hydrochloric acid has been titrated. Although the quality of the titration curve would probably permit determination of the individual concentrations of hydrochloric acid and acetic acid with an error no worse than 1 per cent, the combination of 0.1 $F$ hydrochloric acid and 0.1 $F$ weak acid of $K_a = 10^{-5}$ represents a rough limit for a strong acid-weak acid mixture that can be analyzed with reasonable accuracy.

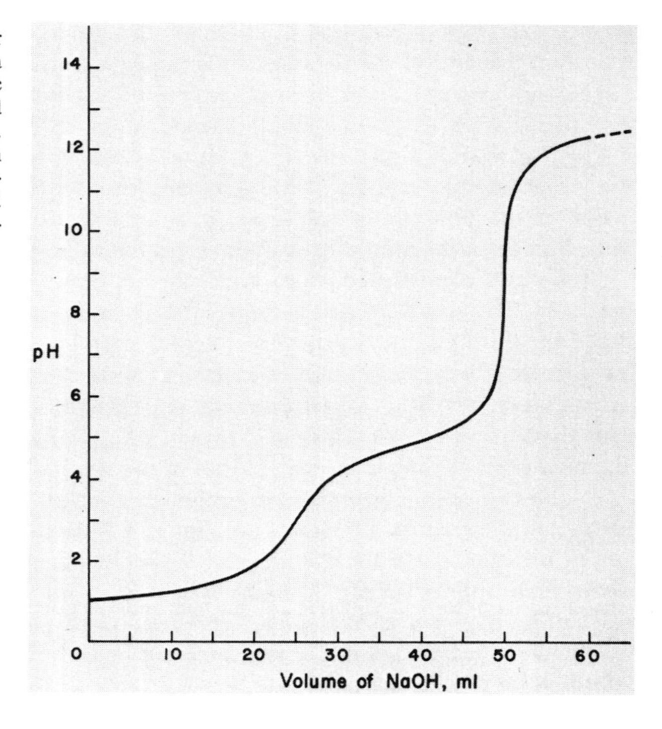


*Figure 5–7.* Titration curve for the titration of 50.00 ml of a mixture of 0.1000 $F$ hydrochloric acid and 0.1000 $F$ acetic acid with 0.2000 $F$ sodium hydroxide. The first step of the titration curve corresponds to the neutralization of the hydrochloric acid and the second step to the neutralization of acetic acid.

Experimentally, the location of the first equivalence point could be best accomplished graphically from a complete plot of the titration curve. Alternatively, an acid-base indicator which changes to the color of its *base* form at pH 3 might be employed, but it would be advisable to prepare a color comparison solution containing the appropriate concentrations of acetic acid and indicator in order to approximate the equivalence-point conditions as closely as possible. Phenolphthalein could be used to signal the second equivalence point.

**Titration of Phosphoric Acid with Sodium Hydroxide Solution.** In the last chapter, the stepwise dissociation of phosphoric acid was represented by the following set of proton-transfer equilibria:

$$H_3PO_4 + H_2O \rightleftharpoons H_2PO_4^- + H_3O^+; \qquad K_{a1} = 7.5 \times 10^{-3}$$
$$H_2PO_4^- + H_2O \rightleftharpoons HPO_4^{2-} + H_3O^+; \qquad K_{a2} = 6.2 \times 10^{-8}$$
$$HPO_4^{2-} + H_2O \rightleftharpoons PO_4^{3-} + H_3O^+; \qquad K_{a3} = 4.8 \times 10^{-13}$$

A titration curve for the titration of 25.00 ml of 0.5000 *F* phosphoric acid with 0.5000 *F* sodium hydroxide solution is shown in Figure 5–8. We shall not pursue the detailed calculations required to construct this curve, but will emphasize the key features of the titration. It is essential to recognize that the present system is identical to an aqueous solution containing a mixture of three weak monoprotic acids, each at an initial concentration of 0.5 *F*, with dissociation constants of $K_{a1}$, $K_{a2}$, and $K_{a3}$, respectively. A mixture of 0.5 *F* concentrations of a weak diprotic acid and a weak monoprotic acid would likewise mimic the behavior of the phosphoric acid-sodium hydroxide titration if the dissociation constants had the same numerical values as those for phosphoric acid.

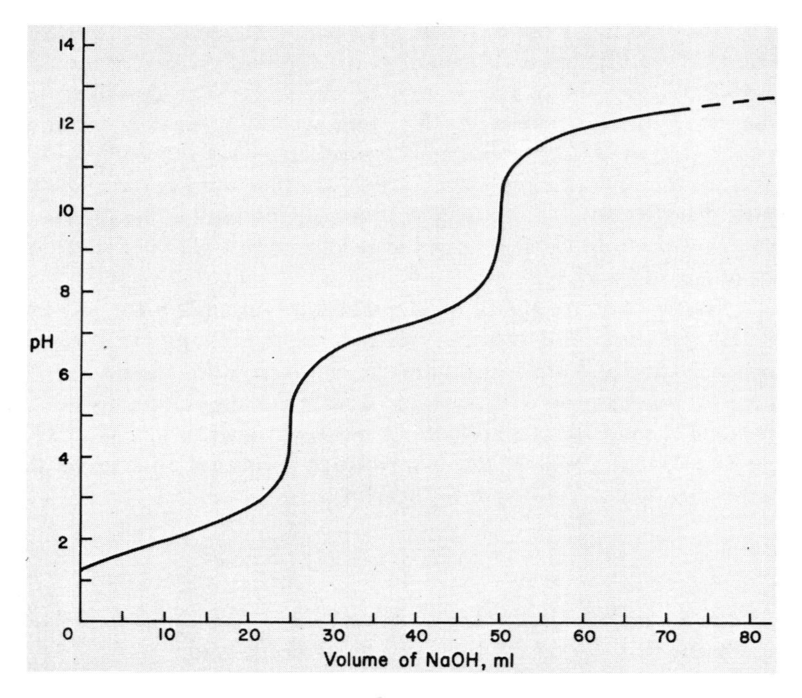


*Figure 5–8.* Titration curve for the titration of 25.00 ml of 0.5000 *F* phosphoric acid with a 0.5000 *F* sodium hydroxide solution. Although the stepwise titration of $H_3PO_4$, first to $H_2PO_4^-$ and then to $HPO_4^{2-}$, is evident, the titration of the third proton of phosphoric acid cannot be observed because $HPO_4^{2-}$ is too weak an acid in water.

Only two equivalence points are observed, corresponding to the successive titrations of $H_3PO_4$ and $H_2PO_4^-$; because $HPO_4^{2-}$ is an exceedingly weak acid, the titration of the third proton of phosphoric acid does not proceed to any significant extent. For the case of phosphoric acid, the curve showing the titration of the first and second protons is resolved into two well-defined steps because $pK_{a1}$ and $pK_{a2}$ differ by slightly more than five $pK$ units ($pK_{a1} = 2.13$ and $pK_{a2} = 7.21$).

As a general rule, if the $pK_a$ values of two weak acids differ by four or more $pK$ units, two distinct equivalence points for the stepwise titration of a mixture of these acids can be obtained. As revealed by Figure 5–5, in order to observe a satisfactory titration curve for the weaker of the two acids, $pK_a$ for the latter must be *smaller* than approximately 8. If the two weak acids in a mixture have $pK_a$ values which differ by less than four units, the reaction of the weaker acid will be well underway before the titration of the stronger one is complete, and the first equivalence point will be poorly defined.

Because the acidities of $H_3PO_4$ and $H_2PO_4^-$ are substantially different, construction of the titration curve in Figure 5–8 is straightforward. First, the initial pH of the solution is governed essentially by dissociation of the first proton from phosphoric acid. Second, up to the first equivalence point, the addition of titrant produces a buffer whose pH varies with the ratio of $[H_3PO_4]$ to $[H_2PO_4^-]$. Third, at the first equivalence point, the solution consists of the amphiprotic species $NaH_2PO_4$, the pH of which can be determined as described in the previous chapter. Fourth, between the first and second equivalence points, when $H_2PO_4^-$ is converted to $HPO_4^{2-}$, the pH is controlled by the ratio of $[H_2PO_4^-]$ to $[HPO_4^{2-}]$. Fifth, at the second equivalence point, another amphiprotic substance ($Na_2HPO_4$) exists. Finally, beyond the second equivalence point, it is necessary to consider both the dissociation of $HPO_4^{2-}$ and the autoprotolysis of water in order to calculate the pH of the solution.

**Titrimetric Analysis of a Carbonate-Bicarbonate Mixture.** An analysis of a sample containing two weak bases is exemplified by the determination of sodium carbonate and sodium bicarbonate in a mixed solid. Figure 5–9 shows a titration curve for the titration with standard 0.3500 $F$ hydrochloric acid of a solution of 1.200 grams of a solid sample containing unknown amounts of $Na_2CO_3$, $NaHCO_3$, and NaCl in 100 ml of water. Two equivalence points are observable, the first pertaining to the titration of carbonate to bicarbonate

$$CO_3^{2-} + H^+ \rightleftharpoons HCO_3^-$$

and the second corresponding to conversion of bicarbonate (including that formed during the first step of the titration) to carbonic acid:

$$HCO_3^- + H^+ \rightleftharpoons H_2CO_3$$

Inspection of Figure 5–9 reveals that the pH change at the first equivalence point is inadequately sharp for use of an acid-base indicator, such as phenolphthalein, unless a color comparison solution is prepared in the manner described in our discussion of the feasibility of weak acid-strong base titrations (page 101). Thus, a solution is prepared having a volume and containing concentrations of phenolphthalein and $NaHCO_3$ as close as possible to the sample solution at the first equivalence point. Then, in an actual titration of the sample, hydrochloric acid is introduced until the color of the solution titrated matches that of the comparison solution. One can obtain results accurate to within ±0.5 per cent by means of this technique. Methyl orange provides a good indication of the second equivalence point, the end point being taken as the first perceptible change from the pure yellow color of the base form to an intermediate orange

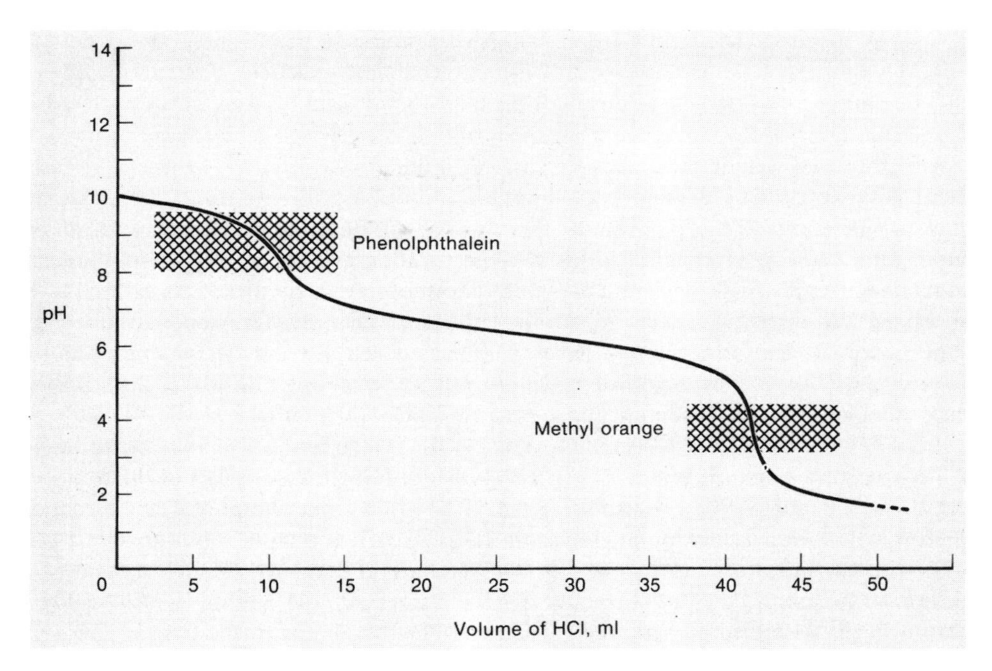


*Figure 5-9.* Titration curve for analysis of a carbonate-bicarbonate mixture with 0.3500 *F* hydrochloric acid. The first step of the titration corresponds to the formation of bicarbonate ion ($HCO_3^-$), and the second step pertains to the conversion of bicarbonate (including the $HCO_3^-$ produced during the first step) to carbonic acid. Phenolphthalein may be employed to indicate the first end point if a special procedure is utilized, whereas methyl orange is used to signal the second end point.

color. Alternatively, it is possible to follow the entire course of the titration with the aid of a glass-electrode pH meter. A complete plot of pH versus volume of titrant can be constructed from the experimental data, and the titration curve can be analyzed graphically to locate the two equivalence points precisely. Furthermore, using a glass-electrode pH meter, one can titrate to the exact pH of the first equivalence point.

In analogy to the titration of a pair of weak acids, the difference between $pK_b$ values for two weak bases must be at least four $pK$ units in order to determine accurately the amounts of the individual components by acid-base titrimetry. For carbonate and bicarbonate, the equilibria which govern their behavior as proton acceptors are

$$CO_3^{2-} + H_2O \rightleftharpoons HCO_3^- + OH^-; \qquad K_{b1} = \frac{K_w}{K_{a2}} = 2.14 \times 10^{-4}$$

and

$$HCO_3^- + H_2O \rightleftharpoons H_2CO_3 + OH^-; \qquad K_{b2} = \frac{K_w}{K_{a1}} = 2.24 \times 10^{-8}$$

Thus, for the present system, $pK_{b1}$ and $pK_{b2}$ differ by almost exactly four units. However, the presence of $HCO_3^-$ in the original sample solution shifts the titration curve for $CO_3^{2-}$ to lower pH values, thereby causing the pH change near the first equivalence point to be smaller and less pronounced than predicted from the magnitudes of $pK_{b1}$ and $pK_{b2}$; this effect will not occur if the weaker base is *not* the conjugate acid of the stronger base. Between the first and second equivalence points, the shape of the titration curve for the conversion of $HCO_3^-$ to $H_2CO_3$ is exactly that expected for the titration of the conjugate base of any weak monoprotic acid.

To calculate the percentage of $Na_2CO_3$ in the mixed solid, note that the amount of carbonate in the sample solution is measured by the volume and concentration of the standard hydrochloric acid added up to the first equivalence point. Since the volume of titrant is close to 10.6 ml, the quantity of carbonate – and, therefore, of sodium carbonate in the original solid sample – is (10.6)(0.350) or 3.71 millimoles; and, inasmuch as the formula weight of $Na_2CO_3$ is 106, the sodium carbonate content of the 1200-milligram sample is

$$\frac{3.71 \text{ millimoles} \cdot 106 \dfrac{\text{milligrams}}{\text{millimole}}}{1200 \text{ milligrams}} \times 100 = 32.8 \text{ per cent}$$

Although 31.2 ml of hydrochloric acid is added between the first and second equivalence points to convert $HCO_3^-$ to $H_2CO_3$, 10.6 ml of titrant reacts with the $HCO_3^-$ formed from the original $CO_3^{2-}$, so only 20.6 ml of hydrochloric acid is equivalent to the $NaHCO_3$ originally present. Thus, the number of millimoles of $NaHCO_3$ is (20.6)(0.350) or 7.21 and, since the formula weight of $NaHCO_3$ is 84.0, the amount of $NaHCO_3$ in the solid sample is

$$\frac{7.21 \text{ millimoles} \cdot 84.0 \dfrac{\text{milligrams}}{\text{millimole}}}{1200 \text{ milligrams}} \times 100 = 50.5 \text{ per cent}$$

It follows that there must be 16.7 per cent of sodium chloride in the mixed solid.

## ACID-BASE TITRATIONS IN NONAQUEOUS SOLVENTS

Earlier, we identified the hydronium ion, $H_3O^+$, as the form of the hydrated proton and as the strongest acid which can exist in water. In nonaqueous solvents, such as ethanol or glacial acetic acid, the proton is not well characterized, but it is clear that the proton is solvated. By analogy to the hydronium ion in water, we shall represent the solvated proton in ethanol and in glacial acetic acid by the formulas $C_2H_5OH_2^+$ and $CH_3COOH_2^+$, respectively. It follows that $C_2H_5OH_2^+$ is the strongest acid to be encountered in ethanol and that $CH_3COOH_2^+$ is the strongest proton donor found in glacial acetic acid.

One of the virtues of the Brønsted-Lowry concept of acids and bases is that proton-transfer reactions in water and in nonaqueous solvents may be discussed in identical terms. If we consider the titration of ammonia ($NH_3$) with a strong acid in water and in glacial acetic acid, the two reactions may be represented by the following equilibria:

$$H_3O^+ + NH_3 \rightleftharpoons H_2O + NH_4^+ \tag{1}$$

$$CH_3COOH_2^+ + NH_3 \rightleftharpoons CH_3COOH + NH_4^+ \tag{2}$$

As each of these reactions proceeds toward equilibrium from left to right, $NH_3$ is accepting a proton, whereas a solvent molecule (water or acetic acid) accepts a proton as the reverse reaction occurs. For the present examples, the final positions of equilibrium are determined by the relative strengths as Brønsted-Lowry bases of $NH_3$ and $H_2O$ in reaction 1 and of $NH_3$ and $CH_3COOH$ in reaction 2. Since water is considerably more basic than acetic acid, the equilibrium position of reaction 2 lies much farther to the right than that of reaction 1. In other words, $NH_3$ appears to be a stronger base (or a better proton acceptor) in glacial acetic acid than it is in water because $CH_3COOH_2^+$ is a much better proton donor than $H_3O^+$.

This discussion leads us to the important conclusion that many bases too weak to be titrated with a strong acid in water can be titrated successfully if glacial acetic acid is chosen as the solvent. In glacial acetic acid, such weak bases appear to be much stronger. Other solvents which, like glacial acetic acid, are less basic or more acidic than water could be similarly employed. Furthermore, weak acids which cannot be titrated in water with strong bases may be determined if the titrations are performed in solvents more basic than water. A solvent which is more basic than water will enhance the strength of a weak acid because such a solvent has a greater affinity for protons. Another reason why nonaqueous solvents are of analytical interest is that many organic compounds having acidic or basic functional groups are insoluble in water and can be titrated only if they are dissolved in organic solvents or perhaps in mixtures of water and an organic solvent.

## Some Properties of Solvents

In general, the acid-base behavior of any solute dissolved in a given solvent will depend on the acid-base properties of the solvent relative to the solute.

**Classification of Solvents.** An **amphiprotic solvent** is one capable of acting as either a Brønsted-Lowry acid or base. Perhaps the most familiar example is water, its amphiprotic character being shown by the reaction between one water molecule (acting as an acid) and a second molecule of water (acting as a base):

$$H_2O + H_2O \rightleftharpoons H_3O^+ + OH^-$$

This proton-transfer process is the so-called **autoprotolysis** of water. Most alcohols, such as ethanol, undergo analogous autoprotolysis reactions and have acid-base properties similar to water:

$$C_2H_5OH + C_2H_5OH \rightleftharpoons C_2H_5OH_2^+ + C_2H_5O^-$$

Glacial acetic acid represents a somewhat different kind of amphiprotic solvent. Although glacial acetic acid does undergo the usual type of autoprotolysis reaction,

$$CH_3COOH + CH_3COOH \rightleftharpoons CH_3COOH_2^+ + CH_3COO^-$$

it is distinctly more acidic than water because all bases appear to be stronger in acetic acid than in water. Other solvents such as liquid ammonia

$$NH_3 + NH_3 \rightleftharpoons NH_4^+ + NH_2^-$$

and ethylenediamine

$$H_2NCH_2CH_2NH_2 + H_2NCH_2CH_2NH_2 \rightleftharpoons H_2NCH_2CH_2NH_3^+ + H_2NCH_2CH_2NH^-$$

exhibit amphiprotic behavior but, compared to water, these are much more basic since they cause all acids to be stronger.

There is another class of solvents, called **inert** or **aprotic solvents**, which do not show any appreciable acid or base properties. Benzene, chloroform, and carbon tetrachloride are typical aprotic solvents, for they have no ionizable protons and they have little or no tendency to accept protons from other substances. In general, it is unrealistic to propose an autoprotolysis reaction for such a solvent.

Still another group of solvents includes those which possess definite base properties but which are without acid properties. Consequently, for these solvents an autoprotolysis reaction cannot be written. Such solvents are frequently included in the list of aprotic or inert solvents. For example, pyridine ($C_5H_5N$) can accept a proton from a Brønsted-Lowry acid such as water,

$$C_5H_5N: + H_2O \rightleftharpoons C_5H_5N{:}H^+ + OH^-$$

but the acid properties of pyridine are virtually nonexistent. Ethers and ketones are capable of accepting protons from strong Brønsted-Lowry acids such as sulfuric acid, but like pyridine they have no ionizable protons.

**Leveling Effect and Differentiating Ability of a Solvent.** In water, the so-called mineral acids – perchloric acid, sulfuric acid, hydrochloric acid, hydrobromic acid, hydriodic acid, and nitric acid – all appear to be equally strong; it is impossible to decide whether any real differences in acid strength exist for these acids. Thus, if we consider the acid strengths of perchloric acid and hydrochloric acid in water, the following acid-base equilibria are involved:

$$HClO_4 + H_2O \rightleftharpoons ClO_4^- + H_3O^+$$

$$HCl + H_2O \rightleftharpoons Cl^- + H_3O^+$$

Since $HClO_4$ and HCl are both much stronger acids than the hydronium ion, $H_3O^+$ (or because water is a stronger base than either $ClO_4^-$ or $Cl^-$), the position of equilibrium in each case lies so far to the right that it is impossible experimentally to distinguish any difference between the two equilibrium positions. Because the strengths of perchloric and hydrochloric acids appear to be identical in water, we speak of water as exerting a **leveling effect** on these two acids.

If the same acids, $HClO_4$ and HCl, are compared in glacial acetic acid as solvent, the pertinent acid-base reactions are

$$HClO_4 + CH_3COOH \rightleftharpoons ClO_4^- + CH_3COOH_2^+$$

and

$$HCl + CH_3COOH \rightleftharpoons Cl^- + CH_3COOH_2^+$$

Earlier, we stated that glacial acetic acid is a much weaker base than water and that the protonated acetic acid molecule ($CH_3COOH_2^+$) is a stronger acid than the hydronium ion ($H_3O^+$). Therefore, neither of these acid-base reactions proceeds so far toward the right in acetic acid as in water. However, the superior acid strength of perchloric acid shows up in the fact that its reaction with the glacial acetic acid solvent attains a greater degree of completion than the reaction of hydrochloric acid with the solvent. Thus, glacial acetic acid possesses the capability to differentiate the acid strengths of $HClO_4$ and HCl, so we call it a **differentiating solvent.**

Clearly, the *leveling* or *differentiating* effect of a solvent depends to a large extent on the acid-base properties of the solvent relative to the dissolved solutes. Although water is a very poor choice of solvent to differentiate the acid strengths of $HClO_4$ and HCl, it is a good solvent to differentiate a mineral acid such as $HClO_4$ or HCl from the much weaker acetic acid. However, a strongly basic solvent such as liquid ammonia would fail to differentiate a mineral acid from acetic acid because the reactions

$$HClO_4 + NH_3 \rightleftharpoons ClO_4^- + NH_4^+$$

and

$$CH_3COOH + NH_3 \rightleftharpoons CH_3COO^- + NH_4^+$$

would both proceed virtually to completion because of the great base strength of $NH_3$.

### Titrations in Basic Solvents

In general, the titration of weakly acidic substances in aqueous media is limited to species having acid dissociation constants not smaller than approximately $10^{-7}$. Acids which possess $K_a$ values below about $10^{-7}$ in water are too weak to provide adequately sharp titration curves for titrations in an aqueous medium with strong bases such as sodium or potassium hydroxide. On the other hand, the use of solvents more basic than water can greatly improve the sharpness of titration curves and the accuracy of the titrations themselves by enhancing the strengths of the solute acids.

**Typical Solvents and Titrants.** Among the common basic solvents are butylamine, ethylenediamine, pyridine, and liquid ammonia. Methanol-benzene mixtures are sometimes used for nonaqueous acid-base titrations. In addition, dimethylformamide (though not as basic as amine solvents) is excellent in titrations of all except very weak organic acids.

Selection of a titrant base for a nonaqueous acid-base titration is as important as the choice of solvent. For best results, a salt of the conjugate base of the solvent should be employed as the titrant, because this is the strongest base capable of existence in a given solvent. A wide variety of titrant bases has been utilized for nonaqueous titrations. When simple alcohols such as methanol or ethanol are used as solvents, the common titrants are sodium methoxide or sodium ethoxide in the corresponding alcoholic solvent. Such titrants are prepared by the reaction between pure sodium metal and the anhydrous alcohol. Occasionally, it is possible to use alcoholic solutions of sodium or potassium hydroxide as titrants. Another class of titrant bases which have gained in popularity in recent years is the tetraalkylammonium hydroxides. For example, tetrabutylammonium hydroxide, $(C_4H_9)_4N^+OH^-$, may be prepared in a variety of solvents, including alcohols and benzene-alcohol mixtures, if a solution of tetrabutylammonium iodide is passed through an anion-exchange column in the hydroxide form.

Strongly basic solvents and titrant solutions quickly absorb carbon dioxide and other acidic impurities, and so they should be protected from the atmosphere. Absorption of moisture can be equally detrimental to the proper behavior of nonaqueous titrants and solvents because the presence of even a trace of water in these nonaqueous systems can exert a pronounced leveling effect on the strengths of solute acids – and perhaps prevent them from being titrated at all.

**End-Point Detection.** Several of the more familiar acid-base indicators, including methyl violet, methyl red, phenolphthalein, and thymolphthalein, function satisfactorily in certain nonaqueous titrations. Frequently, however, the pH ranges over which acid-base indicators change color in nonaqueous solvents are different from those in water, and the color changes are not the same.

It is usually preferable to follow the progress of a nonaqueous acid-base titration potentiometrically. Briefly stated, a potentiometric acid-base titration consists of measuring, versus the volume of titrant added, the potential difference between a reference electrode and a suitable indicator electrode, the latter being sensitive to the hydrogen ion activity of the solution. Design and construction of suitable reference and indicator electrodes for nonaqueous potentiometric titrations can pose some difficult problems. When immersed in strongly basic, nonaqueous solvents, the familiar glass-membrane electrode used for pH measurements and titrations in aqueous media may not function properly, because the presence of water molecules within the surface layers of the glass membrane is essential to the operation of a glass electrode, and this water is removed by the desiccating action of nonaqueous solvents to which the electrode is exposed. Therefore, the antimony oxide electrode and the stainless steel electrode have been employed instead of the glass electrode. Reference electrodes for nonaqueous titrations include the aqueous saturated calomel electrode (SCE) as well as several nonaqueous versions of the calomel electrode.

**Examples of Substances Titrated.** Two classes of weakly acidic organic compounds which can be titrated in basic solvents are carboxylic acids and phenols. Carboxylic acids are sufficiently strong, having $pK_a$ values in water in the neighborhood of 5 or 6, that only modestly basic solvents such as methanol or ethanol are needed to obtain nicely defined titration curves with sharp end points. On the other hand, the determination of phenols, which have $pK_a$ values of approximately 10 in water, requires a much more basic solvent than an alcohol. Anhydrous ethylenediamine is an excellent choice, provided that care is taken to remove and exclude traces of water from the system.

Several other types of weak acids that can be titrated in basic solvents are sulfonamides, barbituric acids, amino acids, and certain salts of amines. Sulfonamides, known more familiarly as the sulfa drugs, possess the $-SO_2NH-$ group whose acidity is considerably enhanced in basic solvents such as butylamine or ethylenediamine. For example, when sulfanilamide is dissolved in butylamine, a proton is transferred from the solute to the solvent, and an ion pair is formed

$$H_2N{-}C_6H_4{-}SO_2NH_2 + C_4H_9NH_2 \rightleftharpoons C_4H_9NH_3^{+-}NH{-}SO_2{-}C_6H_4{-}NH_2$$

which can be titrated with sodium methoxide:

$$C_4H_9NH_3^{+-}NH{-}SO_2{-}C_6H_4{-}NH_2 + CH_3O^-Na^+ \rightleftharpoons H_2N{-}C_6H_4{-}SO_2NH^-Na^+ + CH_3OH + C_4H_9NH_2$$

Mixtures of sulfonamides can be analyzed by taking advantage of the different acid strengths of these compounds in various solvents.

A variety of amine salts can be determined by titration in basic solvents. One representative example of such a titration involves butylamine hydrochloride, $C_4H_9NH_3^+Cl^-$. Although the acid strength of $C_4H_9NH_3^+$ is indeed slight in a solvent such as water, the addition of ethylenediamine to an aqueous suspension of butylamine hydrochloride not only dissolves the amine salt but increases the apparent acid strength of $C_4H_9NH_3^+$ to the point where it can be titrated with sodium methoxide:

$$C_4H_9NH_3^+Cl^- + CH_3O^-Na^+ \rightleftharpoons C_4H_9NH_2 + CH_3OH + Na^+Cl^-$$

One requirement for the successful titration of amine salts is that the anion of the salt cannot be too strong a base. If it is, it may compete so well with the titrant base for the proton that a poorly defined end point is obtained. Among the other amine salts which have been successfully titrated are methylamine hydrochloride ($CH_3NH_3^+Cl^-$), pyridinium perchlorate ($C_5H_5NH^+ClO_4^-$), and quinine sulfate.

### Titrations in Glacial Acetic Acid

In attempting to titrate a variety of bases of widely differing strength with a strong acid in an aqueous solution, one discovers that only those bases which have considerably greater affinity for protons than the solvent (water) can be satisfactorily determined. Fundamentally, the titration reaction involves the transfer of a proton from the conjugate acid of the solvent $HSH^+$ to the solute base B:

$$HSH^+ + B \rightleftharpoons SH + BH^+$$

If the affinity of the solvent SH for protons happens to be comparable to or greater than that of the base B, the titration reaction can never attain any substantial degree of completion and the titration will fail. Therefore, to achieve success in the titration of weak and very weak bases, one must select an acidic solvent with little affinity for protons to enhance as much as possible the meager basicity of the compounds of interest. Note that the solvent requirement is just the opposite of that for the titrimetric determination of weakly acidic substances.

Glacial acetic acid has received overwhelming attention as a titration medium for weak bases, and perchloric acid dissolved in glacial acetic acid is almost universally used as a titrant. When the exclusion or removal of water from the titrant or solvent is desirable in order to prevent leveling effects and improve the sharpness of titration curves, the common practice is to add approximately stoichiometric quantities of acetic anhydride, which consumes water to form more acetic acid by the reaction

$$(CH_3CO)_2O + H_2O \rightarrow 2\ CH_3COOH$$

End-point detection can be accomplished on an empirical basis with the aid of acid-base indicators. However, a better approach is to employ the previously mentioned technique of potentiometric titration. In the case of glacial acetic acid, both the glass electrode and the aqueous saturated calomel reference electrode have been found to function very well.

There are several classes of weak bases which can be determined titrimetrically in glacial acetic acid – amines, amino acids, alkaloids, antihistamines, and anions of weak acids. Primary, secondary, and tertiary amines can be titrated in anhydrous acetic acid with perchloric acid. As an example, the titration of aniline, a primary amine, proceeds according to the reaction

$$C_6H_5{-}NH_2 + CH_3COOH_2^+ \rightleftharpoons C_6H_5{-}NH_3^+ + CH_3COOH$$

where $CH_3COOH_2^+$ is the solvated proton in glacial acetic acid.

In the preceding chapter, the amphiprotic character of alanine (a typical amino acid) was described. Alanine may accept or lose a proton as shown by the equilibria

$$CH_3{-}CH(NH_2){-}COO^- \underset{-H^+}{\rightleftharpoons} \underset{\text{alanine}}{CH_3{-}CH(NH_3^+){-}COO^-} \overset{+H^+}{\rightleftharpoons} CH_3{-}CH(NH_3^+){-}COOH$$

However, in an acidic solvent such as glacial acetic acid, the carboxylic acid group (– COOH) is rendered unreactive, and so the titration of an acetic acid solution of alanine with perchloric acid may be represented as

$$CH_3{-}CH(NH_2){-}COOH + CH_3COOH_2^+ \rightleftharpoons CH_3{-}CH(NH_3^+){-}COOH + CH_3COOH$$

Thus, the titration of amino acids in glacial acetic acid is analogous to the determination of amines. It is interesting to note that alanine exists as a dipolar ion or *zwitterion* in a solvent of high dielectric constant such as water. However, the low dielectric constant of glacial acetic acid does not favor the separation of charge, so the form of alanine shown in the above reaction is predominant.

Anions of weak acids are sufficiently basic in glacial acetic acid to be titrated. Just a few of the anions which can be determined by titration with standard perchloric acid are bicarbonate, carbonate, acetate, bisulfite, cyanide, sulfate, and barbiturate derivatives.

## QUESTIONS AND PROBLEMS

1. Suppose that a solution of sodium hydroxide is permitted to absorb large amounts of carbon dioxide after originally being standardized. Would the original value for the concentration of the sodium hydroxide titrant be high, low, or correct? Explain.
2. If constant-boiling hydrochloric acid, at a pressure of 745 mm of mercury, contains 20.257 per cent HCl by weight, what weight of the constant-boiling solution must be distilled at this pressure for the preparation of exactly 1 liter of 1.000 $F$ hydrochloric acid?
3. If 48.37 ml of a sodium hydroxide solution was required to titrate a 3.972-g sample of pure potassium acid phthalate, what was the concentration of the titrant?
4. Construct the titration curve for each of the following systems:
   (a) 50.00 ml of a 0.5000 $F$ sodium hydroxide solution titrated with 0.5000 $F$ hydrochloric acid

(b) 50.00 ml of a 0.1000 $F$ sodium borate solution titrated with 0.1000 $F$ perchloric acid solution
(c) 50.00 ml of a 0.1500 $F$ aqueous ethanolamine solution titrated with 0.1500 $F$ hydrochloric acid
(d) 50.00 ml of 0.05000 $F$ acetic acid titrated with 0.1000 $F$ aqueous ammonia solution
(e) 50.00 ml of a mixture of 0.1000 $F$ perchloric acid and 0.05000 $F$ propanoic acid titrated with 0.2000 $F$ potassium hydroxide solution
(f) 50.00 ml of a 0.2000 $F$ disodium hydrogen phosphate solution titrated with 0.5000 $F$ perchloric acid

5. Calculate and plot the theoretical titration curve for the titration of 25.00 ml of 0.05000 $F$ formic acid with 0.1000 $F$ sodium hydroxide solution. Include in your plot the points that correspond to the following volumes in milliliters of titrant added: 0, 10.00, 12.45, 12.50, 13.00, and 20.00. What acid-base indicator would you select for this titration?
6. Consider the titration of 25.00 ml of 0.2500 $F$ aqueous ammonia solution with 0.2500 $F$ hydrochloric acid. Calculate and plot the complete titration curve, including the points that correspond to the following volumes in milliliters of titrant added: 0, 5.00, 10.00, 15.00, 20.00, 25.00, 30.00, and 40.00. What acid-base indicator would you select for this titration?
7. Suppose that one desires to determine the concentration $C$ of unknown hydrochloric acid solutions by titration with a standard sodium hydroxide solution of the same $C$ concentration. The titration could be followed with a glass-electrode pH meter. If the end point is taken when the pH of the solution exactly reaches the equivalence-point pH, if the uncertainty in the pH meter readings is ±0.02 pH unit, and if one should not want to incur a titration error greater than ±0.10 per cent, what would be the minimum concentration $C$ of hydrochloric acid which could be titrated by means of this technique?
8. A 25.00-ml aliquot of a benzoic acid solution was titrated with standard 0.1000 $F$ sodium hydroxide, methyl red being used erroneously as the end-point indicator. The observed end point occurred at pH 6.20, after the addition of 20.70 ml of the titrant.
   (a) Calculate the titration error in per cent with the proper sign.
   (b) Calculate the equivalence-point pH for this titration.
   (c) Calculate the concentration of the original benzoic acid solution.
9. A 1.250-g sample of a weak monoprotic acid (HZ) was dissolved in 50 ml of water. The volume of a 0.0900 $F$ sodium hydroxide solution required to reach the equivalence point was 41.20 ml. During the course of the titration, it was observed that, after the addition of 8.24 ml of titrant, the solution pH was 4.30.
   (a) Calculate the formula weight of HZ.
   (b) Calculate the acid dissociation constant of HZ.
   (c) Calculate the theoretical equivalence-point pH for the titration.
10. Suggest an experimental procedure whereby the dissociation constant for an unknown weak monoprotic acid can be determined from the titration curve for that acid, assuming the availability of a pure sample of the unknown acid.

11. Consider the titration of 50.00 ml of $1.000 \times 10^{-4}$ $F$ hydrochloric acid with $1.000 \times 10^{-3}$ $F$ sodium hydroxide solution. Calculate the pH of the solution after the addition of the following volumes of titrant: (a) 0 ml, (b) 1.000 ml, (c) 2.500 ml, (d) 4.900 ml, (e) 4.999 ml, (f) 5.000 ml, (g) 5.555 ml, and (h) 6.000 ml. What means would you employ to locate the equivalence point of the titration?
12. What weight of vinegar should be taken for analysis so that the observed buret reading will be exactly 10 times the percentage of acetic acid in the vinegar if a 0.1000 $F$ sodium hydroxide solution is used as the titrant?
13. A 1.000-g sample of pure oxalic acid dihydrate, $H_2C_2O_4 \cdot 2\,H_2O$, required 46.00 ml of a sodium hydroxide solution for its neutralization. Calculate the concentration of the sodium hydroxide solution.
14. A 2.500-g crystal of calcite, $CaCO_3$, was dissolved in excess hydrochloric acid, the carbon dioxide removed by boiling, and the excess acid titrated with a standard base solution. The volume of hydrochloric acid used was 45.56 ml, and only 2.25 ml of the standard base was required to titrate the excess acid. In a separate titration, 43.33 ml of the base solution neutralized 46.46 ml of the hydrochloric acid solution. Calculate the concentrations of the acid and base solutions.
15. A 0.6300-g sample of a pure organic diprotic acid was titrated with 38.00 ml of a 0.3030 $F$ sodium hydroxide solution, but 4.00 ml of a 0.2250 $F$ hydrochloric acid solution was needed for back titration, at which point the original organic acid was completely neutralized. Calculate the formula weight of the organic acid.
16. A sample of vinegar, weighing 11.40 g, was titrated with a 0.5000 $F$ sodium hydroxide solution, 18.24 ml being required. Calculate the percentage of acetic acid in the vinegar.
17. A hydrochloric acid solution was standardized against sodium carbonate. Just 37.66 ml of the acid was required to titrate a 0.3663-g sample of the sodium carbonate. Later, it was discovered that the sodium carbonate was really the hydrated compound, $Na_2CO_3 \cdot 10\,H_2O$, and not the anhydrous compound as expected. What were the reported and the corrected concentrations of the hydrochloric acid solution?
18. A 1.500-g sample of impure calcium oxide (CaO) was dissolved in 40.00 ml of 0.5000 $F$ hydrochloric acid. Exactly 2.50 ml of a sodium hydroxide solution was required to titrate the excess, unreacted hydrochloric acid. If 1.00 ml of the hydrochloric acid corresponds to 1.25 ml of the sodium hydroxide solution, calculate the percentage of calcium oxide in the sample.
19. The sulfur in a 1.000-g steel sample was converted to sulfur trioxide, which was, in turn, absorbed in 50.00 ml of a 0.01000 $F$ sodium hydroxide solution. The excess, unreacted sodium hydroxide was titrated with 0.01400 $F$ hydrochloric acid, 22.65 ml being required. Calculate the percentage of sulfur in the steel.
20. Malonic acid, HOOC—$CH_2$—COOH (abbreviated $H_2M$), is a diprotic acid which undergoes the following dissociation reactions:

$$H_2M \rightleftharpoons HM^- + H^+; \; pK_{a1} = 2.86$$

$$HM^- \rightleftharpoons M^{2-} + H^+; \; pK_{a2} = 5.70$$

A 20.00-ml aliquot of a solution containing a mixture of disodium malonate ($Na_2M$) and sodium hydrogen malonate (NaHM) was titrated with 0.01000 $F$ hydrochloric acid. The progress of the titration was followed with a glass-electrode pH meter. Two specific points on the titration curve were as follows:

| Volume of HCl added, ml | pH |
|---|---|
| 1.00 | 5.70 |
| 10.00 | 4.28 |

Calculate the total volume in milliliters of hydrochloric acid required to reach the malonic acid ($H_2M$) equivalence point.

21. A solution was known to contain only sodium carbonate and sodium bicarbonate, and its initial pH was found to be 10.63. When a 25.00-ml aliquot of this solution was titrated with 0.1000 $F$ hydrochloric acid, the pH of the solution after the addition of 21.00 ml of the acid was 6.35. What were the molar concentrations of sodium carbonate and sodium bicarbonate in the original solution?
22. Classify each of the following solvents as amphiprotic or aprotic. In addition, indicate (1) if each amphiprotic solvent is predominantly acidic or basic in character and (2) if each aprotic solvent can or cannot accept a proton from a Brønsted-Lowry acid. (a) glacial acetic acid, (b) dioxane, (c) ethylenediamine, (d) methyl isobutyl ketone, (e) benzene, (f) water, (g) diethyl ether, (h) isopropanol, (i) acetone, and (j) butylamine
23. Suggest a solvent and a titrant which would be suitable for the nonaqueous acid-base titration of each of the following substances:

(a) $CH_3$ ... OH — *m*-cresol (used in disinfectants, fumigants, photographic developers, and explosives)

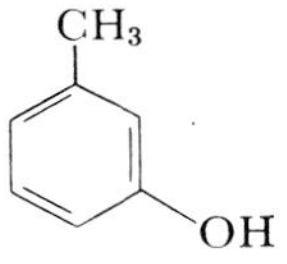

(b) $H_5C_2$ N $C_2H_5$ — diethylaniline (an intermediate for the synthesis of dyestuffs)

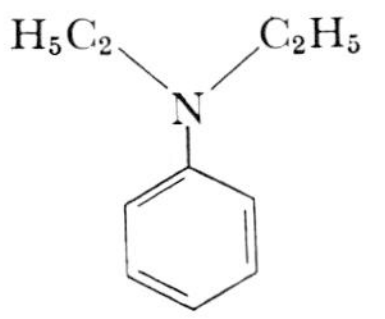

(c) $CH_2CH_2CH_2N(CH_3)_2$, N, Cl, S — chlorpromazine (a sedative and anti-emetic)

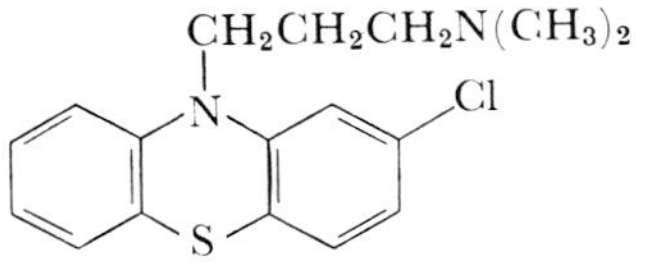

(d) HO—C$_6$H$_4$—C($C_2H_5$)=C($C_2H_5$)—C$_6$H$_4$—OH — diethylstilbestrol (a growth hormone for cattle)

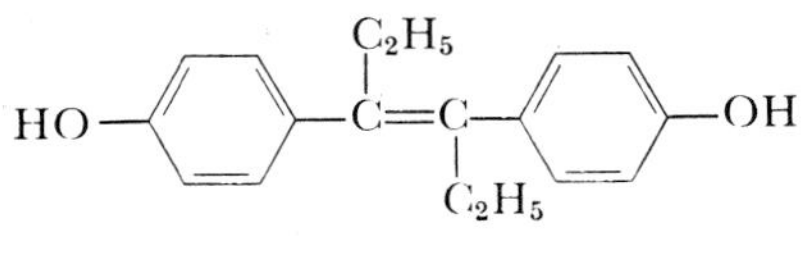

(e) $(H_3C)_2$CH—$CH_2$—CH($NH_2$)—COOH — leucine (an amino acid)

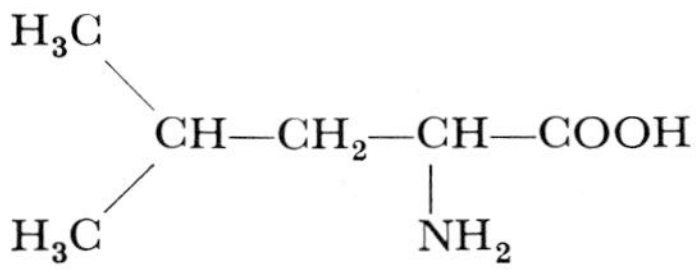

(f) 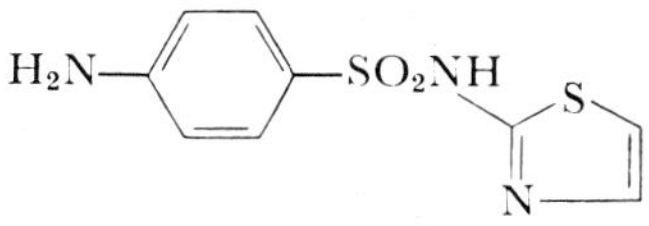 $H_2N-C_6H_4-SO_2NH-$(thiazolyl) sulfathiazole (a sulfa drug effective against diseases due to streptococci)

24. A mixture of ethylamine, diethylamine, and triethylamine may be differentiated and analyzed by means of a set of nonaqueous acid-base reactions:
    (a) Treatment of a mixture of ethylamine, diethylamine, and triethylamine with acetic anhydride in glacial acetic acid converts the primary and secondary amines into essentially neutral acetylation products:

$$C_2H_5NH_2 + (CH_3CO)_2O \rightarrow C_2H_5NHCOCH_3 + CH_3COOH$$
$$(C_2H_5)_2NH + (CH_3CO)_2O \rightarrow (C_2H_5)_2NCOCH_3 + CH_3COOH$$

However, the tertiary amine is unaffected. After the acetylation reaction is complete, the solution is titrated with a standard solution of perchloric acid in glacial acetic acid to determine the quantity of triethylamine.
    (b) If a sample of the three amines is dissolved in a one-to-one mixture of ethylene glycol and isopropanol which contains an excess of salicylaldehyde, the latter reagent reacts with the ethylamine to form a Schiff base:

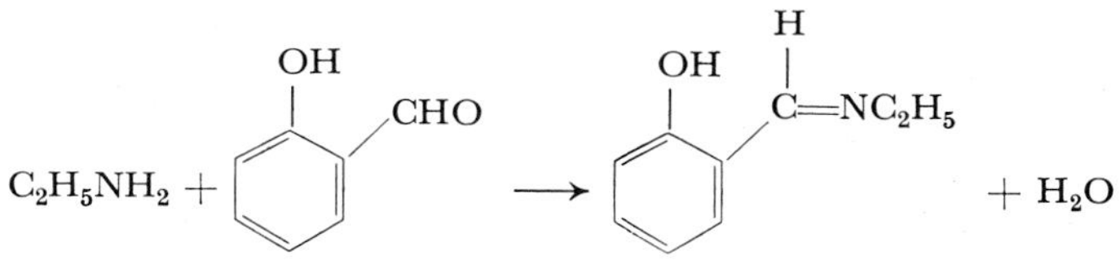

$$C_2H_5NH_2 + C_6H_4(OH)CHO \rightarrow C_6H_4(OH)CH{=}NC_2H_5 + H_2O$$

Since the Schiff base is too weak to react with hydrochloric acid, titration of the mixture with a standard solution of hydrochloric acid in the glycol-alcohol solvent yields the sum of the quantities of diethylamine and triethylamine.

In the analysis of a pure mixture of ethylamine, diethylamine, and triethylamine, separate 1.154-g samples were treated according to each of the two procedures described above. For procedure (a), the final titration required 13.28 ml of 0.4781 *F* perchloric acid titrant. For procedure (b), the titration required 10.69 ml of 0.9536 *F* hydrochloric acid titrant. Calculate the weight percentages of each of the amines in the mixture.

25. Sulfonamides, known more familiarly as sulfa drugs, contain the $-SO_2NH-$ functional group. Sulfanilamide

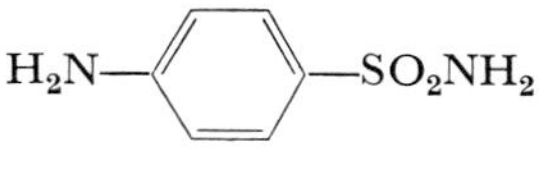

$$H_2N-C_6H_4-SO_2NH_2$$

is the simplest member of this class of compounds and is a relatively weak acid. Replacement of one of the hydrogen atoms of the sulfonamide group with any of a variety of heterocyclic groups results in the formation of many derivatives of sulfanilamide. These more complex sulfa drugs are usually more effective than sulfanilamide for the treatment of cocci and bacterial infections. In addition, the acidity of the remaining hydrogen atom of the sulfonamide group is considerably enhanced by the presence of the heterocyclic group. Thus, sulfathiazole

$H_2N$—⬡—$SO_2NH$
S
N

is a much stronger acid than sulfanilamide. This difference in acidity can be exploited to analyze mixtures of sulfanilamide and sulfathiazole. First, a sample is treated with dimethylformamide, and the resulting solution is titrated with sodium methoxide in benzene-methanol, thymol blue being used as indicator; in dimethylformamide, only the more acidic proton of sulfathiazole can be titrated. Second, another sample is treated with butylamine, and the solution is titrated with the same titrant as before to an azo violet end point; in butylamine, sulfanilamide behaves as a sufficiently strong acid that both it and sulfathiazole can be titrated. In the analysis of a sulfa ointment containing sulfathiazole and sulfanilamide, a 1.500-g sample was extracted with dimethylformamide; the extracts were combined and titrated with 0.1033 $F$ sodium methoxide in benzene-methanol, 2.893 ml of titrant being needed to reach a thymol blue end point. Another 1.500-g sample was treated with butylamine; the butylamine extract was titrated to an azo violet end point with 7.250 ml of the sodium methoxide reagent. Calculate the weight percentages of sulfanilamide and sulfathiazole in the ointment.

## SUGGESTIONS FOR ADDITIONAL READING


1. C. Ayers: Acidimetry and alkalimetry. *In* C. L. Wilson and D. W. Wilson, eds.: *Comprehensive Analytical Chemistry.* Volume IB, Elsevier, New York, 1960, pp. 203–221.
2. R. G. Bates: *Determination of pH.* John Wiley and Sons, New York, 1964.
3. S. Bruckenstein and I. M. Kolthoff: Acid-base strength and protolysis curves in water. *In* I. M. Kolthoff and P. J. Elving, eds.: *Treatise on Analytical Chemistry.* Part I, Volume 1, Wiley-Interscience, New York, 1959, pp. 421–474.
4. J. S. Fritz: *Acid-Base Titrations in Nonaqueous Solvents.* Allyn and Bacon, Inc., Boston, 1973.
5. I. Gyenes: *Titrations in Non-Aqueous Media,* translated by D. Cohen and I. T. Millar. Van Nostrand, Princeton, New Jersey, 1967.
6. W. Huber: *Titrations in Nonaqueous Solvents.* Academic Press, New York, 1967.
7. E. J. King: *Acid-Base Equilibria.* The Macmillan Company, New York, 1965.
8. I. M. Kolthoff: Concepts of acids and bases. *In* I. M. Kolthoff and P. J. Elving, eds.: *Treatise on Analytical Chemistry.* Part I, Volume 1, Wiley-Interscience, New York, 1959, pp. 405–420.
9. I. M. Kolthoff and S. Bruckenstein: Acid-base equilibria in nonaqueous solutions. *In* I. M. Kolthoff and P. J. Elving, eds.: *Treatise on Analytical Chemistry.* Part I, Volume 1, Wiley-Interscience, New York, 1959, pp. 475–542.
10. I. M. Kolthoff and V. A. Stenger: *Volumetric Analysis.* Second edition, Volume 2, Wiley-Interscience, New York, 1947, pp. 49–235.
11. J. Kucharsky and L. Safarik: *Titrations in Non-Aqueous Solvents.* Elsevier, Amsterdam, 1965.

12. H. H. Sisler: *Chemistry in Non-Aqueous Solvents.* Reinhold Publishing Corporation, New York, 1961.
13. C. A. VanderWerf: *Acids, Bases, and the Chemistry of the Covalent Bond.* Reinhold Publishing Corporation, New York, 1961.
14. R. A. Zingaro: *Nonaqueous Solvents.* Heath, Lexington, Massachusetts, 1968.

# COMPLEXOMETRIC TITRATIONS

A **complex ion** or **coordination compound** is a species in which a metal atom or cation is covalently bonded to one or more electron-donating groups. Usually, the metal atom or cation is called the **central atom**, and the term **ligand** is used to designate the electron-donating group. In order for a complex ion to be formed from one or more ligands and a central atom, each ligand must possess at least one unshared pair of electrons, and the central atom must be able to accept an electron pair from each ligand. Thus, a ligand shares a pair of electrons with the central atom in the formation of a covalent bond.

To serve as the basis for a volumetric method of analysis, a complex-formation reaction must be rapid, must proceed according to exact stoichiometry, and must have an equilibrium constant large enough that the titration curve has a sharply defined equivalence point.

Formation of a complex ion by titration of a solution of a metal cation with a standard solution of a complexing agent or ligand has gained importance as a method of volumetric analysis in the last twenty-five years. This is because of the advent of a unique class of ligands – the aminopolycarboxylic acids – that have several electron-donating groups on the same molecule and that form unusually stable one-to-one complexes with many metal ions. To begin this chapter, we shall examine some of the characteristics of complex-formation reactions. Then, the theory and practice of complexometric titrations with ethylenediaminetetraacetic acid (EDTA) will be discussed.

## METAL ION COMPLEXES

An example of the formation of a metal complex is the reaction between the hexaaquochromium(III) ion and cyanide to form the hexacyanochromate(III) anion:

$$Cr(H_2O)_6{}^{3+} + 6\,CN^- \rightleftharpoons Cr(CN)_6{}^{3-} + 6\,H_2O$$

Though the product species, canary-yellow $Cr(CN)_6{}^{3-}$, is very stable, the rate of reaction between violet $Cr(H_2O)_6^{3+}$ and $CN^-$ to form this complex is very slow under conditions suitable for a titration – that is, at room temperature and for stoichiometric quantities of the reactants. This reaction is slow because cyanide ions cannot readily displace the water molecules bound to chromium(III). Therefore, although the formation of the $Cr(CN)_6{}^{3-}$ ion might be expected to be suitable as a titrimetric method on the basis of its stability, the slow rate of reaction prohibits the use of this reaction for analytical purposes. A complex ion characterized by a slow rate of ligand exchange is called an inert or **nonlabile** species. Among the metal ions whose complexes frequently display relatively slight reactivity are chromium(III), cobalt(III), and platinum(IV). There is another group of metals that characteristically form reactive or **labile** complexes, including cobalt(II), copper, lead, bismuth, silver, cadmium, nickel, zinc, mercury, and aluminum. Many complexes of iron(II) and iron(III) are labile, but the cyanide complexes $Fe(CN)_6{}^{4-}$ and $Fe(CN)_6{}^{3-}$ are familiar examples of nonlabile species.

Metal cations have several available orbitals for bond formation with complexing agents; for example, the zinc ion has four coordination sites. However, numerous common ligands, including chloride, bromide, iodide, cyanide, thiocyanate, hydroxide, and ammonia, can occupy only one coordination position of a metal ion. In other words, each of these ligands donates one unshared pair of electrons to the central atom. Such a species is called a **monodentate ligand**, from the Latin word *dentatus,* meaning "toothed." A single zinc ion can react with a maximum of four chlorides, four cyanides, four hydroxides, or four ammonia molecules to form $ZnCl_4{}^{2-}$, $Zn(CN)_4{}^{2-}$, $Zn(OH)_4{}^{2-}$, or $Zn(NH_3)_4{}^{2+}$. Complexes containing more than one kind of ligand, such as $Zn(NH_3)_2(H_2O)_2{}^{2+}$ and $Zn(H_2O)_2(CN)_2$, can be formed.

Other ligands, known as **multidentate ligands**, can contribute two or more electron pairs to a central atom in forming a complex. To identify multidentate species that donate two or three electron pairs to the central atom, we speak of *bidentate* or *terdentate* ligands, respectively. In some instances, a ligand may be *quadridentate, quinquidentate,* or *sexadentate.* Ethylenediamine, $NH_2CH_2CH_2NH_2$, is a bidentate ligand because each nitrogen atom possesses one unshared pair of electrons. Copper(II) can be complexed by two ethylenediamine ligands in much the same way as it is complexed by four ammonia molecules:

$$\left[\begin{array}{ccccc} & H_2N: & & :NH_2 & \\ H_2C & & & & CH_2 \\ | & & Cu & & | \\ H_2C & & & & CH_2 \\ & H_2N: & & :NH_2 & \end{array}\right]^{2+} \qquad \left[\begin{array}{ccc} H_3N: & & :NH_3 \\ & Cu & \\ H_3N: & & :NH_3 \end{array}\right]^{2+}$$

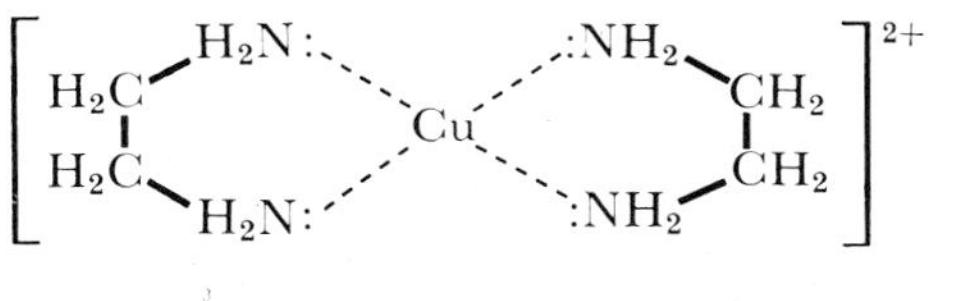

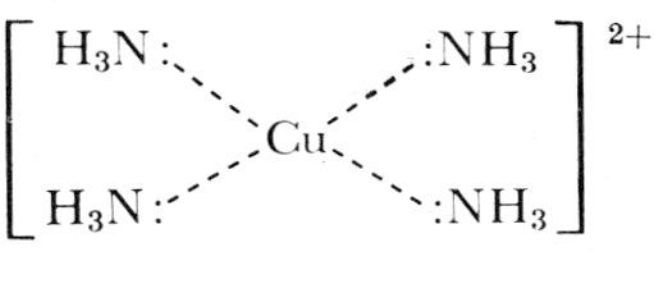

A complex composed of a central metal atom and one or more multidentate ligands is called a **chelate**, or chelate compound, after a Greek word meaning "claw." In a manner of speaking, the two or more electron-donating groups of the ligand act as a claw in bonding to the central atom.

### Stepwise and Overall Formation Constants

Let us consider the interaction between the hydrated zinc ion and ammonia in an aqueous medium:

$$Zn(H_2O)_4{}^{2+} + NH_3 \rightleftharpoons Zn(NH_3)(H_2O)_3{}^{2+} + H_2O$$

$$Zn(NH_3)(H_2O)_3{}^{2+} + NH_3 \rightleftharpoons Zn(NH_3)_2(H_2O)_2{}^{2+} + H_2O$$

$$Zn(NH_3)_2(H_2O)_2{}^{2+} + NH_3 \rightleftharpoons Zn(NH_3)_3(H_2O)^{2+} + H_2O$$

$$Zn(NH_3)_3(H_2O)^{2+} + NH_3 \rightleftharpoons Zn(NH_3)_4{}^{2+} + H_2O$$

These reactions represent the stepwise formation of the monoammine, diammine, triammine, and tetraammine complexes of zinc(II).

A **stepwise formation constant** is associated with each equilibrium and is designated by $K_n$, where, in the present example, the subscript $n$ takes an integral value to indicate the addition of the $n$th ammonia ligand to a complex containing $(n - 1)$ ammonia molecules. Thus, the *first stepwise formation constant,* $K_1$, for the formation of the $Zn(NH_3)(H_2O)_3{}^{2+}$ complex pertains to the equilibrium expression

$$K_1 = \frac{[Zn(NH_3)^{2+}]}{[Zn^{2+}][NH_3]}$$

*Notice that the concentration (activity) of water does not appear in this relation. Furthermore, although water frequently functions as a ligand in aqueous solutions of metal ions, it is customary for simplicity in writing reactions and equilibrium expressions to omit water from the chemical formula of a hydrated cation or complex ion.* We will follow this practice throughout the remainder of this chapter. For the *fourth stepwise formation constant*, we can write

$$K_4 = \frac{[Zn(NH_3)_4{}^{2+}]}{[Zn(NH_3)_3{}^{2+}][NH_3]}$$

For the zinc(II)-ammonia system at 25°C, the numerical values for $K_1$, $K_2$, $K_3$, and $K_4$ are, respectively, 186, 219, 251, and 112.

Sometimes **overall formation constants** are used to characterize the equilibria in systems containing complex ions. Omitting the water molecules from the chemical formulas of all complexes, we can write the equilibria corresponding to the overall reactions for the zinc(II)-ammonia system as follows:

$$Zn^{2+} + NH_3 \rightleftharpoons Zn(NH_3)^{2+}$$

$$Zn^{2+} + 2\,NH_3 \rightleftharpoons Zn(NH_3)_2{}^{2+}$$

$$Zn^{2+} + 3\,NH_3 \rightleftharpoons Zn(NH_3)_3{}^{2+}$$

$$Zn^{2+} + 4\,NH_3 \rightleftharpoons Zn(NH_3)_4{}^{2+}$$

Overall formation constants are designated by $\beta_n$, where the subscript $n$ gives the total number of ligands added to the original aquated cation. For example, the *third overall formation constant*, $\beta_3$, refers to the addition of three ammonia ligands to $Zn(H_2O)_4{}^{2+}$ to form $Zn(NH_3)_3(H_2O)^{2+}$, and the pertinent equilibrium expression is

$$\beta_3 = \frac{[Zn(NH_3)_3{}^{2+}]}{[Zn^{2+}][NH_3]^3}$$

There is a simple relationship between the stepwise formation constants and the overall formation constants for any particular system. This may be illustrated if we multiply the numerator and denominator of the equation defining $\beta_3$ by the quantity $[Zn(NH_3)^{2+}][Zn(NH_3)_2{}^{2+}]$

$$\beta_3 = \frac{[Zn(NH_3)_3{}^{2+}]}{[Zn^{2+}][NH_3]^3} \cdot \frac{[Zn(NH_3)^{2+}][Zn(NH_3)_2{}^{2+}]}{[Zn(NH_3)^{2+}][Zn(NH_3)_2{}^{2+}]}$$

and rearrange the expression as follows:

$$\beta_3 = \frac{[Zn(NH_3)^{2+}]}{[Zn^{2+}][NH_3]} \cdot \frac{[Zn(NH_3)_2{}^{2+}]}{[Zn(NH_3)^{2+}][NH_3]} \cdot \frac{[Zn(NH_3)_3{}^{2+}]}{[Zn(NH_3)_2{}^{2+}][NH_3]}$$

Note that the three terms on the right side of this equation correspond, in order, to $K_1$, $K_2$, and $K_3$. Consequently,

$$\beta_3 = K_1K_2K_3$$

In similar fashion it is possible to derive the relationship between each of the overall formation constants and the stepwise formation constants; the results are

$$K_1 = \beta_1$$
$$K_1K_2 = \beta_2$$
$$K_1K_2K_3 = \beta_3$$
$$K_1K_2K_3K_4 = \beta_4$$
$$K_1K_2K_3K_4 \cdots K_n = \beta_n$$

For the zinc(II)-ammonia system, the overall formation constants have the following values: $\beta_1 = 186, \beta_2 = 4.08 \times 10^4, \beta_3 = 1.02 \times 10^7$, and $\beta_4 = 1.15 \times 10^9$

### Distribution of Metal Among Several Complexes Involving Monodentate Ligands

For most metal ions, and the majority of monodentate ligands, the relative stabilities of the various complexes within a family are such that several species coexist in the same solution for certain concentrations of the ligand. Let us see for the zinc(II)-ammonia system how the fraction of each zinc-ammine complex varies with the concentration of free (uncomplexed) ammonia. We can employ the same approach used in Chapter 4 (pages 79 to 82) for polyprotic acid systems.

In an ammoniacal solution, the sum of the concentrations of all zinc-bearing species will have a fixed value:

$$C_{Zn} = [Zn^{2+}] + [Zn(NH_3)^{2+}] + [Zn(NH_3)_2^{2+}] + [Zn(NH_3)_3^{2+}] + [Zn(NH_3)_4^{2+}]$$

This equation can be rewritten if we utilize the equilibrium expressions involving overall formation constants:

$$C_{Zn} = [Zn^{2+}] + \beta_1[Zn^{2+}][NH_3] + \beta_2[Zn^{2+}][NH_3]^2 + \beta_3[Zn^{2+}][NH_3]^3 + \beta_4[Zn^{2+}][NH_3]^4$$

Now, the concentration of each complex will be some fraction ($\alpha$) of the total concentration, $C_{Zn}$:

$$\alpha_{Zn^{2+}} = \frac{[Zn^{2+}]}{C_{Zn}}$$

$$\alpha_{Zn(NH_3)^{2+}} = \frac{[Zn(NH_3)^{2+}]}{C_{Zn}} = \frac{\beta_1[Zn^{2+}][NH_3]}{C_{Zn}}$$

$$\alpha_{Zn(NH_3)_2^{2+}} = \frac{[Zn(NH_3)_2^{2+}]}{C_{Zn}} = \frac{\beta_2[Zn^{2+}][NH_3]^2}{C_{Zn}}$$

$$\alpha_{Zn(NH_3)_3^{2+}} = \frac{[Zn(NH_3)_3^{2+}]}{C_{Zn}} = \frac{\beta_3[Zn^{2+}][NH_3]^3}{C_{Zn}}$$

$$\alpha_{Zn(NH_3)_4^{2+}} = \frac{[Zn(NH_3)_4^{2+}]}{C_{Zn}} = \frac{\beta_4[Zn^{2+}][NH_3]^4}{C_{Zn}}$$

Note that the sum of the five $\alpha$ values must be unity. Substitution of the relation for $C_{Zn}$ into each of the five preceding equations, followed by cancellation of $[Zn^{2+}]$ terms, yields

$$\alpha_{Zn^{2+}} = \frac{1}{1 + \beta_1[NH_3] + \beta_2[NH_3]^2 + \beta_3[NH_3]^3 + \beta_4[NH_3]^4}$$

$$\alpha_{Zn(NH_3)^{2+}} = \frac{\beta_1[NH_3]}{1 + \beta_1[NH_3] + \beta_2[NH_3]^2 + \beta_3[NH_3]^3 + \beta_4[NH_3]^4}$$

$$\alpha_{Zn(NH_3)_2{}^{2+}} = \frac{\beta_2[NH_3]^2}{1 + \beta_1[NH_3] + \beta_2[NH_3]^2 + \beta_3[NH_3]^3 + \beta_4[NH_3]^4}$$

$$\alpha_{Zn(NH_3)_3{}^{2+}} = \frac{\beta_3[NH_3]^3}{1 + \beta_1[NH_3] + \beta_2[NH_3]^2 + \beta_3[NH_3]^3 + \beta_4[NH_3]^4}$$

$$\alpha_{Zn(NH_3)_4{}^{2+}} = \frac{\beta_4[NH_3]^4}{1 + \beta_1[NH_3] + \beta_2[NH_3]^2 + \beta_3[NH_3]^3 + \beta_4[NH_3]^4}$$

It is evident that the fraction of each species is dependent upon the ammonia concentration but not the total concentration ($C_{Zn}$) of zinc-containing complexes. Using the five final equations for the $\alpha$ values, we computed the fraction of each zinc(II) complex as a function of the logarithm of the concentration of ammonia. Results of these calculations are plotted in Figure 6–1. For ammonia concentrations below $10^{-3}$ $F$, the predominant species is the hydrated cation, $Zn(H_2O)_4{}^{2+}$, whereas the tetraammine-zinc(II) complex, $Zn(NH_3)_4{}^{2+}$, is the major component when the concentration of ammonia exceeds $10^{-2}$ $F$. For ammonia concentrations between $10^{-3}$ and $10^{-2}$ $F$, significant concentrations of all five zinc(II) species are present.

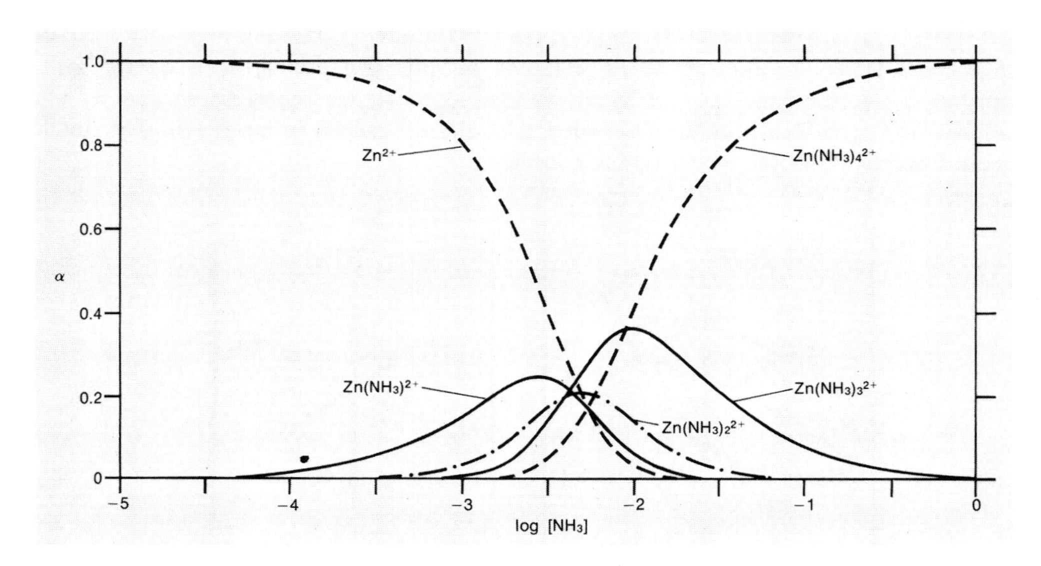


*Figure 6–1.* Distribution of $Zn(H_2O)_4{}^{2+}$ and the various zinc-ammine complexes as a function of the logarithm of the free (uncomplexed) ammonia concentration.

### Advantages of Multidentate Ligands for Complexometric Titrations

There are two reasons why complexes formed from monodentate ligands are unsuitable for complexometric titrations. To understand the difficulties, let us refer again to the zinc(II)-ammonia system. What problems are encountered if we attempt to titrate a solution of $Zn(H_2O)_4{}^{2+}$ with a standard solution of ammonia in water? First, because no single zinc-ammine species is more stable than the others – *the stepwise formation constants are similar in magnitude* – the addition of ammonia produces a changing mixture of complexes, and no simple stoichiometric ratio between ammonia and zinc(II) ever exists. Second, because of the modest stabilities of the zinc ammines – *the stepwise formation constants are relatively small* – the quantitative formation of even the most extensively complexed $Zn(NH_3)_4{}^{2+}$ ion requires a large excess of free ammonia, which can only be achieved well beyond the theoretical equivalence point. Thus, during the addition of titrant, the concentrations of free ammonia and the various zinc-ammine complexes change so gradually that the resulting titration curve never exhibits an equivalence point.

When a complexometric titration is performed with a standard solution of a multidentate ligand, problems associated with the use of monodentate ligands are avoided – other than the possibility of a slow rate of reaction. Most multidentate ligands employed in titrimetry coordinate at several or all positions around a central atom; therefore, only one-to-one metal-ligand complexes are formed and the stoichiometry of the titration reaction is simple. In addition, multidentate ligands usually form much more stable complexes than do chemically similar monodentate ligands, so that well-defined titration curves are obtainable.

## TITRATIONS WITH ETHYLENEDIAMINETETRAACETIC ACID (EDTA)

Today complexometric titrations are performed almost exclusively with a standard solution of one of the family of aminopolycarboxylic acids, ethylenediaminetetraacetic acid (EDTA)

```
HOOC—CH2                              CH2—COO⁻
        \    +              +        /
         HN—CH2—CH2—NH
        /                            \
⁻OOC—CH2                              CH2—COOH
```

being by far the most popular choice. Quite commonly, the parent acid (EDTA) is written as $H_4Y$ in order to emphasize that the compound is tetraprotic. Infrared and nuclear magnetic resonance studies have revealed that in an aqueous environment $H_4Y$ exists as a zwitterion, the protons from two of the four carboxylic acid groups having been transferred to the nitrogen atoms. We can formulate the stepwise dissociation of the parent acid as

$$H_4Y \rightleftharpoons H^+ + H_3Y^-; \qquad K_{a1} = 1.00 \times 10^{-2}$$

$$H_3Y^- \rightleftharpoons H^+ + H_2Y^{2-}; \qquad K_{a2} = 2.16 \times 10^{-3}$$

$$H_2Y^{2-} \rightleftharpoons H^+ + HY^{3-}; \qquad K_{a3} = 6.92 \times 10^{-7}$$

$$HY^{3-} \rightleftharpoons H^+ + Y^{4-}; \qquad K_{a4} = 5.50 \times 10^{-11}$$

where $Y^{4-}$ represents the ethylenediaminetetraacetate anion. Throughout the remainder of this chapter, the abbreviation EDTA will be employed for general purposes, whereas $H_4Y$, $H_3Y^-$, $H_2Y^{2-}$, and so on, will be used in chemical reactions and equilibrium expressions to designate specific forms of EDTA. In solutions with pH values less than 2, the two negatively charged carboxylate sites of $H_4Y$ become protonated to yield the relatively strong acids $H_5Y^+$ and $H_6Y^{2+}$:

$$H_6Y^{2+} \rightleftharpoons H^+ + H_5Y^+; \qquad K = 1.26 \times 10^{-1}$$

$$H_5Y^+ \rightleftharpoons H^+ + H_4Y; \qquad K' = 2.51 \times 10^{-2}$$

However, under the conditions of acidity used for the majority of complexometric titrations with EDTA, the $H_6Y^{2+}$ and $H_5Y^+$ species are not present in significant amounts.

Inasmuch as the parent acid ($H_4Y$) is only sparingly soluble in water, it is not ordinarily used for complexometric titrations. On the other hand, the disodium salt is relatively soluble and, being commercially available in the form $Na_2H_2Y \cdot 2\,H_2O$, serves as the starting material for the preparation of the standard EDTA solutions employed in titrimetry. In practice, the disodium salt is dried in an oven at 80°C for two hours; next, an appropriate amount of the material is accurately weighed and then dissolved in distilled water, and the resulting solution is transferred to a volumetric flask and diluted.

### Composition of EDTA Solutions as a Function of pH

Changes in the hydrogen ion concentration cause enormous variations in the composition of EDTA solutions. Figure 6–2 shows how the fraction ($\alpha$) of EDTA present as each of the seven possible species, $H_6Y^{2+}$ through $Y^{4-}$, varies as a function of pH.* From an examination of Figure 6–2, it is apparent that $H_6Y^{2+}$ is the predominant form of EDTA below pH 1 and that $H_5Y^+$ is the major species between pH 1 and 1.6. Only between pH 1.6 and 2 is $H_4Y$ the most important component of the solution, whereas $H_3Y^-$ is the major constituent from pH 2 to 2.7. For pH values greater than 2.7, there are wide pH ranges within which one species predominates – $H_2Y^{2-}$ between pH 2.7 and 6.2, $HY^{3-}$ between pH 6.2 and 10.2, and $Y^{4-}$ above pH 10.2.

---

*To obtain the data needed to plot Figure 6–2, the total concentration of EDTA species is defined:

$$C_{EDTA} = [H_6Y^{2+}] + [H_5Y^+] + [H_4Y] + [H_3Y^-] + [H_2Y^{2-}] + [HY^{3-}] + [Y^{4-}]$$

Next, the fraction of EDTA present as each species is written, *e.g.*,

$$\alpha_{Y^{4-}} = \frac{[Y^{4-}]}{C_{EDTA}}$$

Then, the concentration of each species other than $H_6Y^{2+}$ is expressed in terms of $[H^+]$, $[H_6Y^{2+}]$, and the six equilibrium constants identified in the preceding section ($K$, $K'$, $K_{a1}$, $K_{a2}$, $K_{a3}$, and $K_{a4}$). Proceeding as outlined on page 80, one obtains the $\alpha$ value for each species as a function only of $[H^+]$ and the equilibrium constants, *e.g.*,

$$\alpha_{Y^{4-}} = \frac{KK'K_{a1}K_{a2}K_{a3}K_{a4}}{[H^+]^6 + K[H^+]^5 + KK'[H^+]^4 + KK'K_{a1}[H^+]^3 + KK'K_{a1}K_{a2}[H^+]^2 + KK'K_{a1}K_{a2}K_{a3}[H^+] + KK'K_{a1}K_{a2}K_{a3}K_{a4}}$$

Finally, each $\alpha$ is computed for many different pH values and the results are plotted.

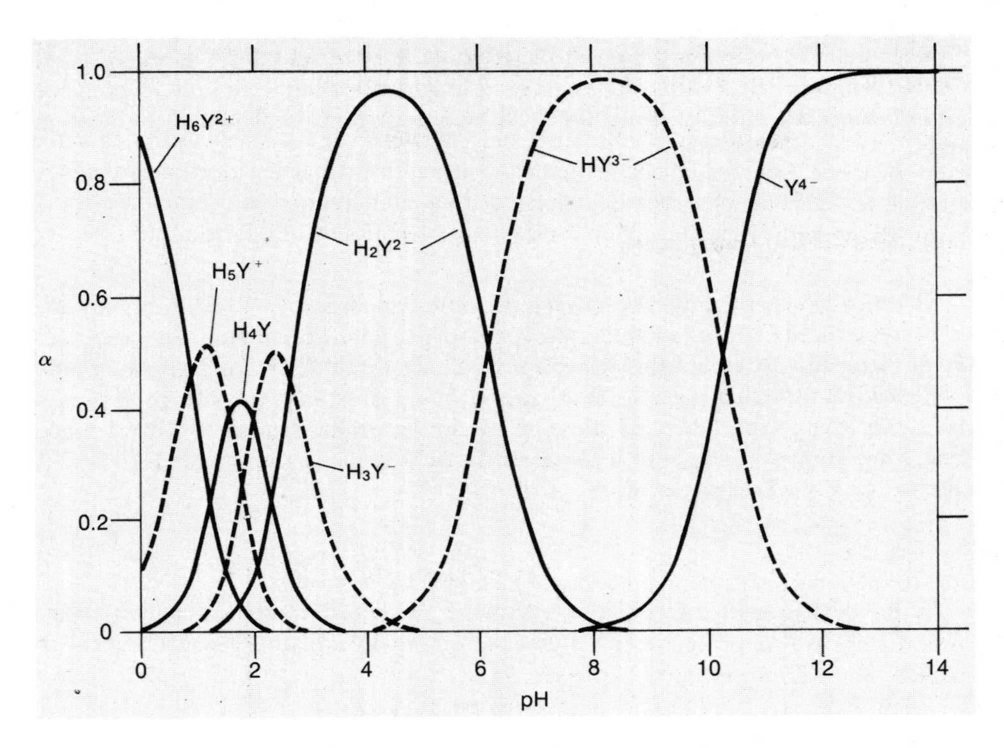


*Figure 6–2.* Distribution of EDTA species as a function of pH.

## Metal Ion-EDTA Complexes

Practically every metal cation in the periodic table forms very stable, one-to-one complexes with the ethylenediaminetetraacetate ion, $Y^{4-}$. This one-to-one stoichiometry of metal-EDTA complexes arises from the fact that the $Y^{4-}$ ion possesses a total of six electron-donating groups – four carboxylate groups and two amine groups – which can occupy four, five, or six coordination positions around a central metal atom. One known example of a six-coordinated species is the cobalt(III)-EDTA complex, $CoY^-$, whose structure is

Metal-EDTA complexes gain stability from the five-membered chelate rings which are formed. In the cobalt(III)-EDTA complex, there are five such five-membered rings. A five-membered ring is especially stable because the bond angles allow all five atoms in the ring to lie in a plane. In some instances, only four or five of the six electron-donating groups of the $Y^{4-}$ anion are bound to a metal ion, and the remaining positions around the metal are occupied by monodentate ligands such as water, hydroxide, or ammonia.

We can write the general reaction for the formation of metal-EDTA complexes as

$$M^{n+} + Y^{4-} \rightleftharpoons MY^{(n-4)+}$$

where $M^{n+}$ denotes a hydrated metal cation. For the preceding process the equilibrium expression is

$$K_{MY} = \frac{[MY^{(n-4)+}]}{[M^{n+}][Y^{4-}]}$$

in which $K_{MY}$ is the formation constant for the $MY^{(n-4)+}$ complex. In Table 6–1 are listed formation constants for a number of metal-EDTA complexes.

As the data presented in Table 6–1 show, metal-EDTA complexes exhibit a great range of stability. An obvious trend is that the formation constants become larger as the charge of the cation increases. However, the stability of a metal-EDTA species can be significantly *decreased* by changes in the composition of a solution. For example, if an acid is added to a solution containing the nickel-EDTA complex, the dissociation of the latter is enhanced because protons combine with the $Y^{4-}$ anion to yield $HY^{3-}$ or $H_2Y^{2-}$ according to the equilibria

$$NiY^{2-} + H^+ \rightleftharpoons Ni^{2+} + HY^{3-}$$

**Table 6–1. Formation Constants of Metal-EDTA Complexes***

| Element | Cation | log $K_{MY}$ | Element | Cation | log $K_{MY}$ |
|---|---|---|---|---|---|
| Aluminum | $Al^{3+}$ | 16.1 | Nickel | $Ni^{2+}$ | 18.6 |
| Barium | $Ba^{2+}$ | 7.8 | Scandium | $Sc^{3+}$ | 23.1 |
| Bismuth | $Bi^{3+}$ | 27.9 | Silver | $Ag^+$ | 7.3 |
| Cadmium | $Cd^{2+}$ | 16.5 | Strontium | $Sr^{2+}$ | 8.6 |
| Calcium | $Ca^{2+}$ | 10.7 | Thallium | $Tl^{3+}$ | 21.5 |
| Cobalt | $Co^{2+}$ | 16.3 | Thorium | $Th^{4+}$ | 23.2 |
| Copper | $Cu^{2+}$ | 18.8 | Titanium | $Ti^{3+}$ | 21.3 |
| Gallium | $Ga^{3+}$ | 20.3 | | $TiO^{2+}$ | 17.3 |
| Indium | $In^{3+}$ | 24.9 | Uranium | $U^{4+}$ | 25.5 |
| Iron | $Fe^{2+}$ | 14.3 | Vanadium | $V^{2+}$ | 12.7 |
| | $Fe^{3+}$ | 25.1 | | $V^{3+}$ | 25.9 |
| Lead | $Pb^{2+}$ | 18.0 | | $VO^{2+}$ | 18.8 |
| Magnesium | $Mg^{2+}$ | 8.7 | | $VO_2^+$ | 15.6 |
| Manganese | $Mn^{2+}$ | 13.8 | Yttrium | $Y^{3+}$ | 18.1 |
| Mercury | $Hg^{2+}$ | 21.8 | Zinc | $Zn^{2+}$ | 16.5 |

* Constants are valid for a temperature of 20°C and an ionic strength of 0.1. Data are from G. Schwarzenbach and H. Flaschka: *Complexometric Titrations*. Translated by H. M. N. H. Irving, second English edition, Methuen, London, 1969, p. 10.

and

$$NiY^{2-} + 2\,H^+ \rightleftharpoons Ni^{2+} + H_2Y^{2-}$$

Alternatively, to the solution of the nickel-EDTA species can be added a ligand that forms a different complex with nickel(II). Thus, in the presence of ammonia, the reaction

$$NiY^{2-} + 6\,NH_3 \rightleftharpoons Ni(NH_3)_6^{2+} + Y^{4-}$$

causes more dissociation of the nickel-EDTA complex than occurs in the absence of ammonia – although the formation constant for the nickel-EDTA species is much larger than that for the hexaamminenickel(II) complex. In each case the effective stability of the nickel-EDTA complex is diminished by a reagent that reacts with either nickel(II) or the $Y^{4-}$ anion.

## Conditional Formation Constants

Let us extend the preceding considerations to the subject of complexometric titrations with EDTA. Most titrations are performed in neutral or in mildly acidic or alkaline media that are well buffered. Typically, the solution to be titrated is buffered with acetic acid and sodium acetate (pH 4 to 5) or with ammonium chloride and ammonia (pH 9 to 10); citrate and tartrate buffers are occasionally employed. Since the predominant form of EDTA under these conditions is $H_2Y^{2-}$ or $HY^{3-}$ (Figure 6–2), the titration process leading to the formation of a metal-EDTA complex might be formulated as

$$M^{n+} + H_2Y^{2-} \rightleftharpoons MY^{(n-4)+} + 2\,H^+$$

or

$$M^{n+} + HY^{3-} \rightleftharpoons MY^{(n-4)+} + H^+$$

However, the production of free hydrogen ion can be deleterious; if the pH becomes too low, the formation of the metal-EDTA complex will be incomplete. One of the purposes of the buffer is to consume the hydrogen ion liberated during the titration. In a number of instances, this buffer plays a second role. For example, cadmium, copper, nickel, and zinc ions are precipitated as insoluble hydroxides near pH 7 or 8. Fortunately, these metal cations form stable ammine complexes, so that the use of an ammonium chloride-ammonia buffer prevents the undesired precipitation of these hydroxides. A buffer that performs these special functions for metal-EDTA titrations is called an **auxiliary complexing agent.**

Although an auxiliary complexing agent helps to ensure the success of a complexometric titration, such a buffer system decreases the stability of the metal-EDTA complex. Consider the formation of the zinc-EDTA complex in a buffer consisting of ammonium chloride and ammonia. For such a system, the total concentration of zinc(II) species *not complexed by EDTA* is given by the expression presented on page 123; that is,

$$C_{Zn} = [Zn^{2+}] + [Zn(NH_3)^{2+}] + [Zn(NH_3)_2^{2+}] + [Zn(NH_3)_3^{2+}] + [Zn(NH_3)_4^{2+}]$$

In addition, the concentration of hydrated zinc(II) can be written as

$$[Zn^{2+}] = \alpha_{Zn^{2+}} C_{Zn}$$

where $\alpha_{Zn^{2+}}$ can be computed as described earlier in this chapter from the concentration of free ammonia and the formation constants for the zinc-ammine species. This latter relation stipulates that, whatever the total concentration of zinc(II) *not complexed by EDTA,* the concentration of $Zn^{2+}$ is *a fixed fraction of the total.* Next, we can formulate an equation* for the total concentration of EDTA *not complexed by zinc(II)*

$$C_{EDTA} = [H_6Y^{2+}] + [H_5Y^{+}] + [H_4Y] + [H_3Y^{-}] + [H_2Y^{2-}] + [HY^{3-}] + [Y^{4-}]$$

and we can state that, *in an ammonium chloride-ammonia buffer of known pH, a constant fraction of the uncomplexed EDTA exists as* $Y^{4-}$:

$$[Y^{4-}] = \alpha_{Y^{4-}} C_{EDTA}$$

Let us substitute the above relations for $[Zn^{2+}]$ and $[Y^{4-}]$ into the expression for the formation constant for the zinc-EDTA complex:

$$K_{ZnY^{2-}} = \frac{[ZnY^{2-}]}{[Zn^{2+}][Y^{4-}]} = \frac{[ZnY^{2-}]}{(\alpha_{Zn^{2+}} C_{Zn})(\alpha_{Y^{4-}} C_{EDTA})}$$

Inasmuch as $\alpha_{Zn^{2+}}$ and $\alpha_{Y^{4-}}$ are constants, we can include them on the left side of the equation:

$$K_{ZnY^{2-}} \alpha_{Zn^{2+}} \alpha_{Y^{4-}} = \frac{[ZnY^{2-}]}{C_{Zn} C_{EDTA}}$$

It is customary to call the product of the three terms on the left side of the last equation a **conditional formation constant,** $K'_{ZnY^{2-}}$:

$$K'_{ZnY^{2-}} = K_{ZnY^{2-}} \alpha_{Zn^{2+}} \alpha_{Y^{4-}}$$

This conditional formation constant is a measure of the effective stability of the zinc-EDTA complex for a particular set of conditions – that is, $K'_{ZnY^{2-}}$ has a specific value for every pH and free ammonia concentration. Obviously, $K'_{ZnY^{2-}}$ will change if a buffer having a different pH or a different ammonia concentration is employed or if a new auxiliary complexing agent is used. Since $\alpha_{Zn^{2+}}$ and $\alpha_{Y^{4-}}$ are both less than unity under most conditions, $K'_{ZnY^{2-}}$ is almost always smaller than $K_{ZnY^{2-}}$, which confirms our previous statement that a metal-EDTA complex has a lower stability in the presence than in the absence of an auxiliary complexing agent.

### Titration Curves for the Zinc(II)-EDTA Reaction in an Ammoniacal Solution

Let us consider the titration of a 50.00-ml sample of 0.001000 *F* zinc nitrate with a 0.1000 *F* $Na_2H_2Y$ solution. We will assume that the zinc(II) solution contains 0.10 *F* ammonia and 0.10 *F* ammonium chloride. Note that the EDTA titrant is 100 times more concentrated than the zinc(II) solution. Therefore, dilution of the sample medium during the titration can be ignored. Because only 0.5000 ml of titrant is required for the titration, it would be performed with the aid of a microburet.

---

*See footnote on page 126.

**Evaluation of the Conditional Formation Constant.** To calculate the conditional formation constant, defined by the relation

$$K'_{ZnY^{2-}} = K_{ZnY^{2-}}\alpha_{Zn^{2+}}\alpha_{Y^{4-}}$$

we must first determine $\alpha_{Zn^{2+}}$ and $\alpha_{Y^{4-}}$. Since the original concentration of ammonia is 0.10 $F$ and the total zinc(II) concentration is only 0.001000 $M$, it is reasonable to assert that the free ammonia concentration remains essentially constant at 0.10 $F$ despite the formation of zinc-ammine complexes. Figure 6–1 shows that the predominant zinc(II) species in this solution is $Zn(NH_3)_4{}^{2+}$ and that very little $Zn^{2+}$ is present. Taking the concentration of ammonia to be 0.10 $F$ and using the relationship derived on page 124, we can show that

$$\alpha_{Zn^{2+}} = \frac{1}{1 + \beta_1[NH_3] + \beta_2[NH_3]^2 + \beta_3[NH_3]^3 + \beta_4[NH_3]^4}$$

$$\alpha_{Zn^{2+}} = \frac{1}{1 + 186(0.1) + 4.08 \times 10^4(0.1)^2 + 1.02 \times 10^7(0.1)^3 + 1.15 \times 10^9(0.1)^4}$$

$$\alpha_{Zn^{2+}} = \frac{1}{1.26 \times 10^5} = 7.94 \times 10^{-6}$$

If the procedure outlined on page 73 is employed to compute the pH and hydrogen ion concentration for a buffer consisting of 0.10 $F$ concentrations of ammonia and ammonium chloride, we find that pH = 9.26 and $[H^+] = 5.55 \times 10^{-10}$ $M$. Substitution of the latter result, along with values for appropriate equilibrium constants, into the expression* for $\alpha_{Y^{4-}}$ yields

$$\alpha_{Y^{4-}} = \frac{KK'K_{a1}K_{a2}K_{a3}K_{a4}}{[H^+]^6 + K[H^+]^5 + KK'[H^+]^4 + KK'K_{a1}[H^+]^3 + KK'K_{a1}K_{a2}[H^+]^2 + KK'K_{a1}K_{a2}K_{a3}[H^+] + KK'K_{a1}K_{a2}K_{a3}K_{a4}}$$

$$\alpha_{Y^{4-}} = \frac{2.57 \times 10^{-24}}{\left\{\begin{array}{l}(2.92 \times 10^{-56}) + (6.64 \times 10^{-48}) + (3.00 \times 10^{-40}) + (5.40 \times 10^{-33}) \\ \quad + (2.08 \times 10^{-26}) + (2.59 \times 10^{-23}) + (2.57 \times 10^{-24})\end{array}\right\}}$$

$$\alpha_{Y^{4-}} = \frac{2.57 \times 10^{-24}}{2.86 \times 10^{-23}} = 8.99 \times 10^{-2}$$

Combining the values of $\alpha_{Zn^{2+}}$ and $\alpha_{Y^{4-}}$ with $K_{ZnY^{2-}}$ found in Table 6–1, we can evaluate the conditional formation constant for the zinc-EDTA complex:

$$\begin{aligned} K'_{ZnY^{2-}} &= K_{ZnY^{2-}}\alpha_{Zn^{2+}}\alpha_{Y^{4-}} \\ &= (3.16 \times 10^{16})(7.94 \times 10^{-6})(8.99 \times 10^{-2}) \\ &= 2.25 \times 10^{10} \end{aligned}$$

Although the conditional formation constant is more than one million times smaller than the true formation constant, the former is still large enough to yield a well defined titration curve.

---

*See footnote on page 126.

**Construction of the Titration Curve.** To assess the feasibility of the titration of zinc(II) with EDTA, it is desirable to construct the complete titration curve – that is, a plot of pZn (the negative logarithm of the concentration of $Zn^{2+}$) as a function of the volume of titrant. Such a curve enables us to predict the accuracy of the titration and to select an appropriate method for locating the equivalence point.

At the Start of the Titration. Before any of the EDTA titrant has been added, the *total* concentration of zinc(II), $C_{Zn}$, is $1.00 \times 10^{-3}$ *M*. It follows that

$$[Zn^{2+}] = \alpha_{Zn^{2+}}C_{Zn} = (7.94 \times 10^{-6})(1.00 \times 10^{-3}) = 7.94 \times 10^{-9}\ M$$

and that pZn is 8.10.

Before the Equivalence Point. When the EDTA titrant is introduced into the zinc(II) solution, the principal reaction may be written as

$$Zn(NH_3)_4{}^{2+} + HY^{3-} \rightleftharpoons ZnY^{2-} + NH_4{}^{+} + 3\ NH_3$$

Because the conditional formation constant for the zinc-EDTA complex is large ($K'_{ZnY^{2-}} = 2.25 \times 10^{10}$), each increment of titrant can be assumed to react completely with an equivalent amount of zinc(II). Furthermore, in calculating pZn, we can safely neglect the tiny amount of zinc(II) produced by dissociation of the zinc-EDTA complex. After the addition of 0.0500 ml of 0.1000 *F* $Na_2H_2Y$ titrant, or 0.00500 millimole, there will remain 0.04500 millimole of zinc(II) *uncomplexed by EDTA* in approximately 50 ml of solution. Therefore, the *total* concentration of zinc(II) uncomplexed by EDTA, $C_{Zn}$, is essentially $9.00 \times 10^{-4}$ *M*. Proceeding as before, we can write

$$[Zn^{2+}] = \alpha_{Zn^{2+}}C_{Zn} = (7.94 \times 10^{-6})(9.00 \times 10^{-4}) = 7.15 \times 10^{-9}\ M$$

and

$$pZn = 8.15$$

Results for six more similar calculations are listed in Table 6–2.

**Table 6–2. Variation of pZn During the Titration of 50.00 ml of 0.001000 *M* Zinc(II) in a 0.1 *F* $NH_3$-0.1 *F* $NH_4Cl$ Buffer with 0.1000 *F* EDTA Solution**

| Volume of EDTA, ml | $[Zn^{2+}]$, *M* | pZn |
|---|---|---|
| 0 | $7.94 \times 10^{-9}$ | 8.10 |
| 0.050 | $7.15 \times 10^{-9}$ | 8.15 |
| 0.100 | $6.35 \times 10^{-9}$ | 8.20 |
| 0.200 | $4.76 \times 10^{-9}$ | 8.32 |
| 0.300 | $3.18 \times 10^{-9}$ | 8.50 |
| 0.400 | $1.59 \times 10^{-9}$ | 8.80 |
| 0.450 | $7.94 \times 10^{-10}$ | 9.10 |
| 0.490 | $1.59 \times 10^{-10}$ | 9.80 |
| 0.500 | $1.68 \times 10^{-12}$ | 11.78 |
| 0.510 | $1.76 \times 10^{-14}$ | 13.75 |
| 0.550 | $3.52 \times 10^{-15}$ | 14.45 |
| 0.600 | $1.76 \times 10^{-15}$ | 14.75 |
| 0.700 | $8.80 \times 10^{-16}$ | 15.06 |

At the Equivalence Point. A calculation of pZn at the equivalence point requires use of the conditional formation constant for the zinc-EDTA complex:

$$K'_{ZnY^{2-}} = \frac{[ZnY^{2-}]}{C_{Zn}C_{EDTA}} = 2.25 \times 10^{10}$$

At the equivalence point, the concentration of the zinc-EDTA complex is $1.00 \times 10^{-3}$ *M* minus the total concentration of zinc(II), $C_{Zn}$, not complexed by EDTA:

$$[ZnY^{2-}] = (1.00 \times 10^{-3}) - C_{Zn}$$

Because the stoichiometry of the complexation process is one-to-one and because no reaction is ever 100 per cent complete, we can state that the total concentration of zinc(II) not complexed by EDTA at the equivalence point is, by definition, equal to the total concentration of EDTA not complexed by zinc(II):

$$C_{Zn} = C_{EDTA}$$

If the two preceding relations are substituted into the expression for $K'_{ZnY^{2-}}$, we have

$$K'_{ZnY^{2-}} = \frac{[ZnY^{2-}]}{C_{Zn}C_{EDTA}} = \frac{(1.00 \times 10^{-3}) - C_{Zn}}{(C_{Zn})^2} = 2.25 \times 10^{10}$$

In solving this equation, let us neglect provisionally the $C_{Zn}$ term in the numerator because the zinc-EDTA complex is so stable. Accordingly,

$$(C_{Zn})^2 = \frac{1.00 \times 10^{-3}}{2.25 \times 10^{10}} = 4.45 \times 10^{-14}; \qquad C_{Zn} = 2.11 \times 10^{-7}\,M$$

Note that $C_{Zn}$ is indeed negligible in comparison to $1.00 \times 10^{-3}$ *M*. Finally, we obtain the result

$$[Zn^{2+}] = \alpha_{Zn^{2+}}C_{Zn} = (7.94 \times 10^{-6})(2.11 \times 10^{-7}) = 1.68 \times 10^{-12}; \qquad pZn = 11.78$$

After the Equivalence Point. Consider the calculation of pZn after the addition of 0.510 ml of the EDTA titrant. This corresponds to an *excess* of (0.010)(0.1000) or $1.00 \times 10^{-3}$ millimole of EDTA in a volume that we shall take to be 50 ml. Therefore, the total concentration of EDTA not complexed by zinc(II), $C_{EDTA}$, is $2.00 \times 10^{-5}$ *M*. Since dilution of the solution is ignored, the concentration of the zinc-EDTA complex, $[ZnY^{2-}]$, is essentially $1.00 \times 10^{-3}$ *M*. When these two conditions are inserted into the relation for $K'_{ZnY^{2-}}$, the result is as follows:

$$K'_{ZnY^{2-}} = \frac{[ZnY^{2-}]}{C_{Zn}C_{EDTA}} = \frac{1.00 \times 10^{-3}}{C_{Zn}(2.00 \times 10^{-5})} = 2.25 \times 10^{10}$$

$$C_{Zn} = \frac{1.00 \times 10^{-3}}{(2.00 \times 10^{-5})(2.25 \times 10^{10})} = 2.22 \times 10^{-9}\,M$$

$$[Zn^{2+}] = \alpha_{Zn^{2+}}C_{Zn} = (7.94 \times 10^{-6})(2.22 \times 10^{-9}) = 1.76 \times 10^{-14}\,M$$

$$pZn = 13.75$$

Table 6–2 includes data obtained from three more calculations of this type, and the complete titration curve is portrayed in Figure 6–3.

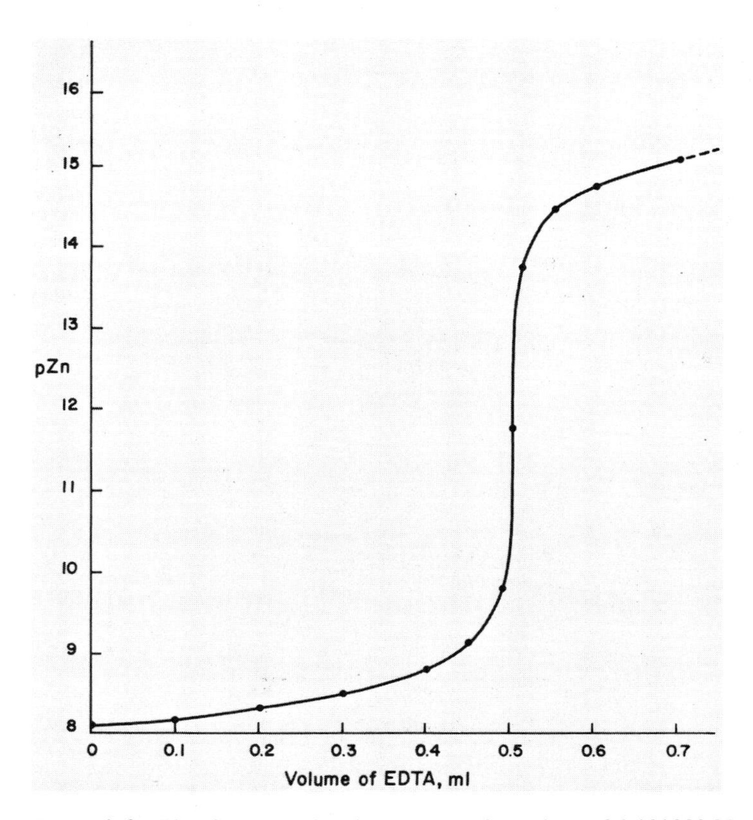


*Figure 6–3.* Titration curve for the titration of 50.00 ml of 0.001000 *M* zinc(II) solution with 0.1000 *F* EDTA. In addition, the zinc(II) solution is 0.1 *F* in ammonia and 0.1 *F* in ammonium chloride.

**Effect of Solution Composition on Titration Curves.** Since the composition of a solution governs the value of the conditional formation constant for the zinc-EDTA complex – because $\alpha_{Zn^{2+}}$ and $\alpha_{Y^{4-}}$ are affected – the success of a complexometric titration with EDTA rests in part on the auxiliary complexing agent.

Influence of Ammonia. How does the concentration of ammonia affect the sharpness of the titration curve? Figure 6–4 shows three titration curves for the reaction of 0.001 *M* zinc(II) with 0.1 *F* EDTA in the presence of ammonia-ammonium chloride buffers. In each case, the ammonia-ammonium ion concentration ratio is unity, so that a constant pH of 9.26 is maintained, but the absolute concentrations differ. According to calculations performed in constructing the titration curve in Figure 6–3, values of pZn *before the equivalence point* depend only upon $\alpha_{Zn^{2+}}$ for a specified initial concentration of zinc(II). As seen in Figure 6–4, variations in the ammonia concentration shift the position of the titration curve before the equivalence point, but have no effect on the curve after the equivalence point. An increase in the concentration of ammonia causes a decrease in $\alpha_{Zn^{2+}}$ and diminishes the change in pZn in the region of the equivalence point, making the equivalence point more difficult to locate accurately. For an ammonia concentration of 1 *F*, the change in pZn near the equivalence point is inadequate for a direct titration.

*Figure 6–4.* Effect of ammonia concentration on the titration curve for the titration of 50.00 ml of 0.001000 *M* zinc(II) with 0.1000 *F* EDTA in ammonia-ammonium chloride buffers. For each curve, the ammonia concentration is as shown and the ammonium chloride concentration has an identical value so that the pH is 9.26.

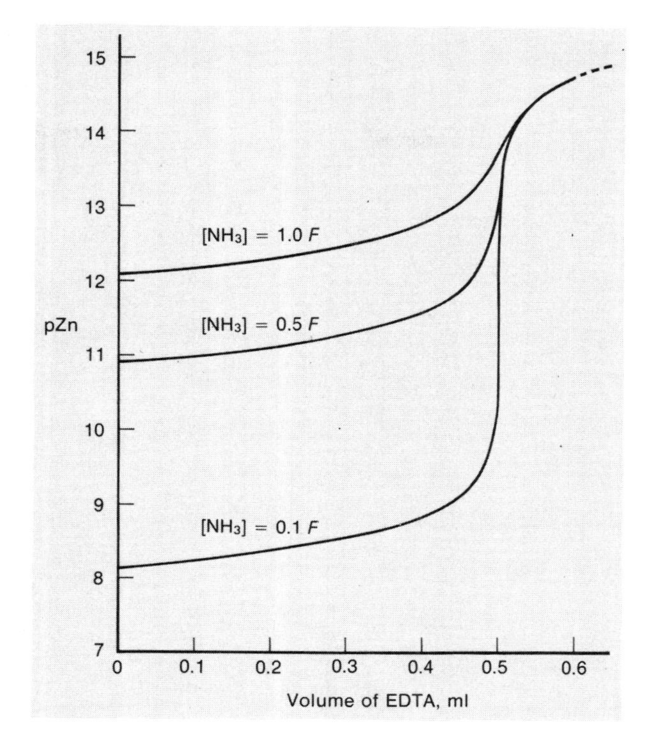


A decrease in the ammonia concentration below 0.1 *F* would improve the quality of the titration curve. However, lowering the concentration of ammonia indiscriminately will lead to the precipitation of zinc hydroxide because, as shown in Figure 6–1, the relative amount of hydrated zinc(II) increases very rapidly for ammonia concentrations less than 0.01 *F*. Therefore, in employing an auxiliary complexing agent, one must seek a compromise – the concentration of the complexing agent must be high enough to prevent precipitation of the metal ion being determined but not so high that a sharp end point cannot be obtained.

**Influence of pH.** Figure 6–5 illustrates the effect of pH on the titration curve for the zinc(II)-EDTA reaction in various solutions containing the same ammonia concentration of 0.1 *F*. In contrast to the influence of the ammonia concentration (Figure 6–4), the pZn values are independent of pH before the equivalence point. On the other hand, the titration curve *after the equivalence point* is displaced toward higher values of pZn as the pH increases from 7 to 10 because more and more EDTA exists in the form of the $Y^{4-}$ anion.

In the presence of 0.1 *F* ammonia, some zinc(II) might be precipitated as insoluble $Zn(OH)_2$ at pH values above 10, although the latter compound is converted to the zinc-EDTA complex during the titration. This difficulty can be overcome through the use of a larger initial concentration of ammonia, but a price must be paid because the titration curve before the equivalence point is shifted upward if extra ammonia is added. For any ligand that forms complexes with zinc(II) as stable as the zinc-ammine species, the change of pZn near the equivalence point becomes undesirably small below pH 7.

### Techniques of Complexometric Titrations

There are four different methods employed to perform complexometric titrations with EDTA – direct titration, back titration, displacement titration, and alkalimetric titration.

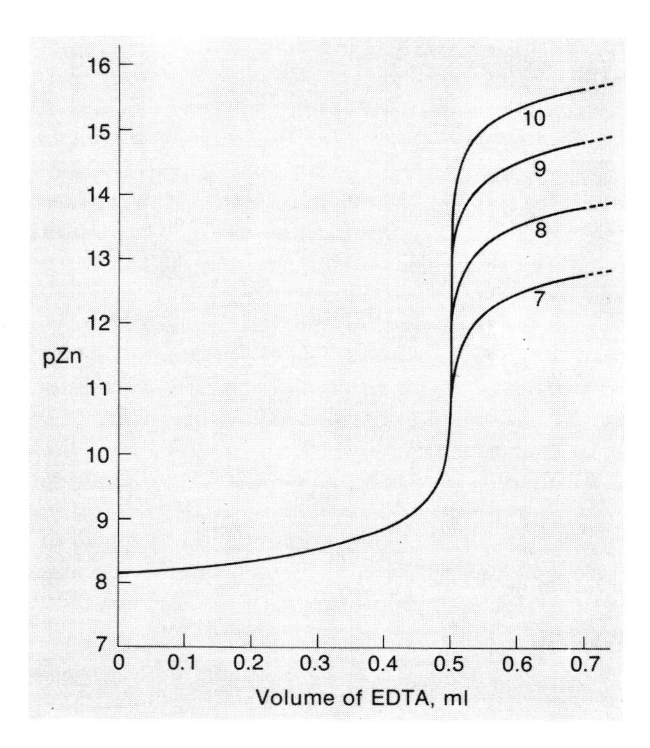


*Figure 6–5.* Effect of pH on the titration curve for the reaction of 0.001000 *M* zinc(II) with 0.1000 *F* EDTA in the presence of 0.100 *F* ammonia. Numbers on the titration curves denote the pH of the ammonia-ammonium chloride buffer. Different pH values are obtained in practice through the use of a range of ammonium chloride concentrations.

**Direct Titration.** A direct determination of a metal ion involves addition of standard EDTA titrant to the sample solution until an appropriate end-point signal is observed. Usually, an auxiliary complexing agent is included in the sample medium to control the pH and to prevent precipitation of the metal hydroxide.

Direct titrations are feasible for a large number of metal cations, including aluminum, barium, bismuth, cadmium, calcium, cerium, cobalt, copper, gallium, indium, iron, lead, magnesium, manganese, mercury, nickel, scandium, strontium, thallium, thorium, zinc, and zirconium as well as the lanthanides. Only those cations that react instantaneously with EDTA are suitable candidates for a direct titration.

**Back Titration.** Certain metal cations react too slowly with EDTA to be titrated directly, whereas others form precipitates within the pH range normally chosen for a direct titration. Moreover, satisfactory end-point indicators are not available for some metal ions that might be determined by means of a direct-titration procedure. When any of these difficulties arises, the technique of back titration is often successful.

In a back titration, one adds to the sample solution a known volume of standard EDTA in excess of that required to combine with the desired metal cation, $M_1^{n+}$. It may be necessary to heat the solution or to alter the solution conditions so that the reaction between $M_1^{n+}$ and EDTA is rapid. Finally, the excess EDTA is back titrated under appropriate conditions with a standard solution of a second metal ion, $M_2^{n+}$. A solution of magnesium(II), zinc(II), or copper(II) is commonly employed as the titrant, and the quantity of metal cation ($M_1^{n+}$) in the sample is calculated by difference from the volumes of the standard solutions of EDTA and metal ion ($M_2^{n+}$) used.*

In the back-titration method, the EDTA complex of $M_2^{n+}$ must be *less* stable than the EDTA complex of $M_1^{n+}$; otherwise, $M_2^{n+}$ will displace $M_1^{n+}$ from its EDTA complex. On the other hand, if the EDTA complex of $M_1^{n+}$ is nonlabile, one need not be concerned

*There need be no restriction on the charges of the two metal cations.

about the relative stabilities of the two metal-EDTA species because no displacement reaction will occur. Back titrations may be performed for the determination of aluminum, cobalt, lead, manganese, mercury, nickel, and thallium.

**Displacement Titration.** Another alternative to the direct-titration method entails addition of a known excess of the magnesium-EDTA complex to a solution of the metal cation, $M^{n+}$, to be determined. Because the magnesium-EDTA species is less stable than most other metal-EDTA complexes (Table 6–1), the displacement reaction

$$MgY^{2-} + M^{n+} \rightleftharpoons Mg^{2+} + MY^{(n-4)+}$$

occurs, and an amount of magnesium ion equivalent to the quantity of $M^{n+}$ is liberated. Titration of the free magnesium ion with a standard EDTA solution permits the determination of $M^{n+}$. Sometimes the zinc-EDTA complex is used instead of the magnesium-EDTA species.

**Alkalimetric Titration.** Since the reaction between a metal cation, $M^{n+}$, and $H_2Y^{2-}$ produces hydrogen ion,

$$M^{n+} + H_2Y^{2-} \rightleftharpoons MY^{(n-4)+} + 2\ H^+$$

titration with a standard sodium hydroxide solution provides a means to determine how much metal ion is present. Care must be taken to ensure that the solution of the metal ion contains no buffering agents. In practice, one uses two burets containing standard EDTA and sodium hydroxide solutions, respectively, introduces an acid-base indicator and adjusts the initial pH of the sample solution, and adds first one and then the other titrant back and forth until a final increment of EDTA solution causes no change in the color of the indicator.

### Selectivity in Complexometric Titrations

As mentioned earlier, EDTA forms stable complexes with most cations. Therefore, this reagent frequently lacks selectivity if one is interested in determining a single metal cation in a mixture or in performing a successive titration of several species in the same solution. What can be done to improve the selectivity of complexometric titrations with EDTA?

**Chemical Separations.** An obvious answer is to separate the desired species from the other components of the sample. Among the techniques that can be employed are precipitation, liquid-liquid extraction, and ion-exchange chromatography – all of which are discussed in later chapters. These methods have the disadvantages, however, that the composition of the sample solution may be altered, perhaps drastically, and that the various components of the sample, once separated, must be recovered in a suitable form before a complexometric titration can be performed.

**Control of Acidity.** Simply by adjusting the pH of a sample solution containing several metal cations, one may cause only a single species to react with EDTA. This is due to the fact that conditional formation constants for metal-EDTA complexes depend in part on the hydrogen ion concentration.

Let us consider the determination of nickel(II). In a solution having an ammonia concentration of 1 *F* and a pH of 10, the conditional formation constant ($K'_{NiY^{2-}}$) for the nickel-EDTA complex is $1.72 \times 10^8$, so that nickel(II) by itself can be accurately titrated with EDTA. However, if the sample solution contains any of the alkaline-earth cations – magnesium, calcium, strontium, and barium – these species (which form no stable ammine complexes) will be co-titrated with nickel(II) because their conditional formation constants at pH 10 are comparable ($1.77 \times 10^8$, $1.77 \times 10^{10}$, $1.41 \times 10^8$, and $2.23 \times 10^7$, respectively) to that for the nickel-EDTA complex. Fortunately, the interference caused by these elements can be prevented if the pH of the sample solution is kept at 3.5 with an appropriate buffer. Ignoring the possible formation of other complexes of the various metal cations and recalculating the conditional formation constants for the EDTA complexes of nickel, magnesium, calcium, strontium, and barium at pH 3.5, we obtain values of $1.00 \times 10^9$, 0.126, 12.6, 0.100, and $1.58 \times 10^{-2}$, respectively. Thus, the EDTA complexes of the alkaline-earth ions have become too unstable to interfere with the titration of nickel(II).

**Use of Masking Agents.** Nickel(II) can be titrated with EDTA in the company of alkaline-earth cations if the pH of the sample solution is properly adjusted. On the other hand, simple pH control yields unsatisfactory results for the determination of manganese(II) in a solution containing other divalent heavy-metal ions such as cadmium, cobalt, copper, mercury, nickel, and zinc. In the latter case, selectivity for the titration of manganese(II) requires addition to the sample solution of a **masking agent**, some substance that reacts with the interfering cations to prevent them from combining with EDTA. Cyanide ion is a widely used masking agent and is an ideal choice for the determination of manganese(II) with EDTA because the other cations form exceedingly stable cyano complexes – $Cd(CN)_4{}^{2-}$, $Co(CN)_6{}^{2-}$, $Cu(CN)_3{}^{2-}$, $Hg(CN)_4{}^{2-}$, $Ni(CN)_4{}^{2-}$, and $Zn(CN)_4{}^{2-}$ – which do not react with the titrant. In performing this determination, one adds potassium cyanide (to mask the other divalent cations), potassium sodium tartrate (an auxiliary complexing agent to prevent precipitation of manganese(II) hydroxide), ascorbic acid (a reducing agent to stop atmospheric oxidation of manganese(II) to manganese dioxide), and an ammonia-ammonium chloride buffer at pH 10. Then, at a temperature of 70 to 80°C, manganese(II) is titrated with a standard EDTA solution.

A large number of masking agents have been utilized for complexometric titrations. Most of these reagents function as complexing ligands, although some serve as precipitants or as substances that produce a change in the oxidation state of a likely interferent. Table 6–3 lists four common masking agents and a few of their uses.

### Metallochromic Indicators for End-Point Detection

For some complexometric titrations with EDTA, especially those used for the determination of iron(III), a few highly specific end-point indicators are available. Thiocyanate ion which forms blood-red-colored complexes with iron(III) – $FeSCN^{2+}$ and $Fe(SCN)_2{}^+$ – can be employed to signal the end point of a complexometric titration of iron(III) in a strongly acidic medium, a sudden disappearance of color occurring at the equivalence point. Salicylic acid

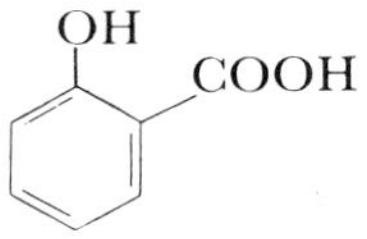

OH
COOH

reacts with iron(III) to yield very stable, reddish-brown complexes having one, two, or three bidentate salicylate anions coordinated to the central atom. When a solution containing iron(III) and a small quantity of salicylic acid is titrated with EDTA at pH 3, the end point is marked by a distinct change from the red-brown color of the iron(III)-salicylate species to the colorlessness of the iron-EDTA complex.

**Table 6–3. Some Masking Agents Used in Complexometric Titrations with EDTA**

| Masking Agent | Typical Applications |
|---|---|
| Cyanide, $CN^-$ | (a) Titrate Pb in ammoniacal tartrate medium in presence of Cu masked as $Cu(CN)_3^{2-}$ or Co masked as $Co(CN)_6^{4-}$<br>(b) Titrate Ca at pH 10 in presence of divalent heavy metals masked as cyanide complexes<br>(c) Titrate In in ammoniacal tartrate medium in presence of Cd, Co, Cu, Hg, Ni, and Zn masked as cyanide complexes |
| Triethanolamine, $N(CH_2CH_2OH)_3$ | (a) Titrate Ni in ammoniacal medium in presence of Al, Fe, and Mn masked as triethanolamine complexes<br>(b) Titrate Mg in ammoniacal medium of pH 10 in presence of Al masked as a triethanolamine complex<br>(c) Titrate Zn or Cd in ammoniacal medium in presence of Al masked as a triethanolamine complex |
| Fluoride, $F^-$ | (a) Titrate Zn in ammoniacal medium in presence of Al masked as $AlF_6^{3-}$ or Mg and Ca masked by precipitation as $MgF_2$ and $CaF_2$, respectively<br>(b) Titrate Ga in glacial acetic acid at pH 2.8 in presence of Al masked as $AlF_6^{3-}$ |
| 2,3-Dimercaptopropanol, $CH_2(OH)—CH(SH)—CH_2(SH)$ | (a) Titrate Mg in ammoniacal medium of pH 10 in presence of Bi, Cd, Cu, Hg, and Pb masked as complexes with 2,3-dimercaptopropanol<br>(b) Titrate Th at pH 3 in the presence of Bi and Pb masked as complexes with 2,3-dimercaptopropanol |

Far more important is the large number of organic dyes, called **metallochromic indicators**, which have been utilized for end-point detection in complexometric titrations. Metallochromic indicators not only form stable, brightly colored complexes with most metal ions, but are acid-base indicators as well. Table 6–4 summarizes information about five typical metallochromic indicators and lists some of the complexometric titrations for which these substances are used.

**Eriochrome Black T.** Perhaps the most widely employed metallochromic indicator is Eriochrome Black T, which has the structure shown in Table 6–4. Eriochrome Black T is a triprotic acid which may be abbreviated as $H_3In$, but loss of a proton from the $-SO_3H$ group, like the first acid dissociation of sulfuric acid, is virtually complete in water. However, the second and third steps of dissociation can be represented by the following equilibria:

$$H_2In^- \rightleftharpoons H^+ + HIn^{2-}; \qquad K_{a2} = 5.01 \times 10^{-7}$$

$$HIn^{2-} \rightleftharpoons H^+ + In^{3-}; \qquad K_{a3} = 2.51 \times 10^{-12}$$

**Table 6-4. Properties and Uses of Some Selected Metallochromic Indicators**

| Indicator | Structure | Dissociation Constants and Colors of Free Indicator Species | Colors of Metal-Indicator Complexes | Applications |
|---|---|---|---|---|
| Eriochrome Black T | 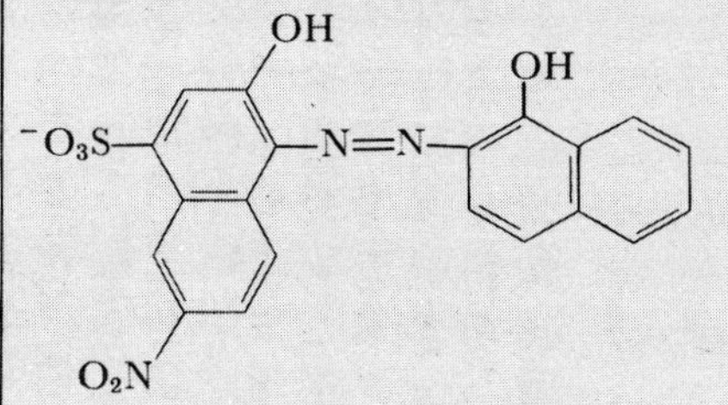 $(H_2In^-)$ | $H_2In^-$ (red); $pK_{a2} = 6.3$<br>$HIn^{2-}$ (blue); $pK_{a3} = 11.6$<br>$In^{3-}$ (orange) | wine-red | Direct titration of Ba, Ca, Cd, In, Pb, Mg, Mn, Sc, Sr, Tl, Zn, and lanthanides<br>Back titration of Al, Ba, Bi, Ca, Co, Cr, Fe, Ga, Pb, Mn, Hg, Ni, Pd, Sc, Tl, V<br>Displacement titration of Ba, Ca, Cu, Au, Pb, Hg, Pd, Sr |
| Calmagite | 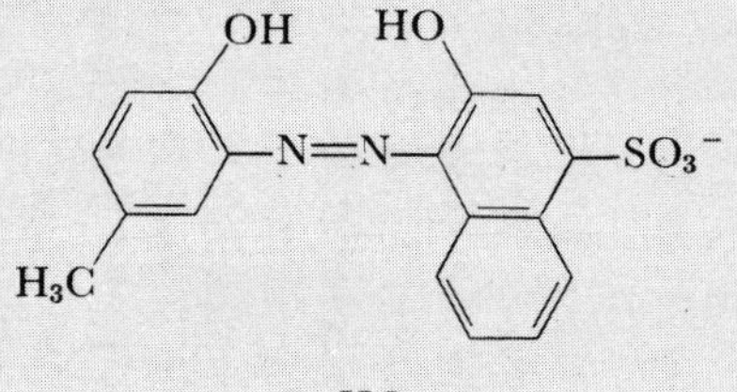 $(H_2In^-)$ | $H_2In^-$ (red); $pK_{a2} = 8.1$<br>$HIn^{2-}$ (blue); $pK_{a3} = 12.4$<br>$In^{3-}$ (orange) | wine-red | Titrations performed with Eriochrome Black T as indicator may be carried out equally well with Calmagite |
| PAN | 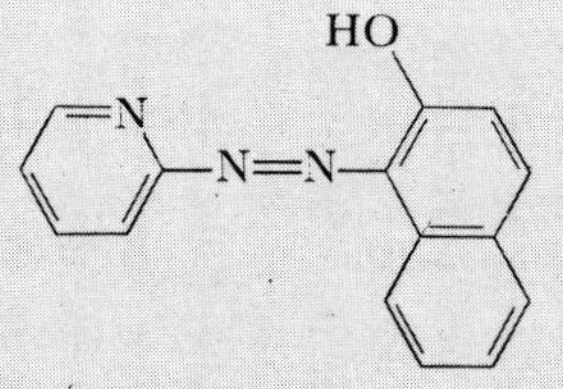 $(HIn)$ | HIn (orange-red); $pK_a = 12.3$<br>$In^-$ (pink) | red | Direct titration of Cd, Cu, In, Sc, Tl, Zn<br>Back titration of Cu, Ga, Fe, Pb, Ni, Sc, Sn, Zn<br>Displacement titration of Al, Ca, Co, Ga, In, Fe, Pb, Mg, Mn, Hg, Ni, V, Zn |

| | | | | |
|---|---|---|---|---|
| Murexide | 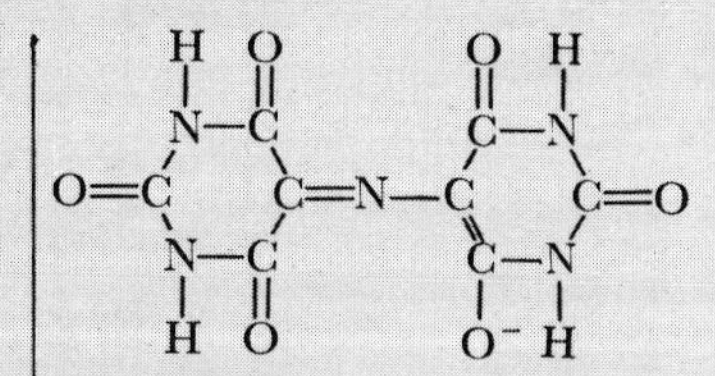 ($H_4In^-$) | $H_4In^-$ (red-violet); $pK_{a2} = 9.2$<br>$H_3In^{2-}$ (violet); $pK_{a3} = 10.9$<br>$H_2In^{3-}$ (blue) | red with $Ca^{2+}$<br>yellow with $Co^{2+}$, $Ni^{2+}$, and $Cu^{2+}$ | Direct titration of Ca, Co, Cu, Ni<br>Back titration of Ca, Cr, Ga<br>Displacement titration of Au, Pd, Ag |
| Pyrocatechol Violet | 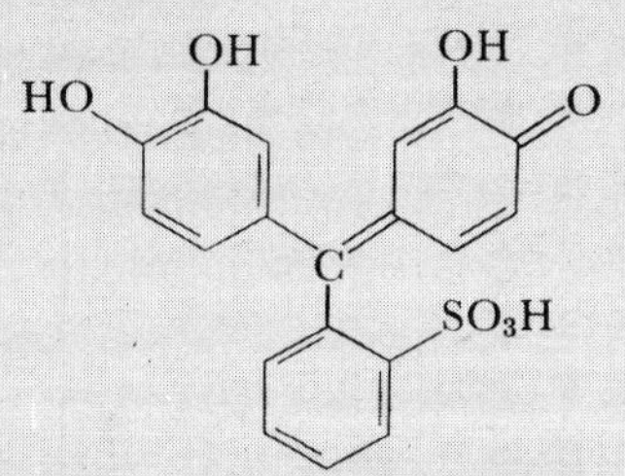 ($H_4In$) | $H_4In$ (red); $pK_{a1} = 0.2$<br>$H_3In^-$ (yellow); $pK_{a2} = 7.8$<br>$H_2In^{2-}$ (violet); $pK_{a3} = 9.8$<br>$HIn^{3-}$ (red-purple); $pK_{a4} = 11.7$ | blue, except red with Th(IV) | Direct titration of Al, Bi, Cd, Co, Ga, Fe, Pb, Mg, Mn, Ni, Th, Zn<br>Back titration of Al, Bi, Ga, In, Fe, Ni, Pd, Sn, Th, Ti |

In an aqueous medium, the colors of the species $H_2In^-$, $HIn^{2-}$, and $In^{3-}$ are red, blue, and orange, respectively. Because most EDTA titrations utilizing Eriochrome Black T are performed in the presence of a buffer such as ammonia-ammonium chloride whose pH is between 8 and 10, the predominant form of the free indicator is usually the bright-blue-colored $HIn^{2-}$ anion. In addition, Eriochrome Black T combines with most of the metal cations listed in Table 6–1 to yield stable, one-to-one, wine-red-colored complexes.

In the direct titration of zinc(II) with EDTA in an ammonia-ammonium ion buffer containing $10^{-6}$ to $10^{-5}$ *F* Eriochrome Black T, the solution has a wine-red color before the equivalence point due to the zinc-indicator complex. Suddenly, after the first tiny excess of EDTA titrant is added, the color of the solution becomes bright blue because formation of the zinc-EDTA complex liberates the $HIn^{2-}$ indicator ion according to the reaction

$$\underset{\text{(wine-red)}}{ZnIn^-} + HY^{3-} \rightleftharpoons ZnY^{2-} + \underset{\text{(blue)}}{HIn^{2-}}$$

For direct titrations of other cations, the same color change is observed at the end point, whereas just the opposite change in color takes place in a back titration. Note that successful end-point detection places restrictions on the equilibrium constants for the competing reactions. With reference to the direct titration of zinc(II), the stability of the zinc-Eriochrome Black T complex must be *less* than that of the zinc-EDTA species but *greater* than that of any zinc ammine under the extant solution conditions.

**Some Other Metallochromic Indicators.** One of the chief difficulties in using Eriochrome Black T is that solutions of it decompose rather rapidly. Only with relatively freshly prepared solutions can the proper end-point color change be realized. This instability appears to stem from the fact that the Eriochrome Black T molecule possesses an oxidizing substituent, the nitro group ($-NO_2$), and two different types of reducing groups, azo ($-N{=}N-$) and phenolic hydroxyl, which are probably involved in intermolecular redox reactions leading to destruction of the indicator.

In recent years, Calmagite has tended to replace Eriochrome Black T as an end-point indicator for complexometric titrations. Calmagite has chemical and physical properties that are remarkably similar to those of Eriochrome Black T, but its structural formula (Table 6–4) reveals the absence of a nitro group. A solution of Calmagite is exceedingly stable, having a shelf life of years. More important is that Calmagite may be used routinely as a substitute for Eriochrome Black T in all titrations requiring the latter indicator without any modification of the procedures.

Certain metal cations form such stable complexes with Eriochrome Black T or Calmagite that EDTA cannot displace the indicator from the metal-indicator species. Therefore, no color change is observed at the equivalence point of the titration or, if a color change does occur, it takes place only gradually after an excess of EDTA has been added. Nickel(II), copper(II), and titanium(IV) behave in this manner. However, nickel(II) and copper(II) can be determined by direct titration with EDTA in an ammoniacal medium if murexide is used as the metallochromic indicator, the end point corresponding to conversion of the yellow metal-murexide complex to the free violet-colored indicator anion.

Another indicator for the direct titration of copper(II) is PAN, the properties of which are compiled in Table 6–4. Usually, the determination is performed in an acetic acid-acetate buffer of pH 5 that contains approximately 50 per cent ethanol. This alcoholic solution sharpens the end point by raising the solubility of the copper-PAN complex so that the rate of reaction between EDTA and the copper-PAN species is increased.

Unlike the majority of metallochromic indicators, pyrocatechol violet is not a hydroxyazo dyestuff but a derivative of triphenylmethane. Pyrocatechol violet can be employed in acidic, neutral, and alkaline solutions for the direct titration of numerous metal cations, although the end-point color change in an ammoniacal medium is a somewhat unsatisfactory blue-to-violet transition.

### Scope of Complexometric Titrations with EDTA

Complexometric titrations with EDTA as well as other aminopolycarboxylic acids have been utilized for the determination of practically every metal cation, both singly and in mixtures. In Table 6–5 is a list of some of the individual metal ions that can be determined and of the kinds of samples that can be analyzed. Needless to say, real samples such as alloys or pharmaceutical products often require considerable pretreatment before a final titration with EDTA is performed. When, as usually occurs, mixtures of several metal ions are present in a sample, it may be necessary to separate the species, for example, by precipitation, extraction, or ion-exchange chromatography. Alternatively, appropriate masking agents can be added to the sample solution, or the pH of the sample solution can be adjusted, so that one metal cation can be titrated in the presence of others.

**Determination of Water Hardness.** Without doubt, the most important early application of titrations with EDTA was the determination of water hardness. In talking about water hardness, we can distinguish between *total* water hardness and *individual* water hardness due separately to calcium and magnesium. To determine total water hardness, one measures the *sum* of the amounts of calcium and magnesium, but usually reports the result in terms of calcium. Generally, the final titration is performed in a solution, buffered at pH 10 with ammonia and ammonium chloride, containing Eriochrome Black T indicator and appropriate masking agents to prevent traces of heavy metals from interfering with the action of the indicator. Furthermore, the EDTA titrant frequently contains a small concentration of the magnesium-EDTA complex, because the latter improves the sharpness of the end-point color change.

To evaluate the hardness of water due to calcium and magnesium individually, it is common practice to follow a procedure with two separate aliquots of the water sample. For one aliquot, the sum of the amounts of calcium and magnesium is established as described in the preceding paragraph. For the second aliquot, only calcium is determined, so that the magnesium content of the water is obtained by difference. In order to determine calcium in the presence of magnesium, the former is titrated with EDTA in a solution having a pH greater than 12; under such conditions, magnesium cannot interfere because it is precipitated as insoluble $Mg(OH)_2$. Murexide is employed as the end-point indicator, and suitable reagents can be added to the solution to mask heavy-metal impurities.

**Determination of Organic Compounds.** Methods have been devised for the determination of organic compounds and functional groups, particularly those found in pharmaceutical materials. These techniques are based upon the reaction between a known excess of a metal ion and the substance to be determined. Depending upon the nature of the chemical system, one can calculate the quantity of organic compound by measuring either the amount of metal ion that has reacted with the organic substance or the quantity of metal ion that remains unreacted.

A number of compounds form precipitates with mercury(II), after which the excess, unreacted metal in the filtrate is determined; these substances include nicotinamide, theobromine, phenobarbital, and ethyl gallate (an antioxidant). Caffeine, quinine, antihistamines, codeine, strychnine, thiamine, and morphine can each be precipitated

**Table 6–5. Some Representative Analyses Involving EDTA Titrations***

| Species Determined | Procedure | Typical Materials Analyzed |
|---|---|---|
| Al(III) | Add slight excess of standard EDTA; adjust pH to 6.5; dissolve salicylic acid indicator in the solution; back-titrate unreacted EDTA with standard iron(III) solution | alloys, cryolite ores, cracking catalysts, clays |
| Bi(III) | Adjust pH of sample solution to 2.5; add pyrocatechol violet indicator; titrate with EDTA | alloys, pharmaceuticals |
| Ca(II) | Add NaOH to neutral sample solution to obtain pH $> 12$; add murexide; titrate with EDTA | pharmaceuticals, phosphate rocks, water, biological fluids |
| Cd(II) or Zn(II) | To neutral sample solution, add $NH_3$–$NH_4^+$ buffer; add Eriochrome Black T indicator and titrate with EDTA | plating baths, alloys |
| Co(II) | Adjust pH of sample solution to 6; add murexide indicator; add ammonia to obtain orange color of cobalt-murexide complex; titrate with EDTA | paint driers, magnet alloys, cemented carbides |
| Cu(II) | Adjust pH of sample solution to 8 with $NH_3$ and $NH_4Cl$; add murexide indicator and titrate with EDTA | alloys, ores, electroplating baths |
| Fe(III) | Adjust pH of sample solution to 4; dissolve salicylic acid indicator in the solution; titrate with EDTA | hemoglobin, limestone, cement, tungsten alloys, paper pulp, boiler scale |
| Ga(III) | To neutral sample solution, add $NH_3$ and $NH_4^+$ to reach pH 9; add Eriochrome Black T indicator and a known excess of standard EDTA; back-titrate the unreacted EDTA with a standard manganese(II) solution | |
| Hf(IV) or Zr(IV) | [same procedure as for Al(III)] | alloys, ores, paint driers, sand |
| Hg(II) | To the sample solution, add a known excess of standard magnesium(II)-EDTA solution; neutralize the solution; add $NH_3$–$NH_4^+$ buffer and Eriochrome Black T indicator; perform substitution titration of free magnesium(II) with EDTA | mercury-containing pharmaceuticals, organomercury compounds, ores |
| In(III) | Add tartaric acid to an acidic sample solution; neutralize the solution; add $NH_3$–$NH_4^+$ buffer and Eriochrome Black T indicator; heat solution to boiling and titrate with EDTA | |
| Mg(II) | [same procedure as for Cd(II)] | pharmaceuticals, aluminum alloys, gun powder, soil, plant materials, water, biological fluids |
| Mn(II) | Add triethanolamine and ascorbic acid to acidic sample solution; neutralize the solution; add $NH_3$–$NH_4^+$ buffer; add Eriochrome Black T and titrate with EDTA | metallurgical slags, silicate rocks, alloys, ferromanganese |
| Ni(II) | Add murexide indicator to neutral sample solution; add $NH_3$ to obtain orange color of nickel-murexide complex; titrate with EDTA | electroplating baths, manganese catalysts, Alnico, ferrites |

**Table 6–5. Some Representative Analyses Involving EDTA Titrations (*continued*)**

| Species Determined | Procedure | Typical Materials Analyzed |
|---|---|---|
| Pb(II) | [same procedure as for In(III), except that a room-temperature titration is performed] | gasoline, ores, paints, alloys, pharmaceuticals |
| Sn(IV) | To strongly acidic sample solution, add known excess of standard EDTA; buffer the solution at pH 5; heat solution to 75°C, add pyrocatechol violet indicator, and titrate unreacted EDTA with standard Zn(II) solution | alloys, electroplating baths |
| Th(IV) | Adjust pH of sample solution to 2; add pyrocatechol violet indicator; warm solution to 40°C and titrate with EDTA | reactor fuels, ores, minerals, glasses, alloys |
| Tl(III) | [same procedure as for Hg(II)] | alloys |

* Information excerpted from G. Schwarzenbach and H. Flaschka: *Complexometric Titrations*. Second English edition, translated by H. M. N. H. Irving, Methuen, London, 1969.

from an acidic solution as a salt of the anion $BiI_4^-$, and the excess bismuth(III) can be determined by titration with EDTA. Biological fluids, chocolates, and sweetened and fermented drinks may be analyzed for sugar if one treats the sample with an alkaline solution of copper(II); the copper(II) is reduced to insoluble $Cu_2O$ by the sugar, and one can determine either how much cuprous oxide is formed or how much copper(II) remains.

Carbon-carbon double bonds can be determined if the sample material is treated with an excess of mercury(II) acetate in the presence of a $BF_3$-etherate catalyst; one measures the mercury(II), which does not add to the double bond, by first adding a known excess of EDTA and then back-titrating the unreacted ligand with a standard solution of zinc(II). If a solution containing amino acids or peptides is shaken with freshly precipitated copper(II) phosphate, the copper(II) that becomes complexed by the amino acid group may be found by a titration with EDTA after the solution is centrifuged to remove the excess precipitate.

## QUESTIONS AND PROBLEMS

1. In view of the fact that ethylenediaminetetraacetate (EDTA) forms stable complexes with so many cations, discuss the approaches by means of which reactions between EDTA and metal ions can be made selective.
2. Suggest an analytical procedure, involving an EDTA titration, for the direct or indirect determination of sulfate ion.
3. Define or identify each of the following terms: labile complex, nonlabile complex, bidentate ligand, chelate, stepwise formation constant, overall formation constant, ammine complex, auxiliary complexing agent, metallochromic indicator, conditional formation constant, masking agent.
4. Estimate from Figure 6–1 the fractions of the various zinc-ammine complexes in a solution containing a free ammonia concentration of 0.00316 *F*. If the total concentration of zinc-ammine complexes in such

a solution is 0.00100 *M*, calculate the concentration of each individual zinc-ammine complex.

5. Write the equilibrium reaction and the equilibrium expression corresponding to the fourth stepwise formation constant for the $Ni(NH_3)_4(H_2O)_2{}^{2+}$ complex.
6. Write the equilibrium reaction and the equilibrium expression corresponding to the third overall formation constant for the $Cd(NH_3)_3(H_2O)_3{}^{2+}$ complex.
7. Calculate the molar concentration of each of the species, $Cu(NH_3)^{2+}$, $Cu(NH_3)_2{}^{2+}$, $Cu(NH_3)_3{}^{2+}$, and $Cu(NH_3)_4{}^{2+}$, in a solution in which the equilibrium concentrations of ammonia and $Cu^{2+}$ ion are $2.50 \times 10^{-3}$ *F* and $1.50 \times 10^{-4}$ *M*, respectively.
8. What is the free ammonia concentration in a system containing the nickel-ammine complexes, if the concentration of $Ni(NH_3)_4{}^{2+}$ is exactly ten times the $Ni(NH_3)_3{}^{2+}$ concentration?
9. Calculate the equilibrium concentrations of $Zn(NH_3)_2{}^{2+}$ and $Zn(NH_3)_3{}^{2+}$ in a solution prepared by the mixing of 10.0 ml of 0.00200 *F* zinc nitrate solution with 40.0 ml of 0.200 *F* aqueous ammonia.
10. Calculate the concentration of each of the three silver(I) species in a solution initially containing 0.400 *F* ammonia and a total silver(I) concentration of 0.00300 *M*.
11. Calculate the equilibrium concentrations of the three species, $Ag^+$, $CN^-$, and $Ag(CN)_2{}^-$, resulting from the mixing of 35.0 ml of 0.250 *F* sodium cyanide solution with 30.0 ml of 0.100 *F* silver nitrate solution.
12. Calculate the equilibrium concentrations of the three species, $Ag^+$, $CN^-$, and $Ag(CN)_2{}^-$, resulting from the mixing of 50.0 ml of 0.200 *F* sodium cyanide solution with 70.0 ml of 0.100 *F* silver nitrate solution.
13. Calculate the concentration of each of the five EDTA species (ignoring $H_6Y^{2+}$ and $H_5Y^+$) in a 0.0250 *F* EDTA solution of pH 9.50.
14. What is the concentration of uncomplexed magnesium ion in a solution prepared by the mixing of equal volumes of 0.200 *F* EDTA and 0.100 *F* magnesium nitrate solutions, assuming that the pH is 9.00?
15. Calculate the concentration of uncomplexed nickel(II) in a solution prepared by the mixing of equal volumes of 0.150 *F* EDTA solution and 0.100 *F* nickel(II) nitrate solution. Assume that the solution is buffered at pH 10.50, but that the buffer constituents do not complex nickel(II).
16. Calculate the conditional formation constant for the nickel(II)-EDTA complex in a solution containing 0.500 *M* ammonium ion and 0.500 *F* free (uncomplexed) ammonia.
17. Calculate the conditional formation constant for the mercury(II)-EDTA complex in a solution of pH 11.00 containing 0.0100 *M* free (uncomplexed) cyanide ion.
18. Construct the titration curve for the titration of 50.00 ml of a 0.001000 *M* copper(II) solution with a 0.1000 *F* EDTA solution. Assume that the copper(II) sample solution contains 0.400 *F* ammonia and 0.200 *F* ammonium nitrate. Successive formation constants for the copper(II)-ammine complexes are listed in Appendix 3.
19. Construct the titration curve for the titration of 100.0 ml of a 0.002000 *M* magnesium ion solution with a 0.1000 *F* EDTA solution at pH 10.

20. A standard solution of calcium chloride was prepared by dissolution of 0.2000 g of pure calcium carbonate in hydrochloric acid. Then the solution was boiled to remove carbon dioxide and was diluted to 250.0 ml in a volumetric flask. When a 25.00-ml aliquot of the calcium chloride solution was used to standardize an EDTA solution by titration at pH 10, 22.62 ml of the EDTA solution was required. Calculate the concentration of the EDTA solution.
21. A 1.000-ml aliquot of a nickel(II) solution was diluted with distilled water and an ammonia-ammonium chloride buffer; it was then treated with 15.00 ml of a 0.01000 $F$ EDTA solution. The excess EDTA was back-titrated with a standard 0.01500 $F$ magnesium chloride solution, of which 4.37 ml was required. Calculate the concentration of the original nickel(II) solution.
22. To a solution containing gallium(III) ion was added an excess of an ammonia-ammonium nitrate buffer of pH 10 and exactly 25.00 ml of a 0.05862 $F$ solution of the magnesium-EDTA complex. Then a few drops of Eriochrome Black T indicator was introduced. Finally, the solution was titrated with 0.07010 $F$ EDTA until a color change from red to blue was observed, 5.91 ml of titrant being used. Calculate the weight in milligrams of gallium present in the original solution.
23. Total water hardness is usually expressed in terms of the number of milligrams of calcium carbonate ($CaCO_3$) per liter of water (in other words, parts per million, ppm, of calcium carbonate), although the titrimetric procedure actually entails a determination of the sum of the amounts of calcium(II) and magnesium(II) in solution.

    A 100.0-ml sample of water was checked for total hardness by means of the procedure described on page 143. In the titration, which was performed in an ammonia-ammonium ion buffer of pH 10 with Calmagite as metallochromic indicator, a volume of 12.58 ml of 0.008826 $F$ EDTA was needed to reach the desired end point. Calculate the total water hardness in terms of parts per million (ppm) of calcium carbonate.

    For another 100.0-ml sample of the water, the calcium content was determined by means of a titration with EDTA at pH 13. Murexide was employed as indicator, and 10.11 ml of the previous EDTA titrant was required. Using the titration data, calculate the calcium and magnesium content of the water in terms of parts per million (ppm) of $Ca^{2+}$ and $Mg^{2+}$, respectively.
24. Impure sodium phenobarbital ($NaC_{12}H_{11}N_2O_3$) was assayed by means of a procedure involving a complexometric titration. A 0.2438-g sample of the powder was dissolved in 100 ml of 0.02 $F$ sodium hydroxide solution at 60°C. After being cooled, the solution was acidified with acetic acid and was transferred to a 250.0-ml volumetric flask. Then a 25.00-ml aliquot of a 0.02031 $F$ mercury(II) perchlorate solution was added to the flask. Next, the solution was diluted to the mark on the volumetric flask and was allowed to stand until the precipitate formed according to the reaction

$$Hg^{2+} + 2\,C_{12}H_{11}N_2O_3^- \rightleftharpoons Hg(C_{12}H_{11}N_2O_3)_2$$

Finally, the solution was filtered to remove the precipitate and the filtrate was analyzed for unreacted mercury(II). A 50.00-ml aliquot of the filtrate was treated with 10.00 ml of a 0.01128 *M* solution of the magnesium-EDTA complex, and the liberated magnesium ion was titrated at pH 10 to an Eriochrome Black T end point with 5.89 ml of 0.01212 *F* EDTA solution. Calculate the percentage of sodium phenobarbital in the powder.

25. Acrylonitrile ($H_2C{=}CHCN$), which is a toxic and flammable liquid, polymerizes to form Orlon fibers. A partially polymerized sample of acrylonitrile, weighing 0.2795 g, was dissolved in 10 ml of a 0.05 *F* solution of $BF_3 \cdot O(C_2H_5)_2$ in methanol. A 0.1772-g portion of pure, anhydrous mercury(II) acetate was dissolved in the methanol solution, and the reaction between monomeric acrylonitrile and mercury(II) acetate

$$H_2C{=}CHCN + Hg(CH_3COO)_2 + CH_3OH \longrightarrow \begin{array}{cc} H_2C & \!\!\!-\!\!\!\; CHCN \\ | & | \\ H_3CO & Hg(CH_3COO) \end{array} + CH_3COOH$$

was allowed to reach completion. Then, 10 ml of an ammonia-ammonium nitrate buffer, 5.000 ml of a 0.1016 *M* solution of the zinc(II)-EDTA complex, 20 ml of water, and a few drops of Eriochrome Black T indicator were added. Zinc(II), released by the reaction between unconsumed mercury(II) and the zinc(II)-EDTA complex, was titrated with 0.05121 *F* EDTA solution, a volume of 2.743 ml being needed to reach the end point. Calculate the weight per cent of monomeric acrylonitrile in the sample.

## SUGGESTIONS FOR ADDITIONAL READING


1. F. Basolo and R. Johnson: *Coordination Chemistry.* W. A. Benjamin, New York, 1964, pp. 114–140.
2. H. Flaschka: *EDTA Titrations.* Second edition, Pergamon Press, New York, 1964.
3. H. Flaschka and A. J. Barnard, Jr.: Titrations with EDTA and related compounds. *In* C. L. Wilson and D. W. Wilson, eds.: *Comprehensive Analytical Chemistry.* Volume IB, Elsevier, New York, 1960, pp. 288–385.
4. G. Schwarzenbach and H. Flaschka: *Complexometric Titrations.* Second English edition, translated by H. M. N. H. Irving, Methuen, London, 1969.
5. F. J. Welcher: *The Analytical Uses of Ethylenediaminetetraacetic Acid.* Van Nostrand, Princeton, New Jersey, 1958.
6. T. S. West: *Complexometry.* Third edition, British Drug Houses Chemicals Limited, Poole, England, 1969.

# 7 APPLICATIONS OF PRECIPITATION REACTIONS

There are two classes of analytical methods based on the formation of precipitates – gravimetric determinations and precipitation titrations. In the usual gravimetric precipitation method, a solution of the species to be determined is mixed with a second solution containing an excess of some reagent that forms a precipitate with the substance sought. Then the resulting precipitate is separated from the solution, washed free of excess precipitant, dried at an appropriate temperature, and finally weighed so that the quantity of substance sought can be calculated.

In a precipitation titration, a solution containing the species to be determined is titrated with a standard solution of a reagent that forms a precipitate with the desired substance. If some technique is available to signal when an amount of titrant stoichiometrically equivalent to the substance sought has been added, the quantity of that substance can be computed from the concentration and volume of titrant used.

Quite obviously, a precipitation titration differs from a gravimetric determination in that the former does not require separation of a precipitate from the solution in which it is formed. However, regardless of which of these two analytical methods is employed, three requirements must be fulfilled by the precipitation process.

First, *the desired constituent must be precipitated quantitatively.* In other words, the quantity of the desired substance left in solution must be a negligible fraction of the total amount of that species originally present.

Second, *the precipitate must be pure.* In a gravimetric determination, the precipitate must not contain significant amounts of impurities unless these substances are readily removable in the washing and drying steps of the procedure. In a precipitation titration, the precipitate must not contain impurities that significantly alter the stoichiometric relationship between the titrant and the constituent being titrated.

Third, *the precipitate must be in a suitable physical form.* In a gravimetric determination, the precipitate must consist of particles that are large enough to be retained by the medium used for filtration. In a titrimetric procedure, the particles of precipitate must either remain dispersed or settle rapidly as particular circumstances require, so that, for example, the indicator color change marking the end point of the titration is not obscured.

In this chapter, we will examine the formation and dissolution of precipitates as well as some typical gravimetric determinations and precipitation titrations.

## FORMATION OF PRECIPITATES

### Nucleation and Crystal Growth

Formation of a precipitate involves two phenomena, **nucleation** and **crystal growth**. Nucleation proceeds through the formation within a supersaturated solution of the smallest particles of a precipitate (nuclei) capable of spontaneous growth. Crystal growth consists of the deposition of ions from the solution upon the surfaces of solid particles which have been nucleated.

As a rule, the particle size of a precipitate is determined by the number of nuclei formed in the nucleation step. In turn, the number of nuclei that form is governed by the extent of supersaturation in the immediate environment where nucleation occurs and by the number and effectiveness of sites upon which nuclei may form. In general, the *larger* the extent of supersaturation, the *smaller* will be the size of individual particles of the precipitate. For this reason, it is desirable to have solutions quite dilute at the time of mixing, particularly if the precipitate is extremely insoluble, in order to promote the formation of large crystals. Furthermore, it is important to form a precipitate under conditions for which it is not too insoluble.

Theoretically, it is possible for a sufficiently large cluster of ions to come together in a supersaturated solution to form a nucleus by the process known as **spontaneous nucleation**. In practical situations, however, it is probable that purely spontaneous nucleation is far less frequent than **induced nucleation**, in which the initial clustering of ions is aided by the presence in the solution of sites which can attract and hold ions. Surfaces of the container in which precipitation occurs provide many nucleation sites, as indicated by the fact that the particle size of a precipitate is strongly influenced by the type of container, how scratch-free it is, and how it was cleaned prior to use. Insoluble impurities in the reagents and in the solvent used to prepare solutions can serve as nucleation sites.

Crystal growth, once a nucleus has been formed, consists of *diffusion* of ions to the surface of the growing crystal and *deposition* of these ions on the surface. Either process can be rate-determining. In general, the diffusion rate is influenced by the specific nature of the ions and their concentrations, by the rate of stirring, and by the temperature of the solution, whereas the rate of deposition of ions is affected by concentration, by impurities on the surface, and by the growth characteristics of the particular crystal.

### Purity of a Precipitate

A precipitate can be contaminated with one or more extraneous substances because those substances are themselves insufficiently soluble in the mother liquid. It is not feasible, for example, to separate chloride from bromide by precipitation of the latter with silver nitrate because the solubilities of the two silver halides are not adequately different from each other. Such situations can be avoided if some other means of separation is utilized. In addition, a precipitate can become contaminated with substances from its mother liquid even when the solubilities of those other substances are not exceeded.

**Behavior of Colloidal Precipitates.** When first formed, a precipitate usually exists as colloidal particles dispersed in the solution phase. Because particles of a colloidal precipitate range from approximately $1 \times 10^{-7}$ to $2 \times 10^{-5}$ cm in diameter, a colloidal precipitate exhibits an enormous ratio of surface area to mass.

Most substances of analytical interest are ionic solids consisting of cations and anions arranged in a crystal lattice. In Figure 7–1 is depicted a row of alternating silver and chloride ions on the surface of a colloidal particle of silver chloride. Ions from the mother liquid which are adsorbed at the negative or positive charge centers on the surface of colloidal particles impart an electrical charge to those surfaces. Generally, one particular ion will be preferentially adsorbed and will impart its charge, either positive or negative, to all the surfaces. If an excess of silver ions is present in the mother liquid, the primary adsorbed layer will consist of silver ions, and the surfaces will acquire a net positive charge. This adsorbed ion layer imparts stability to the colloidal dispersion because the charged particles repel each other.

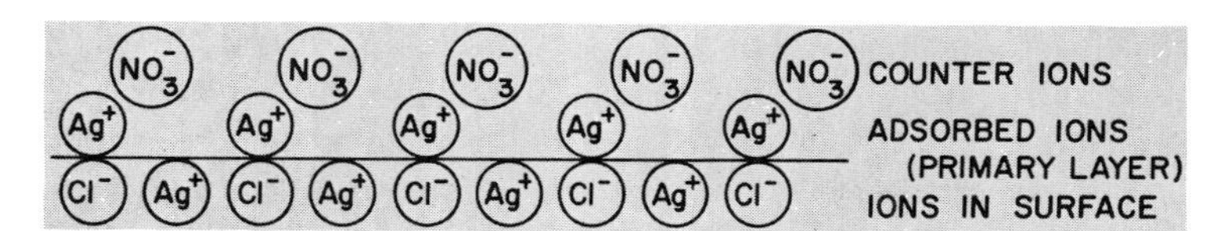


*Figure 7–1.* Schematic representation of adsorption of nitrate counter ions onto a primary adsorbed layer of silver ions at the surface of a colloidal silver chloride particle.

It is undesirable for a precipitate to be colloidal at the time of filtration, because colloidal particles are so fine that they pass through ordinary filtering media. In order for colloidal particles to coagulate and settle to the bottom of a container, the electrical charge imparted to their surfaces by the primary adsorbed ion layer must be removed or neutralized. Removal of this charge is unlikely, because the forces of attraction between the surface and the adsorbed ion layer are quite intense. Fortunately, there is a tendency for adsorption of a second layer of ions, called the counter-ion layer, which in Figure 7–1 consists of nitrate ions. In large part, the completeness with which counter ions cover the primary adsorbed ion layer determines the stability of the colloidal dispersion. If the secondary layer is sufficient to neutralize the charge due to the primary adsorbed ion layer, the particles coagulate rather than repel each other. Generally, the greater the ion content of the mother liquid, the more completely will the secondary layer neutralize the charge imparted to the surface of the particles by the primary layer.

Coagulation of colloidal particles through counter-ion adsorption is reversible, and **peptization** is the process whereby coagulated particles pass back into the colloidal state. Precautions must be taken during the washing of a coagulated precipitate to prevent peptization. When coagulation is accomplished through charge neutralization, peptization will occur if the precipitate is washed with pure water. Instead the wash liquid must contain an electrolyte, such as nitric acid, which will be volatile upon subsequent drying of the precipitate, so that it will not contribute to lack of purity of the precipitate at the time of weighing.

**Coprecipitation.** Precipitation of an otherwise soluble substance along with an insoluble one is termed coprecipitation. These two substances might precipitate simultaneously or one might follow the other. There are several mechanisms by which coprecipitation can occur: surface adsorption, occlusion, post-precipitation, and mechanical entrapment.

Surface Adsorption. As already described, ions are adsorbed from the mother liquid onto the surfaces of precipitated particles. This adsorption involves a primary adsorbed ion layer, which is held very tightly, and a counter-ion layer, which is held more or less loosely. These ions are carried down with the precipitate, causing it to be impure.

Impurities coprecipitated by surface adsorption cause an error in a gravimetric determination only if they are present during the final weighing of a precipitate. During washing it is sometimes possible to replace initially adsorbed ions with species that will be subsequently volatile in the drying of the precipitate. To minimize surface adsorption, it is advantageous to ensure that the solution in which the precipitate forms is dilute; that large crystals are obtained by stirring the reagent solutions as they are mixed; that the substance of interest is precipitated from a hot solution to increase the solubility of all components; and that foreign ions, which form relatively insoluble compounds with the ions of the precipitate, are replaced by other ions forming more readily soluble compounds.

Mechanical Entrapment. Mechanical entrapment is the physical enclosure of a small portion of the mother liquid within tiny hollows or flaws which form during the rapid growth and coalescence of the crystals. These pockets remain filled with the mother liquid and eventually become completely enclosed by the precipitate. Ordinary washing is of no aid in removing entrapped material. When a precipitate is dried at high temperature, the internal pressure in the pockets may rupture the particles with resultant release of the trapped solvent, but nonvolatile solutes present in the trapped mother liquid remain as impurities. To minimize mechanical entrapment, procedures identical to those used to lessen surface adsorption are employed.

Post-Precipitation. Another kind of precipitate contamination closely associated with surface adsorption is post-precipitation. Post-precipitation is encountered in the separation of calcium ion from magnesium ion by precipitation with oxalate. Calcium oxalate is a moderately insoluble compound which can be precipitated quantitatively. Since it precipitates slowly, it is permitted to remain in contact with the mother liquid for some time prior to filtration. Magnesium oxalate is too soluble to precipitate by itself under ordinary conditions. However, if calcium oxalate is precipitated from a solution containing magnesium ion and if the precipitate is allowed to remain in contact with the mother liquid for an excessive time, magnesium oxalate coprecipitates. Apparently, oxalate ion, present in excess in the solution, comprises the primary adsorbed ion layer. This produces a relatively high concentration of oxalate ion localized on the calcium oxalate surface, so that precipitation of some magnesium oxalate ensues. Post-precipitation can be minimized if the desired precipitate is brought to a filterable condition as soon as possible after its first formation.

Occlusion. In the process of occlusion, one ion within a crystal is replaced in the crystal lattice by another ion of similar size and structure. Thus, the impurity becomes permanently incorporated into the crystal lattice, and it cannot be removed by washing. Accordingly, the only way to eliminate this type of contamination is to remove the offending ion prior to precipitation of the desired compound or to dissolve the precipitate and reform it under more favorable conditions.

### Precipitation from Homogeneous Solution

To obtain a precipitate in the form of relatively large individual crystals, supersaturation should be held to a minimum. To accomplish this goal, the method of **precipitation from homogeneous solution** has been developed. In this method, the precipitating agent is not added directly but rather is generated slowly by means of a homogeneous chemical reaction within the solution at a rate comparable to the rate of crystal growth. Thus, the extent of supersaturation does not reach as high a value as would exist if the two reagent solutions were simply mixed directly. This technique is applicable to any precipitation process in which the necessary reagent can be generated slowly by some chemical reaction occurring within the solution of the unknown.

**Precipitation by Means of Urea Hydrolysis.** Urea is an especially useful reagent for the homogeneous precipitation or separation of any substance whose solubility is affected by pH. It hydrolyzes slowly in hot aqueous solutions according to the reaction

$$(NH_2)_2CO + H_2O \rightarrow CO_2 + 2\ NH_3$$

Slow generation of ammonia within an initially acidic or neutral solution serves to raise the pH gradually and uniformly. By adjusting the initial concentration of urea and the pH of the sample solution, and by controlling the temperature at which the solution is heated, one can obtain any desired final pH value or rate of increase of pH.

One of the classic gravimetric procedures entails the precipitation of hydrous aluminum oxide through the addition of aqueous ammonia to a sample solution. However, the voluminous and gelatinous character of the resulting precipitate causes numerous complications including difficulty in the filtering and washing steps and serious coprecipitation of other cations and anions. These problems are largely overcome if the homogeneous precipitation technique is employed. Thus, the pH of an aluminum ion solution can be adjusted to a value at which hydrous aluminum oxide is soluble, an appropriate quantity of urea added, and the solution heated to hydrolyze the urea and to raise the pH so that hydrous aluminum oxide will precipitate quantitatively. This precipitate possesses more desirable physical characteristics – high density and crystallinity – than that obtained by direct mixing of reagents. Upon subsequent ignition, the hydrous aluminum oxide is converted into aluminum oxide, $Al_2O_3$, an excellent weighing form for the determination of aluminum.

**Homogeneous Generation of Sulfate Ions.** Sulfate ion can be generated homogeneously if one heats a solution containing sulfamic acid

$$HSO_3NH_2 + H_2O \rightarrow H^+ + SO_4^{2-} + NH_4^+$$

or if one causes the slow hydrolysis of diethyl sulfate:

$$(C_2H_5)_2SO_4 + 2\ H_2O \rightarrow 2\ H^+ + SO_4^{2-} + 2\ C_2H_5OH$$

Barium ion, or any cation which forms an insoluble sulfate, can be precipitated homogeneously by means of these reactions. By using mixtures of alcohol and water as the solvent along with the homogeneous precipitation technique, one can effect sharper separations of the alkaline earth elements – barium, strontium, and calcium – than is possible by direct addition of a solution containing sulfate ion.

**Precipitation of Sulfides.** Many metal ions form insoluble sulfides, the solubilities of which are very much influenced by pH because hydrogen sulfide is an extremely weak acid. When sulfides are precipitated by the bubbling of hydrogen sulfide gas into a solution, the resulting precipitates have undesirable physical characteristics, not to mention the unpleasant and toxic nature of hydrogen sulfide itself. Metal sulfides can be precipitated homogeneously by means of the slow acid-or-base-catalyzed hydrolysis of thioacetamide:

$$CH_3CSNH_2 + H_2O \rightarrow CH_3CONH_2 + H_2S$$

This homogeneous precipitation method yields metal sulfides which are distinctly granular and much easier to handle in subsequent washing and filtering operations. However, application of this method of generating hydrogen sulfide to analytical procedures is complicated by the fact that some metal ions react directly with thioacetamide to form metal sulfides.

## DISSOLUTION OF PRECIPITATES

Completeness of precipitation of a desired species is generally determined by the equilibrium solubility of that substance under the conditions prevailing either at the time of precipitation or at the time of the subsequent filtration and washing operations. Let us begin our consideration of equilibrium solubility with the dissolution of silver chloride in pure water. It is proper to represent the solubility behavior of silver chloride in water as

$$AgCl(s) \rightleftharpoons AgCl(aq) \rightleftharpoons Ag^+ + Cl^-$$

where solid silver chloride first dissolves to form an aqueous, undissociated molecule which, in turn, dissociates into silver ion and chloride ion.

If we add the equilibria corresponding to each step in the dissolution process, namely

$$AgCl(s) \rightleftharpoons AgCl(aq)$$

and

$$AgCl(aq) \rightleftharpoons Ag^+ + Cl^-$$

the result is

$$AgCl(s) \rightleftharpoons Ag^+ + Cl^-$$

Because the standard state of silver chloride is defined to be the pure solid at 25°C, we can conclude that the activity of solid silver chloride is unity, *i.e.*,

$$(AgCl(s)) = 1$$

if the silver chloride is not contaminated in any way with impurities and if the temperature is 25°C. Throughout the remainder of this chapter, the activities of solid phases will always be assumed to be unity. Therefore, the thermodynamic equilibrium-constant expression for the last of the preceding reactions is

$$(Ag^+)(Cl^-) = K_{ap} = 1.78 \times 10^{-10}$$

where the product of the activities of silver ion and chloride ion is defined as the thermodynamic activity product, $K_{ap}$. This relation is a statement of the fact that, in a solution saturated with solid silver chloride, the product of the activities of silver ion and chloride ion is always constant.

In analytical chemistry, we are often concerned with the concentration of a species instead of its activity, because the former provides information about the quantity of material present in a particular phase or system. In addition, activity coefficients for ions cannot be reliably calculated for electrolyte solutions other than very dilute ones. Moreover, equilibrium calculations are generally done in order to predict the feasibility of a method of separation or analysis. Therefore, it is usually satisfactory to omit activity coefficients and to write relations involving only analytical concentrations, such as

$$[Ag^+][Cl^-] = K_{sp}$$

This latter equation is the familiar **solubility-product expression** for silver chloride, and the equilibrium constant ($K_{sp}$) is called the **solubility product** or the **solubility-product constant**. For almost all equilibrium calculations concerned with the solubility of precipitates, we shall use analytical concentrations and we will assume that the activity product ($K_{ap}$) and the solubility product ($K_{sp}$) are synonymous. Appendix 1 contains a compilation of solubility products.

Let us consider two examples of solubility-product calculations. Suppose that the solubility of nickel hydroxide is known to be 0.00011 g per liter of water and that we wish to calculate the solubility product of this compound. The simplest possible representation of the solubility reaction is

$$Ni(OH)_2(s) \rightleftharpoons Ni^{2+} + 2\ OH^-$$

and the solubility-product expression is

$$[Ni^{2+}][OH^-]^2 = K_{sp}$$

To obtain the molar solubility of nickel hydroxide, we divide the weight solubility by the molecular weight of $Ni(OH)_2$, the latter being 92.7. Thus, the molar solubility, $S$, is given by

$$S = \frac{0.00011 \text{ g}}{1 \text{ liter}} \frac{1 \text{ mole}}{92.7 \text{ g}} = 1.2 \times 10^{-6}\ M$$

Since one nickel ion is formed for each nickel hydroxide molecule which dissolves,

$$[Ni^{2+}] = 1.2 \times 10^{-6}\ M$$

whereas the hydroxide ion concentration is twice that of nickel(II):

$$[OH^-] = 2.4 \times 10^{-6}\ M$$

It follows that the solubility product is given by

$$(1.2 \times 10^{-6})(2.4 \times 10^{-6})^2 = K_{sp}$$

$$6.9 \times 10^{-18} = K_{sp}$$

Starting with the solubility-product expression for a compound, we can calculate the molar solubility. For example, the solubility-product expression for cadmium(II) hydroxide is

$$[Cd^{2+}][OH^-]^2 = 5.9 \times 10^{-15}$$

corresponding to the equilibrium

$$Cd(OH)_2(s) \rightleftharpoons Cd^{2+} + 2\ OH^-$$

Since we obtain one $Cd^{2+}$ ion and two $OH^-$ ions from each cadmium(II) hydroxide molecule which dissolves, the concentration of cadmium ion should be equal to the molar solubility ($S$) of cadmium(II) hydroxide

$$[Cd^{2+}] = S$$

whereas the concentration of the hydroxide ion will be twice the molar solubility:

$$[OH^-] = 2\ S$$

Substituting these results into the solubility-product expression, we obtain

$$(S)(2S)^2 = 5.9 \times 10^{-15}$$

$$4S^3 = 5.9 \times 10^{-15}$$

$$S = 1.1 \times 10^{-5}\ M$$

### Effect of Ionic Strength

Formation and dissolution of a precipitate in analytical procedures invariably occur in the presence of relatively large concentrations of foreign electrolytes. Therefore, we must be concerned about the solubility of a precipitate in the presence of a foreign electrolyte. However, we shall restrict this discussion to systems wherein the desired compound is in equilibrium with a solution containing no ions common to the precipitate – that is, no ions of which the solid is composed, except those produced when the precipitate itself dissolves.

Quantitative predictions of the effect of a foreign electrolyte on the solubility of silver chloride can be based upon the thermodynamic equilibrium-constant expression

$$(Ag^+)(Cl^-) = f_{\pm}^2[Ag^+][Cl^-] = K_{ap}$$

where $f_{\pm}$ is the mean activity coefficient for silver chloride and $[Ag^+]$ and $[Cl^-]$ represent the analytical concentrations of silver ion and chloride ion, respectively. If the small concentration of aqueous molecular silver chloride is neglected, the molar solubility $S$ of silver chloride is measured by the concentration of silver ion. Since the concentrations of silver ion and chloride ion are equal, we can write

$$f_{\pm}^2 S^2 = K_{ap}$$

or

$$S = \frac{(K_{ap})^{1/2}}{f_{\pm}}$$

It is evident from the latter equation that the solubility of silver chloride should increase as the mean activity coefficient decreases; and, of course, the converse is true.

Let us compare the solubility of silver chloride in water and in 0.1 $F$ nitric acid. For water, the ionic strength $I$ is nearly zero because the concentrations of dissolved silver ion and chloride ion are very small. Thus, we shall assume that the mean activity coefficient is essentially unity. Therefore, the solubility $S$ of silver chloride is given by the simple relation

$$S = (K_{ap})^{1/2} = (1.78 \times 10^{-10})^{1/2}$$

$$S = 1.33 \times 10^{-5}\,M$$

A calculation of the solubility of silver chloride in 0.1 $F$ nitric acid requires that we determine the mean activity coefficient of silver chloride. Although the ionic strength of a solution does depend on the concentrations of all ionic species, the contribution by dissolved silver chloride is negligible. Thus, the ionic strength is essentially equal to the concentration of nitric acid. To obtain the mean activity coefficient of silver chloride, we can consult Table 3–2 (page 36). In the table, the single-ion activity coefficients for $Ag^+$ and $Cl^-$ are both found to be 0.755 at an ionic strength of 0.1, so the mean activity coefficient for silver chloride is 0.755 as well. If we now combine this result with the equation for the solubility of silver chloride, the value of $S$ can be calculated:

$$S = \frac{(K_{ap})^{1/2}}{f_{\pm}} = \frac{(1.78 \times 10^{-10})^{1/2}}{0.755}$$

$$S = 1.77 \times 10^{-5}\,M$$

Therefore, the solubility of silver chloride is about 33 per cent higher in 0.1 $F$ nitric acid than in water.

### Common Ion Effect

Quite often, the solubility of a precipitate is lower in an aqueous solution containing a common ion – that is, one of the ions comprising the compound – than in pure water alone. We can illustrate this so-called **common ion effect** by calculating and comparing the solubility of lead iodate, $Pb(IO_3)_2$, in water and in 0.03 $F$ potassium iodate solution.

For lead iodate, the solubility reaction and solubility-product constant may be written as

$$Pb(IO_3)_2(s) \rightleftharpoons Pb^{2+} + 2\ IO_3^-; \qquad K_{sp} = 2.6 \times 10^{-13}$$

In water, each dissolved lead iodate molecule yields one lead ion and two iodate ions. If the solubility of lead iodate is denoted by the symbol $S$, we can write

$$[Pb^{2+}] = S \qquad \text{and} \qquad [IO_3^-] = 2\,S$$

When these mathematical statements are substituted into the solubility-product expression, we obtain

$$[Pb^{2+}][IO_3^-]^2 = (S)(2S)^2 = 2.6 \times 10^{-13}$$

$$4S^3 = 2.6 \times 10^{-13}$$

$$S = 4.0 \times 10^{-5}\ M$$

for the solubility of lead iodate in water.

In the 0.03 $F$ potassium iodate medium, the situation is somewhat more complicated, for there are two sources of iodate ion. At equilibrium, the concentration of iodate ion will be the sum of the contributions from potassium iodate and lead iodate: 0.03 $M$ from potassium iodate and $2S$ from the dissolution of lead iodate, where $S$ represents the equilibrium concentration of lead ion (which is different from that found in the preceding calculation). Thus, the appropriate solubility-product expression is

$$(S)(2S + 0.03)^2 = 2.6 \times 10^{-13}$$

This relation could be solved in a completely rigorous manner, but the manipulations would involve a third-order equation. Therefore, it is worthwhile to consider a suitable simplifying assumption. Since the solubility of lead iodate in water is only $4.0 \times 10^{-5}\ M$ and since the presence of potassium iodate should repress the dissolution of lead iodate, we will neglect the $2S$ term in the preceding expression:

$$(S)(0.03)^2 = 2.6 \times 10^{-13}$$

This equation can be readily solved:

$$S = \frac{2.6 \times 10^{-13}}{9 \times 10^{-4}} = 2.9 \times 10^{-10}\,M$$

Certainly our neglect of the $2S$ term was justified, because the solubility of lead iodate is substantially decreased by the presence of 0.03 $F$ potassium iodate.

### Complexation of Cation with Foreign Ligand

If a foreign complexing agent or ligand is available which can react with the cation of a precipitate, the solubility of a compound can be markedly enhanced. As examples, we can cite the increase in the solubility of the silver halides, AgCl, AgBr, and AgI, in ammoniacal solutions.

For silver bromide in aqueous ammonia solutions, the position of equilibrium for the reaction

$$AgBr(s) \rightleftharpoons Ag^+ + Br^-; \qquad K_{sp} = 5.25 \times 10^{-13}$$

will be shifted toward the right because of the formation of stable silver-ammine complexes. Silver ion reacts with ammonia in a stepwise manner, according to the equilibria

$$Ag^+ + NH_3 \rightleftharpoons AgNH_3^+; \qquad K_1 = 2.4 \times 10^3$$

and

$$AgNH_3^+ + NH_3 \rightleftharpoons Ag(NH_3)_2^+; \qquad K_2 = 6.9 \times 10^3$$

Let us calculate the solubility of silver bromide in a 0.1 $F$ ammonia solution. On the basis of the preceding reactions, we can conclude that each silver bromide molecule which dissolves will yield one bromide ion and either one $Ag^+$, one $AgNH_3^+$, or one $Ag(NH_3)_2^+$. Therefore, the solubility $S$ of silver bromide is measured by the concentration of bromide ion, or alternatively, by the *sum* of the concentrations of all soluble silver species, *i.e.*,

$$S = [Ag^+] + [AgNH_3^+] + [Ag(NH_3)_2^+] = [Br^-]$$

This relation can be transformed into one equation with a single unknown if we introduce the equilibrium-constant expressions for the silver-ammine species. For the concentration of the monoamminesilver(I) complex, it can be shown that

$$[AgNH_3^+] = K_1[Ag^+][NH_3]$$

and the concentration of the diamminesilver(I) complex may be written as

$$[Ag(NH_3)_2^+] = K_2[AgNH_3^+][NH_3] = K_1K_2[Ag^+][NH_3]^2$$

If we substitute the relations for the concentrations of the two silver-ammine complexes into the solubility equation, the result is

$$S = [Ag^+] + K_1[Ag^+][NH_3] + K_1K_2[Ag^+][NH_3]^2 = [Br^-]$$

In a solution saturated with silver bromide, the solubility-product expression must be obeyed; that is,

$$[Ag^+] = \frac{K_{sp}}{[Br^-]}$$

so it follows that

$$S = \frac{K_{sp}}{[Br^-]} + \frac{K_1 K_{sp}[NH_3]}{[Br^-]} + \frac{K_1 K_2 K_{sp}[NH_3]^2}{[Br^-]} = [Br^-]$$

However, since $[Br^-] = S$, we can write

$$S = \frac{K_{sp}}{S} + \frac{K_1 K_{sp}[NH_3]}{S} + \frac{K_1 K_2 K_{sp}[NH_3]^2}{S}$$

Finally, solving for the solubility $S$, we obtain

$$S^2 = K_{sp} + K_1 K_{sp}[NH_3] + K_1 K_2 K_{sp}[NH_3]^2$$
$$S = (K_{sp})^{1/2}(1 + K_1[NH_3] + K_1 K_2[NH_3]^2)^{1/2}$$

Before we insert values for the ammonia concentration and the various constants into the latter equation, note that this relation is a general one for the solubility of *any* sparingly soluble silver salt in an ammonia solution. For the solubility of silver bromide in a 0.1 $F$ ammonia solution, we have

$$S = (5.25 \times 10^{-13})^{1/2}\,[1 + (2.4 \times 10^3)(0.1) + (2.4 \times 10^3)(6.9 \times 10^3)(0.1)^2]^{1/2}$$
$$S = 2.9 \times 10^{-4}\ M$$

In addition, we can calculate the concentration of each soluble silver species. For $Ag^+$:

$$[Ag^+] = \frac{K_{sp}}{[Br^-]} = \frac{5.25 \times 10^{-13}}{2.9 \times 10^{-4}} = 1.78 \times 10^{-9}\ M$$

For $AgNH_3^+$:

$$[AgNH_3^+] = K_1[Ag^+][NH_3] = (2.4 \times 10^3)(1.78 \times 10^{-9})(0.1)$$
$$[AgNH_3^+] = 4.3 \times 10^{-7}\ M$$

For $Ag(NH_3)_2^+$:

$$[Ag(NH_3)_2^+] = K_2[AgNH_3^+][NH_3] = (6.9 \times 10^3)(4.3 \times 10^{-7})(0.1)$$
$$[Ag(NH_3)_2^+] = 2.9 \times 10^{-4}\ M$$

Therefore, the only species which contributes significantly to the solubility of silver bromide is the diamminesilver(I) complex. One of the possible complications which we overlooked in solving this problem was that the formation of silver-ammine complexes might alter the concentration of ammonia. However, the solubility of silver bromide is small enough that neglect of this effect was justified. If the solubility of silver bromide were of comparable magnitude to the ammonia concentration, it would be necessary to consider the change in the concentration of ammonia.

**Reaction of Anion with Acid**

In certain instances, the anion of a slightly soluble compound reacts with one or more of the constituents of a solution phase. For example, the solubility of lead sulfate in nitric acid media is greater than in pure water and becomes larger as the concentration of nitric acid increases.

In pure water, the solubility of lead sulfate is governed by the equilibrium

$$PbSO_4(s) \rightleftharpoons Pb^{2+} + SO_4^{2-}; \qquad K_{sp} = 1.6 \times 10^{-8}$$

and, if no other equilibria prevail, the solubility of lead sulfate should essentially be the square root of the solubility product, or approximately $1.3 \times 10^{-4}$ *M.*

If nitric acid is added to the system, the solubility of lead sulfate increases because sulfate ion reacts with hydrogen ion to form the hydrogen sulfate anion:

$$H^+ + SO_4^{2-} \rightleftharpoons HSO_4^-$$

However, the addition of another proton to the hydrogen sulfate ion does not occur to any appreciable extent because sulfuric acid is a very strong acid. Therefore, when a lead sulfate molecule dissolves in nitric acid, one lead ion is formed, but the sulfate ion from the precipitate may be present as either $SO_4^{2-}$ or $HSO_4^-$. Accordingly, we can express the solubility $S$ of lead sulfate in nitric acid by means of the equation

$$S = [Pb^{2+}] = [SO_4^{2-}] + [HSO_4^-]$$

A relationship between the concentration of sulfate ion and hydrogen sulfate ion can be based on the equilibrium expression for the *second* dissociation of sulfuric acid, namely,

$$HSO_4^- \rightleftharpoons H^+ + SO_4^{2-}; \qquad K_{a2} = 1.2 \times 10^{-2}$$

By substituting the latter information into the solubility equation, we obtain

$$S = [Pb^{2+}] = [SO_4^{2-}] + \frac{[H^+][SO_4^{2-}]}{K_{a2}}$$

and, by utilizing the solubility-product expression for lead sulfate, we can write

$$S = [Pb^{2+}] = \frac{K_{sp}}{[Pb^{2+}]} + \frac{K_{sp}[H^+]}{K_{a2}[Pb^{2+}]}$$

$$[Pb^{2+}]^2 = K_{sp} + \frac{K_{sp}[H^+]}{K_{a2}}$$

$$[Pb^{2+}] = \left(K_{sp} + \frac{K_{sp}[H^+]}{K_{a2}}\right)^{1/2}$$

If the hydrogen ion concentration is known, we can use the latter relation to calculate the concentration of lead ion in equilibrium with solid lead sulfate and, therefore, the solubility of lead sulfate itself.

Suppose that we wish to predict the solubility of lead sulfate in 0.10 $F$ nitric acid. One question to be answered is whether the formation of the hydrogen sulfate anion, through the reaction of $SO_4^{2-}$ with $H^+$, will consume enough hydrogen ion to change significantly the concentration of the latter. We can either assume that the change will be important – in which case the computations are difficult – or that this change can be neglected as a first approximation. If the second alternative is chosen, it is necessary to justify the assumption after the calculations are completed. Let us assume that the concentration of hydrogen ion remains essentially constant at 0.1 *M*. Thus, the lead ion concentration is given by the equation

$$[Pb^{2+}] = \left(1.6 \times 10^{-8} + \frac{(1.6 \times 10^{-8})(0.1)}{(1.2 \times 10^{-2})}\right)^{1/2}$$

which, when solved, yields

$$[Pb^{2+}] = 3.9 \times 10^{-4}\,M$$

Even if all the sulfate derived from the dissolution of lead sulfate were converted to hydrogen sulfate anion, the change in the hydrogen ion concentration would be negligibly small. Therefore, by making a simplifying assumption and later justifying its validity, we obtain the desired result in a straightforward way. If, on the other hand, the lead ion concentration is shown to be comparable to the original concentration of nitric acid, we should immediately suspect that our assumption was false and return to a more rigorous solution to the problem. It should be stressed that the primary reason for making the assumption was the smallness of the solubility product of lead sulfate compared to the initial nitric acid concentration.

## GRAVIMETRIC DETERMINATIONS

In this section of the chapter, we will explore briefly typical procedures employed in gravimetric analysis. Some compounds of interest are strictly inorganic, but others are substances formed by addition of an organic precipitant to a solution containing an inorganic cation or anion. Table 7–1 presents a short compilation of reagents used for gravimetric determinations of various elements. Most of these reagents can form precipitates with a number of species, so it is usually necessary to separate the desired ion from interfering species before the precipitant is added to the sample solution. One can often gain selectivity in the use of these reagents by adjusting the pH of the sample solution or by adding a masking agent. In the discussions which follow, we will consider a few of these determinations in more detail.

### Determination of Chloride

Chloride ion in a wide variety of samples can be determined if an excess of silver nitrate solution is added to a solution of the unknown and the resulting silver chloride precipitate is collected, dried, and weighed. Silver ion forms a number of sparingly soluble salts. However, the specificity of the method for chloride ion is much improved in the presence of a small amount of acid, such as nitric acid. Under these conditions, most anions of weak acids that form insoluble silver salts will not produce precipitates. For example, silver phosphate precipitates in an alkaline medium. However, in an acidic solution, phosphate ion forms $H_2PO_4^-$ and $H_3PO_4$, so that no precipitate is obtained when silver nitrate solution is added. On the other hand, bromide and iodide ions yield

**Table 7–1. Some Reagents for Gravimetric Determinations[a]**

| Reagent | Species Precipitated Quantitatively |
|---|---|
| Ammonia, $NH_3$ | Al(III), Be(II), Cr(III), Cu(II), Fe(III), In(III), La(III), Pb(II), Sc(III), Sn(IV), Th(IV), Ti(IV), U(VI), Y(III), Zr(IV) |
| Anthranilic acid | Cd(II), Co(II), Cu(II), Hg(II), Mn(II), Ni(II), Pb(II), Zn(II) |
| Benzidine | $PO_4^{3-}$, $SO_4^{2-}$ |
| Cupferron | Al(III), Bi(III), Cu(II), Fe(III), Ga(III), Nb(III), Sn(IV), Th(IV), Ti(IV), U(IV), V(V), Zr(IV) |
| Dimethylglyoxime, $H_3C-C(=NOH)-C(=NOH)-CH_3$ | Ni(II), Pd(II) |
| Hydrogen sulfide, $H_2S$ | As(III), Cu(II), Ge(IV), Hg(II) |
| 8-Hydroxyquinoline (Oxine) | Al(III), Cd(II), Co(II), Cu(II), Fe(III), In(III), Mn(II), Mo(VI), Ni(II), Ti(IV), U(VI), Zn(II) |
| 2-Mercaptobenzothiazole | Au(III), Bi(III), Cd(II), Cu(II), Pb(II) |
| Nitron | $ClO_3^-$, $ClO_4^-$, $NO_3^-$, $ReO_4^-$ |

Table 7–1. Some Reagents for Gravimetric Determinations[a] *(continued)*

| Reagent | Species Precipitated Quantitatively |
|---|---|
| α-Nitroso-β-naphthol, (structure: NO, OH) | Co(II), Cu(II), Fe(III) |
| Phenylarsonic acid, $C_6H_5$—As(=O)(OH)$_2$ | Bi(III), Hf(IV), Nb(III), Sn(IV), Ta(V), Th(IV), Zr(IV) |
| Salicylaldoxime, (structure: CH=NOH, OH) | Bi(III), Cu(II), Pb(II), Pd(II) |
| Silver nitrate, $AgNO_3$ | $AsO_4^{3-}$, $Br^-$, $Cl^-$, $CN^-$, $I^-$, $IO_3^-$, $MoO_4^{2-}$, $N_3^-$, $OCN^-$, $S^{2-}$, $SCN^-$, $VO_3^-$ |
| Sodium tetraphenylborate, $NaB(C_6H_5)_4$ | Cs(I), K(I), $NH_4^+$, Rb(I), Tl(I) |
| Tetraphenylarsonium chloride, $(C_6H_5)_4AsCl$ | $ReO_4^-$ |

[a]Information excerpted from L. Meites, ed.: *Handbook of Analytical Chemistry*. McGraw-Hill Book Company, New York, 1963, p. 3-4. No attempt has been made to present an exhaustive list of species precipitated by a given reagent, but all species listed are precipitated quantitatively.

insoluble precipitates with silver ion even in dilute nitric acid, so these anions must be entirely absent or removed before a successful determination of chloride can be undertaken.

Silver halides are decomposed into their constituent elements on exposure to light. Such decomposition is particularly grave during the early stages of the precipitation when the milky suspension presents an enormous surface for photochemical reaction. Therefore, the precipitation should not be conducted in bright light. Silver chloride is almost pure white, but the precipitate is likely to acquire a purplish hue by the time it is weighed, owing to tiny grains of metallic silver on the surface of the white silver chloride. Even a small amount of photodecomposition results in an appreciable purplish coloration, so that one should not be alarmed at the coloration if reasonable caution has been taken to prevent excessive decomposition.

Bromide ion can be determined by means of the same procedure. Iodide determinations are possible, but the method is less satisfactory for three reasons – silver iodide tends to peptize more readily than silver chloride, silver iodide adsorbs certain impurities quite tenaciously, and silver iodide is extremely photosensitive. Thiocyanate and cyanide ions can also be determined in this fashion.

Several different anions of weak acids can be determined by precipitation of their silver salts from a nearly neutral medium. For a successful determination, the sample solution must contain only one anion which can form an insoluble silver salt. For example, phosphate ion can be determined by precipitation and weighing of silver phosphate.

Several cations can be determined by a reversal of the procedure. Thus, silver may be determined quite accurately by the precipitation and weighing of either silver chloride or silver bromide. Mercury(I) and lead(II) may also be precipitated as their chlorides, but solubility losses are considerable with lead chloride. Thallous ion may likewise be determined by precipitation and weighing of thallous chloride.

### Determination of Sulfate

Sulfur is commonly determined by conversion of all the sulfur-containing material in a weighed sample to sulfate ion, after which barium sulfate is precipitated, washed, filtered, dried, and weighed. Barium ion forms insoluble precipitates with a variety of anions other than sulfate. Yet most of them are anions of weak acids so that their barium salts are soluble in acidic media. In dilute acid solutions, only a trace of the sulfate is lost as hydrogen sulfate ion, but essentially all anions of much weaker acids are effectively removed from the scene through formation of their undissociated acids. Fluoride is the only anion that remains troublesome under these conditions. Barium fluoride is quite insoluble in dilute acid solutions, and so fluoride must be removed prior to precipitation of barium sulfate. Such removal can be accomplished readily through volatilization of hydrogen fluoride or through the complexation reaction between boric acid and fluoride.

Of the more common cations other than barium, only lead, calcium, and strontium form nominally insoluble sulfates. Interference from lead may be prevented if acetate is added to complex the lead. Calcium and strontium must be removed prior to precipitation of barium sulfate.

Precipitated barium sulfate tends to retain many extraneous substances from its mother liquid, and herein lies the chief deterrent to the highly accurate determination of sulfate. Many species can and do coprecipitate with barium sulfate. When foreign anions coprecipitate with barium sulfate in a gravimetric sulfate determination, the results tend to be high, whereas coprecipitation of foreign cations leads to results that are too low. One fortunate aspect of these coprecipitation phenomena is clear – when both foreign cations and foreign anions are coprecipitated, the errors tend to compensate and fairly accurate results are obtained.

By using the sulfate procedure in reverse, one can determine barium by adding an excess of sulfate ion to a solution of the unknown. Coprecipitation errors may be serious. In a barium determination, anion coprecipitation leads to low results and cation coprecipitation to high results – just opposite to the situation in sulfate determinations.

Lead can be determined by means of a similar procedure, but lead sulfate is soluble enough (about 4 mg per 100 ml of water at room temperature) to make solubility losses somewhat more significant than in barium sulfate precipitations. Strontium and calcium sulfates are nominally insoluble, but their solubilities (15 mg of $SrSO_4$ and 100 mg of $CaSO_4$, respectively, per 100 ml of water) are too great for good quantitative results. These compounds are rendered less soluble in a mixed alcohol-water solvent, but even then the determinations are not entirely satisfactory.

### Organic Precipitants

A large number of organic reagents react with metal cations to form insoluble precipitates suitable for gravimetric determinations. Most of these precipitants interact with metal ions to produce covalently bonded chelate compounds. Some of these substances combine with cations to yield insoluble salts. In the discussions which follow, we will examine briefly the analytical uses of three representative organic precipitants – dimethylglyoxime, 8-hydroxyquinoline (oxine), and sodium tetraphenylborate. More comprehensive summaries of the applications of organic reagents to gravimetric analysis may be found in the references listed at the end of this chapter, and a selected group of organic precipitants is included in Table 7–1.

**Dimethylglyoxime.** One of the most selective of all organic precipitants is dimethylglyoxime. This compound behaves as a very weak monoprotic acid in water, having a dissociation constant of $2.2 \times 10^{-11}$ at 25°C. Since the solubility of dimethylglyoxime in water is very low, a solution of the compound in ethanol is ordinarily employed as precipitating agent, although an aqueous solution of the sodium salt of dimethylglyoxime can be used.

In weakly acidic or ammoniacal media, two dimethylglyoxime molecules combine with nickel(II) to yield the familiar scarlet-colored bis(dimethylglyoximato) nickel(II) chelate compound:

$$\mathrm{Ni^{2+} + 2\ H_3C{-}C({=}N{-}OH){-}C({=}N{-}OH){-}CH_3 \rightleftharpoons Ni[O{-}N{=}C(CH_3){-}C(CH_3){=}N{-}O{-}H]_2 + 2\,H^+}$$

Palladium(II) forms a yellow precipitate with dimethylglyoxime in dilute acid solutions, and bismuth(III) reacts with dimethylglyoxime at approximately pH 11 to yield an insoluble compound. However, these are the only species that are precipitated by dimethylglyoxime. Therefore, it has become a routine procedure to separate and determine nickel(II) through the use of dimethylglyoxime as a precipitant in solutions with pH values between approximately 5 and 10. This method has been applied extensively to the determination of nickel in steel.

Palladium(II) can be determined by means of a similar procedure if the precipitation is performed in a dilute acid medium, whereas bismuth(III) can be separated as a precipitate from an alkaline solution.

Various other dioximes have been recommended for analytical use as precipitants, principally because they are more soluble in water than dimethylglyoxime. Among these compounds are methylbenzoyldioxime and 1,2-cyclohexanedionedioxime.

**8-Hydroxyquinoline (Oxine).** Whereas dimethylglyoxime has unusual selectivity for just a few species, 8-hydroxyquinoline forms insoluble chelate compounds with practically every metal in the periodic table except the alkali metal elements. Trivalent metal cations, including aluminum, bismuth, cerium, iron, scandium, and thallium, react with three oxine ligands according to the generalized equilibrium

$$\mathrm{M^{3+} + 3\ C_9H_6N(OH) \rightleftharpoons M(C_9H_6NO)_3 + 3\,H^+}$$

whereas dipositive cations, such as cadmium, copper, magnesium, nickel, lead, strontium, and zinc, combine with oxine in a one-to-two metal-to-oxine ratio. Oxine is amphiprotic in aqueous media because it can either gain or lose a proton, as shown by the following equilibria:

N, OH $+ H^+ \rightleftharpoons$ $N^+$, OH H

N, OH $\rightleftharpoons H^+ +$ N, $O^-$

Therefore, depending on the stability of the desired metal oxinate, one can adjust the pH of the solution to achieve a certain degree of selectivity in the precipitation. However, the selectivity of the method can be increased considerably through the use of masking agents. These masking agents are other complexing species, such as cyanide, tartrate, and EDTA, which are added to a sample solution and which prevent the reaction between oxine and interfering metal ions.

**Sodium Tetraphenylborate.** In contrast to the two previous reagents, the precipitates formed by reaction between cations and sodium tetraphenylborate are salts rather than chelate compounds. Sodium tetraphenylborate is especially useful for the quantitative precipitation of ammonium ion and alkali metal cations (potassium, rubidium, and cesium).

For the determination of potassium ion, the sample solution is usually acidified with nitric acid or acetic acid. Then, an appropriate volume of a 3 per cent aqueous solution of sodium tetraphenylborate is added, whereupon potassium tetraphenylborate is immediately precipitated:

$$K^+ + B(C_6H_5)_4^- \rightarrow KB(C_6H_5)_4$$

This compound can be collected on a sintered-glass filter crucible, washed several times with a dilute aqueous solution of sodium tetraphenylborate, washed with cold water, and dried at approximately 100°C until constant weight is attained.

A mixture of ammonium ion and potassium ion can be analyzed. First, both cations are precipitated together as tetraphenylborate salts, and the sum of the weights of the two precipitates is determined. Second, the mixed precipitate is dissolved in acetone, sodium hydroxide solution is added, and the alkaline solution is evaporated to expel ammonia. Third, the residue is dissolved in water, and potassium tetraphenylborate alone is collected, washed, and dried. Using the two weight measurements, one can calculate the individual amounts of ammonium and potassium ions in the original sample.

## PRECIPITATION TITRATIONS

To be suitable for a precipitation titration, a reaction leading to formation of a slightly soluble compound must satisfy three requirements. First, the rate of precipitation, as well as the attainment of solubility equilibrium, must be rapid in order to permit the progress of a titration, as reflected by changes in concentrations, to be followed accurately. If the rate of precipitation or attainment of solubility equilibrium is slow, the technique of back-titration can sometimes be employed. In this method, a measured excess of the precipitant is added; then, the unreacted precipitant is back-titrated with a standard solution of another reagent. Second, the precipitation reaction should be quantitative and should proceed according to well-defined stoichiometry. Unfortunately, few precipitates of analytical interest are formed from a complex aqueous solution with the high purity and definiteness of composition necessary for a successful titration. Third, a suitable means must be available for locating or identifying the equivalence point of the titration. A wide variety of end-point detection methods have been developed, several of which will be discussed subsequently. Other methods, based on the measurement of electromotive force or light absorption, are described in later chapters.

Generally, the silver halides, silver thiocyanate, and a few mercury, lead, and zinc salts are the compounds most frequently involved in precipitation titrations. Information about a number of practical precipitation titrations is summarized in Table 7–2.

### Titration of Chloride with Silver Ion

We shall now consider the titration of chloride with a standard solution of silver nitrate in some detail by constructing the titration curve and by examining methods used to locate the equivalence point.

**Construction of the Titration Curve.** Let us construct the titration curve for the titration of 50.00 ml of 0.1000 *F* sodium chloride solution with 0.1000 *F* silver nitrate solution. This curve will consist of a plot of the negative base-ten logarithm of the chloride concentration (pCl) versus the volume of titrant.

Before the Equivalence Point. Before introduction of silver nitrate solution, the chloride concentration is 0.1000 *M*, so that pCl is 1.00. However, as soon as titrant is added, silver chloride precipitates. Prior to the equivalence point, the *total* chloride concentration is the *sum* of the chloride ion remaining untitrated plus the chloride ion resulting from dissolution of silver chloride:

$$[Cl^-] = [Cl^-]_{untitrated} + [Cl^-]_{from\ AgCl}$$

Now, the concentration of chloride remaining untitrated is given by the following expression:

$$[Cl^-]_{untitrated} = \frac{\text{millimoles } Cl^- \text{ taken} - \text{millimoles } Ag^+ \text{ added}}{\text{total solution volume in milliliters}}$$

To calculate the chloride concentration resulting from dissolution of silver chloride, let us first assume that *all* the added silver nitrate reacts to precipitate silver chloride. Then, for each silver chloride molecule that dissolves, one silver ion and one chloride ion are formed. Therefore,

$$[Cl^-]_{from\ AgCl} = [Ag^+]$$

However, the silver ion concentration is, in turn, given by the solubility product of silver chloride divided by the *total* concentration of chloride ion in the solution, which is $[Cl^-]$. Thus,

$$[Cl^-]_{from\ AgCl} = [Ag^+] = \frac{K_{sp}}{[Cl^-]}$$

**Table 7-2. Precipitation Titrations[a]**

| Species Determined | Titrant | Precipitate Formed | Indicator | Procedure |
|---|---|---|---|---|
| $Ag^+$ | KSCN | AgSCN | Fe(III) | Direct titration in presence of 0.8 *F* $HNO_3$ |
| $AsO_4^{3-}$ | KSCN | AgSCN | Fe(III) | Precipitate $Ag_3AsO_4$ at pH 8; dissolve precipitate in 0.8 *F* $HNO_3$; titrate $Ag^+$ |
| $Br^-$ | $AgNO_3$ and KSCN | AgBr and AgSCN | Fe(III) | Precipitate AgBr with known slight excess of $AgNO_3$ in 0.8 *F* $HNO_3$; add indicator; back-titrate with KSCN |
| | $AgNO_3$ | AgBr | eosin | Direct titration at pH $\geqslant$ 1 |
| $Cl^-$ | $AgNO_3$ and KSCN | AgCl and AgSCN | Fe(III) | Precipitate AgCl with known slight excess of $AgNO_3$ in 0.8 *F* $HNO_3$; remove AgCl by filtration; add indicator; back-titrate with KSCN |
| | $AgNO_3$ | AgCl | $K_2CrO_4$ | Direct titration at pH 7 |
| $F^-$ | $AgNO_3$ and KSCN | AgCl and AgSCN | Fe(III) | Precipitate PbClF at pH 3.6–5.6; dissolve precipitate in 0.8 *F* $HNO_3$; precipitate AgCl with known slight excess of $AgNO_3$; remove AgCl by filtration; add indicator; back-titrate with KSCN |
| $Hg_2^{2+}$ | NaCl | $Hg_2Cl_2$ | bromophenol blue | Direct titration in acid solution |
| $I^-$ | $AgNO_3$ and KSCN | AgI and AgSCN | Fe(III) | Precipitate AgI with known slight excess of $AgNO_3$ in 0.8 *F* $HNO_3$; add indicator; back-titrate with KSCN |
| | $AgNO_3$ | AgI | dichloro-fluorescein | Direct titration in 0.01 *F* $HNO_3$ |
| $SO_4^{2-}$ | $BaCl_2$ | $BaSO_4$ | tetrahydroxy-quinone | To sample solution at pH 7–8, add equal volume of ethanol; titrate with vigorous stirring |
| $Zn^{2+}$ | $K_4Fe(CN)_6$ | $K_2ZnFe(CN)_6$ | Fe(III) or diphenyl-benzidine | Direct titration in solution containing 0.18 *F* HCl and 0.75 *F* $NH_4Cl$ |

[a]Information excerpted from L. Meites, ed.: *Handbook of Analytical Chemistry.* McGraw-Hill Book Company, New York, 1963, p. 3-51.

If the preceding equations are combined, we obtain a general relation for the *total* chloride concentration for any point, except the initial point, up to and including the equivalence point:

$$[Cl^-] = \frac{\text{millimoles } Cl^- \text{ taken} - \text{millimoles } Ag^+ \text{ added}}{\text{total solution volume in milliliters}} + \frac{K_{sp}}{[Cl^-]}$$

Using the above expression, let us calculate pCl values corresponding to the addition of various volumes of titrant. After 10.00 ml of titrant is added,

$$[Cl^-] = \frac{5.000 \text{ millimoles} - 1.000 \text{ millimole}}{60.00 \text{ milliliters}} + \frac{1.78 \times 10^{-10}}{[Cl^-]}$$

$$[Cl^-] = 0.0667 + \frac{1.78 \times 10^{-10}}{[Cl^-]}$$

This equation can be solved by means of the quadratic formula. However, the second term on the right-hand side of the equation can be neglected, as shown by the following reasoning. If the value of $[Cl^-]$ is provisionally taken to be 0.0667 *M*, the second term is only $2.67 \times 10^{-9}$ *M*, which is negligible compared to 0.0667 *M*. Thus, 0.0667 *M* is an excellent approximation for $[Cl^-]$, and so pCl = 1.18. Neglect of the second term is valid in the present example until one is very close to the equivalence point. However, if the original chloride concentration is less than $10^{-4}$ *M*, then, even at the beginning of the titration, the dissolution of silver chloride must be considered and the equation solved exactly.

After 20.00 ml of titrant is added, $[Cl^-]$ = 0.0429 *M*; pCl = 1.37. After 30.00 ml of titrant is added, $[Cl^-]$ = 0.0250 *M*; pCl = 1.60. After 40.00 ml of titrant is added, $[Cl^-]$ = 0.0111 *M*; pCl = 1.95. After 49.00 ml of titrant is added, $[Cl^-]$ = 0.00101 *M*; pCl = 3.00. After 49.90 ml of titrant is added, $[Cl^-] = 1.00 \times 10^{-4}$ *M*; pCl = 4.00. Note that the dissolution of silver chloride has been neglected up to this point.

After 49.99 ml of titrant is added, neglect of the dissolution of silver chloride is no longer justifiable, and so the exact equation must be used to calculate $[Cl^-]$:

$$[Cl^-] = \frac{5.000 \text{ millimoles} - 4.999 \text{ millimoles}}{99.99 \text{ milliliters}} + \frac{1.78 \times 10^{-10}}{[Cl^-]}$$

$$[Cl^-] = 1.000 \times 10^{-5} + \frac{1.78 \times 10^{-10}}{[Cl^-]}$$

Using the quadratic formula, one calculates that $[Cl^-] = 1.92 \times 10^{-5}$ *M*; pCl = 4.72.

**At the Equivalence Point.** When 50.00 ml of titrant has been added, the exact equation simplifies to

$$[Cl^-] = \frac{K_{sp}}{[Cl^-]}$$

so that $[Cl^-]^2 = K_{sp}$; $[Cl^-] = (K_{sp})^{1/2} = 1.33 \times 10^{-5}$ *M*; pCl = 4.87.

**After the Equivalence Point.** Beyond the equivalence point, an excess of silver ion is present in the solution; the *total* silver ion concentration is the *sum* of the contributions from the excess titrant and the dissolution of silver chloride:

$$[Ag^+] = [Ag^+]_{excess} + [Ag^+]_{from\ AgCl}$$

To obtain the concentration of silver ion contributed by the excess titrant, we can use the expression

$$[Ag^+]_{excess} = \frac{\text{millimoles of } AgNO_3 \text{ added beyond equivalence point}}{\text{total volume of solution in milliliters}}$$

Beyond the equivalence point, dissolved silver chloride is the only source of chloride, so that the silver ion concentration from the dissolution of silver chloride is equal to the chloride concentration:

$$[Ag^+]_{\text{from AgCl}} = [Cl^-]$$

In turn, the chloride concentration is related to the *total* silver ion concentration through the solubility product:

$$[Ag^+]_{\text{from AgCl}} = [Cl^-] = \frac{K_{sp}}{[Ag^+]}$$

When these relations are combined, the following equation for the *total* silver ion concentration is obtained:

$$[Ag^+] = \frac{\text{millimoles of } AgNO_3 \text{ added beyond equivalence point}}{\text{total volume of solution in milliliters}} + \frac{K_{sp}}{[Ag^+]}$$

Once $[Ag^+]$ is calculated from this relationship, the value of $[Cl^-]$, and hence pCl, can be determined from the solubility-product expression. Except for points very close to the equivalence point or for the titration of very dilute solutions, the second term on the right-hand side of this last equation can be neglected.

After 60.00 ml of titrant is added, there is an excess of 1.000 millimole of silver ion in a total volume of 110.0 ml of solution:

$$[Ag^+] = \frac{1.000 \text{ millimole}}{110.0 \text{ milliliters}} = 9.09 \times 10^{-3}\,M$$

$$[Cl^-] = \frac{1.78 \times 10^{-10}}{9.09 \times 10^{-3}} = 1.96 \times 10^{-8}\,M; \qquad pCl = 7.71$$

After 70.00 ml of titrant is added, $[Ag^+] = 0.0167\ M$, $[Cl^-] = 1.07 \times 10^{-8}\ M$, and pCl = 7.97. Figure 7–2 shows the titration curve constructed on the basis of these calculations.

**End-Point Detection.** An important characteristic of a titration curve is the large and abrupt change in the concentration of the substance being titrated in the vicinity of the equivalence point, and the purpose of any method of end-point detection is to signal that this point has been reached. Ideally, the end point should coincide with the equivalence point. In the following discussions, we will examine three end-point detection methods useful for the titrimetric determination of chloride.

Mohr Titration. In the Mohr method, the chloride ion is precipitated by titration with standard silver nitrate solution in the presence of a small concentration of chromate ion, and the end point is signaled by the first detectable and permanent appearance of a precipitate of brick-red-colored silver chromate throughout the solution. Silver chromate forms upon addition of a slight excess of silver nitrate only after virtually all the chloride

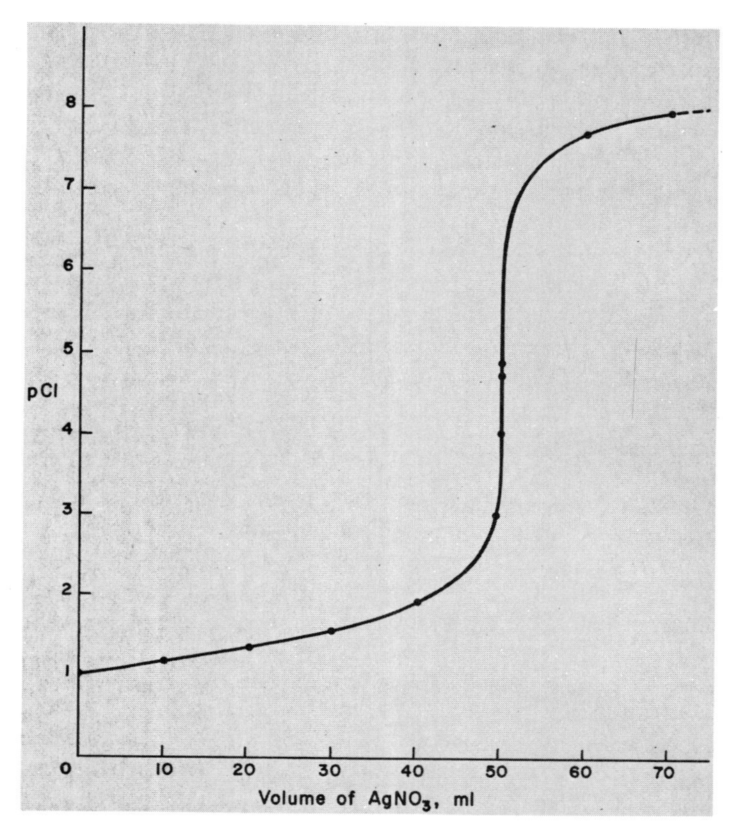


*Figure 7–2.* Titration curve for the titration of 50.00 ml of 0.1000 $F$ sodium chloride solution with 0.1000 $F$ silver nitrate solution.

ion has been precipitated as silver chloride. Frequently, a blank titration is performed with a solution containing no chloride but the same concentration of chromate as an actual experiment. A suspension of white-colored calcium carbonate is used to simulate the silver chloride precipitate, the mixture is titrated with silver nitrate, and the volume of titrant required to produce the same end-point color is measured. This small volume, when subtracted from the total volume of titrant used in an actual titration, provides an end-point correction.

It is essential that the Mohr titration be carried out in a solution which is neither too acidic nor too basic. For pH values higher than 10, there is danger that silver hydroxide or oxide may precipitate:

$$Ag^+ + OH^- \rightleftharpoons AgOH$$

$$2\,AgOH \rightleftharpoons Ag_2O + H_2O$$

In slightly acidic solutions, chromate ion becomes protonated

$$CrO_4^{2-} + H^+ \rightleftharpoons HCrO_4^-; \qquad K = 3.2 \times 10^6$$

and the resulting hydrogen chromate anion dimerizes to form dichromate:

$$2\,HCrO_4^- \rightleftharpoons Cr_2O_7^{2-} + H_2O; \qquad K = 33$$

Below pH 7 the chromate ion concentration decreases quite rapidly, and the titration error increases correspondingly, because more and more titrant must be added beyond the equivalence point to exceed the solubility product of silver chromate.

Both bromide and chloride can be determined by means of the Mohr titration. However, iodide and thiocyanate do not give satisfactory results because precipitates of silver iodide and silver thiocyanate adsorb chromate ions, so that indistinct end points are obtained.

Volhard Titration. In the Volhard titration for the direct determination of silver ion, or for the indirect determination of a number of anions that form insoluble silver salts, the end point is detected by means of the formation of a soluble, colored complex ion. Advantage is taken of the fact that iron(III) in an acidic medium forms an intense blood-red complex with thiocyanate:

$$Fe^{3+} + SCN^- \rightleftharpoons FeSCN^{2+}; \qquad K_1 = 138$$

Although $FeSCN^{2+}$ is not especially stable, a sensitive end-point signal can be obtained because this complex is detectable at very low concentration levels.

It is important that the Volhard titration be performed in an acidic solution. In aqueous media, the parent form of iron(III) is the colorless hexaaquo species, $Fe(H_2O)_6^{3+}$. However, $Fe(H_2O)_6^{3+}$ is a fairly strong acid and undergoes dissociations such as

$$Fe(H_2O)_6^{3+} \rightleftharpoons Fe(H_2O)_5(OH)^{2+} + H^+$$

and

$$Fe(H_2O)_5(OH)^{2+} \rightleftharpoons Fe(H_2O)_4(OH)_2^{+} + H^+$$

Because $Fe(H_2O)_5(OH)^{2+}$ and $Fe(H_2O)_4(OH)_2^{+}$ are orange and brown, their presence would mask the appearance of red-colored $FeSCN^{2+}$. To repress the acid dissociation of $Fe(H_2O)_6^{3+}$, the pH of the solution must be between 0 and 1.

*Direct Determination of Silver.* In the Volhard method for direct determination of silver, an acidic solution containing silver ion and a small concentration of iron(III) is titrated with standard potassium thiocyanate solution, and the end point is marked by the appearance of $FeSCN^{2+}$ upon addition of the first slight excess of titrant.

*Indirect Determination of Chloride.* There has been extensive use of the Volhard method for the determination of chloride and other anions. A carefully measured excess of standard silver nitrate is added to the chloride sample solution, the proper amounts of acid and iron(III) indicator having been introduced. An amount of silver nitrate stoichiometrically equivalent to the chloride reacts to precipitate silver chloride; then, the excess silver nitrate is back-titrated with standard potassium thiocyanate solution to the appearance of the red $FeSCN^{2+}$ complex. To compute the quantity of chloride (or other anion) in the sample solution, the millimoles of silver ion equivalent to the thiocyanate used in the back-titration are subtracted from the total number of millimoles of silver ion originally added to the sample solution.

A serious difficulty occurs when the excess silver ion is back-titrated with potassium thiocyanate in the presence of the silver chloride precipitate. Because silver chloride is more soluble than silver thiocyanate, the addition of thiocyanate causes *metathesis* of silver chloride according to the reaction

$$AgCl + SCN^- \rightleftharpoons Cl^- + AgSCN$$

In effect, some of the silver nitrate reacts both with chloride and with thiocyanate, so the resulting titration data are useless. An obvious way to eliminate metathesis is to remove the silver chloride by filtration after the excess of silver nitrate is added. Care must be taken to wash the silver chloride completely free of the excess silver ions. After removal of the silver chloride, the titration of silver ion with thiocyanate is identical to the direct determination of silver. A second way to overcome metathesis is based on decreasing the rate of the heterogeneous reaction between solid silver chloride and thiocyanate ion. This can be accomplished if the surface area of the silver chloride precipitate is decreased. There are two possibilities – first, boil the solution to coagulate the silver chloride, cool the solution, then titrate it with thiocyanate and, second, coat the silver chloride particles with a water-immiscible substance such as nitrobenzene.

It is possible to perform the indirect Volhard determination of other anions. In particular, for the determinations of bromide and iodide, the metathesis of the silver halide by excess thiocyanate does not occur because silver bromide and silver iodide both have smaller solubility products than silver thiocyanate. None of the special precautions used in the determination of chloride has to be employed for bromide or iodide. However, because iodide does react with iron(III), it is necessary to add the excess of silver nitrate before the iron(III) indicator. A successful determination of anions which form silver salts *more soluble* than silver chloride requires that an excess of silver nitrate be added and that the silver salt be removed by filtration before the back-titration with standard potassium thiocyanate solution.

**Use of Adsorption Indicators (Fajans Method).** Certain organic dyes are adsorbed by colloidal precipitates more strongly on one side of the equivalence point than on the other. In certain cases, the dye undergoes an abrupt color change in the process of being adsorbed, and it can serve as a sensitive end-point indicator in titrations.

One important family of adsorption indicators is derived from fluorescein. Quite often, the sodium salt of this compound, sodium fluoresceinate, is used as an indicator for the titration of chloride with silver nitrate in a neutral or slightly basic solution. This indicator is ionized in solution to form sodium ions and fluoresceinate ions, which will be designated as the indicator anions, $In^-$. We can understand the behavior of the indicator by noting that a colloidally dispersed particle of precipitate attracts to its surface a primary adsorbed ion layer consisting preferentially of the lattice ion present in excess in the surrounding solution. In addition, a counter-ion layer is attracted more or less strongly, the sign of the charges of the counter ions being opposite to those of the primary layer. As soon as the titration is begun, solid silver chloride is formed in the titration vessel. At all points prior to the equivalence point, chloride ion is in excess, so that the primary ion layer consists of adsorbed chloride ions and the counter-ion layer of any available cations, such as sodium or hydrogen ions. Few negatively charged indicator ions, $In^-$, are adsorbed because chloride displaces them. However, beyond the equivalence point, excess silver nitrate is present, so the primary adsorbed ion layer is composed of silver ions and the secondary ion layer consists of negative ions, an appreciable number of which will be the indicator anions. Sodium fluoresceinate imparts a fluorescent yellow-green color to the solution, but when the indicator anions are adsorbed as counter ions on the precipitate, a dramatic color change occurs and the precipitate particles become bright pink. Thus, the end point is marked by the change from a green solution to a pink precipitate. In practice, the particles of precipitate are kept very small and well dispersed through the use of dextrin as a protective colloid, so that the observed color change appears to be one of yellow-green to pink throughout the entire solution.

Two other adsorption indicators are the sodium salts of dichlorofluorescein and eosin. Whereas fluorescein is applicable to silver halide precipitations only if the pH is within the approximate limits of 6 to 10, a pH as low as 4 is acceptable with dichlorofluorescein. In practice, the use of dichlorofluorescein rather than fluorescein is preferable for the titration of halides. Eosin is suitable as an indicator for the precipitation titrations of bromide, iodide, and thiocyanate (but not chloride) in solutions as acidic as pH2. Chloride cannot be determined because the eosinate anion is more strongly adsorbed than chloride even at the start of the titration.

## QUESTIONS AND PROBLEMS

1. Formulate solubility-product expressions for $CaF_2$, $Fe(OH)_3$, $Ca_3(PO_4)_2$, $MgNH_4PO_4$, and $Ag_2CrO_4$.
2. If the solubility product for silver chloride is $1.78 \times 10^{-10}$ and the solubility product for silver chromate is $2.45 \times 10^{-12}$, which compound is more soluble in water?
3. Given the following solubility data in water, evaluate the solubility-product constant for each of the compounds:
   (a) TlCl, 0.32 gram per 100 ml
   (b) $Pb(IO_3)_2$, $3.98 \times 10^{-5}$ mole per liter
   (c) AgI, $1.40 \times 10^{-6}$ gram per 500 ml
   (d) $Mg(OH)_2$, 0.0085 gram per liter
4. From the solubility-product constants tabulated in Appendix 1, calculate the solubility of each of the following compounds in formula weights per liter:
   (a) CdS, (b) $BaF_2$, (c) $BiI_3$, (d) $Cu(IO_3)_2$, (e) $SrSO_4$.
5. Which solution contains a higher concentration of silver ion, pure water saturated with silver iodate or pure water saturated with silver chromate?
6. What is the maximum concentration of calcium ion that can be present in one liter of solution containing 3.00 moles of fluoride ion?
7. What hydroxide ion concentration is needed just to begin precipitation of magnesium hydroxide from a 0.01 $F$ solution of magnesium sulfate?
8. Concentrated potassium iodide (KI) solution is slowly added to a solution that is initially 0.02 $M$ in lead ion and 0.02 $M$ in silver ion. Which cation will precipitate first? What will be its concentration when the second cation starts to be precipitated? Neglect dilution of the sample solution by the potassium iodide medium.
9. Given the following equilibrium data

$$Ag^+ + 2\,NH_3 \rightleftharpoons Ag(NH_3)_2^+; \qquad \beta_2 = 1.62 \times 10^7$$
$$AgBr(s) \rightleftharpoons Ag^+ + Br^-; \qquad K_{sp} = 5.25 \times 10^{-13}$$

   calculate the concentration of ammonia required to prevent precipitation of silver bromide from a solution which is 0.025 $M$ in bromide and 0.045 $M$ in $Ag(NH_3)_2^+$.
10. A solution is saturated with respect to a compound of the general formula $AB_2C_3$:

$$AB_2C_3(s) \rightleftharpoons A^+ + 2\,B^+ + 3\,C^-$$

If this solution is found to contain ion $C^-$ at a concentration of 0.003 *M*, calculate the solubility product of the compound $AB_2C_3$.

11. If the concentration of rubidium ion, $Rb^+$, is decreased by a factor of 111 when 3.00 moles of lithium perchlorate, $LiClO_4$, is added to 1 liter of a saturated rubidium perchlorate solution, what is the solubility product of $RbClO_4$?
12. Calculate the solubility of cuprous iodide in a 0.35 *F* ammonia solution.

$$CuI(s) \rightleftharpoons Cu^+ + I^-; \qquad K_{sp} = 5.1 \times 10^{-12}$$
$$Cu^+ + NH_3 \rightleftharpoons CuNH_3^+; \qquad K_1 = 8.5 \times 10^5$$
$$CuNH_3^+ + NH_3 \rightleftharpoons Cu(NH_3)_2^+; \qquad K_2 = 8.5 \times 10^4$$

13. Excess solid barium sulfate was shaken with 3.6 *F* hydrochloric acid until equilibrium was attained. Calculate the equilibrium solubility of barium sulfate and the concentrations of $SO_4^{2-}$ and $HSO_4^-$.

$$BaSO_4(s) \rightleftharpoons Ba^{2+} + SO_4^{2-}; \qquad K_{sp} = 1.1 \times 10^{-10}$$
$$H_2SO_4 \rightleftharpoons H^+ + HSO_4^-; \qquad K_{a1} \gg 1$$
$$HSO_4^- \rightleftharpoons H^+ + SO_4^{2-}; \qquad K_{a2} = 1.2 \times 10^{-2}$$

14. Equilibria involving the solubilities of lead sulfate and strontium sulfate and the dissociation of sulfuric acid are as follows:

$$PbSO_4(s) \rightleftharpoons Pb^{2+} + SO_4^{2-}; \qquad K_{sp} = 1.6 \times 10^{-8}$$
$$SrSO_4(s) \rightleftharpoons Sr^{2+} + SO_4^{2-}; \qquad K_{sp} = 3.8 \times 10^{-7}$$
$$H_2SO_4 \rightleftharpoons H^+ + HSO_4^-; \qquad K_{a1} \gg 1$$
$$HSO_4^- \rightleftharpoons H^+ + SO_4^{2-}; \qquad K_{a2} = 1.2 \times 10^{-2}$$

If an excess of both lead sulfate and strontium sulfate is shaken with a 0.60 *F* nitric acid solution until equilibrium is attained, what will be the concentrations of $Pb^{2+}$, $Sr^{2+}$, $HSO_4^-$, and $SO_4^{2-}$?

15. Excess barium sulfate was shaken with 0.25 *F* sulfuric acid until equilibrium was attained. What was the concentration of barium ion in the solution?

$$BaSO_4(s) \rightleftharpoons Ba^{2+} + SO_4^{2-}; \qquad K_{sp} = 1.1 \times 10^{-10}$$
$$H_2SO_4 \rightleftharpoons H^+ + HSO_4^-; \qquad K_{a1} \gg 1$$
$$HSO_4^- \rightleftharpoons H^+ + SO_4^{2-}; \qquad K_{a2} = 1.2 \times 10^{-2}$$

16. Solubility-product data for cupric iodate, lanthanum iodate, and silver iodate are as follows:

$$Cu(IO_3)_2(s) \rightleftharpoons Cu^{2+} + 2\,IO_3^-; \qquad K_{sp} = 7.4 \times 10^{-8}$$
$$La(IO_3)_3(s) \rightleftharpoons La^{3+} + 3\,IO_3^-; \qquad K_{sp} = 6.0 \times 10^{-10}$$
$$AgIO_3(s) \rightleftharpoons Ag^+ + IO_3^-; \qquad K_{sp} = 3.0 \times 10^{-8}$$

Suppose that you prepare a separate saturated solution of each salt in pure water. Calculate the equilibrium concentration of iodate ion in each of these three solutions.

17. If the solubility of solid barium iodate, $Ba(IO_3)_2$, in a 0.000540 $F$ potassium iodate ($KIO_3$) solution is 0.000540 formula weight per liter, what is the solubility-product constant for $Ba(IO_3)_2$?
18. Solubility-product data for silver chloride and silver chromate are as follows:

$$AgCl(s) \rightleftharpoons Ag^+ + Cl^-; \qquad K_{sp} = 1.78 \times 10^{-10}$$

$$Ag_2CrO_4(s) \rightleftharpoons 2\,Ag^+ + CrO_4^{2-}; \qquad K_{sp} = 2.45 \times 10^{-12}$$

(a) Suppose that you have an aqueous solution containing 0.0020 $M$ $CrO_4^{2-}$ and 0.000010 $M$ $Cl^-$. If *concentrated* silver nitrate solution (so that volume changes may be neglected) is added gradually with good stirring to this solution, which precipitate ($Ag_2CrO_4$ or AgCl) will form first? Justify your conclusion with a calculation.
(b) Eventually, the silver ion concentration should increase enough to cause precipitation of the second ion. What will be the concentration of the *first* ion when the second ion just begins to precipitate?
(c) What *percentage* of the *first* ion is already precipitated when the second ion just begins to precipitate?

19. The solubility equilibrium and the solubility-product constant for zinc arsenate are as follows:

$$Zn_3(AsO_4)_2(s) \rightleftharpoons 3\,Zn^{2+} + 2\,AsO_4^{3-}; \qquad K_{sp} = 1.3 \times 10^{-28}$$

Calculate the final concentration of $AsO_4^{3-}$ required to precipitate all but 0.1 per cent of the $Zn^{2+}$ in 250 ml of a solution that originally contains 0.2 g of $Zn(NO_3)_2$. Zinc nitrate is very soluble in water.

20. In pure water, the solubility of cerium(III) iodate, $Ce(IO_3)_3$, is 124 mg per 100 ml.
(a) Calculate the solubility-product constant for cerium(III) iodate, which dissolves according to the reaction

$$Ce(IO_3)_3(s) \rightleftharpoons Ce^{3+} + 3\,IO_3^-$$

(b) Calculate the solubility (moles per liter) of cerium(III) iodate in a 0.050 $F$ potassium iodate ($KIO_3$) solution.

21. A 0.8046-g sample of impure barium chloride dihydrate weighed 0.7082 g after it was dried at 200°C for two hours. Calculate the percentage of water in the sample.
22. A silver chloride precipitate weighing 0.3221 g was obtained from 0.4926 g of a soluble salt mixture. Calculate the percentage of chloride in the sample.

23. The sulfur content of a sample is usually expressed in terms of the percentage of $SO_3$ in that sample. Calculate the percentage of $SO_3$ in a 0.3232-g sample of a soluble salt mixture which yielded a barium sulfate precipitate weighing 0.2982 g.
24. What is the percentage of iron in a 0.9291-g sample of an acid-soluble iron ore which yields a ferric oxide ($Fe_2O_3$) precipitate weighing 0.6216 g?
25. From a 0.6980-g sample of an impure, acid-soluble magnesium compound, a precipitate of magnesium pyrophosphate ($Mg_2P_2O_7$) weighing 0.4961 g was obtained. Calculate the magnesium content of the sample in terms of the percentage of magnesium oxide (MgO) present in that sample.
26. A 1.0000-g sample of a mixture of $K_2CO_3$ and $KHCO_3$ yielded 0.4000 g of carbon dioxide ($CO_2$) upon ignition. What weight of each compound was present in the mixture?
27. A 1.1374-g sample contains only sodium chloride and potassium chloride. Upon dissolution of the sample and precipitation of the chloride as silver chloride, a precipitate weighing 2.3744 g was obtained. Calculate the percentage of sodium chloride in the sample.
28. A 0.5000-g sample of a clay was analyzed for sodium and potassium by precipitation of both as chlorides. The combined weight of the sodium and potassium chlorides was found to be 0.0361 g. The potassium in the chloride precipitate was reprecipitated as $K_2PtCl_6$, which weighed 0.0356 g. Calculate the sodium and potassium content of the clay in terms of the percentages of hypothetical $Na_2O$ and $K_2O$ present in the clay.
29. A 0.2000-g sample of an alloy containing only silver and lead was dissolved in nitric acid. Treatment of the resulting solution with cold hydrochloric acid gave a mixed chloride precipitate (AgCl and $PbCl_2$) weighing 0.2466 g. When this mixed chloride precipitate was treated with hot water to dissolve all the lead chloride, 0.2067 g of silver chloride remained. Calculate the percentage of silver in the alloy and calculate what weight of lead chloride was not precipitated by the addition of cold hydrochloric acid.
30. Calculate the percentage of potassium iodide (KI) in a 2.145-g sample of a mixture of potassium iodide and potassium carbonate that, when analyzed according to the Volhard procedure, required 3.32 ml of a 0.1212 $F$ potassium thiocyanate solution after the addition of 50.00 ml of a 0.2429 $F$ silver nitrate solution.
31. For the analysis of a commercial solution of silver nitrate, a 2.075-g sample of the solution was weighed out and diluted to 100.0 ml in a volumetric flask. A 50.00-ml aliquot of the solution was then titrated with 35.55 ml of a potassium thiocyanate solution, of which 1.000 ml corresponds to exactly 5.000 mg of silver. Calculate the weight percentage of silver nitrate in the original solution.
32. An analytical chemist analyzed 0.5000 g of an arsenic-containing sample by oxidizing the arsenic to arsenate, precipitating silver arsenate ($Ag_3AsO_4$), dissolving the precipitate in acid, and titrating the silver with a 0.1000 $F$ potassium thiocyanate solution – 45.45 ml being required. Calculate the percentage of arsenic in the sample.

33. A solid sample is known to contain only sodium hydroxide, sodium chloride, and water. A 6.700-g sample was dissolved in distilled water and diluted to 250.0 ml in a volumetric flask. A one-tenth aliquot required 22.22 ml of 0.4976 *F* hydrochloric acid for titration to a phenolphthalein end point. When another one-tenth aliquot of the sample solution was titrated according to the Volhard procedure, 35.00 ml of a 0.1117 *F* silver nitrate solution was added and 4.63 ml of 0.0962 *F* potassium thiocyanate solution was required for the back-titration. The water content of the sample was determined by difference. What was the percentage composition of the original sample?
34. A method for the determination of low concentrations of chloride ion takes advantage of the common ion effect and the solubility-product principle. The "concentration solubility product" of AgCl(s) in 0.0250 *F* nitric acid at 25°C is $2.277 \times 10^{-10}$. The concentration of chloride in an unknown was determined as follows: A 0.0250 *F* nitric acid solution containing chloride ion at an unknown concentration was shaken with excess pure AgCl(s) at 25°C until solubility equilibrium was attained. A 100.0-ml sample of the equilibrated solution was withdrawn by suction through a sintered-glass filter stick in order to remove any solid AgCl which might have been dispersed in the solution. The 100.0-ml sample was then titrated with a standard $1.020 \times 10^{-4}$ *F* potassium iodide (KI) solution. The end point was reached when 7.441 ml of the standard KI had been added. Calculate the concentration of chloride ion in the original solution to four significant figures. Neglect activity effects, the presence of molecular AgCl(aq), and the possible formation of anionic complexes between silver and chloride ions.
35. Construct the titration curve for the titration of 50.00 ml of a 0.01000 *F* silver nitrate solution with 0.1000 *F* potassium thiocyanate solution in a 1 *F* nitric acid medium.

## SUGGESTIONS FOR ADDITIONAL READING

1. C. Ayers: Argentometric methods. *In* C. L. Wilson and D. W. Wilson, eds.: *Comprehensive Analytical Chemistry.* Volume IB, Elsevier, New York, 1960, pp. 222-237.
2. A. J. Bard: *Chemical Equilibrium.* Harper & Row, New York, 1966, pp. 46-62.
3. T. R. Blackburn: *Equilibrium.* Holt, Rinehart and Winston, New York, 1969, pp. 93-112.
4. J. N. Butler: *Ionic Equilibrium.* Addison-Wesley Publishing Company, Reading, Massachusetts, 1964, pp. 174-205.
5. J. F. Coetzee: Equilibria in precipitation reactions. *In* I. M. Kolthoff and P. J. Elving, eds.: *Treatise on Analytical Chemistry.* Part I, Volume 1, Wiley-Interscience, New York, 1959, pp. 767-809.
6. L. Gordon, M. L. Salutsky, and H. H. Willard: *Precipitation from Homogeneous Solution.* John Wiley and Sons, New York, 1959.
7. W. F. Hillebrand, G. E. F. Lundell, H. A. Bright, and J. I. Hoffman: *Applied Inorganic Analysis.* Second edition, John Wiley and Sons, New York, 1953, pp. 44-46, 711-723, 822-835.
8. I. M. Kolthoff, E. B. Sandell, E. J. Meehan, and S. Bruckenstein: *Quantitative Chemical Analysis.* Fourth edition, The Macmillan Company, New York, 1969, pp. 125-154, 198-247, 248-285, 565-677, 716-725, 795-802.
9. I. M. Kolthoff and V. A. Stenger: *Volumetric Analysis.* Second edition, Volume II, Wiley-Interscience, New York, 1947, pp. 239-344.

10. M. L. Salutsky: Precipitates: Their formation, properties, and purity. *In* I. M. Kolthoff and P. J. Elving, eds.: *Treatise on Analytical Chemistry.* Part I, Volume 1, Wiley-Interscience, New York, 1959, pp. 733-766.
11. H. F. Walton: *Principles and Methods of Chemical Analysis.* Second edition, Prentice-Hall, Englewood Cliffs, New Jersey, 1964, pp. 80-114, 373-389.
12. C. L. Wilson, F. E. Beamish, W. A. E. McBryde, and L. Gordon: Inorganic gravimetric analysis. *In* C. L. Wilson and D. W. Wilson, eds.: *Comprehensive Analytical Chemistry.* Volume IA, Elsevier, New York, 1959, pp. 430-547.

# 8 PRINCIPLES OF OXIDATION-REDUCTION METHODS

Oxidation-reduction methods of analysis, commonly called **redox methods**, probably rank as the most widely used volumetric procedures. Although acid-base, precipitation, and complexometric titrations have usefulness, many substances cannot be determined satisfactorily by means of these kinds of titrations. Redox methods are, however, of much broader applicability for the determination of a wide variety of chemical species, and additional emphasis has been placed on these techniques through the large amount of research in the field of electroanalytical chemistry.

## ELECTROCHEMICAL CELLS

An electrochemical cell consists essentially of two electrodes which are immersed either into the same solution or into two different solutions in electrolytic contact with one another. Across the interface where an electrode meets a solution, a difference in electrical potential develops – which we shall refer to simply as a **potential**. Likewise, across the interface between two solutions of differing composition, a **liquid-junction potential** arises, although it really is a potential difference across the boundary where the two media come into contact. In addition, every electrochemical cell exhibits an **electromotive force** or **emf**, expressed in volts, which is the *sum* of various potential differences within the cell, including the two electrode potentials and perhaps one or more liquid-junction potentials.

There are two kinds of electrochemical cells. A **galvanic cell** is an electrochemical cell in which the *spontaneous* occurrence of electrode reactions produces electrical energy which can be converted into useful work. In an **electrolytic cell**, nonspontaneous electrode reactions are forced to proceed when an external source of emf is connected across the two electrodes. In the operation of an electrolytic cell, electrical energy or work must be expended in causing the electrode reactions to occur.

Oxidation is the loss of electrons, and reduction is the gain of electrons. A substance which is reduced, that is, a substance which causes another chemical species to be oxidized, is an **oxidizing agent** or **oxidant**. Conversely, a substance which causes another species to be reduced, thereby becoming oxidized itself, is a **reducing agent** or **reductant**. Oxidation cannot take place without a corresponding reduction, and no substance can be reduced without some other substance simultaneously being oxidized. In an electrochemical cell, the electrode at which oxidation occurs is called the **anode**, whereas the **cathode** is the electrode at which reduction takes place.

An example of a galvanic cell is depicted in Figure 8–1. One beaker contains a solution of some zinc salt into which a metallic zinc electrode is immersed, and the other beaker contains a solution of a cupric salt into which an electrode of metallic copper is placed. Connecting the two beakers is a **salt bridge**, which provides a pathway for the migration of ions (flow of current) from one beaker to the other when an electrical circuit is completed, yet prevents mixing of the solutions in the two beakers as well as

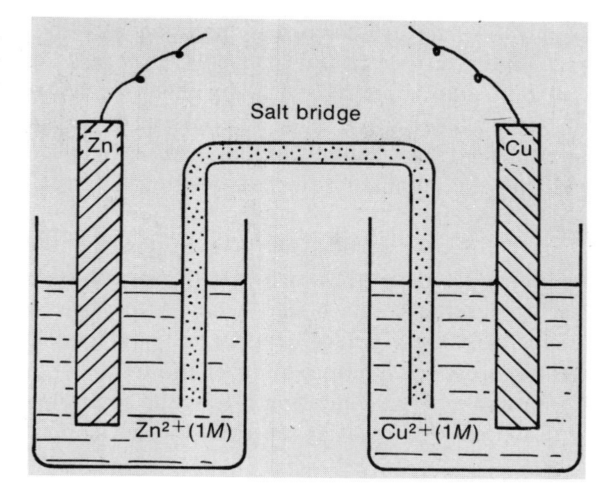


*Figure 8–1.* A simple galvanic cell involving a zinc metal-zinc ion half-cell and a cupric ion-copper metal half-cell connected by a potassium chloride salt bridge. For this cell the concentration (activity) of each ionic species, $Zn^{2+}$ and $Cu^{2+}$, is unity, although in general any values are possible.

any direct electron-transfer reaction between one electrode and the solution in the opposite beaker. In its simplest form, a salt bridge consists of an inverted U-tube filled with a mixture of potassium chloride solution and agar which forms a gel to minimize the leakage of potassium chloride into the two beakers.

If the two metal electrodes are connected by means of a conducting wire, the zinc electrode dissolves according to the reaction

$$Zn \rightleftharpoons Zn^{2+} + 2\ e$$

and cupric ions are reduced to copper atoms which deposit upon the copper electrode:

$$Cu^{2+} + 2\ e \rightleftharpoons Cu$$

We can represent the overall process occurring in this galvanic cell as the combination of the two individual electrode reactions; that is,

$$Zn + Cu^{2+} \rightleftharpoons Zn^{2+} + Cu$$

Electrons produced at the zinc electrode, as zinc atoms are oxidized to zinc ions, flow through the external wire to the copper electrode, where they are available to combine with incoming cupric ions to form more copper metal. Along with the flow of electrons in the external circuit, negatively charged ions in the solutions migrate from the copper half-cell through the salt bridge in the direction of the zinc half-cell. Migration of ions in solution is not all one way, for the production of zinc ions and the consumption of cupric ions cause the movement of cations from the zinc half-cell toward the copper half-cell.

In keeping with previous definitions, the zinc electrode is the *anode*, because it is the electrode at which oxidation occurs, and the copper electrode is the *cathode* since reduction takes place there.

## SHORTHAND REPRESENTATION OF CELLS

It is both time-consuming and space-consuming to draw a complete diagram of an electrochemical cell – electrodes, salt bridge, solutions, and containers – each time a cell is discussed. As a result, a set of rules for representing cells in a shorthand fashion has been developed.

1. Conventional chemical symbols are used to indicate the ions, molecules, elements, gases, and electrode materials involved in a cell. Concentrations of ions or molecules, as well as the partial pressure of each gaseous species, are enclosed in parentheses.

2. A single vertical line | is employed to designate the fact that a boundary between an electrode phase and a solution phase or between two different solution phases exists and that the potential difference developed across this interface is included in the total emf of the cell.

3. A double vertical line || indicates that the liquid-junction potential developed across the interface between two different solutions is ignored or that it is minimized by placing a suitable salt bridge between the two solutions. A liquid-junction potential originates at the interface between two nonidentical solutions because anions and cations diffuse across this interface at different rates.

Several examples will demonstrate the application of these rules.

*Example 8–1.* We can represent the zinc-copper cell depicted in Figure 8–1 as

$$Zn \mid Zn^{2+}\ (1\ M) \parallel Cu^{2+}\ (1\ M) \mid Cu$$

This shorthand abbreviation indicates that a zinc electrode in contact with a 1 $M$ solution of zinc ion gives rise to a potential and, similarly, that a 1 $M$ cupric ion solution in contact with a copper electrode produces another potential – the sum of these two potentials being the emf of the galvanic cell. A double vertical line indicates the presence of a salt bridge and reminds us that any liquid-junction potential is neglected in considering the emf of the cell.

*Example 8–2.* If, for the zinc-copper cell, we wished to specify 1 $F$ solutions of particular salts, such as $Zn(NO_3)_2$ and $CuSO_4$, and if instead of a salt bridge we allowed the two solutions to contact each other directly through a thin porous membrane (to prevent mixing of the two solutions), the cell would be correctly indicated as

$$Zn \mid Zn(NO_3)_2\ (1\ F) \mid CuSO_4\ (1\ F) \mid Cu$$

*Example 8–3.* Suppose we construct an entirely new galvanic cell. Let one electrode consist of a platinum wire in a mixture of 0.2 $M$ ferric ion and 0.05 $M$ ferrous ion in a 1 $F$ hydrochloric acid medium, and let the other electrode be a second platinum wire immersed in a 2 $F$ hydrochloric acid solution saturated with chlorine gas at a pressure of 0.1 atm. A salt bridge can be used to prevent mixing of the two solutions. We can represent this cell as

$$Pt \mid Fe^{3+}\ (0.2\ M),\ Fe^{2+}\ (0.05\ M),\ HCl\ (1\ F) \parallel HCl\ (2\ F) \mid Cl_2\ (0.1\ \text{atm}),\ Pt$$

This example brings up two other significant points. First, when several soluble species are present in the same solution, no special order of listing these is needed. Second, although electron transfer between chloride ion and chlorine molecules involves dissolved species, it is customary to indicate gaseous substances as part of the electrode phase along with the electrode material, *e.g.*, platinum metal.

## ELECTROMOTIVE FORCE AND ITS MEASUREMENT

As discussed in Chapter 3, the free-energy change ($\Delta G$) for a chemical process measures the driving force or tendency for reaction to occur. In addition, $\Delta G$ represents the *maximal amount of useful energy or work* (aside from work due to expansion) that can be obtained from the process.

A quantity which is characteristic of any galvanic cell is $E$, the emf expressed in volts. In turn, the free-energy change for a process which occurs or can be made to occur in a galvanic cell is related to the emf by the equation

$$\Delta G = -nFE$$

where $n$ is the number of faradays of electricity generated by the cell reaction and $F$ is the Faraday constant (96,487 coulombs or 23,060 calories per faraday). By measuring the emf of a galvanic cell, we can determine the free-energy change $\Delta G$ for a given process. Let us see how this can be done for the reaction

$$Zn + Cu^{2+} \rightleftharpoons Zn^{2+} + Cu$$

A galvanic cell in which the desired reaction can occur is shown in Figure 8–1.

A **potentiometer** is an instrument used to determine the emf of a galvanic cell. As a brief introduction to the principles of operation of a potentiometer, let us consider the following simple concepts. Figure 8–2A shows the zinc-copper cell with an external conducting wire connecting the zinc and copper electrodes. In addition, a current-measuring galvanometer, $G$, has been inserted into the external circuit.

Suppose we introduce into the external circuit a source of known emf, as shown in Figure 8–2B, in such a way that it opposes the zinc-copper cell. If the known emf is now varied until it becomes just equal to the emf of the zinc-copper cell, the galvanometer will register a current of zero. When no current flows in the circuit, it follows that the emf values of the zinc-copper cell and the known source are identical, and hence the emf of

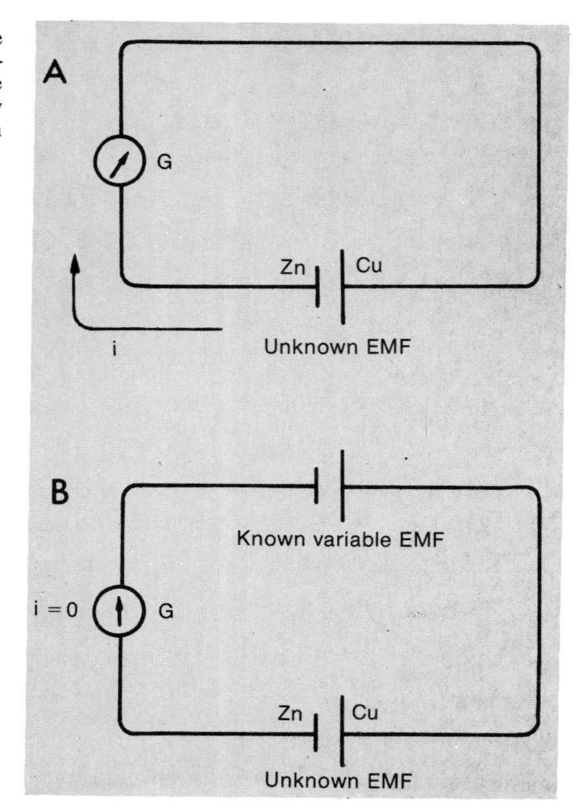


*Figure 8–2.* Diagram to illustrate the principles of the potentiometric measurement of electromotive force. It should be noted that the direction of current flow shown in A is the direction of electron flow. (See text for discussion.)

the zinc-copper cell is determined. In practice, an actual potentiometric measurement consists of varying the known emf until the galvanometer indicates zero current.

Unfortunately, the circuit discussed above lacks the sophistication necessary for the accurate measurement of emf. Introduction of several improvements provides the precision potentiometer shown in Figure 8–3. There must be a source of emf (*V*), which is typically a 3-v battery, but the value of which need not be known exactly. This emf source is connected across the ends of a uniform or linear slidewire resistance (*AB*) accurately calibrated from, say, 0 to 1.5 v in 0.1-mv increments. Most potentiometers have a large fraction of the slidewire replaced by fixed, precision resistors. In the use of such an instrument, the first step is standardization of the potentiometer circuit. To accomplish this standardization, one throws switch *S* to connect a standard Weston cell (Figure 8–4) across the slidewire. A *saturated* Weston cell, whose shorthand representation is

$$-\mathrm{Cd(Hg)} \mid \mathrm{CdSO_4 \cdot \tfrac{8}{3}H_2O(s),\ Hg_2SO_4(s)} \mid \mathrm{Hg}+$$

provides an internationally accepted standard for calibration purposes. Next, the movable contact (*C*) is set to a scale reading along slidewire *AB* equal to the known emf of the Weston cell. This emf will be precisely known for any particular cell and will have a value of about 1.0186 v at 20°C. Then, by tapping the key (*K*) and observing the deflections of the galvanometer needle (*G*), one carefully adjusts the variable resistance (*R*) until the galvanometer indicates that no *net* current is flowing. If the Weston cell is replaced with an unknown cell, all other adjustments are kept unchanged, and no net current-flow is observed when contact *C* is moved to 1.0186 v on slidewire *AB*, the emf of the unknown cell is 1.0186 v. Furthermore, since slidewire *AB* is linear, any other position to which *C* must be moved will give directly the emf of the unknown cell. Therefore, an actual potentiometric measurement involves four steps – (1) the potentiometer is standardized,

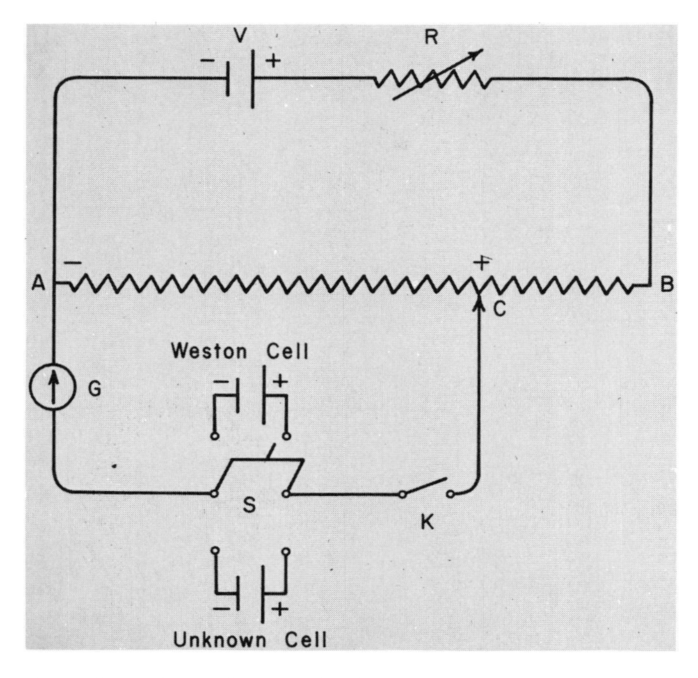


*Figure 8–3.* Circuit diagram of a simple potentiometer with a linear voltage divider.

*Figure 8–4.* Saturated Weston standard cell.

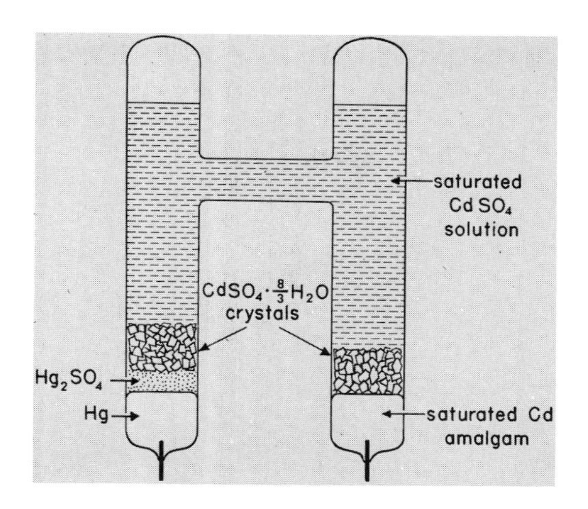


(2) the unknown cell is switched into the circuit, (3) contact *C* and key *K* are, respectively, moved and tapped alternately until potentiometric balance is achieved, and (4) the final position of the sliding contact (*C*) is recorded.

## THERMODYNAMIC AND ELECTROCHEMICAL SIGN CONVENTIONS

We shall now explore some of the relationships between thermodynamics and electrochemistry. In the galvanic cell shown in Figure 8–5, the left-hand compartment contains a platinum electrode dipped into a hydrochloric acid solution of unit activity and bathed with hydrogen gas at a pressure of 1 atm. For convenience, the platinum electrode can be sealed into an inverted tube, with an enlarged open bottom to admit the hydrochloric acid solution and with exit ports for the hydrogen gas to escape. Hydrogen gas at 1 atm pressure enters through a side-arm tube. Hydrogen gas must come into intimate contact with the platinum electrode. This requirement can be fulfilled if the platinum electrode is coated prior to use with a layer of finely divided, spongy platinum metal (**platinum black**). Hydrogen gas can permeate the platinum black, and electron transfer between hydrogen ions and hydrogen molecules (or atoms) is greatly facilitated. This combination of a platinum-black electrode, hydrochloric acid of unit activity, and hydrogen gas at 1 atm pressure is called the **standard hydrogen electrode** (SHE) or the **normal hydrogen electrode** (NHE).

In the right-hand compartment is a silver electrode coated with silver chloride and placed into a hydrochloric acid solution of unit activity which is saturated with silver chloride. A porous membrane can be inserted between the two compartments to prevent direct mixing of the solutions. However, the liquid-junction potential is virtually zero because the solutions on each side of the membrane are nearly identical in composition.

This word and picture description of the cell can be portrayed by means of the shorthand cell representation

$$\mathrm{Pt, H_2\ (1\ atm)}\ |\ \mathrm{HCl}\ (a = 1\ M), \mathrm{AgCl(s)}\ |\ \mathrm{Ag}$$

Associated with every shorthand cell representation is a so-called **conventional cell reaction**, which may or may not proceed as written when the electrodes are short-circuited. To write this cell reaction, we must adopt the following rule:

**Looking at the shorthand cell representation, we combine as *reactants* the reductant (reducing agent) of the left-hand electrode and the oxidant (oxidizing agent) of the**

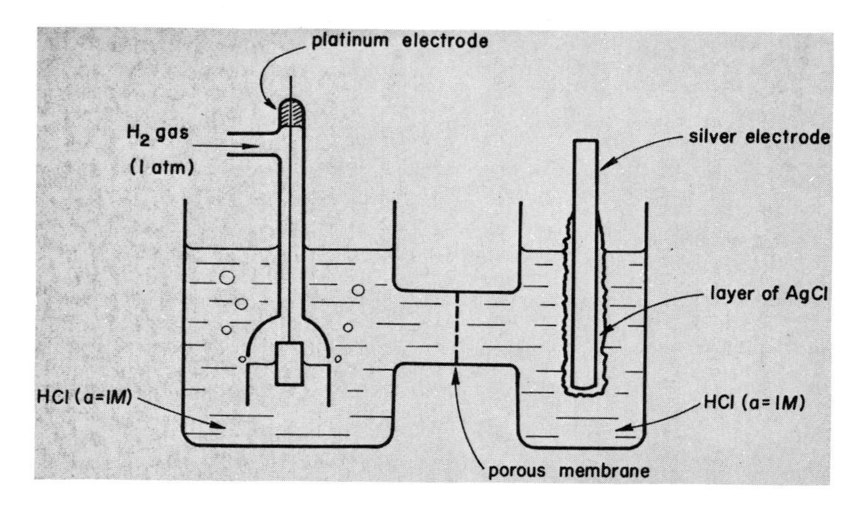


*Figure 8–5.* A galvanic cell consisting of a standard or normal hydrogen electrode (NHE) and a silver-silver chloride electrode.

**right-hand electrode. As *products* we obtain the oxidant of the left-hand electrode and the reductant of the right-hand electrode.**

If we now apply this rule, hydrogen gas (reductant of the left-hand electrode) reacts with silver chloride (oxidant of the right-hand electrode) to yield hydrogen ions (oxidant of the left-hand electrode) and elemental silver (reductant of the right-hand electrode). Chloride ions, freed by the reduction of silver chloride, are also formed. Thus, the cell reaction is

$$H_2 + 2\ AgCl \rightleftharpoons 2\ H^+ + 2\ Ag + 2\ Cl^-$$

Information about the tendency for this cell reaction to occur as written (when the electrodes are short-circuited) is furnished by the free-energy change ($\Delta G$) for the process. According to thermodynamic conventions, the cell reaction will be spontaneous if $\Delta G$ is a negative quantity. If $\Delta G$ is positive, the reaction is nonspontaneous and, in fact, will proceed in the reverse direction. In the event that $\Delta G$ is zero, the system is at equilibrium and no net cell reaction will occur. From the relationship between free-energy change $\Delta G$ and electromotive force $E$,

$$\Delta G = -nFE$$

it can be seen that $E$ is positive for spontaneous processes, negative for nonspontaneous reactions, and zero for systems at equilibrium.*

If the emf for the cell is measured potentiometrically as described previously, it is found to have an absolute magnitude of 0.2222 v. In addition, the cell reaction proceeds

*Strictly speaking, a chemical process can have a free-energy change, but not an emf; only an electrochemical cell possesses an emf. However, because of the fundamental relationship between the free-energy change for an overall cell reaction and the emf for the cell in which that reaction takes place, namely $\Delta G = -nFE$, it is common practice to speak of the emf of the overall cell reaction. When we talk about the emf of an overall cell reaction, we will actually mean the free-energy change for the reaction computed from the equation in the preceding sentence. In addition, the emf of a cell properly has no sign, whereas the emf of a cell reaction is a thermodynamic quantity and can have a + or − sign in accordance with the direction in which the process is written.

from left to right as written; this fact can be demonstrated if we connect the electrodes by a conducting wire and examine their behavior. To indicate that the cell reaction is a spontaneous process and that the driving force for this reaction is 0.2222 v, we can write

$$H_2 + 2\,AgCl \rightleftharpoons 2\,H^+ + 2\,Ag + 2\,Cl^-; \qquad E = +0.2222 \text{ v}$$

Let us now focus our attention more closely on the galvanic cell itself. Since the cell reaction should proceed spontaneously from left to right, we can conclude that there is a tendency for hydrogen gas to be oxidized to hydrogen ions and a tendency for silver chloride to be reduced to silver metal. Therefore, the hydrogen gas-hydrogen ion electrode is the *anode*, and the silver-silver chloride electrode functions as the *cathode.* Likewise, there is a tendency for an excess of electrons to accumulate at the left-hand electrode and for a deficiency of electrons to exist at the right-hand electrode. We can speak only of a *tendency* because, under the conditions of an ideal potentiometric emf measurement, there is no flow of electrons. An excess of electrons at the hydrogen gas-hydrogen ion electrode causes the anode to be the *negative* electrode of the galvanic cell, and a deficiency of electrons at the silver-silver chloride electrode means that the cathode is the *positive* electrode. On the shorthand cell representation, the signs of the electrodes are indicated as follows:

$$-\text{Pt}, H_2\ (1 \text{ atm}) \mid \text{HCl}\ (a = 1\ M), \text{AgCl(s)} \mid \text{Ag}+$$

We come now to one of the most important and fundamental conclusions of this discussion:

**There is no direct relationship between the sign of the emf for a cell reaction and the signs (+ or −) of the electrodes of the cell. An emf for a cell reaction is a *thermodynamic* concept; its sign may be positive (+) when a process occurs spontaneously or negative (−) when a process is nonspontaneous. On the other hand, the sign of any particular electrode of a galvanic cell is an *electrochemical* concept and is invariant.**

To illustrate the significance of these statements, the following galvanic cell may be considered:

$$\text{Ag} \mid \text{AgCl(s)}, \text{HCl}\ (a = 1\ M) \mid H_2\ (1 \text{ atm}), \text{Pt}$$

This is the same galvanic cell we have been discussing, except for the fact that we are viewing it from the other side of the bench top. According to our rule for writing the reaction corresponding to a shorthand cell representation, we obtain

$$2\,H^+ + 2\,Ag + 2\,Cl^- \rightleftharpoons H_2 + 2\,AgCl$$

This reaction is exactly the reverse of the process we considered earlier, so it follows that the present reaction is nonspontaneous. Since the galvanic cell is the same as before, the absolute magnitude of the emf is identical to the previous value. To indicate these facts, we may write

$$2\,H^+ + 2\,Ag + 2\,Cl^- \rightleftharpoons H_2 + 2\,AgCl; \qquad E = -0.2222 \text{ v}$$

From a *thermodynamic* viewpoint, whenever we reverse the direction of a process, the magnitude of the emf remains the same, whereas the sign of the emf is changed to denote whether the process is spontaneous or nonspontaneous. Thus, the reaction *actually would proceed from right to left if the electrodes were short-circuited.* There is a tendency for hydrogen gas to be oxidized to hydrogen ions and for silver chloride to be reduced to silver metal, so the *electrochemical* conclusions made previously about the galvanic cell,

$$-\text{Pt}, \text{H}_2\ (1\ \text{atm})\ |\ \text{HCl}\ (a = 1\ M), \text{AgCl(s)}\ |\ \text{Ag}+$$

are completely valid for the cell

$$+\text{Ag}\ |\ \text{AgCl(s)}, \text{HCl}\ (a = 1\ M)\ |\ \text{H}_2\ (1\ \text{atm}), \text{Pt}-$$

Accordingly, the hydrogen gas-hydrogen ion electrode is the *anode* and the *negative* electrode, whereas the silver-silver chloride electrode is the *cathode* and the *positive* electrode.

## STANDARD POTENTIALS AND HALF-REACTIONS

We can consider every overall cell reaction as being composed of two separate half-reactions. A **half-reaction** is a balanced chemical equation showing the transfer of electrons between two different oxidation states of the same element. For the cell

$$-\text{Pt}, \text{H}_2\ (1\ \text{atm})\ |\ \text{HCl}\ (a = 1\ M), \text{AgCl(s)}\ |\ \text{Ag}+$$

the overall reaction is

$$\text{H}_2 + 2\ \text{AgCl} \rightleftharpoons 2\ \text{H}^+ + 2\ \text{Ag} + 2\ \text{Cl}^-$$

and the two half-reactions are

$$\text{H}_2 \rightleftharpoons 2\ \text{H}^+ + 2\ \text{e}$$

and

$$2\ \text{AgCl} + 2\ \text{e} \rightleftharpoons 2\ \text{Ag} + 2\ \text{Cl}^-$$

Each half-reaction contains the same number of electrons, so when these half-reactions are added together the electrons cancel and the overall reaction is obtained.

**Definition and Determination of Standard Potentials.** We can regard the emf for the cell

$$-\text{Pt}, \text{H}_2\ (1\ \text{atm})\ |\ \text{HCl}\ (a = 1\ M), \text{AgCl(s)}\ |\ \text{Ag}+$$

as the sum of two potentials, one associated with the half-reaction occurring at each electrode of the cell.* Unfortunately, it is impossible to determine the *absolute* value for

---

*Recall that we customarily ascribe an emf to an electrochemical cell as well as to the corresponding overall cell reaction. A similar practice is followed to describe the properties of individual electrodes and half-reactions. An electrode in contact with a solution has a potential, but the half-reaction (or electron-transfer process) which takes place at that electrode cannot truly be said to possess a potential. Instead, the half-reaction has a free-energy change ($\Delta G$) associated with it, and this free-energy change may be related mathematically through the expression, $\Delta G = -nFE$, to the potential ($E$) of the electrode at which that half-reaction occurs. Provided that we recognize these distinctions (and the fact that the potentials and free-energy changes to which we refer are relative values with respect to the normal [or standard] hydrogen electrode), *it is convenient to discuss the potential of either an electrode or a half-reaction.* However, the potential of any individual electrode is a sign-invariant quantity, whereas the potential of a half-reaction is a thermodynamic parameter and may have either a + or − sign, depending upon the direction in which the half-reaction is written. Finally, note that $E$ can symbolize either emf or potential, the context of usage indicating which meaning is intended.

the potential of any individual electrode. All that can be measured is the *sum* of no less than two potentials (actually, potential differences) within a cell; from this sum, the potential of just one electrode cannot be extracted. Therefore, it is necessary to define arbitrarily the value for the potential of some electrode chosen as a reference; then, the potentials for all other electrodes can be quoted *relative* to this standard.

Accordingly, the **normal hydrogen electrode** (NHE) has been adopted as the standard. This electrode (Figure 8–5) involves the hydrogen ion-hydrogen gas half-reaction with hydrogen ion and hydrogen gas in their standard states of unit activity and 1 atm pressure, respectively. Under these conditions, the potential for this system is assigned the value of *zero* at all temperatures and is called the *standard potential, $E^0$, for the hydrogen ion-hydrogen gas electrode*:

$$2\,H^+ + 2\,e \rightleftharpoons H_2; \qquad E^0 = 0.0000\text{ v}$$

One can determine the standard potential ($E^0$), *relative to the standard hydrogen electrode,* for an electrode at which any other half-reaction occurs, if a suitable galvanic cell is constructed in which one electrode is the standard hydrogen electrode and the second electrode involves the half-reaction of interest. Although $E^0$ for the standard hydrogen electrode is *defined* to be independent of temperature, standard potentials for electrodes involving all other half-reactions do vary with temperature. By convention, standard potentials are always determined at 25°C. Conditions necessary for the successful measurement of a standard potential are included in the definition of this quantity – **the *standard potential* for any redox couple is the potential (sign and magnitude) of an electrode consisting of that redox couple under standard-state conditions measured in a galvanic cell against the standard hydrogen electrode at 25°C.**

Notice that the cell

$$-\text{Pt, } H_2\text{ (1 atm)} \mid \text{HCl } (a = 1\ M), \text{AgCl(s)} \mid \text{Ag}+$$

fulfills all requirements of the definition of standard potential, inasmuch as all species are present in their standard states. Since the magnitude of the emf of the galvanic cell is 0.2222 v, and since the silver chloride-silver metal electrode is always the positive electrode (regardless of the bench-top orientation of the cell), we can conclude that *the standard potential, $E^0$, of the silver chloride-silver metal electrode is +0.2222 v (relative to the normal hydrogen electrode).*

**Table of Standard Potentials.** Table 8–1 presents a short list of standard potentials along with the half-reaction to which each pertains, whereas a more complete compilation is provided in Appendix 4. All the half-reactions are written as *reductions.* This practice follows the recommendation set down by the International Union of Pure and Applied Chemistry (IUPAC) in 1953. Each half-reaction in a table of standard potentials automatically implies an overall cell reaction in which the half-reaction of interest (written as a reduction) is combined with the *hydrogen gas-hydrogen ion* half-reaction. Although only half-reactions appear in the table, standard potentials are always derived from a galvanic cell which consists of a standard hydrogen electrode and an electrode involving the desired half-reaction under standard-state conditions. For example, the half-reaction

$$AgCl + e \rightleftharpoons Ag + Cl^-$$

should be thought of as the overall reaction

$$\tfrac{1}{2}\,H_2 + AgCl \rightleftharpoons H^+ + Ag + Cl^-$$

Table 8–1. **Standard and Formal Potentials***

| $E^0$ | Half-reaction |
|---|---|
| 1.685 | $PbO_2 + SO_4^{2-} + 4\,H^+ + 2\,e \rightleftharpoons PbSO_4 + 2\,H_2O$ |
| 1.51 | $MnO_4^- + 8\,H^+ + 5\,e \rightleftharpoons Mn^{2+} + 4\,H_2O$ |
| *1.44* | $Ce^{4+} + e \rightleftharpoons Ce^{3+}$ (1 $F$ $H_2SO_4$) |
| 1.3595 | $Cl_2 + 2\,e \rightleftharpoons 2\,Cl^-$ |
| 1.33 | $Cr_2O_7^{2-} + 14\,H^+ + 6\,e \rightleftharpoons 2\,Cr^{3+} + 7\,H_2O$ |
| 1.15 | $SeO_4^{2-} + 4\,H^+ + 2\,e \rightleftharpoons H_2SeO_3 + H_2O$ |
| 1.000 | $VO_2^+ + 2\,H^+ + e \rightleftharpoons VO^{2+} + H_2O$ |
| 0.7995 | $Ag^+ + e \rightleftharpoons Ag$ |
| 0.771 | $Fe^{3+} + e \rightleftharpoons Fe^{2+}$ |
| 0.740 | $H_2SeO_3 + 4\,H^+ + 4\,e \rightleftharpoons Se + 3\,H_2O$ |
| *0.70* | $Fe^{3+} + e \rightleftharpoons Fe^{2+}$ (1 $F$ HCl) |
| *0.68* | $Fe^{3+} + e \rightleftharpoons Fe^{2+}$ (1 $F$ $H_2SO_4$) |
| 0.559 | $H_3AsO_4 + 2\,H^+ + 2\,e \rightleftharpoons HAsO_2 + 2\,H_2O$ |
| 0.5355 | $I_3^- + 2\,e \rightleftharpoons 3\,I^-$ |
| 0.361 | $VO^{2+} + 2\,H^+ + e \rightleftharpoons V^{3+} + H_2O$ |
| 0.337 | $Cu^{2+} + 2\,e \rightleftharpoons Cu$ |
| 0.334 | $UO_2^{2+} + 4\,H^+ + 2\,e \rightleftharpoons U^{4+} + 2\,H_2O$ |
| 0.2415 | $Hg_2Cl_2 + 2\,K^+ + 2\,e \rightleftharpoons 2\,Hg + 2\,KCl(s)$ (saturated calomel electrode) |
| 0.2222 | $AgCl + e \rightleftharpoons Ag + Cl^-$ |
| 0.153 | $Cu^{2+} + e \rightleftharpoons Cu^+$ |
| 0.10 | $TiO^{2+} + 2\,H^+ + e \rightleftharpoons Ti^{3+} + H_2O$ |
| 0.0000 | $2\,H^+ + 2\,e \rightleftharpoons H_2$ |
| −0.126 | $Pb^{2+} + 2\,e \rightleftharpoons Pb$ |
| −0.255 | $V^{3+} + e \rightleftharpoons V^{2+}$ |
| −0.3563 | $PbSO_4 + 2\,e \rightleftharpoons Pb + SO_4^{2-}$ |
| −0.41 | $Cr^{3+} + e \rightleftharpoons Cr^{2+}$ |
| −0.763 | $Zn^{2+} + 2\,e \rightleftharpoons Zn$ |

* Formal potentials are indicated by italicized entries. Only those standard and formal potentials utilized in this chapter are listed here; a much more complete tabulation is presented in Appendix 4.

Similarly, the entry

$$Zn^{2+} + 2\,e \rightleftharpoons Zn$$

may be considered as

$$H_2 + Zn^{2+} \rightleftharpoons 2\,H^+ + Zn$$

In turn, the first of these overall reactions pertains to the galvanic cell

$$-\text{Pt}, H_2\ (1\ \text{atm})\ |\ \text{HCl}\ (a = 1\ M), \text{AgCl(s)}\ |\ \text{Ag}+$$

and the second reaction is correlated with the galvanic cell

$$+\text{Pt}, H_2\ (1\ \text{atm})\ |\ H^+\ (a = 1\ M)\ \|\ Zn^{2+}\ (a = 1\ M)\ |\ \text{Zn}-$$

When any half-reaction is written as a reduction (either by itself or combined in an overall reaction with the hydrogen gas-hydrogen ion half-reaction under standard-state conditions), the potential of this half-reaction (or the emf of the overall reaction, since the potential for the standard hydrogen electrode is zero) is identical in *sign* and *magnitude* to the potential for the actual physical electrode at which this half-reaction occurs. Specifically, if the reactants and products are present in their standard states, the potential of a reduction half-reaction is the same as the standard potential for the electrode at which the half-reaction takes place. For example, under standard-state conditions, the potential of the lead(II)-lead metal half-reaction written as a reduction

$$Pb^{2+} + 2\ e \rightleftharpoons Pb$$

and the standard potential for the lead(II)-lead metal electrode are both -0.126 v.

Standard-potential data are utilized for many types of thermodynamic calculations. In cases where reduction processes are involved, one may employ the data exactly as they appear in the table of standard potentials. Thus, for the reduction of dichromate to chromic ion, we obtain from the table

$$Cr_2O_7^{2-} + 14\ H^+ + 6\ e \rightleftharpoons 2\ Cr^{3+} + 7\ H_2O; \qquad E^0 = +1.33\ v$$

A positive sign for the potential indicates that, relative to the hydrogen ion-hydrogen gas half-reaction, the reduction is spontaneous. However, if we wish to consider the oxidation of chromic ion to dichromate, both the direction of the half-reaction and the sign of the potential must be reversed; that is,

$$2\ Cr^{3+} + 7\ H_2O \rightleftharpoons Cr_2O_7^{2-} + 14\ H^+ + 6\ e; \qquad E^0 = -1.33\ v$$

Because the potential of a half-reaction is a thermodynamic quantity, we shall use subscripts to indicate the direction of the half-reaction to which each potential corresponds, particularly when we write the potential value separately from the half-reaction. Thus, $E^0_{Cr_2O_7^{2-},Cr^{3+}}$ or, more generally, $E_{Cr_2O_7^{2-},Cr^{3+}}$ denotes the potential for the half-reaction

$$Cr_2O_7^{2-} + 14\ H^+ + 6\ e \rightleftharpoons 2\ Cr^{3+} + 7\ H_2O$$

$E_{Zn,Zn^{2+}}$ would represent the potential for the half-reaction

$$Zn \rightleftharpoons Zn^{2+} + 2\ e$$

and $E_{I_3^-,I^-}$ would stand for the potential of the half-reaction

$$I_3^- + 2\ e \rightleftharpoons 3\ I^-$$

Although the normal hydrogen electrode is the international standard, it is seldom used in practice because it is cumbersome as well as hazardous. Instead, secondary standard electrodes (whose potentials have been previously measured against the hydrogen electrode) are employed. Two especially good secondary standards are the saturated calomel electrode and the silver chloride-silver metal electrode.

Figure 8–6 shows the relative positions on the scale of standard potentials of the chlorine gas-chloride ion, saturated calomel, titanium(IV)-titanium(III), hydrogen ion-hydrogen gas, lead(II)-lead metal, and zinc(II)-zinc metal electrodes. It can be seen that the standard potentials for the chlorine gas-chloride ion, titanium(IV)-titanium(III), lead(II)-lead metal, and zinc(II)-zinc metal electrodes are +1.3595, +0.10, -0.126, and

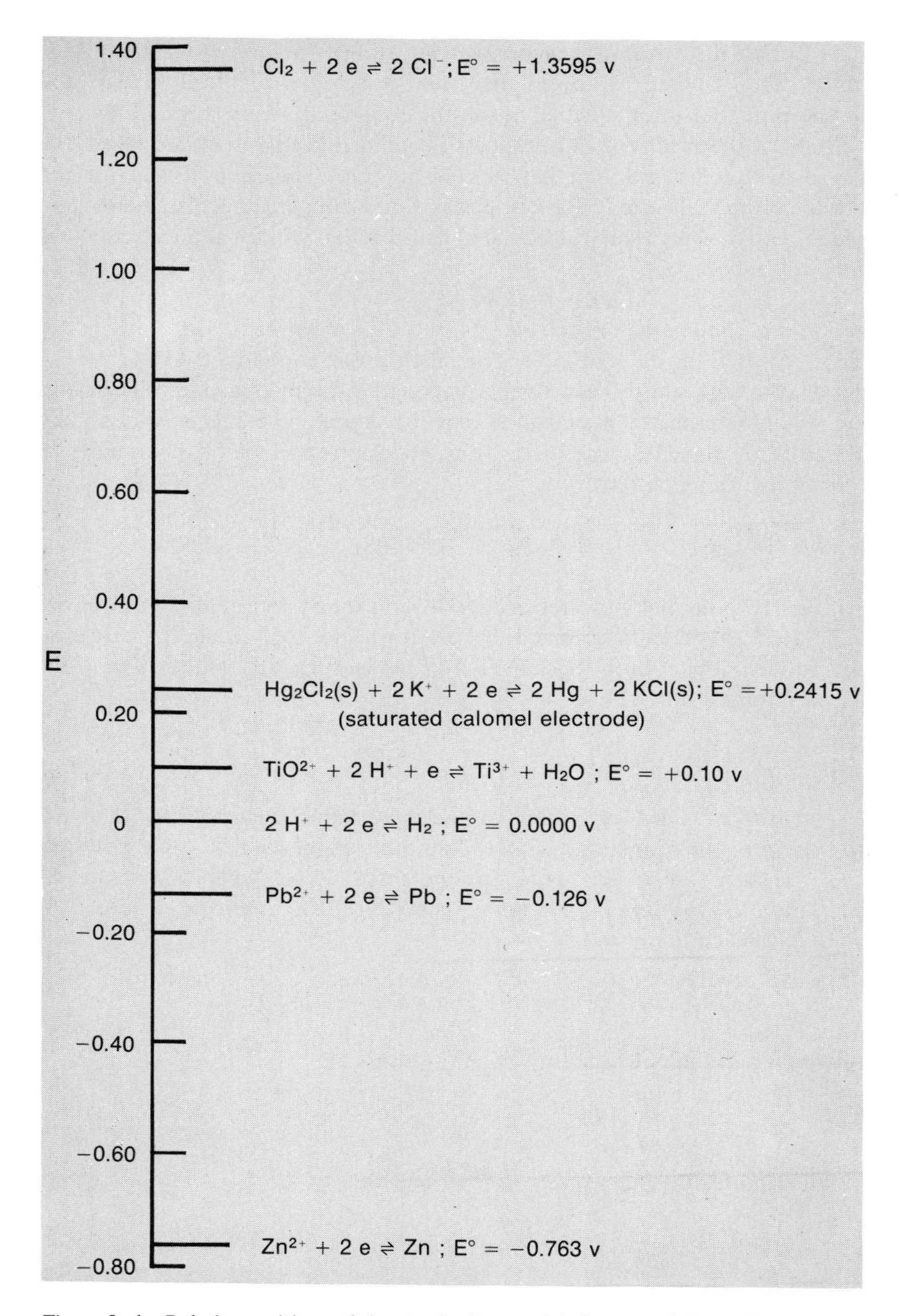


*Figure 8–6.* Relative positions of the standard potentials for several electrode systems.

−0.763 v, respectively, relative to the normal hydrogen electrode. However, the same four electrode systems have standard potentials of +1.1180, −0.14, −0.368, and −1.005 v with respect to the saturated calomel electrode – these values being 0.2415 v *more negative* than the standard potentials on the hydrogen-electrode scale. Thus, if the potential of some electrode is measured against the saturated calomel electrode, it is only necessary to *add* 0.2415 v to obtain the electrode potential on the hydrogen scale.

Note that the emf of the galvanic cell composed of any pair of electrode systems included in Figure 8–6 is the vertical distance expressed in volts between the two electrodes of interest. A cell consisting of chlorine gas-chloride ion and zinc(II)-zinc metal electrodes has an emf under standard-state conditions of 1.3595 – (–0.763) or 2.123 v; the cell constructed from saturated calomel and titanium(IV)-titanium(III) electrodes should exhibit an emf of 0.2415 – (0.10) or 0.14 v; and a cell composed of the lead(II)-lead metal and zinc(II)-zinc metal electrodes would have an emf of –0.126 – (–0.763) or 0.637 v. In addition, the positive electrode in each of these galvanic cells is the one for which the corresponding reduction half-reaction has the more positive potential. Thus, lead metal is the positive electrode in a cell consisting of lead(II)-lead and zinc(II)-zinc metal electrodes.

**Formal Potentials.** Frequently, the use of standard potentials to make chemical and thermodynamic predictions is unjustified and may even lead to serious error. This is because aqueous solutions of analytical interest are usually so concentrated that activity coefficients and activities of species involved in half-reactions cannot be determined. Another problem is that the simple ionic and molecular species written in half-reactions do not necessarily represent the true state of affairs in aqueous solution. For example, the entry

$$Fe^{3+} + e \rightleftharpoons Fe^{2+}; \qquad E^0 = +0.771 \text{ v}$$

in Table 8–1 applies to an aqueous solution in which the activities of the species $Fe(H_2O)_6{}^{3+}$ and $Fe(H_2O)_6{}^{2+}$ are both unity. However, it is virtually impossible to prepare an aqueous solution containing just these species, so the validity of using this standard potential may be questioned. Most cations form complexes with anions present, and some highly charged cations also form polymeric species. In a 1 $F$ hydrochloric acid medium, the species of iron(III) include $Fe(H_2O)_6{}^{3+}$, $Fe(H_2O)_5OH^{2+}$, $Fe(H_2O)_3Cl_3$, $Fe(H_2O)_2Cl_4{}^-$, $Fe(H_2O)_4Cl_2{}^+$, and $Fe_2(H_2O)_8(OH)_2{}^{4+}$, whereas $Fe(H_2O)_6{}^{2+}$, $Fe(H_2O)_5Cl^+$, and $Fe(H_2O)_4Cl_2$ are likely forms of iron(II). Unfortunately, the identities of the various species and the equilibria and equilibrium constants which describe this and most other systems are difficult to ascertain.

To overcome these uncertainties, the use of formal potentials has been suggested. A **formal potential** is defined as the potential (relative to the standard hydrogen electrode) of an electrode involving the half-reaction of interest in a specified electrolyte solution when the *formal concentrations* of the oxidant and reductant are both unity. A formal potential is usually denoted by the symbol $E^{0\prime}$. As an example, the formal potential for the iron(III)-iron(II) half-reaction is +0.70 v in 1 $F$ hydrochloric acid, which differs significantly from the standard potential of +0.771 v for the iron(III)-iron(II) couple.

Some formal potentials are included in Table 8–1, as well as in Appendix 4; the specific electrolyte solutions to which these data pertain, *e.g.*, 2 $F$ hydrochloric acid or 0.1 $F$ potassium chloride, are indicated in parentheses following the half-reactions. When available, a formal potential should be used for thermodynamic calculations *if experimental conditions correspond closely* to those used in the original measurement of the formal potential.

## THE NERNST EQUATION

For a chemical reaction, the driving force, and thus the emf, depends on the activities (concentrations) of reactants and products. For the generalized half-reaction

$$aA + ne \rightleftharpoons bB$$

the expression relating emf to the concentrations of the reactants and products is the familiar **Nernst equation**, which can be written in simplified form as

$$E = E^0 - \frac{0.05915}{n} \log \frac{[B]^b}{[A]^a}$$

In addition, the Nernst equation can be formulated for an overall cell reaction

$$aA + bB \rightleftharpoons cC + dD$$

as

$$E = E^0 - \frac{0.05915}{n} \log \frac{[C]^c[D]^d}{[A]^a[B]^b}$$

For most calculations the numerical constant 0.05915, valid only for 25°C, is shortened to 0.059.

When we utilize the latter form of the Nernst equation, molar concentrations of dissolved ionic and molecular species are employed, pressure in atmospheres is used for gases, the concentration of water is taken to be unity, and concentrations of pure solids are assumed to be unity.

*Example 8–4.* Write Nernst equations for each of the following half-reactions:

(1)

$$2\,H^+ + 2\,e \rightleftharpoons H_2$$

$$E_{H^+,H_2} = E^0_{H^+,H_2} - \frac{0.059}{n} \log \frac{p_{H_2}}{[H^+]^2} = 0.0000 - \frac{0.059}{2} \log \frac{p_{H_2}}{[H^+]^2}$$

(2)

$$Cr_2O_7^{2-} + 14\,H^+ + 6\,e \rightleftharpoons 2\,Cr^{3+} + 7\,H_2O$$

$$E_{Cr_2O_7^{2-},Cr^{3+}} = E^0_{Cr_2O_7^{2-},Cr^{3+}} - \frac{0.059}{n} \log \frac{[Cr^{3+}]^2}{[Cr_2O_7^{2-}][H^+]^{14}}$$

$$= +1.33 - \frac{0.059}{6} \log \frac{[Cr^{3+}]^2}{[Cr_2O_7^{2-}][H^+]^{14}}$$

(3)

$$MnO_4^- + 8\,H^+ + 5\,e \rightleftharpoons Mn^{2+} + 4\,H_2O$$

$$E_{MnO_4^-,Mn^{2+}} = E^0_{MnO_4^-,Mn^{2+}} - \frac{0.059}{n} \log \frac{[Mn^{2+}]}{[MnO_4^-][H^+]^8}$$

$$= +1.51 - \frac{0.059}{5} \log \frac{[Mn^{2+}]}{[MnO_4^-][H^+]^8}$$

Since the potential of a half-reaction is independent of the quantity of chemical reaction, each of the preceding relations could be multiplied or divided by an integer without altering the value of $E$ obtained from the Nernst equation. For example, the dichromate-chromic ion half-reaction could be written as

$$\tfrac{1}{2}\,Cr_2O_7^{2-} + 7\,H^+ + 3\,e \rightleftharpoons Cr^{3+} + 3\tfrac{1}{2}\,H_2O$$

and the corresponding Nernst equation would be

$$E_{Cr_2O_7^{2-},Cr^{3+}} = E^0_{Cr_2O_7^{2-},Cr^{3+}} - \frac{0.059}{n} \log \frac{[Cr^{3+}]}{[Cr_2O_7^{2-}]^{1/2}[H^+]^7}$$

$$= +1.33 - \frac{0.059}{3} \log \frac{[Cr^{3+}]}{[Cr_2O_7^{2-}]^{1/2}[H^+]^7}$$

Sometimes it is necessary or desirable to write a half-reaction as an oxidation rather than a reduction. If we are interested in calculating $E$ for the half-reaction

$$Mn^{2+} + 4\,H_2O \rightleftharpoons MnO_4^- + 8\,H^+ + 5\,e$$

the proper form for the Nernst equation is

$$E_{Mn^{2+},MnO_4^-} = E^0_{Mn^{2+},MnO_4^-} - \frac{0.059}{n} \log \frac{[MnO_4^-][H^+]^8}{[Mn^{2+}]}$$

$$= -1.51 - \frac{0.059}{5} \log \frac{[MnO_4^-][H^+]^8}{[Mn^{2+}]}$$

Notice that the order of the subscripts for $E$ and $E^0$ has been reversed to show the direction of the half-reaction being considered. Also, the sign of $E^0$ has been changed and the concentration terms in the logarithmic part of the equation have been inverted.

## APPLICATIONS OF STANDARD POTENTIALS AND THE NERNST EQUATION

This section of the chapter is concerned with specific examples of the various types of thermodynamic calculations which make use of standard-potential data and the Nernst equation.

**Calculation of the Electromotive Force for an Overall Cell Reaction.** Since knowledge of the emf for an overall cell reaction is often desirable, we shall first consider how to calculate this quantity.

*Example 8–5.* Calculate the emf for the overall reaction corresponding to the galvanic cell

$$Pb \mid Pb^{2+}\ (a = 1\ M) \parallel Cr^{2+}\ (a = 1\ M),\ Cr^{3+}\ (a = 1\ M) \mid Pt$$

First, we should write the overall cell reaction according to the rule established earlier:

$$Pb + 2\,Cr^{3+} \rightleftharpoons Pb^{2+} + 2\,Cr^{2+}$$

This overall reaction can be viewed in terms of its component half-reactions,

$$Pb \rightleftharpoons Pb^{2+} + 2\,e$$

and

$$2\,Cr^{3+} + 2\,e \rightleftharpoons 2\,Cr^{2+}$$

Since the activities of all species which comprise the galvanic cell are unity, the value for the potential of each half-reaction may be obtained directly from Table 8–1. Thus, we can write

$$E_{Pb,Pb^{2+}} = E^0_{Pb,Pb^{2+}} = +0.126 \text{ v}$$

and

$$E_{Cr^{3+},Cr^{2+}} = E^0_{Cr^{3+},Cr^{2+}} = -0.41 \text{ v}$$

Although we are considering two units of the chromium(III)-chromium(II) half-reaction, we do *not* multiply $E^0_{Cr^{3+},Cr^{2+}}$ by a factor of two – emf or potential is an intensity factor and does not depend on the quantity of reaction.

Now, if we combine two half-reactions to obtain an overall reaction, what relationships exist between the properties of the half-reactions and the properties of the net reaction? It is rigorously true that the free-energy change for the overall reaction is equal to the *sum* of the free-energy changes for the two individual half-reactions:

$$\Delta G_{overall} = \Delta G_{Pb,Pb^{2+}} + \Delta G_{Cr^{3+},Cr^{2+}}$$

However, since

$$\Delta G = -nFE$$

we can write

$$-nFE_{overall} = -nFE_{Pb,Pb^{2+}} - nFE_{Cr^{3+},Cr^{2+}} = -nFE^0_{Pb,Pb^{2+}} - nFE^0_{Cr^{3+},Cr^{2+}}$$

Inspection of the two half-reactions, which when added give the overall cell reaction, shows that each involves *two electrons*, since the electrons must cancel to produce the net reaction. In addition, the overall cell reaction must itself involve *two electrons*. Accordingly, the value of $n$ is the *same* for all terms of the preceding equation. Dividing each term by the common factor $-nF$, we obtain

$$E_{overall} = E^0_{Pb,Pb^{2+}} + E^0_{Cr^{3+},Cr^{2+}} = +0.126 - 0.41 = -0.28 \text{ v}$$

Although this result is indistinguishable from that obtained by straightforward addition of the potentials for the two half-reactions, in general it is not permissible to add *intensive* thermodynamic properties such as potential or emf. Only because $n$ happens to be identical for the overall cell reaction and the two half-reactions does the sum of the potentials yield the same result as the sum of the free-energy changes.

*Example 8–6.* Calculate the emf for the overall reaction pertaining to the galvanic cell

$$Cu \mid Cu^{2+} (0.01\ M) \parallel Fe^{2+} (0.1\ M), Fe^{3+} (0.02\ M) \mid Pt$$

First, the overall cell reaction and two half-reactions are as follows:

$$Cu + 2\ Fe^{3+} \rightleftharpoons Cu^{2+} + 2\ Fe^{2+}$$

$$Cu \rightleftharpoons Cu^{2+} + 2\ e$$

$$2\ Fe^{3+} + 2\ e \rightleftharpoons 2\ Fe^{2+}$$

Again, the overall reaction and the half-reactions involve the same number of electrons, so

$$E_{\text{overall}} = E_{\text{Cu,Cu}^{2+}} + E_{\text{Fe}^{3+},\text{Fe}^{2+}}$$

For the present example, the concentrations of the various ions differ from unity, and so it is necessary to use the Nernst equation to calculate the potential for each half-reaction:

$$E_{\text{Cu,Cu}^{2+}} = E^0_{\text{Cu,Cu}^{2+}} - \frac{0.059}{n} \log [\text{Cu}^{2+}] = -0.337 - \frac{0.059}{2} \log (0.01)$$

$$= -0.337 - \frac{0.059}{2} (-2) = -0.278 \text{ v}$$

$$E_{\text{Fe}^{3+},\text{Fe}^{2+}} = E^0_{\text{Fe}^{3+},\text{Fe}^{2+}} - \frac{0.059}{n} \log \frac{[\text{Fe}^{2+}]}{[\text{Fe}^{3+}]}$$

$$= +0.771 - \frac{0.059}{1} \log \frac{(0.1)}{(0.02)} = +0.771 - 0.059 \log 5$$

$$= +0.771 - 0.059\,(0.70) = +0.730 \text{ v}$$

$$E_{\text{overall}} = -0.278 + 0.730 = +0.452 \text{ v}$$

**Calculation of Equilibrium Constants.** One of the most important applications of emf measurements is the evaluation of equilibrium constants for chemical reactions.

A relation between the equilibrium constant and the emf for an overall reaction can be derived. Recall from Chapter 3 that the standard free-energy change for a reaction is related to the equilibrium constant (written in terms of *concentrations*) by the equation

$$\Delta G^0 = -RT \ln K$$

In addition, the standard free-energy change and the emf for a reaction under standard-state conditions may be equated through the expression

$$\Delta G^0 = -nFE^0$$

By combining these two equations, we obtain

$$RT \ln K = nFE^0$$

If we convert this relation for use with base-ten logarithms and substitute appropriate values for the various physical constants ($F$ = 23,060 calories per faraday, $R$ = 1.987 calories mole$^{-1}$ degree$^{-1}$, and $T$ = 298°K), the following result is obtained:

$$\log K = \frac{nE^0}{0.059}$$

This equation is valid only for 25°C or 298°K. It should be noted that $E^0$ is the emf for the reaction of interest under *standard-state conditions*; this value can be easily determined according to the procedure developed in the two preceding examples.

*Example 8–7.* Compute the equilibrium constant for the reaction

$$\text{HAsO}_2 + \text{I}_3^- + 2\,\text{H}_2\text{O} \rightleftharpoons \text{H}_3\text{AsO}_4 + 2\,\text{H}^+ + 3\,\text{I}^-$$

We can express the equilibrium constant as

$$K = \frac{[H_3AsO_4][H^+]^2[I^-]^3}{[HAsO_2][I_3^-]}$$

Next, we can show that the overall reaction is composed of the half-reactions

$$HAsO_2 + 2\,H_2O \rightleftharpoons H_3AsO_4 + 2\,H^+ + 2\,e$$

and

$$I_3^- + 2\,e \rightleftharpoons 3\,I^-$$

which appear side by side in Table 8–1. In accord with previous calculations, we can write

$$E^0_{\text{overall}} = E^0_{\text{HAsO}_2,\text{H}_3\text{AsO}_4} + E^0_{\text{I}_3^-,\text{I}^-} = -0.559 + 0.536 = -0.023\text{ v}$$

If this value is substituted into the expression for log $K$ along with $n = 2$, we find that

$$\log K = \frac{nE^0}{0.059} = \frac{(2)(-0.023)}{0.059} = -0.78; \qquad K = 0.17$$

*Example 8–8.* From standard-potential data, calculate the solubility-product constant for silver chloride, which dissolves according to the reaction

$$AgCl \rightleftharpoons Ag^+ + Cl^-$$

A search of Table 8–1 shows that the two half-reactions

$$Ag \rightleftharpoons Ag^+ + e$$

and

$$AgCl + e \rightleftharpoons Ag + Cl^-$$

can be added together to yield the desired overall reaction. It follows that

$$E^0_{\text{overall}} = E^0_{\text{Ag},\text{Ag}^+} + E^0_{\text{AgCl},\text{Ag}} = -0.7995 + 0.2222 = -0.5773\text{ v}$$

Substituting this latter value for $E^0_{\text{overall}}$ into our equation for log $K$, we obtain

$$\log K = \frac{nE^0}{0.05915} = \frac{(1)(-0.5773)}{0.05915} = -9.760$$

$$K = K_{sp} = 1.74 \times 10^{-10}$$

Using this approach, we can calculate equilibrium constants for redox reactions, solubility products, formation constants for complex ions, and dissociation constants for acids and bases. All we need do is represent the desired reaction as the sum of appropriate half-reactions. If standard-potential data are available, the calculations can be performed as in the two preceding examples.

## THE CERIUM(IV)-IRON(II) TITRATION

One of the classic oxidation-reduction or redox procedures is the determination of iron(II) by titration with a standard solution of cerium(IV):

$$Fe^{2+} + Ce^{4+} \rightleftharpoons Fe^{3+} + Ce^{3+}$$

Usually, the titration of iron(II) is performed in a 0.5 to 2 $F$ sulfuric acid medium, and the titrant consists of a solution of cerium(IV) in 1 $F$ sulfuric acid. Although the preceding simple representation implies the reaction between $Fe(H_2O)_6{}^{2+}$ and $Ce(H_2O)_6{}^{4+}$ to form $Fe(H_2O)_6{}^{3+}$ and $Ce(H_2O)_6{}^{3+}$, the chemistry of both cerium(IV) and iron(II) in aqueous sulfuric acid is modified by formation of complexes such as $Ce(H_2O)_5SO_4{}^{2+}$ and $Fe(H_2O)_4(SO_4)_2{}^{2-}$.

In the present discussion, we shall consider the cerium(IV)-iron(II) titration in a 1 $F$ sulfuric acid medium. Therefore, it is appropriate to utilize the *formal* potentials for the iron(III)-iron(II) and cerium(IV)-cerium(III) half-reactions in 1 $F$ sulfuric acid, which are listed in Table 8–1 as

$$Ce^{4+} + e \rightleftharpoons Ce^{3+}; \qquad E^{0\prime} = +1.44 \text{ v}$$

and

$$Fe^{3+} + e \rightleftharpoons Fe^{2+}; \qquad E^{0\prime} = +0.68 \text{ v}$$

If the equilibrium constant for the titration reaction is computed, the result is

$$\log K = \frac{nE^{0\prime}}{0.059} = \frac{(1)(-0.68 + 1.44)}{0.059} = +12.9; \qquad K = 8 \times 10^{12}$$

Such a large equilibrium constant indicates that the reaction between iron(II) and cerium(IV) is strongly favored and that the quantitative determination of iron(II) should be successful. However, a thermodynamically favorable reaction may not proceed at a conveniently rapid rate. Fortunately, the iron(II)-cerium(IV) reaction is fast, partly because it involves the transfer of only one electron.

**Construction of the Titration Curve.** Calculation and construction of a titration curve provide insight into the probable success of the proposed titration and give information which aids in the selection of a method of end-point detection. We shall discuss the titration of 25.00 ml of 0.1000 $M$ iron(II) with 0.1000 $M$ cerium(IV) in a 1 $F$ sulfuric acid medium.

According to the Nernst equation, the potential of an electrode depends on the concentrations (activities) of reducing and oxidizing agents in solution. Therefore, the variation of potential as a function of the volume of added titrant serves to show the progress of a redox titration. In order to observe continuously the variation of potential during the course of the titration, it is necessary to make potentiometric measurements of the emf of a galvanic cell which consists of a reference electrode and an indicator electrode. A **reference electrode** is one whose potential is fixed with respect to the standard hydrogen electrode. In principle, the reference electrode may be the standard hydrogen electrode, but the saturated calomel electrode is usually employed. An **indicator electrode** is one whose potential varies in accord with the Nernst equation as the concentrations of reactant and product change during a titration. For example, a platinum wire dipped into an iron(II)-iron(III) mixture in 1 $F$ sulfuric acid exhibits a potential which depends on the ratio of concentrations (activities) of iron(II) and iron(III). When a reference electrode and indicator electrode are combined to make a galvanic cell, any change in the overall cell emf is due solely to a change in the potential of the indicator electrode, which reflects in turn a variation in the concentrations of oxidized and reduced species. For convenience, we shall take the standard hydrogen electrode (with a potential defined to be zero) as the reference electrode and a platinum wire electrode as the indicator electrode. Then the emf of the complete galvanic cell consisting of these two electrodes will be identical in magnitude to the potential of the indicator electrode alone.

It should be recalled that the *sign* and *magnitude* of the potential of an electrode in a galvanic cell do not vary if the bench-top orientation of a cell is altered, whereas the potential of a half-reaction as computed from the Nernst equation is a thermodynamic quantity and does depend on the direction in which the half-reaction is written. For the purpose of constructing a redox titration curve, *the potential for the half-reaction of interest, when it is written as a reduction process, is always identical in both sign and magnitude to the potential of the indicator electrode versus the standard hydrogen electrode*. Therefore, as we construct the titration curve, we shall apply the Nernst equation specifically to half-reactions written as reductions.

**At the Start of the Titration.** It is not possible to predict what the potential of the indicator electrode is at the beginning of the titration, because the iron(II)-iron(III) concentration ratio is unknown. Although the sample solution is specified to contain only iron(II), a very tiny amount of iron(III) will inevitably exist because dissolved atmospheric oxygen slowly oxidizes iron(II). Also, in the determination of iron(II) by titration with a standard solution of cerium(IV) or some other oxidant, the original sample invariably consists either partly or wholly of iron(III). Prior to the actual titration, the iron(III) must be converted quantitatively to iron(II) with the aid of some reducing agent. Regardless of the strength of this reducing agent, at least a trace of unreduced iron(III) will remain. Under the usual titration conditions, the quantity of iron(III) is too small to cause a significant error in the determination.

**Prior to the Equivalence Point.** As the first definite point on the titration curve, let us calculate the potential of the platinum indicator electrode after the addition of 1.00 ml of 0.1000 $M$ cerium(IV) titrant to the 25.00-ml sample of 0.1000 $M$ iron(II) in 1 $F$ sulfuric acid.

Initially, 2.500 millimoles of iron(II) are present. Addition of 1.00 ml of 0.1000 $M$ cerium(IV), or 0.100 millimole, results in the formation of 0.100 millimole of iron(III) and 0.100 millimole of cerium(III) as reaction products, and 2.400 millimoles of iron(II) remain. These conclusions follow from the stoichiometry of the titration reaction, because the equilibrium constant is so large ($K = 8 \times 10^{12}$) that the reaction between cerium(IV) and iron(II) is essentially complete. For the same reason, the quantity of cerium(IV) titrant remaining unreacted is extremely small, but can be computed from the known quantities of cerium(III), iron(II), and iron(III) and the equilibrium constant. It should be emphasized that the potential of the platinum indicator electrode can have only one value which is governed in principle by either the iron(II)-iron(III) concentration ratio or the cerium(III)-cerium(IV) concentration ratio. However, the iron(II)-iron(III) ratio prior to the equivalence point can be easily determined from the reaction stoichiometry, whereas the determination of the cerium(III)-cerium(IV) ratio requires a calculation involving the use of the equilibrium expression. Therefore, it is more straightforward to employ the Nernst equation for the iron(III)-iron(II) half-reaction

$$E_{Fe^{3+},Fe^{2+}} = E^{0\prime}_{Fe^{3+},Fe^{2+}} - \frac{0.059}{n} \log \frac{[Fe^{2+}]}{[Fe^{3+}]}$$

for all points on the titration curve *prior* to the equivalence point. Since the logarithmic term involves the ratio of two concentrations, each raised to the first power, it is not necessary to convert the quantities of iron(II) and iron(III) expressed in millimoles into concentration units. After the addition of 1.00 ml of cerium(IV) titrant, there are 0.100 millimole of iron(III) and 2.400 millimoles of iron(II) present. Thus,

$$E_{Fe^{3+},Fe^{2+}} = +0.68 - 0.059 \log \frac{(2.400)}{(0.100)} = +0.68 - 0.059 \log 24$$

$$= +0.68 - 0.08 = +0.60 \text{ v}$$

which is the potential of the platinum indicator electrode with respect to the standard hydrogen electrode.

Similar calculations were performed for other titrant volumes prior to the equivalence point up to 24.99 ml, the results of which are listed in Table 8–2. Even at a point on the titration curve as close to the equivalence point as 24.99 ml, it is valid to determine the quantities of iron(III) and iron(II) from the simple reaction stoichiometry. For points much closer to the equivalence point, it is necessary to use the equilibrium expression for the titration reaction to obtain accurate values for the concentrations of iron(III) and iron(II).

**At the Equivalence Point.** To calculate the potential of the platinum indicator electrode at the equivalence point, recall that the potential can be viewed as being determined by either the iron(III)-iron(II) or the cerium(IV)-cerium(III) half-reaction. Let us symbolize the potential for the indicator electrode as well as for each half-reaction (written as a reduction) by $E_{ep}$ at the equivalence point. Then the Nernst equation for each half-reaction may be written:

$$E_{ep} = E^{0\prime}_{Fe^{3+},Fe^{2+}} - 0.059 \log \frac{[Fe^{2+}]}{[Fe^{3+}]}$$

$$E_{ep} = E^{0\prime}_{Ce^{4+},Ce^{3+}} - 0.059 \log \frac{[Ce^{3+}]}{[Ce^{4+}]}$$

**Table 8–2. Variation of Potential of a Platinum Indicator Electrode During Titration of 25.00 ml of 0.1000 *M* Iron(II) with 0.1000 *M* Cerium(IV)**

| Volume of Cerium(IV), ml | E *vs.* NHE |
|---|---|
| 0 | — |
| 1.00 | +0.60 |
| 2.00 | 0.62 |
| 5.00 | 0.64 |
| 10.00 | 0.67 |
| 12.50 | 0.68 |
| 15.00 | 0.69 |
| 20.00 | 0.72 |
| 24.00 | 0.76 |
| 24.50 | 0.78 |
| 24.90 | 0.82 |
| 24.99 | 0.88 |
| 25.00 | 1.06 |
| 26.00 | 1.36 |
| 30.00 | 1.40 |
| 40.00 | 1.43 |
| 50.00 | 1.44 |

An inspection of the titration reaction

$$Fe^{2+} + Ce^{4+} \rightleftharpoons Fe^{3+} + Ce^{3+}$$

provides us with two useful relations. By definition, the *equivalence point* of any titration is the point at which the quantity of added titrant is stoichiometrically equivalent to the substance being titrated. Since we are titrating 2.500 millimoles of iron(II), the equivalence point corresponds to the addition of 2.500 millimoles of cerium(IV). No reaction is ever 100 per cent complete, but the iron(II)-cerium(IV) reaction has a large equilibrium constant and does attain a high degree of completion. At the equivalence point, the small concentration of unreacted iron(II) is equal to the small concentration of unreacted cerium(IV):

$$[Fe^{2+}] = [Ce^{4+}]$$

Similarly, the concentrations of the products, iron(III) and cerium(III), are equal:

$$[Fe^{3+}] = [Ce^{3+}]$$

If we rewrite the first of the preceding two Nernst equations,

$$E_{ep} = E^{0\prime}_{Fe^{3+},Fe^{2+}} - 0.059 \log \frac{[Fe^{2+}]}{[Fe^{3+}]}$$

substitute the concentration relationships just derived into the second Nernst equation,

$$E_{ep} = E^{0\prime}_{Ce^{4+},Ce^{3+}} - 0.059 \log \frac{[Fe^{3+}]}{[Fe^{2+}]}$$

and add these two equations together, the result is

$$2E_{ep} = E^{0\prime}_{Fe^{3+},Fe^{2+}} + E^{0\prime}_{Ce^{4+},Ce^{3+}} - 0.059 \log \frac{[Fe^{2+}]}{[Fe^{3+}]} - 0.059 \log \frac{[Fe^{3+}]}{[Fe^{2+}]}$$

However, the two logarithmic terms cancel, so at the equivalence point the potential of the platinum indicator electrode with respect to the standard hydrogen electrode is

$$E_{ep} = \frac{E^{0\prime}_{Fe^{3+},Fe^{2+}} + E^{0\prime}_{Ce^{4+},Ce^{3+}}}{2} = \frac{0.68 + 1.44}{2} = +1.06 \text{ v}$$

In the present case, the potential at the equivalence point is the simple average of the formal potentials for the two half-reactions. However, in general the equivalence-point potential is a *weighted* average of the formal or standard potentials for the two half-reactions involved in the titration. In some situations the potential at the equivalence point may depend on the pH and on the concentration of one or more of the reactants or products.

**Beyond the Equivalence Point.** Points on the titration curve after the equivalence point can be most readily determined from the Nernst equation for the cerium(IV)-cerium(III) reaction. It is reasonable to assert that 2.500 millimoles of cerium(III) is present at and beyond the equivalence point, because this is the quantity formed upon oxidation of 2.500 millimoles of iron(II) by cerium(IV). Addition of cerium(IV) after the equivalence point simply introduces excess unreacted cerium(IV). For example, after the addition of a *total* of 26.00 ml of 0.1000 $M$ cerium(IV) titrant, or 1.00 ml in excess of the volume needed to reach the equivalence point, the solution contains 2.500 millimoles of cerium(III) and 0.100 millimole of cerium(IV). Thus we can write

$$E_{Ce^{4+},Ce^{3+}} = E^{0\prime}_{Ce^{4+},Ce^{3+}} - \frac{0.059}{n} \log \frac{[Ce^{3+}]}{[Ce^{4+}]}$$

$$E_{Ce^{4+},Ce^{3+}} = +1.44 - 0.059 \log \frac{(2.500)}{(0.100)} = +1.44 - 0.059 \log 25$$

$$= +1.44 - 0.08 = +1.36 \text{ v}$$

Results of other calculations of the potential of the indicator electrode beyond the equivalence point are included in Table 8–2.

After the addition of 50.00 ml of cerium(IV) titrant, $E_{Ce^{4+},Ce^{3+}} = +1.44$ v. This is a special point on the titration curve, for the quantities of cerium(III) and cerium(IV) present are equal. Thus, the potential of the indicator electrode is identical to the formal potential for the cerium(IV)-cerium(III) half-reaction. There is another special point on the titration curve which occurs halfway to the equivalence point. After the addition of exactly 12.50 ml of cerium(IV) titrant, the quantity of iron(III) produced and the amount of untitrated iron(II) are the same, so the potential of the indicator electrode is identical to the formal potential (+0.68 v) for the iron(III)-iron(II) couple in 1 $F$ sulfuric acid. In Figure 8–7 is plotted the complete titration curve for the iron(II)-cerium(IV) reaction.

## OXIDATION-REDUCTION INDICATORS

There are many ways in which the end point of an oxidation-reduction titration can be detected. As the well-defined titration curve of Figure 8–7 suggests, one approach is to record the complete titration curve. Then the end point can be taken as the midpoint of the steeply rising portion of the curve. Any titration which is followed by measurement of the potential of an indicator electrode as a function of added titrant is termed a **potentiometric titration**. Potentiometric titrations will be discussed in detail in Chapter 10.

A few titrants have such intense colors that they may serve as their own end-point indicators. For example, even a small amount of potassium permanganate imparts a distinct purple tinge to an otherwise colorless solution, so the appearance of the color due to the first slight excess of permanganate in a titration vessel marks the end point of the titration. However, this method of end-point detection may not be applicable if the sample solution contains other highly colored substances.

For some titrations, it is possible to add a species to the sample solution which reacts either with the substance being titrated or with the titrant to produce a sharp color change at the end point. Thus, starch is frequently used as an indicator in titrations involving iodine. Starch forms an intensely colored, dark blue complex with iodine which can serve to indicate the disappearance of the last trace of iodine in the titration of iodine with thiosulfate or the appearance of the first small excess of iodine in the titration of arsenic(III) with triiodide solution.

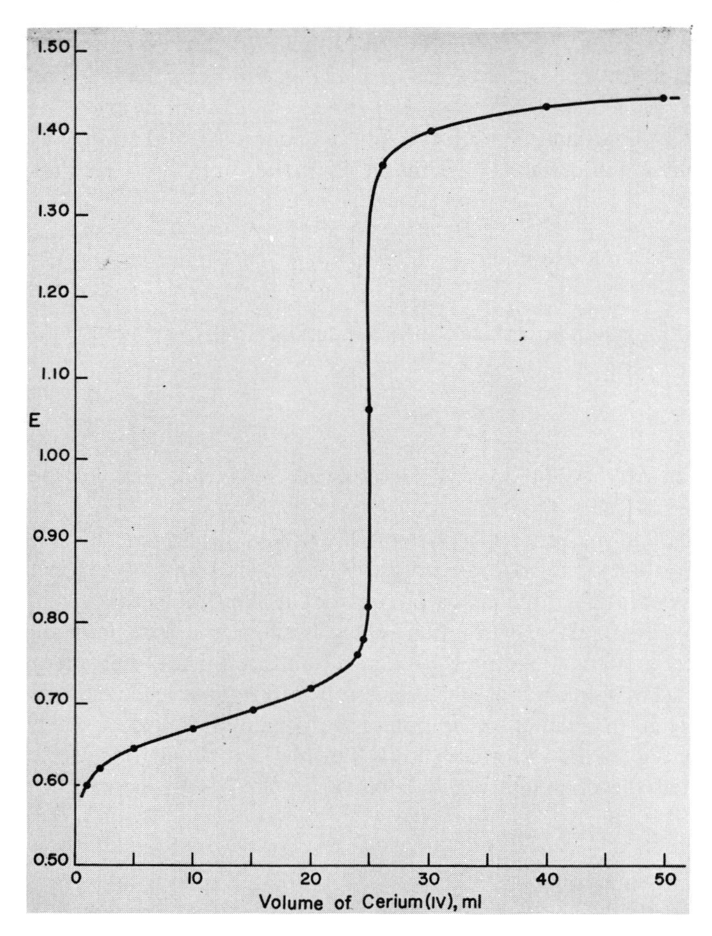


*Figure 8–7.* Titration curve for the titration of 25.00 ml of 0.1000 *M* iron(II) with 0.1000 *M* cerium(IV) in a 1 *F* sulfuric acid medium.

Redox indicators comprise another class of end-point indicators for oxidation-reduction titrations. A **redox indicator** is a substance which can undergo an oxidation-reduction reaction and whose oxidized and reduced forms differ in color. We may represent the behavior of a redox indicator by the half-reaction

$$\mathrm{Ox} + n\mathrm{e} \rightleftharpoons \mathrm{Red}$$

where Ox and Red are the oxidized and reduced forms of the indicator, respectively, and the Nernst equation corresponding to the half-reaction is

$$E_{\mathrm{Ox,Red}} = E^0_{\mathrm{Ox,Red}} - \frac{0.059}{n} \log \frac{[\mathrm{Red}]}{[\mathrm{Ox}]}$$

where $E^0_{\mathrm{Ox,Red}}$ is the standard potential for the indicator half-reaction. As in the case of acid-base indicators, a redox indicator will appear to be the color of the *oxidized* form if [Ox] is at least 10 times greater than [Red]. However, the indicator will have the color of the *reduced* form if [Red] is 10 or more times greater than [Ox]. From the preceding Nernst equation, we can conclude that the color of the *oxidized* form will be seen when

$$E_{\mathrm{Ox,Red}} = E^0_{\mathrm{Ox,Red}} - \frac{0.059}{n} \log \frac{1}{10} = E^0_{\mathrm{Ox,Red}} + \frac{0.059}{n}$$

and that the color of the *reduced* form will be observed if

$$E_{\mathrm{Ox,Red}} = E^0_{\mathrm{Ox,Red}} - \frac{0.059}{n} \log 10 = E^0_{\mathrm{Ox,Red}} - \frac{0.059}{n}$$

At intermediate potentials, the indicator will appear to be a mixture of colors. Thus, a redox indicator requires a potential change of $2 \times (0.059/n)$ or a maximum of about 0.12 v to change from one color form to the other. Theoretically, the range over which this color change occurs is centered on the standard-potential value ($E^0_{\mathrm{Ox,Red}}$) of the indicator. For many redox indicators, the number of electrons ($n$) involved in the half-reaction is 2, so a potential change of only 0.059 v causes the indicator to change color.

In Table 8–3 are listed some common oxidation-reduction indicators along with their standard potentials and the colors of the oxidized and reduced forms. Oxidation-reduction reactions of most redox indicators involve hydrogen ions, and so the potentials at which these indicators change color depend on pH. Standard potentials listed in Table 8–3 pertain to a hydrogen ion concentration (activity) of 1 *M*.

Most frequently used as redox indicator for the iron(II)-cerium(IV) titration in 1 *F* sulfuric acid is tris(1,10-phenanthroline)iron(II) sulfate or **ferroin.** As shown in Figure 8–8, three 1,10-phenanthroline (or *o*-phenanthroline) ligands are coordinated to iron(II) through the lone electron pairs on the nitrogen atoms to form the octahedral indicator

**Table 8–3. Selected Oxidation-Reduction Indicators**

| Indicator | Color of Reduced Form | Color of Oxidized Form | Standard Potential, $E^0$ |
|---|---|---|---|
| Indigo monosulfate | colorless | blue | 0.26 |
| Methylene blue | colorless | blue | 0.36 |
| 1-naphthol-2-sulfonic acid indophenol | colorless | red | 0.54 |
| Diphenylamine | colorless | violet | 0.76 |
| Diphenylbenzidine | colorless | violet | 0.76 |
| Barium diphenylamine sulfonate | colorless | red-violet | 0.84 |
| Sodium diphenylbenzidine sulfonate | colorless | violet | 0.87 |
| Tris(2,2′-bipyridine)-iron(II) sulfate | red | pale blue | 0.97 |
| Erioglaucine A | green | red | 1.00 |
| Tris(5-methyl-1,10-phenanthroline)iron(II) sulfate (methyl ferroin) | red | pale blue | 1.02 |
| Tris(1,10-phenanthroline)-iron(II) sulfate (ferroin) | red | pale blue | 1.06 |
| N-phenylanthranilic acid | colorless | pink | 1.08 |
| Tris(5-nitro-1,10-phenanthroline)iron(II) sulfate (nitroferroin) | red | pale blue | 1.25 |
| Tris(2,2′-bipyridine)-ruthenium(II) nitrate | yellow | pale blue | 1.25 |

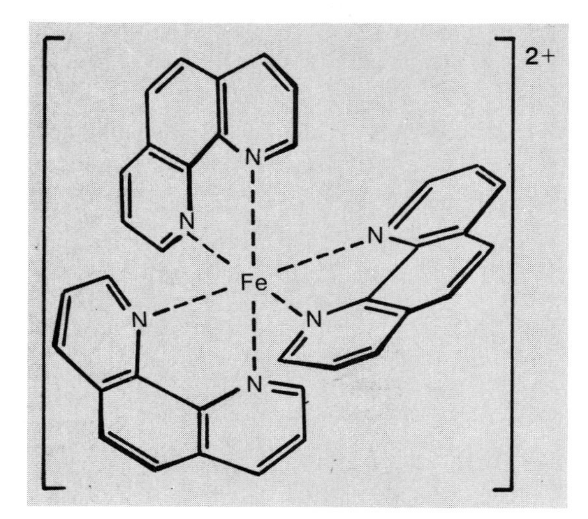


*Figure 8–8.* Structure of the tris(1,10-phenanthroline)iron(II) or ferroin redox indicator.

complex. In its reduced iron(II) form, the indicator has an intense red color, whereas the oxidized iron(III) indicator complex is pale blue.

In the titration of iron(II) with cerium(IV) in a sulfuric acid medium, the ferroin indicator is initially in its reduced, red form, and the end point is signaled by the sudden change to the pale blue color of the oxidized form. Actually, the combination of the blue color of the oxidized indicator and the yellow color of the first slight excess of cerium(IV) titrant imparts a green tinge to the solution at the end point.

Another family of redox indicators is derived from the colorless compound diphenylamine. In the presence of an oxidant, such as potassium dichromate, diphenylamine undergoes an irreversible oxidative coupling reaction to yield diphenylbenzidine, which is also colorless:

$$2\, C_6H_5{-}NH{-}C_6H_5 \rightarrow C_6H_5{-}NH{-}C_6H_4{-}C_6H_4{-}NH{-}C_6H_5 + 2\,H^+ + 2\,e$$

diphenylamine　　　diphenylbenzidine

However, upon further oxidation, diphenylbenzidine is converted to a bright purple compound, known as diphenylbenzidine violet, which owes its deep color to a system of

$$C_6H_5{-}N{=}C_6H_4{=}C_6H_4{=}N{-}C_6H_5$$

diphenylbenzidine violet

conjugated double bonds. This color is sufficiently intense to be easily discernible even in the presence of green-colored chromium(III), the reduction product of dichromate ion. Diphenylbenzidine and diphenylbenzidine violet are interconvertible, and it is this reversible reaction which provides the desired indicator behavior. Either diphenylamine or diphenylbenzidine may be added to the solution being titrated. Both diphenylamine and diphenylbenzidine suffer from the disadvantage that they are relatively insoluble in aqueous media. These substances can be rendered water soluble, however, if the readily ionizable sulfonic acid derivatives are prepared. Diphenylamine sulfonate and diphenylbenzidine sulfonate ions, usually in the form of barium or sodium salts, are soluble in acidic media and undergo the same color change as the unsulfonated species.

## FEASIBILITY OF TITRATIONS

A titration curve provides information concerning whether a titration is feasible and, if so, what method of end-point detection should be used. An important factor which determines the success of any proposed redox titration is the difference between the standard or formal potentials for the titrant and the substance being titrated. For example, the difference between the formal potentials in 1 *F* sulfuric acid for the cerium(IV)-cerium(III) couple ($E^{0\prime} = +1.44$ v) and the iron(III)-iron(II) couple ($E^{0\prime} = +0.68$ v) is large enough to yield a very well defined titration curve (Figure 8–7).

Let us ask the following question: In order to obtain a sufficiently well-defined titration curve, what is the minimum allowable difference in the standard or formal potentials for the titrant and the species to be titrated?

We can gain some insight into this matter by inspecting the titration curves shown in Figure 8–9. These curves show what happens when cerium(IV) is used to titrate reducing agents of various strengths. For our discussion let us assume, first, that the titrant is 0.1 *M* cerium(IV) and, second, that each of the different reductants is at an initial concentration of 0.1 *M* and also that each undergoes a one-electron oxidation. If we start with 50 ml of 0.1 *M* reducing agent, exactly 50 ml of 0.1 *M* cerium(IV) solution will be needed in each titration. We have calculated and plotted the titration curves of Figure 8–9 on the assumption that the solutions involved in these titrations are 1 *F* in sulfuric acid, and so the formal potential for the cerium(IV)-cerium(III) half-reaction ($E^{0\prime} = +1.44$ v versus NHE) is applicable. Standard potentials for the reductants being titrated are printed on the titration curves.

One fact emerges immediately from an inspection of Figure 8–9. That is, the magnitude of the emf change at the equivalence point of a titration is proportional to the *difference* between the formal potential for the cerium(IV)-cerium(III) half-reaction and the standard potential for the substance being titrated. If a species with a standard potential of +0.60 v (versus the hydrogen electrode) is titrated, the emf jumps at least 500 mv near the equivalence point. In the case of a substance with a standard potential of +1.00 v, the emf changes only about 150 mv near the equivalence point. With a species having a standard potential of +1.20 v, there is no very steep portion of the titration curve. Obviously, the size and steepness of the vertical portion of a titration curve govern the accuracy and the precision with which the equivalence point can be located, and thereby govern the feasibility of a titration itself. When the vertical portion is large and steep, as it is for the two bottom titration curves in Figure 8–9, the equivalence point can be easily determined. On the other hand, if the vertical portion is small and not sharply defined, as in the uppermost curve of Figure 8–9, the choice of colored redox indicator or other method for the location of the equivalence point is severely limited.

It is generally recognized that a redox titration can be successfully performed if the *difference* between the standard or formal potentials for the titrant and the substance being titrated is 0.20 v or greater. For differences between 0.20 and 0.40 v, it is preferable to follow the progress of a redox titration by means of potentiometric measurement techniques. On the other hand, when the difference between the formal or standard potentials is larger than about 0.40 v, the use of either colored redox indicators or instrumental methods for the location of the equivalence point gives good results.

Sometimes the shape of a titration curve and thus the feasibility of a titration can be markedly improved by changes in the solution conditions. For example, the formal

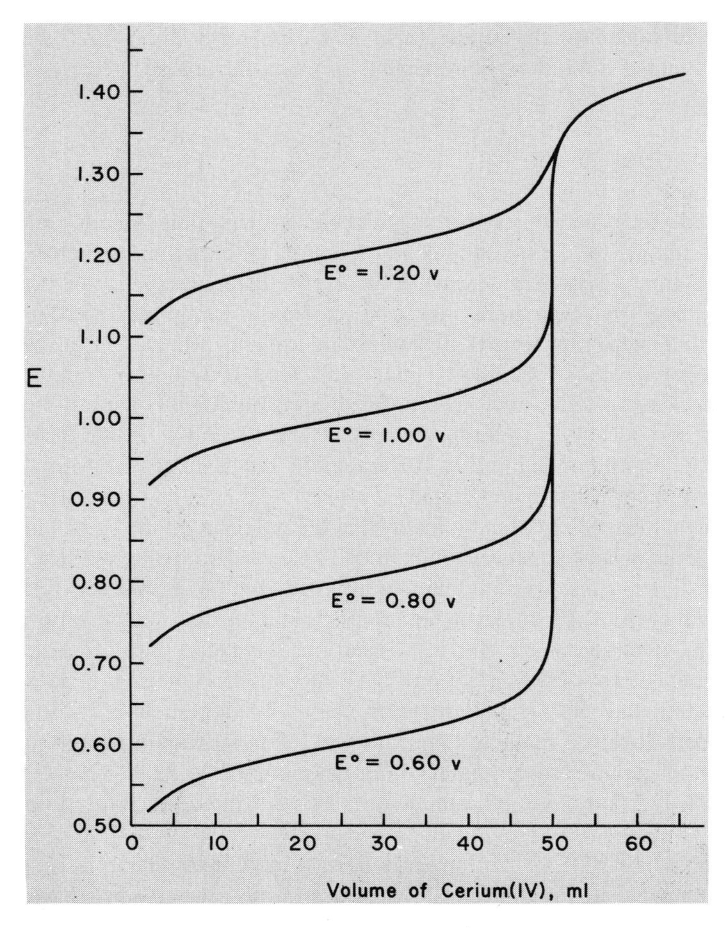


*Figure 8–9.* Titration curves for reducing agents of various strengths with 0.10 $M$ cerium(IV) solution in a 1 $F$ sulfuric acid medium. In each case, it is assumed that 50 ml of 0.10 $M$ reducing agent is titrated and that each reducing agent undergoes a one-electron oxidation. The standard potential for each species titrated is printed on the titration curve. The formal potential for the cerium(IV)-cerium(III) couple is +1.44 v versus NHE in a 1 $F$ sulfuric acid solution.

potentials for many half-reactions are altered when the concentration of acid or of some complexing agent is varied. In addition, the color-change intervals of redox indicators are affected by changes in solution composition. Any prediction concerning the feasibility of a proposed titration must take account of the effects of the concentrations of solution components.

## QUESTIONS AND PROBLEMS

1. Write the shorthand cell representation for each of the following galvanic cells:
   (a) A metallic silver electrode in a 0.015 $F$ silver nitrate solution connected, through a potassium nitrate salt bridge to eliminate the liquid-junction potential, to a 0.028 $F$ nickel chloride solution into which a nickel metal rod is immersed.

(b) A platinum wire in a mixture of 0.10 $M$ cerium(IV) and 0.05 $M$ cerium(III) in a 1 $F$ sulfuric acid solution which is in contact through a permeable membrane with a 10 $F$ sodium hydroxide solution containing 0.05 $M$ permanganate ion ($MnO_4^-$) and 0.001 $M$ manganate ion ($MnO_4^{2-}$) in which a gold wire electrode is immersed.

(c) A palladium wire in a 0.025 $F$ hydrochloric acid solution which is saturated with hydrogen gas at a pressure of 0.5 atm and which is connected, through a potassium chloride salt bridge to minimize the liquid-junction potential, to another half-cell consisting of a zinc metal rod immersed in a 0.04 $F$ zinc nitrate solution.

2. Write the Nernst equation which corresponds to each of the following half reactions:

(a) $Cd + 4\,CN^- \rightleftharpoons Cd(CN)_4^{2-} + 2\,e$
(b) $Cu^{2+} + I^- + e \rightleftharpoons CuI(s)$
(c) $TiO^{2+} + 2\,H^+ + e \rightleftharpoons Ti^{3+} + H_2O$
(d) $Mn^{2+} + 2\,H_2O \rightleftharpoons MnO_2 + 4\,H^+ + 2\,e$
(e) $2\,Hg^{2+} + 2\,e \rightleftharpoons Hg_2^{2+}$
(f) $Hg_2SO_4(s) + 2\,e \rightleftharpoons 2\,Hg + SO_4^{2-}$

3. Calculate the actual emf for the overall reaction occurring in each of the following galvanic cells:

(a) Ni | $Ni^{2+}$ (0.200 $M$) ‖ $Ag^+$ (0.00500 $M$) | Ag
(b) Pt | $I^-$ (0.500 $M$), $I_3^-$ (0.0300 $M$) ‖ $Cd^{2+}$ (0.100 $M$) | Cd
(c) Pb | $Pb^{2+}$ (0.0250 $M$) ‖ $Cu^{2+}$ (0.300 $M$) | Cu
(d) Ag | AgCl(s), $Cl^-$ (0.100 $M$) ‖ $Hg_2SO_4$(s), $SO_4^{2-}$ (2.00 $M$) | Hg
(e) Pt | $VO_2^+$ (0.0200 $M$), $VO^{2+}$ (0.100 $M$), $H^+$ (0.500 $M$) ‖ $Tl^+$ (0.0400 $M$) | Tl

4. For each of the following galvanic cells: (a) Write the two half-reactions and the overall cell reaction, (b) calculate the actual emf for each cell from the Nernst equation, (c) calculate the equilibrium constant for each cell reaction, and (d) identify the anode and cathode and label the electrodes appropriately with either a + or − sign.

Cu | $CuSO_4$ (0.02 $F$) ‖ $Fe^{2+}$ (0.2 $M$), $Fe^{3+}$ (0.01 $M$), HCl (1 $F$) | Pt
Pt | $Pu^{4+}$ (0.1 $M$), $Pu^{3+}$ (0.2 $M$) ‖ AgCl(s), HCl (0.03 $F$) | Ag
Pt | KBr (0.2 $F$), $Br_3^-$ (0.03 $M$) ‖ $Ce^{4+}$ (0.01 $M$), $Ce^{3+}$ (0.002 $M$), $H_2SO_4$ (1 $F$) | Pt
Pt, $Cl_2$ (0.1 atm) | HCl (2.0 $F$) ‖ HCl (0.1 $F$) | $H_2$ (0.5 atm), Pt
Zn | $ZnCl_2$ (0.02 $F$) ‖ $Na_2SO_4$ (0.1 $F$), $PbSO_4$(s) | Pb
Pt | $UO_2^{2+}$ (0.005 $M$), $U^{4+}$ (0.1 $M$), $HClO_4$ (0.2 $F$) ‖ HBr (0.3 $F$), AgBr(s) | Ag

5. Evaluate the equilibrium constant for each of the following reactions:

(a) $Ag + Fe^{3+} \rightleftharpoons Ag^+ + Fe^{2+}$
(b) $2\,Cr^{2+} + Co^{2+} \rightleftharpoons 2\,Cr^{3+} + Co$
(c) $Br_2(aq) + 3\,I^- \rightleftharpoons 2\,Br^- + I_3^-$
(d) $2\,V^{3+} + UO_2^{2+} \rightleftharpoons 2\,VO^{2+} + U^{4+}$
(e) $2\,Co^{3+} + Co \rightleftharpoons 3\,Co^{2+}$

6. Compute the equilibrium constant for the reaction

$$AuCl_4^- + 2\ Au + 2\ Cl^- \rightleftharpoons 3\ AuCl_2^-$$

from the standard potentials for the following half-reactions:

$$AuCl_2^- + e \rightleftharpoons Au + 2\ Cl^-; \qquad E^0 = +1.154\ v$$
$$AuCl_4^- + 2\ e \rightleftharpoons AuCl_2^- + 2\ Cl^-; \qquad E^0 = +0.926\ v$$

7. Given the information that

$$I_2(aq) + 2\ e \rightleftharpoons 2\ I^-; \qquad E^0 = +0.6197\ v$$

and

$$I_2(s) + 2\ e \rightleftharpoons 2\ I^-; \qquad E^0 = +0.5345\ v$$

where the symbol (aq) indicates unit activity of iodine in the aqueous phase and (s) denotes pure solid iodine at unit activity, calculate the solubility (moles/liter) of solid iodine in water.

8. Evaluate the equilibrium constant for the formation of the triiodide ion,

$$I_2(aq) + I^- \rightleftharpoons I_3^-$$

from the knowledge that

$$I_2(aq) + 2\ e \rightleftharpoons 2\ I^-; \qquad E^0 = +0.6197\ v$$

and

$$I_3^- + 2\ e \rightleftharpoons 3\ I^-; \qquad E^0 = +0.5355\ v$$

9. Calculate the solubility-product constant for copper(I) iodide,

$$CuI(s) \rightleftharpoons Cu^+ + I^-$$

given the following information:

$$Cu^{2+} + e \rightleftharpoons Cu^+; \qquad E^0 = +0.153\ v$$

and

$$Cu^{2+} + I^- + e \rightleftharpoons CuI; \qquad E^0 = +0.86\ v$$

10. Given that

$$Ag + 2\ CN^- \rightleftharpoons Ag(CN)_2^- + e; \qquad E^0 = +0.31\ v$$

and

$$Ag \rightleftharpoons Ag^+ + e; \qquad E^0 = -0.80\ v$$

calculate the equilibrium (dissociation) constant for the following reaction:

$$Ag(CN)_2^- \rightleftharpoons Ag^+ + 2\ CN^-$$

11. Consider the following galvanic cell:

$$Pt \mid PuO_2^{2+}\ (0.01\ M),\ Pu^{4+}\ (0.001\ M),\ H^+\ (0.1\ M) \parallel Cu^{2+}\ (0.001\ M) \mid Cu$$

(a) Write the two pertinent half-reactions and the overall cell reaction.
(b) Calculate the $E^0$ and the equilibrium constant for the overall cell reaction.
(c) Calculate the actual emf of the galvanic cell.
(d) Which electrode is the negative electrode?
(e) A bar of pure copper metal weighing 127 g was placed into 1 liter of a 0.0500 *F* $PuO_2(ClO_4)_2$ solution. The pH of this solution was maintained constant at exactly 3.00. What was the concentration of $PuO_2^{2+}$ at equilibrium?

12. The following galvanic cell was constructed for the measurement of the dissociation constant of a weak monoprotic acid, HA:

$$-Pt, H_2\ (0.8\ atm)\ |\ HA\ (0.5\ F),\ NaCl\ (1.0\ F),\ AgCl(s)\ |\ Ag+$$

If the observed electromotive force of this cell was 0.568 v, what is the dissociation constant for the weak acid?

13. Calculate the equilibrium concentrations of cobalt(II) and thallium(I) ions resulting when a 0.250 *F* cobalt(II) sulfate, $CoSO_4$, solution is treated with an excess of pure thallium metal.

14. Titanium(III) may be oxidized by iron(III) in an acidic solution, as shown by the reaction

$$Ti^{3+} + Fe^{3+} + H_2O \rightleftharpoons TiO^{2+} + Fe^{2+} + 2\ H^+$$

Calculate the final concentration of $Ti^{3+}$ ion in a solution prepared by mixing 25.0 ml each of 0.0200 *M* $Ti^{3+}$ and $Fe^{3+}$ solutions, both being 1 *F* in sulfuric acid.

15. As described earlier in this chapter, cerium(IV) and iron(II) react to form cerium(III) and iron(III):

$$Ce^{4+} + Fe^{2+} \rightleftharpoons Ce^{3+} + Fe^{3+}$$

(a) Calculate the equilibrium concentrations of all four species in a solution prepared by mixing 5.0 ml of 0.0500 *F* cerium(IV) sulfate with 25.0 ml of 0.0150 *F* iron(II) sulfate. Assume that the solutions contain 1 *F* sulfuric acid.
(b) Calculate the equilibrium concentrations of all four species in a solution prepared by mixing 25.0 ml of 0.0200 *F* cerium(IV) sulfate with 17.0 ml of 0.0200 *F* iron(II) sulfate. Assume that the solutions contain 1 *F* sulfuric acid.

16. The hypothetical galvanic cell

$$-A\ |\ A^{2+}\ ||\ B^{2+}\ |\ B+$$

has an emf of 0.360 v when the concentrations of $A^{2+}$ and $B^{2+}$ are equal. What will be the observed emf of the cell if the concentration of $A^{2+}$ is 0.100 *M* and the concentration of $B^{2+}$ is $1.00 \times 10^{-4}$ *M*?

17. Given the following standard-potential data,

$$Cd^{2+} + 2\ e \rightleftharpoons Cd; \qquad E^0 = -0.403\ v$$

$$Fe^{2+} + 2\ e \rightleftharpoons Fe; \qquad E^0 = -0.440\ v$$

calculate the equilibrium concentration of cadmium ion when a 0.0500 $M$ cadmium ion solution is shaken with an excess of pure iron filings.

18. Cadmium metal reacts with vanadium(III) according to the reaction

$$Cd + 2\ V^{3+} \rightleftharpoons Cd^{2+} + 2\ V^{2+}$$

If a 0.0750 $M$ vanadium(III) solution is shaken with an excess of cadmium metal, what will be the concentrations of $V^{3+}$, $V^{2+}$, and $Cd^{2+}$ at equilibrium?

19. If the electromotive force of the galvanic cell

$$+Pb \mid Pb^{2+}\ (0.0860\ M) \parallel Tl^{+} \mid Tl-$$

is 0.280 v, what must be the concentration of thallium ion in the right half-cell?

20. If the electromotive force of the galvanic cell

$$-Ag \mid AgCl(s),\ Cl^{-}\ (0.100\ M) \parallel Fe^{3+}\ (0.0200\ M),\ Fe^{2+} \mid Pt+$$

is 0.319 v, what must be the concentration of $Fe^{2+}$ in the right half-cell? Assume that the solution in the right half-cell is 1 $F$ sulfuric acid.

21. If the electromotive force of the galvanic cell

$$-Pt,\ H_2\ (0.250\ \text{atm}) \mid \text{solution of unknown pH} \parallel AgCl(s),\ Cl^{-}\ (1.00\ M) \mid Ag+$$

is 0.621 v, what is the pH of the unknown sample?

22. A 20.00-ml sample of 0.1000 $M$ uranium(IV) in a 1 $F$ sulfuric acid medium is titrated with 0.1000 $M$ cerium(IV) in 1 $F$ sulfuric acid, according to the reaction

$$2\ Ce^{4+} + U^{4+} + 2\ H_2O \rightleftharpoons 2\ Ce^{3+} + UO_2^{\ 2+} + 4\ H^{+}$$

Suppose that the progress of the titration is followed by means of the potentiometric measurement of the potential of a platinum indicator electrode versus a saturated calomel electrode (SCE).

(a) Calculate the potential of the platinum indicator electrode versus SCE at the equivalence point of the titration.

(b) What concentration of uranium(IV) remains unoxidized at the equivalence point?

(c) Calculate what volume of cerium(IV) titrant solution must be added to cause the potential of the platinum indicator electrode to become +0.334 v versus SCE.

(d) Calculate the potential of the platinum electrode versus SCE after the addition of 50.00 ml of cerium(IV) titrant.

23. Consider the titration of 20.00 ml of 0.05000 $M$ iron(III) in a 1 $F$ sulfuric acid medium with 0.02000 $M$ titanium(III), also in 1 $F$ sulfuric acid:

$$Fe^{3+} + Ti^{3+} + H_2O \rightleftharpoons Fe^{2+} + TiO^{2+} + 2\ H^{+}$$

(a) Calculate and plot the complete titration curve, taking enough points to define the titration curve smoothly.

(b) Calculate the concentration of iron(III) unreduced at the equivalence point of the titration.

(c) If methylene blue indicator (see Table 8–3), which undergoes a sharp color change from blue to colorless at a potential of +0.33 v versus the normal hydrogen electrode (NHE), is employed to signal the end point of this titration, what volume of titanium(III) titrant would be required to reach this end point?

24. Calculate and plot the titration curves for each of the following:

(a) The titration of 20.00 ml of 0.02000 *M* tin(II) in a 1 *F* hydrochloric acid medium with 0.02000 *M* vanadium(V) in 1 *F* hydrochloric acid:

$$SnCl_4^{2-} + 2\ VO_2^{+} + 4\ H^{+} + 2\ Cl^{-} \rightleftharpoons SnCl_6^{2-} + 2\ VO^{2+} + 2\ H_2O$$

(b) The titration of 25.00 ml of 0.1000 *M* iron(III) in 1 *F* sulfuric acid with 0.05000 *M* chromium(II) in a 1 *F* sulfuric acid medium:

$$Fe^{3+} + Cr^{2+} \rightleftharpoons Fe^{2+} + Cr^{3+}$$

25. A 25.00-ml sample of 0.02000 *M* $VO_2^{+}$ in 1 *F* perchloric acid was titrated with 0.05000 *M* titanium(III) in 1 *F* perchloric acid according to the following reaction:

$$VO_2^{+} + Ti^{3+} \rightleftharpoons VO^{2+} + TiO^{2+}$$

The titration was followed potentiometrically by measuring the potential between a platinum indicator electrode and a saturated calomel reference electrode (SCE). Determine the following information:

(a) The equilibrium constant for the titration reaction.

(b) The potential of the platinum indicator electrode versus the normal hydrogen electrode (NHE) at the equivalence point of the titration.

(c) The concentration of $VO_2^{+}$ at the equivalence point.

(d) The volume of titanium(III) titrant added at the point at which the potential of the platinum indicator electrode is +1.00 v versus the normal hydrogen electrode (NHE).

(e) The potential of the platinum indicator electrode versus the saturated calomel electrode (SCE) after 8.00 ml of the titanium(III) had been added.

## SUGGESTIONS FOR ADDITIONAL READING

1. R. G. Bates: Electrode potentials. *In* I. M. Kolthoff and P. J. Elving, eds.: *Treatise on Analytical Chemistry*. Part I, Volume 1, Wiley-Interscience, New York, 1959, pp. 319-359.
2. H. A. Laitinen: *Chemical Analysis*. McGraw-Hill Book Company, New York, 1960, pp. 276-297, 326-341.
3. W. M. Latimer: *Oxidation Potentials*. Second edition, Prentice-Hall, Englewood Cliffs, New Jersey, 1952.
4. J. J. Lingane: *Electroanalytical Chemistry*. Second edition, Wiley-Interscience, New York, 1958, pp. 1-70, 129-157.
5. S. Wawzonek: Potentiometry: Oxidation-reduction potentials. *In* A. Weissberger and B. W. Rossiter, eds.: *Physical Methods of Chemistry*. Volume I, Part IIA, Wiley-Interscience, New York, 1971, pp. 1-60.

# 9 APPLICATIONS OF OXIDATION-REDUCTION REACTIONS

Because oxidation-reduction reactions provide a foundation for innumerable methods of chemical analysis, it is appropriate to explore the properties and uses of some of the most popular titrants. Three of the strongest oxidizing agents used in redox titrimetry – potassium permanganate, potassium dichromate, and cerium(IV) – will be examined first. Then, the great versatility of the triiodide-iodide system will be demonstrated through a discussion of reactions in which triiodide ion serves as an oxidizing agent and of processes in which iodide ion acts as a reductant for many oxidizing agents. Finally, the uses of periodate for the determination of organic substances will be considered.

Before the individual reagents are discussed, it is worthwhile to outline what characteristics one seeks in choosing a titrant and what preliminary steps must be taken to prepare the substance being determined for reaction with the titrant.

## REQUIREMENTS FOR SELECTION OF TITRANTS

Although examination of a table of standard potentials reveals the existence of species covering a wide range of oxidizing and reducing strengths, relatively few substances qualify for selection as titrants for redox methods of analysis. For any particular titration, the titrant must satisfy several requirements. First, the titrant must be strong enough to react to practical completion with the substance being titrated. Second, the titrant must not be so powerful that it is able to react with some component of the solution being titrated other than the desired species. Strong oxidants such as silver(II) and cobalt(III) easily satisfy the first requirement, but fail to meet the second requirement because they quickly oxidize the solvent (water) in which they are dissolved. Third, the titrant must react rapidly with the substance being determined. A certain reaction may appear to be favorable from a thermodynamic point of view, but the reaction may not occur at a convenient rate. This is often true if the redox reaction involves multiple electron transfer, or the formation or rupture of chemical bonds. Finally, a simple and precise method for location of the equivalence point must be available.

## PRELIMINARY ADJUSTMENT OF OXIDATION STATES OF TITRATED SUBSTANCES

For the performance of a successful redox titration, the species to be determined must exist in a single oxidation state which can react stoichiometrically and rapidly with the titrant. However, preparation of a sample solution often leaves the desired substance in an oxidation state which is unreactive toward the titrant or perhaps in a mixture of oxidation states. In these instances, the sample solution must be treated with a suitable oxidizing or reducing agent to adjust the oxidation state of the substance prior to the final titration. What characteristics should this oxidizing or reducing agent possess? Clearly, the reagent must convert the desired substance quickly and quantitatively to the proper oxidation state. In addition, it must be possible, as well as convenient, to separate or remove the excess oxidant or reductant from the sample solution, lest the reagent interact later with the titrant. Finally, the oxidizing or reducing agent should display a certain degree of selectivity. Let us consider some of the more widely employed oxidizing and reducing agents used to adjust the oxidation state of a species.

### Oxidizing Agents

**Sodium Bismuthate.** A powerful oxidizing agent, this sparingly soluble compound has an indefinite composition, but consists largely of $NaBiO_3$. In the presence of excess sodium bismuthate in a nitric acid medium, manganese(II) is converted to permanganate, chromium(III) becomes dichromate, and cerium(III) is oxidized to cerium(IV). Unreacted sodium bismuthate may be separated from the solution by filtration.

**Potassium Peroxodisulfate.** In boiling acidic solutions containing a trace of silver(I) as catalyst, potassium peroxodisulfate ($K_2S_2O_8$) quantitatively oxidizes cerium(III) to cerium(IV), manganese(II) to permanganate, chromium(III) to dichromate, and vanadium(IV) to vanadium(V). After the oxidation reactions are complete, continued boiling of the solution for a few minutes decomposes the excess peroxodisulfate:

$$2\,S_2O_8^{2-} + 2\,H_2O \rightleftharpoons 4\,SO_4^{2-} + O_2 + 4\,H^+$$

**Silver(II) Oxide.** Dark brown silver(II) oxide, AgO, is a strong oxidant which smoothly converts manganese(II) to permanganate, chromium(III) to dichromate, cerium(III) to cerium(IV), and vanadium(IV) to vanadium(V) as small portions of the reagent are dissolved in cold perchloric acid or nitric acid. It is easy to recognize when enough silver(II) oxide has been added due to the chocolate-brown color of unreacted $Ag^{2+}$ ion; the excess silver(II) is easily reduced by water if the solution is warmed:

$$4\,Ag^{2+} + 2\,H_2O \rightleftharpoons 4\,Ag^+ + O_2 + 4\,H^+$$

**Hydrogen Peroxide.** In an alkaline medium, hydrogen peroxide oxidizes chromium(III) to chromate, manganese(II) to manganese dioxide, arsenic(III) to arsenic(V), antimony(III) to antimony(V), and vanadium(IV) to vanadium(V). In acidic solutions, this reagent quantitatively converts iron(II) to iron(III) and iodide ion to molecular iodine, but reduces dichromate to chromium(III) and permanganate to manganese(II). Excess hydrogen peroxide decomposes by disproportionation when the acidic or alkaline solution is boiled for a few minutes:

$$2\,H_2O_2 \rightleftharpoons O_2 + 2\,H_2O$$

### Reducing Agents

**Hydrogen Sulfide and Sulfur Dioxide.** These gases are readily soluble in aqueous media and are relatively mild reducing agents, being widely employed for the reduction of iron(III) to iron(II) in acid solutions prior to titrations with standard solutions of oxidants. In addition, hydrogen sulfide and sulfur dioxide reduce vanadium(V) to vanadium(IV), as well as stronger oxidizing agents such as permanganate, cerium(IV), and dichromate. One need only boil the solution to remove the excess of either gas. Besides the unpleasant and toxic nature of these gases, reductions with sulfur dioxide tend to be slow, and the use of hydrogen sulfide leads to the formation of colloidal sulfur which may react with strongly oxidizing titrants.

**Tin(II) Chloride.** Hydrochloric acid solutions of this reagent are utilized almost exclusively for the quantitative reduction of iron(III) to iron(II) prior to titration of the latter with a standard solution of permanganate, dichromate, or cerium(IV).

**Metals and Metal Amalgams.** Perhaps the most versatile method to accomplish reduction of a species to a definite oxidation state is treatment of the sample solution with a metal. Among the metallic elements employed as reductants are zinc, aluminum, cadmium, silver, mercury, copper, nickel, bismuth, lead, tin, and iron.

In the form of wire or rod, the metal can be inserted directly into the solution and, when the reduction is complete, the unused reductant may be washed carefully and withdrawn. Usually, one finds it more convenient to prepare a *reductor*, as shown in Figure 9–1, by packing granules of the appropriate metal reductant into a glass column; then the sample solution is percolated through the column and is collected in a receiving flask. Generally, a solution treated with a metal reductant must be blanketed with nitrogen or carbon dioxide to prevent exposure to the atmosphere before the final titration, since lower oxidation states of many species are quickly oxidized by oxygen.

Aluminum, zinc, and cadmium are potent reductants, which means that they lack specificity. As a consequence of their reducing power, these substances react vigorously with hydrogen ion to yield hydrogen gas. This side-reaction wastes the metallic reductant and introduces unwanted amounts of the corresponding metal ion into the sample solution. To eliminate this difficulty, zinc and cadmium may be amalgamated with mercury. For zinc, this process involves thorough agitation of a solution of mercury(II) nitrate in contact with zinc granules for a few minutes until each metal particle becomes coated with a thin film of elemental mercury. Amalgamated zinc and cadmium have little tendency to reduce hydrogen ion in acidic solutions and, therefore, make ideal packings for reductor columns.

Without doubt, the **Jones reductor** – a glass tube approximately 2 cm in diameter packed with a 30- to 40-cm column of amalgamated zinc (Figure 9–1) – is more extensively used than any other amalgam reductor. If a column similar to that employed for the Jones reductor is filled with pure silver granules, one obtains the well known **silver reductor**. By itself, elemental silver is a poor reducing agent. However, in the presence of hydrochloric acid, metallic silver becomes an effective, though mild, reductant. For example, when a 1 $F$ hydrochloric acid solution of uranium(VI) is passed through the

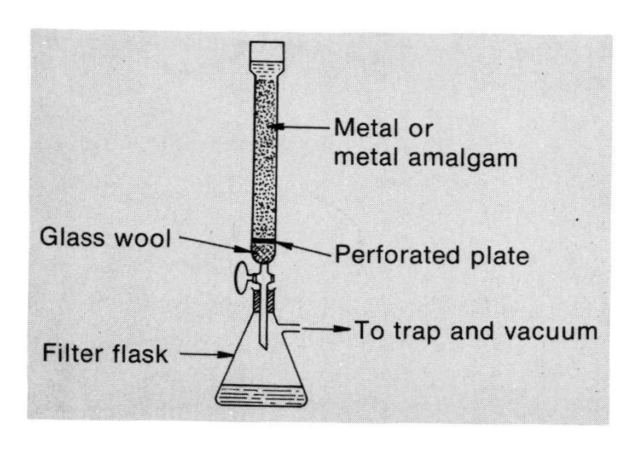


*Figure 9–1.* A metal or metal amalgam reductor.

reductor, uranium(IV) is formed quantitatively and metallic silver is oxidized to silver chloride:

$$UO_2^{2+} + 4\,H^+ + 2\,Ag + 2\,Cl^- \rightleftharpoons U^{4+} + 2\,AgCl + 2\,H_2O$$

Since silver metal in hydrochloric acid is not as strong a reductant as amalgamated zinc, use of a silver reductor may permit greater selectivity in the preferential adjustment of the oxidation state of one substance in a mixture of several species. A comparison of the capabilities of the Jones reductor and the silver reductor is presented in Table 9–1.

**Table 9–1. Comparison of Reductions Accomplished with the Jones Reductor and the Silver Reductor***

| Jones Reductor (1 *F* $H_2SO_4$ Medium) | Silver Reductor (1 *F* HCl Medium) |
|---|---|
| $Cr^{3+} + e \rightleftharpoons Cr^{2+}$ | $Cr^{3+}$ is not reduced |
| $Cu^{2+} + 2\,e \rightleftharpoons Cu$ | $Cu^{2+} + 2\,Cl^- + e \rightleftharpoons CuCl_2^-$ |
| $Fe^{3+} + e \rightleftharpoons Fe^{2+}$ | $Fe^{3+} + e \rightleftharpoons Fe^{2+}$ |
| $MoO_2^{2+} + 4\,H^+ + 3\,e \rightleftharpoons Mo^{3+} + 2\,H_2O$ | $MoO_2^{2+} + e \rightleftharpoons MoO_2^+$ |
| $TiO^{2+} + 2\,H^+ + e \rightleftharpoons Ti^{3+} + H_2O$ | $TiO^{2+}$ is not reduced |
| $\left\{\begin{matrix} UO_2^{2+} + 4\,H^+ + 2\,e \rightleftharpoons U^{4+} + 2\,H_2O \\ UO_2^{2+} + 4\,H^+ + 3\,e \rightleftharpoons U^{3+} + 2\,H_2O \end{matrix}\right\}$† | $UO_2^{2+} + 4\,H^+ + 2\,e \rightleftharpoons U^{4+} + 2\,H_2O$ |
| $VO_2^+ + 4\,H^+ + 3\,e \rightleftharpoons V^{2+} + 2\,H_2O$ | $VO_2^+ + 2\,H^+ + e \rightleftharpoons VO^{2+} + H_2O$ |

* Taken with permission from I. M. Kolthoff and R. Belcher: *Volumetric Analysis.* Volume 3, Interscience Publishers, Inc., New York, 1957, p. 12.

† A mixture of uranium(III) and uranium(IV) is obtained from the Jones reductor; however, the uranium(III) can be converted to uranium(IV) if the solution is shaken in contact with oxygen for several minutes.

## POTASSIUM PERMANGANATE

### Half-Reactions Involving Permanganate Ion

Manganese may exist in a number of stable oxidation states. However, for the majority of oxidation-reduction titrations, the important states are manganese(VII), manganese(IV), and manganese(II). In Appendix 4, one finds the two half-reactions

$$MnO_4^- + 8\,H^+ + 5\,e \rightleftharpoons Mn^{2+} + 4\,H_2O; \qquad E^0 = +1.51\text{ v}$$

and

$$MnO_4^- + 4\,H^+ + 3\,e \rightleftharpoons MnO_2 + 2\,H_2O; \qquad E^0 = +1.695\text{ v}$$

Factors that govern which of the two reduction products, $Mn^{2+}$ or $MnO_2$, is actually formed are quite complex and involve consideration of both kinetics and thermodynamics. In neutral and alkaline media as well as in weakly acidic solutions, manganese dioxide is the product. For titrations in acidic media having hydrogen ion concentrations of 0.1 *M* or greater, manganous ion is the reduction product.

Bright green manganate ion, $MnO_4^{2-}$, is the reduction product of permanganate in strongly alkaline media:

$$MnO_4^- + e \rightleftharpoons MnO_4^{2-}; \qquad E^0 = +0.564\text{ v}$$

In sodium hydroxide solutions more concentrated than 2 $F$, organic compounds reduce permanganate according to this half-reaction.

### Preparation and Stability of Permanganate Solutions

Potassium permanganate is rarely available in a sufficiently high state of purity to permit its direct use as a primary standard substance. Even reagent-grade potassium permanganate is invariably contaminated with small quantities of manganese dioxide. In addition, ordinary distilled water, from which potassium permanganate solutions are prepared, contains organic matter which can reduce permanganate to manganese dioxide.

Permanganate solutions are inherently unstable because $MnO_4^-$ is capable of oxidizing water spontaneously:

$$4\,MnO_4^- + 2\,H_2O \rightleftharpoons 4\,MnO_2 + 3\,O_2 + 4\,OH^-; \qquad E^0 = +0.187\text{ v}$$

Fortunately, the rate of this reaction is exceedingly slow if proper precautions are taken in the original preparation of the solution. Heat, light, acids, bases, manganese(II) salts, and especially manganese dioxide catalyze the permanganate-water reaction. Any manganese dioxide present in the potassium permanganate solution must be removed by filtration if a stable titrant is to be obtained. Solutions of potassium permanganate should be stored in dark bottles and kept out of bright light and away from dust as much as possible.

### Standardization of Permanganate Solutions

Among the substances suitable for the standardization of potassium permanganate solutions are arsenious oxide ($As_2O_3$), sodium oxalate ($Na_2C_2O_4$), and pure iron wire.

Arsenious oxide, commercially available as a primary-standard-grade solid, is initially dissolved in a sodium hydroxide solution, which is subsequently acidified with hydrochloric acid and titrated with potassium permanganate:*

$$2\,MnO_4^- + 5\,HAsO_2 + 6\,H^+ + 2\,H_2O \rightleftharpoons 2\,Mn^{2+} + 5\,H_3AsO_4$$

However, this reaction does not proceed rapidly without a catalyst. Iodine monochloride is an excellent catalyst for this titration although, in a hydrochloric acid solution, iodine monochloride actually exists as the species $ICl_2^-$.

Sodium oxalate is another substance used for the standardization of potassium permanganate solutions. It is available commercially in a very pure state, and it dissolves in sulfuric acid media with the resultant formation of oxalic acid. We may represent the stoichiometry of the permanganate-oxalic acid reaction as

$$2\,MnO_4^- + 5\,H_2C_2O_4 + 6\,H^+ \rightleftharpoons 2\,Mn^{2+} + 10\,CO_2 + 8\,H_2O$$

Usually, the titration is carried out at solution temperatures near 70°C so that the rate of reaction is conveniently fast.

---

*Some authors use $H_3AsO_3$ as the formula of arsenious acid. We prefer to write $HAsO_2$ as the predominant form of arsenic(III) because arsenious acid behaves as a weak monoprotic acid in aqueous media.

Specially purified iron wire may be dissolved in acid, converted to iron(II), and the iron(II) titrated to iron(III) with the permanganate solution to be standardized.

### End-Point Detection for Permanganate Titrations

Solutions of potassium permanganate are so intensely colored that a single drop of a 0.02 $F$ titrant imparts a perceptible color to 100 ml of water, the reduction product of permanganate in an acid medium, $Mn^{2+}$, being practically colorless. Therefore, if the solution being titrated is colorless, the appearance of the pale pink color due to the first trace of excess titrant may be taken as the equivalence point. If the titrant is too dilute to serve as its own indicator, tris(1,10-phenanthroline)iron(II) and other substituted ferroin-type indicators can be used.

### Analytical Uses of Potassium Permanganate in Acidic Solutions

As already mentioned, except for the oxidation of some organic compounds with permanganate in sodium hydroxide media, most applications involve titrimetry under strongly acidic conditions, for which permanganate is reduced to manganous ion. Information about a number of useful direct titrations with a standard potassium permanganate solution is summarized in Table 9–2. In the following paragraphs, we shall examine two of these applications in more detail.

**Determination of Iron in an Ore.** Hematite ($Fe_2O_3$), limonite ($2\ Fe_2O_3 \cdot 3\ H_2O$), and magnetite ($Fe_3O_4$), which are the important iron ores, can usually be dissolved in hot hydrochloric acid. After the oxide is dissolved, the iron will exist wholly or partly as iron(III). Since the titration with standard potassium permanganate solution requires that all iron be present as iron(II), the iron(III) formed during dissolution of the sample must be quantitatively reduced.

A preferred method for the reduction of iron(III) entails addition of a very small excess of tin(II) chloride to the hot hydrochloric acid solution of the iron sample,

$$2\ Fe^{3+} + SnCl_4^{2-} + 2\ Cl^- \rightleftharpoons 2\ Fe^{2+} + SnCl_6^{2-}$$

followed by destruction of the excess tin(II) with mercury(II):

$$SnCl_4^{2-} + 2\ HgCl_4^{2-} \rightleftharpoons SnCl_6^{2-} + Hg_2Cl_2(s) + 4\ Cl^-$$

Enough tin(II) must be introduced to cause complete reduction of iron(III). However, if the excess of tin(II) is too great, elemental mercury rather than mercurous chloride (calomel) may form:

$$SnCl_4^{2-} + HgCl_4^{2-} \rightleftharpoons SnCl_6^{2-} + Hg(l) + 2\ Cl^-$$

Although solid mercurous chloride does not interfere with a successful titration, metallic mercury ruins the determination because, being present in a finely divided colloidal state, it reacts to a significant extent with permanganate.

**Determination of Calcium in Limestone.** A useful indirect oxidation-reduction method of analysis is the determination of calcium in limestone. Preparation of the sample solution entails dissolution of a weighed portion of powdered limestone in hydrochloric acid:

$$CaCO_3 + 2\ H^+ \rightarrow Ca^{2+} + CO_2 + H_2O$$

Table 9–2. Some Titrations Performed with Standard Potassium Permanganate in Acidic Media

| Substance Determined | Half-Reaction for Substance Titrated | Procedure and Conditions for Titration |
|---|---|---|
| Sb(III) | $SbCl_4^- + 2\,Cl^- \rightleftharpoons SbCl_6^- + 2\,e$ | Titrate in 2 *F* HCl |
| As(III) | $HAsO_2 + 2\,H_2O \rightleftharpoons H_3AsO_4 + 2\,H^+ + 2\,e$ | Titrate in 1 *F* HCl with ICl as catalyst |
| $Br^-$ | $2\,Br^- \rightleftharpoons Br_2 + 2\,e$ | Titrate in boiling 2 *F* $H_2SO_4$ to expel $Br_2$ |
| $Fe(CN)_6^{4-}$ | $Fe(CN)_6^{4-} \rightleftharpoons Fe(CN)_6^{3-} + e$ | Titrate in 0.2 *F* $H_2SO_4$ with erioglaucine A as indicator |
| $H_2O_2$ | $H_2O_2 \rightleftharpoons O_2 + 2\,H^+ + 2\,e$ | Titrate in 1 *F* $H_2SO_4$ |
| $I^-$ | $I^- + HCN \rightleftharpoons ICN + H^+ + 2\,e$ | Titrate in presence of 0.1 *F* HCN and 1 *F* $H_2SO_4$; ferroin indicator |
| Fe(II) | $Fe^{2+} \rightleftharpoons Fe^{3+} + e$ | Reduce iron(III) with tin(II) or in Jones reductor; titrate in 1 *F* $H_2SO_4$ or in 1 *F* HCl with Zimmermann-Reinhardt reagent added |
| Mo(III) | $Mo^{3+} + 2\,H_2O \rightleftharpoons MoO_2^{2+} + 4\,H^+ + 3\,e$ | Reduce molybdenum to molybdenum(III) in Jones reductor; treat with excess iron(III) in 1 *F* $H_2SO_4$ and titrate iron(II) formed |
| $H_2C_2O_4$ | $H_2C_2O_4 \rightleftharpoons 2\,CO_2 + 2\,H^+ + 2\,e$ | Titrate in 1 *F* $H_2SO_4$ at 70°C |
| Ca(II), Mg(II), Zn(II), Co(II), La(III), Th(IV), Ba(II), Sr(II), Pb(II) | $H_2C_2O_4 \rightleftharpoons 2\,CO_2 + 2\,H^+ + 2\,e$ | Metal oxalates are precipitated, washed, and redissolved in 1 *F* $H_2SO_4$; liberated oxalic acid titrated at 70°C |
| Sn(II) | $SnCl_4^{2-} + 2\,Cl^- \rightleftharpoons SnCl_6^{2-} + 2\,e$ | Reduce tin(IV) to tin(II) with bismuth amalgam in 5 to 12 *F* HCl; exclude oxygen during titration |
| Ti(III) | $Ti^{3+} + H_2O \rightleftharpoons TiO^{2+} + 2\,H^+ + e$ | Reduce titanium(IV) to titanium(III) with zinc metal, lead amalgam, or in Jones reductor; exclude oxygen; titrate in 1 *F* $H_2SO_4$ or HCl |
| W(III) | $W^{3+} + 2\,H_2O \rightleftharpoons WO_2^{2+} + 4\,H^+ + 3\,e$ | Reduce tungsten to tungsten(III) with lead amalgam at 50°C; titrate in 2 *F* HCl |
| U(IV) | $U^{4+} + 2\,H_2O \rightleftharpoons UO_2^{2+} + 4\,H^+ + 2\,e$ | Reduce uranium to uranium(III) in Jones reductor; expose solution to air to obtain uranium(IV); titrate in 1 *F* $H_2SO_4$ |
| V(IV) | $VO^{2+} + H_2O \rightleftharpoons VO_2^+ + 2\,H^+ + e$ | Reduce vanadium to vanadium(IV) with bismuth amalgam; titrate in 1 *F* $H_2SO_4$ |

Next, the calcium is precipitated as calcium oxalate monohydrate under carefully adjusted conditions:

$$Ca^{2+} + C_2O_4^{2-} + H_2O \rightleftharpoons CaC_2O_4{\cdot}H_2O$$

Then the precipitate is separated from its mother solution by filtration and is washed free of excess oxalate. Finally, the calcium oxalate monohydrate is dissolved in sulfuric acid, the oxalate species being converted into oxalic acid,

$$CaC_2O_4{\cdot}H_2O + 2\,H^+ \rightleftharpoons Ca^{2+} + H_2C_2O_4 + H_2O$$

and the oxalic acid solution is titrated with a standard potassium permanganate solution. Although calcium ion is not involved directly in the final titration, it is stoichiometrically related to the amount of oxalic acid titrated.

### Determination of Organic Compounds with Alkaline Permanganate

A major difficulty encountered in the determination of organic substances is their slow rate of reaction with permanganate. Such behavior is understandable, because most organic compounds are eventually degraded to carbon dioxide and water, a process which usually involves the rupture of at least several carbon-carbon and carbon-hydrogen bonds. Slow reactions can be overcome if the desired organic compound is treated with an excess of permanganate and if the reaction is allowed to proceed for extended periods of time.

Earlier, we mentioned that permanganate is reduced to the green-colored manganate ion by organic substances in concentrated sodium hydroxide solutions. This reaction serves as the basis of a quantitative method for the determination of a number of organic compounds, if the desired substance is dissolved in a known excess of strongly alkaline standard permanganate solution and allowed to react at room temperature for approximately 30 minutes. For example, glycerol which is oxidized to carbonate

$$\underset{\displaystyle OH}{\underset{|}{H_2C}}\!-\!\!-\!\underset{\displaystyle OH}{\underset{|}{CH}}\!-\!\!-\!\underset{\displaystyle OH}{\underset{|}{CH_2}} + 14\,MnO_4^- + 20\,OH^- \rightarrow 3\,CO_3^{2-} + 14\,MnO_4^{2-} + 14\,H_2O$$

may be determined by acidification of the solution after the alkaline oxidation, followed by titrimetric reduction of all higher oxidation states of manganese to manganese(II); the glycerol content of the sample is computed from the *difference* between the equivalents of permanganate originally taken and the equivalents of higher oxidation states of manganese found in the final titration.

Other organic compounds which have been determined by means of this procedure are glycolic acid, tartaric acid, citric acid, ethylene glycol, phenol, salicylic acid, formaldehyde, glucose, and sucrose.

## POTASSIUM DICHROMATE

### Reduction of Dichromate Ion

When we turn to the dichromate ion, we find that the only important reduction involving this species is

$$Cr_2O_7^{2-} + 14\,H^+ + 6\,e \rightleftharpoons 2\,Cr^{3+} + 7\,H_2O; \qquad E^0 = +1.33\text{ v}$$

Potassium dichromate is invariably employed as a titrant in acidic media, for in neutral and alkaline solutions chromic ion, $Cr^{3+}$ or $Cr(H_2O)_6{}^{3+}$, forms an insoluble hydrous oxide, and dichromate is converted to the chromate ion:

$$Cr_2O_7{}^{2-} + 2\ OH^- \rightleftharpoons 2\ CrO_4{}^{2-} + H_2O$$

### Preparation and Stability of Standard Dichromate Solutions

Potassium dichromate may be purchased as a high-purity, reagent-grade solid which is suitable for the direct preparation, by weight, of standard solutions. Furthermore, potassium dichromate is readily soluble in water, and the resulting solutions are stable for many years if protected against evaporation.

### End-Point Detection for Titrations with Dichromate

Dichromate solutions have an orange-yellow color, whereas chromium(III) may be either green or violet, depending upon the composition of the medium in which it is formed as a reduction product. Not surprisingly, the color due to excess dichromate is masked by that of chromium(III), so dichromate cannot serve as its own end-point indicator. Sodium diphenylbenzidine sulfonate and barium diphenylamine sulfonate (Table 8–3) are well suited as end-point indicators for titrations with potassium dichromate.

### Analytical Uses of Potassium Dichromate

Because potassium dichromate is a less potent oxidant than potassium permanganate, it has fewer analytical applications. Undoubtedly, the most important use of dichromate is the titration of iron(II),

$$Cr_2O_7{}^{2-} + 6\ Fe^{2+} + 14\ H^+ \rightleftharpoons 2\ Cr^{3+} + 6\ Fe^{3+} + 7\ H_2O$$

since the method is highly precise in the presence of hydrochloric acid, the most common solvent for iron-containing alloys and ores. Copper(I) and uranium(IV), which react quantitatively with iron(III) to produce a stoichiometric amount of iron(II), can be determined indirectly by means of the preceding reaction. Although hydroquinone in 2 $F$ hydrochloric acid at 50°C may be titrated directly with a standard potassium dichromate solution, the oxidation of most organic compounds is much too slow to be of practical value.

## CERIUM(IV)

### Reactions of Cerium(IV) and Cerium(III)

Although the half-reaction

$$Ce^{4+} + e \rightleftharpoons Ce^{3+}$$

implies the one-electron reduction of $Ce(H_2O)_6^{4+}$ to $Ce(H_2O)_6^{3+}$, these simple aquo ions do not predominate in an aqueous medium. In the case of $Ce(H_2O)_6^{4+}$, proton-transfer reactions such as

$$Ce(H_2O)_6^{4+} + H_2O \rightleftharpoons Ce(H_2O)_5(OH)^{3+} + H_3O^+$$

and

$$Ce(H_2O)_5(OH)^{3+} + H_2O \rightleftharpoons Ce(H_2O)_4(OH)_2^{2+} + H_3O^+$$

occur regardless of what anions are present in solution. Analogous reactions prevail with cerium(III), although the extent of such processes is less, owing to the smaller positive charge of cerium(III). These reactions depend on the hydrogen ion concentration of the solution and on the concentrations of cerium(IV) and cerium(III).

In addition, the appearance of three different *formal* potentials in Appendix 4 for the cerium(IV)-cerium(III) couple (+1.70 v in 1 *F* perchloric acid, +1.61 v in 1 *F* nitric acid, and +1.44 v in 1 *F* sulfuric acid) indicates that both cerium(IV) and cerium(III) form complexes with anions of the mineral acids. Perchlorate ion coordinates with cerium(III) and cerium(IV) to yield, respectively, $Ce(H_2O)_5(ClO_4)^{2+}$ and $Ce(H_2O)_5(ClO_4)^{3+}$. In nitric acid solutions, there is formation of $Ce(H_2O)(NO_3)_5^-$ and $Ce(NO_3)_6^{2-}$; and, in sulfuric acid, species such as $Ce(H_2O)_5(SO_4)^{2+}$, $Ce(H_2O)_3(SO_4)_3^{2-}$, and $Ce(H_2O)_5(SO_4)^+$ have been identified.

## Preparation and Stability of Cerium(IV) Titrants

Ammonium hexanitratocerate(IV), $(NH_4)_2Ce(NO_3)_6$, is a good primary standard for the direct preparation of cerium(IV) solutions. However, cerium(IV) solutions are more commonly prepared from cerium(IV) bisulfate, $Ce(HSO_4)_4$, or ammonium sulfatocerate(IV), $(NH_4)_4Ce(SO_4)_4 \cdot 2\,H_2O$. Moreover, the solution must contain a high concentration of acid, usually 1 *M* or greater, to prevent the precipitation of hydrous cerium(IV) oxide.

Sulfuric acid solutions of cerium(IV) are stable indefinitely and, if necessary, can be heated for short periods of time. Although solutions of cerium(IV) in nitric acid and in perchloric acid possess greater oxidizing strength, they undergo slow decomposition because of the reduction of cerium(IV) by water, a process which is light-induced. Solutions of cerium(IV) in hydrochloric acid are decidedly unstable, owing to the oxidation of chloride ion to chlorine gas by cerium(IV).

## Standardization of Cerium(IV) Solutions

Procedures used in the standardization of cerium(IV) titrants differ little from those used to standardize potassium permanganate solutions. Substances suitable for standardizing cerium(IV) solutions include arsenious oxide, sodium oxalate, and iron wire.

As in the standardization of potassium permanganate, a catalyst is required for rapid reaction between cerium(IV) and arsenic(III). Usually, after the solid arsenious oxide is initially dissolved in a sodium hydroxide medium, sulfuric acid rather than hydrochloric acid is used to acidify the solution, osmium tetroxide ($OsO_4$) is added as the catalyst, and the mixture is titrated with the cerium(IV) solution to be standardized.

Pure sodium oxalate may be dissolved in perchloric acid and employed for the standardization at room temperature of cerium(IV) solutions in either nitric acid or perchloric acid. If one desires to standardize a cerium(IV) sulfate titrant by an analogous procedure, it is necessary to heat the oxalic acid solution to approximately 70°C so that the rate of the cerium(IV)-oxalic acid reaction will be adequately fast.

In addition to pure iron wire, other substances for the standardization of cerium(IV) titrants include Oesper's salt, $FeC_2H_4(NH_3)_2(SO_4)_2 \cdot 2\,H_2O$, and Mohr's salt, $Fe(NH_4)_2(SO_4)_2 \cdot 6\,H_2O$.

### End-Point Detection for Cerium(IV) Titrations

Although cerium(IV) is yellow in all mineral acid solutions, whereas its reduced form, cerium(III), is colorless, the color of the former is not sufficiently intense to serve as an end-point signal, nor does it interfere with recognition of the color changes of redox indicators. Some common indicators, listed in Table 8–3, for cerium(IV) titrations are tris(1,10-phenanthroline)iron(II), tris(5-nitro-1,10-phenanthroline)iron(II), and tris(5-methyl-1,10-phenanthroline)iron(II).

### Analytical Uses of Cerium(IV) Solutions

Many volumetric analyses involving direct titration with a standard potassium permanganate solution can be performed equally well with cerium(IV). As a matter of fact, all of the procedures listed in Table 9–2 are adaptable with little or no modification for use with cerium(IV) titrants. There are, however, some unique applications of cerium(IV) for the determination of organic substances.

**Determination of Organic Compounds.** Cerium(IV) in 4 *F* perchloric acid quantitatively oxidizes many organic substances, frequently within about 15 minutes at room temperature. Perhaps the most striking feature about the reactions between cerium(IV) and organic compounds is that the oxidation products depend on the nature of the organic substance:

1. A compound with *hydroxyl* groups on two or more adjacent carbon atoms has the bond between each of these carbon atoms cleaved, and each fragment is oxidized to a saturated carboxylic acid:

$$\underset{\text{glycerol}}{\underset{\text{OH}}{\underset{|}{H_2C}}-\underset{\text{OH}}{\underset{|}{CH}}-\underset{\text{OH}}{\underset{|}{CH_2}}} + 3\,H_2O \rightarrow 3\,HCOOH + 8\,H^+ + 8\,e$$

2. A compound with an *aldehyde* group bound to a carbon atom also having a ketone group or a hydroxyl group has the bond between the two carbon atoms cleaved, and each fragment is oxidized to a saturated carboxylic acid:

$$\underset{\text{glyceraldehyde}}{H-\underset{\text{O}}{\underset{\|}{C}}-\underset{\text{OH}}{\underset{|}{CH}}-\underset{\text{OH}}{\underset{|}{CH_2}}} + 3\,H_2O \rightarrow 3\,HCOOH + 6\,H^+ + 6\,e$$

3. A compound with a *carboxyl* group bound to a carbon atom which also has a hydroxyl group, or a *carboxyl* group bound to an aldehyde group or another carboxyl group, has the bond between the two carbon atoms broken; each carboxyl group is oxidized to carbon dioxide, and the other fragments are oxidized to saturated carboxylic acids:

$$HOOC{-}CH(OH){-}CH(OH){-}COOH + 2\,H_2O \rightarrow 2\,CO_2 + 2\,HCOOH + 6\,H^+ + 6\,e$$

tartaric acid

4. A compound with a *ketone* group adjacent to a carboxyl group, an aldehyde group, or another ketone group – or a *ketone* group bound to a carbon atom with a hydroxyl group – has the bond between the two carbon atoms cleaved; each fragment with a ketone group is oxidized to a saturated carboxylic acid, and the other fragments are oxidized in accordance with previous rules:

$$H{-}C(=O){-}C(=O){-}CH_2{-}CH_3 + 2\,H_2O \rightarrow HCOOH + CH_3CH_2COOH + 2\,H^+ + 2\,e$$

1,2-butanedione

5. A compound with an active *methylene* group ($-CH_2-$), that is, a methylene group bound to two carboxyl groups, two aldehyde groups, two ketone groups, or combinations of these groups, has each of the two carbon-carbon bonds broken; each methylene fragment is oxidized to formic acid, and each other fragment is oxidized in accordance with the preceding rules:

$$HOOC{-}CH_2{-}COOH + 2\,H_2O \rightarrow 2\,CO_2 + HCOOH + 6\,H^+ + 6\,e$$

malonic acid

In order to determine an organic compound, such as one of those just mentioned, the sample is added to a known excess of cerium(IV) in 4 $F$ perchloric acid and the reaction mixture is allowed to stand (or perhaps is heated at 50 to 60°C) for up to 20 minutes. Unreacted cerium(IV) can be back-titrated with a standard oxalic acid solution, with tris(5-nitro-1,10-phenanthroline)iron(II) as the end-point indicator.

## IODINE (TRIIODIDE)

Iodine exists in a number of analytically important oxidation states, represented by such familiar species as iodide, molecular iodine (or triiodide ion), iodine monochloride, iodate, and periodate. Of particular interest are oxidation-reduction processes involving the two lowest oxidation states, namely, iodide and iodine (triiodide ion). Although three different half-reactions listed in Appendix 4 pertain to these substances

$$I_2(aq) + 2\,e \rightleftharpoons 2\,I^-; \qquad E^0 = +0.6197\text{ v}$$

$$I_2(s) + 2\,e \rightleftharpoons 2\,I^-; \qquad E^0 = +0.5345\text{ v}$$

$$I_3^- + 2\,e \rightleftharpoons 3\,I^-; \qquad E^0 = +0.5355\text{ v}$$

the third half-reaction provides the most realistic picture of the redox behavior of the iodine-iodide system because it includes the two predominant species, triiodide and iodide ions, encountered in practical situations.

Triiodide ion is a sufficiently good oxidizing agent to react quantitatively with a number of reductants. On the other hand, iodide ion is easily enough oxidized to permit its quantitative reaction with many strong oxidizing agents. Accordingly, two major classifications of redox methods involving the use of the triiodide-iodide half-reaction have been developed. First, there are the *direct* methods in which a standard solution of triiodide, or iodine dissolved in a potassium iodide medium, is utilized for the direct titration of the substance to be determined. Second, many *indirect* procedures are employed in which triiodide is formed through the reaction of excess iodide ion with some oxidizing agent, followed by the titration of the triiodide ion with a standard solution of sodium thiosulfate. In a direct method, the end point is marked by the first permanent appearance of free iodine (triiodide) in the titration vessel; in an indirect procedure, the final disappearance of triiodide ion signals the end point.

### Preparation and Stability of Iodine (Triiodide) Solutions

Although solid iodine is sparingly soluble in water (0.00133 *M*), its solubility is considerably enhanced in an aqueous potassium iodide medium, owing to the formation of triiodide ion:

$$I_2(aq) + I^- \rightleftharpoons I_3^-; \qquad K = 708$$

Thus, a so-called *iodine* titrant is actually a solution of potassium iodide in which the requisite weight of solid molecular iodine has been dissolved; typically, such a solution contains 0.05 $M\ I_3^-$, 0.20 $M\ I^-$, and only $3.5 \times 10^{-4}$ $M\ I_2$, and might be more appropriately termed a *triiodide* titrant. Nevertheless, triiodide ion and molecular iodine behave identically in oxidation-reduction processes.

Standard triiodide solutions may be prepared directly from an accurately weighed portion of pure solid iodine, but it is simpler to prepare an iodine (triiodide) solution of approximately the desired concentration and to standardize it against pure arsenious oxide.

Triiodide solutions are inherently unstable for at least two reasons. One cause of this instability is the volatility of iodine. Loss of iodine can occur in spite of the facts that excess potassium iodide is present and that most of the dissolved iodine really exists as the triiodide ion. However, if the standard solution is stored in a tightly stoppered bottle, there should be no significant loss of iodine for several weeks.

Another reason why triiodide solutions undergo changes in concentration is that dissolved atmospheric oxygen causes the oxidation of iodide ion to iodine (triiodide):

$$6\,I^- + O_2 + 4\,H^+ \rightleftharpoons 2\,I_3^- + 2\,H_2O; \qquad E^0 = +0.693\text{ v}$$

Fortunately, this oxidation proceeds very slowly, even though the forward reaction is strongly favored. Since hydrogen ion is a reactant, the oxidation of iodide becomes more significant as the pH decreases, but the problem is not at all serious for essentially neutral triiodide solutions.

### End-Point Detection for Titrations Involving Iodine

Three different techniques are commonly employed for end-point detection in titrations involving iodine. Provided that iodine (triiodide) is the only colored substance in the system, the appearance or disappearance of the yellow-brown triiodide color is itself an extremely sensitive way to locate the end point of a titration. In a perfectly colorless solution, it is possible to detect visually a concentration of triiodide as low as $5 \times 10^{-6}$ *M* which, in a solution volume of 100 ml, corresponds to only 0.01 ml of 0.05 *F* iodine (triiodide) titrant.

Triiodide ion forms a very intense blue-colored complex with colloidally dispersed starch, which serves well as an indication of the presence of iodine. Thus, the end point of a titration with a triiodide solution is marked by the appearance of the blue starch-iodine color, whereas the blue color disappears at the end point of an indirect iodine procedure. As little as $2 \times 10^{-7}$ *M* iodine (triiodide) gives a detectable blue color with starch under optimum conditions.

Another form of end-point detection involves the addition of a few milliliters of carbon tetrachloride or chloroform to the titration flask. Since molecular iodine is much more soluble in the nonpolar organic layer than in the aqueous phase, iodine will concentrate itself in the heavier, lower layer and will impart its violet color to this layer. As the end point is approached, the flask is thoroughly shaken after each addition of titrant, and the organic phase is examined for the presence or absence of the iodine color.

### Standardization of Iodine (Triiodide) Solutions

Iodine or triiodide solutions are almost always standardized by titration against a solution of arsenious oxide. We may write the pertinent reaction* as

$$HAsO_2 + I_3^- + 2\,H_2O \rightleftharpoons H_3AsO_4 + 3\,I^- + 2\,H^+;\ K = 0.17$$

Due to the small magnitude of the equilibrium constant, the titration reaction is adequately complete only if the concentrations of the product species remain low. In particular, the hydrogen ion concentration can be adjusted to and controlled at almost any suitably small value through the use of an appropriate buffer; pH 7 to 9 is the optimum range for the arsenic(III)-triiodide reaction. Accordingly, after a weighed portion of arsenious oxide is dissolved in sodium hydroxide medium, enough hydrochloric acid is added to neutralize or slightly acidify the solution. Next, several grams of sodium bicarbonate are added to provide a carbonic acid-bicarbonate buffer of pH 7 to 8; then starch indicator is added and the solution is titrated with the iodine (triiodide) solution to be standardized.

---

*See footnote at the bottom of page 218.

### Reaction Between Iodine (Triiodide) and Thiosulfate

Whereas most other oxidizing agents convert thiosulfate to sulfite or sulfate, the reaction between thiosulfate and iodine (or triiodide) to yield the tetrathionate ion

$$I_3^- + 2\ S_2O_3^{2-} \rightleftharpoons 3\ I^- + S_4O_6^{2-}$$

is unusual. In general, the iodine-thiosulfate reaction proceeds rapidly according to the preceding well-defined stoichiometry at all pH values between 0 and 7. In mildly alkaline media, however, even iodine oxidizes thiosulfate to sulfate,

$$4\ I_3^- + S_2O_3^{2-} + 10\ OH^- \rightleftharpoons 2\ SO_4^{2-} + 12\ I^- + 5\ H_2O$$

although the reaction is by no means quantitative until the pH becomes very high. Above a pH of approximately 8 or 9, triiodide disproportionates into iodide and hypoiodous acid,

$$I_3^- + OH^- \rightleftharpoons 2\ I^- + HOI$$

the latter appearing to be the substance which most readily oxidizes thiosulfate to sulfate:

$$4\ HOI + S_2O_3^{2-} + 6\ OH^- \rightleftharpoons 4\ I^- + 2\ SO_4^{2-} + 5\ H_2O$$

However, the chemistry of iodine in alkaline media is complicated further by the disproportionation of hypoiodous acid into iodate and iodide:

$$3\ HIO + 3\ OH^- \rightleftharpoons IO_3^- + 2\ I^- + 3\ H_2O$$

Since such side reactions are not unique to the iodine-thiosulfate system, the successful use of iodine or triiodide titrants is generally restricted to solutions having pH values less than about 8.

### Preparation and Stability of Sodium Thiosulfate Solutions

Pure sodium thiosulfate pentahydrate, $Na_2S_2O_3 \cdot 5\ H_2O$, can be crystallized and stored under carefully controlled conditions, and an accurately weighed portion of the solid may be dissolved in water for the direct preparation of a standard solution. However, it is more straightforward to prepare a solution of approximately the desired strength and to standardize it against potassium dichromate, potassium iodate, or a previously standardized iodine (triiodide) solution.

There are at least two factors which influence the stability of a thiosulfate solution – the pH of the solution and the presence of sulfur-consuming bacteria.

In a very dilute acid medium, thiosulfate slowly decomposes into elemental sulfur and hydrogen sulfite ion,

$$S_2O_3^{2-} + H^+ \rightleftharpoons HS_2O_3^- \rightarrow S + HSO_3^-$$

but, in a 1 *F* hydrochloric acid medium, decomposition of thiosulfate is extensive within only a minute or two. This decomposition changes the effective concentration of thiosulfate titrants because the $HSO_3^-$ ion reduces twice the amount of iodine (triiodide) that thiosulfate does, as can be seen from the equilibria

$$HSO_3^- + I_3^- + H_2O \rightleftharpoons HSO_4^- + 3\,I^- + 2\,H^+$$

and

$$2\,S_2O_3^{2-} + I_3^- \rightleftharpoons S_4O_6^{2-} + 3\,I^-$$

Although thiosulfate is unstable in acidic media, nothing prevents its use as a titrant for iodine or triiodide in even relatively concentrated acid solutions, if the titration is carried out in such a manner that no significant local excess of thiosulfate is present.

Sulfur-metabolizing bacteria present in distilled water can convert thiosulfate to a variety of products, including elemental sulfur, sulfite, and sulfate. One usually boils the distilled water in which the solid reagent is to be dissolved as a means of destroying the bacteria. Furthermore, it is common to introduce 50 to 100 mg of sodium bicarbonate into a liter of sodium thiosulfate titrant because bacterial action is minimal near pH 9 or 10.

### Standardization of Sodium Thiosulfate Solutions

All procedures employed for the standardization of thiosulfate solutions ultimately involve the reaction between iodine (triiodide) and thiosulfate. A previously standardized iodine (triiodide) solution suffices well for the standardization of a sodium thiosulfate titrant. In addition, strong oxidizing agents, such as potassium dichromate and potassium iodate, may serve for the standardization through the indirect method of analysis involving iodine.

Potassium dichromate is an excellent reagent for the standardization of thiosulfate solutions. If excess potassium iodide is dissolved in an acidic solution containing an accurately known amount of dichromate, the latter oxidizes iodide to triiodide

$$Cr_2O_7^{2-} + 9\,I^- + 14\,H^+ \rightleftharpoons 2\,Cr^{3+} + 3\,I_3^- + 7\,H_2O$$

and the liberated iodine may be titrated with the sodium thiosulfate solution to a starch-iodine end point.

Another substance which quantitatively oxidizes iodide to triiodide in an acid medium is potassium iodate:

$$IO_3^- + 8\,I^- + 6\,H^+ \rightleftharpoons 3\,I_3^- + 3\,H_2O$$

One may titrate the iodine (triiodide) produced by this reaction with the sodium thiosulfate solution.

### Analytical Uses of the Triiodide-Iodide System

Because triiodide ion is a relatively mild oxidant, it can react quantitatively only with substances which are easily oxidizable. Some of the substances which can be determined by direct titration with a standard iodine (triiodide) solution are listed in Table 9–3, along with the half-reaction which each species undergoes, plus information about the experimental procedure.

Table 9–3. Some Titrations Performed with Standard Potassium Permanganate in Acidic Media

| Substance Determined | Half-Reaction for Substance Titrated | Procedure and Conditions for Titration |
|---|---|---|
| Sb(III) | $SbCl_4^- + 2\,Cl^- \rightleftharpoons SbCl_6^- + 2\,e$ | Titrate in 2 *F* HCl |
| As(III) | $HAsO_2 + 2\,H_2O \rightleftharpoons H_3AsO_4 + 2\,H^+ + 2\,e$ | Titrate in 1 *F* HCl with ICl as catalyst |
| $Br^-$ | $2\,Br^- \rightleftharpoons Br_2 + 2\,e$ | Titrate in boiling 2 *F* $H_2SO_4$ to expel $Br_2$ |
| $Fe(CN)_6^{4-}$ | $Fe(CN)_6^{4-} \rightleftharpoons Fe(CN)_6^{3-} + e$ | Titrate in 0.2 *F* $H_2SO_4$ with erioglaucine A as indicator |
| $H_2O_2$ | $H_2O_2 \rightleftharpoons O_2 + 2\,H^+ + 2\,e$ | Titrate in 1 *F* $H_2SO_4$ |
| $I^-$ | $I^- + HCN \rightleftharpoons ICN + H^+ + 2\,e$ | Titrate in presence of 0.1 *F* HCN and 1 *F* $H_2SO_4$; ferroin indicator |
| Fe(II) | $Fe^{2+} \rightleftharpoons Fe^{3+} + e$ | Reduce iron(III) with tin(II) or in Jones reductor; titrate in 1 *F* $H_2SO_4$ or in 1 *F* HCl with Zimmermann-Reinhardt reagent added |
| Mo(III) | $Mo^{3+} + 2\,H_2O \rightleftharpoons MoO_2^{2+} + 4\,H^+ + 3\,e$ | Reduce molybdenum to molybdenum(III) in Jones reductor; treat with excess iron(III) in 1 *F* $H_2SO_4$ and titrate iron(II) formed |
| $H_2C_2O_4$ | $H_2C_2O_4 \rightleftharpoons 2\,CO_2 + 2\,H^+ + 2\,e$ | Titrate in 1 *F* $H_2SO_4$ at 70°C |
| Ca(II), Mg(II), Zn(II), Co(II), La(III), Th(IV), Ba(II), Sr(II), Pb(II) | $H_2C_2O_4 \rightleftharpoons 2\,CO_2 + 2\,H^+ + 2\,e$ | Metal oxalates are precipitated, washed, and redissolved in 1 *F* $H_2SO_4$; liberated oxalic acid titrated at 70°C |
| Sn(II) | $SnCl_4^{2-} + 2\,Cl^- \rightleftharpoons SnCl_6^{2-} + 2\,e$ | Reduce tin(IV) to tin(II) with bismuth amalgam in 5 to 12 *F* HCl; exclude oxygen during titration |
| Ti(III) | $Ti^{3+} + H_2O \rightleftharpoons TiO^{2+} + 2\,H^+ + e$ | Reduce titanium(IV) to titanium(III) with zinc metal, lead amalgam, or in Jones reductor; exclude oxygen; titrate in 1 *F* $H_2SO_4$ or HCl |
| W(III) | $W^{3+} + 2\,H_2O \rightleftharpoons WO_2^{2+} + 4\,H^+ + 3\,e$ | Reduce tungsten to tungsten(III) with lead amalgam at 50°C; titrate in 2 *F* HCl |
| U(IV) | $U^{4+} + 2\,H_2O \rightleftharpoons UO_2^{2+} + 4\,H^+ + 2\,e$ | Reduce uranium to uranium(III) in Jones reductor; expose solution to air to obtain uranium(IV); titrate in 1 *F* $H_2SO_4$ |
| V(IV) | $VO^{2+} + H_2O \rightleftharpoons VO_2^+ + 2\,H^+ + e$ | Reduce vanadium to vanadium(IV) with bismuth amalgam; titrate in 1 *F* $H_2SO_4$ |

Many oxidizing agents are capable of converting iodide quantitatively to free iodine, which in the presence of excess iodide forms the triiodide ion. Since a stoichiometric relationship exists between the original quantity of oxidant and the amount of triiodide produced, a determination of triiodide, by titration with a standard solution of sodium thiosulfate, provides data from which the quantity of oxidizing agent may be calculated. In applying the indirect method of analysis, one must always be sure that the reaction between the strong oxidant and excess iodide has reached completion before beginning the titration with thiosulfate. Otherwise, any strong oxidizing agent present at the start of the titration will probably oxidize thiosulfate to sulfur and sulfate as well as tetrathionate, thereby upsetting the stoichiometry of the triiodide-thiosulfate reaction. Included in Table 9–4 are many of the species which can be determined by means of the indirect method.

## PERIODIC ACID

To the chemist interested in the determination of organic compounds, the oxidizing properties of periodic acid have special significance because this reagent undergoes a number of highly selective reactions with organic functional groups.

### Nature of Periodic Acid Solutions

Detailed studies have demonstrated that paraperiodic acid, $H_5IO_6$, is the parent compound in aqueous periodic acid solutions. In an aqueous medium, paraperiodic acid behaves as a moderately weak diprotic acid:

$$H_5IO_6 \rightleftharpoons H^+ + H_4IO_6^-; \qquad K_{a1} = 2.3 \times 10^{-2}$$

$$H_4IO_6^- \rightleftharpoons H^+ + H_3IO_6^{2-}; \qquad K_{a2} = 4.4 \times 10^{-9}$$

Furthermore, the product of the first acid dissociation, $H_4IO_6^-$, may dehydrate to form the metaperiodate ion, $IO_4^-$:

$$H_4IO_6^- \rightleftharpoons IO_4^- + 2\,H_2O; \qquad K = 40$$

Paraperiodic acid is an extremely powerful oxidant, the standard potential for the half-reaction

$$H_5IO_6 + H^+ + 2\,e \rightleftharpoons IO_3^- + 3\,H_2O$$

being in the neighborhood of +1.6 to +1.7 v versus NHE.

### Preparation and Standardization of Periodate Solutions

Periodate solutions may be prepared from paraperiodic acid, potassium metaperiodate, or sodium metaperiodate, but the latter compound is preferred because it exhibits high solubility and is easily purified. Despite the oxidizing strength of periodic acid and periodate salts, aqueous solutions containing these species are remarkably stable, and water is only slowly oxidized to oxygen and ozone. All organic matter must be rigorously excluded from periodate solutions.

**Table 9–4. Some Indirect Iodine Methods**

| Substance Determined | Half-Reaction for Substance Determined | Experimental Conditions |
|---|---|---|
| $IO_4^-$ | $IO_4^- + 8\,H^+ + 7\,e \rightleftharpoons \frac{1}{2}\,I_2 + 4\,H_2O$ | 0.5 $F$ HCl |
| $IO_3^-$ | $IO_3^- + 6\,H^+ + 5\,e \rightleftharpoons \frac{1}{2}\,I_2 + 3\,H_2O$ | 0.5 $F$ HCl |
| $BrO_3^-$ | $BrO_3^- + 6\,H^+ + 6\,e \rightleftharpoons Br^- + 3\,H_2O$ | 0.5 $F$ $H_2SO_4$ or HCl |
| $Br_2$ | $Br_2 + 2\,e \rightleftharpoons 2\,Br^-$ | Dilute acid |
| $Cl_2$ | $Cl_2 + 2\,e \rightleftharpoons 2\,Cl^-$ | Dilute acid |
| HOCl | $HOCl + H^+ + 2\,e \rightleftharpoons Cl^- + H_2O$ | 0.5 $F$ $H_2SO_4$ |
| $O_3$ | $O_3 + 2\,H^+ + 2\,e \rightleftharpoons O_2 + H_2O$ | Absorb $O_3$ in slightly alkaline KI solution; add $H_2SO_4$ and titrate liberated $I_3^-$ |
| $O_2$ | $(O_2 + 4\,Mn(OH)_2 + 2\,H_2O \rightleftharpoons 4\,Mn(OH)_3)$ $Mn(OH)_3 + 3\,H^+ + e \rightleftharpoons Mn^{2+} + 3\,H_2O$ | To $O_2$ containing sample, add solutions of $Mn^{2+}$ and of NaOH–KI; after 1 minute, acidify with $H_2SO_4$ and titrate liberated $I_3^-$ |
| $H_2O_2$ | $H_2O_2 + 2\,H^+ + 2\,e \rightleftharpoons 2\,H_2O$ | 1 $F$ $H_2SO_4$ with $NH_4MoO_3$ catalyst |
| $MnO_4^-$ | $MnO_4^- + 8\,H^+ + 5\,e \rightleftharpoons Mn^{2+} + 4\,H_2O$ | 0.1 $F$ HCl or $H_2SO_4$ |
| $MnO_2$ | $MnO_2 + 4\,H^+ + 2\,e \rightleftharpoons Mn^{2+} + 2\,H_2O$ | 0.5 $F$ $H_3PO_4$ or HCl |
| $PbO_2$ | $PbO_2 + 4\,H^+ + 2\,e \rightleftharpoons Pb^{2+} + 2\,H_2O$ | 2 $F$ HCl |
| $Cr_2O_7^{2-}$ | $Cr_2O_7^{2-} + 14\,H^+ + 6\,e \rightleftharpoons 2\,Cr^{3+} + 7\,H_2O$ | 0.4 $F$ HCl; wait 5 minutes before titration |
| Ba(II), Pb(II) | $Cr_2O_7^{2-} + 14\,H^+ + 6\,e \rightleftharpoons 2\,Cr^{3+} + 7\,H_2O$ | Precipitate $BaCrO_4$ or $PbCrO_4$; dissolve washed precipitate in 1 $F$ HCl |
| Ce(IV) | $Ce^{4+} + e \rightleftharpoons Ce^{3+}$ | 1 $F$ $H_2SO_4$ |
| $S_2O_8^{2-}$ | $S_2O_8^{2-} + 2\,e \rightleftharpoons 2\,SO_4^{2-}$ | Mix $S_2O_8^{2-}$ and $I^-$ in neutral medium; after 15 minutes, acidify solution and titrate $I_3^-$ |
| Cu(II) | $Cu^{2+} + I^- + e \rightleftharpoons CuI(s)$ | pH 1–2 |
| $Fe(CN)_6^{3-}$ | $Fe(CN)_6^{3-} + e \rightleftharpoons Fe(CN)_6^{4-}$ | 1 $F$ HCl |
| As(V) | $H_3AsO_4 + 2\,H^+ + 2\,e \rightleftharpoons HAsO_2 + 2\,H_2O$ | 5 $F$ HCl |
| Sb(V) | $SbCl_6^- + 2\,e \rightleftharpoons SbCl_4^- + 2\,Cl^-$ | 5 $F$ HCl |

For the standardization of a periodate solution, an aliquot is added to an excess of potassium iodide solution buffered at pH 8 to 9, whereupon periodate is reduced to iodate and an equivalent quantity of triiodide is liberated:

$$IO_4^- + 3\,I^- + H_2O \rightleftharpoons IO_3^- + I_3^- + 2\,OH^-$$

A standard solution of arsenic(III) can be employed to titrate the triiodide to a starch-iodine end point.

### Reactions of Periodate with Organic Compounds

Since oxidations of organic molecules by periodic acid are slow, an excess of this reagent, as well as a reaction time ranging from a few minutes to several hours, is needed to achieve quantitative results. Most periodate oxidations are performed in an aqueous medium, although the limited solubility of the organic compound in water may require that ethanol or glacial acetic acid be used as solvent. Oxidations are usually carried out at or below room temperature because undesired side reactions occur at elevated temperatures.

Organic compounds with hydroxyl (–OH) groups on two adjacent carbon atoms are oxidized by periodic acid at room temperature. This oxidation process, known as the **Malaprade reaction**, involves scission of the bond between these carbon atoms followed by conversion of each hydroxyl group to a carbonyl group. For example, 1,2-butanediol is oxidized to formaldehyde and propionaldehyde:

$$\underset{\displaystyle OH}{\overset{}{H_2C}}—\underset{\displaystyle OH}{CH}—CH_2—CH_3 + H_4IO_6^- \rightarrow \underset{\displaystyle O}{\overset{}{H_2C}}\!\!=\; + \;\underset{\displaystyle O}{HC}\!\!=—CH_2—CH_3 + IO_3^- + 3\,H_2O$$

Many organic substances have a carbonyl group (>C=O) attached either to another carbonyl group or to a carbon atom with a hydroxyl group. Periodic acid reacts with these compounds to cleave the carbon-carbon bond and to oxidize a carbonyl group to a carboxyl group (–COOH) and a hydroxyl group to a carbonyl group. Thus, the oxidation of one mole of glyoxal yields two moles of formic acid,

$$H—\underset{\displaystyle O}{\overset{}{C}}\!(=)—\underset{\displaystyle O}{C}\!(=)—H + H_4IO_6^- \rightarrow H—\underset{\displaystyle O}{C}\!(=)—OH + HO—\underset{\displaystyle O}{C}\!(=)—H + IO_3^- + H_2O$$

and the oxidation of 2-hydroxy-3-pentanone results in the formation of propanoic acid and acetaldehyde:

$$H_3C—CH_2—\underset{\displaystyle O}{C}\!(=)—\underset{\displaystyle OH}{CH}—CH_3 + H_4IO_6^- \rightarrow H_3C—CH_2—\underset{\displaystyle O}{C}\!(=)—OH + \underset{\displaystyle O}{HC}\!(=)—CH_3 + IO_3^- + 2\,H_2O$$

It is of considerable significance to the success of periodate oxidations that organic compounds with single or isolated hydroxyl or carbonyl groups are unreactive *at room temperature*. At elevated temperatures, the selectivity of periodate oxidations is lost because carboxyl groups as well as the isolated functional groups will react, and the rules for predicting the course of these reactions become invalid.

Situations may be encountered in which three or more adjacent functional groups are present in the same molecule, as, for example, in glycerol. To establish what products are formed in the oxidation of glycerol by periodate, we can apply two rules: (1) assume that chemical attack begins at two adjacent functional groups on one end of the molecule and progresses one carbon-carbon bond at a time toward the other end of the molecule, and (2) use information from previous examples to determine the nature of the products. Accordingly, the first step of the oxidation,

$$\underset{\text{glycerol}}{\underset{\text{OH}}{\overset{|}{H_2C}}\text{—}\underset{\text{OH}}{\overset{|}{CH}}\text{—}\underset{\text{OH}}{\overset{|}{CH_2}}} + H_4IO_6^- \rightarrow \underset{\text{O}}{\overset{\|}{H_2C}} + \underset{\text{O}}{\overset{\|}{HC}}\text{—}\underset{\text{OH}}{\overset{|}{CH_2}} + IO_3^- + 3\,H_2O$$

is followed by the reaction

$$\underset{\text{O}}{\overset{\|}{HC}}\text{—}\underset{\text{OH}}{\overset{|}{CH_2}} + H_4IO_6^- \rightarrow \underset{\text{O}}{\overset{\|}{HC}}\text{—OH} + \underset{\text{O}}{\overset{\|}{CH_2}} + IO_3^- + 2\,H_2O$$

so the overall oxidation of 1 mole of glycerol yields 1 mole of formic acid plus 2 moles of formaldehyde.

### Analytical Uses of Periodate Solutions

Two procedures are employed in the application of the Malaprade reaction to the determination of organic compounds. In the first method, a titrimetric measurement of the quantity of periodate consumed in the oxidation process is performed and, in the second, more elaborate technique, a detailed analysis of the identities and quantities of the reaction products is involved.

**Determination of Ethylene Glycol.** Consider the analysis of an aqueous ethylene glycol solution. If a known excess of a standard sodium metaperiodate solution is added to the sample and the mixture set aside for one hour at room temperature, the ethylene glycol will be quantitatively oxidized to formaldehyde, and an equivalent amount of periodate will be reduced to iodate. As described earlier, one can adjust the pH of the reaction mixture to 7 (employing a carbonic acid-bicarbonate buffer), add an excess of potassium iodide, and titrate the resulting triiodide with a standard arsenic(III) solution; the *difference* between the amount of periodate originally used in the reaction mixture and that found by means of the arsenic(III) titration is proportional to the quantity of ethylene glycol in the sample.

**Analysis of a Mixture of Ethylene Glycol, 1,2-Propylene Glycol, and Glycerol.** Treatment of a solution containing ethylene glycol, 1,2-propylene glycol, and glycerol with excess periodic acid causes the following reactions to occur:

$$\underset{\text{ethylene glycol}}{\underset{\text{OH}}{\overset{|}{H_2C}}\text{—}\underset{\text{OH}}{\overset{|}{CH_2}}} + H_5IO_6 \rightarrow 2\,HCHO + IO_3^- + H^+ + 3\,H_2O$$

$$\underset{\text{1,2-propylene glycol}}{\underset{\text{OH}}{\underset{|}{\text{H}_2\text{C}}}\text{—}\underset{\text{OH}}{\underset{|}{\text{CH}}}\text{—CH}_3} + H_5IO_6 \rightarrow HCHO + CH_3CHO + IO_3^- + H^+ + 3\,H_2O$$

$$\underset{\text{glycerol}}{\underset{\text{OH}}{\underset{|}{\text{H}_2\text{C}}}\text{—}\underset{\text{OH}}{\underset{|}{\text{CH}}}\text{—}\underset{\text{OH}}{\underset{|}{\text{CH}_2}}} + 2\,H_5IO_6 \rightarrow 2\,HCHO + HCOOH + 2\,IO_3^- + 2\,H^+ + 5\,H_2O$$

Formic acid arises only from oxidation of glycerol. Therefore, if the reaction mixture is titrated with standard sodium hydroxide solution after the periodate oxidation is complete, the quantity of formic acid can be related to the concentration of glycerol in the sample. However, one must correct for the amount of periodic acid in the reaction mixture by titration with sodium hydroxide of a blank solution containing all the reagents except the three unknowns.

That acetaldehyde originates only from 1,2-propylene glycol is utilized in the determination of the latter substance. However, it is essential to separate acetaldehyde from formaldehyde, which is an oxidation product of all three compounds in the mixture. To perform this separation, a stream of carbon dioxide is passed through the solution resulting from the periodic acid treatment. Highly volatile formaldehyde and acetaldehyde are both carried by the carbon dioxide gas through a closed system which includes specific collecting solutions for the aldehydes. Formaldehyde is preferentially absorbed by a glycine solution,

$$HCHO + H_2NCH_2COOH \rightarrow H_2C{=}NCH_2COOH + H_2O$$

and acetaldehyde is trapped in a sodium bisulfite medium:

$$CH_3CHO + NaHSO_3 \rightarrow H_3C\text{—}\overset{\overset{\text{SO}_3\text{H}}{|}}{\underset{\underset{\text{H}}{|}}{\text{C}}}\text{—}O^-Na^+$$

Then, the excess of the sodium bisulfite collecting solution is destroyed, the acetaldehyde-bisulfite addition compound is decomposed, and the liberated bisulfite (which is stoichiometrically equivalent to acetaldehyde) is titrated with a standard iodine solution.

Since oxidation of ethylene glycol does not lead to any unique products, it is necessary to establish the ethylene glycol content of the mixture indirectly. Accordingly, the sum of all three components in the sample can be determined from the periodic acid consumed in the oxidation reactions; then the previously measured amounts of glycerol and 1,2-propylene glycol are subtracted from this sum to obtain the quantity of ethylene glycol.

## QUESTIONS AND PROBLEMS

1. Calculate the formality of a potassium permanganate solution, 35.00 ml of which is equivalent to 0.2500 g of 98.00 per cent pure calcium oxalate.

2. If 40.32 ml of a solution of oxalic acid can be titrated with 37.92 ml of a 0.4736 *F* sodium hydroxide solution and if 30.50 ml of the same oxalic acid solution reacts with 47.89 ml of a potassium permanganate solution, what is the formality of the permanganate solution?
3. Calculate the per cent purity of a sample of impure $H_2C_2O_4 \cdot 2\,H_2O$ if 0.4006 g of this material requires a titration with 28.62 ml of a potassium permanganate solution, 1.000 ml of which contains 5.980 mg of potassium permanganate.
4. A slag sample is known to contain all of its iron in the forms FeO and $Fe_2O_3$. A 1.000-g sample of the slag was dissolved in hydrochloric acid according to the usual procedure, reduced with stannous chloride, and eventually titrated with 28.59 ml of a 0.02237 *F* potassium permanganate solution. A second slag sample, weighing 1.500 g, was dissolved in a nitrogen atmosphere in order to prevent atmospheric oxidation of iron(II) during the dissolution process and, without any further adjustment of the oxidation state of iron, it was immediately titrated with the same potassium permanganate solution. If 15.60 ml of the permanganate solution was required in the second experiment, calculate (a) the total percentage of iron in the slag and (b) the individual percentages of FeO and $Fe_2O_3$.
5. The quantity of glycerol in a sample solution was determined by reacting the glycerol with 50.00 ml of a strongly alkaline 0.03527 *F* potassium permanganate solution at room temperature for 25 minutes. After the reaction was complete, the mixture was acidified with sulfuric acid, and the excess unreacted permanganate was reduced with 10.00 ml of a 0.2500 *F* oxalic acid solution. Then the excess oxalic acid was back-titrated with a 0.01488 *F* potassium permanganate solution, 26.18 ml being required. Calculate the weight of glycerol in the original sample.
6. In the determination of the manganese dioxide content of a pyrolusite ore, a 0.5261-g sample of the pyrolusite was treated with 0.7049 g of pure sodium oxalate, $Na_2C_2O_4$, in an acid medium. After the reaction had gone to completion, 30.47 ml of 0.02160 *F* potassium permanganate solution was needed to titrate the excess, unreacted oxalic acid. Calculate the percentage of manganese dioxide in the pyrolusite.
7. A 25.00-ml aliquot of a stock hydrogen peroxide solution was transferred to a 250.0-ml volumetric flask, and the solution was diluted and mixed. A 25.00-ml sample of the diluted hydrogen peroxide solution was acidified with sulfuric acid and titrated with 0.02732 *F* potassium permanganate, 35.86 ml being required. Calculate the number of grams of hydrogen peroxide per 100.0 ml of the original stock solution.
8. What weight of an iron ore should be taken for analysis so that the volume of 0.1046 *F* cerium(IV) sulfate solution used in the subsequent titration will be numerically the same as the percentage of iron in the ore?
9. A silver reductor was employed to reduce 0.01000 *M* uranium(VI) to uranium(IV) in a 2.00 *F* hydrochloric acid medium according to the reaction

$$UO_2^{2+} + 2\,Ag + 2\,Cl^- + 4\,H^+ \rightleftharpoons U^{4+} + 2\,AgCl + 2\,H_2O$$

Calculate the fraction of uranium(VI) which remains unreduced after this treatment.

10. Cerium(IV) in 4 *F* perchloric acid is an effective oxidant for many organic compounds. Write balanced half-reactions for the oxidation of each of the following organic substances:

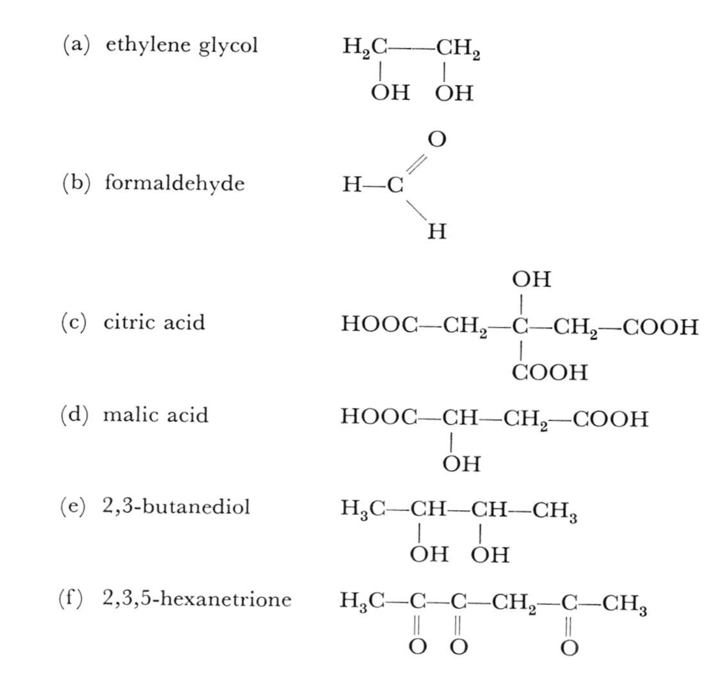


11. A 20.00-ml aliquot of a 1 *F* hydrochloric acid solution containing iron(III) and vanadium(V) was passed through a silver reductor. The resulting solution was titrated with a 0.01020 *F* potassium permanganate solution, 26.24 ml being required to reach a visual end point. A second 20.00-ml aliquot of the sample solution was passed through a Jones reductor, proper precautions being taken to ensure that air oxidation of the reduced species did not occur. When the solution was titrated with the 0.01020 *F* permanganate solution, 31.80 ml was required for the titration. Calculate the concentrations of iron(III) and vanadium(V) in the original sample solution.
12. A potassium permanganate solution was standardized against 0.2643 g of pure arsenious oxide. The arsenious oxide was dissolved in a sodium hydroxide solution; the resulting solution was acidified and finally titrated with the permanganate solution, of which 40.46 ml was used. Calculate the formality of the potassium permanganate solution.
13. A cerium(IV) sulfate solution was standardized against 0.2023 g of 97.89 per cent pure iron wire. If 43.61 ml of the cerium(IV) solution was used in the final titration, calculate its formality.
14. If 36.81 ml of a 0.1206 *F* sodium hydroxide solution is needed to neutralize the acid formed when 34.76 ml of a potassium permanganate solution is treated with sulfur dioxide,

$$2\ MnO_4^- + 5\ H_2SO_3 \rightleftharpoons 2\ Mn^{2+} + 4\ H^+ + 5\ SO_4^{2-} + 3\ H_2O$$

the excess sulfur dioxide being removed by boiling, what volume of 0.1037 $M$ iron(II) solution would be oxidized by 38.41 ml of the potassium permanganate solution?

15. If 5.000 ml of a commercial hydrogen peroxide solution, having a density of 1.010 g/ml, requires 18.70 ml of a 0.1208 $F$ cerium(IV) solution for titration in an acid medium, what is the percentage by weight of hydrogen peroxide in the commercial solution?
16. If 37.12 ml of an iodine (triiodide) solution is required to titrate the sample solution prepared from 0.5078 g of pure arsenious oxide, calculate the formality of the iodine (triiodide) solution.
17. A 0.1936-g sample of primary-standard potassium dichromate was dissolved in water, the solution was acidified, and, after the addition of potassium iodide, the titration of the liberated iodine required 33.61 ml of a sodium thiosulfate solution. Calculate the formality of the thiosulfate solution.
18. The sulfur from a 5.141-g steel sample was evolved as hydrogen sulfide and collected in an excess of an ammoniacal cadmium solution. The resulting cadmium sulfide precipitate was washed and suspended in water to which a few drops of acetic acid had been added. Then 25.00 ml of a 0.002027 $F$ potassium iodate solution was added to the mixture, followed by 3 g of potassium iodide and 10 ml of concentrated hydrochloric acid. The liberated iodine oxidized the hydrogen sulfide gas to sulfur, and the excess unreacted iodine was titrated with 0.1127 $F$ sodium thiosulfate solution from a microburet, 1.085 ml being used. Calculate the percentage of sulfur in the steel.
19. A certain oxidizing agent has a molecular weight of 250.0. A 0.3125-g sample of this compound was treated with excess potassium iodide in an acidic medium, and the liberated iodine (triiodide) was titrated with 20.00 ml of a 0.1250 $F$ sodium thiosulfate solution. How many electrons per molecule are gained by the oxidizing agent in its reaction with iodide?
20. Write balanced equations showing all reactants and products for the oxidation of each of the following compounds with periodic acid at room temperature:

(a) acetoin

$$\begin{array}{cccccccc} H_3C & - & C & - & CH & - & CH_3 \\ & & \| & & | & & \\ & & O & & OH & & \end{array}$$

(b) biacetyl

$$\begin{array}{ccccccc} H_3C & - & C & - & C & - & CH_3 \\ & & \| & & \| & & \\ & & O & & O & & \end{array}$$

(c) glycolic aldehyde

$$\begin{array}{ccccc} H_2C & - & C & - & H \\ | & & \| & & \\ OH & & O & & \end{array}$$

(d) gluconic acid

$$\begin{array}{ccccccccccc} H_2C & - & CH & - & CH & - & CH & - & CH & - & COOH \\ | & & | & & | & & | & & | & & \\ OH & & OH & & OH & & OH & & OH & & \end{array}$$

(e) fructose

$$\begin{array}{ccccccccccc} H_2C & - & CH & - & CH & - & CH & - & C & - & CH_2 \\ | & & | & & | & & | & & \| & & | \\ OH & & OH & & OH & & OH & & O & & OH \end{array}$$

(f) dihydroxyacetone

$$\begin{array}{ccccc} H_2C & - & C & - & CH_2 \\ | & & \| & & | \\ OH & & O & & OH \end{array}$$

21. A sample solution containing glycerol and ethylene glycol was transferred to a 100-ml volumetric flask, 50.00 ml of a standard 0.01913 $F$ paraperiodic acid ($H_5IO_6$) solution was added, and the mixture was diluted to the calibration mark with distilled water. The reaction solution was allowed to stand at room temperature for one hour; then the following analyses were performed on aliquots of this reaction mixture:

    A 25.00-ml aliquot was titrated with a 0.02041 $F$ sodium hydroxide solution, 14.39 ml being required to reach a phenolphthalein end point. In a blank titration, 24.90 ml of the sodium hydroxide solution was required to titrate a 25.00-ml portion of the *original* 0.01913 $F$ paraperiodic acid to the same end point.

    A second 25.00-ml aliquot of the reaction mixture was neutralized with sodium bicarbonate to a pH of approximately 7, and potassium iodide was added. The liberated iodine was titrated with a 0.01029 $F$ arsenic(III) solution to a starch-iodine end point, 5.47 ml being needed.

    Calculate the total number of milligrams of glycerol and of ethylene glycol in the original sample.

22. A 0.4191-g sample of Paris green, an arsenic-containing insecticide, was treated with hydrochloric acid and a reducing agent; and the arsenic was distilled as arsenic trichloride ($AsCl_3$) into a receiving vessel which contained distilled water. The hydrogen chloride which accompanied the arsenic trichloride was neutralized with an excess of solid sodium bicarbonate and the solution was titrated with a 0.04489 $F$ iodine (triiodide) solution, 37.06 ml being required. Calculate the percentage of arsenious oxide, $As_2O_3$, in the sample.

23. Ascorbic acid (vitamin C) in a juice drink was determined by means of a redox procedure involving iodine. A 100.0-ml sample of the drink was acidified with 20 ml of 6 $F$ sulfuric acid, and exactly 25.00 ml of 0.005204 $F$ iodine (triiodide) solution was pipetted into the sample. During a one-minute waiting period, ascorbic acid was oxidized to dehydroascorbic acid according to the half-reaction

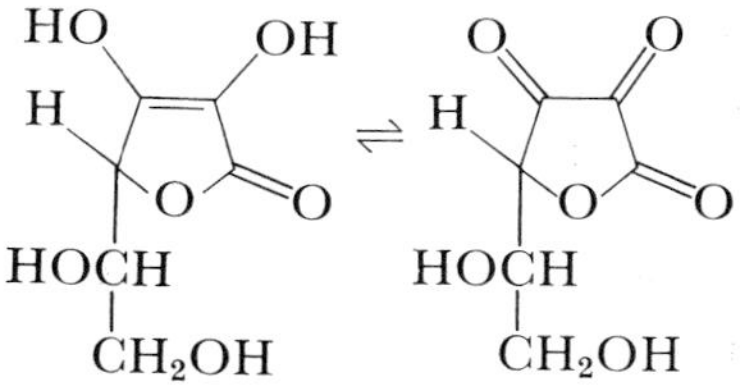

```
HO        OH            O       O
H  \____/               H  \\___//
    |    |       ⇌         |    |        + 2 H+ + 2 e
    \ O /=O                \ O /=O
HOCH                    HOCH
  |                       |
 CH2OH                   CH2OH
```

Next, 3 ml of 0.5 per cent starch indicator was added. Excess, unreacted iodine (triiodide) was reduced through the addition of a known small excess of standard sodium thiosulfate solution; 5.00 ml of 0.009744 $F$ thiosulfate solution was used. Finally, the sample was titrated with the 0.005204 $F$ iodine solution to the first permanent appearance of the starch-iodine color, 2.23 ml being required. Calculate the weight of ascorbic acid in 100 ml of the juice drink. How many fluid ounces of the drink must be consumed to provide the minimum daily requirement (30 mg) of vitamin C?

24. Of the amino acids generally present in protein materials, threonine ($\alpha$-amino-$\beta$-hydroxybutyric acid) is the only one which yields acetaldehyde when treated with sodium metaperiodate:

$$CH_3CH(OH)CH(NH_2)COOH + H_4IO_6^- \rightarrow$$
$$CH_3CHO + CHOCOOH + NH_3 + IO_3^- + 2\,H_2O$$

A protein food tablet, weighing 660 mg, was subjected to hydrolysis in the presence of 6 *F* hydrochloric acid, after which the excess acid was removed by means of vacuum evaporation. The protein hydrolysate was treated at room temperature with sodium metaperiodate in a solution containing excess sodium bicarbonate. A gas train was set up so that acetaldehyde could be separated from formaldehyde, the latter resulting from the periodate oxidation of serine and hydroxylysine. Acetaldehyde was collected in a solution of sodium bisulfite, in which a stable acetaldehyde-bisulfite addition compound is formed. After the excess sodium bisulfite had been decomposed (by oxidation of $HSO_3^-$ with $I_3^-$), borax and sodium carbonate were added to the solution to decompose the acetaldehyde-bisulfite addition compound. Then the liberated bisulfite was titrated with a standard 0.01125 *F* iodine (triiodide) solution to a starch-iodine end point:

$$HSO_3^- + I_3^- + 3\,OH^- \rightleftharpoons SO_4^{2-} + 3\,I^- + 2\,H_2O$$

If 17.92 ml of the iodine titrant was used, calculate the percentage by weight of threonine in the protein tablet.

25. The reaction between hypochlorite and iodide ions in an acidic medium

$$OCl^- + 3\,I^- + 2\,H^+ \rightleftharpoons Cl^- + I_3^- + H_2O$$

is the basis of a method for the determination of "available chlorine" in bleaching powders and bleach solutions. A 2.622-g sample of a household bleach (sodium hypochlorite) solution was treated with excess potassium iodide, and the solution was acidified with 15 ml of 2 *F* sulfuric acid. The liberated iodine (triiodide) was immediately titrated in the presence of starch indicator with 0.1109 *F* sodium thiosulfate solution, 35.58 ml being required. Calculate the "available chlorine" in the bleach solution.

The strength of a chlorine-containing bleaching agent is expressed in terms of "available chlorine." Available chlorine is a measure of the oxidizing power of the chlorine present in the bleaching agent. For example, pure, solid sodium hypochlorite (NaOCl) contains 47.62 weight per cent chlorine. However, the oxidizing strength of chlorine in $OCl^-$ is twice that of chlorine in $Cl_2$ because the reduction of $OCl^-$ to $Cl^-$ is a two-electron reaction whereas the reduction of $Cl_2$ to $Cl^-$ is a one-electron process *per* chlorine. Thus, the "available chlorine" in pure NaOCl is said to be (2)(47.62) or 95.24 per cent.

## SUGGESTIONS FOR ADDITIONAL READING

1. A. Berka, J. Vulterin, and J. Zýka: *Newer Redox Titrants.* Pergamon Press, London, 1965.
2. I. M. Kolthoff and R. Belcher: *Volumetric Analysis.* Volume 3, Wiley-Interscience, New York, 1957.
3. B. Kratochvil: Analytical oxidation-reduction reactions in organic solvents. In *Critical Reviews in Analytical Chemistry.* Volume 1, Chemical Rubber Company, Cleveland, Ohio, 1971, pp. 415–454.
4. H. A. Laitinen: *Chemical Analysis.* McGraw-Hill Book Company, New York, 1960, pp. 326–451.
5. G. F. Smith: *Analytical Applications of Periodic Acid and Iodic Acid.* G. F. Smith Chemical Company, Columbus, Ohio, 1950.
6. G. F. Smith: *Cerate Oxidimetry.* G. F. Smith Chemical Company, Columbus, Ohio, 1942.

# 10 ION-SELECTIVE ELECTRODES AND POTENTIOMETRIC TITRATIONS

We observed in Chapter 8 that the emf of a galvanic cell depends on the activities of reducing and oxidizing agents in equilibrium with the electrodes. Such behavior can be exploited analytically in two ways. First, the magnitude of the emf of a galvanic cell can be related by means of the Nernst equation to the activity of a single species, providing us with the procedure known as **direct potentiometry**. Second, the variation of the emf of a galvanic cell in response to changes in the activities of chemical species can serve as a valuable method for following the progress of a titration, commonly referred to as the technique of **potentiometric titration**.

In the past, the use of direct potentiometry in chemical analysis had been largely restricted to pH measurements. However, within the last decade or so, the development of numerous ion-selective electrodes has revived interest in the subject of direct potentiometry. Potentiometric titrimetry has enjoyed widespread popularity because it is applicable to every kind of chemical reaction and because the required instrumentation is simple and inexpensive.

## INSTRUMENTATION FOR POTENTIOMETRIC METHODS OF ANALYSIS

Every potentiometric measurement requires that a galvanic cell, consisting of a sensitive indicator electrode and a stable reference electrode, be constructed with the solution to be studied in direct contact with the indicator electrode. A reference electrode can either be inserted into the sample solution or be brought into contact with the sample solution through an appropriate salt bridge. To determine the emf of the galvanic cell, a potential-measuring instrument is needed.

### Indicator Electrodes

Although we shall consider a wide variety of indicator electrodes throughout this chapter, three criteria govern the selection of such electrodes. First, the potential of an indicator electrode should be related through the Nernst equation to the activity of the species being determined. Second, an indicator electrode should respond rapidly and reproducibly to variations in the activity of the substance of interest. Third, an indicator electrode should possess a physical form which permits measurements to be performed conveniently.

### Reference Electrodes

A reference electrode is one whose potential remains constant during potentiometric measurements. Preparation of a reference electrode should be simple, and the potential of a reference electrode should be reproducible and should undergo no significant change when a small current passes through the electrode during potentiometric measurements.

Without doubt, the most commonly used reference electrode is the saturated calomel electrode (SCE), composed of metallic mercury and solid mercurous chloride (calomel) in contact with, and in equilibrium with, a saturated aqueous solution of potassium chloride. We can represent the electrochemical equilibrium which characterizes the behavior of a calomel electrode by the half-reaction

$$Hg_2Cl_2(s) + 2\,e \rightleftharpoons 2\,Hg(l) + 2\,Cl^-$$

At 25°C, the potential of the saturated calomel electrode is +0.2415 v versus NHE. Saturated calomel electrodes are commercially available, one of the most familiar forms appearing in Figure 10–1. Another useful reference electrode is the silver-silver chloride electrode. It is analogous to the calomel electrode, except that silver metal and silver chloride replace the mercury and mercurous chloride, respectively, and the pertinent half-reaction is

$$AgCl(s) + e \rightleftharpoons Ag(s) + Cl^-$$

A silver-silver chloride electrode prepared with saturated potassium chloride solution has a potential of +0.197 v versus NHE at 25°C.

### Potential-Measuring Instruments

A reliable value for the emf of a galvanic cell can often be obtained with a simple potentiometer-galvanometer arrangement, as described in Chapter 8. However, if the electrical resistance of a cell is extraordinarily large – the resistance of a glass membrane electrode utilized for pH measurements may itself be 50 megohms or more – an ordinary

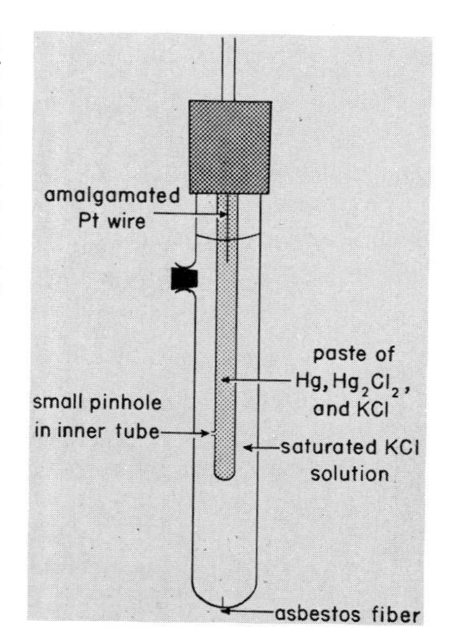


*Figure 10–1.* Saturated calomel reference electrode. An inner tube – the electrode proper – contains a paste of metallic mercury, mercurous chloride, and potassium chloride in contact with an amalgamated platinum wire which leads to a potential-measuring instrument. An outer tube – actually a salt bridge – is filled with a saturated potassium chloride solution. Electrolytic contact between the inner and outer compartments is obtained by means of a pinhole in the side of the inner tube. A porous asbestos fiber, sealed into the tip of the outer (salt bridge) tube, provides contact to the sample solution.

potentiometer cannot be used, because a galvanometer is not sensitive enough to indicate accurately when a known, variable emf is balanced against the emf of the galvanic cell.

Two kinds of potential-measuring devices are commonly employed to measure emf values of high-resistance galvanic cells. A *direct-reading* instrument is essentially a high input impedance vacuum-tube or field-effect-transistor voltmeter. For some purposes, the voltage is converted to a proportional current which is passed through an ammeter whose dial face is calibrated in terms of pH or the activity (concentration) of some desired species. A *potentiometric* or *null-detector* meter incorporates a relatively straightforward potentiometer circuit, except that the usual galvanometer is replaced with a vacuum-tube or solid-state amplifier to improve the sensitivity and accuracy of the emf measurements; these instruments commonly provide a direct readout of the activity (concentration) of the species of interest. Direct-reading instruments are especially convenient for use in potentiometric titrations, where the variation of emf as a function of added titrant is more important than any individual emf value. Null-detector meters are more accurate than direct-reading instruments; consequently, they are preferable for precise, single measurements with ion-selective electrodes.

## DIRECT POTENTIOMETRY

### *PRINCIPLES AND TECHNIQUES OF DIRECT POTENTIOMETRY*

Suppose that we wish to analyze a dilute aqueous solution of silver nitrate by means of direct potentiometry. One approach is to construct a galvanic cell consisting of a saturated calomel reference electrode and a silver wire indicator electrode:

$$\mathrm{Hg} \mid \mathrm{Hg_2Cl_2(s), KCl(s)} \mid \mathrm{KNO_3}\ (1\ F) \mid \mathrm{AgNO_3} \mid \mathrm{Ag}$$

$$E_{\mathrm{Hg,Hg_2Cl_2}} \qquad E_{\mathrm{j}} \qquad E_{\mathrm{j}}' \qquad E_{\mathrm{Ag^+,Ag}}$$

To prevent the precipitation of silver chloride, the calomel electrode is separated from the silver nitrate solution by a salt bridge containing 1 $F$ potassium nitrate solution.

Now, the overall emf of the galvanic cell is the sum of the potentials of the reference and indicator electrodes, plus two liquid-junction potentials – one at the boundary between the saturated calomel electrode and the potassium nitrate salt bridge, and the other at the interface between the salt bridge and the silver nitrate solution:

$$E_{\mathrm{cell}} = E_{\mathrm{Hg,Hg_2Cl_2}} + E_{\mathrm{j}} + E_{\mathrm{j}}' + E_{\mathrm{Ag^+,Ag}}$$

By utilizing a potentiometer, we can determine the overall emf of the galvanic cell with an error not exceeding 1 mv. If we extract from this measurement the value for the potential of the silver indicator electrode, the activity of silver ion can be computed from the relation

$$E_{\mathrm{Ag^+,Ag}} = E^0_{\mathrm{Ag^+,Ag}} - 0.059 \log \frac{1}{(\mathrm{Ag^+})}$$

To obtain $E_{\mathrm{Ag^+,Ag}}$, it is necessary to subtract individual values for $E_{\mathrm{Hg,Hg_2Cl_2}}$, $E_{\mathrm{j}}$, and $E_{\mathrm{j}}'$ from $E_{\mathrm{cell}}$. Obviously, any uncertainties or variations in the potential of the reference electrode or the liquid-junction potentials will produce corresponding errors in the values of $E_{Ag^+,Ag}$ and the silver ion activity calculated from the Nernst equation. It is relatively easy to obtain a saturated calomel reference electrode whose potential is known to within 1 mv or less. What do we know about the liquid-junction potentials?

### Liquid-Junction Potentials

When two solutions containing dissimilar ions – or the same ions at different concentrations – are brought into contact with each other, a potential difference develops at the boundary between these two media. This liquid-junction potential arises because the various ionic species diffuse or migrate across the interface at characteristically different rates.

As a simple illustration of the phenomenon of a liquid-junction potential, let us examine what happens at the interface between a saturated (4.2 $F$) potassium chloride solution and a 1 $F$ potassium chloride medium. Schematically, this phase boundary can be depicted as

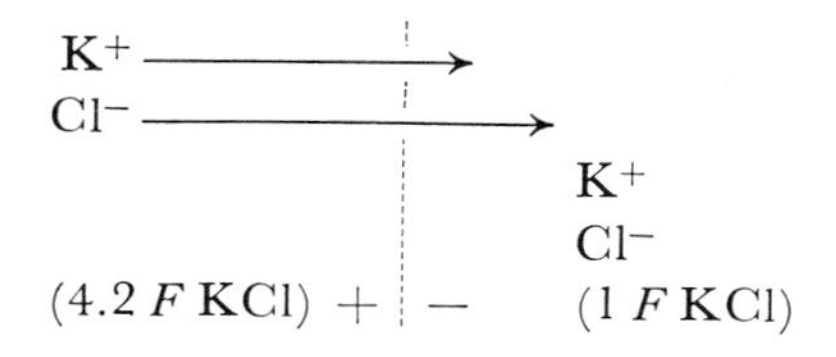


Under the influence of a concentration gradient, potassium ions and chloride ions both diffuse from left to right across the boundary, but chloride ion is approximately 4 per cent more mobile than potassium ion. Therefore, the former outraces the latter, and the right side of the boundary acquires a small negative charge, whereas the left side of the liquid junction becomes slightly positive. Thus, it is the separation of charge which causes the liquid-junction potential. However, this charge separation cannot increase beyond a certain equilibrium state.

Liquid-junction potentials encompass a large range of values, depending upon the charges, mobilities, and concentrations of the ionic species and the nature of the solvent on each side of the boundary. At best, a liquid-junction potential may be only 1 or 2 mv. More often than not, liquid-junction potentials reach 10 mv, if not higher values. Unavoidable and unknown liquid-junction potentials constitute a major drawback to the analytical use of direct potentiometry.

### Calibration Techniques in Direct Potentiometry

Most analytical applications of direct potentiometry rely upon an empirical calibration technique which minimizes the effects of uncertainties in the potential of the reference electrode and in the liquid-junction potentials. Assume we are interested in the determination of silver ion in unknown silver nitrate solutions, and let us prepare a series of solutions containing known concentrations of silver ion. Each known solution must have the same composition and ionic strength as the anticipated unknown samples so that activity coefficients will be constant. Next, the known solutions are transferred, one by one, into a galvanic cell such as that described earlier, and the overall emf of the cell is measured potentiometrically for each solution. Finally, we construct a calibration curve, which is essentially a straight line, by plotting the overall emf of the galvanic cell as a function of the logarithm of the silver ion concentration.

When an unknown silver ion solution is subsequently placed in the galvanic cell, the observed emf can be translated into the concentration of silver ion by interpolation of the calibration curve. Although the reference-electrode potential and the liquid-junction potentials need not be known in this procedure, the values of these parameters must remain constant during the establishment of the calibration curve and the measurement of the emf of the galvanic cell containing the unknown solution. Note that the calibration technique permits direct determination of the analytical concentration of a species, whereas the potential of an indicator electrode is governed by the activity of that species.

### Method of Standard Addition

Another way to overcome the effects of uncertain liquid-junction potentials and changing activity coefficients on direct potentiometric measurements is the method of standard addition.

Suppose that we wish to determine the concentration of silver ion in a solution by using the electrochemical cell described earlier. First, we transfer an accurately known volume of the sample solution into the cell. Second, we measure the potential of the silver indicator electrode ($E_1$) relative to the saturated calomel reference electrode. Third, we pipet a known volume of a standard silver ion solution into the sample solution contained in the cell. This standard solution should be relatively concentrated so that only a tiny volume needs to be added; in this way, there is almost no change at all in the composition and ionic strength of the original sample medium, so that the liquid-junction potential and activity coefficients remain essentially constant. Fourth, we measure the new potential of the silver indicator electrode ($E_2$) versus the reference electrode. Then, knowing the change in the potential of the indicator electrode ($\Delta E = E_2 - E_1$), the volume ($V_X$) of the original sample solution, and the volume ($V_S$) and concentration ($C_S$) of the standard silver ion solution, we can calculate the concentration ($C_X$) of silver ion in the sample by means of the relation

$$C_X = C_S\left(\frac{V_S}{V_X + V_S}\right)\left[10^{n(\Delta E)/0.059} - \left(\frac{V_X}{V_X + V_S}\right)\right]^{-1}$$

Of the various terms in the preceding expression, it is $\Delta E$ (or the difference between $E_2$ and $E_1$) that is subject to the most uncertainty, because $C_S$, $V_S$, and $V_X$ can be determined very accurately. Errors in the measurement of $E_1$ and $E_2$ become more serious as $\Delta E$ decreases. Therefore, it is desirable to make $\Delta E$ as large as possible by the addition of a reasonably large volume ($V_S$) of the standard solution. However, if too great a volume of standard solution is added, the composition of the unknown medium may be altered and there may be uncertain changes in the liquid-junction potential and activity coefficients. To obtain the best results, one must seek a compromise.

### A Fundamental Limitation of Direct Potentiometry

Even when the greatest possible care is exercised, it is practically impossible to measure the potential of an indicator electrode (or the emf of a galvanic cell) with an accuracy better than 1 mv. An uncertainty of 1 mv or more may arise from a number of sources, including a change in the potential of the reference electrode, an unknown liquid-junction potential, or a variation in temperature.

If we wish to determine the activity (concentration) of a species by means of direct potentiometry, how seriously will the result be affected by a 1-mv error in the potential of the indicator electrode or in the overall emf of a galvanic cell? To answer this question, let us perform two calculations. Suppose that the *true* potential of a silver indicator electrode immersed in a silver ion solution at 25°C should be +0.592 v versus NHE. Taking the standard potential for the silver ion-silver metal half-reaction to be +0.800 v versus NHE, we may write

$$E_{Ag^+,Ag} = E^0_{Ag^+,Ag} - 0.05915 \log \frac{1}{(Ag^+)}$$

$$+0.592 = +0.800 - 0.05915 \log \frac{1}{(Ag^+)}$$

$$\log \frac{1}{(Ag^+)} = \frac{0.208}{0.05915} = 3.51_6$$

$$\log (Ag^+) = -3.51_6 = 0.48_4 - 4.000$$

$$(Ag^+) = 3.05 \times 10^{-4}\ M$$

Now let us assume that the potential of the silver indicator electrode immersed in the same sample solution was *observed* to be +0.593 v versus NHE, an error of only 1 mv. If the silver-ion activity is calculated as before, we obtain $(Ag^+) = 3.16 \times 10^{-4}$ *M*. Thus, a 1-mv uncertainty in the indicator-electrode potential leads to an error of just under 4 per cent in the activity of silver ion. Since possible uncertainties due to unknown liquid-junction potentials may easily exceed 1 mv, the attainment of accurate analytical results from direct potentiometry usually requires the use of an empirical calibration technique or the method of standard addition.

## *APPLICATIONS OF DIRECT POTENTIOMETRY*

During the last ten years, the commercial availability of a family of ion-selective electrodes responding to more than twenty different cations and anions – and the hundreds of novel and diverse applications of these ion sensors to many scientific and technological disciplines – has created a spectacular renaissance in the field of direct potentiometry. Among the virtues of direct potentiometry are that measurements can be performed on exceedingly small samples – much less than a milliliter – and that the sample solution is not altered or destroyed during the experimental measurements.

### Determination of pH

In biochemical processes, in the rates of many organic and inorganic reactions, and in a wide variety of separations and measurements in analytical chemistry, the activity or concentration of hydrogen ion in a solution plays a critical role. Consequently, the determination of pH is one of the most important practical applications of direct potentiometry.

**Glass Membrane Electrode.** Of all the electrodes sensitive to hydrogen ions, the glass membrane electrode, or simply the glass electrode, is unique, because the mechanism of its response to hydrogen ion involves an ion-exchange process rather than electron transfer.

Glass electrodes may be purchased in a wide variety of sizes and shapes for such diverse purposes as the determination of the pH of blood or other biological fluids, the continuous measurement and recording of the hydrogen ion activity in flowing solutions, or the evaluation of the pH of a single drop or less of solution. However, the most familiar form of the glass membrane electrode is depicted in Figure 10–2. A thin-walled bulb, fabricated from a special glass highly sensitive to the hydrogen ion activity of a solution, is sealed to the bottom of an ordinary glass tube. Inside the glass bulb is a dilute aqueous hydrochloric acid solution, usually 0.1 $F$ in concentration. A silver wire coated with a layer of silver chloride is immersed into the hydrochloric acid medium, and the silver wire is extended upward through the resin-filled tube to provide electrical contact to the external circuit.

Most glasses employed for the construction of pH-sensitive membrane electrodes contain at least 60 weight per cent $SiO_2$, along with smaller amounts of oxides of the alkali metal and alkaline earth elements. For many years soda-lime glass, consisting by weight of 72 per cent $SiO_2$, 22 per cent $Na_2O$, and 6 per cent CaO, was widely used. However, this formulation has been replaced by lithia glasses containing $SiO_2$, $Li_2O$, and CaO, or $SiO_2$, $Li_2O$, and BaO, whose advantages will be discussed later.

In order to function as a pH-sensitive electrode, the surfaces of a glass membrane must be hydrated. A water-soaked glass membrane consists of a middle layer of dry glass, perhaps 0.5 mm in thickness, sandwiched between inner and outer hydrated gel layers which are typically $10^{-4}$ mm thick. As hydration of a pH-sensitive membrane takes place, lithium ions (or other singly charged cations) in the surface layers of the glass become aquated and are exchanged for protons from the adjacent solution

$$\mathrm{Li^+_{glass} + H^+_{solution} \rightleftharpoons Li^+_{solution} + H^+_{glass}}$$

by the same kind of mechanism that occurs with ordinary cation-exchange resins. So complete is the ion-exchange process at the surfaces of the inner and outer hydrated layers of the glass that all sites originally held by lithium ions are occupied by hydrogen ions; thus, the exterior surfaces of the glass membrane consist of a gel of hydrated silicic acid. Later, when a sample solution of unknown pH is brought into contact with the outer hydrated gel layer of a glass membrane electrode, the magnitude of the resulting

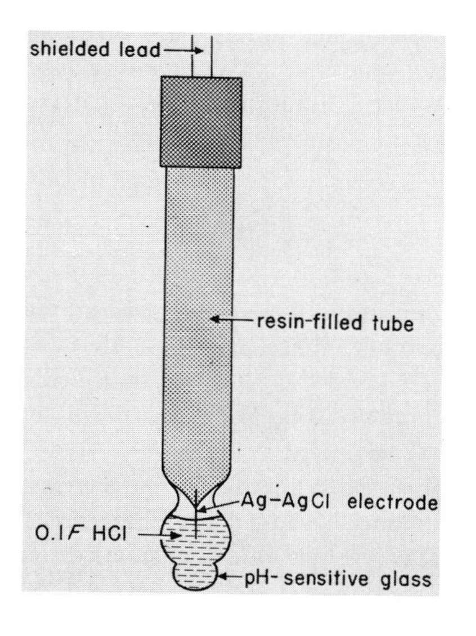


*Figure 10–2.* Glass membrane electrode.

phase-boundary potential can be correlated with a numerical value for the pH of that sample.

**Practical pH Measurements.** A glass membrane electrode in combination with the saturated calomel reference electrode provides the galvanic cell usually employed for practical pH measurements. We can describe such a cell in shorthand fashion as

| Ag | AgCl(s), HCl (0.1 *F*) | glass membrane | sample solution of unknown pH | $Hg_2Cl_2(s)$, KCl(s) | Hg |
|---|---|---|---|---|---|

Each vertical line in the cell representation denotes a phase boundary across which a potential develops. Therefore, the emf for this galvanic cell is composed of five parts: (1) the potential of the silver-silver chloride electrode, (2) the potential between the hydrochloric acid solution inside the glass electrode and the inner wall of the glass membrane, (3) the potential between the outer wall of the glass membrane and the solution of unknown pH, (4) the liquid-junction potential between the solution of unknown pH and the saturated potassium chloride solution of the calomel electrode, and (5) the potential of the saturated calomel electrode.

Many careful experimental studies have demonstrated that the overall emf, $E_{cell}$, for such a cell is related to the pH of the sample solution by the expression

$$E_{cell} = K + 0.059 \text{ pH}$$

In this equation, $K$ includes the first, second, fourth, and fifth sources of potential listed above. However, the value of $K$ can never be precisely known because the liquid-junction potential is uncertain. *Therefore, all practical* pH *determinations necessarily involve a calibration procedure in which the* pH *of an unknown solution is compared to the* pH *of a standard buffer.* If we transfer a standard buffer into the galvanic cell and measure the emf of the cell, it follows that

$$(E_{cell})_s = K + 0.059 \text{ (pH)}_s$$

where the subscript $s$ pertains to the standard buffer. Similarly, if an unknown solution is present in the cell, we obtain

$$(E_{cell})_x = K + 0.059 \text{ (pH)}_x$$

where the subscript $x$ refers to the unknown solution. When the latter two relationships are combined, the result is

$$\text{(pH)}_x = \text{(pH)}_s + \frac{(E_{cell})_x - (E_{cell})_s}{0.059}$$

which has been adopted at the National Bureau of Standards as the *operational definition* of pH. Implicit in this definition are two assumptions, namely, that the pH of the standard buffer is accurately known and that $K$ has the same value when either the standard buffer or the unknown sample is present in the galvanic cell.

A list of accepted pH values for several standard buffer solutions is presented in Table 10–1. In practice, one selects a standard buffer with a pH as close as possible to that of the unknown sample. For example, if the unknown solution has a pH close to 4, it is preferable that the standard buffer be potassium acid phthalate, the latter having a pH of

**Table 10–1. pH Values of Standard Buffers***

| Temperature, °C | Potassium Tetroxalate, 0.05 *M* | Potassium Hydrogen Tartrate, Saturated at 25°C | Potassium Acid Phthalate, 0.05 *M* | $KH_2PO_4$, 0.025 *M*; $Na_2HPO_4$, 0.025 *M* | Borax, 0.01 *M* |
|---|---|---|---|---|---|
| 0 | 1.666 | — | 4.003 | 6.984 | 9.464 |
| 10 | 1.670 | — | 3.998 | 6.923 | 9.332 |
| 20 | 1.675 | — | 4.002 | 6.881 | 9.225 |
| 25 | 1.679 | 3.557 | 4.008 | 6.865 | 9.180 |
| 30 | 1.683 | 3.552 | 4.015 | 6.853 | 9.139 |
| 40 | 1.694 | 3.547 | 4.035 | 6.838 | 9.068 |
| 50 | 1.707 | 3.549 | 4.060 | 6.833 | 9.011 |

* Values taken with permission from R. G. Bates: *Determination of pH.* John Wiley and Sons, New York, 1964, p. 76. The uncertainties in these values are about ±0.005 unit at 25°C, but somewhat larger at other temperatures.

4.01 at 25°C (Table 10–1). For optimum conditions, when the pH values of the standard buffer and unknown sample are essentially identical, the uncertainty in the measured pH of an unknown solution is about ±0.02 pH unit. There are pH meters available commercially from which one can read pH values with a *precision* of ±0.003 unit or better, a feature which is frequently useful for studying *changes* in the pH of a system under carefully controlled conditions. However, such precision must not be misconstrued, for the *accuracy* of the measurements is still no better than one or two hundredths of a pH unit.

**Errors in Glass-Electrode pH Measurements.** In strongly basic media, or in moderately alkaline solutions containing high concentrations of alkali metal ions, glass membrane electrodes show a so-called **alkaline error**, in which the observed pH is *lower* than the true value by an amount which increases as the pH becomes greater. This error arises because cations other than hydrogen ion – particularly, sodium ion and, to a lesser extent, lithium and potassium ions – compete for the ion-exchange sites at the surface of the outer hydrated gel layer. Studies have shown that the size of the alkaline error varies with the identity and concentration of the extraneous cation, with the temperature, and with the glass composition.

Soda-lime glass (22 per cent $Na_2O$, 6 per cent CaO, and 72 per cent $SiO_2$) exhibits a correct hydrogen-ion response at room temperature between pH 1 and 9. However, in the presence of 1 *M* sodium ion at 25°C, the observed pH is *low* by 0.2 unit at pH 10, by 1.0 unit at pH 12, and by 2.5 units at pH 14. In general, the corresponding alkaline errors are less for lithium ion and smaller still for solutions containing potassium ion. If sodium oxide in a glass is replaced with lithium oxide, the alkaline error is greatly reduced. A glass composed of 10 per cent $Li_2O$, 10 per cent CaO, and 80 per cent $SiO_2$ by weight gives an alkaline error of less than 0.2 pH unit in the presence of 1 *M* sodium ion at pH 13. Nowadays, almost all commercially available glass electrodes for pH measurements contain lithium oxide, providing quite accurate hydrogen-ion response up to nearly pH 13.

At the other end of the pH scale, the glass electrode exhibits an **acid error** which is opposite from that observed in alkaline solutions. In very strongly acidic media, the observed pH is *higher* than the true value. Fortunately, the magnitude of the acid error is negligibly small down to at least pH 1.

Finally, a glass electrode does not respond rapidly when inserted into solutions which are poorly buffered, presumably because the rate of attainment of ion-exchange equilibrium in the outer hydrated gel layer is slow.

### Ion-Selective Electrodes for the Determination of Other Cations and Anions

For a long time, the analytical uses of direct potentiometry were limited to the determination of pH, primarily because of the lack of sensitive and selective indicator electrodes for other ions. However, recent years have witnessed the development of a number of interesting ion-selective electrodes. Such electrodes are finding important applications in the monitoring of industrial operations, in water analysis, in oceanography, in medical diagnosis, in the measurement of pollution, and in the study of biochemical systems.

There are several classes of ion-selective indicator electrodes – glass membrane electrodes for cations other than hydrogen ion, liquid ion-exchange membrane electrodes, and solid-state electrodes. All these electrodes respond, as does the pH-sensitive glass electrode, to changes in the activity of a species in the same oxidation state but in two different phases separated by a membrane. Contrast this behavior to that of a platinum indicator electrode responding to changes in the ratio of activities of two different oxidation states of some element in a single phase (*e.g.*, a solution containing $Fe^{2+}$ and $Fe^{3+}$) or to that of a silver indicator electrode responding to changes in the activity of silver ion in solution; these systems involve electron transfer between two oxidation states of the same element.

**Glass Electrodes for the Determination of Cations.** As long ago as 1925, it was discovered that the addition of small percentages of $Al_2O_3$ to pH-sensitive glass results in enhanced response to alkali metal ions. Yet, thirty years elapsed before careful investigations confirmed that electrodes made from sodium aluminosilicate glass composed of $Na_2O$, $Al_2O_3$, and $SiO_2$ are sensitive to sodium ion as well as other cations. By suitable variations in the composition of sodium aluminosilicate glasses, electrodes have been developed for the determination of $Li^+$, $Rb^+$, $Cs^+$, $Tl^+$, $NH_4^+$, $Na^+$, $K^+$, and $Ag^+$. Table 10–2 presents information about some of the most useful glass membrane electrodes. Included in the data are selectivity ratios,* which measure the seriousness of interferences. For example, the electrode for the determination of lithium ion has a selectivity constant $K_{Li^+/Na^+}$ of approximately 3, so that lithium ion is only three times more effective than sodium ion in governing the electrode potential. However, the same glass electrode has better than a 1000-fold preference for lithium ion over potassium ion; that is, $K_{Li^+/K^+} > 1000$.

Analytical uses of cation-selective glass electrodes are impressive in both number and scope. Due to the availability of glass electrodes sensitive to sodium and potassium ions, the perfection of special miniature and flow-through electrodes for the determination of these two cations, and the physiological importance of these two species, some of the most interesting applications are in the field of biomedical analysis. Thus, it is feasible to

*Selectivity ratios depend on the nature and composition of the solution being investigated. Values quoted in this chapter are primarily intended to give semiquantitative information about the preferential response of an electrode to equal concentrations of two different ions. Some authors prefer to report the selectivity ratio as a number less than 1; for example, the selectivity ratio ($K_{Li^+/Na^+}$) for the lithium electrode in Table 10–2 might be written as 1/3 or 0.33. For any particular situation, it is necessary to identify which form of expression is meant. Most likely, the source of ambiguity will be resolved in the near future by one of the committees of the International Union of Pure and Applied Chemistry (IUPAC).

Table 10–2. **Compositions and Characteristics of Some Cation-Selective Glasses***

| Cation of Interest | Glass Composition (mole per cent) | Selectivity Ratio** | Characteristics |
|---|---|---|---|
| $Li^+$ | 15% $Li_2O$, 25% $Al_2O_3$, 60% $SiO_2$ | $K_{Li^+/Na^+} \approx 3$ | Best for $Li^+$ in presence of $H^+$ and $Na^+$ |
| | | $K_{Li^+/K^+} > 1000$ | |
| $Na^+$ | 11% $Na_2O$, 18% $Al_2O_3$, 71% $SiO_2$ | $K_{Na^+/K^+} \approx 2800$ at pH 11 | Nernstian response down to approximately $10^{-5}$ $M$ $Na^+$ |
| | | $K_{Na^+/K^+} \approx 300$ at pH 7 | |
| $K^+$ | 27% $Na_2O$, 5% $Al_2O_3$, 68% $SiO_2$ | $K_{K^+/Na^+} \approx 20$ | Nernstian response down to somewhat less than $10^{-4}$ $M$ $K^+$ |
| $Ag^+$ | 11% $Na_2O$, 18% $Al_2O_3$, 71% $SiO_2$ | $K_{Ag^+/Na^+} > 1000$ | Less selective but more reproducible than glass containing 28.8% $Na_2O$, 19.1% $Al_2O_3$, and 52.1% $SiO_2$ with $K_{Ag^+/H^+} \approx 100{,}000$ |

* Data taken with permission from G. A. Rechnitz: Chem. Eng. News *45*:146, June 12, 1967.

** See footnote on page 251.

measure the activities of sodium and potassium ions in urine, serum, cerebrospinal fluid, whole blood, plasma, bile, brain cortex, kidney tubules, and muscle fibers. Cystic fibrosis, characterized by abnormally high sodium levels in sweat, can be diagnosed through determination of the sodium ion activity on the skin surface. There are numerous practical uses of sodium- and potassium-selective glass electrodes for the analysis of water and soil extracts.

One recent development entails the use of a glass electrode (27 mole per cent $Na_2O$, 4 mole per cent $Al_2O_3$, and 69 mole per cent $SiO_2$) responsive to ammonium ion for the determination of enzymes and their corresponding substrates. For example, urease is an enzyme which catalyzes the hydrolysis of urea to ammonium ion,

$$CO(NH_2)_2 + H_2O + 2\,H^+ \xrightarrow{\text{urease}} CO_2 + 2\,NH_4^+$$

the latter being measured potentiometrically to obtain the original quantity of urea. A similar procedure can be followed for the analysis of solutions containing glutamine, asparagine, glutamic acid, as well as various amines. Alternatively, the ammonium-ion electrode may be employed as an enzyme-sensing device. A cellophane membrane is sealed around the glass bulb of the electrode, and a solution of urea is injected into the annular space between the cellophane and glass bulb. If this electrode system is immersed into an unknown urease solution, urea diffuses outward through the cellophane and reacts with urease to form ammonium ion. Then, some of the ammonium ions diffuse through the cellophane membrane and are detected by the electrode. A calibration curve can be constructed so that the response of the electrode can be translated into the amount of urease present.

**Liquid Ion-Exchange Membrane Electrodes.** A liquid ion exchanger consists of a water-immiscible solvent containing a high-molecular-weight organic solute with acidic or basic functional groups, or perhaps chelating properties, which interacts strongly and rather selectively with the ion to be determined. If a porous membrane is impregnated with a liquid ion-exchange substance, and if this membrane is placed between two different solutions containing the desired species, a potential develops across the membrane which depends on the activities of the desired ion in the solutions on either side of the membrane.

Figure 10–3 illustrates a commercially available electrode designed for the determination of calcium ion. A porous, plastic filter-membrane is held in contact with a reservoir filled with the liquid ion exchanger – a solution of calcium didecylphosphate in di-*n*-octylphenyl phosphonate – so that the membrane becomes permeated by the liquid cation exchanger. This membrane separates the solution to be analyzed from the internal compartment of the electrode. In the internal compartment is a calcium chloride solution of fixed concentration into which dips a silver-silver chloride reference electrode. It can be shown that the overall emf of a galvanic cell in which the calcium-selective electrode and a saturated calomel reference electrode are in electrolytic contact with a calcium-ion solution follows the equation

$$E_{\text{cell}} = K + \frac{0.059}{2}\,\text{pCa}$$

where $K$ includes the potentials of the calomel electrode and the inner silver-silver chloride electrode. A calcium-selective electrode obeys the preceding relation for calcium-ion activities as low as $10^{-5}\,M$ and exhibits marked specificity for calcium ion in the presence of strontium, magnesium, barium, sodium, and potassium ions. Calcium-ion activity in living organisms is known to affect many physiological processes, including blood coagulation, enzyme function, nerve conduction, bone formation, intestinal secretion and absorption, hormonal release from endocrine glands, and cerebral function. Thus, some of the most important applications of the calcium-selective electrode are in the areas of biomedical research and clinical medicine.

A liquid ion-exchange membrane electrode has been developed for the measurement of potassium-ion activity. Closely resembling the calcium-selective electrode in design, this electrode utilizes a dilute solution of valinomycin in diphenyl ether as the liquid ion exchanger; valinomycin (an antibiotic) is an uncharged, cyclic macromolecule with a high affinity for potassium ion, but not sodium ion. In addition, a Nernstian response to potassium-ion activities ranging from $10^{-6}$ to $0.1\,M$ is observed. An electrode similar to the potassium-selective valinomycin sensor has been prepared for the measurement of ammonium ion in the presence of calcium ion and the alkali metal cations. It employs a

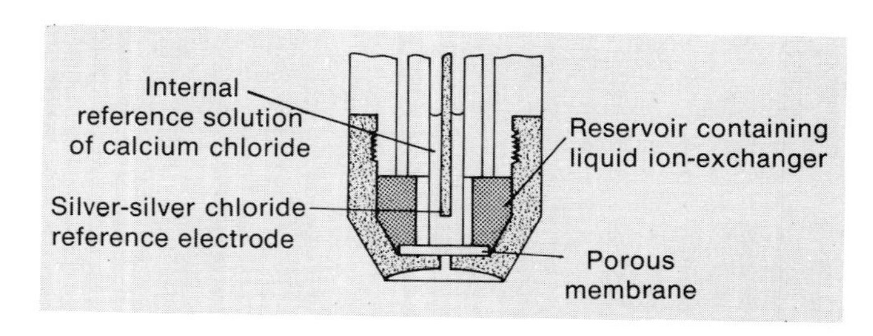


*Figure 10–3.* Schematic diagram of calcium-selective liquid ion-exchange membrane electrode.

mixture of nonactin and monactin (which are macrocyclic antibiotics) dissolved in tris(2-ethylhexyl)phosphate as the liquid ion exchanger.

A liquid ion-exchange membrane electrode, responding almost equally well to calcium and magnesium ions, has been devised for the monitoring of water hardness. Other electrodes which contain chelating ion-exchange substances are sensitive to copper(II) and lead(II). Liquid membrane electrodes employing substituted 1,10-phenanthroline complexes of iron(III) and nickel(II) as anion exchangers can be purchased for the determination of $BF_4^-$, $ClO_4^-$, and $NO_3^-$. An electrode utilizing the salt of a high-molecular-weight tetraalkylammonium cation serves as a chloride sensor. Properties of these liquid ion-exchange membrane electrodes are compiled in Table 10–3.

**Solid-State Membrane Electrodes.** Ionically conducting materials – single crystals, mixed crystals, and polycrystalline solids – have been employed for the construction of a variety of solid-state membrane electrodes for the direct potentiometric determination of cations and anions. A comparison of the design features of these solid-state sensors with the more familiar glass membrane electrode is provided in Figure 10–4.

A synthetic, *single crystal* of lanthanum fluoride, doped with europium(II) to enhance its conductivity, makes an excellent solid-state membrane electrode which responds selectively to fluoride ion over a range of activity from $10^{-6}$ to 1 *M* and which exhibits at least a thousand-to-one preference for fluoride over chloride, bromide, iodide, nitrate, bicarbonate, and sulfate ions. There are a number of applications for the fluoride-selective electrode. It has been utilized for continuous monitoring of fluoride

**Table 10–3. Characteristics of Liquid Ion-Exchange Membrane Electrodes***

| Ion Determined | Measurement Range | pH Range for Use | Interferences | Applications |
|---|---|---|---|---|
| $Ca^{2+}$ | pCa = 0 to 5 | 5.5 to 11 | $Zn^{2+}$, $Fe^{2+}$, $Pb^{2+}$ | biological fluids, EDTA titrations, water analysis |
| $Cu^{2+}$ | pCu = 1 to 5 | 4 to 7 | $Fe^{2+}$, $Ni^{2+}$, $Zn^{2+}$ | |
| $Pb^{2+}$ | pPb = 2 to 5 | 3.5 to 7.5 | $Cu^{2+}$, $Fe^{2+}$, $Zn^{2+}$, $Ni^{2+}$ | |
| $BF_4^-$ | $pBF_4$ = 1 to 5 | 2 to 12 | $I^-$, $NO_3^-$, $Br^-$ | analysis of alloys, semiconductors, paints, and pigments after conversion of boron to $BF_4^-$ by treatment with HF |
| $Cl^-$ | pCl = 0 to 5 | 2 to 10 | $ClO_4^-$, $I^-$, $NO_3^-$, $Br^-$, $OH^-$, $HCO_3^-$, $SO_4^{2-}$, $CH_3COO^-$ | foods, pharmaceuticals, water analysis, soaps, especially useful in solutions of sulfide or strong reductants |
| $ClO_4^-$ | $pClO_4$ = 1 to 5 | 4 to 10 | $OH^-$, $I^-$ | potentiometric titration of $ClO_4^-$ with tetraphenylarsonium chloride |
| $NO_3^-$ | $pNO_3$ = 1 to 5 | 2 to 12 | $ClO_4^-$, $I^-$, $ClO_3^-$, $Br^-$, $S^{2-}$ | fertilizers, soils, foods, plants, explosives, water supplies |

* Data taken in part from Orion Research, Inc., *Analytical Methods Guide*, fourth edition, October, 1972.

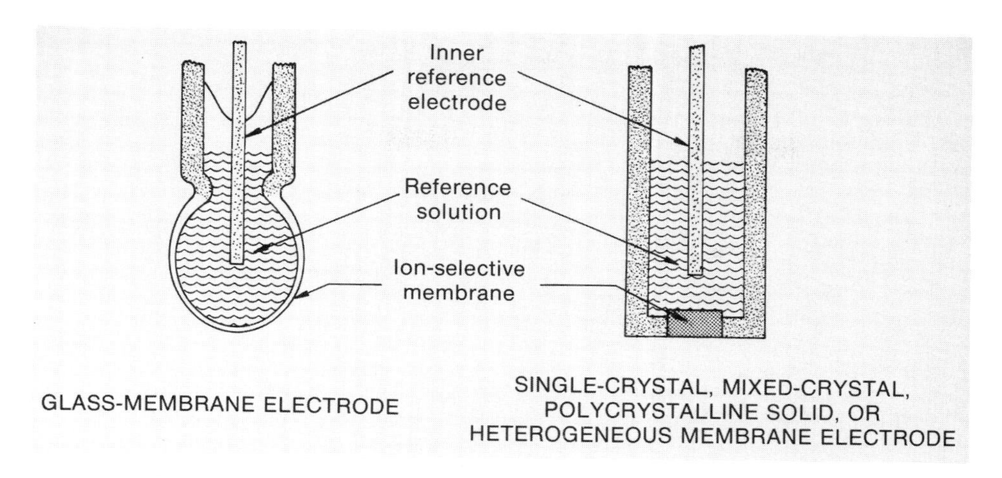


*Figure 10–4.* Comparison between the construction of a glass-membrane electrode and the various solid-state and heterogeneous membrane electrodes.

levels in drinking water. In pollution control, the electrode is useful for the determination of fluoride in air and stack-gas samples. Fluoride in toothpaste, pharmaceuticals, urine, saliva, dissolved bone and teeth, and multiple vitamins is measurable. Among the industrial uses of the electrode are fluoride analyses of chromium electroplating baths, plastics, pesticides, and fertilizers.

Solid-state electrodes responsive to cadmium, copper, and lead ions have been prepared from *mixed-crystal* membranes consisting of silver sulfide to which is added CdS, CuS, or PbS, respectively. Electrodes selective toward thiocyanate, chloride, bromide, and iodide can be fabricated if silver sulfide containing finely divided and well dispersed AgSCN, AgCl, AgBr, or AgI is pressed into the form of a disc or pellet, and mounted into the bottom of a glass tube as shown in Figure 10–4. Moreover, the silver iodide-silver sulfide mixture provides a solid-state membrane electrode suitable for the measurement of cyanide ion. By itself, *polycrystalline* silver sulfide can be pressed into pellet form for the preparation of a solid-state electrode which responds to both sulfide and silver ions. In addition, the latter is a useful indicator electrode for potentiometric titrations of halide mixtures or of cyanide with standard silver nitrate solution. Some analytical applications of these various electrodes, as well as substances that interfere with their response, are listed in Table 10–4.

## POTENTIOMETRIC TITRATIONS

We may characterize a potentiometric titration as one in which the change of the emf of a galvanic cell during a titration is recorded as a function of added titrant. Although the major aim of this procedure is the precise location of the equivalence point, thermodynamic information, including dissociation constants for weak acids and formation constants for complex ions, can be obtained from potentiometric titration curves.

Compared to other methods for the location of equivalence points, the technique of potentiometric titrimetry offers a number of advantages. It is applicable to systems which are so brightly colored that visual methods of end-point detection are useless, and it is especially valuable when no internal chemical indicator is available. Moreover, it

**Table 10–4. Characteristics of Solid-State Membrane Electrodes***

| Ion Determined | Measurement Range | Interferences | Applications |
|---|---|---|---|
| $F^-$ | pF = 0 to 6 | $OH^-$ (pH < 8.5 for pF = 6, pH < 11 for pF = 2) | water supplies, plating baths, toothpaste, pharmaceuticals, teeth and bone |
| $Cd^{2+}$ | pCd = 1 to 7 | $Fe^{3+}$, $Pb^{2+}$, $Ag^+$, $Hg^{2+}$, $Cu^{2+}$ | water, industrial wastes, plating baths, paper, and pigments |
| $Cu^{2+}$ | pCu = 0 to 8 | $S^{2-}$, $Ag^+$, $Hg^{2+}$, $Cl^-$, $Br^-$, $Fe^{3+}$, $Cd^{2+}$ | plating baths, water supplies, sewage, ore refining, fungicides |
| $Pb^{2+}$ | pPb = 1 to 7 | $Cd^{2+}$, $Ag^+$, $Hg^{2+}$, $Cu^{2+}$, $Fe^{3+}$ | petroleum products, foods, air pollution, EDTA titration of lead, titration of $SO_4^{2-}$ with lead |
| $CN^-$ | pCN = 2 to 6 | $S^{2-}$, $I^-$ | industrial wastes, plating baths, determination of amygdalin |
| $SCN^-$ | pSCN = 0 to 5 | $I^-$, $Cl^-$, strong reductants, species that form insoluble silver salts or stable silver complexes | |
| $Cl^-$ | pCl = 0 to 4.3 | $S^{2-}$, $I^-$, $Br^-$, $CN^-$ | foods, pharmaceuticals, soaps, plastics, glass, water analysis |
| $Br^-$ | pBr = 0 to 5.3 | $S^{2-}$, $I^-$, $CN^-$ | titration of epoxy groups with HBr |
| $I^-$ | pI = 0 to 7.3 | $S^{2-}$, $CN^-$ | |
| $S^{2-}$ | pS = 0 to 17 | (none) | industrial wastes, paper and pulp industries |
| $Ag^+$ | pAg = 0 to 17 | $Hg^{2+}$ | titration of mixed halides or of $CN^-$ with silver ion |

* Data taken from Orion Research, Inc., *Analytical Methods Guide*, fourth edition, October, 1972, and from A. K. Covington: Chem. Brit., *5*:388, 1969.

eliminates subjective decisions concerning color changes of end-point indicators as well as the need for indicator blank corrections.

## *TECHNIQUES OF POTENTIOMETRIC TITRIMETRY*

Equipment required for the performance of potentiometric titrations is relatively simple, as shown in Figure 10–5. A titration vessel, a buret containing standard titrant, and suitable indicator and reference electrodes are the necessities, and there should be provision for efficient stirring of the solution. Usually, the titrant can be added rather rapidly at the beginning of the titration until the equivalence point is approached. Thereafter, until the equivalence point is passed, one should introduce the titrant in small

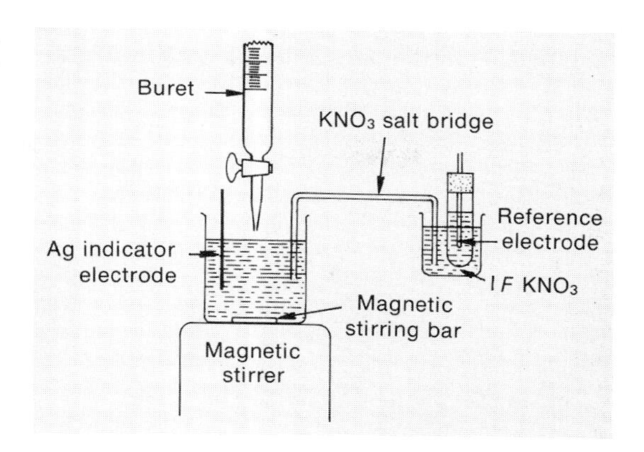


*Figure 10–5.* Apparatus for potentiometric titration of chloride with silver nitrate solution.

increments and should wait for equilibrium to be established before adding more reagent. After the addition of each increment, sufficient time must be allowed for the indicator-electrode potential to become reasonably steady, so that it exhibits a drift of perhaps not more than a few millivolts per minute.

In contrast to direct potentiometry, the measurement of the emf of a galvanic cell does not require the highest possible accuracy, because only the rather large *changes* in the potential of an indicator electrode near the equivalence point of a titration have significance.

### Determination of Equivalence Points

A typical set of potentiometric-titration data for the precipitation reaction involving chloride and silver ions is presented in Table 10–5. In the first column is a list of buret readings, whereas the second column shows the potential of a silver indicator electrode, measured with respect to the saturated calomel electrode (SCE), corresponding to each volume reading. Notice that the volume increments are large near the beginning of the titration when the potential of the indicator electrode varies only slightly. However, in the region of the equivalence point, *small* and *equal* volume increments are introduced. Previous experience with a particular titration alerts one to the fact that the equivalence point is being approached. Any of several methods may be employed for determination of the equivalence point of a potentiometric titration.

**Graphical Methods.** One procedure involves visual inspection of the complete titration curve. If the potential of the indicator electrode is plotted versus the volume of titrant, the resulting titration curve exhibits a maximal slope – that is, a maximal value of $\Delta E/\Delta V$ – which may be taken as the equivalence point. Figure 10–6A illustrates this approach with the experimental data in Table 10–5, although only that portion of the titration curve near the equivalence point is pictured.

An extension of the preceding technique entails preparation of a plot of $\Delta E/\Delta V$, the change in potential per volume increment of titrant, as a function of the volume of titrant. Such a graph, derived from the titration data of Table 10–5, is shown in Figure 10–6B. If the volume increments near the equivalence point are equal, one can simply take the difference in millivolts between each successive pair of potential readings as $\Delta E/\Delta V$; however, if the volume increments are unequal, it is necessary to normalize the data. To obtain the volume, $V$, pertaining to each value of $\Delta E/\Delta V$, one must *average* the two volumes corresponding to the successive potential readings. For example, in Table 10–5 the change in potential, $\Delta E$, is 29 mv as one proceeds from 16.00 to 16.10 ml of

**Table 10–5. Potentiometric Titration Data for Titration of 3.737 Millimoles of Chloride with 0.2314 *F* Silver Nitrate**

| Volume of $AgNO_3$, ml | *E vs.* SCE, volts | $\Delta E/\Delta V$, mv/0.1 ml | $\Delta^2 E/\Delta V^2$ |
|---|---|---|---|
| 0 | +0.063 | | |
| 5.00 | 0.071 | | |
| 10.00 | 0.086 | | |
| 15.00 | 0.138 | | |
| 15.20 | 0.145 | | |
| 15.40 | 0.153 | | |
| 15.60 | 0.161 | | |
| | | 5 | |
| 15.70 | 0.166 | | |
| | | 7 | |
| 15.80 | 0.173 | | |
| | | 9 | |
| 15.90 | 0.182 | | +5 |
| | | 14 | |
| 16.00 | 0.196 | | +15 |
| | | 29 | |
| 16.10 | 0.225 | | +35 |
| | | 64 | |
| 16.20 | 0.289 | | −34 |
| | | 30 | |
| 16.30 | 0.319 | | −14 |
| | | 16 | |
| 16.40 | 0.335 | | −5 |
| | | 11 | |
| 16.50 | 0.346 | | |

titrant added. Accordingly, $\Delta E/\Delta V$ is 29 mv per 0.1 ml, and $V$ should be taken as 16.05 ml. Figure 10–6B consists of two curves extrapolated to a point of intersection whose position along the abscissa indicates the equivalence-point volume. It is only a happenstance that one of the experimental points lies so close to this point of intersection.

**Analytical Method.** A rapid, accurate, and convenient procedure for the location of an equivalence point relies upon the mathematical definition that the second derivative ($\Delta^2 E/\Delta V^2$) of the titration curve is zero at the point where the first derivative ($\Delta E/\Delta V$) is a maximum. In the fourth column of Table 10–5 are listed the individual values of $\Delta^2 E/\Delta V^2$ – the difference between each pair of $\Delta E/\Delta V$ values – and Figure 10–6C presents a graph of these results as a function of the volume of titrant. Since the second derivative ($\Delta^2 E/\Delta V^2$) changes sign and passes through zero between volume readings of 16.10 and 16.20 ml, the equivalence-point volume must lie within this interval. If the volume increments are small enough, one may assume that the ends of the upper and lower curves in Figure 10–6C can be joined by a straight line. Provided that this approach is valid, it can be shown by mathematical interpolation that the equivalence-point volume should be

$$16.10 + 0.10\left(\frac{35}{35 + 34}\right) = 16.15 \text{ ml}$$

Note that the analytical method requires only experimental data similar to those listed in Table 10–5; no graphical constructions are necessary.

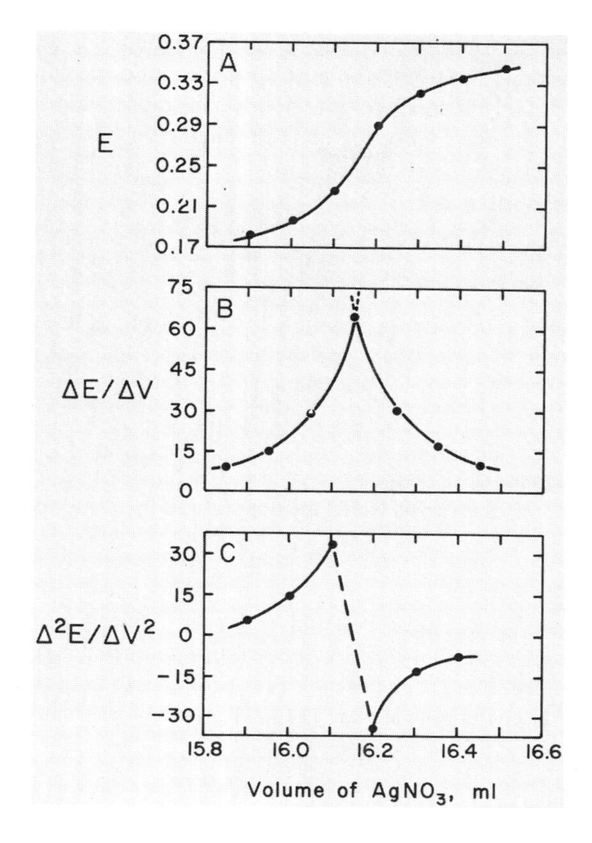


*Figure 10–6.* Titration curves for the titration of 3.737 millimoles of chloride with 0.2314 $F$ silver nitrate. Curve A: normal titration curve, showing the region near the equivalence point. Curve B: first derivative titration curve. Curve C: second derivative titration curve. All data points are taken from Table 10–5.

## *APPLICATIONS OF POTENTIOMETRIC TITRIMETRY*

Potentiometric titrations can be performed for precipitation, complex-formation, acid-base, and oxidation-reduction reactions. Titration accuracies are fully as good as with conventional end-point indicators and may be even better.

### Precipitation Titrations

Potentiometric end-point detection is applicable to precipitation titrations involving silver(I), mercury(II), lead(II), and zinc(II), as well as anions such as chloride, bromide, iodide, thiocyanate, and ferrocyanide.

Indicator electrodes for potentiometric precipitation titrations may be fabricated from the metal whose cation comprises the precipitate. Alternatively, a membrane electrode which responds to an appropriate ion is often useful. In some instances, the indicator electrode may be one whose potential is governed by the anion of the precipitate. For example, the progress of a titration of chloride ion with standard silver nitrate solution can be monitored if the potential of a silver indicator electrode versus a saturated calomel reference electrode is measured. Figure 10–5 illustrates the physical appearance of the titration apparatus, whereas the shorthand representation of the galvanic cell shown in this diagram is

$$\text{Hg} \,|\text{Hg}_2\text{Cl}_2(\text{s}), \text{KCl}(\text{s})|\, \text{KNO}_3\ (1\ F)\ |\ \text{chloride sample solution}\ |\ \text{Ag}$$

A potassium nitrate salt bridge is inserted between the sample solution and the saturated calomel electrode to prevent diffusion of potassium chloride solution into the sample.

Once a small volume of silver nitrate titrant is added, the solution becomes saturated with silver chloride and the indicator electrode behaves as a silver-silver chloride electrode. If we neglect liquid-junction potentials, the overall emf of the galvanic cell can be written as

$$E_{cell} = E_{Hg,Hg_2Cl_2} + E_{AgCl,Ag}$$

and, if the Nernst equation for the silver chloride-silver metal half-reaction is substituted into the latter relation, it becomes

$$E_{cell} = E_{Hg,Hg_2Cl_2} + E^0_{AgCl,Ag} - 0.059 \log (Cl^-)$$

When the respective values for $E_{Hg,Hg_2Cl_2}$ and $E^0_{AgCl,Ag}$ (namely, -0.2415 and +0.2222 v versus NHE) are substituted into this equation, the result is

$$E_{cell} = -0.0193 - 0.059 \log (Cl^-) = -0.0193 + 0.059 \text{ pCl}$$

Thus, a linear relationship exists between the value of pCl and the emf of the galvanic cell. Accordingly, if one performs a potentiometric titration of chloride with silver nitrate solution and constructs a plot of emf as a function of the volume of titrant, the titration curve would be similar to that shown in Figure 7–2.

### Complexometric Titrations

Most present applications of complexometric titrations center on the reactions of metal ions with various members of the family of aminopolycarboxylic acids, notably ethylenediaminetetraacetic acid (EDTA). A large number of metal ions form stable one-to-one complexes with the ethylenediaminetetraacetate anion, so there exists considerable interest in the use of potentiometric methods to monitor the progress of metal-EDTA titrations.

Elemental mercury in the presence of a small concentration of the very stable mercury(II)-EDTA complex provides a versatile and unique indicator-electrode system for many metal-EDTA titrations. A convenient form of mercury indicator electrode can be prepared if a gold wire is amalgamated by immersion into mercury.

Suppose we wish to determine calcium ion in an unknown sample solution by means of a potentiometric EDTA titration. Into the titration vessel containing the sample, let us insert a mercury indicator electrode and a saturated calomel reference electrode. We shall then introduce a small concentration, approximately $10^{-4}$ *M*, of the mercury(II)-EDTA complex. Finally, let us titrate the mixture with a standard solution of EDTA, recording the potential of the mercury indicator electrode as a function of the titrant volume. It can be shown, although we will not present the derivation here, that the potential of the mercury indicator electrode in the vicinity of the equivalence point is given by the relation

$$E_{Hg^{2+},Hg} = K - \frac{0.059}{2} \text{ pCa}$$

Thus, the mercury indicator electrode exhibits a linear response to pCa, the negative logarithm of the calcium ion concentration (activity). If the potential of the indicator electrode is plotted versus the volume of titrant, a titration curve similar to that shown in Figure 6–3 is obtained.

Wide applicability of the potentiometric procedure stems from the fact that the mercury(II)-EDTA complex is unusually stable. Indeed, only metal cations whose EDTA complexes are *less* stable than the mercury(II)-EDTA species can be determined satisfactorily. However, Table 6–1 on page 128 provides evidence that this requirement is met by many cations of analytical importance and, in accord with this observation, several dozen elements may be titrated potentiometrically with EDTA.

### Acid-Base Titrations

Availability of glass electrodes, saturated calomel electrodes, and pH meters has rendered potentiometric acid-base titrations more convenient to perform than practically any other method of volumetric analysis. Particularly impressive is the increasing number of applications in the field of nonaqueous acid-base titrimetry, some of which were described in Chapter 5.

Potentiometric acid-base titrimetry offers several additional advantages which other neutralization methods of analysis lack. First, the former technique allows the recording of complete titration curves, which is especially valuable for the study of mixtures of acids or bases in either aqueous or nonaqueous solvents. Second, the potentiometric procedure permits one to apply graphical or analytical methods for the location of equivalence points, a necessity in the analysis of multicomponent systems. Third, quantitative information concerning the relative acid or base strengths of solute species can be obtained. For example, as shown in Figure 5–8, the titration of phosphoric acid with a standard sodium hydroxide solution produces a titration curve with two steps, corresponding to the consecutive reactions

$$H_3PO_4 + OH^- \rightleftharpoons H_2PO_4^- + H_2O$$

and

$$H_2PO_4^- + OH^- \rightleftharpoons HPO_4^{2-} + H_2O$$

Halfway to the first equivalence point, the concentrations of phosphoric acid and dihydrogen phosphate are essentially equal. Substitution of the latter condition – namely, $[H_3PO_4] = [H_2PO_4^-]$ – into the equilibrium expression

$$K_{a1} = \frac{[H^+][H_2PO_4^-]}{[H_3PO_4]}$$

leads to the conclusion that the measured pH is numerically equal to $pK_{a1}$. In a similar fashion, the value of $pK_{a2}$ may be obtained directly from the pH reading midway between the first and second equivalence points.

### Oxidation-Reduction Titrations

Principles and applications of oxidation-reduction methods of analysis have been considered in detail in earlier chapters, so it suffices to mention here that a large majority of the determinations outlined previously can be accomplished by means of potentiometric titrimetry.

Most potentiometric redox titrations involve the use of a platinum indicator electrode, a saturated calomel reference electrode, and a suitable potential-measuring instrument. Unless chloride ion has a deleterious effect upon the desired oxidation-reduction reaction, the calomel reference electrode can be immersed directly into the titration vessel; otherwise, an appropriate salt bridge connecting the reference electrode to the solution in the titration vessel is required. Although a platinum wire or foil is almost invariably employed as the indicator electrode, gold, palladium, and carbon are occasionally used.

If an indicator electrode responds reversibly to the various oxidants and reductants in the chemical system, titration curves recorded in practice exhibit excellent agreement with theoretical curves derived according to procedures discussed in Chapter 8.

## QUESTIONS AND PROBLEMS

1. Suppose that the activity of copper(II) in an aqueous medium is to be determined through a direct potentiometric measurement with a copper metal indicator electrode, the pertinent half-reaction being

$$Cu^{2+} + 2\,e \rightleftharpoons Cu$$

   (a) If an error of exactly 1 mv in the observed potential of the copper electrode is incurred, calculate the corresponding per cent uncertainty in the copper(II) activity.
   (b) If the error in the copper(II) activity is not to exceed 1 per cent, calculate the maximal tolerable uncertainty in the potential of the copper indicator electrode.

2. Silver ion in 100.0 ml of a dilute nitric acid medium was titrated potentiometrically with a standard $2.068 \times 10^{-4}$ $F$ potassium iodide solution, according to the reaction

$$Ag^{+} + I^{-} \rightleftharpoons AgI(s)$$

A silver iodide indicator electrode and a saturated calomel reference electrode were used, and the following titration data resulted:

| Volume of KI, ml | $E$ versus SCE, volts |
|---|---|
| 0 | +0.2623 |
| 1.00 | 0.2550 |
| 2.00 | 0.2451 |
| 3.00 | 0.2309 |
| 3.50 | 0.2193 |
| 3.75 | 0.2105 |
| 4.00 | 0.1979 |
| 4.25 | 0.1730 |
| 4.50 | 0.0316 |
| 4.75 | −0.0066 |

   (a) Calculate the concentration of silver ion in the original sample solution.
   (b) Compute the titration error, in per cent with the proper sign, if potassium iodide titrant is added until the potential of the indicator electrode reaches +0.0600 v versus SCE.

3. A 0.2479-g sample of anhydrous sodium hexachloroplatinate(IV), $Na_2PtCl_6$, was analyzed for its chloride content by means of a potentiometric titration with standard silver nitrate solution. The weighed sample was decomposed in the presence of hydrazine sulfate, the platinum(IV) being reduced to the metal, and the liberated chloride ion was titrated with 0.2314 *F* silver nitrate. A silver indicator electrode, a saturated calomel reference electrode, and a nitric acid salt bridge were used in the performance of the titration. The resulting experimental data were as follows:

| Volume of $AgNO_3$, ml | *E* versus SCE, volts |
|---|---|
| 0 | +0.072 |
| 13.00 | 0.140 |
| 13.20 | 0.145 |
| 13.40 | 0.152 |
| 13.60 | 0.160 |
| 13.80 | 0.172 |
| 14.00 | 0.196 |
| 14.20 | 0.290 |
| 14.40 | 0.326 |
| 14.60 | 0.340 |

   Calculate the apparent percentage of chloride in the sample, the true percentage of chloride in sodium hexachloroplatinate(IV), and the per cent relative error between these two results.

4. Derive the equation, shown on page 246, pertaining to the method of standard addition as employed in direct potentiometry.

5. Calcium ion in a water sample was determined by means of a direct potentiometric measurement. Exactly 100.0 ml of the water was placed in a beaker, and a saturated calomel reference electrode (SCE) and a calcium-selective membrane electrode were immersed in the solution. The potential of the calcium electrode was found to be -0.0619 volt versus SCE. When a 10.00-ml aliquot of 0.00731 *F* calcium nitrate was pipetted into the beaker and mixed thoroughly with the water sample, the new potential of the calcium electrode was observed to be -0.0483 volt versus SCE. Calculate the molar concentration of calcium ion in the original water sample.

6. When the emf of the galvanic cell

$$-Ag \mid Ag(S_2O_3)_2^{3-}\ (0.001000\ M),\ S_2O_3^{2-}\ (2.00\ M) \parallel Ag^+\ (0.0500\ M) \mid Ag+$$

   was measured, a value of 0.903 volt was obtained. Calculate the overall formation constant, $\beta_2$, for the $Ag(S_2O_3)_2{}^{3-}$ complex, which is formed according to the reaction

$$Ag^+ + 2\ S_2O_3^{2-} \rightleftharpoons Ag(S_2O_3)_2^{3-}$$

7. If the electromotive force of the galvanic cell

$$-Pt,\ H_2\ (0.500\ atm) \mid \text{solution of unknown pH} \mid Hg_2Cl_2(s),\ KCl(s) \mid Hg+$$

   is 0.366 volt, calculate the pH of the unknown solution, neglecting the liquid-junction potential.

8. A liquid ion-exchange membrane electrode useful for the determination of ammonium ion ($NH_4^+$) was prepared by impregnation of a Millipore filter with a saturated solution of 72 per cent nonactin and 28 per cent monactin in tris(2-ethylhexyl)phosphate. A silver-silver chloride electrode was placed behind the ammonium-selective membrane; a saturated calomel reference electrode was combined with the ammonium ion electrode to obtain a complete electrochemical cell:

| $-$Ag | $AgCl(s)$, $LiCl$ (0.0100 $F$) | nonactin-monactin membrane | solution of unknown $NH_4^+$ activity | ‖ | $Hg_2Cl_2(s)$, $KCl(s)$ | Hg + |
|---|---|---|---|---|---|---|

When two different solutions having ammonium ion activities of $3.19 \times 10^{-3}$ and $1.08 \times 10^{-5}$ $M$ were placed in the cell, the observed emf values were 0.2105 and 0.0644 volt, respectively, at 25.00°C. What is the ammonium ion activity of a sample solution that yields an emf of 0.1128 volt? Assume that the liquid-junction potentials remain the same throughout the series of measurements.

9. A lanthanum fluoride ($LaF_3$) solid-state electrode can be used to follow the progress of potentiometric titrations of fluoride ion with standard lanthanum nitrate solution and can be employed for the direct measurement of the fluoride ion concentration in a solution. Exactly 100.0 ml of a 0.03095 $F$ sodium fluoride (NaF) solution was titrated potentiometrically with 0.03318 $F$ lanthanum nitrate solution, the titration reaction being

$$La^{3+} + 3\,F^- \rightleftharpoons LaF_3(s)$$

A solid-state $LaF_3$ membrane electrode, a saturated potassium chloride salt bridge, and a saturated calomel reference electrode (SCE) were employed, and the following titration data were obtained:

| Volume of $La(NO_3)_3$, ml | $E$ versus SCE, volts |
|---|---|
| 0 | −0.1046 |
| 29.00 | −0.0249 |
| 30.00 | −0.0047 |
| 30.30 | +0.0041 |
| 30.60 | 0.0179 |
| 30.90 | 0.0410 |
| 31.20 | 0.0656 |
| 31.50 | 0.0769 |
| 32.50 | 0.0888 |
| 36.00 | 0.1007 |
| 41.00 | 0.1069 |
| 50.00 | 0.1118 |

(a) Using the known concentrations of the sodium fluoride and lanthanum nitrate solutions, calculate the theoretical volume of titrant needed. How does this value compare to that obtained by means of the so-called analytical method for locating the equivalence point?

(b) It has been established that the emf of a cell consisting of a lanthanum fluoride membrane electrode and a saturated calomel reference electrode in contact with a sodium fluoride solution follows the equation

$$E = K + 0.05915 \text{ pF}$$

where $K$ is a constant for the system and where pF denotes the negative logarithm of the fluoride ion concentration. Using the initial point of the titration data, evaluate the constant in the above relation.

(c) Using the result of part (b), determine the concentration of fluoride ion after the addition of 50.00 ml of titrant. Ignore changes in activity coefficients due to variations in solution composition.

(d) Calculate the concentration of free lanthanum ion after the addition of 50.00 ml of titrant.

(e) Using the results of parts (c) and (d), evaluate the solubility-product constant for $LaF_3(s)$.

10. Quinhydrone is a sparingly soluble one-to-one addition compound formed from hydroquinone and quinone. When solid quinhydrone is dissolved in an aqueous medium, equal concentrations of hydroquinone and quinone result. If a platinum wire is dipped into the solution, the potential of the electrode is governed by the reversible reaction

$$\text{quinone} + 2\,H^+ + 2\,e \rightleftharpoons \text{hydroquinone} \quad ; \quad E^0 = +0.6994 \text{ v versus NHE}$$

Because the potential of the quinone-hydroquinone half-reaction depends on the concentration of hydrogen ion, it is possible to utilize this system for the measurement of pH.

A solution of unknown pH was saturated with quinhydrone and a platinum electrode was dipped into it. Then electrolytic contact was made between the sample solution and a saturated calomel reference electrode (SCE) by means of a potassium chloride salt bridge. Schematically, the electrochemical cell may be represented as

$$\text{Pt} \left| \begin{matrix} \text{solution of unknown pH} \\ \text{saturated with quinhydrone} \end{matrix} \right\| Hg_2Cl_2(s), \quad KCl(s) \Big| \text{Hg}$$

If the potential of the platinum electrode was found to be +0.2183 volt versus SCE, what was the pH of the unknown solution?

11. In order to determine the equilibrium constant for the proton-transfer reaction between pyridine and water,

$$C_5H_5N + H_2O \rightleftharpoons C_5H_5NH^+ + OH^-$$

the following electrochemical cell was constructed:

$$-\text{Pt}, \quad H_2\ (0.200 \text{ atm}) \left| \begin{matrix} C_5H_5N\ (0.189\ M), \\ C_5H_5NH^+Cl^-\ (0.0536\ M) \end{matrix} \right\| \begin{matrix} Hg_2Cl_2(s), \\ KCl(s) \end{matrix} \Big| \text{Hg} +$$

If the electromotive force of the cell is 0.563 volt at 25°C, what is the equilibrium constant ($K_b$) for the above reaction?

12. A sodium-sensitive membrane electrode was fabricated from a glass composed of 11 per cent $Na_2O$, 18 per cent $Al_2O_3$, and 71 per cent $SiO_2$. Then, the sodium-selective electrode and a saturated calomel reference electrode were dipped into an aqueous sodium chloride solution of unknown concentration to produce the following electrochemical cell:

$$-\text{Ag} \left| \begin{matrix} \text{AgCl(s)}, \\ \text{NaCl } (1.00\ F) \end{matrix} \right| \begin{matrix} \text{sodium-sensitive} \\ \text{glass membrane} \end{matrix} \left| \begin{matrix} \text{sample solution} \\ \text{of unknown pNa} \end{matrix} \right\| \begin{matrix} \text{Hg}_2\text{Cl}_2\text{(s)}, \\ \text{KCl(s)} \end{matrix} \left| \text{Hg} + \right.$$

When the emf of the cell was measured at 25°C with a potentiometer, a value of 0.2033 volt was obtained. When the unknown solution was replaced with a sodium chloride solution containing a sodium ion activity of $3.23 \times 10^{-3}\ M$, the emf of the cell was observed to be 0.1667 volt at 25°C.

(a) Assuming no changes in the liquid-junction potential or the asymmetry potential of the glass membrane, and assuming a perfect Nernstian response of the glass membrane to sodium ion, calculate the sodium ion activity in the unknown solution.

(b) If the accuracy of each emf measurement is no better than ±1 mv, what is the range of sodium ion activities within which the true value must fall?

## SUGGESTIONS FOR ADDITIONAL READING

1. R. G. Bates: *Determination of pH*. John Wiley and Sons, New York, 1964.
2. R. P. Buck: Potentiometry: pH measurements and ion selective electrodes. *In* A. Weissberger and B. W. Rossiter, eds.: *Physical Methods of Chemistry*. Volume I, Part IIA, Wiley-Interscience, New York, 1971, pp. 61–162.
3. A. K. Covington: Ion-selective electrodes. Chem. Brit., *5*:388, 1969.
4. R. A. Durst, ed.: *Ion-Selective Electrodes*. National Bureau of Standards, Special Publication No. 314, Washington, 1969.
5. R. A. Durst: Ion-selective electrodes in science, medicine, and technology. Amer. Sci., *59*:353, 1971.
6. G. Eisenman, ed.: *Glass Electrodes for Hydrogen and Other Cations*. Dekker, New York, 1967.
7. G. Eisenman: The electrochemistry of cation-sensitive glass electrodes. *In* C. N. Reilley, ed.: *Advances in Analytical Chemistry and Instrumentation*. Volume 4, Wiley-Interscience, New York, 1965, pp. 213–369.
8. G. Eisenman, R. Bates, G. Mattock, and S. M. Friedman: *The Glass Electrode*. Wiley-Interscience, New York, 1966.
9. J. J. Lingane: *Electroanalytical Chemistry*. Second edition, Wiley-Interscience, New York, 1958, pp. 9–35, 71–167.
10. G. A. Rechnitz: Ion-selective electrodes. Chem. Eng. News *45*:146, June 12, 1967.
11. G. A. Rechnitz: Chemical studies at ion-selective membrane electrodes. Accts. Chem. Res., *3*:69, 1970.

# 11 COULOMETRY AND POLAROGRAPHY

A number of methods of chemical analysis are based upon the principle that a desired reaction can take place when an external voltage of appropriate magnitude and polarity is impressed across the electrodes of an electrochemical cell. Using **electrogravimetry**, one can accomplish a number of useful analyses by depositing a metal quantitatively upon a previously weighed electrode and by reweighing the electrode to obtain the amount of the metal. In **controlled-potential coulometry**, the potential of an anode or cathode is maintained constant so that only a single reaction occurs at that electrode. By integrating the current, which flows as a function of time, one can determine the total quantity of electricity produced by the desired reaction and can calculate the amount of the species of interest according to Faraday's law.

Another technique, **coulometric titration**, is a method in which a titrant, electrochemically generated at constant current, reacts with the substance to be determined. Since the magnitude of the constant current is analogous to the concentration of a standard titrant solution, and the time required to complete the titration is equivalent to the volume of titrant solution, the current-time product is directly related to the unknown quantity of substance.

In 1922, Jaroslav Heyrovský published a paper* describing a new electroanalytical technique called **polarography**, a discovery that eventually led to his receiving the Nobel prize in chemistry for 1959. In polarography, one is interested in the current-potential curves resulting from electron-transfer processes that occur at the surface of a dropping mercury electrode only a few square millimeters in area. From such measurements the identity and concentration of the reactant can be determined. In his Nobel lecture,† Heyrovský reviewed some of the virtues of polarography – that solutions as dilute as $10^{-5}$ *M* can be analyzed, that samples as small as 0.05 ml may be handled routinely, and that almost every element, in one form or another, as well as many hundreds of organic compounds can be determined.

## ELECTROGRAVIMETRY

Electrogravimetric methods of analysis usually involve deposition of the desired metallic element upon a previously weighed cathode, followed by subsequent reweighing of the electrode plus deposit to obtain by difference the quantity of the metal. Among the elements that have been determined in this manner are cadmium, copper, nickel, silver, tin, and zinc. A few substances can be oxidized at a platinum anode to form an insoluble and adherent precipitate suitable for gravimetric measurement. An example of this procedure is the oxidation of lead(II) to lead dioxide in a nitric acid medium.

---

*J. Heyrovský: Chem. listy *16*:256, 1922.

†J. Heyrovský: The trends of polarography. In *Nobel Lectures in Chemistry for 1942-1962*. Elsevier Publishing Company, Amsterdam, pp. 564-584.

Representative of most electrogravimetric procedures is the determination of copper. Suppose that we desire to measure the quantity of copper(II) in a 1 *F* sulfuric acid medium by depositing the metal upon a previously weighed platinum electrode. Typical apparatus, consisting of a large platinum gauze cathode and a smaller anode, is depicted in Figure 11–1. Both electrodes are immersed in the solution to be electrolyzed, although some applications may require that the anode and cathode be in different compartments separated by a porous membrane to prevent interaction of the products formed at the anode with the metal deposit on the cathode. A direct-current power supply (*E*) and a rheostat (*R*) enabling one to vary the voltage impressed across the cell constitute the essential electrical circuitry. It is customary to incorporate both a voltmeter (*V*) and an ammeter (*A*) so that the magnitudes of the applied voltage and current can be monitored continuously. Efficient stirring of the solution is mandatory. A magnetic stirrer or a motor-driven propeller is often employed; sometimes, the anode has a design which permits it to be rotated with the aid of a motor.

There are two techniques by which the reduction of copper(II) and deposition of copper metal can be accomplished electrolytically. In the first method the electrolysis is performed at constant applied voltage, whereas the second procedure requires that the current passing through the cell be kept constant.

#### Electrolysis at Constant Applied Voltage

In beginning our discussion of electrolysis at constant applied voltage, we shall assume that the original concentration of copper(II) is 0.01 *M* and that 2.0 to 2.5 v – a voltage sufficient to cause rapid and quantitative deposition of copper – is impressed across the electrochemical cell. As soon as the deposition of copper metal has commenced, the cathodic process is the reduction of cupric ion at a copper-plated platinum electrode

$$Cu^{2+} + 2\,e \rightleftharpoons Cu$$

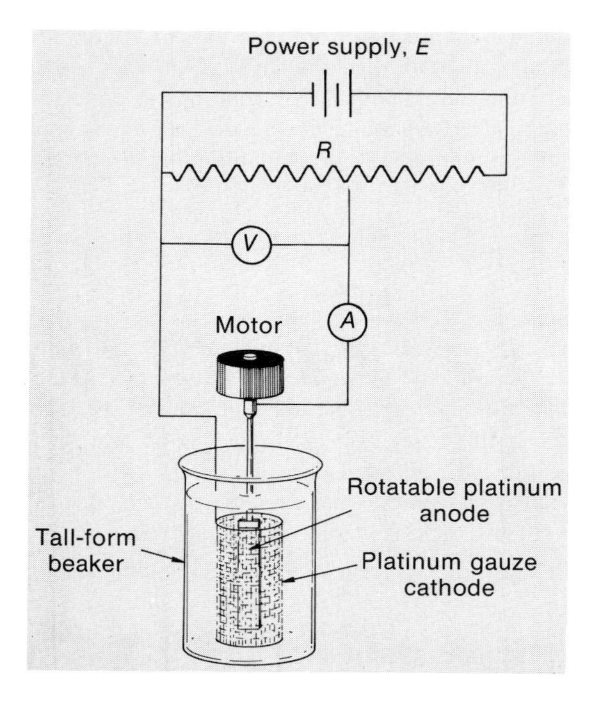


*Figure 11–1.* Simple apparatus for electrodeposition of metals.

whereas oxidation of water

$$2\,H_2O \rightleftharpoons O_2 + 4\,H^+ + 4\,e$$

occurs at the anode.

Figure 11–2 depicts the experimentally observed potential-time behavior of the copper cathode. It can be seen that the potential of the cathode decreases rapidly after about five minutes to approximately −0.3 v and undergoes a slower drop to a final value near −0.44 v. Any further shift of the cathode potential is halted by the steady evolution of hydrogen gas at the copper-plated platinum electrode.

### Electrolysis at Constant Current

A second method by which electrogravimetric determinations are performed involves the use of constant-current electrolysis. In this technique the external voltage impressed across an electrolytic cell is varied so that the current remains constant during the course of the deposition.

For the copper(II)-sulfuric acid system described above, the potential-time behavior of the copper cathode during a constant-current electrolysis will resemble the curve shown in Figure 11–2. Of course, the larger the constant current, the shorter will be the time required to complete the deposition of copper. However, if an unusually high

*Figure 11–2.* Schematic variation of potential of copper-plated platinum cathode during deposition of copper by means of electrolysis at constant applied voltage or at constant current from a solution initially 0.01 $M$ in copper(II) and 1 $F$ in sulfuric acid.

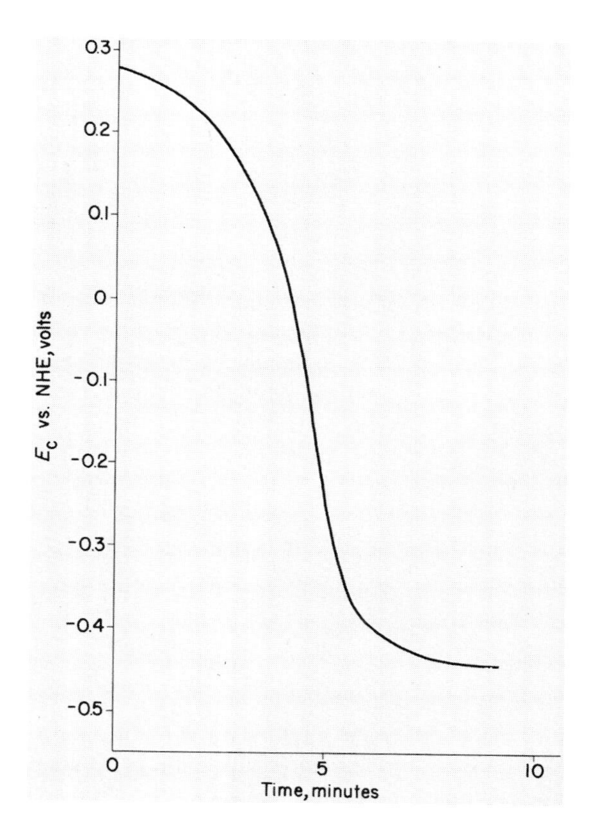

current is employed, it is impossible for cupric ions alone to sustain the desired rate of electron transfer and, almost as soon as electrolysis begins, a significant fraction of the total current will be due to evolution of hydrogen gas. As a result of vigorous gas evolution, metal deposits obtained by means of constant-current electrolysis frequently have poorer physical properties than those resulting from electrolysis at constant applied voltage. Formation of hydrogen gas can be eliminated through the use of a **cathodic depolarizer**, some substance more easily reduced than hydrogen ion but one that does not interfere with the smooth deposition of copper or any other metal. Nitric acid is a common cathodic depolarizer, because nitrate ion undergoes an eight-electron reduction to ammonium ion at a more positive potential than the reduction of hydrogen ion to molecular hydrogen.

### Feasibility of Determinations and Separations

That the deposition of copper described in the preceding sections permits the attainment of accurate analytical results can be shown through a simple calculation. Suppose that no more than one part in 10,000 of the original quantity of copper(II) is to remain in solution at the end of the electrolysis. Since the sulfuric acid medium initially contained cupric ion at a concentration of 0.01 *M*, the copper(II) concentration must be decreased to $10^{-6}$ *M*. Using the Nernst equation

$$E_{\mathrm{Cu^{2+},Cu}} = E^0_{\mathrm{Cu^{2+},Cu}} - \frac{0.059}{2}\log\frac{1}{[\mathrm{Cu^{2+}}]}$$

we can conclude that, if the potential of the cathode has a value of

$$E_{\mathrm{Cu^{2+},Cu}} = 0.337 - \frac{0.059}{2}\log\frac{1}{(10^{-6})} = +0.160 \text{ v versus NHE}$$

or less when the electrolysis is interrupted, copper metal will have been quantitatively deposited. Figure 11–2 reveals that this requirement is easily met, since the cathode potential reaches −0.44 v.

Although the large negative shift of the cathode potential does ensure the quantitative deposition of copper, an uncontrolled change in the potential is undesirable if separation of copper from other elements is to be achieved. Any metal cations present in the sulfuric acid solution which undergo even partial deposition at potentials more positive than −0.44 v versus NHE will be plated upon the cathode along with copper. Among the species that could interfere with the determination of copper are bismuth(III), antimony(III), tin(II), and cobalt(II), the pertinent half-reactions and standard potentials being

$$\mathrm{BiO^+ + 2\,H^+ + 3\,e \rightleftharpoons Bi + H_2O}; \qquad E^0 = +0.32 \text{ v}$$

$$\mathrm{SbO^+ + 2\,H^+ + 3\,e \rightleftharpoons Sb + H_2O}; \qquad E^0 = +0.212 \text{ v}$$

$$\mathrm{Sn^{2+} + 2\,e \rightleftharpoons Sn}; \qquad E^0 = -0.136 \text{ v}$$

$$\mathrm{Co^{2+} + 2\,e \rightleftharpoons Co}; \qquad E^0 = -0.277 \text{ v}$$

Codeposition of tin and cobalt could be avoided if the electrolysis were stopped before the cathode potential became too negative. However, as commonly performed, electrogravimetry is a non-selective, "brute-force" technique in which metallic elements are separated into two groups – those species more easily reduced than hydrogen ion or water and those elements more difficult to reduce than hydrogen ion or the solvent. Therefore, electrogravimetric analysis is usually applicable to systems in which the metal ion of interest is the only reducible substance other than hydrogen ion or water. In some instances, one can improve the selectivity of electrogravimetry by changing the pH of the sample solution by adding appropriate complexing agents to alter the reduction potentials for the species of interest.

### Influence of Experimental Conditions on Electrolytic Deposits

Analyses based on electrodeposition must fulfill the same requirements as other gravimetric methods. Accordingly, the desired substance must be deposited quantitatively and the deposit must be pure. Above all, the deposit must possess physical properties such that subsequent washing, drying, and weighing operations will not cause significant mechanical loss or chemical alteration of the substance.

**Current Density.** Generally, the variable with the greatest effect upon the physical state of an electrolytic deposit is current density, since increases in the latter reduce the size of individual metal crystals. As a rule, current densities ranging from 0.005 to 0.05 amp/cm$^2$ yield metal deposits which are smooth and lustrous. Excessively high current densities are undesirable because the resulting deposit has a tendency to consist of fragile whiskers or dendrites and because simultaneous evolution of hydrogen gas will cause spongy deposits to form.

**Evolution of Gas.** Liberation of a gas during the deposition of a metal invariably produces a rough, spongy solid. In addition, vigorous evolution of a gas may detach fragments of the metal deposit from the electrode. Evolution of gas can be minimized or eliminated through the use of a depolarizer.

**Solution Composition.** It is known empirically that reduction of a stable metal ion complex almost always yields a much smoother and more tightly adherent deposit than is obtainable from a solution of the aquated cation. For example, metallic silver plated from an alkaline cyanide medium containing the species $Ag(CN)_2^-$, or from an ammonia solution in which $Ag(NH_3)_2^+$ is predominant, is characteristically smooth and shiny. On the other hand, a loosely adherent, dendritic deposit of elemental silver is obtained from an aqueous silver nitrate solution in which the simple hydrated argentous ion exists.

## CONTROLLED-POTENTIAL COULOMETRY

In the preceding section of this chapter, we concluded that electrolysis at either constant applied voltage or constant current lacks selectivity for analytical separations and determinations. This is because variations in the potential of the cathode or anode at which the desired reaction is supposed to occur may cause extraneous processes to take place. On the other hand, by controlling the potential of an anode or cathode so that only the desired reaction will proceed and by taking advantage of the influence of solution composition on the standard or formal potentials for half-reactions, one can achieve utmost specificity in electrochemical determinations and separations.

### Feasibility of Separations and Determinations

Suppose that we wish to separate and determine by means of controlled-potential coulometry with a platinum gauze cathode both cadmium(II) and zinc(II) in a 1 *F* ammonia solution initially containing 0.01 $M$ concentrations of the respective metal-ammine complexes, $Cd(NH_3)_4{}^{2+}$ and $Zn(NH_3)_4{}^{2+}$.

Since the half-reaction and standard potential for the two-electron reduction of $Cd(NH_3)_4{}^{2+}$ are

$$Cd(NH_3)_4{}^{2+} + 2\ e \rightleftharpoons Cd + 4\ NH_3; \qquad E^0 = -0.61\ v$$

we can predict from the Nernst equation

$$E_{Cd(NH_3)_4{}^{2+},Cd} = E^0_{Cd(NH_3)_4{}^{2+},Cd} - \frac{0.059}{2} \log \frac{[NH_3]^4}{[Cd(NH_3)_4{}^{2+}]}$$

that cadmium metal should begin to deposit rapidly when the cathode attains a potential of

$$E_{Cd(NH_3)_4{}^{2+},Cd} = -0.61 - \frac{0.059}{2} \log \frac{(1)^4}{(0.01)} = -0.67\ v\ \text{versus NHE}$$

Similarly, from the half-reaction

$$Zn(NH_3)_4{}^{2+} + 2\ e \rightleftharpoons Zn + 4\ NH_3; \qquad E^0 = -1.04\ v$$

we can expect that metallic zinc will start to undergo rapid deposition when the cathode potential reaches

$$E_{Zn(NH_3)_4{}^{2+},Zn} = E^0_{Zn(NH_3)_4{}^{2+},Zn} - \frac{0.059}{2} \log \frac{[NH_3]^4}{[Zn(NH_3)_4{}^{2+}]}$$

$$E_{Zn(NH_3)_4{}^{2+},Zn} = -1.04 - \frac{0.059}{2} \log \frac{(1)^4}{(0.01)} = -1.10\ v\ \text{versus NHE}$$

These calculations reveal that the potential of the electrode must not be more positive than -0.67 v if deposition of cadmium is to occur nor more negative than -1.10 v if codeposition of zinc is to be avoided. Let us control the electrode potential at -0.88 v, midway between the preceding two values, and let us stir the solution very efficiently. What concentration of $Cd(NH_3)_4{}^{2+}$ remains after the solution has been exhaustively electrolyzed? Using the Nernst equation, we can determine that the final concentration of $Cd(NH_3)_4{}^{2+}$ is $7.1 \times 10^{-10}$ $M$ and that very complete deposition of cadmium is obtainable.

At the conclusion of the first part of this procedure – signaled by the fact that the current falls practically to zero – the cathode may be removed from the electrolytic cell, and washed, dried, and weighed for the determination of cadmium. Before the deposition of zinc, the cadmium might be dissolved from the platinum cathode with nitric acid. Alternatively, the cadmium could be left on the electrode, and the latter returned to the electrolytic cell. Then the cathode potential is adjusted so that zinc(II) is reduced to the elemental state.

A prediction of the potential needed to ensure the quantitative deposition of zinc can be based on the Nernst equation for reduction of $Zn(NH_3)_4{}^{2+}$. To deposit 99.9 per cent of the zinc, corresponding to a final $Zn(NH_3)_4{}^{2+}$ concentration of $1 \times 10^{-5}$ $M$, the required cathode potential is -1.19 v versus NHE. If the cathode potential is maintained at -1.19 v, the bulk concentration of $Zn(NH_3)_4{}^{2+}$ decreases with time and the current diminishes until it becomes almost zero – a very small current will continue to flow. At this point, the controlled-potential deposition of zinc may be regarded as complete, the cathode removed from the cell, and the plated zinc washed, dried, and weighed.

However, the controlled-potential deposition of metals does not require any weighing operations. If the working-electrode potential is controlled and only one electrode reaction occurs, the total quantity of electricity produced by that reaction can be related to the weight or concentration of the original electroactive species through Faraday's law.

### Current-Time Behavior and Faraday's Law

A useful characteristic of controlled-potential electrolysis is the variation of current with time. Let us consider the deposition of cadmium metal upon a platinum cathode from a well-stirred 0.01 *M* solution of $Cd(NH_3)_4{}^{2+}$ in 1 *F* aqueous ammonia. Suppose the cathode potential is controlled at a value corresponding to that for $Cd(NH_3)_4{}^{2+}$ reduction – for example, −0.88 v versus NHE. At the moment the controlled-potential deposition begins, the current due to reduction of $Cd(NH_3)_4{}^{2+}$ ion has some initial value, $i_0$. When one half of the cadmium has been deposited, the current, which is proportional to the bulk concentration of $Cd(NH_3)_4{}^{2+}$, will have diminished to $0.5i_0$. After 90 per cent of the $Cd(NH_3)_4{}^{2+}$ ion has undergone reduction, the current would be $0.1i_0$, and the current after 99 and 99.9 per cent of the cadmium is deposited should have a value of $0.01i_0$ and $0.001i_0$, respectively.

For the ideal situation described in the previous paragraph, the current decays in accordance with the exponential expression

$$i_t = i_0 e^{-kt}$$

where $i_0$ is the initial current and $i_t$ is the current at any time $t$. In this relation, the constant $k$ increases if the electrode area or the rate of mechanical stirring or the temperature becomes greater, whereas $k$ decreases if the volume of the solution is increased. To minimize the time required to complete a controlled-potential electrolysis, these experimental parameters should be manipulated to make the value of $k$ as large as possible. Although the form of the exponential decay law prevents the current from ever reaching zero, the magnitude of the final current relative to the initial current does provide the most straightforward means to determine when an electrolysis is quantitatively complete. For example, if an error not exceeding one part per thousand is desired, the electrolysis should be continued until the current decays to 0.1 per cent of its initial value.

Fundamentally, the quantity of electricity, $Q$, in coulombs passed in any electrolysis is given by the relation

$$Q = \int_0^t i_t \, dt$$

where $i_t$ is the current in amperes as a function of time and $t$ is the duration of the electrolysis in seconds. Integration of the current-time curve for a controlled-potential electrolysis may be accomplished in either of two ways. First, a mechanical or electronic *current-time integrator*, with provision for direct readout of the number of coulombs of electricity, can be incorporated into the electrical circuit. Alternatively, the quantity of electricity corresponding to the desired reaction can be determined by means of a *chemical coulometer* placed in series with the electrolysis cell.

Once the quantity of electricity, $Q$, has been measured, the weight, $W$, in grams of the substance being determined is obtainable from Faraday's law of electrolysis

$$W = \frac{QM}{nF}$$

where $M$ is the formula weight of the species that is oxidized or reduced, $n$ is the number of faradays of electricity involved in the oxidation or reduction of one formula weight of substance, and $F$ is the Faraday constant (96,487 coulombs).

### Equipment for Controlled-Potential Coulometry

Figure 11–3 shows the apparatus needed for controlled-potential coulometry – a cell with appropriate working, auxiliary, and reference electrodes; a variable voltage source-potentiostat combination; and a coulometer.

A coulometric cell contains three electrodes. First, there is the *working electrode*, whose potential is controlled to permit only one reaction to occur. Second, the *auxiliary electrode* serves, along with the working electrode, to complete the electrolysis circuit. Third, there must be a *reference electrode* against which the potential of the working electrode is measured and controlled.

Platinum and mercury working electrodes are commonly employed, but gold, carbon, and silver are occasionally useful. Mercury is so easily oxidized that it can be used as an anode only for reactions of readily oxidizable substances; platinum, gold, and carbon are much better anode materials. On the other hand, mercury is superior for cathodic reactions. An auxiliary electrode is quite often fabricated from the same material as the working electrode. There is a danger that substances formed at the auxiliary electrode may be stirred to the working electrode at which they would react, or vice versa.

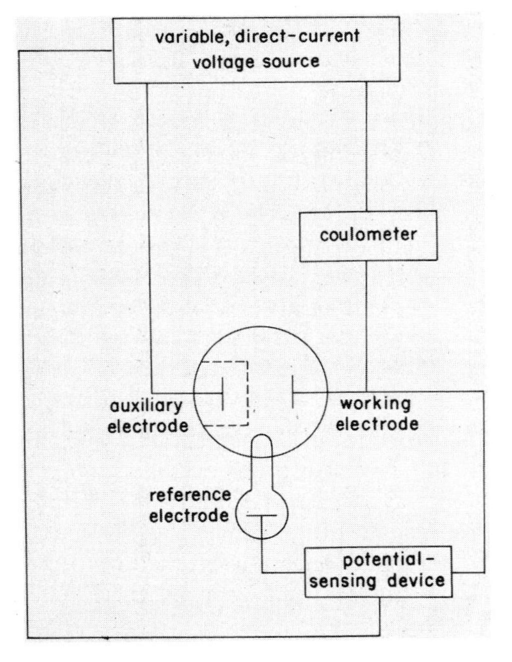


*Figure 11–3.* Apparatus for controlled-potential coulometry.

To prevent this, the auxiliary electrode is situated in a separate compartment of the cell that makes electrolytic contact with the working-electrode compartment through a sintered-glass disk. A saturated calomel or a silver-silver chloride half-cell is usually chosen as the reference electrode; it is brought into electrolytic contact with the sample solution through a salt bridge.

For oxidation or reduction processes at a platinum gauze working electrode, the cell illustrated in Figure 11–1 is suitable. A length of platinum wire placed in a separate compartment of the cell could be employed as the auxiliary electrode, and an appropriate reference electrode must be added. It would be necessary to close the cell with a stopper containing holes for insertion of the electrodes and other accessories used for stirring the solution and removing dissolved oxygen.

Throughout a controlled-potential electrolysis, the total voltage applied to the cell must be continuously decreased to keep the potential of the working electrode constant. Many electronic *potentiostats* are commercially available which automatically monitor the potential of the working electrode and adjust the applied voltage to maintain that potential at the preselected value.

### Applications of Controlled-Potential Coulometry

Controlled-potential coulometry has been applied to the determination of both inorganic and organic substances. In general, analyses based on controlled-potential coulometry are accurate to within 0.5 to 1 per cent, although errors may be as small as 0.1 per cent. Controlled-potential coulometry is best suited for the determination of 0.05 to 1 milliequivalent of the desired substance in 50 to 100 ml of solution.

**Cathodic Separation and Determination of Metals.** An elegant example of the use of controlled-potential coulometry for the separation and determination of metal cations is the analysis of alloys containing copper, bismuth, lead, and tin. A weighed sample is dissolved in a mixture of nitric acid and hydrochloric acid; sodium tartrate, succinic acid, and hydrazine are added to the solution and the pH is adjusted to 6.0. From this solution, copper can be deposited quantitatively upon a platinum cathode whose potential is maintained at −0.30 v versus SCE. Then the cathode is weighed to establish the quantity of copper, the copper-coated cathode is returned to the cell, and bismuth is plated upon the electrode at a potential of −0.40 v. After the gravimetric determination of bismuth, the cathode is again returned to the cell for the deposition of lead at a potential of −0.60 v. Finally, when the weight of the lead deposit has been determined, the cathode is replaced in the cell, the solution is acidified to decompose the tin(IV)-tartrate complex, the tin(IV) is reduced at a potential of −0.60 v, and the weight of tin metal is obtained.

Procedures such as the one just outlined are available for the separation and determination of one or more components of the following mixtures: antimony, lead, and tin; lead, cadmium, and zinc; silver and copper; nickel, zinc, aluminum, and iron; and rhodium and iridium.

A number of separations and determinations have been accomplished with the aid of the mercury pool cathode. Lead(II) can be separated from cadmium(II) by deposition of lead into mercury. Copper(II) and bismuth(III) in an acidic tartrate solution can be separated and determined by means of controlled-potential coulometry with a mercury pool cathode, and uranium(VI) may be determined through controlled-potential reduction to uranium(IV) in a 1 *F* sulfuric acid medium.

**Anodic Reactions at Platinum and Silver Electrodes.** Controlled-potential determinations involving anodic processes at platinum electrodes in 1 *F* sulfuric acid include the oxidation of iron(II) to iron(III), of arsenic(III) to arsenic(V), and of thallium(I) to insoluble thallium(III) oxide.

By using silver electrodes one can carry out the controlled-potential coulometric determination of individual halides as well as the analysis of halide mixtures, the reaction being

$$Ag + Br^- \rightleftharpoons AgBr + e$$

when bromide ion is present.

**Determination of Organic Compounds.** A variety of organic substances have been determined by means of controlled-potential coulometry. Reduction of picric acid to triaminophenol at a mercury pool cathode

$$C_6H_2(OH)(NO_2)_3 + 18\,H^+ + 18\,e \rightarrow C_6H_2(OH)(NH_2)_3 + 6\,H_2O$$

permits solutions of the original acid to be analyzed. Other nitro compounds (including *o*-nitrophenol, *m*-nitrophenol, and *p*-nitrobenzoic acid) as well as azo dyes and nitroso compounds can be quantitatively reduced at mercury. Trichloroacetate ion undergoes a pair of two-electron reductions at a mercury electrode, first to dichloroacetate

$$Cl_3CCOO^- + H^+ + 2\,e \rightarrow Cl_2HCCOO^- + Cl^-$$

and then to monochloroacetate:

$$Cl_2HCCOO^- + H^+ + 2\,e \rightarrow ClH_2CCOO^- + Cl^-$$

Since the potentials for the two reactions differ by 0.8 v, it is possible to measure trichloroacetate in the presence of much larger or much smaller quantities of dichloroacetate and monochloroacetate. Primary, secondary, and tertiary aliphatic amides such as acetamide, N-methylacetamide, and N,N-dipropylpropionamide, respectively, can be determined by oxidation at a platinum electrode in an acetonitrile medium. Substituted hydroquinones undergo a two-electron reaction at platinum anodes which serves as the basis of an analytical method.

## COULOMETRIC TITRATIONS

Coulometric titrimetry is based upon the constant-current electrolytic generation of a titrant which reacts quantitatively with the substance to be determined. By multiplying together the magnitude of the constant current and the time required to generate an amount of titrant stoichiometrically equivalent to the desired species, one obtains the quantity of the desired substance.

### Principles of Coulometric Titrimetry

Several aspects of coulometric titrimetry can be illustrated if we consider the determination of cerium(IV) by reduction to cerium(III) in a 1 *F* sulfuric acid medium.

Assume that an oxygen-free cerium(IV) solution is contained in a coulometric cell equipped with a stirrer, that a platinum generator electrode is immersed in the sample solution, that a platinum auxiliary electrode is situated in a separate compartment filled with 1 $F$ sulfuric acid, and that a constant current is passed through this electrolytic cell to cause the generator electrode to be the cathode.

If a relatively large concentration of iron(III) is added to the sample solution, the half-reaction

$$Fe^{3+} + e \rightleftharpoons Fe^{2+}$$

takes place at the cathode. However, the iron(II) formed at the electrode surface is stirred into the body of the sample solution and chemically reduces cerium(IV) to cerium(III):

$$Ce^{4+} + Fe^{2+} \rightleftharpoons Ce^{3+} + Fe^{3+}$$

Since iron(II) is converted back to iron(III), only one net reaction occurs – *the quantitative reduction of cerium(IV)*. If a method of end-point detection is available, such as a means to determine the first tiny excess of electrogenerated iron(II), the quantity of cerium(IV) can be related to the product of the constant current, $i$, and the electrolysis time, $t$. Accordingly, we speak of the *coulometric titration of cerium(IV) with electrogenerated iron(II)*, where the latter species is called a **coulometric titrant**. Of course, some of the cerium(IV) is reduced through direct electron-transfer at the working electrode.

## Advantages of Coulometric Titrimetry

Compared to conventional volumetric methods of analysis, coulometric titrimetry possesses several advantages.

Perhaps the most important virtue of coulometric titrimetry is that exceedingly small quantities of titrant can be accurately generated. It is not uncommon to perform coulometric titrations in which increments of titrant are generated with a constant current of 10 ma flowing for 0.1 second – which corresponds to only 1 millicoulomb or approximately $1 \times 10^{-5}$ milliequivalent. Such an amount of reagent would require the addition of 1 $\mu$l of 0.01 $N$ titrant from a microburet.

Coulometric titrimetry is inherently capable of much higher accuracy than ordinary volumetric analysis because the two parameters of interest – current and time – can be determined experimentally with exceptional precision. In addition, the preparation, storage, and handling of standard titrant solutions are eliminated. Moreover, titrants which are unstable or otherwise troublesome to preserve can be electrochemically generated in situ. Such uncommon reagents as silver(II), manganese(III), titanium(III), copper(I), tin(II), bromine, and chlorine can be conveniently prepared as coulometric titrants.

## Apparatus and Techniques of Coulometric Titrimetry

Basic equipment needed for the performance of coulometric titrations includes a suitable source of constant current, an accurate timing device, and a coulometric cell, as indicated schematically in Figure 11–4.

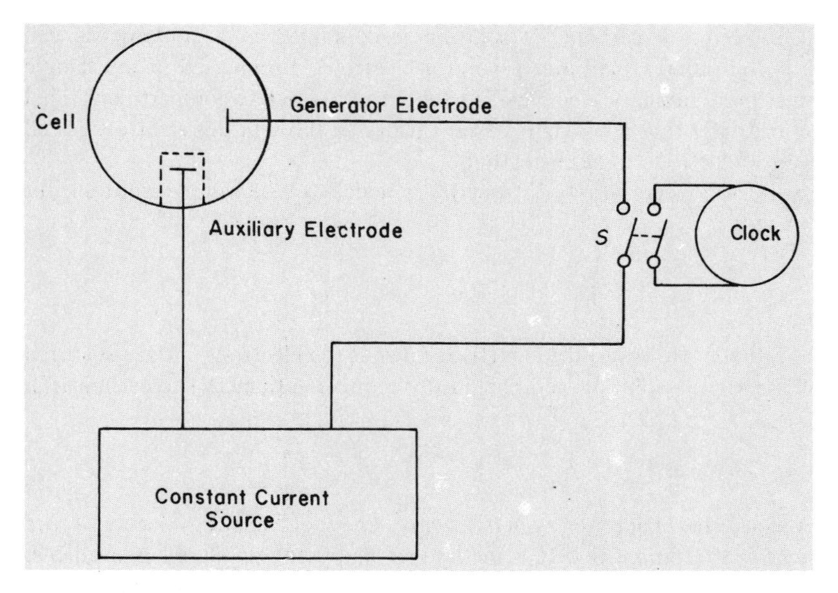


*Figure 11–4.* Diagram of essential apparatus for performance of constant-current coulometric titrations.

**Constant-Current Sources.** For the majority of coulometric titrations, currents ranging from 5 to 20 ma are employed, although the use of constant currents as large as 100 ma or as small as 1 ma is occasionally desirable. Electronic instruments, permitting current regulation to within ±0.01 per cent and delivering currents of 200 ma or more, are commercially available. Some of these constant-current sources provide for direct readout of the number of equivalents of the substance being titrated.

**Measurement of Time.** In the performance of a coulometric titration, it is invariably necessary to open and close switch *S* a number of times as the equivalence point is approached so that small increments of titrant can be generated and permitted to react before the titration is either continued or terminated. This procedure parallels the manual closing and opening of a buret stopcock in classic volumetric analysis. Ordinarily, time is measured with the aid of an electric stopclock powered by a synchronous motor which, in turn, is actuated through the same switch (*S*) used to start and stop the coulometric titration.

**Coulometric Cells.** Every coulometric cell contains a **generator electrode**, at which the titrant is produced, and an auxiliary electrode to complete the electrolysis circuit. Generator electrodes typically range from 10 to 25 $cm^2$ in area and are usually fabricated from platinum, gold, silver, or mercury. In most instances, a length of platinum wire serves as the auxiliary electrode. Frequently, the reaction at the auxiliary electrode yields a substance that interferes with the coulometric process. Therefore, one customarily isolates the auxiliary electrode from the main sample solution by placing the auxiliary electrode in a tube containing electrolyte solution but closed at the bottom with a porous glass disk. As for other coulometric methods of analysis, the sample solution must be stirred. In addition, the cell may be closed with a fitted cover having an inlet tube for the continuous introduction of nitrogen or sometimes carbon dioxide for removal and exclusion of oxygen from the sample solution.

**End-Point Detection.** Other than the mode by which titrant is added to the sample solution, there is little difference between coulometric titrimetry and the more familiar methods of classic volumetric analysis. It is not surprising that all of the techniques discussed earlier in the text for end-point detection are applicable. Visual methods for the location of equivalence points, including the use of colored acid-base and redox indicators, have been widely employed in coulometric titrimetry. Among the instrumental techniques of end-point detection are potentiometric, amperometric, and spectrophotometric methods.

### Analytical Applications of Coulometric Titrimetry

Every kind of titration – acid-base, precipitation, complex-formation, and oxidation-reduction – has been successfully performed coulometrically. Coulometric titrations are best suited to the determination of quantities of a substance ranging from 1 μg or less up to about 100 mg, errors characteristically being 0.1 to 0.3 per cent.

**Acid-Base Titrations.** For the determination of both strong and weak acids, advantage may be taken of the fact that electroreduction of water at a platinum cathode yields hydroxide ion:

$$2\,H_2O + 2\,e \rightleftharpoons H_2 + 2\,OH^-$$

An advantage of the constant-current coulometric titration of acids with electrogenerated hydroxide ion is that one can avoid the nuisance of having to protect a standard base solution from atmospheric carbon dioxide. Hydrogen ion, for the coulometric titration of bases, can be generated from the oxidation of water at a platinum anode.

By adding a small amount of water to a nonaqueous solvent, one can electrogenerate hydrogen ions at a platinum anode for the titration of weak bases without seriously modifying the properties of the solvent system. One-milligram samples of pyridine, benzylamine, and triethylamine have been determined with an error of less than 2 per cent in acetonitrile containing lithium perchlorate trihydrate as both the supporting electrolyte and the source of water.

**Precipitation Titrations.** Individual halide ions can be accurately determined with anodically generated silver ion, and it is possible to analyze mixtures of these species as well. Organic chlorides or bromides may be analyzed if, after combustion of the desired compound, the liberated halide is titrated coulometrically. In addition, mercaptans are determinable with anodically formed silver ion. Electrogenerated mercury(I) can be employed to titrate chloride, bromide, and iodide individually and in mixtures; chromate, oxalate, molybdate, and tungstate form precipitates with electrogenerated lead(II).

**Complexometric Titrations.** Ethylenediaminetetraacetate is generated through reduction of the stable mercury(II)-EDTA-ammine complex at a mercury pool cathode in an ammonia-ammonium nitrate buffer:

$$HgNH_3Y^{2-} + NH_4^+ + 2\,e \rightleftharpoons Hg + 2\,NH_3 + HY^{3-}$$

As the EDTA species ($HY^{3-}$) is released, it reacts rapidly with a metal ion to form the corresponding metal-EDTA complex. This procedure has been applied to the titration of calcium(II), copper(II), lead(II), and zinc(II), but can be extended to the determination of other metal ions which form EDTA complexes.

**Oxidation-Reduction Titrations.** Coulometric titrimetry has been applied more extensively to redox processes than to all other kinds of reactions combined. Table 11-1 lists some of the titrants and applications which have been investigated. One of the interesting uses of coulometric redox titrimetry is the continuous monitoring of hydrogen sulfide and sulfur dioxide in air and gas samples with electrogenerated bromine or

**Table 11–1. Coulometric Oxidation-Reduction Titrations**

| Coulometric Titrant | Generator Electrode Reaction | Applications |
|---|---|---|
| $Cl_2$ | $2\ Cl^- \rightleftharpoons Cl_2 + 2\ e$ | chlorination of unsaturated fatty acids; oxidation of As(III) to As(V), of Tl(I) to Tl(III), and of iodide to iodine |
| $Br_2$ | $2\ Br^- \rightleftharpoons Br_2 + 2\ e$ | continuous monitoring of $H_2S$ and $SO_2$ in gases and air; bromination of olefins, phenols, and aromatic amines ; bromination of oxine and metal oxinates ; determination of mustard gas |
| $OBr^-$ | $Br^- + 2\ OH^- \rightleftharpoons OBr^- + H_2O + 2\ e$ | determination of protein in serum; oxidation of ammonia to nitrogen |
| $I_3^-$ | $3\ I^- \rightleftharpoons I_3^- + 2\ e$ | continuous monitoring of $H_2S$ and $SO_2$ in gases and air; oxidation of As(III) to As(V), of Sb(III) to Sb(V), of hydrogen sulfide to sulfur, and of sulfur dioxide to sulfate |
| $Ce^{4+}$ | $Ce^{3+} \rightleftharpoons Ce^{4+} + e$ | oxidation of Fe(II) to Fe(III), of ferrocyanide to ferricyanide, of Ti(III) to Ti(IV), of As(III) to As(V), of U(IV) to U(VI), of iodide to iodine, and of hydroquinone to quinone |
| $Mn^{3+}$ | $Mn^{2+} \rightleftharpoons Mn^{3+} + e$ | oxidation of As(III) to As(V), of Fe(II) to Fe(III), and of $H_2C_2O_4$ to $CO_2$ |
| $Ag^{2+}$ | $Ag^+ \rightleftharpoons Ag^{2+} + e$ | oxidation of Ce(III) to Ce(IV), of V(IV) to V(V), of As(III) to As(V), and of $H_2C_2O_4$ to $CO_2$ |
| $CuCl_2^-$ | $Cu^{2+} + 2\ Cl^- + e \rightleftharpoons CuCl_2^-$ | reduction of $Cr_2O_7^{2-}$ to Cr(III), of V(V) to V(IV), of Fe(III) to Fe(II), of $IO_3^-$ to $ICl_2^-$, and of Au(III) to Au(0) |
| $Fe^{2+}$ | $Fe^{3+} + e \rightleftharpoons Fe^{2+}$ | reduction of Ce(IV) to Ce(III), of V(V) to V(IV), of $Cr_2O_7^{2-}$ to Cr(III), and of $MnO_4^-$ to Mn(II) |
| $Ti^{3+}$ | $TiO^{2+} + 2\ H^+ + e \rightleftharpoons Ti^{3+} + H_2O$ | reduction of Fe(III) to Fe(II), of Ce(IV) to Ce(III), of U(VI) to U(IV), of Mo(VI) to Mo(V), and of V(V) to either V(IV) or V(III) |

triiodide. In addition, the protein content of serum has been determined with anodically formed hypobromite ion.

## POLAROGRAPHY

Polarography derives its analytical importance from two characteristics of current-potential curves or polarograms obtained with a dropping mercury electrode. A ent-potential curve consists of a plot of the current, which flows as reactions occur at orking electrode, versus the potential of that electrode, measured against an opriate reference electrode. In polarography, the position of a polarogram along the ntial axis may indicate the identity of the substance which undergoes electron sfer. Under experimental conditions that are easily achieved, a polarogram exhibits a ısion-controlled limiting current whose magnitude is governed by the concentration ıe electroactive substance.

## larographic Cell and the Dropping Mercury Electrode

Some of the apparatus used to perform polarographic measurements is pictured in re 11–5. A so-called H-cell, comprising two compartments separated by a red-glass disk and an agar plug saturated with potassium chloride, is commonly oyed. A dropping mercury electrode consists of a length of thin-bore capillary tubing hed to the bottom of a stand tube and a mercury reservoir. Under the influence of ty, mercury issues from the orifice of the capillary in a series of identical droplets, oximately one-half millimeter in diameter and a few square millimeters in area. By ting the height of the mercury column, one can vary the **drop time** – that is, the required for a fresh drop of mercury to emerge, grow, and fall from the tip of the

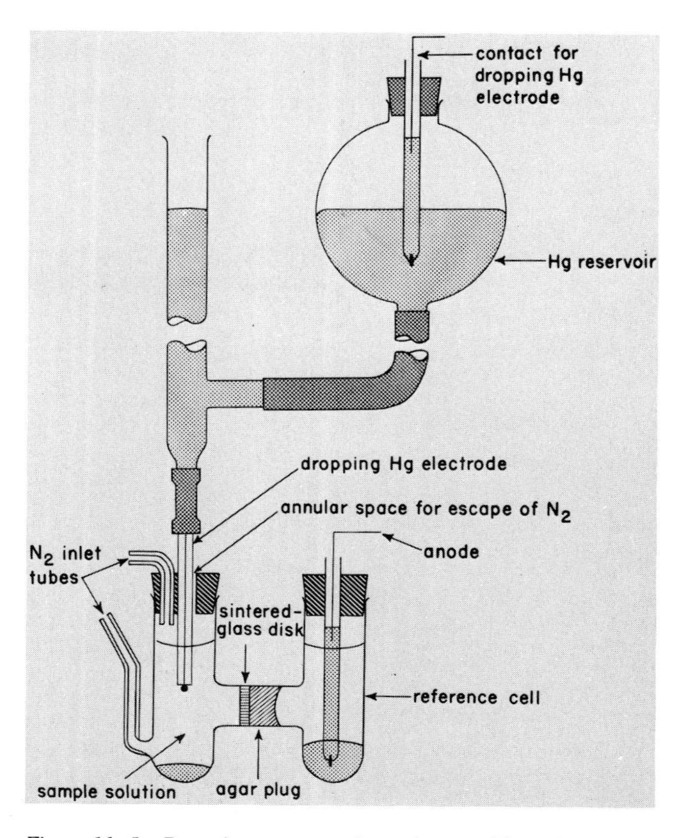


*Figure 11–5.* Dropping mercury electrode assembly and H-cell for polarographic measurements. (Redrawn, with permission, from a paper by J. J. Lingane and H. A. Laitinen: Ind. Eng. Chem., Anal. Ed., *11*:504, 1939.)

capillary. Drop times range from three to six seconds with mercury heights of 30 cm or more. Nitrogen is bubbled through the sample solution to remove oxygen prior to the recording of a polarogram and is passed over the surface of the solution during the measurements.

A reference electrode, most often a saturated calomel electrode, is situated in the second compartment of the H-cell. Because the reference electrode serves simultaneously as the auxiliary electrode, it must be large compared to the area of a mercury drop so that the potential of the reference electrode remains essentially constant during the passage of microampere currents through the polarographic cell.

### Nature of a Polarogram

Suppose that an oxygen-free 0.1 *F* potassium nitrate solution containing $2.00 \times 10^{-3}$ *F* thallium(I) nitrate is transferred into a polarographic cell. If the potential of the dropping mercury electrode is scanned in a cathodic direction from 0 to −1.8 v versus SCE, the polarogram shown as curve *A* in Figure 11–6 is observed. This current-potential curve was obtained with the aid of a **polarograph**, an instrument which automatically increases the voltage impressed across the polarographic cell and simultaneously records the current-potential curve on moving chart paper. According to polarographic convention, currents resulting from cathodic processes are positive, whereas anodic currents are negative. Current oscillations are due to the periodic growth and fall of the mercury drops.

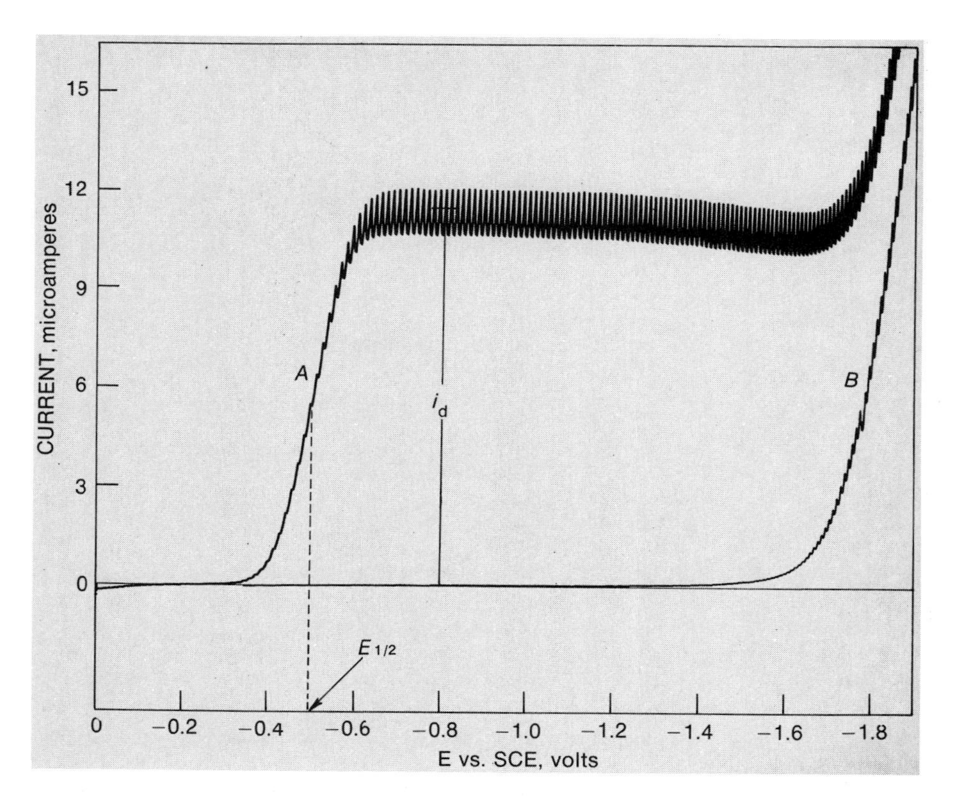


*Figure 11–6.* A representative polarogram. Curve *A*: polarogram for reduction of $2.00 \times 10^{-3}$ *F* thallium(I) nitrate in a 0.1 *F* potassium nitrate medium. Curve *B*: residual-current curve for a 0.1 *F* potassium nitrate solution.

A complete plot of current as a function of the dropping-electrode potential is termed a **polarogram**, and the rapidly ascending portion of the curve, as appears around −0.45 v versus SCE in Figure 11–6, is known as a **polarographic wave**. Reduction of thallium(I) with the resultant formation of a thallium amalgam

$$Tl^{+} + e + Hg \rightleftharpoons Tl(Hg)$$

is responsible for this polarographic wave. A well-defined limiting-current (plateau) region for reduction of thallium(I) appears between −0.7 and −1.6 v; however, the abrupt rise in current beyond the latter potential is caused by reduction of potassium ions to form a potassium amalgam.

Even when thallium(I) is excluded from the potassium nitrate solution, a small **residual current** flows through the polarographic cell, as may be seen in curve *B* of Figure 11–6. In polarography, the residual current arises from charging of the so-called electrical double-layer at the surface of the mercury electrode; the mercury-solution interface behaves essentially as a tiny capacitor. In order for the mercury to acquire the potential demanded by the external applied voltage, electrons move toward or away from the mercury surface, depending on the required potential and on the composition of the solution. Since each mercury drop emerging from the orifice of the capillary must be charged to the proper potential, a continuous flow of current results.

### Polarographic Diffusion Current

There are three mechanisms by which an electroactive substance, such as thallium(I), may reach the surface of an electrode – *migration* of a charged species due to the flow of current through a cell, *convection* caused by stirring or agitation of the solution and by density or thermal gradients, and *diffusion* of a substance from a region of high concentration to one of lower concentration. Successful polarographic measurements demand that the electroactive species reach the surface of a mercury drop solely by diffusion, this being the only process amenable to straightforward mathematical treatment. To eliminate electrical migration of an ion, one introduces into the sample solution a 50- to 100-fold excess of an innocuous **supporting electrolyte** – for example, potassium nitrate. Convection can be prevented if the sample solution is unstirred and if the polarographic cell is mounted in a vibration-free location.

Of particular interest in polarography is the **diffusion layer**, a film of solution perhaps 0.05 mm in thickness at the surface of the electrode. By referring to the reduction of thallium(I) in a potassium nitrate medium (curve *A*, Figure 11–6), let us see what happens at the surface of the dropping mercury electrode as its potential is varied. At approximately −0.1 v versus SCE, little, if any, thallium(I) is reduced. If the potential of the dropping mercury electrode is shifted to about −0.45 v, a significant fraction of the thallium(I) at the electrode surface is reduced to yield a thallium amalgam, and a finite current flows through the polarographic cell. As a consequence of this electrolysis, a concentration gradient is established, causing thallium(I) to diffuse from the bulk of the solution toward the electrode surface. If thallium(I) did not diffuse toward the electrode, the current would fall to zero almost instantaneously because the surface concentration of thallium(I) would drop to the value governed by the electrode potential and would change no more. However, the current is sustained by diffusion of additional thallium(I) toward the electrode under the influence of the concentration gradient.

According to the simplest model for a diffusion-controlled process, the observed current is directly related to the difference between the concentration of the electroactive species in the bulk of the solution, $C$, and at the surface of the electrode, $C_0$. Thus, we can write

$$i = k(C - C_0)$$

where $i$ is the current and $k$ is a proportionality constant. Fundamentally, it is the potential of the dropping mercury electrode that fixes the surface concentration ($C_0$) of the electroactive substance. At potentials more negative than -0.70 v versus SCE for the reduction of thallium(I), essentially all the reactant reaching the electrode surface is immediately reduced. Thus, the surface concentration is so small compared to the bulk concentration that the term ($C - C_0$) becomes virtually equal to $C$, and the current attains a limiting value called the **diffusion current**, which we denote by the symbol $i_d$ (Figure 11–6):

$$i_d = kC$$

Analytical applications of polarography rely upon the Ilkovic equation

$$i_d = 607nD^{1/2}\,Cm^{2/3}\,t^{1/6}$$

which expresses the dependence of the diffusion current on the concentration of the electroactive species as well as other parameters. In this relation $i_d$ is the time-average diffusion current in microamperes that flows during the lifetime of a single mercury drop, $n$ is the number of faradays of electricity per mole of electroactive substance reduced or oxidized, $D$ is the diffusion coefficient (cm$^2$/second) of the reactant, $C$ is the bulk concentration of the electroactive species in millimoles per liter, $m$ is the rate of flow of mercury in milligrams per second, and $t$ is the drop time in seconds. Among other terms, the numerical constant 607 includes the value of the faraday (96,487 coulombs).

### Equation of the Polarographic Wave

If we assume that a species originally present in the sample solution undergoes reversible polarographic reduction to some substance which is soluble either in the solution or in the mercury phase,

$$\text{Ox} + n\,\text{e} \rightleftharpoons \text{Red}$$

the concentrations (activities) of the reactant and product *at the solution-electrode interface* are related to the potential of the dropping mercury electrode through the Nernst equation:

$$E = E^0 - \frac{0.059}{n}\log\frac{[\text{Red}]_{\text{surface}}}{[\text{Ox}]_{\text{surface}}}$$

In this expression, $E$ and $E^0$ represent, respectively, the potential of the dropping mercury electrode and the standard potential for the pertinent half-reaction, each measured with respect to the same reference electrode. In addition, the observed potential of the dropping mercury electrode and the surface concentrations of the oxidized and reduced species are average values during the lifetime of a mercury drop.

**Half-Wave Potential.** Provided that the reactant and product species are completely soluble in either the solution or the mercury drop, it can be demonstrated, although we shall not present the derivation in this text,* that in the region of the polarographic wave the potential of the dropping mercury electrode is related to the observed current, $i$, and the diffusion current, $i_d$, by the expression

$$E = E_{1/2} - \frac{0.059}{n} \log \frac{i}{i_d - i}$$

where the **half-wave potential**, $E_{1/2}$, is defined as the potential of the dropping mercury electrode when the observed current is exactly one half of the diffusion current; that is, $E = E_{1/2}$ for $i = i_d/2$ (Figure 11–6). This is the most familiar form of the equation for a polarographic wave.† Note that the half-wave potential is independent of the concentration of the reactant, but is characteristic of the identity of the electroactive substance in a given supporting electrolyte solution. A list of half-wave potentials for reduction of some common metal ions in various media is presented in Table 11–2. Frequently, one can identify the species responsible for an unknown polarographic wave by searching a table of half-wave potentials.

**Table 11–2. Half-Wave Potentials for Reduction of Metal Ions in Various Media**[a]

| | Supporting Electrolyte Solution | | | | | |
|---|---|---|---|---|---|---|
| **Metal Ion**[b] | **1 *F* $HNO_3$** | **1 *F* KCl** | **1 *F* $NH_3$, 1 *F* $NH_4Cl$** | **1 *F* KCN** | **0.05 *F* EDTA, 0.8 *F* $CH_3COONa$ pH 12** | **1 *F* NaOH** |
| Cd(II) | −0.59 | −0.64 | −0.81 | −1.18 | −1.28 | −0.78 |
| Co(II) | — | −1.20 | −1.29 | −1.45 | NR[d] | −1.46 |
| Cu(II) | −0.01 | +0.04; −0.22[c] | −0.24; −0.51[c] | NR[d] | −0.51 | −0.41 |
| Ni(II) | — | — | −1.10 | −1.36 | NR[d] | — |
| Pb(II) | −0.41 | −0.44 | — | −0.72 | −1.32 | −0.76 |
| Tl(I) | −0.48 | −0.48 | −0.48 | — | −0.60 | −0.48 |
| Zn(II) | −1.00 | −1.00 | −1.35 | NR[d] | NR[d] | −1.53 |

[a] All half-wave potentials are quoted with respect to the saturated calomel reference electrode (SCE); data excerpted from L. Meites: *Polarographic Techniques*. Second edition, Wiley-Interscience, New York, 1965, pp. 615–670.
[b] All species are reduced to form a metal amalgam aside from exceptions noted for copper(II).
[c] Copper(II) gives two waves, corresponding to stepwise reduction to copper(I) and copper (0).
[d] No wave is observed before reduction of the supporting electrolyte-solvent system itself.

---

*Details of the derivation may be found in J. J. Lingane: *Electroanalytical Chemistry*. Second edition, Wiley-Interscience, New York, 1958, p. 261.

†We have considered only the reversible reduction of a substance to yield a product soluble in the solution or in mercury. For the reversible oxidation of a species to give a soluble product, the equation for the polarographic wave has the same form as that for a cathodic process except that a plus (+) sign precedes the logarithmic term. See J. J. Lingane: *Electroanalytical Chemistry*. Second edition, Wiley-Interscience, New York, 1958, p. 263 for the derivation of these relations.

## Experimental Techniques of Polarography

**Electrical Apparatus.** To perform polarographic measurements, one should be able to impress voltages ranging from 0 to 3 v across the cell, and one must know the potential of the dropping mercury electrode versus the reference electrode with a precision of 10 mv. In addition, it is necessary to measure the cell current, which usually encompasses a range from 0.1 to 100 $\mu$a, with an accuracy of ±0.01 $\mu$a.

A simple electrical circuit for polarographic experiments is illustrated in Figure 11–7. Two 1.5-v batteries connected in series provide a source of voltage which can be varied through adjustment of a movable contact along the slidewire $R_1$. If the double-pole double-throw switch is moved to the right-hand position, the potential of the dropping mercury electrode versus the reference electrode can be determined with the aid of a potentiometer. For measurement of the current, the switch is thrown to the left-hand position and the $iR$ drop across a precision 10,000-ohm resistor ($R_2$) is obtained by means of the potentiometer. A galvanometer is often used as a null-point detector in the potentiometer circuit; if so, the galvanometer must be damped by placement of a resistor across its terminals so that the potentiometer can be balanced in spite of the current oscillations caused by the growth and fall of the dropping mercury electrode.

With only slight modification, the circuit of Figure 11–7 can be converted into an automatic recording polarograph. First, the slidewire may be replaced by a motor-driven, 100-ohm, 10-turn potentiometer which varies the applied voltage in a known and linear fashion. Second, the current may be recorded in synchronization with the change in applied voltage, if a strip-chart potentiometer recorder is connected across resistor $R_2$. A large number of all-electronic, recording polarographs are available commercially.

**Measurement of Diffusion Currents.** Undoubtedly, the most reliable procedure for measurement of the diffusion current entails the recording of separate current-potential curves for the sample solution and for a blank solution containing all components except the electroactive substance. Next, one selects a potential at which the diffusion

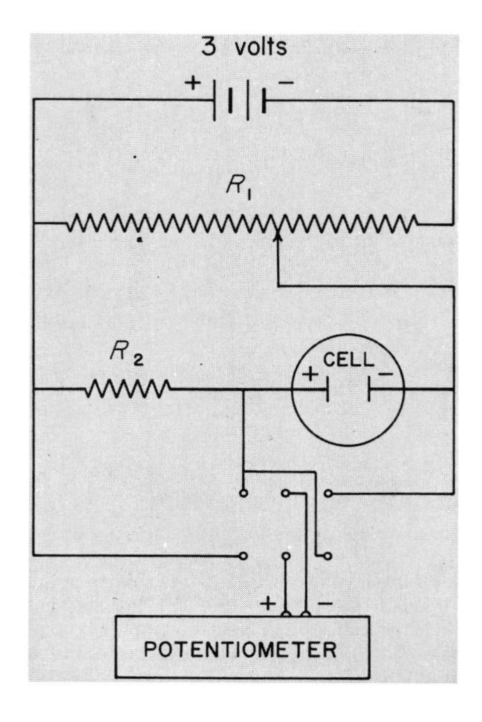


*Figure 11–7.* Schematic diagram of circuitry for manual performance of polarographic measurements. (Redrawn, with permission, from a paper by J. J. Lingane: Anal. Chem., *21*: 45, 1949.)

current is to be evaluated and measures the difference in current between the two curves. Since the limiting-current region of the polarogram for the desired substance and the residual-current curve often have different slopes, the diffusion current may vary slightly as a function of potential. However, the accuracy of the measurement is not impaired if a self-consistent procedure is followed.

**Determination of Concentration.** We shall now examine some procedures used to evaluate the concentration of an electroactive substance. A simple and straightforward approach involves the preparation of a series of standard solutions, the measurement of the diffusion current for each sample, and the construction of a plot of observed diffusion current versus concentration. With the aid of this calibration curve, the concentration of the desired substance in an unknown solution is determined from a single measurement of the diffusion current.

A second procedure for determination of the concentration of an electroactive species is the **method of standard addition.** Suppose that a polarogram is recorded for a solution containing an unknown concentration, $C_u$, of the desired substance and the observed diffusion current is $(i_d)_1$. If an aliquot $V_s$ of a solution of the electroactive substance of known concentration $C_s$ is added to a volume $V_u$ of the unknown solution, the diffusion current will attain a new value $(i_d)_2$. Using data collected from the two experiments, one can compute the concentration of the unknown sample from the relation

$$C_u = \frac{C_s V_s (i_d)_1}{(V_u + V_s)(i_d)_2 - V_u (i_d)_1}$$

Best results are achieved if the diffusion currents, $(i_d)_1$ and $(i_d)_2$, differ by at least a factor of two.

## Analytical Applications of Polarography

Hundreds of inorganic and organic substances have been subjected to polarographic investigation, and many procedures for the analysis of individual species and mixtures of compounds have been developed. For routine determinations, the optimum concentration range for polarography is $10^{-4}$ to $10^{-2}$ *M*, and analytical results accurate to within ±2 per cent are usually attainable. With special care, the uncertainty of a polarographic analysis may be as small as a few tenths of 1 per cent.

**Determinations of Inorganic Ions.** Through an appropriate choice of supporting electrolyte, pH, and complexing ligands, practically any metal ion can be reduced at the dropping mercury electrode to an amalgam or to a soluble lower oxidation state. Divalent cations such as cadmium, cobalt, copper, lead, manganese, nickel, tin, and zinc can be determined in many different complexing and noncomplexing media. Alkaline earth ions exhibit well-defined polarographic waves in solutions containing tetraethylammonium iodide as the supporting electrolyte. Polarographic characteristics of the tripositive states of aluminum, bismuth, chromium, europium, gallium, gold, indium, iron, samarium, uranium, vanadium, and ytterbium in a number of supporting electrolyte solutions have been reported. Polarographic reductions of chromate, iodate, molybdate, selenite, tellurite, and vanadate, as well as the anionic chloride complexes of tungsten(VI), tin(IV), and molybdenum(VI), are included in the list of analytically useful procedures.

**Analysis of Mixtures.** Solutions containing two or more electroactive substances can be analyzed polarographically if the half-wave potentials for the various species differ by at least 0.2 v. As shown in Figure 11–8, a mixture of thallium(I), cadmium(II), and nickel(II) in an ammonia-ammonium chloride buffer can be resolved and analyzed by means of polarography.

Table 11–3. Polarographic Behavior of Organic Functional Groups

| Functional Group | Class of Compounds | Exemplary Reaction |
| --- | --- | --- |
| C—I<br>C—Br | organic halides | $C_6H_5{-}CH_2Br + H^+ + 2\,e \longrightarrow C_6H_5{-}CH_3 + Br^-$ |
| C═C | carbon-carbon double bond conjugated with another C═C bond or with an aromatic ring | $C_6H_5{-}CH{=}CH_2 + 2\,H^+ + 2\,e \longrightarrow C_6H_5{-}CH_2CH_3$ |
| C≡C | phenyl-substituted acetylenes | $C_6H_5{-}C{\equiv}C{-}C_6H_5 + 4\,H^+ + 4\,e \longrightarrow C_6H_5{-}CH_2CH_2{-}C_6H_5$ |
| C═O | quinones | $O{=}C_6H_4{=}O + 2\,H^+ + 2\,e \longrightarrow HO{-}C_6H_4{-}OH$ |
|  | aldehydes, ketones | $2\,C_6H_5{-}C({=}O){-}CH_3 + 2\,H^+ + 2\,e \longrightarrow 2\,C_6H_5{-}\dot{C}(OH){-}CH_3 \longrightarrow \text{dimer}$ |
|  |  | $C_6H_5{-}C({=}O){-}CH_3 + 2\,H^+ + 2\,e \longrightarrow C_6H_5{-}CH(OH){-}CH_3$ |
| C—OH | hydroquinones | $HO{-}C_6H_4{-}OH \longrightarrow O{=}C_6H_4{=}O + 2\,H^+ + 2\,e$ |

Table 11–3. **Polarographic Behavior of Organic Functional Groups** *(continued)*

| Functional Group | Class of Compounds | Exemplary Reaction |
|---|---|---|
| C—SH | mercaptans (thiols) | $HS{-}CH_2COO^- + Hg \longrightarrow HgS{-}CH_2COO^- + H^+ + e$ |
| S—S | disulfides | $H_5C_2{-}S{-}S{-}C_2H_5 + 2\,H^+ + 2\,e \longrightarrow 2\,C_2H_5{-}SH$ |
| S=O | sulfoxides | $H_3C{-}S({=}O){-}CH_3 + 2\,H^+ + 2\,e \longrightarrow H_3C{-}S{-}CH_3 + H_2O$ |
| O—O | peroxides, hydroperoxides | $C_6H_5{-}C({=}O){-}O{-}O{-}C({=}O){-}C_6H_5 + 2\,H^+ + 2\,e \longrightarrow 2\,C_6H_5{-}COOH$ |
| N=O | nitro and nitroso compounds | $C_6H_5{-}C({=}O){-}CH_2NO_2 + 4\,H^+ + 4\,e \longrightarrow C_6H_5{-}C({=}O){-}CH_2NHOH + H_2O$ |
| N=N | azo compounds | $C_6H_5{-}N{=}N{-}C_6H_5 + 2\,H^+ + 2\,e \longrightarrow C_6H_5{-}NH{-}NH{-}C_6H_5$ |
| C≡N | nitriles | $CH_3CO{-}C_6H_4{-}CN + 4\,H^+ + 4\,e \longrightarrow CH_3CO{-}C_6H_4{-}CH_2NH_2$ |
| C=N | imines, oximes | $C_6H_5{-}CH{=}CH{-}C(CH_3){=}NH + 2\,H^+ + 2\,e \longrightarrow C_6H_5{-}CH{=}CH{-}CH(CH_3){-}NH_2$ |

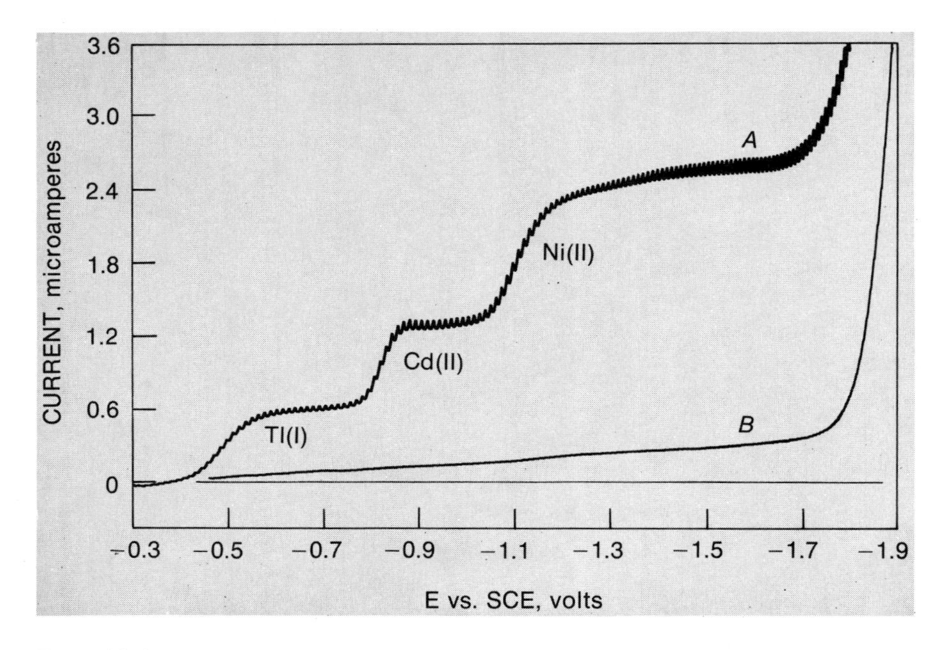


*Figure 11–8.* Curve *A*: polarogram for a mixture of thallium(I), cadmium(II), and nickel(II), each at a concentration of $1.00 \times 10^{-4}$ *M*, in a buffer consisting of 1 *F* ammonia and 1 *F* ammonium chloride. Curve *B*: residual-current curve for the supporting electrolyte solution alone.

**Polarography of Organic Compounds.** In contrast to inorganic substances, the reactions of organic compounds at the dropping mercury electrode are usually irreversible and often proceed in several steps. Polarographic data pertaining to organic molecules are frequently difficult to interpret, because of the effect of pH on the rate, mechanism, and products of an organic electrode process, the influence of solvent on the course of the reaction, the importance of the nature and concentration of the supporting electrolyte, and the occurrence of side reactions involving partially oxidized or reduced species. Despite these problems, there are a number of useful applications of polarography to the determination of organic substances. In Table 11–3 are listed some of the functional groups found in organic molecules that are electroactive at the dropping mercury electrode and that give diffusion currents proportional to concentration.

## QUESTIONS AND PROBLEMS

1. When a silver electrode is polarized anodically in a solution containing chloride, bromide, or iodide ion, the following reaction occurs

$$Ag + X^- \rightleftharpoons AgX + e$$

where $X^-$ denotes the halide ion of interest.

(a) Suppose that a silver electrode is to be polarized anodically in the controlled-potential electrolysis of an aqueous solution initially 0.0200 *F* in sodium bromide (NaBr) and 1 *F* in perchloric acid. What is the minimal potential versus NHE (normal hydrogen electrode) at which the silver anode must be controlled in order to ensure an accuracy of at least 99.9 per cent in the determination of bromide?

(b) Suppose that controlled-potential coulometry with a silver anode is to be utilized for the separation and analysis of a mixture of 0.0500 *M* chloride and 0.0250 *M* iodide in a 1 *F* perchloric acid medium. Within what range of potentials (versus NHE) must the silver anode be maintained to ensure the deposition of at least 99.9 per cent of the iodide as silver iodide without removal of any chloride ion from the sample solution? What is the minimal potential (versus NHE) at which the silver anode must be kept to ensure that at least 99.9 per cent of the chloride ion is deposited as silver chloride?

2. Imagine that you wish to separate copper(II) from antimony(III) by controlled-potential deposition of copper upon a platinum cathode from a 1 *F* perchloric acid medium originally containing 0.0100 *M* copper(II) and 0.0750 *M* antimony(III). At what potential versus SCE (saturated calomel electrode) must the platinum cathode be controlled to cause 99.9 per cent deposition of copper? What fraction of the antimony(III) will be reduced to the metallic state? If the volume of the solution is 100 ml, what will be the total gain in weight of the cathode?

$$Cu^{2+} + 2\,e \rightleftharpoons Cu; \qquad E^0 = +0.337 \text{ v versus NHE}$$

$$SbO^+ + 2\,H^+ + 3\,e \rightleftharpoons Sb + H_2O; \qquad E^0 = +0.212 \text{ v versus NHE}$$

3. An unknown sample of hydrochloric acid was titrated coulometrically with hydroxide ion electrogenerated at a constant current of 20.34 ma, a total generation time of 645.3 seconds being required to reach the equivalence point. Calculate the number of millimoles of hydrochloric acid in the sample.

4. To determine the quantity of chlorobenzene impurity in benzene, a 1.000-$\mu$l sample of the liquid was injected from a microliter syringe into a vaporization chamber at 200°C. Immediately, the vaporized sample was swept with a stream of nitrogen into an adjoining chamber at 800°C and was mixed with oxygen, whereupon rapid combustion of the organic compounds occurred to yield carbon dioxide, water, and hydrogen chloride. These gaseous products were carried by the nitrogen stream into a coulometric titration cell (equipped with a silver generator anode and a platinum auxiliary cathode) and were trapped with a 70 per cent solution of acetic acid in water. Next, a constant current of 64.2 $\mu$a was passed through the electrolysis cell until the potential of a separate silver indicator electrode versus a silver-silver acetate reference electrode showed that the equivalence point had been reached. If the coulometric titration required 48.4 seconds, calculate the weight percentage of chlorobenzene in the benzene sample. Assume that the density of the liquid sample is equal to that of pure benzene.

5. A mixture of nickel(II) and cobalt(II) was analyzed by means of controlled-potential coulometry. An electrolysis cell was assembled having a well-stirred mercury pool cathode and a platinum auxiliary anode. One hundred milliliters of supporting electrolyte solution (1.00 *F* in pyridine, 0.30 *M* in chloride ion, and 0.20 *F* in hydrazine and with a

pH of 6.89) was deaerated with nitrogen gas, and exactly 5.000 ml of the nickel-cobalt sample solution was pipetted into the cell. Quantitative reduction of nickel(II) to the elemental (amalgam) state was accomplished with the potential of the mercury pool held at −0.95 v versus SCE; an electromechanical current-time integrator in series with the cell read 60.14 coulombs when the current had decayed. Then the potential of the mercury cathode was adjusted to −1.20 v versus SCE for the reduction of cobalt(II) to the elemental (amalgam) state; the final reading of the current-time integrator, corresponding to the sum of the quantities of nickel(II) and cobalt(II) was 351.67 coulombs. Calculate the molar concentrations of nickel(II) and cobalt(II) in the original sample solution.

6. An indium(III)-containing sample solution was buffered with a mixture of acetic acid and sodium acetate, was heated to approximately 75°C, and was treated with a slight excess of 8-hydroxyquinoline (oxine) to precipitate quantitatively the insoluble indium(III) oxinate salt:

$$In^{3+} + 3\ HOC_9H_6N \rightleftharpoons In(OC_9H_6N)_3 + 3\ H^+$$

After the precipitate was washed, it was redissolved in a minimal volume of concentrated hydrochloric acid, and the resulting solution was diluted with water to exactly 50.00 ml in a volumetric flask. A 200.0-$\mu$l aliquot of this solution was pipetted into a titration vessel containing 50 ml of 0.2 $F$ sodium bromide solution and equipped with a platinum generator anode and a platinum auxiliary cathode. Bromine, electrogenerated by oxidation of bromide ion at a constant current of 12.53 ma, reacted with oxine according to the reaction

$$HOC_9H_6N + 2\ Br_2 \rightarrow HOC_9H_4NBr_2 + 2\ HBr$$

An end point was reached after a titration time of 186.6 seconds. Calculate the weight of indium(III) in milligrams in the original sample solution.

7. Monoprotonated ethylenediaminetetraacetate ($HY^{3-}$) can be electrogenerated by means of the constant-current reduction of the mercury(II)-EDTA-ammine complex

$$HgNH_3Y^{2-} + NH_4^+ + 2\ e \rightleftharpoons Hg + 2\ NH_3 + HY^{3-}$$

at a mercury pool cathode (page 279), whereupon the free ligand reacts rapidly and quantitatively with many metal ions, including lead(II):

$$Pb^{2+} + HY^{3-} \rightleftharpoons PbY^{2-} + H^+$$

A 50.00-ml sample of gasoline, containing tetraethyllead, was refluxed with concentrated hydrochloric acid, and the resulting solution was evaporated almost to dryness. Enough water was added to dissolve the slightly soluble precipitate of lead(II) chloride. Then the lead(II) solution was added to a titration vessel containing an ammoniacal solution of $HgNH_3Y^{2-}$ and equipped with a mercury pool cathode and a platinum auxiliary anode. A constant current of 93.44 ma was passed through the cell. A mercury indicator electrode (page 260) was employed to follow the progress of the titration potentiometrically, and the end point was reached after a titration time of 578.3 seconds. If the density of tetraethyllead is 1.65 g/ml, calculate the number of milliliters of tetraethyllead per gallon of gasoline.

8. In a solution containing 2.5 *F* ammonia, 1 *F* ammonium chloride, and 2 *F* potassium chloride, trichloroacetate undergoes reduction to dichloroacetate

$$Cl_3CCOO^- + H_2O + 2\,e \rightarrow Cl_2CHCOO^- + OH^- + Cl^-$$

at a mercury pool cathode held at a potential of -0.90 v versus SCE. Dichloroacetate is reduced to monochloroacetate at potentials near -1.65 v versus SCE, whereas monochloroacetate is electroinactive under such conditions. A 0.3339-g sample of impure trichloroacetic acid, known to be contaminated with dichloroacetic and monochloroacetic acids, was assayed by means of controlled-potential coulometry. An electrolysis cell consisting of a mercury pool cathode and a platinum auxiliary anode was employed, the trichloroacetic acid sample was dissolved in the supporting electrolyte solution specified above, and the potential of the cathode was maintained at -0.90 v versus SCE throughout the electrolysis. A current-time integrator in series with the coulometric cell read 352.8 coulombs when the electrolysis current had dropped essentially to zero. Calculate the weight percentage of trichloroacetic acid in the sample material.

9. Vitamin C tablets, advertised to each contain 120 mg of ascorbic acid, were analyzed by means of controlled-potential coulometry. A single tablet was dissolved in approximately 100 ml of a solution consisting of a 0.1 *F* biphthalate-phthalate buffer of pH 6.03 along with 0.25 per cent oxalic acid (to inhibit the autoxidation of ascorbic acid). This solution was transferred to an electrolysis cell equipped with a platinum gauze anode and a platinum auxiliary cathode, the potential of the anode was set at +1.09 v versus SCE, and the electro-oxidation of ascorbic acid

$$C_6H_8O_6 \rightleftharpoons C_6H_6O_6 + 2\,H^+ + 2\,e$$

was allowed to reach completion. A current-time integrator in series with the electrolysis cell had a final reading (corrected for the oxidation of oxalic acid) of 128.9 coulombs. Calculate the weight of ascorbic acid in the tablet.

10. Jolly and Boyle (Anal. Chem., *43*:514, 1971) have devised a method for the constant-current coulometric titration of weak acids dissolved in liquid ammonia. At a platinum cathode immersed in liquid ammonia containing potassium bromide as a supporting electrolyte, electrons are produced which become solvated by the ammonia:

$$e(\text{platinum}) \rightarrow e(\text{ammonia})$$

These ammoniated electrons can react with any weak acid (HA) to yield hydrogen gas and the conjugate base of the acid:

$$e(\text{ammonia}) + HA \rightarrow \tfrac{1}{2}\,H_2 + A^-$$

For coulometric titrations in liquid ammonia, the cell must be cooled with a dry ice-alcohol mixture; a platinum auxiliary anode is employed, and potentiometric end-point detection is utilized.

A *moist* sample of ammonium chloride dissolved in liquid ammonia containing potassium bromide was titrated with ammoniated electrons generated with a constant current of 50.99 ma. One end point, corresponding to the conversion of ammonium ion to ammonia, was detected at a titration time of 811.3 seconds. A second end point, due to the conversion of water to hydroxide ion, was observed at a *total* titration time of 948.6 seconds. Calculate the weight in milligrams of ammonium chloride in the sample, and determine the weight percentage of water in the moist ammonium chloride.

11. In a 1 $F$ sodium hydroxide medium, the reduction of the tellurite anion, $TeO_3^{2-}$, yields a single, well-developed polarographic wave. With a dropping mercury electrode whose $m$ value was 1.50 mg/second and whose drop time was 3.15 seconds, the observed diffusion current was 61.9 $\mu$a for a 0.00400 $M$ solution of the tellurite ion. If the diffusion coefficient of the tellurite anion is $0.75 \times 10^{-5}$ cm$^2$/second, to what oxidation state is tellurium reduced under these conditions?
12. A $2.23 \times 10^{-3}$ $F$ solution of osmium tetroxide ($OsO_4$) in 1 $F$ hydrochloric acid was examined polarographically. For a dropping mercury electrode with a drop time of 4.27 seconds and a mercury flow rate of 3.29 mg/second, a cathodic diffusion current of 62.4 $\mu$a was observed. Although the diffusion coefficient of $OsO_4$ is unknown, it is reasonable to assume that the value is close to that of the chromate ion ($CrO_4^{2-}$), the latter species having a diffusion coefficient of $1.07 \times 10^{-5}$ cm$^2$/second. On the basis of this information, write the half-reaction for the reduction of osmium tetroxide in 1 $F$ hydrochloric acid.
13. An unknown cadmium(II) solution was analyzed polarographically according to the method of standard addition. A 25.00-ml sample of the unknown solution was found to yield a diffusion current of 1.86 $\mu$a. Next, a 5.00-ml aliquot of a standard $2.12 \times 10^{-3}$ $M$ cadmium(II) solution was added to the unknown sample, and the resulting mixture gave a diffusion current of 5.27 $\mu$a. Calculate the concentration of cadmium(II) in the known solution.
14. A trace impurity of nickel(II) in cobalt(II) salts can be measured polarographically. A 3.000-g sample of reagent-grade $CoSO_4 \cdot 7\,H_2O$ was transferred to a 100.0-ml volumetric flask and was dissolved in a small amount of water. Then, 2 ml of concentrated hydrochloric acid, 5 ml of pyridine, and 5 ml of 0.2 per cent gelatin were added to the volumetric flask, and the solution was diluted with water to the calibration mark and was mixed well. Exactly 75.00 ml of this solution was placed in a polarographic cell, and the resulting polarogram exhibited a diffusion current of 1.97 $\mu$a at $-0.96$ v versus SCE. Next, 4.00 ml of a $9.24 \times 10^{-3}$ $F$ nickel(II) chloride solution was pipetted into the polarographic cell. Subsequently, the diffusion current at $-0.96$ v versus SCE was increased to 3.95 $\mu$a. Assuming that the nickel(II) impurity was present as $NiSO_4 \cdot 7\,H_2O$, calculate the weight per cent of this compound in the $CoSO_4 \cdot 7\,H_2O$ reagent.
15. In a buffer solution consisting of 1 $F$ ammonia and 1 $F$ ammonium chloride, copper(II) exists predominantly as the tetraammine complex,

$Cu(NH_3)_4{}^{2+}$. This species undergoes stepwise reduction at the dropping mercury electrode:

$$Cu(NH_3)_4{}^{2+} + e \rightleftharpoons Cu(NH_3)_2{}^{+} + 2\ NH_3; \quad E_{1/2} = -0.24 \text{ v versus SCE}$$

$$Cu(NH_3)_2{}^{+} + e + Hg \rightleftharpoons Cu(Hg) + 2\ NH_3; \quad E_{1/2} = -0.51 \text{ v versus SCE}$$

A series of polarograms was recorded for different concentrations of copper(II) in the ammonia-ammonium chloride buffer, and the limiting current was measured at -0.70 v versus SCE (on the plateau of the polarographic wave) for each concentration.

| Concentration of copper(II), millimoles/liter | Limiting current, $\mu$a |
|---|---|
| 0 | 0.18 |
| 0.489 | 3.24 |
| 0.990 | 6.55 |
| 1.97 | 13.18 |
| 3.83 | 25.2 |
| 8.43 | 56.0 |

When the polarogram for a buffer solution containing an unknown concentration of copper(II) was recorded, a limiting current of 7.49 $\mu$a was observed. Calculate the concentration in moles/liter of copper(II) in the unknown solution.

## SUGGESTIONS FOR ADDITIONAL READING


1. M. Březina and P. Zuman: *Polarography in Medicine, Biochemistry, and Pharmacy*. Wiley-Interscience, New York, 1958.
2. D. R. Crow: *Polarography of Metal Complexes*. Academic Press, New York, 1969.
3. D. D. DeFord and J. W. Miller: Coulometric analysis. *In* I. M. Kolthoff and P. J. Elving, eds.: *Treatise on Analytical Chemistry*. Part I, Volume 4, Wiley-Interscience, New York, 1959, pp. 2475-2531.
4. J. Heyrovský and P. Zuman: *Practical Polarography*. Academic Press, New York, 1968.
5. I. M. Kolthoff and J. J. Lingane: *Polarography*. Second edition, Wiley-Interscience, New York, 1952.
6. A. J. Lindsey: Electrodeposition. *In* C. L. Wilson and D. W. Wilson, eds.: *Comprehensive Analytical Chemistry*. Volume II-A, Elsevier, Amsterdam, 1964, pp. 7-64.
7. J. J. Lingane: *Electroanalytical Chemistry*. Second edition, Wiley-Interscience, New York, 1958.
8. L. Meites: *Polarographic Techniques*. Second edition, Wiley-Interscience, New York, 1965.
9. L. Meites: Voltammetry at the dropping mercury electrode (polarography). *In* I. M. Kolthoff and P. J. Elving, eds.: *Treatise on Analytical Chemistry*. Part I, Volume 4, Wiley-Interscience, New York, 1959, pp. 2303-2379.
10. G. W. C. Milner: *Principles and Practice of Polarography and Other Voltammetric Processes*. Longmans, Green & Company, Ltd., London, 1957.
11. G. W. C. Milner and G. Phillips: *Coulometry in Analytical Chemistry*. Pergamon Press, New York, 1967.
12. O. H. Müller: Polarography. *In* A. Weissberger and B. W. Rossiter, eds.: *Physical Methods of Chemistry*. Volume I, Part IIA, Wiley-Interscience, New York, 1971, pp. 297-421.
13. W. C. Purdy: *Electroanalytical Methods in Biochemistry*. McGraw-Hill Book Company, New York, 1965.

14. G. A. Rechnitz: *Controlled-Potential Analysis.* The Macmillan Company, New York, 1963.
15. W. B. Schaap: Polarography. *In* F. J. Welcher, ed.: *Standard Methods of Chemical Analysis.* Sixth edition, Volume III-A, Van Nostrand, Princeton, New Jersey, 1966, pp. 323-376.
16. W. D. Shults: Coulometric methods. *In* F. J. Welcher, ed.: *Standard Methods of Chemical Analysis.* Sixth edition, Volume III-A, Van Nostrand, Princeton, New Jersey, 1966, pp. 459-492.

# 12 PHASE EQUILIBRIA AND EXTRACTIONS

Modern chemical analysis requires that chemical separations be given heavy emphasis. A few examples make the point clearly:

In water quality studies, a simple determination of "dissolved organic carbon" is relatively uninformative because we need to know something about the nature of the "organic carbon." For example, a few parts in $10^6$ of compound A may be harmless, but several parts in $10^9$ of compound B may be very hazardous. At the same time, these two compounds might be very similar in most of their chemical characteristics, responding identically to all convenient tests. In such cases – and they occur very frequently – analyses for A and B individually can be performed only after A and B have been *separated* from each other and from all possible interferences.

In physiological studies, detailed analytical information is routinely obtained by techniques which directly incorporate chemical separations. To know that a certain body fluid is enriched in lipid material means little until the identities of the excess lipids are ascertained, and this can be learned only after the various lipids present have been *separated* and individually determined.

In organic chemistry, the equivalent weight of an organic acid can be determined, and its empirical formula deduced, if a weighed sample is titrated with standard base. But what useful formula can result if the organic acid happens to be a mixture of two or three compounds? If the acid of primary interest is not *separated* from the accompanying impurities, the observed equivalent weight will be chemically meaningless.

To summarize, chemical separations are important because quantitative analyses, qualitative analyses, and structural analyses all must begin with a step which clearly defines the sample. Computer programmers have given us an "axiom for our times" which describes the situation with chemical systems as delicately and accurately as it does for computer systems: "Garbage in, garbage out!"

In the following sections, we shall consider chemical separations in general, stressing first that separation of compounds implies a separation of phases. Second, we shall consider mechanisms by which phases can be equilibrated and separated. As a final part of this introduction, which is designed to organize and unify all our thinking about separations, we shall construct a table of phase equilibria which amounts to a simple catalog of all possible separation techniques.

## SEPARATION OF COMPOUNDS REQUIRES SEPARATION OF PHASES

It is not possible to purify compounds by selecting individual molecules. We must work on a larger scale, and there is no way around the generalization that, if two compounds are to be separated, we must, somewhere along the line, get them into two different and separable phases. For example, if we wish to separate a carboxylic acid ($pK_a \sim 5$) from an alkane (no functional groups), we can equilibrate an ether solution of the two compounds with aqueous 0.01 $F$ sodium hydroxide in a separatory funnel (see Figure 12–9). The alkane is only slightly soluble in the aqueous phase and remains in the ether. The carboxylic acid, however, can dissociate:

$$RCOOH + H_2O \rightleftharpoons RCOO^- + H_3O^+; \qquad \frac{[RCOO^-]\,[H^+]}{[RCOOH]} = K_a \sim 10^{-5}$$

At the pH of the aqueous phase, about 12, this dissociation is complete:

$$\frac{[RCOO^-]}{[RCOOH]} \sim \frac{10^{-5}}{[H^+]} = \frac{10^{-5}}{10^{-12}} = 10^7$$

The resulting carboxylate anion is far more soluble in the aqueous phase than in the ether, so that the carboxylate is partitioned quantitatively into the aqueous phase. Separation of the phases is accomplished by draining the heavier aqueous phase from the funnel. It is this phase separation which provides the desired separation of compounds – in this case, of the alkane from the acid.

Liquid-liquid extraction is by no means the only convenient method of compound and phase separation. In Chapter 7, precipitation is discussed, and examples are given of the separation of silver by precipitation as the chloride, of nickel by precipitation of the bis(dimethylglyoximato) complex, and others. Brewers and distillers have long been aware of the possibilities of volatility and adsorptivity as tools of chemical separation. A more volatile and purified ethanol-rich phase is separated from fermented mash by distillation, and other compounds with unpleasant tastes and aromas are removed from water-ethanol solutions by adsorption on charcoal.

In precipitations, recrystallizations, and all separations based on solubility, the nature of the phase separation is clear – a solid is separated from a liquid. The separation of two liquid phases, as in the ether-aqueous base system described earlier, is also easily grasped. It is important to extend this concept to more subtly defined phases. For example, in the adsorption on charcoal mentioned above, it is best not to think of the separation of a "charcoal phase" from a liquid phase, but instead to realize that an adsorbed phase (the compounds held on the surface of the charcoal) has been separated from the liquid phase. This is not pedantic nit-picking. If large quantities of liquid phase are passed through the charcoal bed, the adsorbed compounds will be slowly released. In a slightly different situation, the compounds in the adsorbed phase might be *displaced* by substances even more strongly attracted to the charcoal surface. In either case, the result is the same – the adsorbed compounds reappear in the liquid phase, and the "charcoal phase" loses its identity as the host for the desired compounds. The suggested view – separation of adsorbed and liquid phases – is superior because it emphasizes the transitory nature of the desired separation.

For emphasis, we repeat: at the heart of any chemical separation are the processes of (1) phase contact and equilibration, and (2) phase separation. These steps occur in all separation techniques, and a key to understanding a given method is the identification and classification of the steps according to the nature of the phases involved and the mechanism of phase contact and separation.

**Mechanisms of Phase Contact.** Consideration of the ways in which phases can be brought together, equilibrated, and separated reveals four levels of complexity. These levels are summarized in Figure 12–1.

The simplest possibility is shown in Figure 12–1a, and is a single-stage operation like that discussed in the acid-alkane example used above. The phases need not be liquids

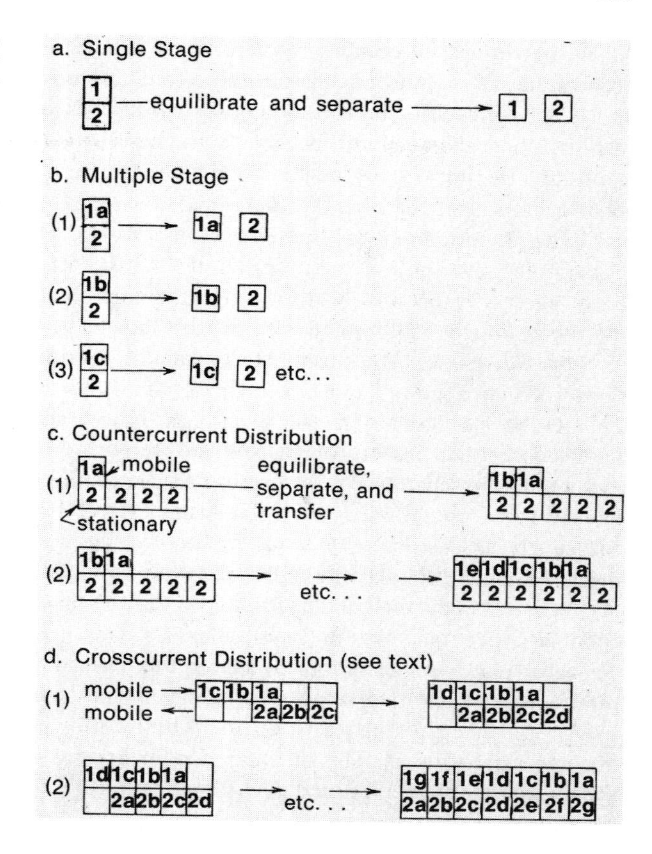


*Figure 12–1.* A schematic representation of the techniques of phase contact. Two types of phases (**1** and **2**) are shown; sequentially used batches of each phase are assigned additional letters (**1a**, **1b**, **1c**, etc.) in the order of their introduction to the system.

only. In terms of complexity, we might deal similarly with a single-stage crystallization or gas-liquid equilibration.

A second level of complexity is reached (with no great strain . . . these distinctions are terribly simple, but well worth keeping in mind when trying to get a general view of separation processes) when fresh batches of one of the phases are repeatedly equilibrated with the other. This is shown in Figure 12–1b, and is a technique commonly employed in extractions when the aim is to increase the yield of some partially extracted product. The process can be carried out automatically in discontinuous steps, as in a Soxhlet extractor (see Figure 12–11 and its accompanying discussion), or continuously, with incremental additions and withdrawals of fresh extracting phase, as in a continuous liquid extractor (see Figure 12–10 and its accompanying discussion).

At the third level of complexity an important new feature is introduced. In countercurrent distribution (CCD), *multiple contacts* occur between *pairs* of phases.* In CCD, the initial equilibration occurs between only one pair of phases, but succeeding transfer operations bring about the creation of many new phase pairs, all of which are equilibrated simultaneously in later operations. As Figure 12–1c shows, it is convenient to regard one phase as stationary and the other as mobile, since it migrates through the apparatus. The mobile phase–stationary phase distinction is also characteristic of chromatographic separations (Chapter 13), which differ from this stepwise process only in that equilibrations occur between incremental regions in continuous phases instead of between individual batches of mobile and stationary phase.

---

*Dr. Lyman C. Craig was the major developer of liquid-liquid stepwise-operation countercurrent distribution. The whole technique bears an abbreviation honoring him, "CCD," for "Craig Countercurrent Distribution."

A fourth level of complexity is recognized but not consistently named. Here, we will use the term "crosscurrent distribution," but this is a name without any official status, and it is by no means uniformly applied by all chemists. As noted in Figure 12–1d, the distinguishing characteristic is that *both* phases are moving through the apparatus in opposite directions. This might seem to differ only trivially from CCD, in which the relative motion of the phases is the same; however, there is an important distinction, and calculations pertaining to the two types of systems differ greatly. An example of crosscurrent distribution is furnished by a distillation column in which the refluxing liquid phase streams downward through the column from the condenser to the still pot, and is continually equilibrated with upward moving vapor phase.

**Types of Phases.** It is useful to make a distinction between "bulk" and "thin-layer" phases. When a liquid and a solid precipitate are separated, or when two liquids are separated in a separatory funnel, each phase has a recognizable volume extending in three dimensions. Such extensive phases are termed "bulk phases." By contrast, when compounds are separated by adsorption on some solid material, the adsorbed "phase" can be viewed as extending in two dimensions only. The third dimension, normal to the surface of the adsorbent, extends for only a few molecular diameters. Such phases are logically termed "thin-layer phases."

Two strikingly different types of thin-layer phases can be distinguished, depending upon whether the "solute" molecules are free to diffuse within the phase. If the thin-layer phase is due simply to adsorption, then the adsorbed molecules are relatively immobile. In the other type of thin layer, the solute molecules are not held to the surface, but instead are free to diffuse within a thin layer of some liquid phase dispersed on some inert support. This distinction is further explained by Figure 12–2. Because thin layers are usually encountered in chromatography, the figure further identifies the phases as stationary and mobile.

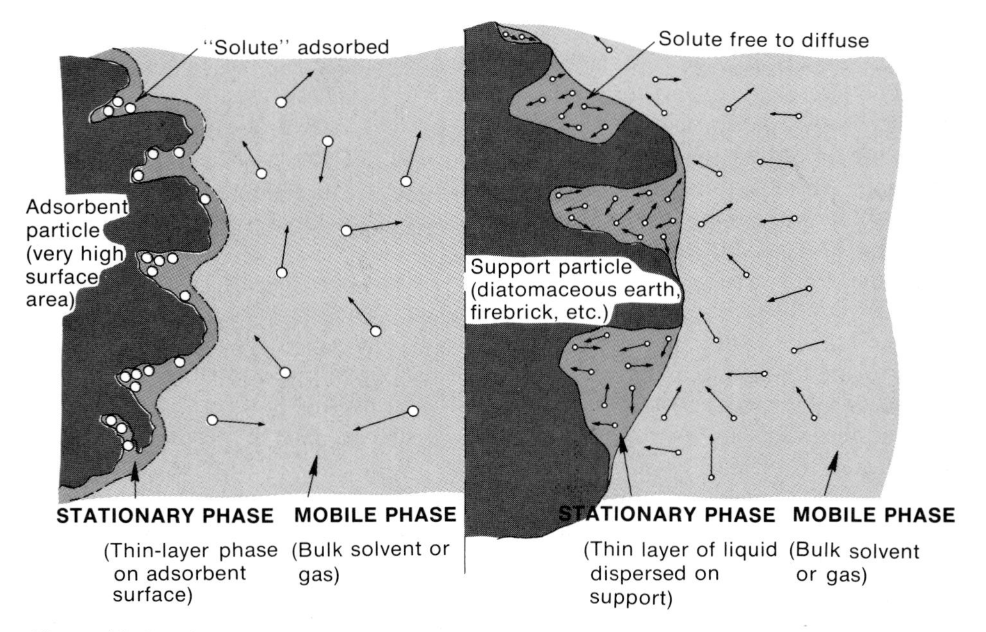


*Figure 12–2.* The two types of thin-layer phases. The small white circles represent solute molecules, and the small arrows on some of them represent diffusion vectors.

To summarize, five different phase types can be listed as shown below:

Bulk
- gas
- liquid
- solid

Thin-layer
- solute free to diffuse
- solute adsorbed

These phases can be arranged in a matrix as shown in Figure 12–3, with at least nine different phase-pairs being possible.

Any chemical separation technique can be succinctly described by noting which phase-pair (Figure 12–3) is used and which mechanism of phase contact (Figure 12–1) is employed. Given nine phase-pairs and four mechanisms of contact, there are, in principle, at least 36 fundamentally different separation techniques. Whether all have been tried or are even possible is uncertain, but it is clearly neither practical nor useful to attempt an exhaustive discussion in this book. Instead, we will first deal with general aspects of phase-contact mechanisms, and then turn to a discussion of representative techniques.

## PHASE EQUILIBRIA

Consider some solute, Z, that is equilibrated between an aqueous phase and an organic solvent phase. The equilibrium can be simply represented as

$$\mathrm{Z(aq)} \rightleftharpoons \mathrm{Z(org)}$$

In a way entirely analogous to the treatment of equilibrium constants in Chapter 3, we can define the **thermodynamic partition coefficient**, $\mathscr{K}_\mathrm{p}$, as

$$\mathscr{K}_\mathrm{p} = \frac{(a_\mathrm{Z})_\mathrm{org}}{(a_\mathrm{Z})_\mathrm{aq}} \tag{12–1}$$

where $a_\mathrm{Z}$ represents the activity of Z.

| | | BULK PHASES | | | THIN-LAYER PHASES | |
|---|---|---|---|---|---|---|
| | | GAS | LIQUID | SOLID | SOLUTE FREE TO DIFFUSE | SOLUTE ADSORBED |
| BULK PHASES | GAS | gaseous diffusion | gas-liquid extraction, distillation | sublimation | gas-liquid chromatography "GLC" | gas-solid chromatography |
| BULK PHASES | LIQUID | | liquid-liquid extraction | crystallization | liquid-liquid chromatography | liquid-solid chromatography<br>ion-exchange chromatography |

*Figure 12–3.* Phase equilibria and chemical separation techniques.

### The Partition Coefficient

As in systems previously discussed, the required activity coefficients are rarely known, so that an approximation based on concentrations alone is commonly employed. This approximate relation is

$$K_p = \frac{[Z]_1}{[Z]_2} \qquad (12\text{–}2)$$

where $K_p$ is called the **partition coefficient**, and the subscripts refer to the concentrations of the solute Z in phases 1 and 2. Because dilute solutions of molecular species more closely approximate ideal behavior than those containing ionic solutes, the approximation given here is better than that made earlier for equilibrium constants. It is best, however, always to remember that we are dealing with a ternary system (two solvents and a solute), the actual behavior of which can be accurately described only by a triangular phase diagram. Our present assumption of ideality will serve well in the great majority of cases; but, whenever high concentrations of solutes are encountered, or whenever the two solvent phases are not completely immiscible, significant deviations can be expected.

### The Distribution Ratio

In the context of practical chemical procedures, it is best to carry our interest yet another step beyond ideal activity ratios. For example, in the extraction of an organic acid, HA, from water by an organic solvent, there is one especially important question. How efficient is the extraction or, in other words, how much acid is left in the water phase? The partition coefficient relates only to the ratio $[HA]_{org}/[HA]_{aq}$, which does not tell the whole story in this case, because, in the aqueous phase, the acid can dissociate. The solute in the aqueous phase is thus present in two different forms, HA and $A^-$, and questions about completeness of extraction must consider the ratio $[HA]_{org}/([HA] + [A^-])_{aq}$. In general, such a ratio of the total concentration of a solute in two phases is termed a **distribution ratio**, and is denoted by $D_c$:

$$D_c = \frac{[\text{total concentration of all forms of Z}]_1}{[\text{total concentration of all forms of Z}]_2} = \frac{c_1}{c_2} \qquad (12\text{–}3)$$

Quantitative treatment of the distribution of a dissociable acid between two phases, such as the one just discussed, is simple and interesting. We can picture the situation as follows:

| | |
|---|---|
| organic phase (1) | HA |
| | $\upharpoonleft\downharpoonright$ |
| aqueous phase (2) | $HA \rightleftharpoons A^- + H^+$ |

The two equilibria are governed by the expressions

$$K_p = \frac{[HA]_1}{[HA]_2} \quad \text{and} \quad K_a = \frac{[A^-]_2[H^+]_2}{[HA]_2} \qquad (12\text{–}4), (12\text{–}5)$$

If we assume that the charged species are essentially insoluble in the organic phase, the distribution ratio is given by

$$D_c = \frac{[HA]_1}{[HA]_2 + [A^-]_2} \tag{12-6}$$

Combination of equations (12–4), (12–5), and (12–6) leads to the equation

$$D_c = \frac{[HA]_1}{\dfrac{[HA]_1}{K_p} + \dfrac{K_a\,[HA]_1}{K_p\,[H^+]_2}} \tag{12-7}$$

which can be further simplified to yield

$$D_c = \frac{K_p[H^+]_2}{[H^+]_2 + K_a} \tag{12-8}$$

When plotted on logarithmic coordinates, equation (12–8) will have two straight-line regions:

first, if $[H^+]_2 \gg K_a$, then $D_c = K_p$, and $\log D_c = \log K_p = (\text{const.})$;

second, if $K_a \gg [H^+]_2$, then

$$D_c = \frac{K_p[H^+]_2}{K_a}$$

and, upon taking logarithms, we obtain

$$\log D_c = \log K_p + \log [H^+]_2 - \log K_a$$

which, because $K_a$ and $K_p$ are constants, is an expression of the form

$$\log D_c = (\text{const.})' - \text{pH}$$

If we take values for $K_a$ and $K_p$ that are typical for the distribution of a fatty acid between ether and water, namely, $K_a = 10^{-5}$ and $K_p = 10^3$, the graph in Figure 12–4 can be constructed, showing clearly the well-known dependence of fatty acid extraction on pH.

Another interesting situation occurs when an extracted compound polymerizes in the organic phase. Here a carboxylic acid may again serve as an example. Acetic acid is almost completely dimerized in non-polar solvents such as benzene. Choosing conditions, *viz.*, $\text{pH} < 2$, such that dissociation of acetic acid in the aqueous phase is negligible, we can represent the overall equilibrium as shown below:

| | |
|---|---|
| organic phase (1) | $HA \rightleftharpoons (HA)_2$ |
| | $\upharpoonleft\downharpoonright$ |
| aqueous phase (2)<br>$\text{pH} < 2$ | $HA$ |

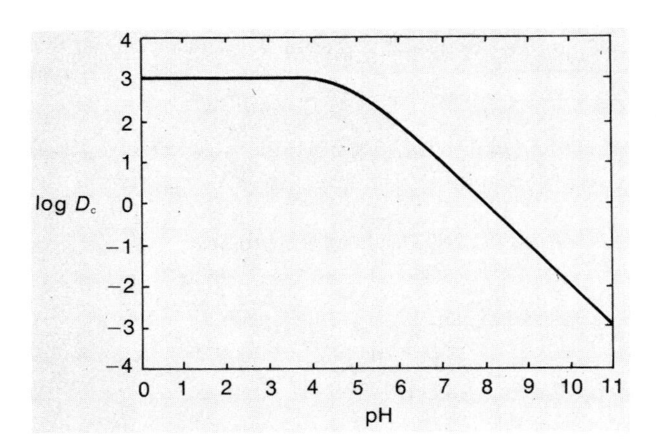


*Figure 12–4.* Distribution ratio ($D_C$) as a function of pH for the partitioning of a carboxylic acid between ether and an aqueous buffer when $K_a = 10^{-5}$ and $K_p = 10^3$.

The two equilibria are governed by the expressions

$$K_p = \frac{[HA]_1}{[HA]_2} \quad \text{and} \quad K_2 = \frac{[(HA)_2]_1}{[HA]_1^2} \qquad (12\text{–}4), (12\text{–}9)$$

The distribution ratio which we might determine, for example, by titrimetrically measuring the total concentration of acetic acid in each phase, is given by

$$D_c = \frac{[HA]_1 + 2[(HA)_2]_1}{[HA]_2} \qquad (12\text{–}10)$$

Substitution of equations (12–4) and (12–9) into equation (12–10) leads to the expression

$$D_c = K_p + 2\, K_2 K_p^2 [HA]_2 \qquad (12\text{–}11)$$

In this case, $D_C$ depends on the total acetic acid concentration in the system.

A variation of $D_C$ causes curvature in plots of $c_1$ (the total concentration of all forms of solute in phase 1) versus $c_2$ (the total concentration of all forms of solute in phase 2). Such plots are called **partition isotherms**, even though they truly represent the distribution ratio, not the partition coefficient. Several are shown in Figure 12–5. The linear relationship depicted by *a* is the behavior expected for the ideal situation when $D_C$ remains constant. Curve *b* shows the variation in $D_C$ which would be observed for solute association (for example, dimer formation) in phase 1. Curve *c* shows an isotherm frequently encountered when phase 1 is an adsorbed phase; the curve levels off as the concentration of solute approaches the value required to cause monolayer coverage of the adsorbent.

## THEORY OF PHASE-CONTACT METHODS

### Single Equilibration

**The Contact Unit.** For the sake of concreteness, consider the typical situation in which an aqueous solution is extracted with an organic solvent having a density less than

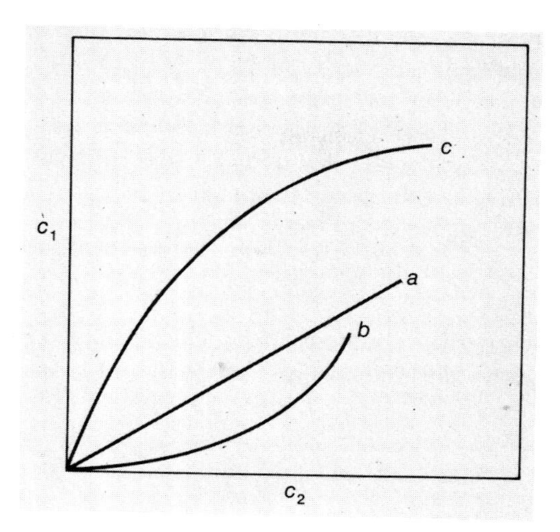


*Figure 12–5.* Three types of partition isotherms.

1.0 g cm$^{-3}$. The organic (upper) phase is termed the "numerator phase," and all parameters associated with it are assigned the subscript 1. The aqueous (lower) phase is termed the "denominator phase" and is denoted by the subscript 2. The letter $V$ logically refers to volumes, and $c$ represents the total concentration of all forms of the specific solute of interest. The situation before and after equilibration is summarized in Figure 12–6. Notice particularly that the initial concentration of solute in the aqueous (sample) phase *prior to equilibration* is denoted as $c_0$.

**Solute Partitioning.** At equilibrium, the ratio of the total concentrations of solute in the two phases is given by

$$\frac{c_1}{c_2} = D_c \qquad (12\text{–}3)$$

where the distribution ratio, $D_c$, will be assumed to be constant throughout the discussion. However, $D_c$ refers only to *concentrations*, and the actual *amounts* of solute present in each phase depend on the phase volumes. For $c$ expressed in moles/liter and $V$ in liters, we can write the following relationships:

$$\text{moles of solute in phase 2 initially} = c_0V_2 \qquad (12\text{–}12)$$

$$\text{moles of solute in phase 1 after equilibration} = c_1V_1 \qquad (12\text{–}13)$$

$$\text{moles of solute in phase 2 after equilibration} = c_2V_2 \qquad (12\text{–}14)$$

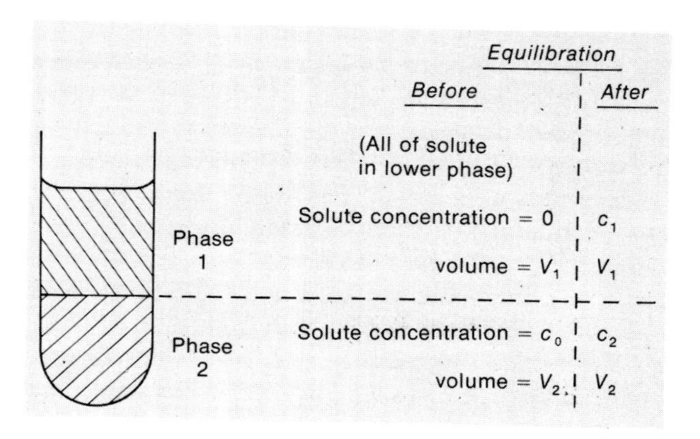


*Figure 12–6.* Phase volumes and solute concentrations in a single equilibration stage. (See text for further discussion.)

It is very useful to define symbols for the *fractional amounts* of solute in each phase after equilibration:

$$p \equiv \text{fraction of solute in phase 1 after equilibration} = \frac{c_1V_1}{c_0V_2} \tag{12–15}$$

$$q \equiv \text{fraction of solute in phase 2 after equilibration} = \frac{c_2V_2}{c_0V_2} \tag{12–16}$$

Using the definitions above and setting $V_1/V_2 = V_r$ (the volume ratio), the reader should show that

$$q = \frac{1}{D_cV_r + 1} \tag{12–17}$$

and

$$p = \frac{D_cV_r}{D_cV_r + 1} \tag{12–18}$$

and should not proceed until these derivations have been completed.

An obvious application for calculations of this sort exists in cases where quantitative extraction is required. If $D_c$ is known or can be measured, the efficiency of a single-stage extraction is simply $p$. Note that, within limits set by the value of $D_c$ and by the size of laboratory apparatus, $p$ can be increased by increasing $V_r$. To calculate the per cent of solute extracted into phase 1, all we need to do is express $p$ in per cent:

$$\text{per cent extracted} = E = 100\,p \tag{12–19}$$

*Example 12–1.* Calculate the concentrations and amounts in each phase after extraction of 100 ml of $10^{-2}$ $M$ acetanilide in water with 100 ml of ether if $D_c = 3.0$. What will be the effect of using 1000 ml of ether?

For the extraction with 100 ml of ether, $V_r = \frac{100}{100} = 1$. Using equations (12–17) and (12–18), we obtain

$$p = \frac{D_cV_r}{D_cV_r + 1} = \frac{3.0}{3.0 + 1} = \frac{3}{4}$$

$$q = \frac{1}{D_cV_r + 1} = \frac{1}{3.0 + 1} = \frac{1}{4}$$

$$\begin{aligned}\text{amount in ether phase} &= c_1V_1 = pc_0V_2\\ &= (0.75)(10^{-2})(0.1)\\ &= 7.5 \times 10^{-4} \text{ mole of acetanilide}\end{aligned}$$

$$\begin{aligned}\text{amount in aqueous phase} = c_2V_2 &= qc_0V_2\\ &= (0.25)(10^{-2})(0.1)\\ &= 2.5 \times 10^{-4} \text{ mole of acetanilide}\end{aligned}$$

$$\text{concentration in ether phase} = c_1 = \frac{pc_0V_2}{V_1}$$

$$= \frac{7.5 \times 10^{-4} \text{ mole}}{0.1 \text{ liter}} = 7.5 \times 10^{-3} M$$

$$\text{concentration in aqueous phase} = c_2 = qc_0$$

$$= (0.25)(10^{-2}) = 2.5 \times 10^{-3} M$$

For the extraction with 1000 ml of ether, $V_r = 10$:

$$p = \frac{D_cV_r}{D_cV_r + 1} = \frac{(3.0)(10)}{(3.0)(10) + 1} = \frac{30}{31} = 0.97$$

Thus the efficiency (completeness) of the extraction is increased from 75 to 97 per cent if the volume of the ether phase is increased from 100 to 1000 ml.

## Repeated Equilibrations

**Stepwise Partitioning of the Solute.** The disappointingly low value for $p$ often obtained in calculations like those above leads to a consideration of repetitive extraction. Multiple extractions with fresh upper phase will remove more material. A constant fractional amount, $p$, of solute is removed from the lower phase for each equilibration. The situation through $n$ extraction steps is summarized in Figure 12–7, which must be carefully studied. The last step of Figure 12–7 gives general expressions for the fractional amounts of solute in each phase after any number, $n$, of extractions. Thus, after $n$ extractions, we have:

*solute amounts*

$$\text{upper phase:} \quad pq^{n-1} \cdot (\text{initial amount of solute}) = pq^{n-1}c_0V_2 \qquad (12\text{–}20)$$

$$\text{lower phase:} \quad q^n \cdot (\text{initial amount of solute}) = q^nc_0V_2 \qquad (12\text{–}21)$$

*solute concentrations*

$$\text{upper phase:} \quad \frac{pq^{n-1}c_0V_2}{V_1} = \frac{pq^{n-1}c_0}{V_r} \qquad (12\text{–}22)$$

$$\text{lower phase:} \quad \frac{q^nc_0V_2}{V_2} = q^nc_0 \qquad (12\text{–}23)$$

The total quantity of solute extracted is, of course, the sum of the amounts of solute in all the successive upper phases or, much more simply, the total amount of solute originally present in the lower phase minus that remaining in the lower phase after the last extraction:

$$(p + pq + pq^2 + \cdots + pq^{n-1})c_0V_2 = (1 - q^n)c_0V_2 \qquad (12\text{–}24)$$

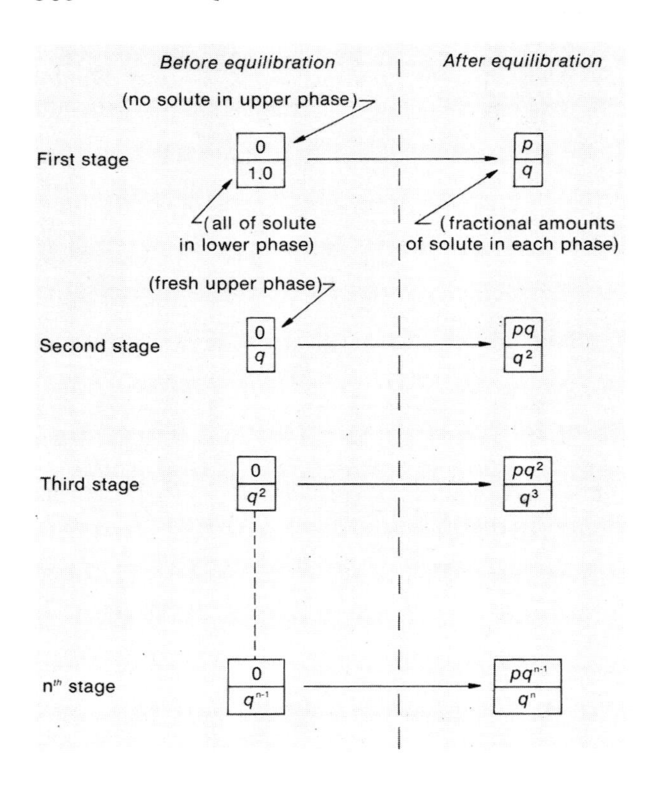


*Figure 12–7.* Solute partitioning in a repetitive extraction scheme.

*Example 12–2.* Consider again the acetanilide-ether-water system used in Example 12–1. Compare the efficiencies of the following three extraction procedures:

I. Single extraction with 100 ml of ether
II. Single extraction with 1000 ml of ether
III. Repetitive extraction with ten 100-ml portions of ether

We have already performed the calculations for alternatives I and II in Example 12–1, and have found efficiencies of 75 and 97 per cent, respectively. It remains to see what improvement results from dividing the 1000 ml of ether into ten equal portions as proposed in alternative III. Our interest lies mainly in the final distribution of material; but first, simply to see how the calculations are done, let us find what the concentrations and amounts of acetanilide are in each phase after the fifth extraction:

upper phase: as noted in Figure 12-7, the fraction of solute remaining in the upper phase is given by $pq^4$; and the amount can be calculated from equation (12–20)

$$\begin{aligned}\text{amount} &= pq^4c_0V_2\\ &= (\tfrac{3}{4})(\tfrac{1}{4})^4(10^{-2})(0.1)\\ &= (0.75)(3.9\times10^{-3})(10^{-3})\\ &= 2.9\times10^{-6}\text{ mole of acetanilide}\end{aligned}$$

The concentration of acetanilide is determined by dividing this amount by the volume (100 ml = 0.1 liter).

$$c_1 = \frac{2.9\times10^{-6}}{0.1} = 2.9\times10^{-5}\ M$$

lower phase: as noted in Figure 12–7, the fraction of solute remaining in the lower phase is given by $q^5$; and the amount can be calculated from equation (12–21)

$$\text{amount} = q^5 c_0 V_2$$
$$= (9.8 \times 10^{-4})(10^{-2})(0.1)$$
$$= 9.8 \times 10^{-7} \text{ mole of acetanilide}$$

The concentration ($c_2$) is determined simply by dividing by the volume of solvent (0.1 liter), and is found to be $9.8 \times 10^{-6}$ $M$.

At the end of ten extractions, the key question is: What fraction of the acetanilide has been removed from the water solution? The lower phase will contain a fraction $q^{10}$. According to equation (12–24), the total fraction of acetanilide extracted will be

$$\sum_{n=1}^{10} pq^{n-1} = 1 - q^{10}$$
$$= 1 - (\tfrac{1}{4})^{10} = 1 - 10^{-6}$$
$$= 0.999999$$
$$\text{or } E = 99.9999 \text{ per cent}$$

Realizing the fanciful aspects of such a number, we see that the efficiency of the third alternative surpasses that of the second by a factor of about $3 \times 10^4$. In fact, the calculation of the distribution of acetanilide after only five 100-ml extractions shows that alternative to be far superior to a single one-liter extraction:

$$E \text{ (five extractions)} = 100\,(1 - q^5)$$
$$= 100\,(1 - 9.8 \times 10^{-4})$$
$$\sim 100\,(0.999) = 99.9 \text{ per cent}$$

**Separation of Two Partitioned Solutes.** Up to now, we have been concerned with obtaining the maximum extent of extraction of just one solute. In a practical situation, it is likely that we would be equally concerned with minimizing the extent of extraction of some other solute, thus obtaining a separation of the two compounds. Clearly, the distribution ratios of the two solutes will have much to do with this, and we might naively suppose that the separability of two compounds A and B will be determined by the ratio $D_{cA}/D_{cB}$ (here, and in the following discussion, the subscripts A and B relate to compounds A and B). The calculations below contrast two arbitrarily chosen systems, both with the ratio $D_{cA}/D_{cB} = 10^3$, and show that the situation is somewhat more complicated.

| SYSTEM 1 | SYSTEM 2 |
|---|---|
| $D_{cA} = 32, D_{cB} = 0.032$ | $D_{cA} = 10^3, D_{cB} = 1$ |
| For a single extraction with $V_r = 1$ | |
| $p_A = 0.97, p_B = 0.03$ | $p_A = 0.999, p_B = 0.50$ |
| For equal initial concentrations of A and B in phase 2 | |
| A obtained 97% pure | A obtained 66% pure |
| 97% of the B removed | only 50% of the B removed |

It happens that system 1 above deals with the most favorable case. For $V_r = 1$, the best separation of A from B will be obtained when

$$\sqrt{D_{cA} D_{cB}} = 1 \tag{12–25}$$

Because the $D_c$ values relate only to concentrations and not to amounts, it is best to broaden our considerations by rewriting equation (12–25) in terms of $p$ and $q$, and in terms of concentrations and volumes [see equations (12–15) and (12–16)]:

$$1 = \left[\left(\frac{p_A}{q_A}\right)\left(\frac{p_B}{q_B}\right)\right]^{1/2} = \left[\left(\frac{c_{1A}V_{1A}}{c_{2A}V_{2A}}\right)\left(\frac{c_{1B}V_{1B}}{c_{2B}V_{2B}}\right)\right]^{1/2} = \sqrt{D_{cA}D_{cB}V_r^2} \tag{12–26}$$

Rearrangement of equation (12–26) shows that it is possible to optimize the conditions for a separation by adjusting $V_r$ so that

$$V_r = (D_{cA}D_{cB})^{-1/2} \tag{12–27}$$

In system 2 above, application of equation (12–27) leads to an optimum $V_r$ of 0.032. Therefore, if the extraction of one liter of lower phase with 32 ml of upper phase is practical, a separation equal to the optimum result obtained in the first case can be achieved.

## Countercurrent Distribution

**Pattern of Equilibrations.** In the preceding section, we have considered the separation of two solutes by selective extraction. Unhappily, even when the $D_c$ values differ by $10^3$, the purity of the extracted material does not exceed 97 per cent in the best case ($D_{cA} = D_{cB}^{-1}$). Any repeated stages of extraction can only offer the same extraction ratios and, although yield can be increased in this way, purity cannot. Consider now a second step in which the upper phase obtained in system 1 above, containing 97 per cent A and 3 per cent B, is equilibrated with a fresh lower phase. In this second step, 97 per cent of the A will stay in the upper phase, giving a yield of $(0.97)^2 100 = 94$ per cent, whereas only 3 per cent of the B will stay in the upper phase, giving a yield of $(0.03)^2 100 = 0.09$ per cent. This process is summarized in Figure 12–8. Noting the quantities shown in the figure, we can observe that the second step produces an upper phase in which nearly all the solute is compound A. In fact, if the upper phase were collected and the solvent removed by evaporation, the purity of the recovered solute would be

$$\frac{0.9409}{0.9409 + 0.0009}(100) = 99.9\%\text{ A}$$

This great increase in the purity of A in the upper phase shows that exposure to fresh lower phase is, to say the least, a very helpful step. Furthermore, if the lower phase from the first step were to be extracted with fresh upper phase, it would be found to contain 99.9 per cent B. Given these remarkable improvements in separation of solutes, we are bound to ask how far this phenomenon can be pushed. As it happens, we have just described a *countercurrent extraction* process; and, although this example is but two steps long and offers high purity only because of the favorable distribution ratios chosen, it is essentially the same as chromatographic processes which involve $10^7$ and more steps and allow separation of, for example, diastereomers.

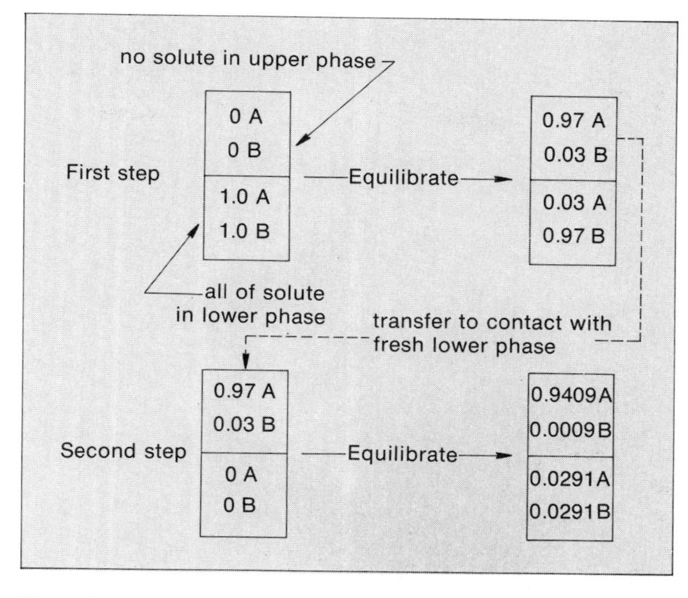


*Figure 12–8.* Schematic representation of the first and second steps in the countercurrent distribution of solutes A and B as defined in System 1 of the text.

It is of interest to study countercurrent distribution more closely, not because it remains a widely applied technique in solvent extraction, but because it forms a useful introduction to the theory of chromatographic separations. Therefore, we shall delay this discussion until the next chapter, in which chromatography is discussed in detail.

## PRACTICAL ASPECTS AND APPLICATIONS

### Apparatus for Extractions

A single-batch extraction can be performed readily in a separatory funnel, an example of which is shown in Figure 12–9. The two immiscible liquids are placed in the separatory funnel, which is then shaken vigorously to ensure intimate contact between the phases. The liquids are allowed to separate into two layers, and the denser liquid is withdrawn through the stopcock.

**Continuous Liquid-Liquid Extraction.** In this technique, the liquid to be extracted has droplets of the extracting solvent passed through it continually. A relatively dense extracting solvent can simply be dropped through the liquid to be extracted, or a relatively light solvent can rise through the liquid buoyantly. Apparatus can be constructed to continue this process indefinitely, without attention, only a small volume of extracting solvent being required. The only energy input is that required to recycle and purify the solvent by a distillation process which also does the necessary work against gravity.

*Figure 12–9.* A separatory funnel, used in single-batch extractions, containing two immiscible liquids that have been shaken together and allowed to separate.

An apparatus adaptable to solvents of any density is shown in Figure 12–10. As shown, extracting solvents of high density can be used. The solvent is distilled from the reservoir, condensed, and dripped through the liquid to be extracted as it returns to the reservoir by way of the tube attached to the bottom of the extractor. In the case of low-density solvents, the return tube can be removed and replaced by two simple plugs, and a funnel tube with a fritted-glass-disk solvent-disperser is placed in the extractor. Condensed solvent falls into the funnel and is forced through the fritted-glass disk by the hydrostatic head of condensate. The light solvent rises through the liquid and returns to the reservoir by overflowing in the extractor sidearm. In either case, if the extractor volume is too large, glass beads can be added to fill the space and, incidentally, to improve liquid-liquid contact by providing a tortuous path for the extracting solvent.

This technique offers the great advantage of unattended operation. However, the same distillation which ensures a constant supply of fresh solvent can cause problems. Volatile solutes may be lost, or, in a few cases, recycled with the solvent. Either way, the apparent recovery will be misleadingly low.

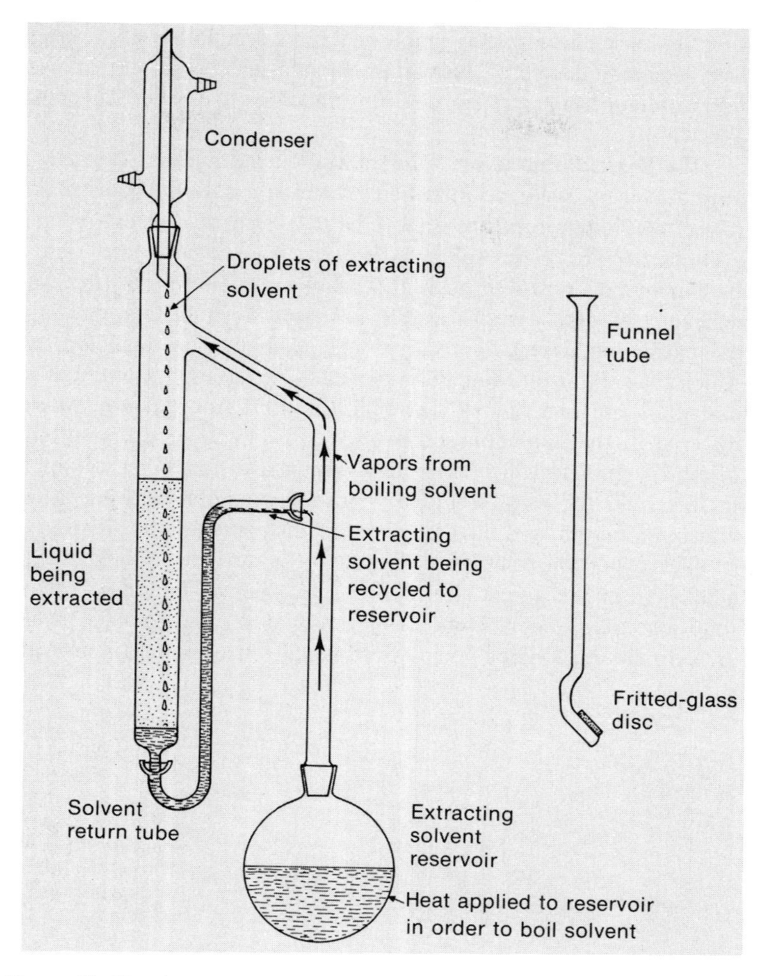


*Figure 12–10.* A continuous liquid-liquid extraction apparatus. As shown, it is arranged for use with an extracting solvent denser than the liquid to be extracted. A lighter solvent can be employed if the solvent-return tube is removed and the fritted-glass disk funnel-tube is placed in the extractor. In that case, the extracting solvent returns to the reservoir by overflowing in the extractor sidearm.

The question of efficiency in this extraction process is rather complex. The attainment of equilibrium requires a certain amount of mass transfer which, in a batch extraction, is substantially aided by the shaking which takes place. In addition, the thorough mixing which takes place in a separatory funnel generates an enormous interfacial area, favoring mass transfer. The quiescent conditions prevailing in continuous extraction seem almost designed to avoid these advantages, although a rotating stirrer might easily be connected to a shaft extending down through the condenser (at least in the case of a dense extracting solvent). Systematic quantitative treatment of extraction efficiency is impossible, but a useful empirical approach is available. A known quantity of an easily detected solute can be added to the extractor and the rate of its extraction into the solvent monitored by analysis of aliquots withdrawn at various times. Determination of the solute concentration in the extracting solvent will allow calculation of the solute remaining unextracted. That amount will decrease with a characteristic half-time which can be determined from a plot of the logarithm of the fraction of solute extracted versus time. In turn, the resulting graph will allow calculation of the length of time required for any necessary degree of extraction, though, of course, due allowance must be made for any difference between the distribution ratios of the test substance and the solute to be extracted.

**The Soxhlet Extractor.** The continuous extraction of solids presents a problem. An apparatus can easily be imagined in which a solvent reservoir is placed below a filter funnel containing solid material to be extracted, the solvent is boiled, and its vapor led to a condenser above the solid material by way of a bypass tube. The condensate could then percolate down through the solid, extracting it, on the way back to the solvent reservoir. In such cases, however, channels inevitably develop in the solid material, and the extraction is very inefficient. This fault can be remedied by arranging a system in which the solid is thoroughly immersed in extracting solvent, at least periodically. Soxhlet extractors are one way of accomplishing this. Another way would be to place the solid material in the apparatus shown in Figure 12–10 such that the top of the material was below the level at which extracting solvent overflowed by way of the solvent-return tube. In this way, the solid would be continually extracted by a flow of solvent, but would remain immersed, and the danger of channeling would be greatly reduced.

The operation of a Soxhlet extractor is not continuous, and, in this regard, it differs greatly from the above alternative. A Soxhlet extractor is shown in Figure 12–11. The solid material is finely ground or powdered in order to offer the greatest possible surface area for liquid contact and is then loaded into an "extraction thimble" made of filter

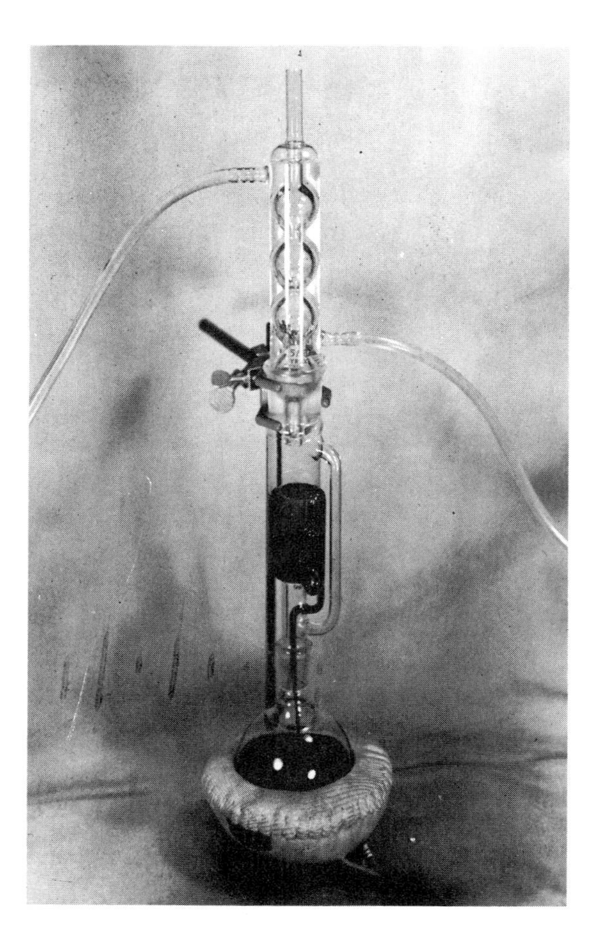

*Figure 12–11.* A Soxhlet extractor for the discontinuous extraction of solids. The solvent-return tube is shown here just in front of the extraction chamber, and is filled with dark fluid which was being siphoned back into the solvent reservoir at the moment this picture was taken.

paper. The thimble fits into the extraction chamber and allows passage of the extracting liquid while it retains the solid. Freshly distilled solvent drips from the condenser and fills the extraction chamber until it reaches the highest level of the solvent-return tube. At that moment, the tube operates as a siphon, draining the extraction chamber completely as it returns the solvent to the reservoir. The operation is thus a sequence of fillings and siphonings, which is characterized as a "discontinuous" extraction process. The operation is amusing to watch, if nothing else. Chemists with small children find that they can occupy them in the laboratory simply by having an operating Soxhlet extractor for them to watch. (Playing with the light spot on a storage oscilloscope is a far better distraction, however.)

Operation of a Soxhlet extractor can become uncomfortably dramatic if mixed solvents are used without proper care. A low-boiling fraction (solvent or azeotrope) might distill into the extraction chamber, and, by the time the siphon dumps it into the reservoir, a substantially higher-boiling mixture may be present. Depending on relative volumes of high- and low-boiling solvents, the low-boiling fraction can become superheated and may be volatilized at a speed high enough to discourage even the best condenser. Another problem which arises, particularly with large extractors, is cessation of boiling when the large volume of cool solvent returns to the reservoir. Severe bumping can be encountered during the reheating, and stable boiling may never be regained. Boiling chips are of little help because they are effective for only a few hours. A magnetically driven stirrer placed in the reservoir is the best solution. Finally, the extraction thimble itself can be a major source of contamination when minute amounts of organic material are sought.

### Extraction of Molecular Species

**Organic Compounds in Natural Samples.** For the organic chemist interested in natural products, or for geochemists interested in organic constituents of rocks and natural water samples, extraction is an everyday procedure. As a general method for the isolation of crude organic fractions, it is the inevitable prelude to a further series of chromatographic separations. Accordingly, the practical examples taken up in Chapter 14 will extend the analyses whose beginnings are discussed here.

Many natural surface waters are now recognized as chronically contaminated by persistent pesticides or their degradation products. The levels of contamination are low, almost always below one part in $10^9$ (abbreviated ppb, for "parts per billion"). However, because of concentrative processes elsewhere in nature, even these low levels of contamination must be monitored and studied in the hope that techniques for their reduction can be found. Due to the terribly dilute solutions in which these compounds present themselves to the analyst, there is an obvious need for concentrating them prior to further analysis. Because of their volatility, chemical fragility, and chances of steam distillation at such very low concentrations, it is not practically possible to isolate these substances by removal of the water through distillation. Instead, it is found that extraction with *n*-hexane offers the best approach. Typically, a one-liter water sample will be extracted with three 25-ml portions of solvent. The extracts are "dried" with anhydrous sodium sulfate, and the 75 ml of solvent is evaporated under a stream of nitrogen gas to whatever extent is required. It is possible to reduce the volume of the *n*-hexane to only a few microliters, thus achieving enrichments approaching a million-fold. The enriched extract is further studied by high-sensitivity gas-liquid chromatography.

Solvent extraction of solid materials is the first step in the examination of organic constituents of dried plant tissues (such as leaves and shredded bark) or crushed rocks and soils. Plant tissues can sometimes be satisfactorily extracted in a separatory funnel, but the denser and finer-grained inorganic materials usually require Soxhlet extraction. With the goal of extracting as much of the organic material as possible, it is necessary to choose a solvent in which both moderately polar compounds and non-polar species such as large alkanes are quite soluble. A non-polar solvent such as hexane would not do because many polar compounds (phenolics, for example) would be inefficiently extracted. On the other hand, long-chain alkanes would be poorly extracted by methanol. Chloroform would be a good compromise, but its purification at levels required for trace analyses is relatively difficult. A good choice turns out to be a benzene-methanol mixture. The choice of solvent is not the only problem – even finely ground solids, if loosely dispersed in the extraction thimble initially, can form a dense mud, offering little chance for good phase contact. For this reason, ultrasonic extraction of solvent-dispersed inorganic material is frequently carried out simply by dipping a beaker containing the slurry into an ultrasonic cleaning bath for a few minutes. This is best done after an hour or so of stirring the solid with the solvent, and care must be taken that the input of acoustical energy does not result in an inadvertent synthesis of trace contaminants by solvent degradation; however, this is not a problem with benzene-methanol mixtures.

**Fractionation of Organic Mixtures.** If a sample preparation procedure has complete extraction as its goal, conditions and solvents are carefully chosen (as noted above) in order to minimize the extent to which the isolation procedure predetermines the makeup of the resulting analytical sample. Yet, it would be foolish to expect that these effects can be completely eliminated, and it often serves better to accentuate them in a way which can produce well-defined sample fractionation. Consider a mixture of organic compounds dissolved in some nonaqueous solvent. Extraction with aqueous base (see page 298) will remove molecules with dissociable acidic functional groups. Careful choice of the pH can influence the selectivity of this extraction. The opposite possibility exists for the extraction of basic species into an acidic aqueous solution. The neutral species remain in the organic solvent.

This kind of fractionation can be coupled with sample isolation. For example, in the investigation of lipid fractions in single-celled organisms, the first step is cell disruption, usually accomplished by pressure gradients or ultrasonic energy. In either case, the broken cells are left in an ethanol or water solution. The solution is made strongly alkaline (10 per cent by weight potassium hydroxide) and is allowed to stand overnight. In this step, most ester linkages (triglycerides) are hydrolyzed, and the freed acids are, of course, completely neutralized by the strong base. Extraction with a non-polar solvent such as *n*-heptane removes only the so-called "non-saponifiable" lipids (sterols) and leaves behind the great bulk of cellular material. Large masses of cell debris can be a problem during the extraction, particularly after the hydrolysis. Before the extraction, it is convenient to introduce a purely mechanical separation step, namely, removal of the debris by centrifugation. A fraction containing fatty acids can be obtained by acidification of the aqueous phase, followed by a second heptane extraction.

### Extraction of Metal Ion Complexes

Ionic species do not have appreciable solubility in non-polar liquids, but ions bound up in neutral complexes with organic ligands offer a hydrophobic exterior and possess substantial solubility in some organic solvents. The distribution ratios for such systems greatly favor extraction of the complexes into the organic phase. If an aqueous solution of two metal ions is equilibrated with an organic solvent containing a ligand which forms a complex with only one of the ionic species, the metal ions can be separated easily and with excellent resolution.

**Complex Equilibria in Two-Phase Systems.** There are a number of individual processes which must be considered, and these are summarized in the scheme below:

$$\begin{array}{lcc} \text{organic phase} & HL & ML_n \\ \hdashline & \rightleftharpoons & \uparrow\downarrow \\ \text{aqueous phase} & HL & \\ & \rightleftharpoons & \\ M^{n+} + (n) & L^- \rightleftharpoons & ML_n \\ & + & \\ & H^+ & \end{array}$$

Because the organic ligand is usually a weak acid, it is denoted by the symbol HL. This species participates in two equilibria: (1) partitioning between phases and (2) dissociation into $H^+$ and $L^-$. Usually, $L^-$ is the form of the ligand involved in complexation, and a third equilibrium in this system is that between $L^-$ and $M^{n+}$, the metal ion to be extracted. Finally, a fourth equilibrium involves partitioning of the metal-ligand complex ($ML_n$) between the organic and aqueous phases. The entire process is controlled by adjustment of the pH of the aqueous phase, which regulates the amount of $L^-$ available and, therefore, determines the extent of complexation. When two metal ions which form metal-ligand complexes of differing stability are to be separated from each other, it is often possible to adjust the pH of the aqueous phase, and thereby control the concentration of $L^-$, at a value such that the more strongly complexed metal ion reacts completely with $L^-$ and is extracted into the organic phase, while the less strongly complexed metal ion does not combine with $L^-$ and remains in the aqueous phase.

This technique is rather general and finds many important applications. A number of the reagents listed in Table 7–1 are commonly employed in such extraction processes. However, the method is by no means all-powerful. It can happen that the metal ions of interest form equally stable complexes with the available ligands and that no pH can be found at which a good separation is possible. More often, we are not dealing with just *two* metal ions, but with *ten* or more. In such instances, interferences can pose a major problem.

## QUESTIONS AND PROBLEMS

1. Define and contrast: (a) thermodynamic partition coefficient, (b) partition coefficient, (c) distribution ratio.
2. For what type of species will the partition coefficient and the distribution ratio usually be equal?
3. Consider a diprotic acid ($H_2A$) with dissociation constants $K_{a1}$ and $K_{a2}$. Derive an expression for the distribution ratio, $D_c$, for partitioning of this acid between an organic solvent and an aqueous phase of controlled pH. Your result will be analogous to equation (12–8) and should give $D_c$ as a function of $K_p$ (the partition coefficient for distribution of $H_2A$ between the aqueous and organic phases), $K_{a1}$, $K_{a2}$, and $[H^+]_{aq}$.

4. Given equation (12–11) on page 304, tell what data you would collect and how you would treat these data in order to determine $K_p$ and $K_2$ for a system such as that described in the text.
5. Which of the partition isotherms in Figure 12–5 represents (a) the distribution of a dissociable acid HA between an organic solvent and an aqueous buffer, (b) the distribution of a dissociable acid between an organic solvent and an aqueous phase whose pH is controlled only by the dissociation of HA itself?
6. Derive equations (12–17) and (12–18).
7. Calculate the fraction of a solute A extracted from 100 ml of an aqueous phase into 50 ml of an originally pure immiscible organic solvent, if the distribution ratio of the solute, $D_c$, is 80 and if A exists as a monomeric species in each phase.
8. Experiment shows that 90 per cent of a substituted phenol is removed from a water sample by extraction with an equal volume of benzene. What percentage of the substituted phenol will be extracted if the volume of benzene is doubled?
9. Consider some solute-water-organic solvent system with a distribution ratio of 10. The volume of water sample from which the solute is to be extracted is 25 ml. Determine the fraction of solute remaining unextracted after (a) one extraction with 250 ml of the organic solvent and (b) three extractions with 25 ml of the organic solvent; (c) express the preceding results in terms of per cent efficiency of extraction.
10. Suppose that you desire to extract a given solute from one solvent into a second, immiscible solvent, the distribution ratio for the solute being only 3.50. If the volume of the first phase, initially containing all the solute, is 10 ml, calculate the number of successive extractions with fresh 10-ml portions of the second solvent needed to extract a minimum of 99 per cent of the solute from the original solvent.
11. Long-chain alkanoic acids have a distribution ratio of about 0.1 for partitioning between hexane and an aqueous buffer with a pH of 10. The distribution ratio for partitioning of some polar lipids in the same solvent system is about 25. (a) Determine the value of the ratio (hexane volume/buffer volume) which will optimize the separation of the lipids from the acids. (b) It is very important to isolate an alkanoic acid-fraction free of the polar lipids mentioned above. Tell how you would increase the purity of the acid fraction after extracting it from the hexane.
12. If five is taken as the maximum convenient number of repeated extractions, and if $\Sigma V_1$, the total volume of extracting solvent, is equal to $V_2$, the volume of the aqueous phase, what is the minimum $D_c$ required for 99.9 per cent extraction efficiency?
13. In a particular continuous liquid-liquid extractor, the following concentrations of propanol were found in the aqueous phase after various total extraction times: $t = 0$ (start of experiment), $[C_3H_7OH] = 0.35\,M$; $t = 1$ hr, $[C_3H_7OH] = 0.088\,M$; $t = 4$ hr, $[C_3H_7OH] = 0.0014\,M$. How long would it take to achieve 99.9 per cent extraction efficiency?
14. Sketch a practical liquid-liquid continuous extraction apparatus which allows stirring of the extracted phase even when a solvent less dense than the sample is being used.

15. A sample of foilage is to be studied for its organic base content. Tell how you would work-up an organic base fraction from plant material.
16. The technique of solvent extraction is a powerful method for the evaluation of equilibrium constants. The equilibrium constant for the reaction

$$I_2 + I^- \rightleftharpoons I_3^-$$

was determined in this manner.
   (a) The partition coefficient for the distribution of molecular iodine between water and carbon tetrachloride was established by measurement of the equilibrium concentrations of $I_2$ in the two phases. Calculate the partition coefficient obtained in a typical experiment, if the titration of 100.0 ml of the aqueous phase required 13.72 ml of 0.01239 *F* sodium thiosulfate solution and if the titration of 2.000 ml of the carbon tetrachloride layer required 23.87 ml of the same sodium thiosulfate solution.
   (b) Next, molecular iodine was partitioned between carbon tetrachloride and an aqueous 0.1000 *F* potassium iodide solution. The titration of a 2.000-ml portion of the carbon tetrachloride phase with a 0.01239 *F* sodium thiosulfate solution required 17.28 ml of the titrant, whereas the titration of a 5.000-ml aliquot of the aqueous phase required 25.93 ml of the sodium thiosulfate solution. Calculate the equilibrium constant for the formation of triiodide ion.
   (c) Why is molecular iodine so much more soluble in carbon tetrachloride than in water? Why is triiodide not extracted into carbon tetrachloride?
17. A study was made of the distribution of formic acid, HCOOH, between water and benzene at 25°C. Various amounts of formic acid were dissolved in water having a pH of 1.0, and the aqueous solutions were shaken in contact with benzene until equilibrium was attained. For each experiment, aliquots of the aqueous and benzene phases were taken, and the total *formal* concentration of formic acid in each phase was determined by means of titration with a standard sodium hydroxide solution. From five different experiments, the following data were collected:

| HCOOH in aqueous phase, *F* | 1.632 | 3.436 | 5.115 | 6.863 | 8.852 |
|---|---|---|---|---|---|
| HCOOH in benzene phase, *F* | 0.003117 | 0.008418 | 0.01514 | 0.02402 | 0.03629 |

   (a) Assuming that monomeric formic acid is the only important species in water at pH 1, determine what is the principal form of formic acid in benzene. Draw its structure.
   (b) Evaluate the equilibrium constant for the association of monomeric formic acid in benzene

$$n\ \text{HCOOH (benzene)} \rightleftharpoons [\text{HCOOH}]_n\ \text{(benzene)}$$

where $[HCOOH]_n$ is the predominant form of formic acid in benzene.

(c) From experiments at 75°C, the equilibrium constant for the preceding association reaction was found to be 2.47. Calculate the value of $\Delta H$ for this reaction, and discuss the significance of the result.

## SUGGESTIONS FOR ADDITIONAL READING

1. E. W. Berg: *Physical and Chemical Methods of Separation.* McGraw-Hill Book Company, New York, 1963.
2. L. C. Craig and D. Craig: Laboratory extraction and countercurrent extraction. *In* A. Weissberger, ed.: *Technique of Organic Chemistry.* Second edition, Part I, Volume III (*Separation and Purification*), Wiley-Interscience, New York, 1956, Chapter II.
3. H. Irving and R. J. P. Williams: Liquid-liquid extraction. *In* I. M. Kolthoff and P. J. Elving, eds.: *Treatise on Analytical Chemistry.* Part I, Volume 3, Wiley-Interscience, New York, 1959, pp. 1309–1365.
4. G. H. Morrison and H. Freiser: *Solvent Extraction in Analytical Chemistry.* John Wiley and Sons, New York, 1957.
5. R. H. Perry, C. H. Chilton, and S. D. Kirkpatrick: *Chemical Engineers' Handbook.* Fourth edition, McGraw-Hill Book Company, New York, 1963.
6. H. Purnell: *Gas Chromatography.* John Wiley and Sons, New York, 1962.
7. R. E. Treybal: *Liquid Extraction.* Second edition, McGraw-Hill Book Company, New York, 1963.

# 13 CHROMATOGRAPHY

In Chapter 12 we discussed phase equilibria and chemical separations by examining extraction processes involving bulk phases and small numbers of equilibrations. In chromatography – the subject of this chapter – substances are separated not by discrete extraction steps but by being blown or washed through a tube in which they are continually exchanged and equilibrated between a *stationary phase* and a *mobile phase.* The mobile phase must be some gas or liquid in which the substances to be separated are at least partially soluble, and the stationary phase is usually a thin-layer phase like one of those described in Figure 12–2. The same fundamental laws of phase equilibria apply to both extractions and chromatography but, instead of equilibria between upper and lower phases in a separatory funnel, we must deal with equilibria between mobile and stationary phases in a chromatographic column. Rather than well-defined equilibration steps in which two bulk phases are deliberately mixed and then allowed to separate, we must deal with whatever equilibration takes place as the mobile phase passes over the stationary phase. Although these contrasts help to "define" chromatography by comparison with extraction processes, it seems best to provide an example for further clarification.

The example chosen here is the simple liquid-solid chromatographic column illustrated in Figure 13–1. In practice, the column might be a glass tube about one centimeter in internal diameter. A length of it, about 15 cm, would be packed with stationary phase; some additional length at the top would be left empty to provide a reservoir for the mobile phase; and a stopcock would be attached to the bottom to regulate the flow of mobile phase. Although the stationary phase can take many different forms, let us assume for the moment that it is some adsorbent (perhaps alumina or silica gel), "activated" by being dried at 150°C and having an average particle diameter of about 100 $\mu$m, which has been packed into the column by pouring in a slurry and drawing off the excess solvent through the stopcock. The mobile phase should be some solvent which easily dissolves the sample without reacting with it, and which is not so polar that it displaces adsorbed sample molecules from the stationary phase.

Consider the separation of a binary mixture on this column. Before the sample is added, the stopcock is opened briefly and any mobile phase above the top of the stationary-phase bed is allowed to drain through the column. Then, the sample (a mixture of components 1 and 2) is dissolved in a minimum volume of mobile phase, and applied to the top of the column. Following this, the stopcock is again opened briefly, this time to draw the portion of mobile phase containing the sample into the very topmost portion of the column, as shown in Figure 13–1A. At this moment, the sample (solute) molecules are distributed between an adsorbed phase and the solution phase in ratios determined by their individual adsorption isotherms. Let us say that component 2 is more strongly adsorbed and, therefore, that a smaller proportion of it is in the solution phase. Next, a volume of mobile phase (the **eluent**) is added to the empty space above the adsorbent bed, and the process of **elution** is begun by opening the stopcock so that this mobile phase slowly passes through the column (say, at 0.1 ml/min).

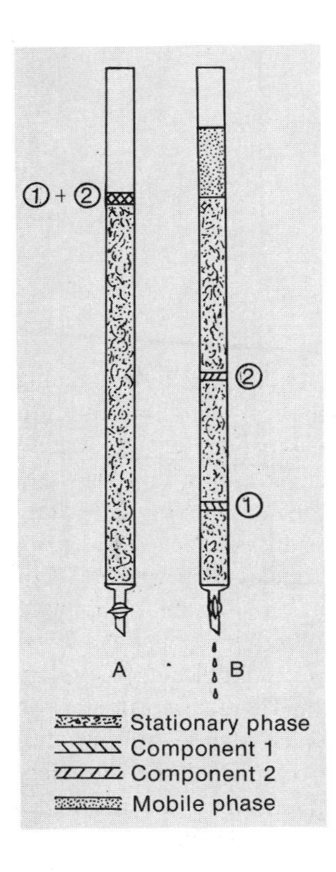


*Figure 13–1.* Diagrammatic representation of the retention and elution of a two-component mixture on a chromatographic column; *A,* just after addition of sample; *B,* during elution. Component 2 is more strongly retained than 1.

From this time onward, chaos prevails. If the reader will pardon childish simplicity and the attribution of eyes and emotions to individual molecules, which, in reality, are responding only to purely physical forces, a memorable description of these events is possible. Imagine a molecule of solute, sitting happily in the solution phase just after the sample has been added and the adsorbed phase-solution phase equilibrium has been established. This molecule cannot get onto the adsorbent because its surface is too crowded, and to do so would disturb the equilibrium. Suddenly, someone opens the stopcock, and the elevator in which this molecule is riding starts to go down. That is the end of equilibrium. Looking around as it descends, our molecule soon sees an uncrowded adsorbent surface and settles down on it, trying to bring the concentration of adsorbed-phase molecules up and into equilibrium with the solution-phase concentration. A different solute molecule, which had been among those adsorbed in the initial equilibration, suddenly finds that all its solution-phase colleagues have gone down on the elevator, and that the solvent around it is "empty." It jumps into solution, striving to get that concentration up into equilibrium with the crowded adsorbent. As it rides along, it finds the adsorbent surface quite crowded by our friend above and others like it which had initially been in the solution phase. Soon, however, this second molecule passes the peak of adsorbent crowding, and the forces which continually strive to establish equilibrium lead it to sit down on the adsorbent again. It is by exactly this sort of molecular leapfrog that solutes pass through the column. As the solvent elevator moves downward at a constant rate, molecules which are strongly adsorbed (component 2) are in the mobile phase for only a small fraction of any increment of time, and make slow progress relative to less strongly adsorbed species (component 1). Thus, we arrive at the situation depicted in Figure 13–1B, which shows that two solute zones have formed in the column.

In the example chosen, the solute molecules are repeatedly adsorbed on and desorbed from the solid stationary phase. Because of this mechanism of interaction, chromatography involving a solid stationary phase is often termed "adsorption chromatography," and is sometimes discussed as though it differed radically from liquid-liquid or gas-liquid chromatography. These latter techniques, involving a thin-layer stationary phase in which the solute is free to diffuse, can be grouped together under the term "partition chromatography." In terms of the picture of solute migration which the example provides, the distinction between adsorption and partition chromatography is unimportant. In the case of partition chromatography, we should think of sample molecules not as looking for uncrowded adsorbent surfaces but, instead, as looking for uncrowded swimming pools – that is, volumes of stationary phase in which the solute concentration is too low to be in equilibrium with the surrounding mobile phase.

If the flow of mobile phase is maintained, and if each drop of mobile phase is analyzed for solute content the instant it emerges from the column, a chromatogram like that in Figure 13–2 is obtained. In practice, such chromatograms are easily obtained with the aid of a **detector**, a special analytical instrument arranged for continuous analysis of a flowing fluid stream and having an electrical output signal which is in some known way proportional to the chemical input. Components of a chemical mixture are characterized by the degree to which they are retained on a given column, and the most common measurement of this is the **retention time**, as noted in Figure 13–2. Equally characteristic is the **retention volume**, which is a measure of the amount of mobile phase required for elution of a given component. Because even columns which are identically prepared usually differ slightly in their retention characteristics, it is convenient to use **relative retention times**, or **relative retention volumes**, in which the retention of any particular component is expressed relative to that of some standard compound co-injected with the sample.

## Solute Migration

No solute can move through the column faster than the eluent which carries it, and the extent to which any compound falls short of this maximum velocity is measured by **R**, the **retention ratio**, or **retardation factor**. If the average molecule of mobile phase

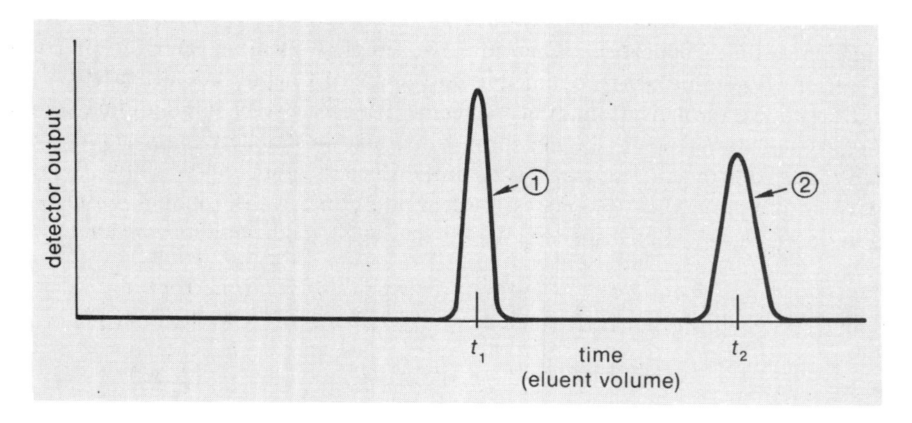


*Figure 13–2.* Graphic representation of output of the analyzing device (or detector) during a chromatographic separation; the retention time for component 1 is $t_1$, and for component 2 it is $t_2$.

requires 15 minutes for passage through a 15-cm column, the eluent velocity ($v$) would be 1.0 cm/min or 0.016 cm/sec. (The instantaneous velocity of any given molecule can vary widely from this value and, because of tortuosity in the flow path, the true distance travelled during passage through the column will be somewhat greater than 15 cm. The "velocity" defined here refers only to the average component parallel to the bulk flow.) If a particular solute zone takes an hour to pass through the column under the same conditions, its velocity would be 0.004 cm/sec or one-fourth the velocity of the mobile phase. The retardation factor is thus 0.25. The retardation factor or retention ratio offers a measure of the fraction of time which the average solute molecule spends in the mobile phase. Thus, we can write

$$\mathbf{R} = \frac{t_M}{t_M + t_S} \tag{13–1}$$

where $t_M$ is the time in the mobile phase and $t_S$ is the time in the stationary phase. In this case, since $\mathbf{R} = 0.25$, a solute molecule spends 25 per cent of its time in the mobile phase and travels 25 per cent as fast as the eluent. Notice that, if a substance is entirely unretained and spends no time in the stationary phase, it will have $\mathbf{R} = 1.0$ and will travel with the same velocity as the mobile phase. If the molecules of a substance spend, on the average, half their time in each phase, we have $t_M = t_S$ and $\mathbf{R} = 0.5$. It is interesting to recall that we have compared the mobile phase to an elevator always moving at a constant speed. Therefore, because all eluted species must travel the same column length, they must spend exactly the same length of time in the mobile phase. It is only $t_S$ which varies from component to component, $t_M$ remaining constant.

In chromatography, the **partition coefficient** ($K$) is defined by the relation

$$K = \frac{c_S}{c_M} \tag{13–2}$$

where $c_S$ and $c_M$ are the solute concentrations in the stationary and mobile phases, respectively. It is not obvious, but this definition differs significantly from that given for $K_p$ in equation (12–2). For any chromatographic system, $K$ (without a subscript) is determined by observation of the actual migration rate of the solute. The migration rate depends on the collective movement of all forms of the solute, and $K$ is, therefore, more closely related to $D_c$ [equation (12–3)] than to $K_p$. Nevertheless, the definition, terminology, and symbol adopted here are the result of international agreement, and do find consistent application. Therefore, we employ them in spite of the unfortunate contrast with some of the definitions and terms of Chapter 12, which are, incidentally, the result of another international agreement by a separate international committee. In order to relate $K$ and $\mathbf{R}$, we can formulate a relationship between $K$ and the phase times ($t_S$ and $t_M$) by noting that the times depend not only on the relative solute concentrations in the two phases, but, in addition, on the volumes within which these concentrations must be dispersed. Within any region of the column, we have

$$\begin{pmatrix}\text{ratio of times spent by}\\ \text{solute in each phase}\end{pmatrix} = \begin{pmatrix}\text{ratio of quantities of}\\ \text{solute in each phase}\end{pmatrix} = \begin{matrix}\text{(partition coefficient)}\\ \times\ \text{(ratio of phase volumes)}\end{matrix}$$

$$\frac{t_S}{t_M} = \frac{c'_S\ V'_S}{c'_M\ V'_M} = K\left(\frac{V_S}{V_M}\right) \tag{13–3}$$

where $V_S$ and $V_M$ are the volumes of stationary and mobile phase, respectively, within the column and where the primed quantities pertain to the region under consideration. In the third expression of equation (13–3), the primes can be dropped by assuming, in the region under consideration, that overall equilibrium prevails ($K = c'_S/c'_M$) and that the *ratio* of phase volumes is characteristic of the column as a whole. Rearrangement and substitution into equation (13–1) yields

$$\mathbf{R} = \frac{1}{1 + \dfrac{t_S}{t_M}} = \frac{1}{1 + K\left(\dfrac{V_S}{V_M}\right)} = \frac{V_M}{V_M + KV_S} \tag{13–4}$$

which expresses the retention ratio in terms of the partition coefficient and column parameters.

A relationship between **R** and the **retention volume**, $V_R$, can be derived if we note that an eluent molecule introduced at the top of the column will appear at the outlet after the passage of a volume $V_M$ of mobile phase. A solute travelling with half the eluent velocity (**R** = 0.5) will take twice as long to traverse the column, so that $V_R = 2V_M$. A substance travelling at one-tenth of the eluent velocity will require ten times longer, and thus $V_R = 10V_M$. Evidently,

$$\frac{V_R}{V_M} = \frac{1}{\mathbf{R}} \tag{13–5}$$

Then, substituting into equation (13–4), we obtain

$$\mathbf{R} = \frac{V_M}{V_R} = \frac{V_M}{V_M + KV_S} \tag{13–6}$$

Therefore,

$$V_R = V_M + KV_S \tag{13–7}$$

In some forms of chromatography, it is easily possible to determine the mobile-phase volume, $V_M$. In such cases, the **adjusted retention volume**, $V'_R$, is sometimes used:

$$V'_R = V_R - V_M = KV_S \tag{13–8}$$

This parameter has the feature of being directly proportional to $K$. In either situation, these expressions are useful because they reveal the simple relationship between $V_R$ and $K$.

## PLATE THEORY OF CHROMATOGRAPHY

Succinct though it may be, the foregoing discussion of solute migration tells nothing about the *shape* of chromatographic zones. Are all the molecules of a given solute transmitted in perfect unison, emerging at the instant an eluent volume $V_R$ has passed through the column? In Figure 13–2 we have correctly indicated that this is not the case, and that the zones are somewhat diffuse, with maximum concentrations in the center of the zone gradually tailing off to zero at either side. The figure further indicates that late-eluting peaks are relatively broad, a characteristic directly attributable to their longer residence time in the column, because the mechanisms of zone-broadening can act for a longer period of time, thus producing greater effects. It is the task of chromatographic theory to consider all the mechanisms of zone-broadening, and to explain in detail the shapes of chromatographic zones, basing these explanations to the greatest possible extent on a consideration of the physical processes actually occurring in the chromatographic system.

The practical significance of zone-breadth lies in its effect on the separating power of any given chromatographic system. Consider, for example, the separation of zones 1 and 2 in Figure 13–2. As shown, the zones are completely separated or "resolved." If the breadth of each zone were increased by a factor of four or more, however, peak overlap would begin to occur and the degree of separation would be very significantly affected. Alternatively, if we consider the number of additional solute zones which will fit between times $t_1$ and $t_2$ in Figure 13–2, it is clear that the number of components the system can separate without overlap depends very strongly on zone-breadth.

It is necessary to develop some quantitative basis for considering zone-breadth. Although it is dominant at present, chromatography is hardly the first successful tool for the separation of chemical compounds. Consequently, much of the terminology in chromatographic theory derives from earlier techniques, and a brief digression is necessary in order to explain these terms. In particular, in the field of organic chemistry, techniques of fractional distillation received very extensive development prior to the advent of gas-liquid chromatography. In fractional distillation, it was recognized that the quality of a separation depends strongly on the construction and use of the "fractionating column." The column is interposed between the "still pot" in which the organic mixture is boiled and the "condenser" in which the vapor from the still pot is condensed prior to being either returned to the still pot or taken off as product. Within the column itself, rising vapors from the still pot are continuously equilibrated with liquid (condensed vapors) returning from the condenser. The efficiency of separation depends on the nature of the in-column equilibration in a way which might be summarized by saying, for example, "this vapor appears to have been equilibrated with the liquid at six different stages within the column." As the vapor from the still pot rises in the column, each stage of equilibration results in an enrichment of the vapor in its more volatile component, thus increasing the purity of the vapor reaching the condenser. In fact, processes in the column are continuous, not stepwise, and if we observed a typical fractionating column in operation we would see little other than vapor bubbling up through descending liquid. Certainly, we could not *see,* for example, six discrete equilibrations. Nevertheless, regardless of which processes actually prevailed, the efficiency of the column could be summarized by noting that the composition of the product corresponded to that which would be obtained after six theoretically perfect equilibrations.

Some, but not all, fractionating columns are constructed so that liquid-vapor equilibrations take place as the vapor bubbles upward through liquid percolating downward through perforated plates placed in the column. Ideally, a complete liquid-vapor equilibration should take place on each plate. Although such perfection is rarely, if ever, achieved, this type of column construction gave rise to the term **theoretical plate.** If a column employed in any chemical separation process appears to provide $n$ complete phase equilibrations during a single passage of material through the column, the efficiency of the column as a fractionating device is expressed by stating that the column possesses $n$ theoretical plates.

#### Relation of Chromatography to Countercurrent Distribution

The concept of discrete stages of equilibration in chromatography can be greatly clarified by a brief reconsideration of countercurrent distribution. Figure 12–8 and its accompanying discussion showed how a two-stage extraction process could greatly improve product purity. The important characteristic of the process was the second extraction step, in which the first upper phase was equilibrated with fresh lower phase. This process is correctly termed a two-stage countercurrent distribution, and is summarized here by the first four charts in Figure 13–3. Each vertical pair of boxes in Figure 13–3 represents an upper-lower phase pair in an extraction tube. At the outset, all the solute is in the lower phase of the "zeroth" tube; for mathematical reasons which will become apparent later, it is convenient to number the extraction tubes from zero, not one. After the first equilibration, fractional amounts $p$ and $q$ are in the upper and lower phases, respectively. In the next step, the upper phase in the zeroth tube is transferred to the first tube, and fresh upper phase is added to the zeroth tube. A second equilibration follows, and the solute becomes distributed among the phases as noted in the fourth chart. Additional transfers and equilibrations, which can continue indefinitely, have the effect of moving the upper phase through the series of extraction tubes. Because of this pattern of movement, the upper phase is analogous to a chromatographic mobile phase, and the lower phase is analogous to a stationary phase. Each extraction tube is analogous to whatever column height is required to achieve one complete stage of equilibration

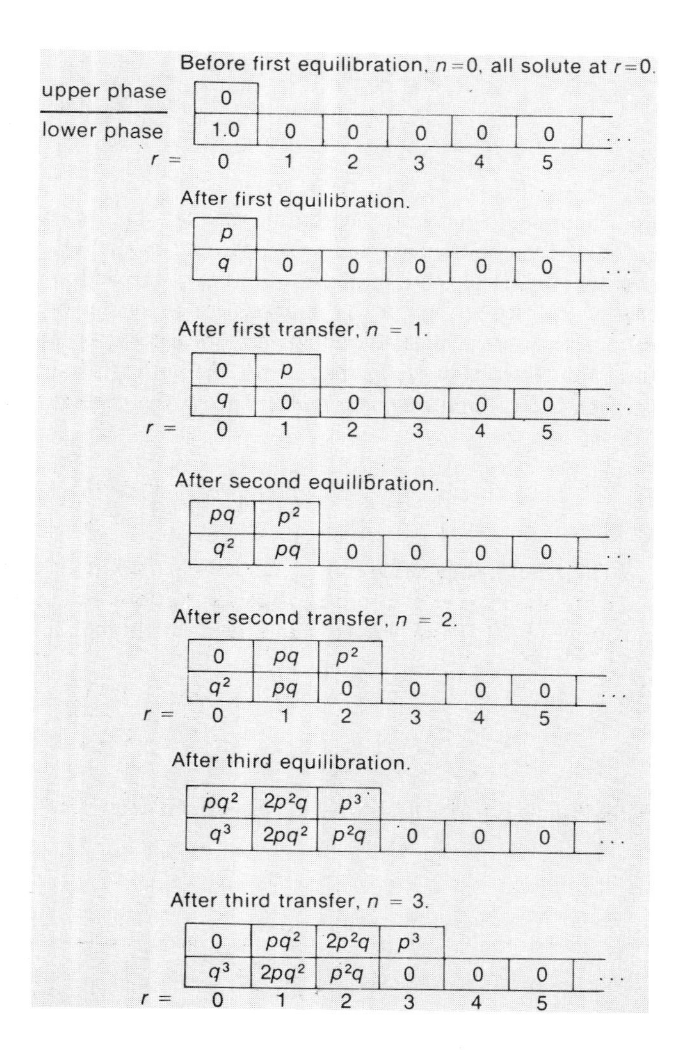


*Figure 13–3.* Fractional amounts of solutes in countercurrent distribution.

between the mobile and stationary phases, and this same column subunit is equivalent to a single theoretical plate.

**Calculation of Zone Profiles.** To trace the evolution of a chromatographic zone, we can consider the distribution of a solute among the theoretical plates of a chromatographic column. Imagine that the column is divided into individual theoretical-plate subunits, each containing an amount of mobile phase $\Delta V_M$ and of stationary phase $\Delta V_S$. A solute is added to the column and enters the zeroth theoretical plate. A fractional amount $p$ remains in the mobile phase, while a fractional amount $q$ enters the stationary phase. Although the flow of a chromatographic mobile phase is continuous, it can be viewed in terms of the addition of successive $\Delta V_M$ units to the column; addition of the next $\Delta V_M$ unit has the effect of transferring the fractional amount $p$ to the first theoretical plate. The distribution continues to develop exactly as outlined in Figure 13–3. The fractional amount of solute in each theoretical plate after the addition of one, two, and three $\Delta V_M$ units can be determined by combining the mobile and stationary phase concentrations shown for each theoretical plate in Figure 13–3:

| Number ($n$) of $\Delta V_M$ units added | Fractional amount of solute in theoretical plate ($r$) | | | |
|---|---|---|---|---|
| | 0 | 1 | 2 | 3 |
| 1 | $q$ | $p$ | | |
| 2 | $q^2$ | $2pq$ | $p^2$ | |
| 3 | $q^3$ | $3pq^2$ | $3p^2q$ | $p^3$ |

For example, in the first theoretical plate ($r = 1$) after two transfers ($n = 2$), an amount $pq$ is in the mobile phase and an amount $pq$ is in the stationary phase, and the total in that theoretical plate, as noted above, is $2pq$. Inspection shows that, after the addition of any given number, $n$, of $\Delta V_M$ increments, the series of total solute fractions in the various theoretical plates is obtained from the expansion of $(q + p)^n$. Conveniently, the total solute fraction ($F_{r,n}$) in the $r$th theoretical plate after $n$ $\Delta V_M$ units have passed through the column can be calculated from the general form of the binomial expansion:

$$F_{r,n} = \frac{n!}{(n - r)!\, r!}\, p^r q^{(n-r)} \tag{13–9}$$

The relationship between $p$, $q$, and the partition coefficient, $K$, can be determined if we write the ratio $p/q$ in terms of concentrations and phase volumes [see, for example, equations (12–15) and (12–16)] and then substitute $K$ for the ratio $c_S/c_M$:

$$\frac{p}{q} = \frac{c_M V_M}{c_S V_S} = \frac{1}{K}\frac{V_M}{V_S} \tag{13–10}$$

Substituting $p = (1 - q)$ or $q = (1 - p)$, we can determine that

$$p = \frac{1}{1 + K\left(\dfrac{V_S}{V_M}\right)} \quad \text{and} \quad q = \frac{K\left(\dfrac{V_S}{V_M}\right)}{1 + K\left(\dfrac{V_S}{V_M}\right)} \tag{13–11, 13–12}$$

To examine quantitatively the development of chromatographic zones and the separation of two solutes with different $K$ values, we can consider the behavior of two solutes, one with $p = 3/4$ and $q = 1/4$, and the other with $p = 1/4$ and $q = 3/4$. In a typical chromatographic column with $V_M/V_S = 10$, these parameters correspond to solutes with $K$ values of 3.33 and 30, respectively. As an example, let us ask what fraction of the solute with $K = 3.33$ will be in the fourth theoretical plate after six $\Delta V_M$ units of mobile phase have passed through the column. In this case, the values to be inserted in equation (13–9) are $r = 4$, $n = 6$, $p = 3/4$, and $q = 1/4$. We obtain $F_{4,6} = 15\,p^4q^2$, or 1215/4096. Carrying out similar calculations for each of the solutes in each theoretical plate gives the results plotted in Figure 13–4. Notice that the stepwise equilibration model which we have used not only indicates the solutes will be separated, but also shows the zones should have finite width.

As elution continues, that is, as more mobile phase is passed through the column and, therefore, $n$ increases, more and more plates are populated with solute molecules. The continuing development of the zone profiles is shown in Figure 13–5, which indicates the shapes and locations of the chromatographic zones after the passage of 25 and 100 $\Delta V_M$ units of mobile phase.

In a typical chromatographic separation, both $n$ and $r$ become very large, and calculations using equation (13–9), which contains factorial terms, would be extremely cumbersome. It is also true, however, that, when $n$ and $r$ are very large, exponential approximations can be substituted for the factorial terms. A modified form of equation (13–9) which is applicable under typical chromatographic conditions ($n \gg r$) is given below:

$$F_{r,n} = \frac{(np)^r e^{-np} e^r}{\sqrt{2\pi r}\, r^r} \quad (\text{for } n \gg r) \qquad (13\text{–}13)$$

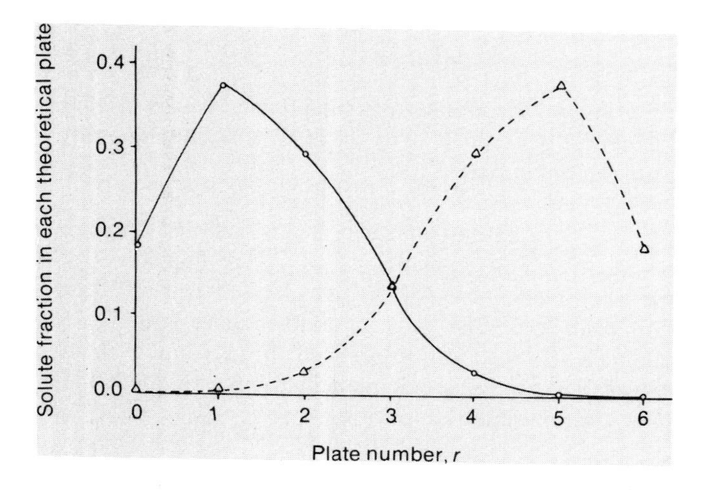


*Figure 13–4.* Fractional amounts of solute in each plate of a chromatographic column after the addition of six $\Delta V_M$ units of mobile phase. In this example, it has been assumed that $V_M/V_S = 10$. The two curves represent two solutes, one with $K = 30$ (—○—), and the other with $K = 3.33$ (—△—).

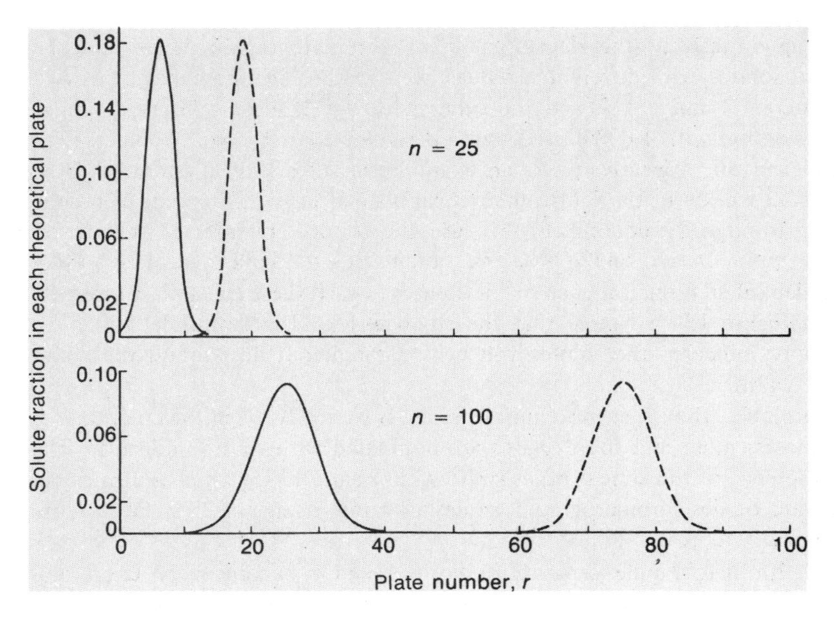


*Figure 13–5.* Fractional amount of solute in each theoretical plate of a chromatographic column after the addition of 25 and 100 $\Delta V_M$ units of mobile phase. Conditions are as specified for Figure 13–4.

If we denote the plate containing the largest solute fraction as $r_{max}$, equation (13–13) can be used to calculate the solute fraction at the zone maximum or "top" of the chromatographic peak. It can be shown that $r_{max} = np$ (in other words, progress of a solute through the column is directly proportional to the fraction of solute in the mobile phase), and substitution of $r = r_{max} = np$ into equation (13–13) yields

$$F_{r_{max},n} = \frac{1}{\sqrt{2\pi np}} = \frac{1}{\sqrt{2\pi r_{max}}} \tag{13–14}$$

Consider the situation in which the peak is at the end of a column containing $N$ theoretical plates, that is, $r_{max} = N$. The absolute quantity of solute in the $N$th theoretical plate can be calculated from

$$\begin{pmatrix}\text{absolute quantity}\\ \text{in plate}\end{pmatrix} = \begin{pmatrix}\text{fractional amount}\\ \text{in plate}\end{pmatrix}\begin{pmatrix}\text{total quantity}\\ \text{in column}\end{pmatrix}$$

For example, if one per cent of the solute is in the $N$th plate ($F_{N,n} = 0.01$) and one millimole of solute was introduced to the column, then 0.01 millimole of solute is in the $N$th plate. Expressing this mathematically, we can define $Q_{N,n}$ as the quantity in the $N$th plate after $n$ transfers and write

$$Q_{N,n} = F_{N,n}(m) = \frac{m}{\sqrt{2\pi N}} \tag{13–15}$$

where $m$ is the moles of solute introduced to the column and we have substituted $r_{max} = N$ in equation (13–14).

If the solute moves through a column of $N$ plates in time $t_R$, its rate of movement (plates per unit time) is $N/t_R$. We have calculated above the quantity of material in the $N$th plate when it contains the solute maximum. As that peak leaves the column, the maximum rate of solute escape, $S_{max}$, will be (dimensions shown for clarity)

$$S_{max} = Q_{N,n}\left(\frac{\text{moles}}{\text{plate}}\right)\frac{N}{t_R}\left(\frac{\text{plates}}{\text{unit time}}\right) = \frac{Q_{N,n}}{t_R}N\left(\frac{\text{moles}}{\text{unit time}}\right) = \frac{Nm}{\sqrt{2\pi N}\,t_R} \qquad (13\text{–}16)$$

This expression can be solved for $N$:

$$N = \frac{2\pi(S_{max})^2 t_R^{\,2}}{m^2} \qquad (13\text{–}17)$$

The quantity of solute $m$ (moles introduced to the column) is proportional to the peak area. Assuming that the peak is triangular (choosing some other shape only changes the factor ½) we can write

$$m = k(\text{area}) = k\tfrac{1}{2}ht_w \qquad (13\text{–}18)$$

where $k$ is a constant of proportionality, $t_w$ is the width at the base of the peak, and $h$ is the peak height. The maximum rate of solute escape, $S_{max}$, is proportional to the peak height; that is,

$$S_{max} = kh \qquad (13\text{–}19)$$

where $k$ and $h$ are just as defined above. Substitution of equations (13–18) and (13–19) into equation (13–17) yields

$$N = \frac{2\pi(kh)^2 t_R^{\,2}}{(\frac{1}{2}kht_w)^2} = 8\pi\left(\frac{t_R}{t_w}\right)^2 \qquad (13\text{–}20)$$

If we replace the assumption of triangular peaks with the more realistic Gaussian peak shape, we obtain

$$N = 16\left(\frac{t_R}{t_w}\right)^2 \qquad (13\text{–}21)$$

where $t_w$ is defined as shown in Figure 13–6. Recasting this expression by defining $\tau = t_w/4$, where $\tau$ is the standard deviation of the peak profile in time units, we obtain

$$N = \left(\frac{t_R}{\tau}\right)^2 \qquad (13\text{–}22)$$

Equation (13–21) provides a way in which the number of theoretical plates in a column can be very easily determined from the basic experimental data. In addition, the relationship between peak width and retention time is specified by these equations. We

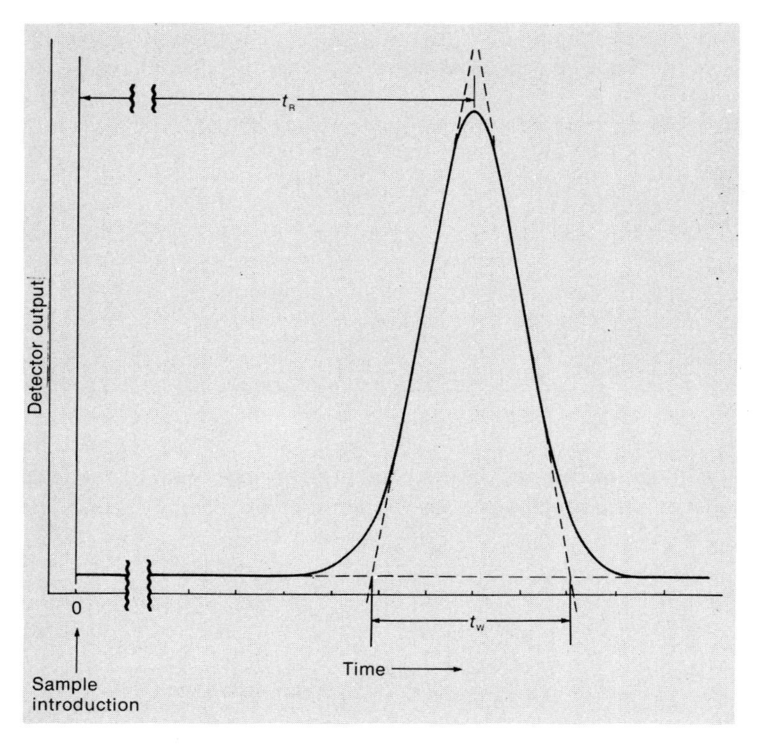


*Figure 13–6.* Definition of peak width, $t_W$, used in determining the efficiency of a chromatographic column. For chromatographic peaks which are perfectly Gaussian in shape, $t_W$ is four standard deviations.

noted earlier that late-eluting zones are broader than early ones, and we now can express this fact quantitatively by rearranging equation (13–21):

$$t_W = \frac{4t_R}{\sqrt{N}} \tag{13–23}$$

The efficiency of any column is best judged by $H$, the height equivalent to a theoretical plate, which is defined by the relation

$$H = \frac{L}{N} \tag{13–24}$$

where $L$ is the length of the column. Substituting for $N$, we obtain

$$H = \frac{\tau^2 L}{t_R^{\,2}} \tag{13–25}$$

Our consideration of the chromatographic column as a series of discrete theoretical plates has provided a very simple expression [equation (13–21)] relating the shape of the chromatographic zone and the number of theoretical plates in the column. Because it provides a highly useful means of describing the performance of a given chromatographic column, it is very widely used. After making a new column, running a test chromatogram, and measuring $t_R$ and $t_W$, an experimenter might substitute the data into equation (13–21), calculate $N$, and proudly announce, "this column has an efficiency equivalent to 24,000 theoretical plates."

### Shortcomings of the Plate Theory

The fact that it provides a rational basis for a convenient measure of column efficiency is almost the only virtue of the theory of chromatography we have discussed thus far. This theory is colloquially termed the "plate theory." Although it does predict that chromatographic zones should be broadened during transmission through the column and that late-eluting peaks should be relatively broad, it achieves these successes in spite of some flagrant deficiencies, which we can briefly enumerate:

1. Built into the theory is the assumption that $K$ is a constant, that is, that the partition isotherm (Figure 12–5) is linear. Particularly in adsorption chromatography, this is often not true.
2. It is assumed that equilibration is rapid compared to the movement of the mobile phase. In fact, it is assumed that, the moment a mobile-phase increment ($\Delta V_M$) is added to the column and all prior mobile-stationary equilibria are upset, diffusion and redistribution of the solute in these phases take place instantaneously. However, diffusion is never instantaneous, and, particularly at high mobile-phase flow-rates, material might be swept along in the mobile phase from one plate to the next before equilibration is complete.
3. It is assumed that spreading of the chromatographic zone by diffusion from one theoretical plate to another ("longitudinal diffusion") does not occur. In effect, it is assumed that diffusion does not occur. To say that this contradicts the second assumption is something of an understatement.
4. The column is assumed to consist of a number of discrete volume elements; and it is assumed that the mobile phase is added in $\Delta V_M$ increments rather than continuously. Both these "assumptions" are, of course, completely untrue.

Because the plate theory assumes a linear isotherm and an ideal diffusion situation, it is termed a "linear ideal" model for the chromatographic process. Other models which overcome various shortcomings of this treatment are termed "non-linear ideal," "linear non-ideal," and "non-linear non-ideal."

**Important Variables Excluded.** From the preceding discussion, it is clear that mobile-phase velocity has an important effect, but nowhere is it considered in the plate theory. Similarly, the dimensions of the phases are of great importance because they determine the distances over which diffusion must occur, and these variables are also excluded. These faults derive from the principal weakness of the treatment, namely, the failure to consider the physical processes which actually occur during zone migration.

## RATE THEORY OF CHROMATOGRAPHY

A treatment which avoids the assumption of instantaneous equilibrium and other shortcomings of the plate theory must pay close attention to the *rates* at which equilibrium can, in reality, be attained under typical chromatographic conditions. In addition, the *rates* of diffusion in the mobile and stationary phases have to be considered. This focus on kinetic aspects has earned for improved chromatographic theories the general designation, "rate theory," though it would be more precise to use the term *linear non-ideal.* The first comprehensive exposition of such a theory was provided by the

Dutch chemists van Deemter, Klinkenberg, and Zuiderweg, in 1956, and the general equation for plate height as a function of mobile-phase velocity is sometimes referred to as the "van Deemter equation." There has been much additional development, in large part due to the work of the American chemist J. Calvin Giddings. The interested reader will find an extensive and highly readable discussion of the details of modern chromatographic theory in his book, cited as supplemental reading at the end of this chapter.

**Organization and Central Idea.** In the rate theory, each of the mechanisms which can contribute to zone-broadening is considered separately. A mathematical expression for the relationship between plate height and variables important in the zone-broadening mechanism is derived and, with due attention to the details of their interaction, these separate expressions are combined to yield a general plate-height equation; that is, total $H$ = ($H$ due to diffusion) + ($H$ due to slow equilibration) + ($H$ due to non-uniform flow patterns) or, more simply,

$$H = H_d + H_e + H_f \tag{13-26}$$

In reality, all these mechanisms act together, and the effects of any one are nearly impossible to isolate. Understanding their combination, however, is not difficult. For example, a solute zone carried along in a moving gas stream will tend to spread out, its sharp boundaries becoming blurred as time passes. A zone which was initially infinitely thin will eventually take on a Gaussian distribution, and the width of the diffusion-spread zone can be characterized in terms of the standard deviation of that distribution. Slow equilibration and flow patterns within the column will also produce spreading into a Gaussian profile. In each case, it is possible to calculate the standard deviation of the resulting zone. Using statistical principles (see Chapter 2), we have a rule for the combination of standard deviations when numbers are added or subtracted. For the sum of three terms

$$\text{sum} = \text{X} + \text{Y} + \text{Z} \tag{13-27}$$

we can write

$$(\sigma_{\text{sum}})^2 = \sigma_{\text{X}}{}^2 + \sigma_{\text{Y}}{}^2 + \sigma_{\text{Z}}{}^2 \tag{13-28}$$

The calculated standard deviation of the sum ($\sigma_{\text{sum}}$), while understandably larger than any of the individual standard deviations ($\sigma_X$, $\sigma_Y$, or $\sigma_Z$), is considerably less than ($\sigma_X + \sigma_Y + \sigma_Z$). This occurs because errors can frequently offset each other. If, for example, the individual X chosen is on the high side of its mean, Y and Z are likely to be on the low side of their means, and the sum not drastically affected. In other words, the chance of observing a large positive or negative deviation *simultaneously* in all three variables is very low. Figure 13–7 depicts the chromatographic parallel. The reader should recognize that the variance of the final zone profile ($\sigma^2$) is the sum of the individual variances, and that the resulting combined standard deviation is $\sqrt{\sigma_d{}^2 + \sigma_e{}^2 + \sigma_f{}^2}$, where the subscripts d, e, and f represent the zone-broadening mechanisms of diffusion, slow equilibration, and flow patterns, respectively. In terms of physical processes, we can easily understand this method of combination by realizing that the individual broadening mechanisms can offset each other, at least partially. For example, a solute molecule speeded up (moved far from the center of the zone) by diffusive spreading might be delayed by slow equilibration or by non-uniform flow effects.

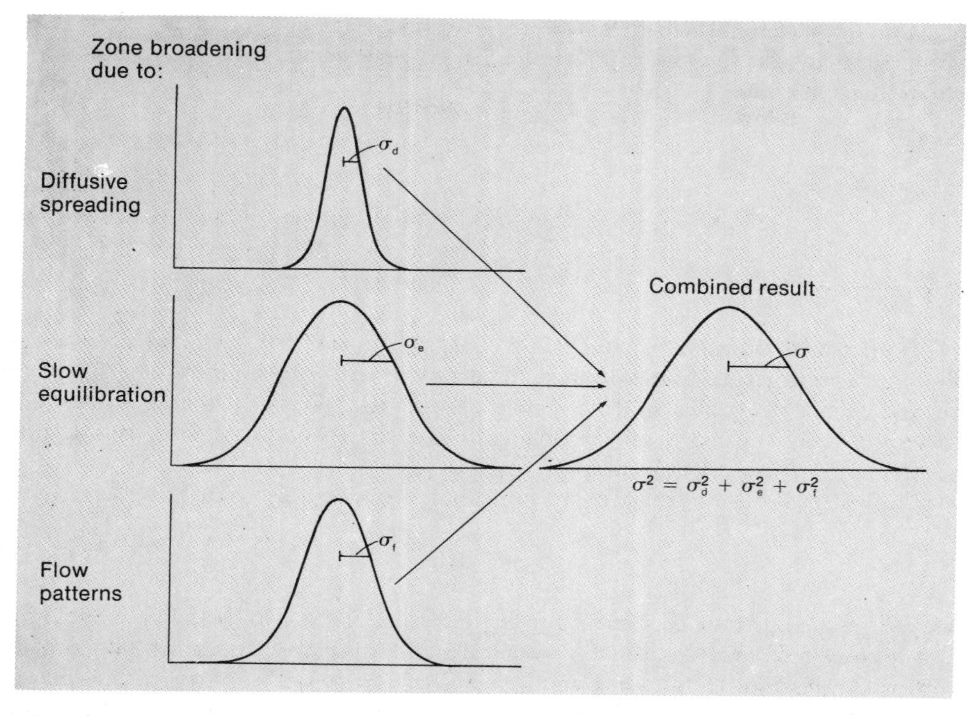


*Figure 13–7.* Combination of the different mechanisms of chromatographic zone-broadening to yield the observed zone profile which results from the simultaneous operation of all these effects.

We can relate the standard deviation, $\sigma$, which represents the size of the zone in the column, to $H$ and $N$ by means of equation (13–25), derived earlier from plate theory:

$$H = \frac{\tau^2 L}{t_R{}^2} \tag{13–25}$$

In this equation, the standard deviation, $\tau$, is expressed in time units, so that $\sigma$ (which has units of length) cannot be directly substituted for $\tau$. To understand the relation between $\tau$ and $\sigma$, we can consider first the relation between retention time and column length. The retention time is simply the column length divided by the rate at which the zone travels through the column (recall that **R**, a ratio, is dimensionless):

$$t_R = \frac{L}{\mathbf{R}v} \tag{13–29}$$

Similarly, the standard deviation in time units is related to the standard deviation in distance units through the expression

$$\tau = \frac{\sigma}{\mathbf{R}v} \tag{13–30}$$

Substituting equations (13–29) and (13–30) into equation (13–25) leads to an uncomplicated equation relating plate height, column length, and the variance of the chromatographic zone:

$$H = \frac{\sigma^2}{L} \tag{13–31}$$

### Mechanisms of Zone-Broadening

**Longitudinal Diffusion.** Figure 13–8 depicts this process in terms of the interdiffusion of two molecular species in some gas stream flowing in an ordinary tube. The same process occurs in chromatographic mobile phases, both gases and liquids. In partition chromatography, diffusion of the solute in the stationary phase also causes zone-broadening, but to a lesser extent.

The variance of a zone broadened by diffusion is given by the general relation

$$\sigma_d{}^2 = 2Dt \tag{13–32}$$

where $D$ is the diffusion coefficient for interdiffusion of the two molecular species (*i.e.*, mobile-phase molecules and solute molecules) and $t$ is the time over which diffusion has occurred. For diffusive spreading in the mobile phase, $(\sigma_{dM})^2 = 2D_M t_M$. It has already been noted that all eluted species spend exactly the same length of time, $t_M$, in the mobile phase and that $t_M = L/v$; thus, $(\sigma_{dM})^2 = 2D_M L/v$. Therefore, the contribution to the plate height due to longitudinal diffusion in the mobile phase ($H_{dM}$) is given by

$$H_{dM} = \frac{(\sigma_{dM})^2}{L} = \frac{2D_M}{v} \tag{13–33}$$

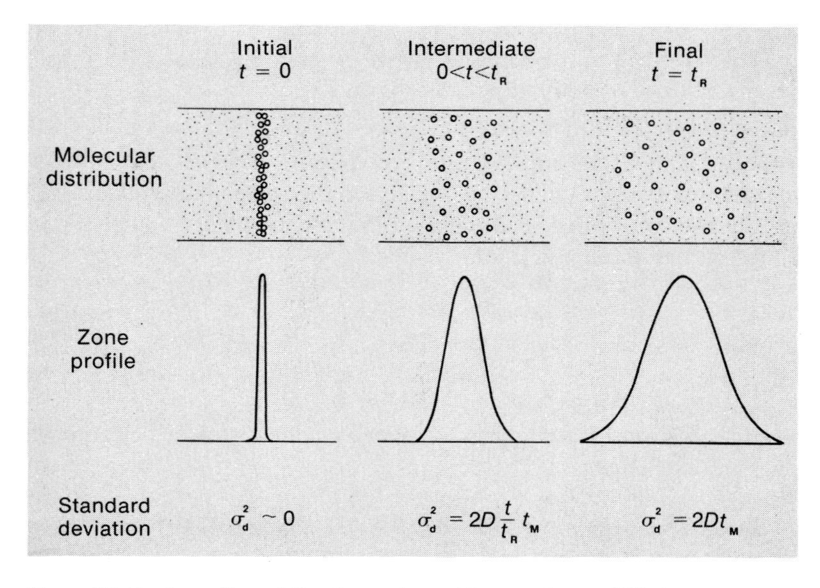


*Figure 13–8.* Spreading of the chromatographic zone due to diffusion, as seen at three different times during elution.

Equation (13–33) shows that, as $v$ decreases, $H_{dM}$ will increase. This happens because a lower velocity gives a longer time for passage through the column and, during that increased time, more zone spreading can occur.

Diffusive spreading in the stationary phase also depends on $v^{-1}$, and, to summarize the relationship between plate height and zone-broadening due to *all* types of longitudinal diffusion, we can write

$$H_d = \frac{B}{v} \tag{13–34}$$

where $B$ is a constant dependent on the diffusion coefficients for the solute in the mobile and stationary phases.

**Slow Equilibration.** In an "ideal" system, the transfer of solute molecules between the mobile and stationary phases is instantaneous. However, such ideality is not achieved for two reasons. First, there is kinetic control of the rate at which molecules can cross the interface, an effect generally termed "sorption-desorption kinetics." Second, there is kinetic control of the rate at which molecules can arrive at the interface and become available for transfer. This second effect is due to the finite rate of diffusion of solute molecules in both the stationary and mobile phases, and is called "diffusion-controlled kinetics." These combined effects force the chromatographic zone in the stationary phase to lag behind that in the mobile phase, trying, as it were, to "catch up," but being prevented from doing so by the rate at which solute molecules can be moved about. The observed chromatographic zone is a combination of the mobile-phase and stationary-phase populations and, when they are offset, the resultant zone is considerably broadened. The degree of offset depends on the velocity of the mobile phase, so that the variance of the chromatographic zone due to slow equilibration must vary directly with $v$. The essential points of this discussion are depicted graphically in Figure 13–9, and we can summarize the relationship between plate height and zone-broadening due to slow equilibration by writing

$$H_e = Cv \tag{13–35}$$

where $C$ is a constant dependent upon the following column characteristics:

1. The rate at which solute exchange can take place across the mobile phase–stationary phase interface.

2. The average time required for a solute molecule to reach the interface from within the stationary phase. This time required for diffusion within the stationary phase depends on the dimensions and nature of the stationary phase itself. In adsorption chromatography, where the "stationary phase" is either non-existent or two-dimensional, depending on your point of view, the diffusion time is zero. In partition chromatography, $C$ depends on $d^2/D_S$, where $d$ is the depth (or thickness) of the stationary-phase layer and $D_S$ is the diffusion coefficient for the solute in the stationary phase.

3. The average time required for a solute molecule to reach the interface from within the mobile phase. By reasoning completely analogous to the preceding arguments, we can conclude that $C$ also depends on $d_p{}^2/D_M$, where $d_p$ is the average diameter of the stationary-phase particles (this dimension, in effect, determines the average size of the channels through which the mobile phase flows and, thus, the average distance which solute molecules must diffuse in the mobile phase), and $D_M$ is the diffusion coefficient for the solute in the mobile phase.

**Flow Patterns.** A packed chromatographic column is a maze through which the mobile phase travels. There are countless pathways by which a molecule might travel

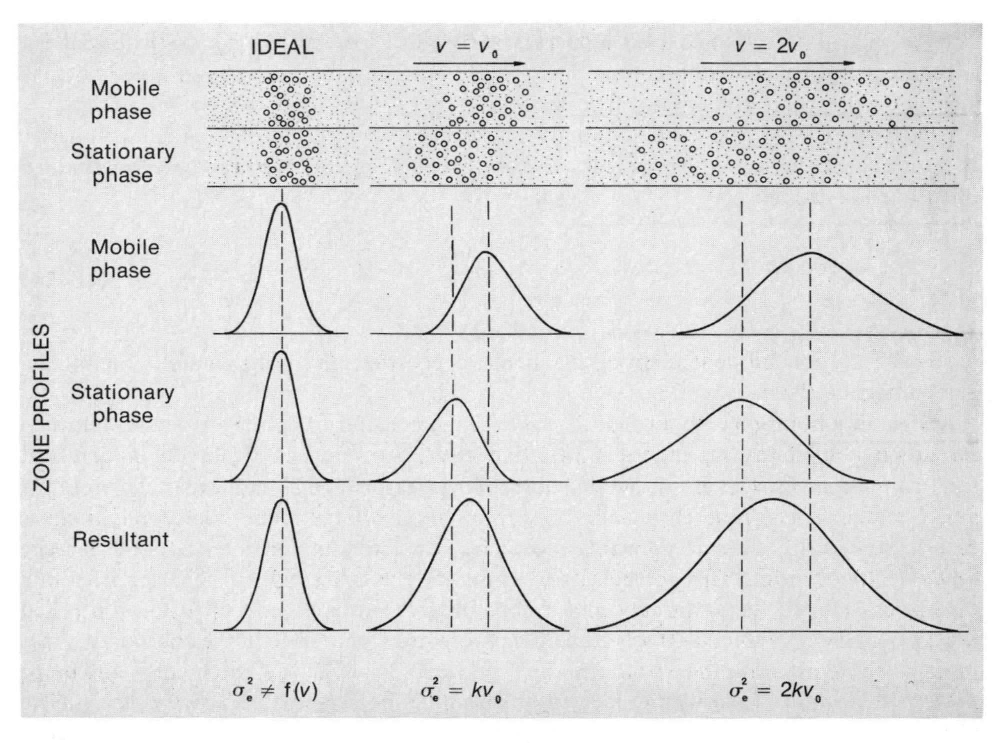


*Figure 13–9.* Spreading of the chromatographic zone due to slow equilibration of solute between the mobile and stationary phases. The first set of profiles depicts the situation assumed by the plate theory, which artificially requires that zone variance be independent of mobile-phase velocity. The second and third sets of profiles realistically depict the situations at some arbitrary mobile-phase velocity $v_0$ and at twice that value. As noted, $\sigma_e^2$ is directly proportional to mobile-phase velocity.

from one end of the column to the other. These pathways are not all the same; some are a bit longer than average and some are a bit shorter. Molecules which travel a short path will emerge, all other factors being equal, a bit sooner than average. If there is a Gaussian distribution of flow-path lengths, this will lead to a Gaussian distribution of travel times. That is, a zone which started as infinitely thin at the column inlet will be dispersed to a Gaussian profile with some characteristic $\sigma_f^2$ at the column outlet. To a first approximation, the relationship between plate height and zone-broadening due to flow patterns can be summarized by writing

$$H_f = A \tag{13–36}$$

where $A$ is a constant dependent upon the structure and arrangement of the column packing.

### Dependence of Plate Height on Mobile-Phase Velocity

**Combination of H Terms.** Substituting equations (13–34), (13–35), and (13–36) into equation (13–26), we can write

$$H = A + \frac{B}{v} + Cv \tag{13–37}$$

In this expression, each contribution to $H$ is expressed in terms of a constant and some power of the mobile-phase velocity. Specifically, $B = 2D_M$; $C$ includes constants depending on sorption-desorption kinetics, $d^2$, $D_S$, $d_p{}^2$, and $D_M$; and $A$ depends on the structure and particle size of the column packing. Figure 13–10 shows a plot of $H$ versus $v$ and indicates how each term contributes to the observed variation of plate height as a function of mobile-phase velocity. Notice that because mobile phase velocity, $v$, is directly proportional to mobile-phase flow-rate, $F$, the variation of $H$ can be fit to an equation having the form

$$H = A' + \frac{B'}{F} + C'F \tag{13–38}$$

where the new set of constants required by the change from $v$ to $F$ is indicated by the primed parameters $A'$, $B'$, and $C'$, which play the same role here as in equation (13–37).

**Optimum Velocity.** Figure 13–10 indicates that there is an optimum velocity or flow rate at which $H$ is minimized. When it is important to obtain the maximum number of theoretical plates in a given separation, the optimum flow rate can be quite accurately found. First, for this purpose, the dependence of plate height upon flow rate can be approximately represented by a relation such as equation (13–38) but containing only two constants

$$H = \frac{X}{F} + YF \tag{13–39}$$

where $X$ and $Y$ are experimentally determined constants. Next, since we desire to minimize $H$ by choosing some optimum flow rate, equation (13–39) can be differentiated and set equal to zero

$$\frac{dH}{dF} = -\frac{X}{F^2} + Y = 0 \tag{13–40}$$

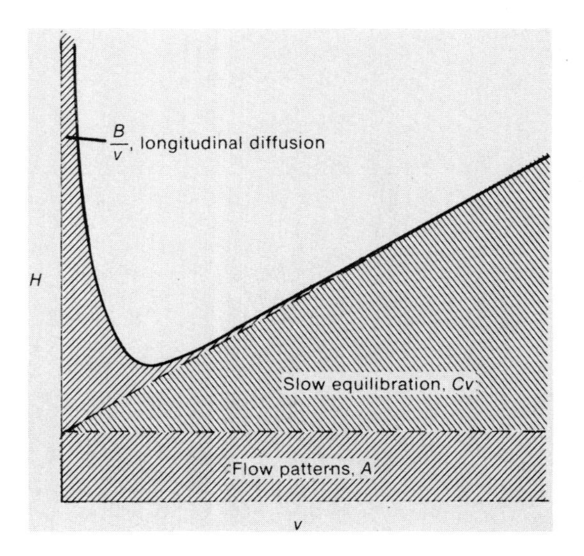


*Figure 13–10.* A graph showing the dependence of plate height on mobile-phase velocity. The contributions of various peak-broadening mechanisms are indicated.

and the resulting expression can be solved for the optimum flow rate:

$$F_{\text{opt}} = \sqrt{\frac{X}{Y}} \tag{13–41}$$

Given two independent sets of observed values for $H$ and $F$, we can determine $X$ and $Y$ by substitution in equation (13–39) and by solution of simultaneous equations. Examples of such calculations will be presented in Chapter 14.

## NON-LINEAR CHROMATOGRAPHY

The rate theory, though it avoids assumptions of instantaneous diffusion, still assumes a linear partition isotherm. We shall now see that non-linearity in this isotherm can exert a major influence on peak shape. It is probably worth repeating that the assumption of linearity is usually well justified for partition chromatography, but is frequently untenable for adsorption chromatography.

Two different types of isotherms and the resulting peak shapes are illustrated in Figure 13–11. Case *a* is the so-called Langmuir adsorption isotherm. The resulting peak shows a substantial "tail" dragging behind the maximum concentration. At the same time, the leading edge of the peak is particularly abrupt. The shape of the isotherm requires that, as the total quantity of solute increases, the fraction in the mobile phase increases and that, accordingly, the areas of highest concentration migrate with the greatest velocity. Thus, a Gaussian peak is transformed into the shape shown in a way that can be remembered thus: "the fast-moving high-concentration center catches up with

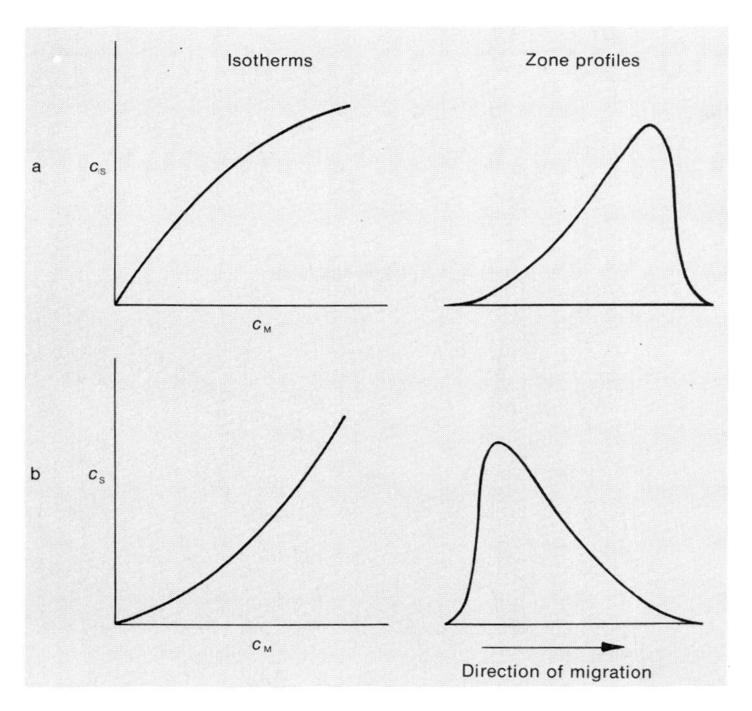


*Figure 13–11.* Non-linear partition isotherms and the resulting chromatographic zone profiles.

the relatively slow-moving front and moves far ahead of the relatively slow-moving tail." A mirror-image argument describes case *b*, which is sometimes observed when a column is overloaded in gas-liquid partition chromatography. The same shape can be attributed to limited solubility of solute in the stationary phase. In either case, the solute apparently acts as its own stationary phase as total concentrations become very high.

## RESOLUTION

Resolution is a measure of the extent of overlap between two adjacent peaks. As such, it is a measure of the success of a given separation. We can quantitatively define resolution ($\mathcal{R}$) by writing

$$\mathcal{R} = \frac{\text{peak separation}}{\text{peak width}} = \frac{\text{distance between zone maxima}}{4\sigma} \qquad (13\text{–}42)$$

We need only be careful that the distance of separation and the zone width are both expressed in the same units. In some forms of chromatography, distance might be convenient; in most cases of elution chromatography performed on columns, time units are most convenient.

The properties of $\mathcal{R}$ are reviewed in Figure 13–12, which shows the appearance of chromatograms representing various degrees of resolution. The extent of cross-contamination which would result in each case (if we assume peaks of equal size) can be easily calculated from the normal curve of error. If a pair of peaks with $\mathcal{R} = 0.5$ (that is, two peaks separated by one-half peak width or $2\sigma$) is split down the middle, and each half is collected and analyzed, it will be found that each peak contains 84.13 per cent of its major component and 15.87 per cent of its minor component. This large degree of overlap is shown by the first pair of peaks in Figure 13–12. For $\mathcal{R} = 1.0$, a resolution which is considered sufficient for most practical purposes, each peak contains 97.73 per cent of the major component and 2.27 per cent of the minor component. The degree of resolution required for quantitative recovery in the usual sense, and the resolution referred to as "complete" by many chromatographers, is $\mathcal{R} = 1.5$. In this instance, the peak maxima are separated by six standard deviation units, and each peak contains 99.87 per cent of the major component and only 0.13 per cent of the minor component. The case in Figure 13–12 for which $\mathcal{R} = 2.0$ makes the interesting point that $(\mathcal{R} - 1)$ peaks can be inserted in the center of any pair resolved with resolution $\mathcal{R}$. For example, if $\mathcal{R} = 4.0$, three peaks can be added between the resolved pair without degrading resolution below $\mathcal{R} = 1.0$.

It is useful to relate $\mathcal{R}$ to experimental variables. The distance traveled by any zone is $L$ or $\mathbf{R}vt$, and so the distance between zone maxima, $\Delta L$, is $(\Delta\mathbf{R})vt$. Making the further substitution from equation (13–31) that $\sigma = \sqrt{HL}$, we obtain

$$\mathcal{R} = \frac{(\Delta\mathbf{R})vt}{4\sqrt{HL}} \qquad (13\text{–}43)$$

which can be simplified by substitution of $vt = L/\mathbf{R}$

$$\mathcal{R} = \frac{\Delta\mathbf{R}}{4\mathbf{R}}\sqrt{\frac{L}{H}} \qquad (13\text{–}44)$$

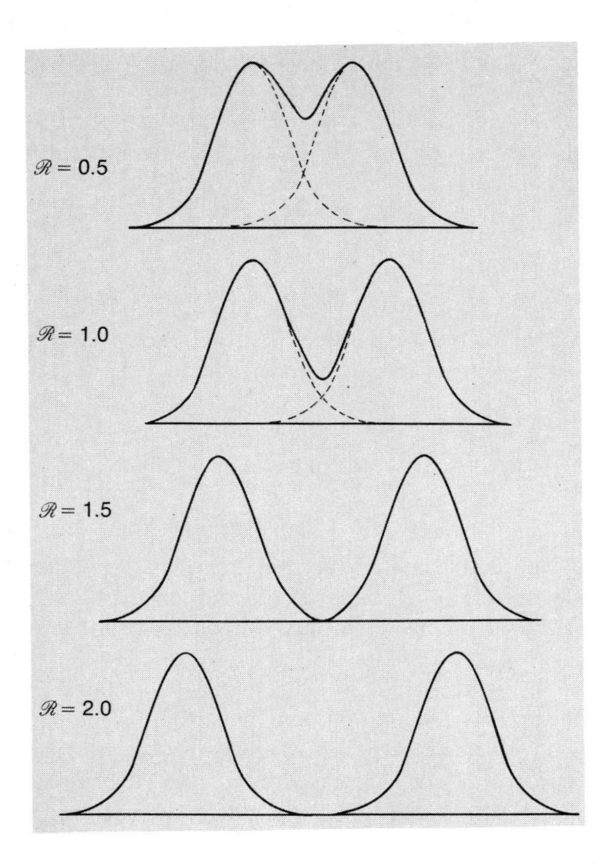


*Figure 13–12.* Pairs of chromatographic peaks showing various degrees of resolution, that is, various values of the parameter $\mathcal{R}$, defined and discussed in the text.

or, since $H = L/N$,

$$\mathcal{R} = \frac{\Delta \mathbf{R}}{4\mathbf{R}} \sqrt{N} \tag{13–45}$$

Because $\mathcal{R}$ is proportional to $\sqrt{N}$, a doubling of resolution will require the number of theoretical plates to be quadrupled. The dependence of $\mathcal{R}$ on the relative retention $(\Delta \mathbf{R}/\mathbf{R})$ is not surprising, but it is important to examine this situation in greater detail. Although the derivation is lengthy and must be omitted here, the preceding equation can be rewritten as

$$\mathcal{R} = \frac{\sqrt{N}\,\Delta K}{4K}(1 - \mathbf{R}) \tag{13–46}$$

Now it can be seen that, for a given relative difference in partition coefficients, $\Delta K/K$, $\mathcal{R}$ depends not only on $\sqrt{N}$, but also on $(1 - \mathbf{R})$. That is, the smaller $\mathbf{R}$ is, the better the resolution. Zones that travel rapidly will be poorly separated; or, to put it differently, zones that travel rapidly require a substantially greater number of theoretical plates for attainment of the same resolution. For example, inserting the values $\mathcal{R} = 1.0$, $\Delta K/K = 0.1$, and $\mathbf{R} = 0.5$ in the expression above, we find that $N = 6400$. For the same $\mathcal{R}$ and $\Delta K/K$, but with $\mathbf{R} = 10^{-2}$, we obtain $N = 1600$. The situation revealed in this discussion is summarized in Figure 13–13, which presents a plot of $N$, in this case the

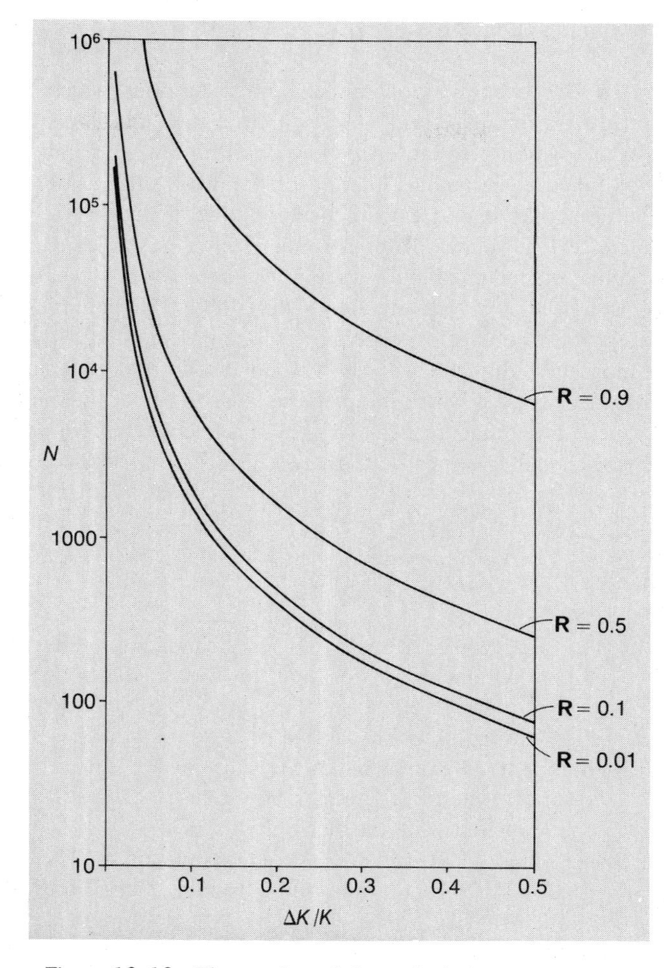


*Figure 13–13.* The number of theoretical plates required for a separation with resolution $\mathcal{R}$ = 1.0 plotted as a function of $\Delta K/K$ and **R**. Notice that $N$ increases greatly as **R** increases, and that, accordingly, maximum resolution can only be obtained in conjunction with substantial retention.

required number of plates for $\mathcal{R}$ = 1.0, as a function of $\Delta K/K$ for various values of **R**. The particularly sharp dependence of $N$ on **R**, as **R** becomes large, leads to a definition of the number of effective theoretical plates which is well worth keeping in mind:

$$\text{number of effective theoretical plates} = N(1 - \mathbf{R})^2$$

It might appear that this situation condemns certain separations to the realm of impossibility. This is not necessarily so. It must be remembered that, while **R** and $K$ are not independent, **R** and the ratio $\Delta K/K$ are. For example, in many cases, **R** can be substantially reduced if the temperature of the chromatographic column is decreased. In other situations, one can decrease **R** by changing the stationary phase (increasing its affinity for the solutes).

### Chromatographic Systems

In this chapter, we have examined theories which account for the spreading of chromatographic zones. These theories are fundamental to any understanding of chromatography. Because they are applicable to all forms of chromatography, they serve the immensely valuable function of impressing the reader with the unity of this field. In addition, they have great practical value in giving a good explanation of the importance and effect of many different experimental variables. However, it is important not to become preoccupied with theoretical aspects, or to take the view that everything of importance in chromatography goes on in the column. We have already seen that a detector and recording system are vital adjuncts in a chromatographic measurement, and a moment's thought will show that the chromatographic process is only part of an analytical *system* which couples the steps of separation and quantitative measurement. It is these integrated systems which find such extensive practical application in modern chemical analysis. Specific examples are discussed in Chapter 14.

## QUESTIONS AND PROBLEMS

1. The partition coefficient, $K$, for substance A in a particular chromatographic column is greater than that for substance B. Which compound is more strongly retained in the chromatographic column?
2. The time required for passage of the mobile phase through a particular column is 25 min. What is the value of **R** for some solute which has a retention time of 261 min? How much time does the solute spend in the mobile phase and in the stationary phase?
3. The value of **R** for a particular solute on a certain chromatographic column is 0.1. The volume of mobile phase in the column, $V_M$, is 2.0 ml. What is the value of $t_S$ for the solute when the flow rate of the mobile phase is 10 ml/min?
4. For the system described in problem 3, what is the value of $K$ for the solute if $V_S$ is 0.5 ml?
5. A column 10 cm long is operated with a mobile-phase velocity of 0.01 cm/sec. Component A requires 40 min for elution. What fraction of the time required for its elution does A spend in the mobile phase? What is the value of **R** for this compound?
6. In gas chromatography, the velocity of the mobile phase can be measured directly if one injects some solute like methane which is entirely unretained by the stationary phase. On a capillary column 50 m long, the retention time of methane is 71.5 sec and the retention time of $n$-heptadecane is 12.6 min. (a) What is the velocity of the mobile phase? (b) What is the value of **R** for the $n$-heptadecane zone? (c) What is the velocity of the $n$-heptadecane zone?
7. Write an expression for the retention time ($t_R$) of some component in terms of the times ($t_M$ and $t_S$) which that component spends in the mobile and stationary phases.
8. Under fixed conditions in a particular gas-liquid partition chromatographic column, substance A is eluted with **R** = 0.5 and $V_R$ = 100 ml. The flow rate of the mobile phase must remain constant, but $V_S$ (the amount of liquid stationary phase) can be changed from its existing

value of 1.5 ml. By what factor must $V_S$ be changed in order to double $V_R$? Is this factor generally applicable any time $V_R$ is to be doubled, or does it apply only in this particular case?

9. Recast equation (13–7) in terms of flow rate of the mobile phase ($F$, ml/sec) and the retention time ($t_R$, sec) instead of the retention volume ($V_R$, ml).
10. Following the example of equation (13–8), write an expression for the adjusted retention time ($t'_R$) in terms of $t_R$ and $t_M$.
11. Two components have adjusted retention times (see problem 10) of 15 and 20 min on a particular chromatographic column. On a different column with a larger stationary-phase volume, the first component has an adjusted retention time of 22 min. What will be the adjusted retention time of the second component on this column?
12. In a particular liquid-liquid chromatographic column, compound A has $K = 10$, and compound B has $K = 15$. The column has $V_S = 0.5$ ml and $V_M = 1.5$ ml, and is operated with a mobile-phase flow-rate of 0.5 ml/min. Calculate $V_R$, $t_R$, and **R** for each component.
13. In the column described in problem 12, the volume of the inert support is 15 ml. Calculate the mobile-phase velocity given a total column length of 30 cm.
14. Make a schematic sketch of a chromatographic column, showing the imaginary divisions between theoretical plates and labeling the incremental phase volumes $\Delta V_M$ and $\Delta V_S$. Use the sketch to show how the mobile phase moves through the column as $n$ is increased.
15. In the derivation of equation (13–20) which relates $N$, $t_R$, and $t_w$, it is assumed that the chromatographic peaks are triangular. Repeat the derivation, assuming instead that the peaks are rectangular with a width $t_w$ centered on retention time $t_R$.
16. A chromatographic column is tested and found to produce a peak having a Gaussian shape and a width of 40 sec at a retention time of 25 min. How many theoretical plates does the column have under the conditions of the test?
17. Some chromatographic columns can be operated at efficiencies corresponding to $10^5$ theoretical plates. Calculate the peak widths obtained from such a column at retention times of 100, 1000, and $10^4$ sec. Assume a Gaussian peak-shape.
18. If the column described in problem 16 is two meters long, what is the height equivalent to a theoretical plate in this case?
19. Explain in your own words the differences between the plate theory and the rate theory of chromatography.
20. Explain in your own words why the rate theory treats the individual zone-broadening mechanisms in terms of their standard deviations, and tell how those standard deviations combine.
21. Some dye is dumped into a well. One month later the dye first appears in another well 12 meters away. In the absence of any possible effects of ground-water flow, and if the soil is assumed to be uniform in texture, how long will it be before the dye appears in a third well 78 meters distant from the first?
22. If the length of a column is doubled, by what factor will zone-broadening due to longitudinal diffusion increase? Why?

23. Two gas-liquid chromatographic columns differ only in that one has a greater amount of liquid stationary phase than the other. Will their optimum mobile-phase velocities differ? If so, how and why?
24. Explain why uniformity of column packing and the use of a narrow range of particle sizes are important in obtaining highest column efficiency.
25. The diffusion coefficient ($D_M$) for *n*-octane in helium at 30°C and 1 atmosphere pressure is 0.248 $cm^2\ sec^{-1}$. In nitrogen, the same constant has the value of 0.0726 $cm^2\ sec^{-1}$. Notice which terms in equation (13–37) are affected by this difference, and sketch two lines on a graph of $H$ versus $v$ (see Figure 13–10) showing how this difference will change the shape and location of the curve representing the dependence of column efficiency on mobile-phase velocity. Which gas will allow faster mobile-phase velocities?
26. A particular gas-liquid chromatographic column two meters in length has an efficiency of 2450 theoretical plates at a flow rate of 15 ml/min and an efficiency of 2200 theoretical plates at a flow rate of 40 ml/min. What is the optimum flow rate, and approximately what efficiency should be obtainable at that flow rate?
27. A particular gas-liquid chromatographic column two meters in length is tested at three different flow rates using helium as the mobile phase and found to have the following performance characteristics:

| methane | *n*-octadecane | |
|---|---|---|
| $t_R$ | $t_R$ | $t_w$ |
| 18.2 sec | 2020 sec | 223 sec |
| 8.0 sec | 888 sec | 99 sec |
| 5.0 sec | 558 sec | 68 sec |

(a) Determine the mobile-phase velocity for each of the runs. (b) Determine the number of theoretical plates and the value of $H$ for each of the runs. (c) By solving simultaneous equations, find the values for the constants in an equation of the form $H = A + B/v + Cv$ and graph the result. (d) Through what range of mobile-phase velocities can 90 per cent of the column efficiency be retained? (e) What is the optimum mobile-phase velocity? (f) A particular separation requires 1100 theoretical plates. What is the fastest mobile-phase velocity at which this can be achieved, and how much time will be saved by running at the maximum flow rate instead of at the optimum?

28. A particular chromatographic column has an efficiency corresponding to 4200 theoretical plates and has retention times for octadecane and 2-methylheptadecane of 15.05 and 14.82 min, respectively. To what degree can these compounds be resolved on this column? How many theoretical plates would be required for unit resolution at those retention times?
29. On a liquid-solid chromatographic column one meter in length, operating with an efficiency of $10^4$ theoretical plates, the retention times of $\alpha$-cholestane and $\beta$-cholestane are 4025 and 4100 sec, respectively. If these two compounds are to be separated with unit resolution, how many theoretical plates will be required? How long a column of this same type will be required in order to achieve this resolution if $H = 0.1$ mm?

## SUGGESTIONS FOR ADDITIONAL READING

1. J. C. Giddings: *Dynamics of Chromatography*. Part I (*Principles and Theory*), Dekker, New York, 1965, Chapters 1, 2, and 7.
2. H. Purnell: *Gas Chromatography*. John Wiley and Sons, New York, 1962, Chapter 7.

# 14 CHROMATOGRAPHIC SYSTEMS

In this chapter we will discuss four techniques that find wide application in modern chemical analysis. The first, thin-layer chromatography (TLC), is close to the ultimate in simplicity and convenience. In terms of its separation principle, thin-layer chromatography furnishes an example of adsorption chromatography. In terms of its method of operation, it is the only representative of plane (as opposed to column) chromatography that we shall discuss. The remaining techniques – gas-liquid chromatography (GLC), ion-exchange chromatography, and molecular-exclusion chromatography – offer three different separation mechanisms.

## THIN-LAYER CHROMATOGRAPHY

### Principles

**Technique of Operation and Separation.** In the earliest days of chromatography, very simple systems were used. Frequently, a rudimentary liquid-solid chromatographic column like that shown in Figure 13–1 was eluted until the bands due to colored sample components were visibly separated on the column. At that stage, one performed the "analysis" simply by noting the sizes and positions of the bands in the column. Crude as this technique might seem, it is also simple and, in chemical analysis, there is no greater virtue save accuracy itself. The chromatographic medium itself furnishes the visual record of the separation – there are no liquid or gas fractions to collect and analyze, and no detector outputs to plot or record – and the chemist sees at a glance what is happening.

In 1944, A. J. P. Martin and co-workers developed an even simpler technique in which the column was replaced by a strip of paper. The cellulose fibers of the paper (which contain a considerable amount of water) serve as a stationary phase, and an organic solvent mobile phase travels across the paper by capillary flow (as a liquid migrates in a blotter). Colored solutes can be seen directly, and many others can be detected if the paper chromatogram is sprayed with some chemical reagent which will mark the zones by color formation. The astonishing simplicity of paper chromatography led to its wide acceptance and application. It has the great disadvantage, however, of offering no choice of stationary phase. If the compounds to be separated are not retained by hydrous cellulose fibers, some other technique must be used.

Thin-layer chromatography allows a reasonable choice of adsorbents and has a few additional advantages (such as speed and freedom from contamination) over paper chromatography. At the same time, it offers comparable simplicity. In this technique, the adsorbent is spread on a glass plate (or sheet of plastic or aluminum foil) in a layer about 0.25 mm thick. The plate is placed on one edge in a chamber to which mobile phase is added to a depth of a few millimeters. From that edge and submerged portion of the plate, the mobile phase slowly covers the rest of the plate by capillary flow. The sample is spotted on the plate a short distance above the submerged edge and is eluted by the flow of mobile phase, which carries sample components upward through the adsorbent bed. Development of the chromatogram is stopped when the leading edge of the mobile phase (known in thin-layer chromatography as the "solvent front") gets near the top of the plate. The individual chromatographic zones are located and characterized by means of their $\mathbf{R_f}$ values:

$$\mathbf{R_f} = \frac{\text{distance traveled by solute}}{\text{distance traveled by mobile phase}} \tag{14–1}$$

Though closely related to **R**, the retardation factor (page 324), $\mathbf{R_f}$ is not exactly equal to it. A ratio of velocities is expressed by **R**, a ratio of distances by $\mathbf{R_f}$. In fact, the solvent front migrates a bit faster than the bulk of the mobile phase, so that $\mathbf{R} \sim 1.15\,\mathbf{R_f}$.

Because a verbal description of the details of a technique like thin-layer chromatography is inevitably unsatisfactory, we have been as brief as possible. As a supplement, Figures 14–1 to 14–3 review the procedure and should be carefully studied.

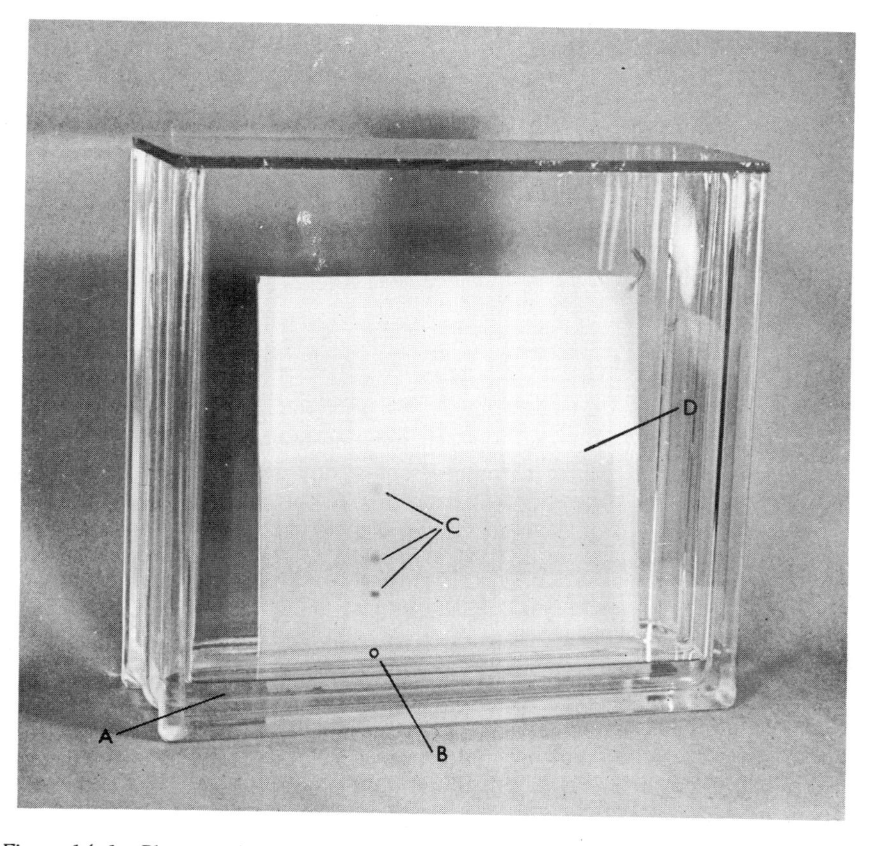


*Figure 14–1.* Photograph of a thin-layer chromatogram in the process of development. The letters indicate: *A*, mobile phase; *B*, point of solute application; *C*, solute zones; and *D*, solvent front. Development is carried out in an enclosed chamber in order to ensure saturation of the atmosphere with solvent vapor. As a further precaution against unsaturation of the gas phase, the chamber should be lined with mobile-phase soaked filter paper, but this has been omitted here for the sake of visibility.

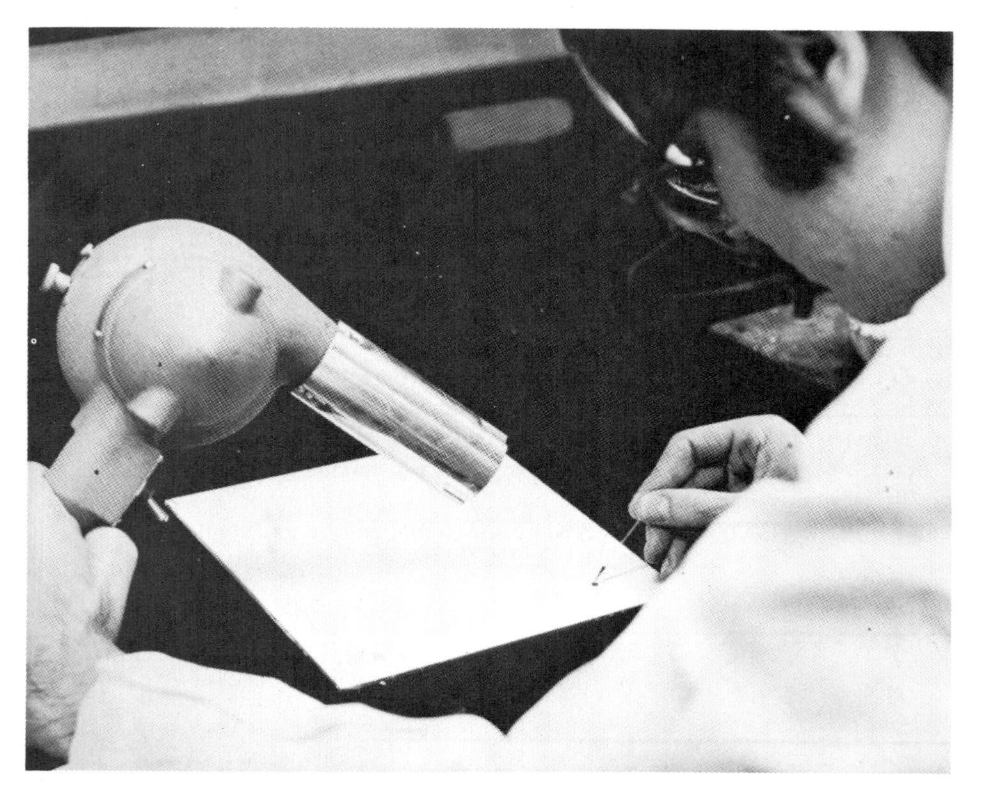

*Figure 14–2.* The process of solute application in thin-layer chromatography. The solutes are dissolved in a suitable solvent which is applied to the layer in tiny drops in order to keep the initial zone as small as possible. In order to get an appreciable quantity of material in one spot, drops are repetitively applied to the plate, the solvent being evaporated with a hot-air blower as shown.

## The Adsorbent Bed

**Properties of Adsorbents.** We can use this opportunity for a general discussion of adsorbents used in liquid chromatography, not necessarily limiting ourselves to the thin-layer technique. An invaluable reference in this more general context is the book by L. R. Snyder, cited at the end of this chapter.

The most important characteristic of any adsorbent is its **activity**, that is, the extent to which it will retain solute species. The activity is a composite of many factors – strength and density of active sites on the adsorbent surface, surface area, and water content of the adsorbent material all play important roles. In addition, the *apparent* activity of an adsorbent in any given chromatographic application depends strongly on the mobile phase. Interactions between solutes and the mobile phase and between the mobile phase and the adsorbent can be as strong as solute-adsorbent interactions, and the degree of retention of any given solute in an adsorption chromatographic separation can accordingly be grossly affected by the nature of the mobile phase. Generalizations are elusive, and it is best to focus on some individual adsorbents. The great majority of adsorption chromatographic separations are performed with either silica or alumina. A third "adsorbent" discussed below is cellulose, included here because of its utility in thin-layer chromatography. Cellulose is not truly an adsorbent, but rather a partitioning medium.

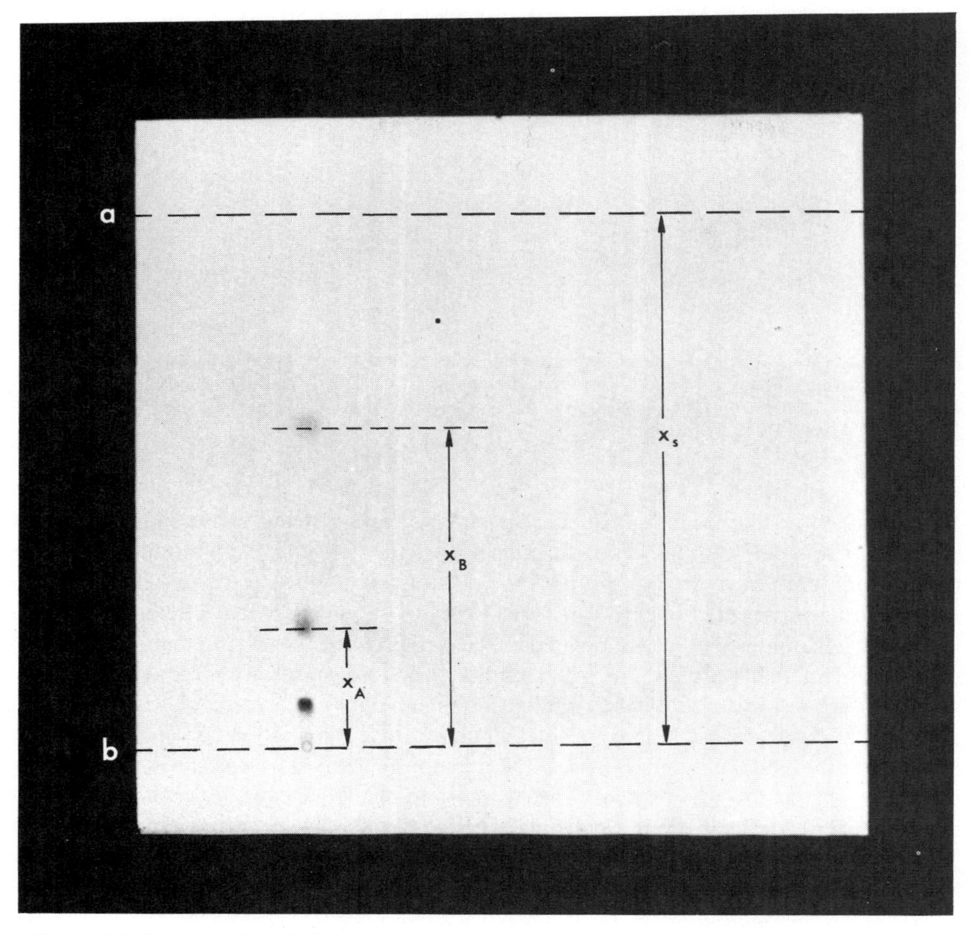


*Figure 14–3.* A developed thin-layer chromatogram. In this case, the solutes happen to be dyes which are visible without further treatment. Other methods for detection of solute zones are reviewed in the text. The indicated distances allow calculation of the $R_f$ values for the solutes. Line *b* marks the starting line, line *a* the solvent front; $R_{fA} = x_A/x_S$ and $R_{fB} = x_B/x_S$.

Silica. Silica, silica gel, and silicic acid are all terms commonly applied to the material produced by acidification of silicate solutions followed by washing and drying of the resulting gel. The product has an enormous surface area (~500 $m^2/g$), and the surface consists of Si–OH groups spaced at intervals of about 5 Å. These groups are the active sites, and the activity of a given batch of silica adsorbent depends on the number and availability of these sites. Unless surface-adsorbed water has been driven off by activation of the silica through heating at 150 to 200°C in air, many of the sites are masked. At higher temperatures, activity is lost because further dehydration occurs due to loss of water from adjacent hydroxyl groups, the product being an inactive Si–O–Si linkage. These steps can be reversed by exposure of the silica to water or water vapor. Above 400°C, surface area is permanently lost by the knitting together of surfaces, as shown in Figure 14–4, and high activity cannot be regained by any degree of rehydration. The active sites interact with polar solutes chiefly by means of hydrogen bonding, with the solute acting as an electron donor. Thus, alkenes are more strongly retained than alkanes and, because of the acidic nature of silica gel, basic substances are particularly strongly held.

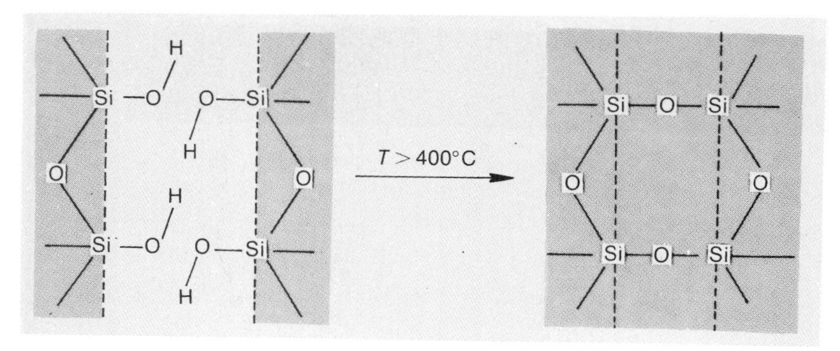


*Figure 14–4.* Mechanism for the dehydration and consequent deactivation of silica at temperatures above 400°C. (Redrawn, with permission, from L. R. Snyder: *Principles of Adsorption Chromatography*. Dekker, New York, 1968, p. 158, by courtesy of Marcel Dekker, Inc.)

Alumina. In sharp contrast to silica, alumina shows a steady increase in activity on heating up to 1000°C. This is good evidence that the active sites on alumina, whatever their nature, are not hydroxyl groups. In fact, alumina exhibits retention characteristics which can be interpreted in terms of three types of solute-adsorbent interactions. In the first type, adsorbates with easily polarizable electrons (Lewis bases) interact with the very strong positive fields around the $Al^{3+}$ ions in the alumina surface. Interactions of the second type result in preferential retention of acidic solutes, apparently by proton donation to basic sites ($O^{2-}$ ions) in the alumina surface. Finally, certain aromatic species are apparently retained by charge-transfer interaction, possibly by electron transfer to the positively charged $Al^{3+}$ sites mentioned earlier. It seems generally true that, when two solutes differ chiefly in their electronic structure, alumina is the preferred adsorbent. Thus, it is superior to silica in its resolution of related aromatic and olefinic hydrocarbons, for example.

Highly active alumina can be produced by overnight air drying of the material at 400°C. It is common practice to produce alumina of lower activity by redistribution of a known amount of water on the adsorbent surface. In practice, this is easily done by adding a known amount of liquid to a weighed amount of alumina and shaking the mixture to speed equilibration. Chromatographic alumina commonly has a surface area of 150 $m^2/g$. A monolayer of water amounts to $3.5 \times 10^{-4}$ $ml/m^2$. Thus, the addition of about 2.5 weight per cent of water to alumina will produce an adsorbent with about half a monolayer of water, and this is found to be a good general-purpose adsorbent. There is, in fact, a widely used scale of alumina activities based on this rehydration technique. Known as the **Brockmann scale** after its developer, it defines fully activated alumina as activity grade I; alumina with 3 per cent added water as activity grade II; with 6 per cent water as grade III; with 10 per cent water as grade IV; and with 15 per cent water as grade V. Notice that grade V corresponds to the presence of about three monolayers of water. The utility of such an adsorbent is somewhat questionable. Certainly the largest effects on adsorbent activity are expected in the range from 0 to 6 per cent water.

Alumina, particularly the fully activated material, is far more likely than silica to catalyze chemical reactions in chromatographed solutes. Since alumina is somewhat alkaline, base-labile molecules are especially likely to be affected; double-bond migrations and even ring-expansions have been reported. This tendency to sabotage is one reason why alumina takes second place to silica in popularity as an adsorbent.

Modified Adsorbents. A technique particularly useful in thin-layer chromatography is the addition of silver nitrate (5 to 15 per cent by weight) to silica gel layers. The resulting mixed adsorbent shows strong selectivity, preferentially retaining species which contain double bonds and are thus able to form $\pi$-complexes with the silver ions. The relative retention even depends on the *number* of double bonds in each molecule, and the adsorbent finds wide application in the characterization of lipid fractions.

Cellulose. The situation for cellulose is complex. It has been commonly regarded as a partitioning medium, the cellulose structure serving chiefly to hold microscopic pools of water into which solutes can be partitioned. It is inescapable, however, that the cellulose must also function to some degree as an adsorbent, and any rationalizations of observed patterns of retention must take this behavior into account. In general, cellulose is useful for the chromatographic separation of substances that are too polar to escape irreversible adsorption on more active media such as silica and alumina. The use of paper chromatography for the separation of such substances has persisted long after thin-layer chromatography has taken over most other applications. Thin-layer plates can be spread with layers of microcrystalline cellulose, and thus mimic the properties of paper, at the same time offering increased speed of separation.

**Preparation of the Adsorbent Bed.** Thin layers are prepared by spreading water slurries of the adsorbent on a carefully cleaned supporting material, usually glass. After the spreading operation is complete, the plates are allowed to dry in air for about 30 minutes before being activated. The time required for drying and activation depends somewhat on the thickness of the layer, which is usually 0.25 mm for analytical separations or 1.0 mm for preparative-scale separations.

The particle size of the adsorbent is very small, generally less than 15 $\mu$m. The adsorbent bed is thus about 30 particles thick. Layers of pure adsorbent have very poor mechanical characteristics and are prone to disaggregation. A binder can be incorporated to cure this problem, and the resulting plates can (in the case of a starch binder) even be written on without flaking. The most common binder is calcium sulfate, which is added to the adsorbent as the hemihydrate, $CaSO_4 \cdot \frac{1}{2} H_2O$ (plaster of Paris), in the amount of 5 to 20 per cent by weight. When mixed with water, the binder is hydrated, and rapidly sets to form the rigid dihydrate, $CaSO_4 \cdot 2 H_2O$ (gypsum). Dehydration of the dihydrate begins at 100°C, a phenomenon that quite severely limits the conditions used for activation of the adsorbent. Drying at 110°C for 30 minutes is commonly employed, but above 130°C the binder is dehydrated so fast that plates are useless after only 30 minutes. Common starch (added to the extent of about 5 per cent by weight) is a superior binder in that it provides more durable layers which, in addition, allow much faster migration of the mobile phase. Starch has the great disadvantage of being an organic material which reacts with many of the more drastic zone-detection reagents.

### The Mobile Phase

**The Role of the Mobile Phase in Adsorption Chromatography.** It is dangerous to regard the mobile phase merely as the solvent which carries the sample through the chromatographic bed. At the same time the mobile phase carries out that function, it also takes a very active role in the adsorption process by competing with the sample components for space in the adsorbed layer of molecules. If the solvent is strongly adsorbed, it will be displaced by the sample only with difficulty. In this case, the solute will spend a relatively great amount of time in the mobile phase and will travel through the chromatographic bed rapidly. Following this example to its logical conclusion, we see that, for a given solute and adsorbent, the rate of solute migration depends on the strength with which the solvent is adsorbed, and *not on the solubility of the sample in the mobile phase*. This fact will often upset the ordinary view of solvent-solute pairing. For example, alkanes are more soluble in heptane than in methanol ("like dissolves like"), but methanol will move alkanes (or anything else, for that matter) through an adsorption chromatographic column much faster than heptane, simply because the methanol is so strongly adsorbed that the alkanes do not get a chance to "sit down."

After this observation, we can add that solvent purity is of great importance. For example, heptane contaminated with 1 per cent methanol would be a *very different* eluent from pure heptane. Even small amounts of a polar impurity can have a large effect owing to their interaction with the adsorbent surface and, thus, can alter migration rates very significantly.

### Sample Application

Figure 14–2 shows how a micropipet is used to apply a spot of sample solution to a thin-layer chromatography plate. The maximum amount of material which should be applied in a single spot is about 200 $\mu$g for 0.25-mm thick layers. For preparative-scale separations, the sample is spread out in a line of closely spaced dots or applied as a streak across the bottom of the plate by some specialized device. In this way, sample loads of about 1 mg per centimeter of streak length can be handled on plates with adsorbent layers that are 1 mm thick. After elution, the zones are scraped from the plate and the separated sample components are recovered by liquid extraction of the adsorbent.

### Detection of Chromatographic Zones

Many compounds either are colored and can be directly seen on the chromatographic plate or are naturally fluorescent and can be seen as bright zones of fluorescence when an ultraviolet lamp is shined on the plate. However, the great majority of compounds are colorless and require some special detection technique. One universally applicable method is the spraying of the plate with concentrated sulfuric acid, followed by heating to 100°C. Any organic compounds can then be seen as black charred zones. (The technique is not applicable to layers with a starch binder.) Chromic or nitric acid can be added to speed the charring. A second nearly universal technique is exposure of the plate to iodine vapor. This is easily accomplished by placing a few iodine crystals along with the plate in an empty development chamber. The iodine vapor *does not* react with the organic compounds, but is preferentially condensed in the solute zones, apparently being bound to the organic molecules purely by van der Waals interactions. Thus, the zones appear as orange spots against a yellow background. The process is entirely reversible – once the plate is removed from the tank, the spots disappear in a few minutes – so that this technique is particularly useful when solute zones must be recovered for further analysis.

Organic compounds which absorb ultraviolet light at wavelengths above 230 nm can be detected through the use of fluorescent indicators. The entire plate is treated with the indicator, and the whole surface is thus fluorescent when exposed to ultraviolet light. However, in regions where some compound absorbs the incident radiation, the fluorescence is greatly reduced and the solute zones are seen as dark spots against the fluorescent background. A very convenient fluorescent indicator is a mixture of zinc silicate, which is excited by light in the wavelength region from 230 to 290 nm, and zinc sulfide, which is excited in the region from 330 to 390 nm. These inorganic compounds are simply mixed (1 per cent or less by weight) with the stationary-phase material prior to the spreading of the layers. Fluorescent organic dyes, which operate on the same principle, can be sprayed onto the plate after development.

For many compounds, there are specific reagents which can be sprayed on the plate to allow zone detection by formation of some distinctive color; an example is the use of ninhydrin to detect α-amino acids. When applicable, these specific reagents offer great advantages. First, unresolved species which do not form colors with the detection reagent will not interfere in the analysis. Second, the process of detection becomes, in addition, one of identification. If the spot does not react properly, the chemist is altered to the fact that something has gone awry.

### Applications

A simple example illustrates the role which thin-layer chromatography often plays in sample isolation. In Chapter 12 (page 316), a procedure for the extraction of a non-saponifiable lipid fraction was outlined. In work in the authors' laboratories, the object of this procedure has been the eventual isolation of a sterol fraction from the non-saponifiable lipids. In a typical experiment, 27.15 g of wet algal cells yielded 94 mg of neutral lipids, including trace amounts of sterols, organic pigments, and alkanes. Next, the 94-mg neutral lipid fraction was further separated by means of chromatography on a silica gel column. Column chromatography was used in place of thin-layer chromatography because only a crude separation was sought and because the large amount of material would require the preparation and development of many TLC plates. The column was eluted first with *n*-heptane in order to remove alkanes. The second eluent, benzene, carried through the sterols and a few pigments, giving a fraction weighing 14.6 mg. This smaller fraction, now stripped of material markedly more polar than sterols, was chromatographed on a silica gel thin-layer plate, a one-to-one mixture of diethyl ether and *n*-heptane serving as the eluent. In this latter step, any sterols present were well resolved from the interfering pigments, and the sterol band could be scraped from the plate for further examination by means of gas-liquid chromatography.

## GAS-LIQUID CHROMATOGRAPHY

### General Aspects

As its name implies, this form of chromatography makes use of a gaseous mobile phase and a liquid stationary phase. It is frequently lumped together with gas-solid chromatography, and there are many books published with the title, "Gas Chromatography," indicating a survey of both techniques.

The essential parts of a gas chromatograph are shown in Figure 14–5. The mobile phase is known as the "carrier gas," and it is convenient to follow the operation of the instrument by tracing the flow of carrier gas. Inertness is the main requirement for a carrier gas. Although air was used as the carrier gas in some of the very earliest experiments, it has been largely superseded by nitrogen and helium; and a typical gas chromatography laboratory is cluttered with high-pressure gas cylinders. It is necessary that the flow of the mobile phase be accurately controlled. This is accomplished with the aid of special valves which can, depending on their type, be adjusted either to provide a constant flow-rate of gas independent of downstream flow resistance or, alternatively, to provide a constant output carrier-gas pressure. The required pressure is seldom more than a few atmospheres. Flow rates depend critically on column size, but are typically about 20 to 50 ml min$^{-1}$.

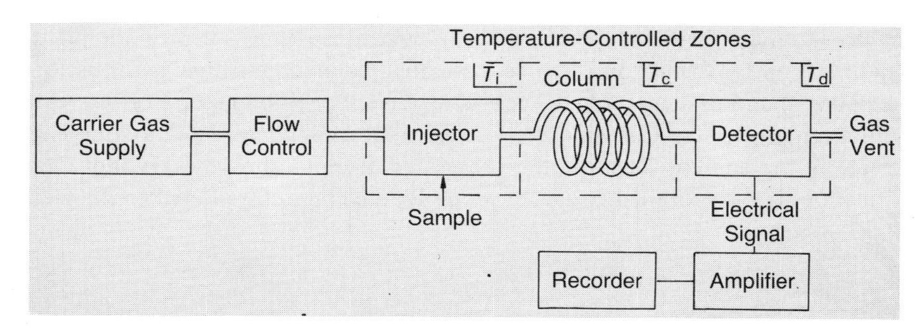


*Figure 14–5.* Block diagram of a gas chromatograph.

On its way to the column, the carrier gas passes through the injector. This small volume is heated to cause rapid vaporization of the sample, which is injected with a syringe through a small silicone rubber disc known as the "septum." The injector is commonly about 50°C hotter than the column. The volume of sample solution injected is usually several microliters or less. The carrier gas sweeps the sample vapor directly onto the column, which is connected to the injector as closely as possible. In the column, the liquid stationary phase is dispersed as a thin layer on an inert supporting material. Columns are ordinarily between 30 cm and 5 m in length, and most have an internal diameter of about 2.5 mm. The inert supporting material has an average particle diameter of about 160 $\mu$m. Because the liquid film is very unevenly distributed on the supporting material, it is meaningless to talk about film thickness. Instead, the extent of liquid phase *loading* is specified by the weight of liquid in comparison to the support; for example, a column with 2 per cent liquid phase has 2 g of liquid for every 100 g of support. Liquid-phase loadings are usually 0.5 to 5 per cent in analytical columns. Column temperatures depend entirely on the volatility of the sample and can range from – 196°C (liquid-nitrogen temperature) to 350°C. Both extremes are rare. A precisely thermostated oven for control of the column temperature is a vital part of the chromatograph. Unfortunately, only a few liquid phases can be taken above 250°C without themselves vaporizing and swamping the detector.

The limitation on column temperature has the effect of setting a lower limit on sample volatility. For example, the largest straight-chain alkane that can be chromatographed without difficulty is *n*-triacontane ($C_{30}H_{62}$). We can gain some idea of the generosity of this limitation by noting that the boiling point of *n*-triacontane is 450°C at atmospheric pressure and that at 304°C its vapor pressure is only 15 torr. If great attention is given to uniform heating of all parts of the chromatograph and if specially designed columns are used, it is possible to chromatograph normal alkanes as large as $C_{60}H_{122}$. This cannot be taken as an indication of the ultimate limit of gas-liquid chromatography, however, because most compounds do not possess the thermal stability of alkanes and are thermally degraded before their vapor pressures become high enough (greater than about 5 torr) to allow successful chromatography.

The detector is an instrument for the continuous measurement of the solute content of the carrier-gas stream. The detector sensitivity (that is, the minimum detectable sample flow) establishes the sensitivity of gas-liquid chromatography as an analytical technique, and it is remarkably high. Detectors in common use can measure a flow of $10^{-10}$ g sec$^{-1}$, whereas the most sensitive types can detect a solute flow of $10^{-14}$ g sec$^{-1}$. From the detector, the carrier gas (and sample, if it has not been consumed by the detector) is simply vented to the atmosphere. The electrical signal from the detector carries the vital information about solute zones. This information is usually displayed on a graphic recorder which automatically presents a graph of solute flow versus time.

### Solute Retention and Column Operation

**Retention Volume.** In Chapter 13, we derived an expression relating retention volume ($V_R$) and the partition coefficient ($K$),

$$V_R = V_M + KV_S \tag{14-2}$$

where $V_M$ and $V_S$ are the volumes of the mobile and stationary phases, respectively. In gas chromatography, a special circumstance prevails – the mobile phase is a compressible fluid with a volume dependent on pressure and temperature. In practice, the carrier-gas flow is measured at room temperature and the prevailing atmospheric pressure and must be corrected to take these factors into account. Correction for the difference in temperature inside and outside the column can be made by means of the equation

$$F_c = F_a \frac{T_c}{T_a} \tag{14-3}$$

where $F$ is the flow rate of carrier gas (ml min$^{-1}$) and $T$ is the absolute temperature (°K), the subscripts a and c referring to *a*mbient laboratory conditions and *c*olumn conditions, respectively. Correction for the difference between atmospheric and column pressures requires first a knowledge of the average pressure within the column. The average column pressure $\bar{p}$ is not simply $(p_i + p_o)/2$, where $p_i$ and $p_o$ are the column inlet and outlet pressures, respectively. The proper expression is*

$$\bar{p} = \frac{2}{3}\left(\frac{p_i^3 - p_o^3}{p_i^2 - p_o^2}\right) \tag{14-4}$$

or

$$\frac{p_o}{\bar{p}} = j = \frac{3}{2}\left[\frac{(p_i/p_o)^2 - 1}{(p_i/p_o)^3 - 1}\right] \tag{14-5}$$

The **corrected retention volume**, $V_R^\circ$, is the retention volume measured at average column pressure and is given by

$$V_R^\circ = \frac{p_o}{\bar{p}} V_R = jV_R = jF_c t_R \tag{14-6}$$

where $t_R$ is the retention time.

We have earlier noted [equation (13–8), page 325] that a direct proportionality exists between $V_R'$, the **adjusted retention volume**, and $K$. For a non-compressible mobile phase

$$V_R' = KV_S = V_R - V_M \quad \text{(non-compressible mobile phase)} \tag{14-7}$$

---

*A derivation of this equation is given in most specialized texts on gas chromatography. See, for example, H. Purnell: *Gas Chromatography*. John Wiley and Sons, New York, 1962, p. 67.

In gas chromatography, the adjusted retention volume also requires correction, and the **net retention volume** ($V_N$) is given by

$$V_N = jV'_R = KV_S \quad \text{(compressible mobile phase)} \qquad (14\text{–}8)$$

In practice, $V'_R$ is easily determined directly, allowing simple calculation of $V_N$. It is only necessary to inject some substance which passes through the column unretained and to watch for the detector signal. For many detectors, a few microliters of air serves this purpose; otherwise, one can use a tiny amount of methane obtained by flushing the injecting syringe in the gas stream from a laboratory gas cock. The time between injection and the appearance of a signal is $t_M$, the time required for passage of the mobile phase through the column. Because the air or methane is unretained and spends no time in the stationary phase, it follows that

$$V_M = t_M F_c \qquad (14\text{–}9)$$

and that

$$V'_R = (t_R - t_M) F_c = t'_R F_c \qquad (14\text{–}10)$$

where $t'_R$ is the **adjusted retention time.**

**Dependence of Retention Volume on Column Temperature.** The partition coefficient, $K$, is a thermodynamic quantity which depends on temperature as do all equilibrium constants:

$$\ln K = -\frac{\Delta G^0}{RT} \qquad (14\text{–}11)$$

where $\Delta G^0$ is the standard free-energy change for the gas-liquid partitioning equilibrium and $R$ is the universal gas constant. Rearranging equation (14–8) and substituting equation (14–10), we can write

$$K = \frac{V_N}{V_S} = j\frac{V'_R}{V_S} = j\frac{t'_R F_c}{V_S} \qquad (14\text{–}12)$$

If we consider the behavior of various compounds under constant chromatographic conditions, so that $V_S$, $j$, and $F_c$ are all constants, we can write equation (14–12) in terms of arbitrary constants and natural logarithms and obtain

$$\ln K = \ln V_N + (\text{const})_1 = \ln V'_R + (\text{const})_2 = \ln t'_R + (\text{const})_3 \qquad (14\text{–}13)$$

Combining equations (14–11) and (14–13), we see that each of the retention variables in equation (14–13) depends logarithmically on the reciprocal of the absolute temperature:

$$\ln V_N \propto \ln V'_R \propto \ln t'_R \propto \frac{1}{T} \qquad (14\text{–}14)$$

A more precise statement of this relationship takes the form

$$\log t'_R = \frac{A}{T} + B \qquad (14\text{–}15)$$

where $A$ and $B$ are arbitrary constants.

*Example 14–1.* For a particular column, the adjusted retention time ($t'_R$) of $n$-pentacosane is 20 min at $T_c = 220°C$ and 30 min at 210°C. If $t'_R$ must not exceed 1 hr, what is the lowest possible column temperature?

To evaluate the constants $A$ and $B$, we must solve two simultaneous equations obtained by substituting the known values of $t'_R$ and $T$ into equation (14–15):

$$\log 20 = \frac{A}{493} + B$$

$$\log 30 = \frac{A}{483} + B$$

$$A = 4140 \qquad \text{and} \qquad B = -7.1$$

Insertion of these values of $A$ and $B$ and of the maximum $t'_R$ of 60 min into equation (14–15) allows calculation of the desired result:

$$\log 60 = \frac{4140}{T} - 7.1$$

$$T = 466°K = 193°C$$

**Dependence of Retention Volume on Vapor Pressure – Homologous Series of Compounds.** Each compound in a *homologous series* differs from its predecessor by one $-CH_2-$group. Thus, the $n$-alkanoic acids, for example, form a homologous series, as do the $n$-alkanes. With any such series, there is a remarkable regularity in vapor pressures such that, at any given $T$,

$$\log p_i^\circ = k_1 + k_2 i \tag{14–16}$$

where $p_i^\circ$ is the saturation vapor pressure of the compound with $i$ carbon atoms and where $k_1$ and $k_2$ are empirical constants. This imposes a regularity on values of $K$ for the members of any homologous series. If we assume ideal solution behavior such that, for any compound, $p_M = X_S p_i^\circ$, where $p_M$ is the solute partial pressure in the mobile phase and $X_S$ is the mole fraction of solute in the stationary phase, then we find that $K$ is inversely proportional to the saturation vapor pressure:

$$K = \frac{c_S}{c_M} \propto \frac{X_S}{p_M} = \frac{1}{p_i^\circ} \tag{14–17}$$

Expressing the relationship shown in equation (14-17) in logarithmic terms and noting that the values of $k_2$ in equation (14–16) are always negative, we obtain

$$\log K \propto -\log p_i^\circ \propto i \tag{14–18}$$

Combining equations (14–13) and (14–18), we can write

$$\log t'_R = Ai + B \tag{14–19}$$

Figure 14–6 demonstrates this relationship for the homologous series of methyl esters of $n$-alkanoic acids. An interesting application of this expression is given in problem 10 at the end of this chapter.

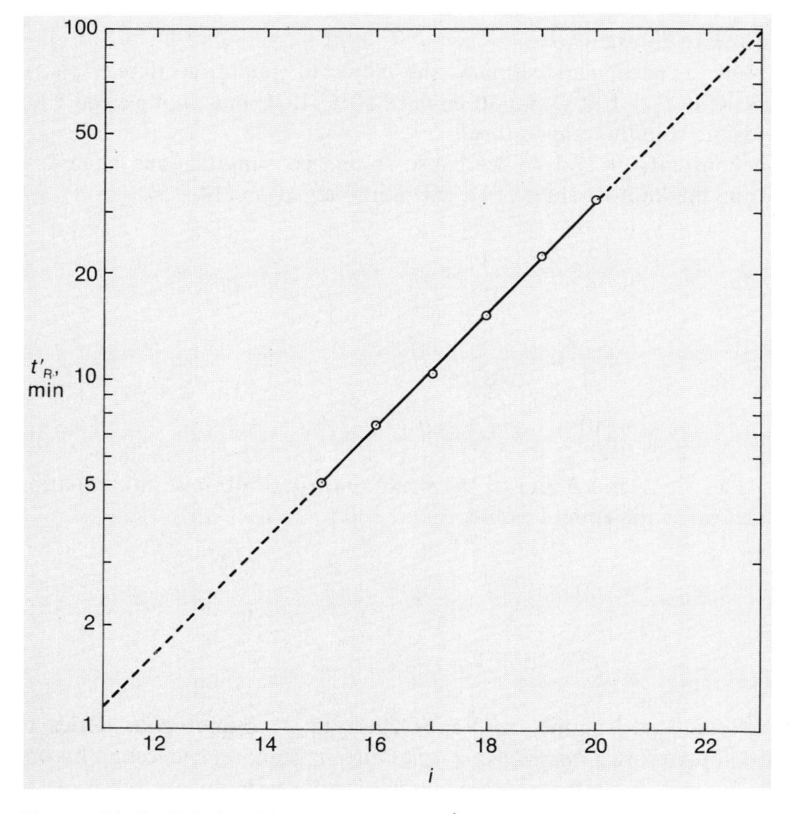


*Figure 14–6.* Relationship between log $t'_R$ and $i$ for compounds in the homologous series $H_3C(CH_2)_iCO_2CH_3$. Although the absolute values of $t'_R$ are dependent on experimental conditions, the logarithmic interdependence shown here will hold for members of the series on any given column.

**Relative Retention and Retention Indices.** When compounds are to be identified by their retention characteristics, one technique is to report **relative retention**, $r$:

$$r_{a,b} = \frac{t'_{Ra}}{t'_{Rb}} = \frac{V'_{Ra}}{V'_{Rb}} = \frac{V_{Na}}{V_{Nb}} = \frac{K_a}{K_b} \tag{14–20}$$

Notice that, although $r_{a,b}$ can be very simply determined from adjusted retention times, it is equal to the fundamental quantity $K_a/K_b$, and is thus independent of $V_M$, $V_S$, $p_i$, $p_o$, and $F_c$. It is, however, dependent on the nature of the liquid phase and on column temperature. The compound $b$ is usually some widely available standard. For example, retention characteristics of sterols are reported relative to cholesterol. One worker in laboratory A might read that chemists working in laboratory B had found some sterol of unknown structure having $r_{x,\text{cholesterol}} = 1.25$ for a specified column type and temperature. Using a column with the same liquid phase at the same temperature, chemists in laboratory A could check to see if, by chance, they had the same new compound. In fact, it would not be too unusual for two different compounds to both have the same $r$. Measurements with two or three different liquid phases would be required before any confidence in the identity is justified.

The **retention index** ($I$) furnishes the best method for reporting any compound's gas chromatographic characteristics. It might be noted, for example, that benzene has a retention index of $I = 650$ on a particular column. This means that, under the conditions employed, benzene behaves like an $n$-alkane with 6.50 carbon atoms. This observation would be made by injecting benzene together with $n$-hexane and $n$-heptane and noting that benzene was eluted between the two alkanes. For a chromatogram run at constant column temperature, it would not be exactly halfway. Remember that, in a homologous series, $i$ (the number of carbon atoms) varies directly with $\log t'_R$ and that, therefore, the interpolation between the $n$-alkane peaks must be *logarithmic.* Notice that, for any $n$-alkane, $I = 100\,i$. The $n$-alkanes provide a convenient measuring scale against which all other compounds can be compared.

*Example 14–2.* An unknown compound has an adjusted retention time ($t'_R$) of 11.23 min. The adjusted retention times of $n$-octadecane ($C_{18}H_{38}$) and $n$-eicosane ($C_{20}H_{42}$), co-injected with the unknown, are 10.50 and 14.65 min, respectively. Calculate the retention index ($I$) of the unknown.

A graphical solution is shown in Figure 14–7. This graph is invaluable in illustrating these calculations, but ordinarily the arithmetic solution is used. According to the law of similar triangles,

$$\frac{a}{b} = \frac{A}{B}$$

Notice that $I_{unknown} = 1800 + b$, and that a, A, and B are all known:

$$a = \log 11.23 - \log 10.50 = 0.0292$$

$$A = \log 14.65 - \log 10.50 = 0.1446$$

$$B = 2000 - 1800 = 200$$

$$b = \frac{aB}{A} = \frac{0.0292}{0.1446}(200) = 40.4$$

$$I_{unknown} = 1800 + b = 1840.4$$

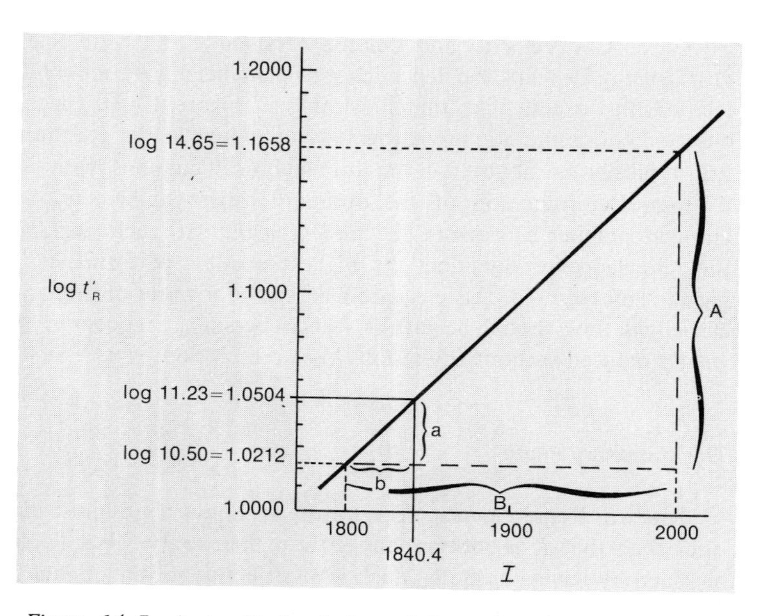


*Figure 14–7.* A graphical solution of Example 14–2. (See text for discussion.)

The retention-index system has many advantages. It is easy to remember the definition of the index in terms of *n*-alkane retention characteristics. The retention index is almost as simple to calculate as an $r$ value, and it conveys some immediate qualitative idea of retention characteristics. Given the $I$ values of a number of unknown compounds, homologous series are obvious. For example, the higher homolog of the compound in Example 14–2 should appear at about $I = 1940$, the lower homolog at about $I = 1740$. Finally, retention indices are less influenced by column temperature than are relative retentions. Equal column temperatures should still be used for comparisons whenever possible, but valuable information can be gained from measurements taken at temperatures as much as 30°C apart.

**Programmed Column Temperature.** A look at Figure 14–6 shows that, for the column temperature chosen, the smallest ester having a retention time greater than one minute is methyl undecanoate, and the largest member of the homologous series having a retention time of less than 100 min is methyl docosanoate. Even if chemical patience extended beyond 100 min, the peak width at such long retention times would be so large that sensitivity would be substantially reduced. From a practical standpoint, not more than nine members of the homologous series could be observed under the conditions used and, even then, the total time required for the separation would be an hour. An excellent alternative mode of operation is available. If the column temperature is increased linearly during a chromatographic analysis, the members of any homologous series will be eluted at approximately equally spaced intervals ($i \propto t'_R$ rather than $i \propto \log t'_R$). This technique, called **programmed-temperature gas chromatography**, offers great advantages for the analysis of mixtures which cover a wide range of compound volatilities. The comparison with isothermal column operation is further illustrated by Figure 14–8, which shows the advantage of programmed-temperature analysis of the homologous series of methyl esters discussed in connection with Figure 14–6. Notice that the retention-time scale is linear, not logarithmic.

Because $t'_R$ is not *exactly* proportional to $i$, the rigorous calculation of retention indices in programmed-temperature chromatography is not possible. However, over any small range (3 carbon numbers), simple non-logarithmic interpolation allows approximate calculation of the retention index.

**Carrier-Gas Velocity and Column Efficiency.** The rate theory of chromatography satisfactorily explains the dependence of plate height, $H$, on mobile-phase velocity, $v$, and relationships exactly like that illustrated in Figure 13–10 (page 339) can generally be observed for gas-liquid chromatography columns. At the optimum carrier-gas velocity, $H$ will typically be about 0.4 mm for analytical columns with an internal diameter of 2.5 mm. Determination of the optimum carrier-gas velocity is easy, and the novice chromatographer carries out the measurement with great zeal, ever thereafter secure in the knowledge of obtaining the highest possible performance. If that performance is higher than required, this elegance can lead to a waste of time. In any practical case, the carrier-gas flow should be just as fast as possible. Frequently, the analysis time can be greatly reduced without any serious loss in resolution.

### The Stationary Phase

**Practical Requirements.** The stationary liquid phase must have molecular characteristics such that $K$ is substantially greater than unity. That is, good retention must be obtained by having a liquid which is an effective solvent for the species to be separated. The liquid phase must have a low volatility so that at high column temperatures it does not "bleed," thereby swamping the detector and eventually ruining the column. The

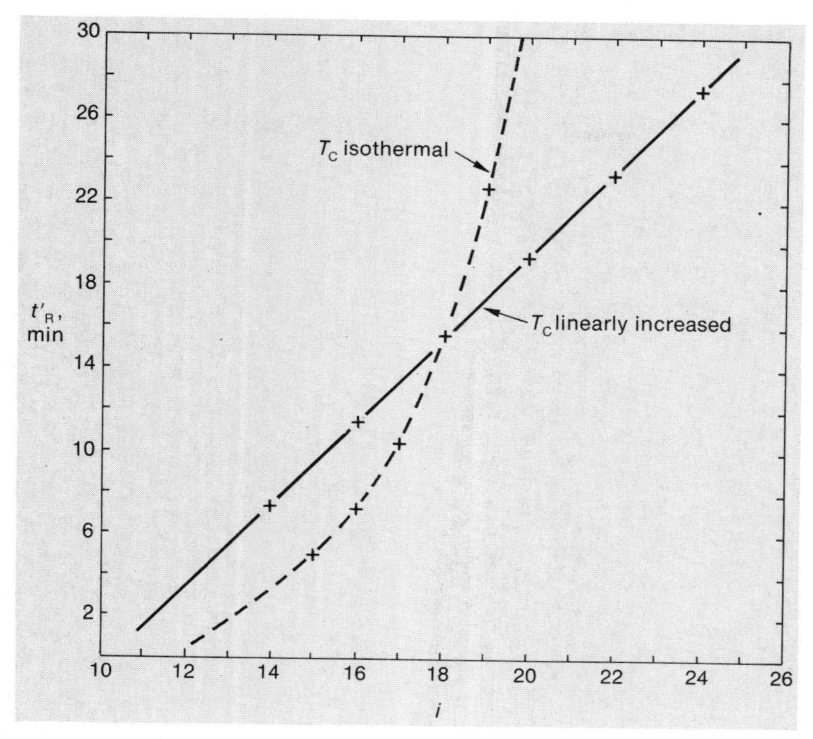


*Figure 14–8.* Adjusted retention time ($t'_R$) as a function of carbon number ($i$) for the homologous series of methyl esters shown in Figure 14–6. In the present situation, the possibility of linear temperature-programming has been introduced, and the graph demonstrates that many more members of the homologous series can be separated in a much shorter time if the column temperature is increased linearly. The dashed line shows the relationship of Figure 14–6 plotted against this linear $t'_R$ scale.

liquid must be dispersed on the solid support as a thin film so that diffusion times of solute molecules within the stationary phase are not excessively long. These rudimentary requirements lead to the common use of high molecular weight polymers dispersed on inorganic supporting particles as gas chromatographic stationary phases.

**Liquid Phases.** A representative selection of examples is shown in Figure 14–9. This is only the tip of the iceberg – at least 600 different materials are presently being sold and used as liquid stationary phases for GLC. However, when the need arises for a liquid phase to accomplish a particular separation, choosing from among this incredible variety is not as difficult as might be expected. First of all, only about 20 of these liquids differ significantly in their practical characteristics. Therefore, the choice is among roughly 20 alternatives, not 600. Secondly, it would (by now) be a rare separation indeed that had not already been accomplished by someone else and reported in the published literature. Such a report is only a starting point; you must ask yourself whether the previous worker's choice was optimal, and learn whether there have been any recent developments which would change the picture, but the information to be gained from prior work is generally extremely helpful.

In the end, the standard rule, "like dissolves like," often forms the best starting point for choice of a liquid phase. Referring to Figure 14–9, for example, we can observe that mixtures of esters are frequently separated on columns employing liquid phase 9, "DEGS" (DEGS is an acronym for diethylene glycol succinate polyester). In this case, *esters* are separated through the use of a poly*ester* liquid phase. Similarly, alkane mixtures are frequently analyzed on columns with an Apiezon L liquid phase (which is, itself,

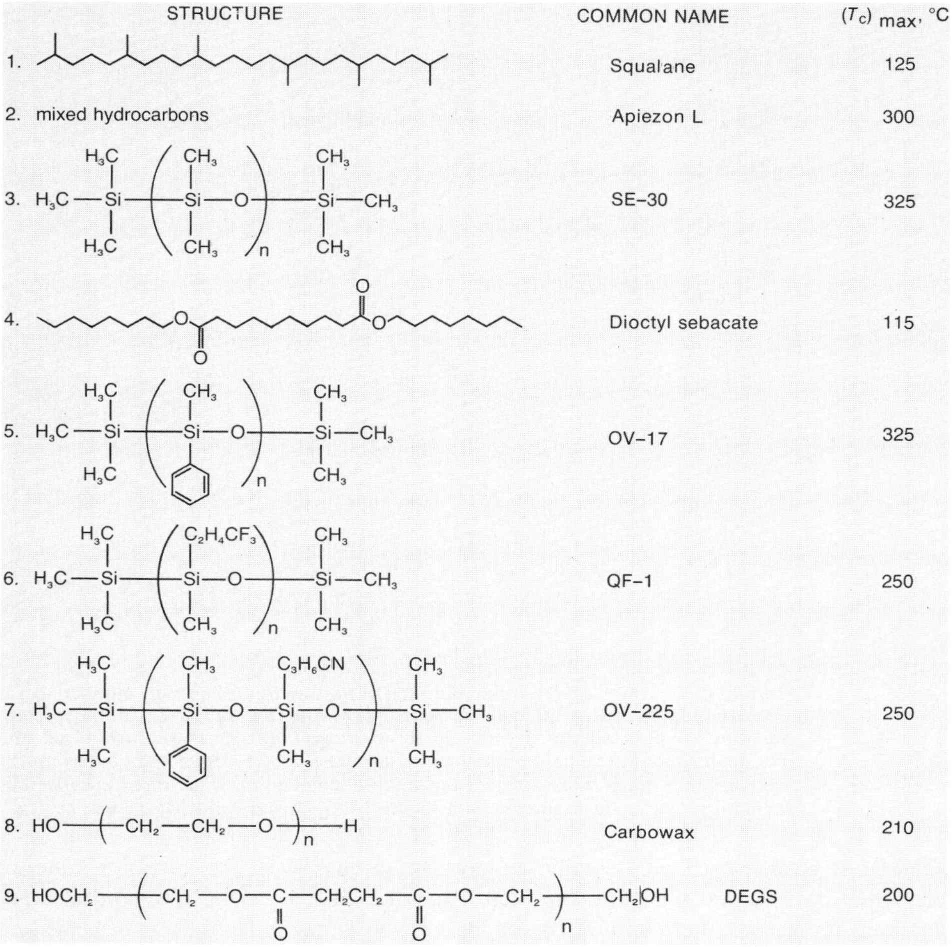

| | STRUCTURE | COMMON NAME | $(T_c)_{max}$, °C |
|---|---|---|---|
| 1. | | Squalane | 125 |
| 2. | mixed hydrocarbons | Apiezon L | 300 |
| 3. | $H_3C—Si(H_3C)(H_3C)—(Si(CH_3)(CH_3)—O)_n—Si(CH_3)(CH_3)—CH_3$ | SE-30 | 325 |
| 4. | | Dioctyl sebacate | 115 |
| 5. | $H_3C—Si(H_3C)(H_3C)—(Si(CH_3)(C_6H_5)—O)_n—Si(CH_3)(CH_3)—CH_3$ | OV-17 | 325 |
| 6. | $H_3C—Si(H_3C)(H_3C)—(Si(C_2H_4CF_3)(CH_3)—O)_n—Si(CH_3)(CH_3)—CH_3$ | QF-1 | 250 |
| 7. | $H_3C—Si(H_3C)(H_3C)—(Si(CH_3)(C_6H_5)—O—Si(C_3H_6CN)(CH_3)—O)_n—Si(CH_3)(CH_3)—CH_3$ | OV-225 | 250 |
| 8. | $HO—(CH_2—CH_2—O)_n—H$ | Carbowax | 210 |
| 9. | $HOCH_2—(CH_2—O—C(=O)—CH_2CH_2—C(=O)—O—CH_2)_n—CH_2OH$ | DEGS | 200 |

*Figure 14–9.* Chemical structures of representative liquid phases used in gas chromatography. The values of $(T_C)_{max}$ indicate the highest temperature at which a given liquid phase can be used without vaporizing to so great an extent that the solute detector is swamped.

made up of alkanes). These examples can be roughly summarized by the observation that the polarity of the liquid phase should match the polarity of the solute. Indeed, it is found that polar substances exhibit broad, poor quality peaks on columns with non-polar liquid phases, whereas non-polar substances give broad peaks in columns with highly polar liquid phases.

Why should obtaining a match between solute polarity and liquid-phase polarity be so important? This amounts to asking "why must the solute dissolve in the liquid phase" and, by restating the question in that way, we have already taken a step toward the answer. Consider the fate of a highly polar solute molecule in the presence of a non-polar liquid phase. The solute will be only very slightly dissolved in the liquid; yet, at the prevailing column temperature, its volatility may be such that it cannot remain completely in the gaseous mobile phase. It will, therefore, condense at least partially, forming an ill-defined third phase. As elution of the column proceeds, the highly polar solute will be moved through the column. However, it will not be cleanly partitioned between the liquid phase and the mobile phase. Instead, the individual molecules will, more-or-less, "rattle" through the column, sticking first to the column wall, then to a bare support particle, then being very reluctantly and occasionally accepted by the liquid phase. This wide variety of starting and stopping mechanisms will give rise to unusually broad chromatographic zones.

The large selection of liquid phases has been developed not because the polarity match between the solute and the liquid phase has to be nearly perfect, but because "polarity" is such a multidimensional property. One pair of "polar" molecules might interact, for example, by means of hydrogen-bonding. Another pair might be equally well termed "polar" on the basis of a dipole-dipole interaction. Many other mechanisms are possible and, for any given class of compounds, several different types of "polarity" might be involved in a unique balance of interaction mechanisms. A single example must suffice. In many naturally occurring steroid mixtures, both ketone and alcohol functional groups are present on the hydrocarbon steroid skeleton. The resulting compounds are of intermediate polarity and might, in principle, be separable on any phase of moderate polarity (*e.g.*, liquid phases 4 through 7 in Figure 14–9). However, with most liquid phases, complete separations are difficult to obtain because the ketones and alcohols are both retarded to the same extent and pass through the column without being resolved. The fluorosilicone, QF-1 (liquid phase 6), finds use because it has the specific property of *selectively* retarding ketones, thus allowing the required ketone-alcohol separation to be accomplished.

**Solid Supports.** Diatomite, the material which goes under the name diatomaceous earth in North America and kieselguhr in Europe – which serves as a filtration aid, inert filler in dynamite, main constituent of firebricks, and even as the abrasive ingredient of toothpaste – is simply silica in another one of its many natural forms. Diatomite is the siliceous residue that remains after the organic constituents of diatoms, a form of microscopic algae, have been stripped away by time and decay. It has a surface area of about $20\ m^2 g^{-1}$ (much lower than silica gel) and a beautiful microscopic structure derived from its biologic origin. Solid supports for gas chromatography are prepared from diatomite by heat treatment.

Heating diatomite to 900°C with a sodium carbonate flux produces a very fragile and glassy particulate material in which the mineral structure has been largely destroyed and the adsorptive activity reduced by a decrease in the surface area to about $0.3\ m^2 g^{-1}$. The material is white and, though sold under many different trade names, it is generally identifiable by the suffix "W." The support can be further treated by being washed with acid and base. Material with very low surface polarity is produced by reaction of the surface–OH groups with "silanizing agents" (see page 373).

An alternate method of heat treatment yields a support which is superior in terms of mechanical strength and, therefore, ease of handling, but which is very inferior in its surface adsorption characteristics. In the manufacture of firebrick, diatomite is mixed with clay, and the mixture is pressed into bricks and fired above 900°C. For use as a chromatographic support, this brick material can be crushed and graded according to size. In the absence of the sodium carbonate flux, much better preservation of the diatomite microstructure is found, and the specific area is about $4\ m^2 g^{-1}$. The material is pink in color, and commercial products generally carry the suffix "P." Deactivation of this support even by careful silanization is essentially impossible, so that the white material is much preferred for use with polar samples.

Zone-broadening due to non-uniform flow patterns depends very much on the regularity of the column packing. For this reason, the support particles are usually size-classified by means of a sieving process so that in any batch the range of particle diameters is small. The size fractions used in chromatography are summarized in Table

14–1, which gives in addition the column diameters for which each size is ordinarily utilized.

**Coating of the Support.** Preparation of stationary phases for gas-liquid chromatography might serve as an example of a laboratory procedure that is simple in theory, but difficult in practice. The liquid phase is taken up in some solvent, the solid support is added to this solution, and the solvent is evaporated, leaving a coating of liquid phase on the solid support. The problem is to avoid uneven distribution of the liquid and, particularly with white diatomite supports, to do this without extensive fracturing of the support by stirring. Procedures which avoid solvent evaporation are more effective. The interested reader can find detailed accounts in the list of references at the end of this chapter.

**Table 14–1. Sizes of Solid Supports Used in Gas-Liquid Chromatography**

| Size Fractions | | | Internal Diameter of Column for which Support is Typically Used, mm |
|---|---|---|---|
| U. S. Standard Screens | | Range of Particle Diameters, μm | |
| Upper mesh size | Lower mesh size | | |
| 45 | 60 | 250–350 | >10 |
| 60 | 80 | 177–250 | 5.5 |
| 80 | 100 | 149–177 | 2.5–5.5 |
| 100 | 120 | 125–149 | 2.5 |

After the carrier solvent for the liquid phase has been removed, the liquid-coated support is packed into a column by pouring in the particles while the column is vibrated and a gas stream is flowing to aid transport of the particles. Then the column is "conditioned" by slow heating to a temperature that is 50°C above its intended maximum temperature of actual use, although the temperature maxima listed in Figure 14–9 should not be exceeded. During this first heating of the column, substantial bleeding of solvent and liquid phase can be expected, so the column must not be attached to the chromatographic detector if gross contamination of the detector is to be avoided.

**Open Tubular Columns.** In columns of very small diameter, the inert supporting material is omitted, and the stationary liquid phase is simply coated on the inside wall of the column tube itself. Band broadening due to flow effects in the column packing is entirely avoided in this way, and the column has, in addition, a much lower resistance to carrier-gas flow. Together, these features substantially increase the speed and efficiency with which separations can be effected. The openness of the column allows enormous lengths (often up to 100 m) and, consequently, astonishing numbers of theoretical plates – as many as $10^6$. An example of open tubular column performance is provided by the chromatogram on the cover of this text. It often happens that high-resolution chromatography reveals the extreme complexity of natural mixtures such as physiological fluids or petroleums. The chromatogram on the cover reveals the presence of over 200 different components and demonstrates dramatically the value of high-resolution open tubular columns. The very small inside diameter of such columns, typically 0.25 mm, has led many workers to call them "capillary" columns, although the aspect of column construction which deserves emphasis is not the smallness of the diameter but, rather, the openness of the bore.

### Injectors

The sample introduction system, like the column, must come into contact with the chemical sample without modifying or degrading it. (Requirements for the design and operation of detectors are not as stringent because, even if the sample is chemically changed by the detector, the separation has already taken place, and useful information on the rate of solute flow can still be obtained). If the sample is injected as a solution and if, inside the injection port, droplets of the solution come in contact with hot metal surfaces, chemical degradation of heat-labile species is virtually assured. In order to avoid this decomposition, many injection ports are glass-lined, the ability of glass to catalyze degradation reactions being substantially less than that of stainless steel. A still better system of "on-column injection" allows for injection of the sample directly into the column packing. The use of glass tubing for columns is very helpful and is mandatory for the successful chromatography of some particularly sensitive samples. The use of glass columns is, however, not as important as the use of glass injectors; the difference is the opportunity for liquid contact in the injector, whereas, within the column, the sample is always in the vapor state.

The silicone rubber septum through which the syringe needle is pushed must withstand temperatures up to 300°C without degradation and, at the same time, must be soft enough to be penetrated easily at much lower temperatures. These are difficult requirements, and the available compromises are all somewhat unsatisfactory. The major problem is that most septa do degrade at least slightly at temperatures above approximately 225°C. The result is "septum bleed," which is at least as troublesome as bleeding of the liquid phase. It is particularly vexing in programmed-temperature operation, where septum bleed appears not as a steady background, but instead as a sequence of highly confusing peaks which often cannot be distinguished from sample peaks. This happens because materials baked off the septum at 300°C are swept by the carrier gas only as far as the first cold region of the column and are "injected" back into the carrier-gas stream through the process of temperature programming.

### Detectors

Detectors are the vital measuring devices in chromatographic analytical systems which combine both separation and measurement. The remarkably high sensitivity which the popular types of detectors possess allows the successful application of gas chromatography to hundreds of interesting chemical problems not otherwise approachable. This high sensitivity also allows the use of samples so small that effective isotherm linearity is virtually guaranteed. On the other hand, detectors generally should be viewed as some of the greatest destroyers of information currently in use. The point is basically philosophical, but of tremendous practical importance. A detector is a "transducer" – the input is a "chemical signal," solute zones in a flowing gas stream; the output is an electrical signal, a single current or voltage proportional to the sample flux. While it is sometimes convenient to think of molecules as billiard balls, the individual identities of the input molecules are of great interest in the present situation. Because it measures only gross sample flux, the detector destroys information about molecular identities. The ultimate detector would give a different kind of output signal for each different form of molecular input.

**Flame-Ionization Detector.** A cross-sectional view of a flame-ionization detector is shown in Figure 14–10. Several design variants having different construction exist, but the principle of operation is the same in all cases. In the base of the detector, the carrier gas is mixed with hydrogen (in some instances, hydrogen is used as the carrier gas, and

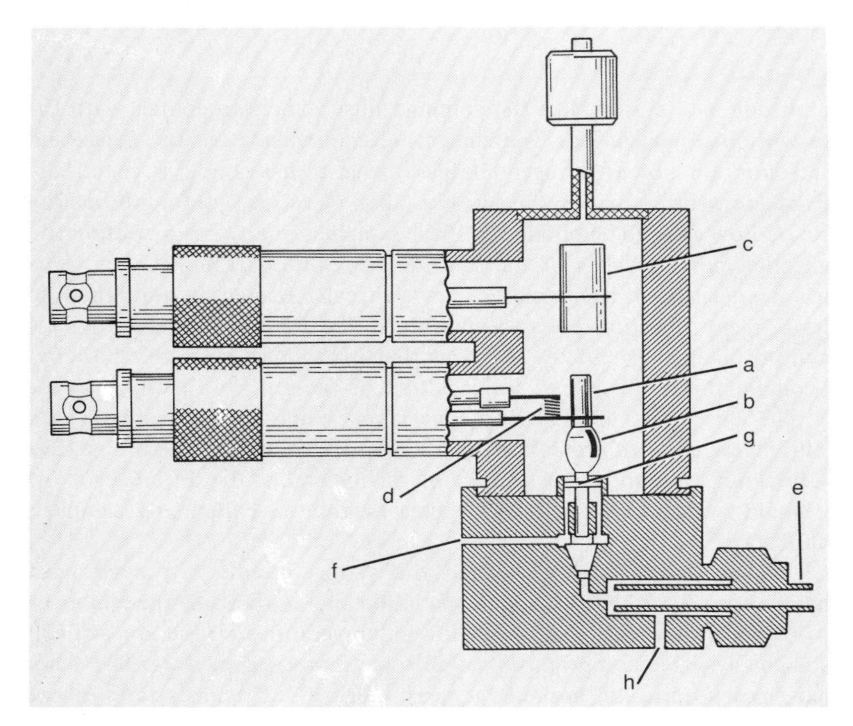


*Figure 14–10.* Cross-sectional view of flame-ionization detector for use in gas chromatography. *a,* Anode, flame tip; *b,* glass section for electrical insulation of flame tip; *c,* cathode; *d,* ignitor coil; *e,* column; *f,* air supply; *g,* air diffuser; *h,* hydrogen inlet. (Courtesy of Varian Aerograph.)

this step is not required). The hydrogen sustains a flame at the flame tip, the required oxidant being supplied by a steady stream of air or oxygen. The usual gas composition flowing into the detector is about 1 part carrier gas, 1 part hydrogen, and 10 parts air. A small heating coil is frequently installed as shown in order to light the flame. The detector operates by measuring the concentration of ions in the flame, the ion concentration being directly proportional to the amount of carbon in the flame.

The ion concentration is determined by measuring the conductivity of the flame between the anode and the cathode. If the electric field between these electrodes is large, 100 V/cm or more, virtually every ion formed in the flame is collected, the ion current is independent of changes in the electric field strength, and the detector is said to be operating in the saturation region. In practice, the voltage applied between the two electrodes is about 300 V. A realistic minimum detectable current with most amplifier systems is about $1 \times 10^{-13}$ amp. Knowing in addition that the efficiency of ion production and collection is about $10^{-6}$ mole of ions collected per mole of carbon introduced to the flame, we can make an interesting calculation concerning detector sensitivity. Let us convert the minimum detectable current of $1 \times 10^{-13}$ amp into the minimum detectable flow-rate of carbon:

$$\begin{aligned}\text{minimum detectable flow-rate of carbon} &= \left(\frac{1 \times 10^{-13}\ \text{coulomb}}{\text{sec}}\right)\left(\frac{1\ \text{ion}}{1.6 \times 10^{-19}\ \text{coulomb}}\right)\left(\frac{1\ \text{atom introduced}}{10^{-6}\ \text{ion collected}}\right)\\ &\quad \times \left(\frac{1\ \text{mole C}}{6 \times 10^{23}\ \text{atoms}}\right)\left(\frac{12\ \text{g C}}{\text{mole C}}\right)\\ &= 1.2 \times 10^{-11}\ \text{g C/sec}\end{aligned}$$

A usable quantitative measurement will require a signal at least ten times this minimum value; and, since the chromatographic zone will be at least ten seconds wide, it follows that one nanogram of carbon is measurable.* The range of linear response extends to very much larger quantities, accurate measurements of as much as 100 $\mu$g being possible.

The flame-ionization detector is sensitive only to compounds which produce ions in the flame; in practice, this means carbon compounds with C–C or C–H bonds. Thus, the detector is insensitive to inorganic gases, such as $O_2$ and $N_2$, as well as to CO, $CO_2$, and $CS_2$. The sensitivity of the detector toward a given compound depends on the number of carbon atoms. Within a homologous series like the *n*-alkanes, this relationship is quite precisely applicable. Structural and chemical effects can play an important role, however. For example, the sensitivity for 2,2,4-trimethylpentane is 15 per cent greater than that for *n*-octane, and 30 per cent greater than that for ethylbenzene, in spite of the fact that all three compounds contain eight carbon atoms. The sensitivity for acetone is only twice the sensitivity for methane, and it seems that, in general, carbonyl carbons "do not count" in the determination of sensitivity.

**Electron-Capture Detector.** A second type of ionization detector is the electron-capture detector. In it, the chromatographic effluent gas is ionized by a stream of particles from some radioactive source, ordinarily either $TiH_2$ containing some ${}^{3}_{1}H$ or nickel foil containing some ${}^{63}_{28}Ni$ (both isotopes are beta-emitters, though alpha-emitters can be used). The ions are collected and their concentration thereby measured by an electrode and amplifier system like that used in the flame-ionization detector. The principle of operation differs greatly, however, in that solute zones are detected by the *decrease* which they cause in the otherwise steady ion current. This decrease occurs because the extent of ionization depends sharply on the concentration of free electrons in the detector, and some chemical species are extremely efficient at capturing free electrons. The minimum sample flux detectable for substances with high electron affinities, for example, halogen-substituted compounds, is on the order of $10^{-13}$ g/sec, and the detector is thus considerably more sensitive than the flame-ionization detector *for these species.* Electron-capture detectors are sensitive to compounds containing halogens, phosphorus, lead, or silicon, in addition to others such as polynuclear aromatics, nitro compounds, and some ketones. It happens that pesticides generally contain phosphorus or chlorine, so the detector is thus ideally suited for the measurement of low levels of these compounds. It is also possible to deliberately introduce halogens into species to which the detector would not ordinarily respond. For example, acids can be esterified with fluorinated alcohols; alcohols and amines can be treated with fluorinated acid anhydrides.

**Thermal-Conductivity Detector.** Until recently, this form of detector was certainly the most widely used, not only because of its simplicity and universal applicability but also because of its relatively low cost. The cost differential between the flame-ionization and thermal-conductivity detectors is now down to about $300; and, since the former offers much higher sensitivity and far better quantitative accuracy, it is definitely to be preferred for the analysis of organic compounds.

A thermal-conductivity detector operates by measuring solute zone-induced changes in the thermal conductivity of the column effluent. Thus, the sensitivity depends only on the difference in thermal conductivity between the pure carrier gas and the carrier gas mixed with some solute. In view of the low solute concentrations encountered, this

*Sensitivities three orders of magnitude better can be achieved by increasing the efficiency of ion production and collection to about $10^{-4}$ and by measuring carbon ion currents as small as 0.1 per cent of the hydrogen flame background. The calculation given above reflect the practical capabilities of commercial instruments in wide use.

difference is inevitably small, yet the appearance of a zone can be sensed quite accurately by measuring the change in resistance of a wire heated by passage of a constant electric current and surrounded by the column effluent. An increase or decrease in the thermal conductivity of the surrounding gas will cause the wire temperature to decrease or increase, with a consequent decrease or increase of the resistance of the wire. The sensitivity is enhanced if one chooses a carrier gas with a thermal conductivity differing from that of the solute to the greatest extent possible, a consideration dictating the use of hydrogen or helium for maximum sensitivity with large organic compounds. Trace constituents of gas samples can be determined by using the major components as carrier gas. For example, the retention characteristics of nitrogen and krypton are very similar, and the determination of trace amounts of krypton in nitrogen (or air) is very difficult because the krypton is "buried" in the tail of the nitrogen peak. If such a sample is analyzed, with nitrogen being used as the carrier gas, the nitrogen in the sample is "invisible" to the thermal-conductivity detector, and the small amount of krypton can be successfully detected.

## Applications of Gas Chromatography

**Analysis of Steroidal Hormones.** Medical biochemistry provides an interesting and important area for the application of modern gas-liquid chromatographic techniques. As an example, we can consider the analysis of steroid hormones. All these compounds contain the basic carbon skeleton shown below:

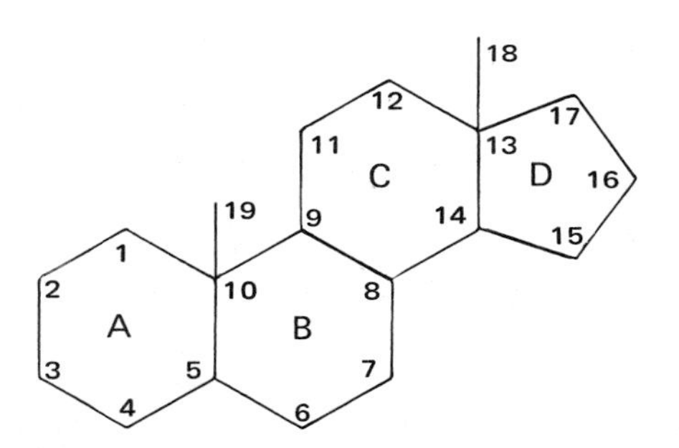


The system of numbers and letters allows convenient designation of structural details in various compounds. An important aspect of the structure of most steroids is the stereochemistry at each asymmetric carbon atom. This almost always has to do with the positioning of a functional group on one of the ring carbons; a dashed line is used to indicate a bond projecting behind the plane of the paper, and a solid line is drawn to indicate a bond projecting in front of the plane of the paper. One stereochemical detail not involving a functional group occurs at the A-B ring junction. When the methyl group attached to carbon-10 projects in front of the plane of the paper, the hydrogen at carbon-5 can project either up (solid line) or down (dashed line). In the former case, the A-B ring junction is *cis*, as in *cis*-decalin. When the hydrogen at carbon-5 lies on the side of the ring system opposite to the methyl group at carbon-10, the ring junction is *trans*, as in *trans*-decalin. Because of its importance in indicating the stereochemistry of the ring junction, the location of the hydrogen at carbon-5 is always specified when that hydrogen is present in a structure. Small differences in the structure of a steroid hormone can have an enormous effect on its physiological properties. Compare, for example, the structures of testosterone, the male sex hormone, and estradiol and progesterone, female sex hormones:

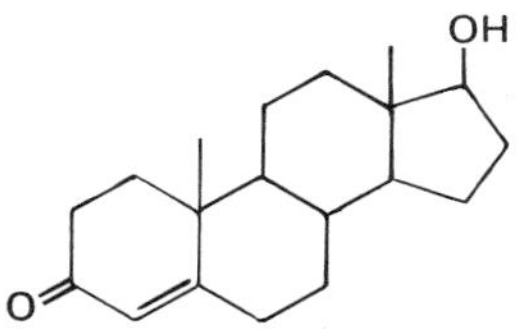

OH O OH

O O HO

testosterone 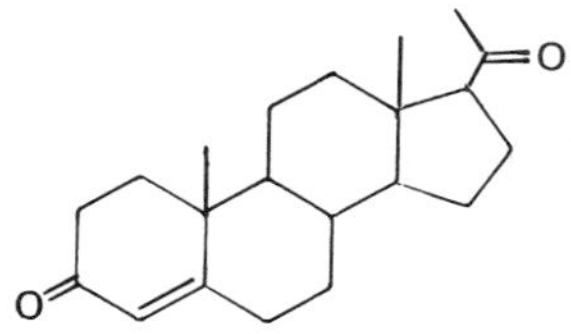 progesterone 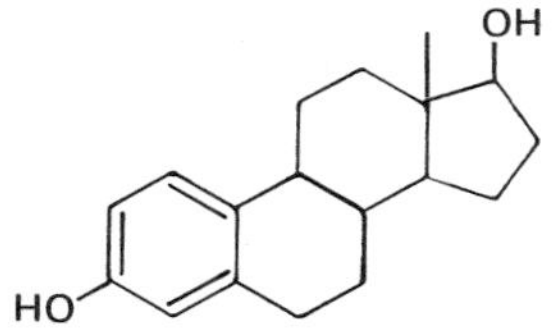 estradiol

Analytical techniques of the highest possible specificity are required if these subtle differences in molecular structure are to be accurately recognized. The full range of complexity is almost unbelievable. Not only must the primary sex hormones be considered; there are, in addition, literally hundreds of different metabolites, therapeutic drugs, anti-fertility agents, and other hormones which can be important in any given situation. Therefore, at the same time that we require *specificity* in the identification and analysis of any particular compound, the scope of the problem requires that practical techniques also possess *broad applicability*. It would be highly useful if a single methodology allowed the simultaneous analysis of many steroidal compounds.

The chromatogram shown in Figure 14–11 demonstrates that high-resolution gas-liquid chromatography provides both specificity and broad applicability. The compounds responsible for the numbered peaks are shown below and on the following page.

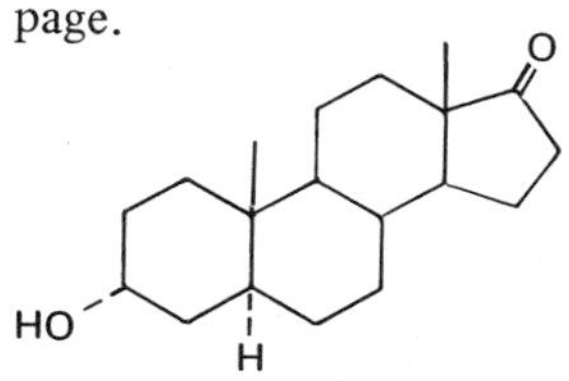

O O O

HO H HO H HO

1. Androsterone 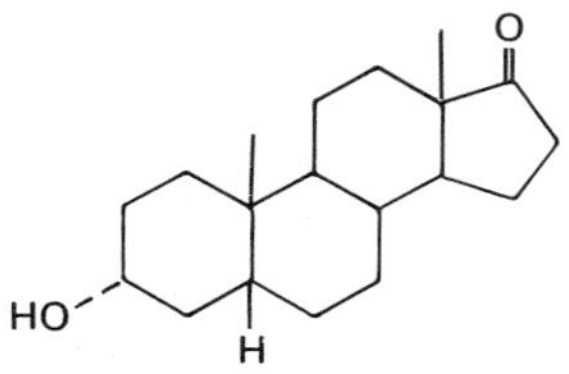 2. Etiocholanolone 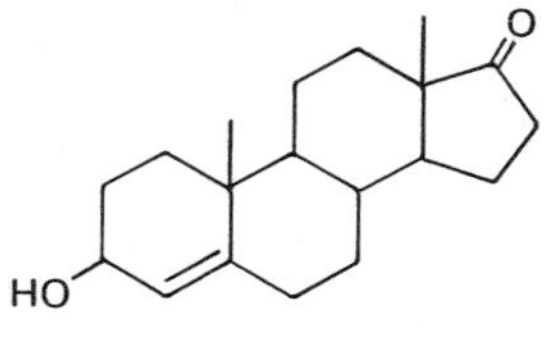 3. Dehydroepiandrosterone

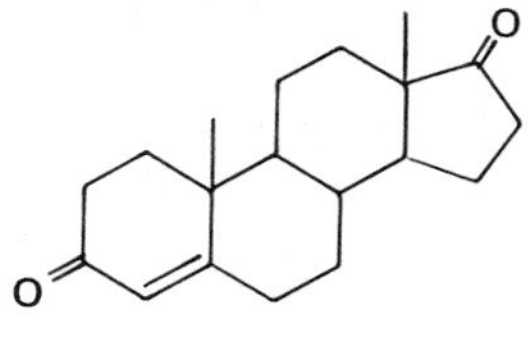

O OH OH

O HO H HO H

4. Androstendione 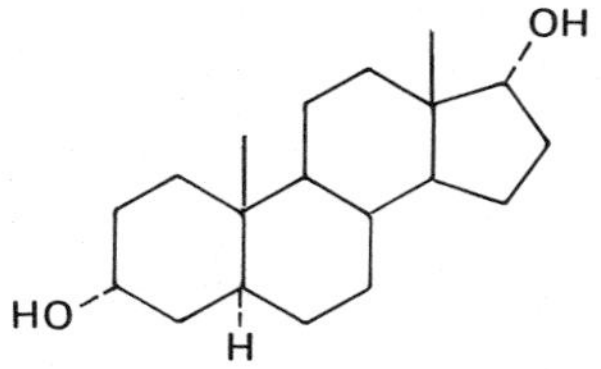 5. Androstanediol 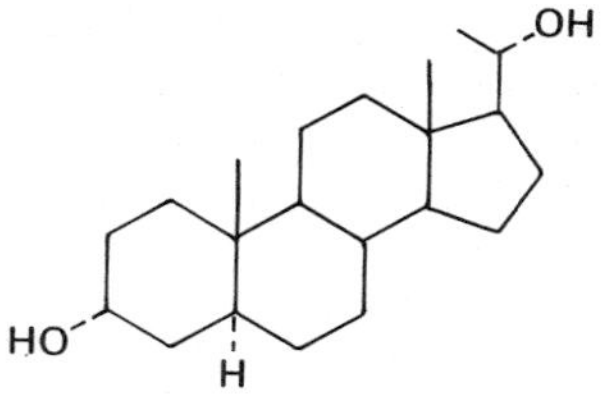 6. Allo-pregnanediol

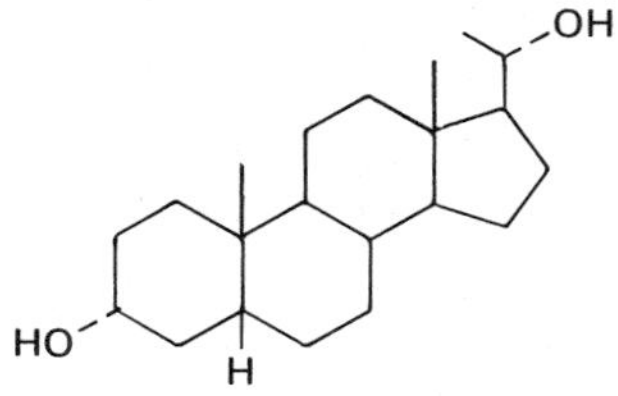

OH O OH OH

HO H HO HO H

7. Pregnanediol 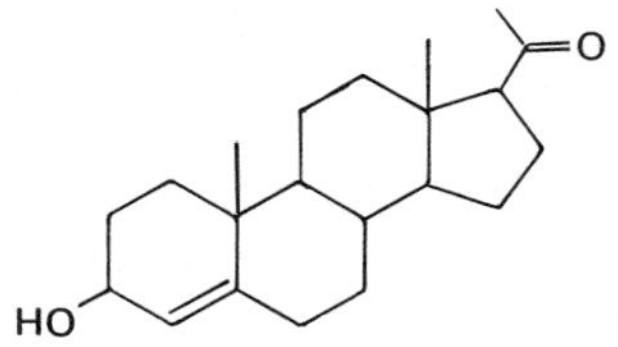 8. Pregnenolone 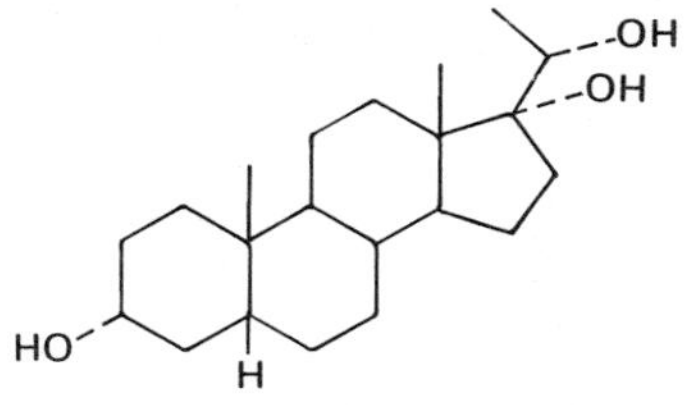 9. Pregnanetriol

10. Progesterone

11. Estriol

12. Tetrahydrocortisol

13. Allo-tetrahydrocortisol

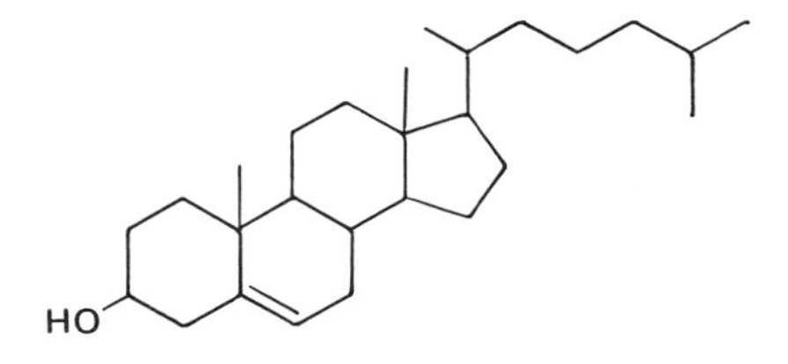


14. Cholesterol

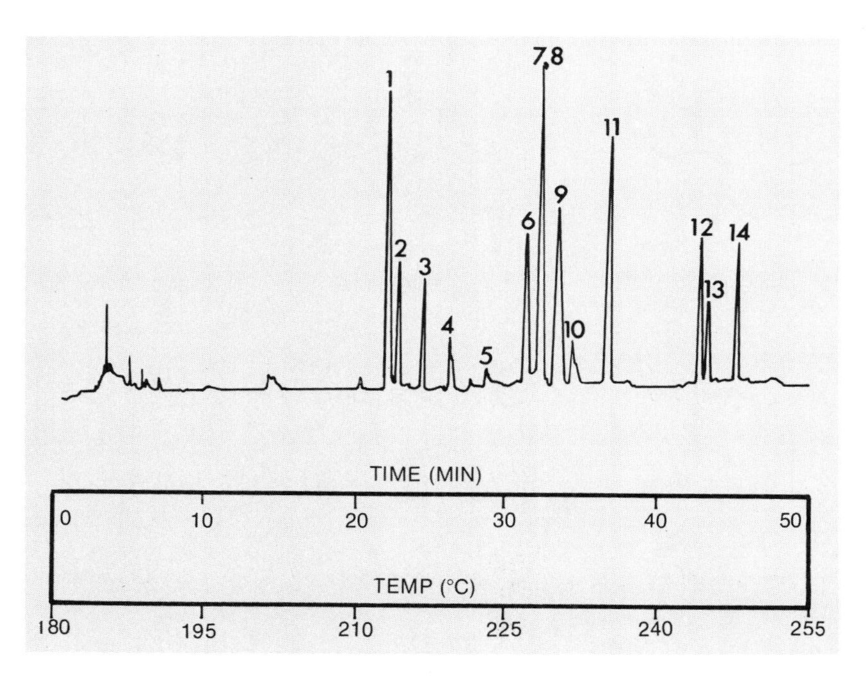


*Figure 14–11.* A chromatogram of the methoxime trimethylsilyl derivatives of the steroids numbered and identified in the text. This chromatogram was obtained with a glass capillary column 0.26 mm (inside diameter) × 20 m (length) coated with SE-30 (liquid phase 3 in Figure 14–9) in a layer 2.5 $\mu$m thick on the inner wall of the column. Helium carrier gas was passed through the column at 2 ml/min, and the column temperature was programmed from 190° to 260°C at 1.5°C/min. A hydrogen flame-ionization detector was used; each larger peak corresponds to approximately 25 ng of compound. The minimum detectable sample flow was about 0.1 ng/sec. (Chromatogram provided by M. V. Novotny and M. P. Maskarinec, Department of Chemistry, Indiana University.)

Direct gas-liquid chromatography of the compounds shown here is extremely difficult because temperatures high enough to volatilize these quite polar molecules are also high enough to cause considerable pyrolytic degradation. To overcome this problem, a *volatile derivative* of each steroid is prepared. The derivatization procedure employed in this instance is typical. First, the mixture of steroids is heated with methoxylamine in order to transform each ketone group into a methoxime group:

$$R{=}O + H_2N{-}OCH_3 \xrightarrow[75^\circ C,\ 30\ min]{pyridine\ solvent} R{=}N{-}OCH_3 + H_2O$$

Next, each hydroxyl group is transformed to a less polar and more thermally stable trimethylsilyl ether by use of a "silanizing reagent" such as N,O-bis(trimethylsilyl)trifluoroacetamide:

$$R{-}OH + CF_3{-}C({=}N{-}Si(CH_3)_3){-}O{-}Si(CH_3)_3 \xrightarrow{60\text{-}75^\circ C,\ 2\ hr} R{-}O{-}Si(CH_3)_3 + CF_3{-}C({=}O){-}NHSi(CH_3)_3$$

By this procedure, a molecule such as tetrahydrocortisol gains 317 mass units of molecular weight (from 366 to 683 g/mole), but the resulting derivative possesses enough thermal stability to allow successful chromatography to be performed. As noted in the figure caption, such compounds are termed "methoxime trimethylsilyl derivatives."

In practice, the order of elution of such compounds can be predicted only very approximately, and individual standards must be prepared and examined for each compound of interest. In the present example, compounds 1 and 2, 6 and 7, and 12 and 13 differ only in the stereochemistry of the A-B ring junction. In the first two cases, the *trans* isomer is eluted before the *cis* compound, but in the last case that order is reversed. Compound 3 differs from 4 only at carbon-3. Interestingly, the larger trimethylsilyl ether derivative of 3 is eluted before the smaller methoxime derivative of 4. Apparently, the latter can interact more strongly with the stationary phase. A similar case occurs when the trimethylsilyl ether derivatives of compounds 6 and 7 are compared to the derivative of compound 8, which contains one trimethylsilyl ether group and one methoxime group.

This example is representative of thousands that might be cited in the general context of biochemical analyses. First, the separation and analysis of closely related compounds is required. Second, derivatization is required in order to provide both thermal stability and volatility. Third, substantial success *can* be obtained.

**Trace Atmospheric Constituents.** There are no stable fluorine compounds released to the atmosphere by natural geochemical equilibria. Halocarbons and sulfur hexafluoride are, however, quite important industrial compounds, and substantial quantities are released into the atmosphere by uncontrolled evaporative losses. For trichlorofluoromethane the estimated amount is about $2 \times 10^5$ tons per year, and for sulfur hexafluoride the amount is about 100 tons per year. J. E. Lovelock, the inventor of the electron-capture detector, has shown that the resultant concentrations in the atmosphere are on the order of $10^{-11}$ part by volume for $CCl_3F$ and $10^{-13}$ part by volume for $SF_6$. These fantastically low concentrations can be measured if one employs an electron-capture detector, and repeated analyses show that the concentration variations reveal patterns of atmospheric circulation. Air which has recently passed over heavily industrialized territory is rich in these compounds, whereas, at the other extreme, air which has just passed over the ocean is particularly clean.

The reader might recognize that this accidental geophysical experiment on the tracing of air masses has become infamous. The chlorofluoromethane, $CCl_3F$ (most common trade name, "Freon-11"), found success as a built-in atmospheric tracer for the same reason that it found wide commercial application; namely, it is so unreactive that it persists in the atmosphere just as it fails to react with other chemicals in aerosol cans (in which a mixture of $CCl_3F$ and $CCl_2F_2$ serves as a propellant). In fact, $CCl_3F$ and $CCl_2F_2$ are so unreactive that they survive for tens of years in the lower atmosphere (the troposphere). Anything which survives that long in the atmosphere will eventually become uniformly distributed, not only horizontally, but also with respect to altitude. As these "unreactive" materials reach the stratosphere, they are increasingly exposed to the ultraviolet portion of the solar spectrum. No material is inert in such an environment – even the $N_2$ and $O_2$ which make up the bulk of the earth's atmosphere are led to participate in a wide variety of photochemical reactions. In fact, one such reaction is responsible for the creation of a relatively stable layer of ozone, $O_3$, which has the effect of protecting the lower atmosphere from incoming ultraviolet radiation. A problem develops because chlorine atoms produced by photochemical degradation of the chlorofluoromethanes catalyze the destruction of ozone, thus allowing the transmission of more far-ultraviolet radiation to the earth's surface. In this way, the uncontrolled release of chlorofluoromethanes poses a hazard to the entire terrestrial ecology, which depends strongly on the stability of the solar radiation spectrum.

We are left to speculate where mankind would be without gas chromatography using electron-capture detection. Without this analytical tool, we would have been much slower to learn about chlorofluoromethanes in the atmosphere *and,* we might as well add, about persistent halogenated pesticide residues in the environment and about carcinogenic chlorohydrocarbons in drinking water.

## ION-EXCHANGE CHROMATOGRAPHY

We have spoken up until now as if all solute separations, and, therefore, all chromatographic techniques, were concerned only with neutral molecules. Of course, this is not true. Many solutes of analytical interest either are inherently ionic (metal ions, for example) or can be made ionic within certain pH ranges (organic acids and bases fall into this category). For the liquid chromatographic separation of such solutes, it is frequently advantageous to employ an "ion-exchange resin" as the stationary phase. These resins are organic polymers incorporating charged sites which can bind solute ions by direct coulombic interaction. The strength of the resin-solute interaction determines the partition coefficient, $K$. Because the nature and strength of the interaction between the resin and any given ion depend strongly on the chemical nature of the solute ion, different ions often have different $K$ values, and separations are frequently possible.

### Preparation and Types of Ion-Exchange Resins

Ion exchangers are of two principal types, cationic and anionic. Cation exchangers possess fixed sites of negative charge and are able, thus, to bind positively charged ions. Anion exchangers reverse this situation.

**Cation Exchangers.** One of the most widely used ion-exchange resins consists of a cross-linked sulfonated polystyrene polymer. Synthesis of this resin begins with the catalyzed polymerization of a mixture of styrene and divinylbenzene.

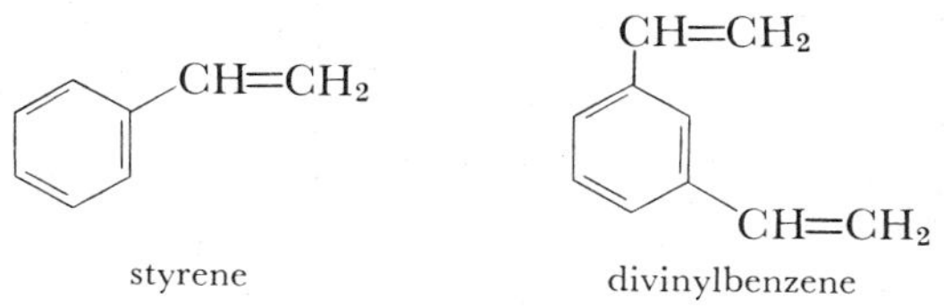

styrene divinylbenzene

This reaction yields spherically shaped beads of cross-linked polystyrene. Next, the polymer beads are sulfonated to produce a substance of the following general structure:

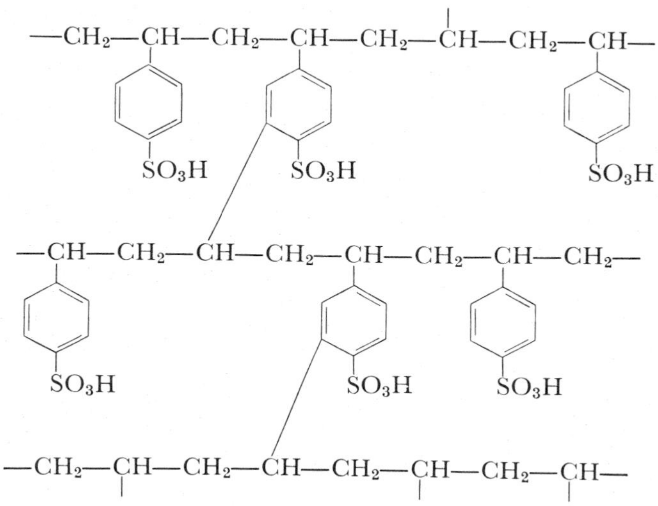

When this resin is brought into contact with an aqueous solution, the sulfonic acid groups can dissociate. The resulting resin-bound $-SO_3^-$ site can bind and exchange cations. Another frequently employed cation exchanger incorporates carboxyl groups, which can dissociate to form $-COO^-$ exchange sites.

Notice that cation-exchange resins incorporate acidic functional groups. If the resin-bound group is strongly acidic (*e.g.*, $-SO_3H$), positively charged solutes can be retained by the resin even at quite low pH values. When the resin-bound group is weakly acidic (*e.g.*, $-CO_2H$), hydrogen ions can displace resin-bound solutes at low pH values.

**Anion Exchangers.** These materials must incorporate positively charged resin-bound sites. Such sites are always basic and range in strength from quaternary amine groups, $-CH_2-\overset{+}{N}(CH_3)_3$, down to tertiary and secondary amine groups, $-CH_2-\overset{+}{N}H(CH_3)_2$ and $-CH_2-\overset{+}{N}H_2(CH_3)$, respectively.

## Ion-Exchange Equilibria

Consider an ion $M^+$ which is being partitioned between a cation-exchanger "resin phase" and the mobile phase. If we depict this process simply by writing

$$(M^+)_{resin} \rightleftharpoons (M^+)_{mobile}$$

we are showing only half the picture. Electrical neutrality *must* be preserved. The exchange site in the resin is *always* bound to some positive ion. In order to be bound by the resin, $M^+$ must displace this other ion, which we shall call $X^+$. A more complete view of the binding of $M^+$ shows clearly that this process is an ion *exchange*:

$$(X^+)_{resin} + (M^+)_{mobile} \rightleftharpoons (X^+)_{mobile} + (M^+)_{resin}$$

In any practical case, partitioning of $M^+$ between the resin and the mobile phase depends very strongly on what other ions are present in the system and how successfully they can compete with $M^+$ for the ion-exchange sites on the resin.

In general, the resin exhibits a preference for

(1) the ion of higher charge,

(2) the ion with the smaller solvated equivalent volume, and

(3) the ion which has greater polarizability.

These criteria are closely analogous to the selectivity rules for the adsorption of ions from solution onto the surface of a colloidal particle, as described in Chapter 7. The first preference noted above is purely electrostatic in origin. Multiply charged ions have stronger electrostatic fields and are more strongly bound to a site of opposite charge. The second preference is related in an interesting way to resin structure. The backbone of the resin is an aromatic hydrocarbon polymer and is strongly hydrophobic. When placed in water, the resin beads swell because water enters to solvate the ionic functional groups. This swelling creates a substantial pressure in the bead, the position of equilibrium being determined by the point of balance between the forces of solvation on one hand and contraction of the stretched-out hydrophobic polymer on the other. Choosing an emotional model again, we might say, "The resin loves to squeeze out a big hydrated ion and substitute a small one." Interestingly, this preference for a small solvated ion diminishes sharply as the degree of cross-linking of the polymer is decreased. The cross-links can be viewed as the springs that create the pressure which "squeezes out" the larger ions.

### Separations by Ion-Exchange Chromatography

**Separation of Rare-Earth Metal Ions.** It is often possible to achieve the separation of two metal cations through the addition of some reagent which will complex one or both ions in the solution. Since the equilibrium constants for the formation of complexes of similar ions, even those of identical charge, may differ significantly, one can alter the relative concentrations of the two adsorbable cations in the solution and the relative affinities of each ion for the resin phase.

A classic example of the application of this approach is the successful separation of the rare-earth (lanthanide) ions. Prior to the time that these elements were first separated by ion-exchange chromatography, the only feasible method for the isolation and purification of the rare earths involved tens or even hundreds of tedious fractional precipitations. Although this precipitation technique presumably did provide pure rare-earth compounds, careful reexamination of these preparations through the application of modern physical-analytical procedures has often revealed these supposedly pure compounds to be relatively impure. If a solution containing rare-earth ions, such as $La^{3+}$,

$Ce^{3+}$, $Eu^{3+}$, $Gd^{3+}$, $Tb^{3+}$, $Er^{3+}$, $Tm^{3+}$, $Yb^{3+}$, or $Lu^{3+}$, is introduced into a cation-exchange column, all these ions will initially be adsorbed on the resin phase. It is impossible to separate these cations by elution of the column with a solution of a simple salt such as sodium chloride or ammonium chloride, because each of these ions has a 3+ charge and nearly the same ionic (solvated) radius. On the other hand, if citrate is added to the eluent, a remarkably sharp separation of the rare-earth ions is achieved; citrate forms a complex of sufficiently different stability with each cation so that, one at a time, the rare earths can be eluted from the column and collected separately in receiving flasks. The separation is still so difficult, however, that the complete elution of the rare-earth ions may require a period of over 100 hours.

**Amino Acid Analysis.** To say that amino acid analyses are important in biochemistry is a generous understatement. Our understanding of everything from genetics to the physical chemistry of enzymes depends upon amino acid analyses, more amino acid analyses, and still more. . . . Because of this central role, many different analytical technologies are available and in widespread use. Of special importance, however, is the ion-exchange chromatographic technique first introduced in 1951 by Stanford Moore and William Stein. This technique is favored because of its very high reliability and quantitative accuracy. Moore and Stein, together with D. H. Spackman, reported in 1958 on the development of fully automated equipment, and since that time the analytical developments introduced by these biochemists have formed the basis of a major instrumental technology which has involved whole companies and the sales of thousands of instruments. Moore and Stein shared the 1972 Nobel Prize for chemistry with Christian B. Anfinsen, the award being made for their individual investigations of the structure and activity of the enzyme ribonuclease. The amino acid composition of this enzyme was determined by Moore, Stein, and C. H. W. Hirs in one of the very first applications of the new ion-exchange chromatographic technique. Appropriately, the chromatogram in Figure 14–12 is from a report on that work.

Amino acids can be separated chromatographically with an ion-exchange stationary phase because the amino acids themselves exist in several ionic forms. In a strongly acidic solution ($pH < 1$), all the amine groups and other basic substituents are protonated, and it is in this form that they are shown in Figure 14–13. Starting with the fully protonated species, we can examine the dissociations which occur as the pH is gradually increased until basic conditions are reached. As a first example, consider aspartic acid. At a pH of approximately 0.85, the carboxyl group with a $pK_a$ of 2.09 will be about 5 per cent dissociated into the $-CO_2^-$ form. As the pH is increased to 2.09, this carboxyl group will become 50 per cent dissociated. Further increases in the pH lead to dissociation of the carboxyl group with a $pK_a$ of 3.86. At a pH of 2.97, dissociation of the more acidic carboxyl group is about 88 per cent complete, whereas dissociation of the less acidic carboxyl group is approximately 12 per cent complete. As a result, the positive charge on the protonated amine group is balanced by one full negative charge distributed between the carboxyl groups, and the net charge on each molecule of aspartic acid is zero. The pH at which an amino acid has zero net charge is termed the **isoelectric point.** Increases in pH above 5.11 lead to greater than 95 per cent dissociation of the second carboxyl group, but dissociation of the protonated amine group will not be even 5 per cent complete until the pH exceeds 8.57. Possessing only two dissociable groups, threonine is somewhat simpler in behavior. Specifically, the isoelectric point lies exactly midway between the two $pK_a$ values, and the molecule has at least a slight net positive charge at any pH below 6.53. Lysine has *two* basic functional groups and, even when dissociation of the carboxyl group is more than 99 per cent complete ($pH > 4.18$), the molecule still bears a net charge of +1.

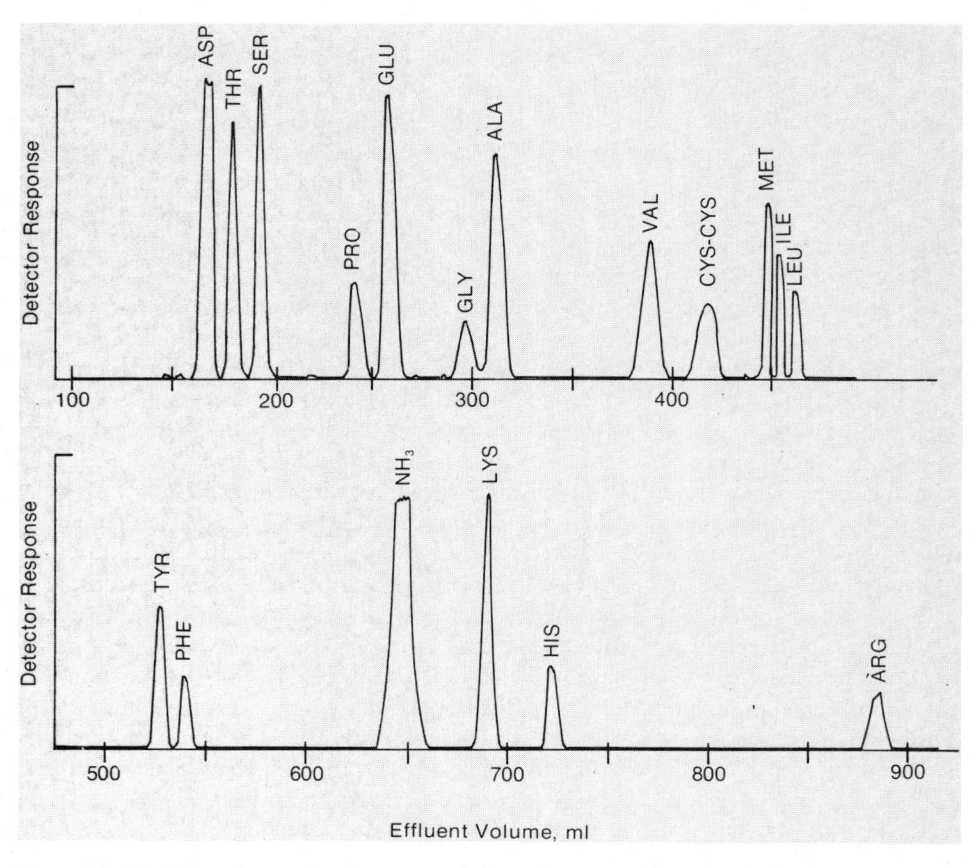


*Figure 14–12.* Ion-exchange chromatogram of the amino acids derived from the total hydrolysis of ribonuclease. The stationary phase was a sulfonated cross-linked (4 per cent divinylbenzene) polystyrene resin with an average particle diameter of approximately 50 μm. The chromatographic column was 150 cm in length and had an internal diameter of 9 mm; its temperature was held constant at 30°C until the elution of serine (SER) was complete, then increased to 50°C and held constant for elution of the remaining amino acids. For the first portion of the chromatogram, the mobile phase was a sodium citrate buffer with pH = 3.1 and $[Na^+] = 0.2\,M$. After the elution of alanine (ALA) was complete, "gradient elution" was employed, with the pH of the mobile phase being gradually increased to 5.1 and its ionic strength slowly increased to correspond to $[Na^+] = 1.4\,M$. The flow rate of the mobile phase was 3 ml/hr for the first 210 ml of effluent, 6 ml/hr in the interval preceding 340 ml of effluent, and 8 ml/hr (12 ml/hr per cm² of column cross-sectional area) thereafter. A total analysis time of 160 hr was required, although in favorable instances it could be reduced to 100 hr. (Redrawn, with permission, from a paper by C. H. W. Hirs, W. H. Stein, and S. Moore: J. Biol. Chem. *211*:941, 1954.)

With this background, we can turn our attention to the chromatogram in Figure 14–12. Note that the initial pH of the eluent is 3.1, a value well below the isoelectric point of all the amino acids shown except aspartic and glutamic acids. With these exceptions, therefore, all the amino acids should have a net positive charge. Because the ion-exchange resin is a sulfonated material, the exchange sites are $-SO_3^-$ groups which can retain the positively charged species. In this case, we would expect the positively charged species to be relatively immobile compared to aspartic and glutamic acids, with these two compounds emerging from the column quite rapidly. Indeed, this is what occurs, even to the extent that aspartic acid, which has the lower isoelectric point, precedes glutamic acid. Strikingly, however, threonine and serine appear as interlopers even though they must carry substantial net positive charges at this pH. Leaving aside, for the moment, any question of why threonine precedes serine, we can observe that both

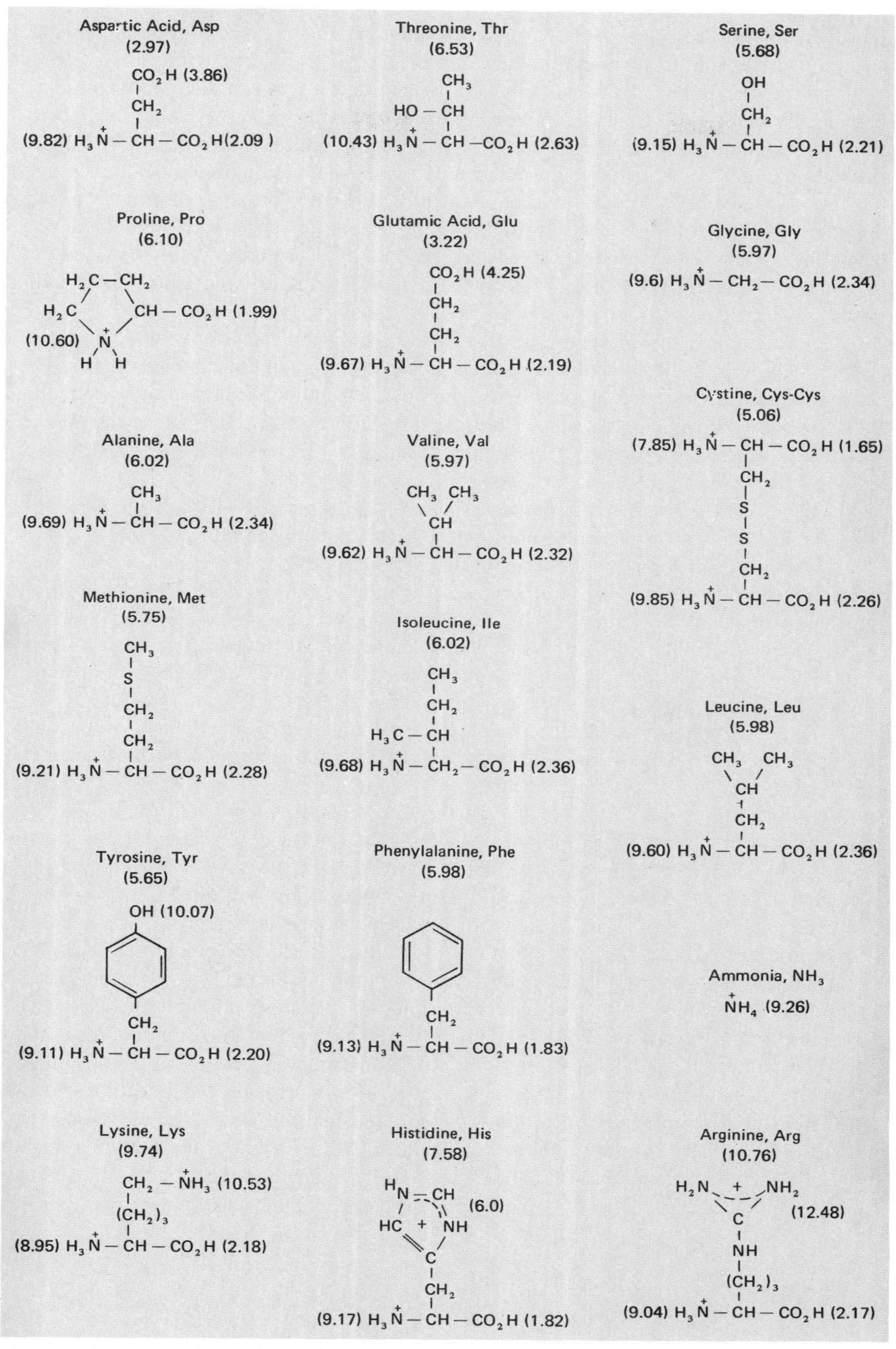


*Figure 14–13.* Structures, standard abbreviations, and acid-base characteristics for the amino acids represented in Figures 14–12 and 14–14. All amino acids are shown in their fully protonated forms. The acid dissociation constant (expressed as $pK_a$) for each functional group is given in parentheses next to the functional group to which it applies. The pH at the isoelectric point (no net charge on the amino acid in aqueous solution) is given in parentheses just below each name.

these amino acids are hydroxylated and that they are, in fact, the only such *aliphatic* acids shown here. From the point of view of ionic charge, these compounds do not differ from glycine, alanine, and others; therefore, we must turn to a non-ionic mechanism to explain the early elution of threonine and serine. Apparently, there is something about the –OH group that is important, and it is helpful to return to the concept, "like dissolves like." The organic material forming the resin beads is an aromatic polymer, and it certainly resembles serine and threonine only slightly. The aqueous mobile phase, however, can interact quite well with the –OH groups on serine and threonine. As a result these compounds are partitioned into the mobile phase to a much greater extent and, therefore, are eluted more rapidly than the non-hydroxylated amino acids with similar isoelectric points.

A peak-by-peak consideration of the chromatogram indicates that non-ionic interactions between the resin and the amino acids play a major role in determining the elution sequence. On a purely ionic basis, compounds would be expected to be eluted in order of increasing isoelectric point. In fact, this defines only the gross aspects of the sequence, and the importance of the following additional factors can be deduced from a study of the chromatogram:

1. Hydroxylated amino acids precede their non-hydroxylated analogs (threonine and serine before alanine, valine, and so on; tyrosine before phenylalanine).

2. Amino acids with aliphatic side-chains interact with the hydrocarbon portion of the resin such that retention generally increases with increasing size of the aliphatic group (glycine before alanine before valine before the leucines).

3. Aromatic amino acids (tyrosine, phenylalanine, and histidine) interact particularly strongly with the resin and are retained much better than would be expected on the basis of their isoelectric points.

4. Steric hindrance around the protonated amine site can reduce the ionic interaction between the resin and the amino acid. Thus, proline appears much earlier than expected, and isoleucine precedes leucine.

As noted in the caption of Figure 14–12, the pH and ionic strength of the mobile phase were both increased as elution of the column continued. If this were not done, very long times would have been required for the appearance of the late-eluting amino acids. Although aspartic and glutamic acids were eluted by a buffer with a pH very near their isoelectric points, the "basic" amino acids (lysine, histidine, and arginine) were eluted with a buffer having a pH substantially below their isoelectric points. This can be explained both by the long times involved and, more importantly, by the increased sodium ion concentration. A higher concentration of $Na^+$ competes more effectively with the amino acids for ion-exchange sites in the stationary phase, and thus increases the migration rate of the amino acids.

In the ion-exchange chromatographic separation of amino acids, the detection system most frequently employed makes use of the fact that all amino acids react with ninhydrin, eventually forming a colored product. In the first step of the reaction, a single mole of amino acid reacts with a single mole of ninhydrin; the amino acid is oxidized to the corresponding aldehyde, $CO_2$, and $NH_3$, whereas ninhydrin is reduced to hydrindantin:

$$\text{ninhydrin} + H_2N\text{–CH(R)–}CO_2H \rightarrow \text{hydrindantin} + RCHO + CO_2 + NH_3$$

ninhydrin hydrindantin

In the second step, the previously formed hydrindantin and $NH_3$ react with an additional mole of ninhydrin to yield a purple product:

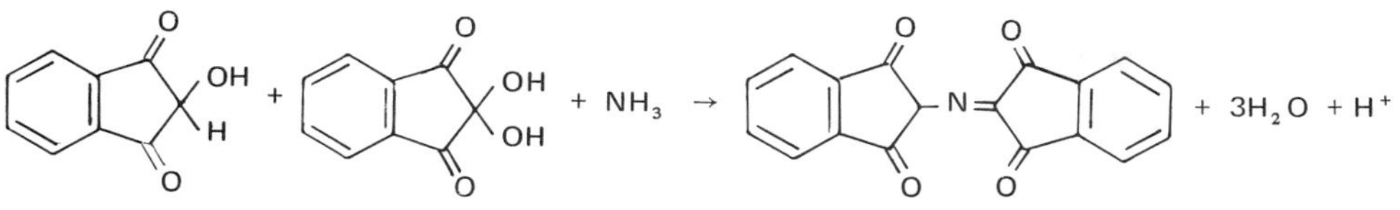

$$+ \; NH_3 \rightarrow \; + \; 3H_2O \; + \; H^+$$

purple product

Notice that no portion of the carbon skeleton of the original amino acid appears in the final product. With the exception of proline and related compounds, all amino acids react to produce the same colored species. Therefore, one can monitor the amino acid concentration in the column effluent simply by mixing ninhydrin with the effluent stream and continuously measuring the color intensity at the single wavelength characteristic of the purple product. Proline reacts with ninhydrin to produce a different colored substance, so that the color intensity at another wavelength must be monitored when proline measurements are required. This colorimetric method of detection is quite sensitive; for example, the glycine peak in Figure 14–12 corresponds to 0.4 $\mu$mole of the amino acid.

Although the fundamental concepts of ion-exchange chromatography have not changed, the passage of more than 22 years since the recording of the chromatogram in Figure 14–12 has brought with it many improvements. A major drawback of the 1954 procedure was the long elution time required (100 to 160 hours as noted in the figure caption). The fully automated procedure introduced in 1958 reduced the total time of analysis to 24 hours, but necessitated the use of two columns, the second to handle the late-eluting basic amino acids. Faster flow rates (47 ml/hr per $cm^2$ of column cross-sectional area) could be employed in the automatic procedure, in part because improvements in the plumbing allowed higher pressures (4 atmospheres) and in part because smaller and more uniform stationary-phase particles were utilized. Size and uniformity of resin beads are particularly important parameters [refer to equations (13–35) and (13–36) with their accompanying discussions]. Large stationary-phase particles require relatively slow flow rates because time must be allowed for diffusion of solute within a large bead which contains a substantial volume of stagnant liquid. When small particles are employed, diffusion times are lowered; it happens that reduction of the particle size "pays off" handsomely because diffusion times depend on the *second power* of the particle diameter. However, the production of small, uniform particles is difficult and their use is complicated by the extraordinarily high pressures required to obtain high flow rates. For these reasons, many forms of liquid chromatography, including ion exchange, have, until recently, been carried out under inherently limiting conditions.

The development of pulseless, high-pressure pumps and new techniques for the production of stationary phases for liquid chromatography, together with improvements in allied areas ranging from chromatographic theory to non-reactive, high-pressure plumbing, have recently resulted in quite incredible further improvements. Ion-exchange chromatography has participated in these advances, as the chromatogram in Figure 14–14 dramatically indicates. Recorded in just over one hour at a flow rate of 382.5 ml/hr per $cm^2$ of column cross-sectional area, the chromatogram includes most of the same amino acids identified in Figure 14–12. The pressure required to obtain this high flow rate is 146 atmospheres. The elution sequence differs slightly due to changes in the column parameters noted in the captions. In particular, the more highly cross-linked resins used in the modern procedure exclude relatively large molecules like glutamic acid and cystine

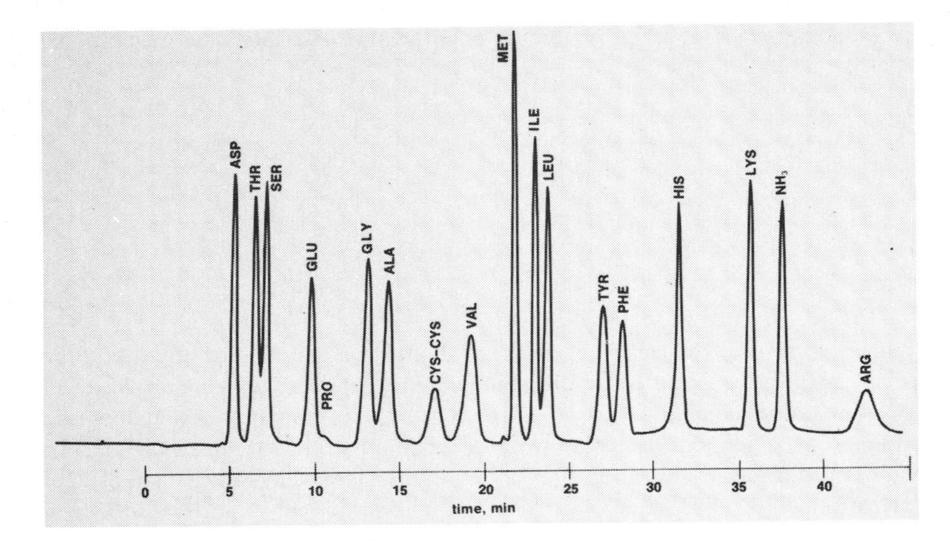


*Figure 14–14.* Ion-exchange chromatogram of the amino acids in a synthetic calibration mixture containing 10 nanomoles of each of 17 amino acids and ammonia. The stationary phase was a sulfonated cross-linked (8 per cent divinylbenzene) polystyrene resin with a particle diameter of 8 ± 1 μm (8 μm) is approximately 20 wavelengths of visible light or 0.0003 inch – *very* small particles, like fine silt). The chromatographic column was 30 cm in length and had an internal diameter of 1.75 mm; its temperature was held constant at 45°C until the elution of serine was complete, then increased to 60°C and held constant for elution of the remaining amino acids. Three different buffers were used sequentially as the eluting mobile phase: 1. (start to 16 min) pH = 3.1, $[Na^+] = 0.2\ M$; 2. (16 to 23 min) pH = 4.1, $[Na^+] = 0.2\ M$; 3. (23 min to end) pH = 6.2, $[Na^+] = 1.0\ M$. The flow rate of the mobile phase was 9.2 ml/hr (382.5 ml/hr per cm² of column cross-sectional area) at a column inlet pressure of 2150 psi (146 atm). (Courtesy of Durrum Instrument Corp.)

from the resin phase to slightly greater extent, causing them to elute more rapidly. Part of the cystine displacement, as well as all the changes at the end of the chromatogram, can be ascribed to the more highly basic final buffer solution in the modern procedure. The sensitivity of the modern procedure is also much higher; each peak in Figure 14–14 corresponds to 0.010 μmole, 40 times less than one of the smallest peaks in Figure 14–12.

## MOLECULAR-EXCLUSION CHROMATOGRAPHY

### Principles

We have observed that ion-exchange resins prefer to retain the ion with the smallest hydrated size. This preference can be explained in terms of cross-links in the resin that act as springs to squeeze out larger particles. An extreme example of ion-selection based on size occurs in some natural (zeolite) ion-exchangers. These materials have pores of rigidly fixed size which are so small that ions must enter "bare," or stripped of their hydration spheres. Large ions such as $Cs^+$ or polyatomic species are completely excluded from the interior of the exchanger. When this exclusion occurs, the large ions pass unretarded through the ion-exchange chromatographic column and are cleanly separated from the smaller ionic species which have access to the interior of the exchanger.

An analogous separation based on size takes place in molecular-exclusion chromatography. A chromatographic column is packed with beads of material having well controlled porosity. There are three subunits of the total volume within the chromatographic column: (1) $V_b$, the volume occupied by the inert matrix of the beads themselves; (2) $V_i$, the volume of mobile phase inside the porous beads; and (3) $V_o$, the volume of mobile phase outside the porous beads. If some molecule has a size small enough that all the mobile phase, both inside and outside of the beads, is accessible to it, then the volume of mobile phase required to effect its elution from the column will be $(V_i + V_o)$. On the other hand, molecules which are so large that they are completely excluded from the porous beads can be eluted with a mobile-phase volume of only $V_o$. Because the pore size cannot be perfectly controlled, there is an intermediate range of molecular sizes which has access to only some of the pores. For these species the retention volume will be greater than $V_o$ but less than $(V_o + V_i)$, with the *larger* molecules being eluted first.

Figure 14–15 illustrates the principle of separation of molecules by means of molecular-exclusion chromatography. A representative chromatogram is shown in Figure 14–16, in which the unusual relationship between molecular size and retention volume is emphasized. In all other chromatographic techniques, larger molecules are more strongly retained and are sometimes eluted only with difficulty. In molecular-exclusion chromatography, not only is the situation reversed, but there is – theoretically at least – an upper bound on retention volume. Because the matrix of the porous material is never

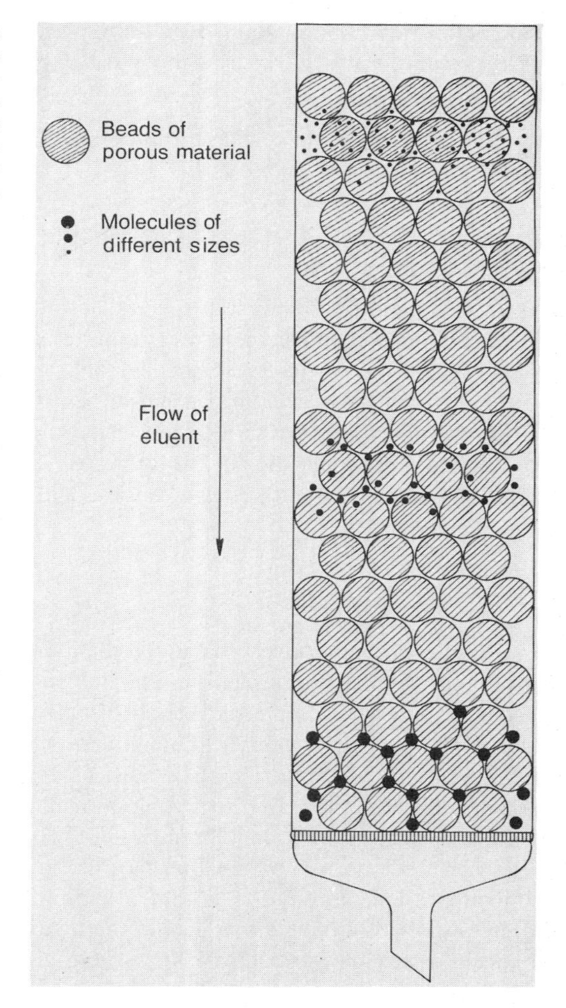


*Figure 14-15.* An intermediate stage in the elution of a molecular-exclusion chromatographic column. The large gray circles represent the beads of porous material. The black circles represent molecules of various sizes. Because they are partially or completely excluded from the mobile phase inside the porous beads, the larger molecules make faster progress through the column.

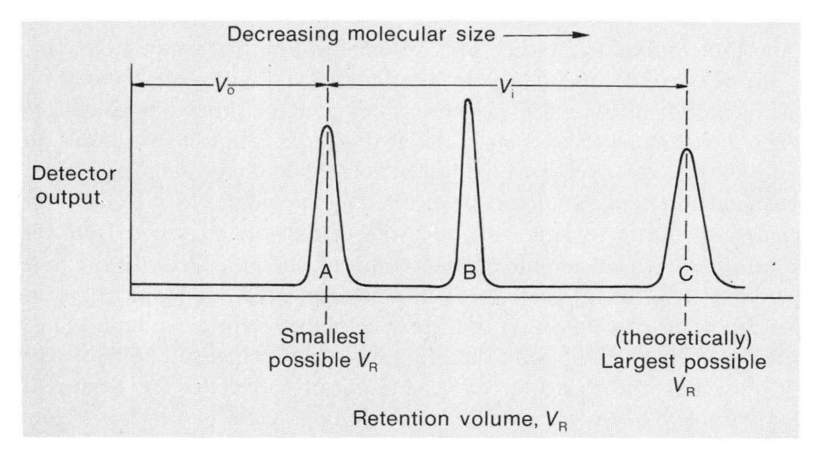


*Figure 14–16.* A representative molecular-exclusion chromatogram in which peak A represents a large molecule completely excluded from the porous beads, peak C represents a small species able to diffuse freely throughout the porous material, and peak B represents some substance of intermediate size having only partial access to the interior of the porous beads.

completely inert, adsorption of solute species can occur and will have the effect of further retarding their progress through the column. When adsorption occurs, retention volumes greater than $(V_o + V_i)$ can sometimes be observed. In general, however, an adequate expression for the retention volume $(V_R)$ is

$$V_R = V_o + KV_i \tag{14–21}$$

where, in molecular-exclusion chromatography, $K$ is the fraction of $V_i$ accessible to a given solute.

A separation of chemical species according to size is often very useful. In the fractionation of complex mixtures from natural sources and in polymer chemistry, this task arises with particular frequency, and specialists in these disciplines have joined chromatographers in creating an entire new field of research in the period since 1960. Here we must be very brief, and can only cover the broadest aspects of column technology and applications. The interested reader will find detailed information in the references cited both in the figure captions and at the end of this chapter.

### Practical Aspects

**Nature of the Porous Material.** To be useful in molecular-exclusion chromatography, a porous material must swell up and imbibe the liquid phase, creating a solvent-filled "sponge" into which molecules can diffuse. Because molecular-exclusion chromatography is carried out with a variety of liquid phases ranging from water to hydrocarbon solvents, a wide variety of porous materials must be available, ranging from hydrophilic materials which will swell in water to lipophilic materials which will imbibe non-polar organic solvents. The most widely used water-loving material is an artificially cross-linked polysaccharide derived by treatment of dextran (a natural glucose polymer) with varying amounts of epichlorohydrin to control the extent of cross-linking. At least eight different degrees of cross-linking are available; the "tightest" gel will exclude polysaccharides with molecular weights as small as 700. A molecular weight in excess of 200,000 is required for complete exclusion from the most "open" gel. Other porous materials include polyacrylamide (ten different degrees of exclusion size) and gels of agarose with exclusion limits ranging up to molecular weights of 150,000,000. Rigid materials such as controlled-porosity glass beads can also be used. Molecular-exclusion chromatography performed with an aqueous mobile phase is sometimes called "gel filtration chromatography."

Porous materials suitable for use with organic mobile phases include exhaustively methylated dextran gels, polystyrene formed from dilute solutions in order to produce "macroreticular" resins, and controlled-porosity glass beads. Like the hydrophilic media, these materials are available with a range of pore size. Polystyrene resins, for example, are available in 12 degrees of porosity with exclusion limits for styrene polymers ranging from 2500 to 410,000,000 in molecular weight. In addition, because the degree of swelling and the resulting pore size depend on the particular organic solvent chosen, conditions can be very carefully tailored to specific requirements. Indeed, one can even make the pore size change during the course of the chromatographic elution by employing a solvent gradient – that is, a gradual change in the composition of the mobile phase. Molecular-exclusion chromatography with an organic mobile phase is often referred to as "gel permeation chromatography."

**Columns and Detectors; Controlling the Flow of Mobile Phase.** Gel filtration chromatography with dextran gels is commonly carried out with a simple glass column 2.5 cm in diameter and 50 cm in length. In such a column, $V_o$ will be 50 to 100 ml and $(V_o + V_i)$ will be 200 to 250 ml. Sample sizes of a few milligrams are placed on the column simply by the addition of solutions to the top of the column. Detection of the solute zones as they emerge from the column can be achieved by spectrophotometric monitoring of the eluate, by measurement of the refractive index of the eluate, or by the collection of aliquots for later analysis. Mobile phase is allowed to flow by gravity through gel filtration columns at a rate of about 3.5 ml per hour for each square centimeter of column cross section. Thus, for a column with a diameter of 2.5 cm, the flow rate is around 16 ml per hour, and the time required for elution of the smallest molecules will be about 16 hours. Faster flow rates cannot be sustained because the soft gel is deformed by the shear forces of the mobile-phase stream; so the column fails, either by extrusion of the gel or by plugging of the column.

In gel permeation chromatography with organic solvents, the technique of high-speed liquid chromatography is much more widely employed. This is possible because the polystyrene resins used in work with organic solvents are far more rigid and have much better mechanical properties than the dextran beads used in gel filtration.

## Applications

**Desalting.** There is frequently very extensive use of buffer solutions during the isolation of biochemical samples in order to control the pH in a way which either maintains enzyme activity or allows certain separations to be made. Consequently, the end result of many isolation schemes is an aqueous solution of the desired molecule *plus* electrolytes from various buffers. If a dextran gel is available which will exclude the sample molecules, "desalting" of the sample solution can be accomplished by means of gel filtration. It is only necessary that the salt-sample solution be applied to the column and that elution with pure water be performed. The relatively small ions of the buffers are not excluded from the gel and are eluted far behind the much larger sample molecules.

**Determination of Molecular Weight.** Although molecular exclusion takes place, strictly speaking, as a function of molecular *size* rather than molecular weight, the use of standards having a chemical nature similar to an unknown allows molecular weights to be determined by interpolation.

For example, the relationship between $V_R$ or $K$ and the molecular weights of proteins transmitted by a certain dextran gel column is shown in Figure 14–17. The dextran gel had a nominal exclusion limit of 800,000 in molecular weight. The diameter of the gel beads was 50 to 75 $\mu$m, the dimensions of the column were 2.5 cm (diameter) $\times$ 50 cm (length), and the buffer (pH 7.5) used as a mobile phase passed through the column at a rate of 15 to 18 ml per hour. The relationship between protein molecular weight and $V_R$ is nicely demonstrated, and it is clear that the molecular weight of an unknown protein could be estimated from its $V_R$ on this column. The retention characteristics of glycoproteins differ systematically, and it can be seen that careful calibration is required if the technique is to be applied to some different class of compounds. In addition to the protein results, Figure 14–17 shows that the experimenter has determined $V_o$ by measurement of the retention volume of a soluble blue dextran (molecular weight, approximately 2,000,000) and has determined $(V_i + V_o)$ by observing the retention volume of sucrose (molecular weight, 342). For "tighter" gels, $V_o$ can be determined through the use of some smaller molecule. Hemoglobin is sometimes used, and frequent practitioners of this art can be recognized by their scarred thumbs.

The use of gel permeation chromatography for determining the distribution of molecular weights for products of a polymerization reactions is demonstrated by Figure 14–18. The porous material used in this case was a soft polystyrene gel cross-linked with 2 per cent divinylbenzene. The column dimensions were 5 cm (diameter) $\times$ 200 cm (length). As can be deduced from the chromatogram, molecules completely excluded from the gel had a retention volume of 830 ml, indicating that $V_o$ was 830 ml. The observed retention volume of benzene was 3050 ml, indicating that $V_i$ was (3050 - 830) or 2220 ml. Tetrahydrofuran was used as an eluent, and a flow rate of 200 ml per hour was obtained when a pressure of about 2.7 atm was applied to the column. A differential refractometer was utilized to measure the difference between the refractive index of the pure eluent and that of the eluate. The sample was a mixture of oligomers having the general formula

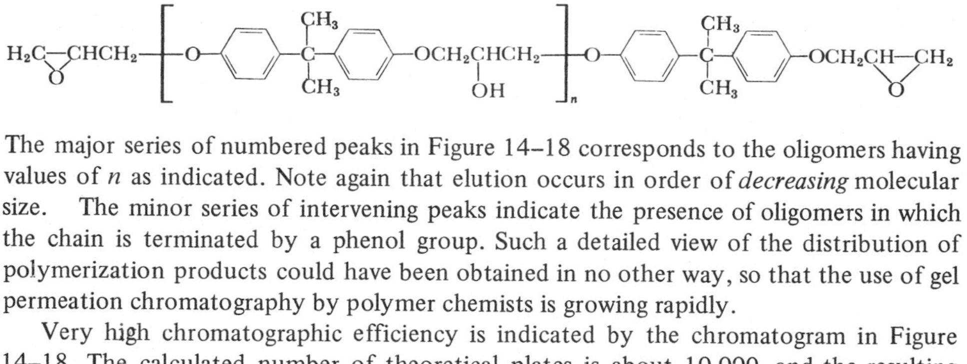

$$\underset{\text{O (epoxide)}}{H_2C{-}CHCH_2}{-}\left[O{-}C_6H_4{-}\overset{CH_3}{\underset{CH_3}{C}}{-}C_6H_4{-}OCH_2\underset{OH}{CH}CH_2\right]_n{-}O{-}C_6H_4{-}\overset{CH_3}{\underset{CH_3}{C}}{-}C_6H_4{-}\underset{\text{O (epoxide)}}{OCH_2CH{-}CH_2}$$

The major series of numbered peaks in Figure 14–18 corresponds to the oligomers having values of $n$ as indicated. Note again that elution occurs in order of *decreasing* molecular size. The minor series of intervening peaks indicate the presence of oligomers in which the chain is terminated by a phenol group. Such a detailed view of the distribution of polymerization products could have been obtained in no other way, so that the use of gel permeation chromatography by polymer chemists is growing rapidly.

Very high chromatographic efficiency is indicated by the chromatogram in Figure 14–18. The calculated number of theoretical plates is about 10,000, and the resulting height equivalent to a theoretical plate is 200 $\mu$m. The average particle diameter in the swollen gel is 74 $\mu$m; thus, the plate height is between 2 and 3 particle diameters, a level of performance difficult to exceed.

## CONCLUDING COMMENT

Scientists engaged in studies of natural systems frequently must choose between alternative explanations of natural phenomena. One guiding principle is "choose the simple alternative"; that is, do not expect nature to be unnecessarily complicated.

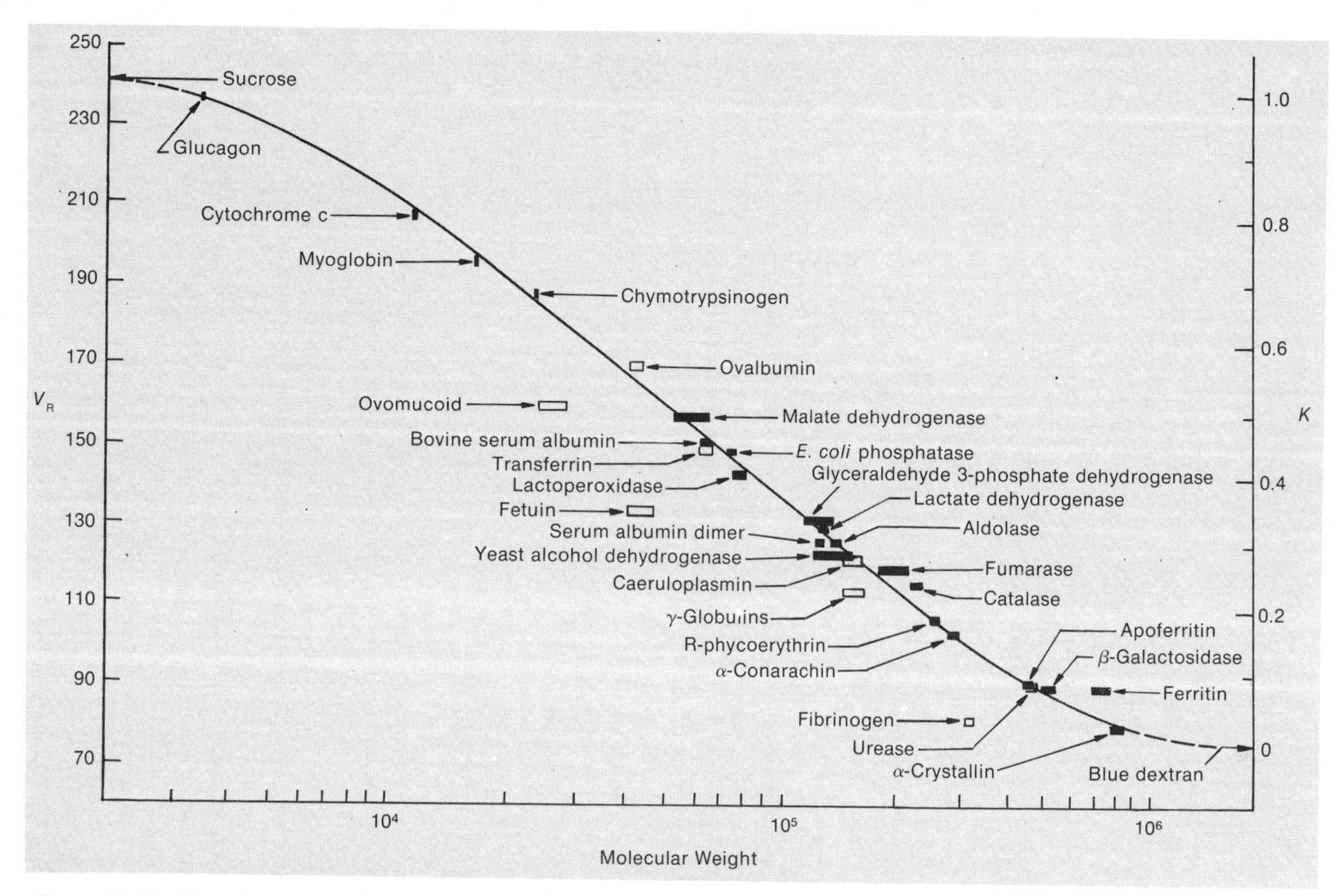


*Figure 14–17.* Plot of retention volume ($V_R$) and partition coefficient ($K$) against the logarithm of molecular weight for a series of proteins (solid rectangles) and some glycoproteins (open rectangles). The lengths of the bars indicate uncertainties in the known molecular weights, and the widths indicate uncertainties of ±1 ml in $V_R$. Experimental details in text. (Redrawn, with permission, from the paper by P. Andrews: Biochem. J. *96*:595, 1965.)

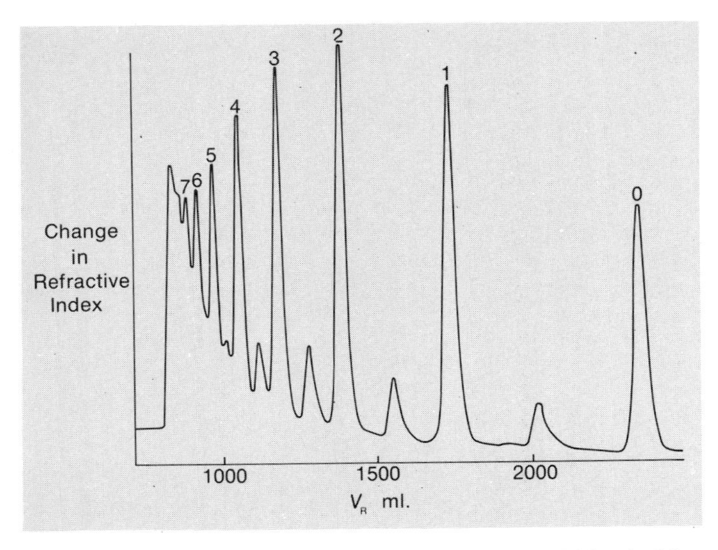


*Figure 14–18.* Gel permeation chromatogram of the epichlorohydrin-2,2-di(4'-hydroxyphenyl) propane copolymer discussed in the text. (Redrawn, with permission, from the paper by W. Heitz, B. Bömer, and H. Ullner: Makromol. Chem. *121*:102, 1969.)

Perhaps we will learn, eventually, to understand the necessity of the complexity which chromatographic techniques are revealing in, for example, biochemistry and organic geochemistry. At the same time, the underlying simplicity will, no doubt, become evident. We can already be sure, however, that understanding will never come without the continuous development of refined techniques for chemical separations. These developments can hardly be regarded as a trivial service, inasmuch as our very existence depends on the maintenance of a safe environment and the continual development of our knowledge of the chemistry of the biosphere.

## QUESTIONS AND PROBLEMS

1. Compound A migrates 7.6 cm from its point of application on a thin-layer chromatographic plate, whereas in the same time the solvent front migrates 16.2 cm beyond the point of sample application. (a) Calculate $\mathbf{R_f}$ for compound A. (b) On an identical plate, the solvent front has moved 14.3 cm beyond the point of sample application; where should compound A be located on this plate?
2. Silica sample A has an $\mathbf{R_f}$ for quinazoline of 0.50 with a one-to-three benzene-methanol solvent. With the same sample and solvent system, silica sample B has an $\mathbf{R_f}$ of 0.40. Which silica is more active?
3. If you want to produce maximum activity in a particular silica sample, how do you treat it?
4. Diatomite is a form of $SiO_2$ with a surface area per unit mass of less than 50 $m^2/g$. Should it be more or less active than silica gel?
5. If you want to produce maximum activity in an $Al_2O_3$ adsorbent, how do you treat it?
6. Consider an alumina sample with a specific surface area of 123 $m^2/g$. Calculate the "number of monolayers" (realizing the fanciful aspects of

such a number) corresponding to each step of the Brockmann activity scale.

7. Given the requirement that a separation is to be performed by means of thin-layer chromatography, choose the adsorbent best suited for each of the following:
   (a) the separation of a mixture of molecules according to their functional groups
   (b) the separation of methyl octadecanoate from methyl octadecenoate
   (c) the fractionation of a mixture of aromatic hydrocarbons
   (d) the separation of

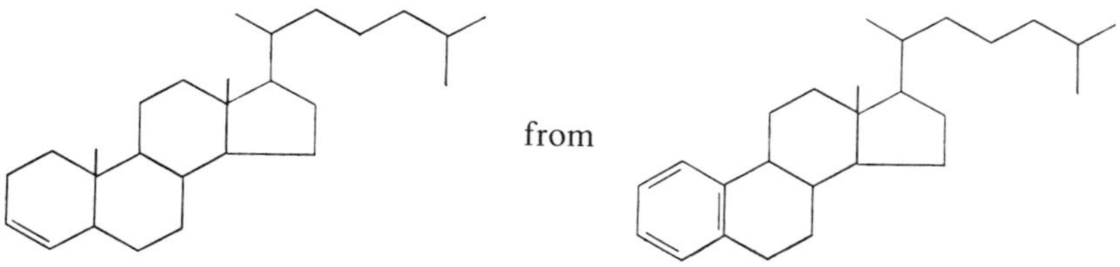

from

   (e) the separation of lysine from "neutral" amino acids
8. A friend complains that he is always losing the alkane fraction from his samples. He is sure that the alkanes are present when he starts his procedure; however, by the time he does thin-layer chromatography on zinc sulfide-impregnated plates, the spot for the alkanes cannot be found. He is making a big mistake – what is it?
9. A particular gas chromatographic column is operated with an inlet pressure of 45 psig (psig = gauge pressure) and an outlet pressure of 1 atm. What is the average pressure inside the column?
10. Given the following retention times for alkanes on an SE-30 column at 150°C, calculate the retention time of heptadecane under the same conditions: $CH_4$, $t_R = 25$ sec; $C_{13}H_{28}$, $t_R = 1.20$ min; $C_{15}H_{32}$, $t_R = 6.35$ min.
11. A certain gas chromatographic column is operated isothermally at 165°C with a flow rate of helium carrier gas of 24 ml/min measured at 23°C and 740 torr pressure. The column inlet pressure is 25 psig. Calculate the corrected retention volume for some compound which has a retention time of 18.72 min.
12. A given gas-liquid chromatographic column is loaded with 1.272 g of a packing which is described as "3 per cent SE-30 on Chromosorb W." The density of SE-30 is about 0.92 g/ml. Calculate $K$ for a compound which has a net retention volume of 273 ml on this column.
13. A particular gas chromatographic peak has adjusted retention times of 24.21 and 17.24 min at 180 and 200°C, respectively. The column in use has an efficiency corresponding to 5000 theoretical plates. What is the highest column temperature which can be used such that the peak width will still be greater than 25 sec? Assume that $t_R$ does not differ greatly from $t'_R$.
14. The methyl ester of the 22-carbon $n$-alkanoic acid has adjusted retention times of 72.0 and 58.3 min on a particular column at temperatures of 180 and 200°C, respectively. Will it be possible to elute this compound

in less than 45 min, if the liquid phase in the column has a maximum operating temperature of 210°C?

15. Compound A is co-injected into a column along with $n$-tetracosane and $n$-hexacosane. The observed retention times are A, 10.20 min; $n$-$C_{24}H_{50}$, 9.81 min; and $n$-$C_{26}H_{54}$, 11.56 min. Calculate the retention index of A.
16. Under the conditions in problem 15 above, a peak appears at an observed retention time of 12.02 min. Might it be a homolog of A? (What is its retention index?)
17. In the identification of cockroach pheromones, a peak was observed to be eluted at a column temperature of 155.2°C during a chromatographic run with temperature programming from 100 to 200°C at 2°C per minute. The elution temperatures of $n$-tetradecane and $n$-hexadecane under the same conditions were 141 and 162°C, respectively. Estimate the retention index of the peak.
18. A colleague is running cholesterol quite satisfactorily on chromatograph X. One day, he moves his column over to chromatograph Y and finds that the cholesterol peak appears much earlier than before. After days of checking flow rates, column temperatures, and so on, he finally traps the peak as it is eluted and discovers that the "cholesterol" has the second structure shown below:

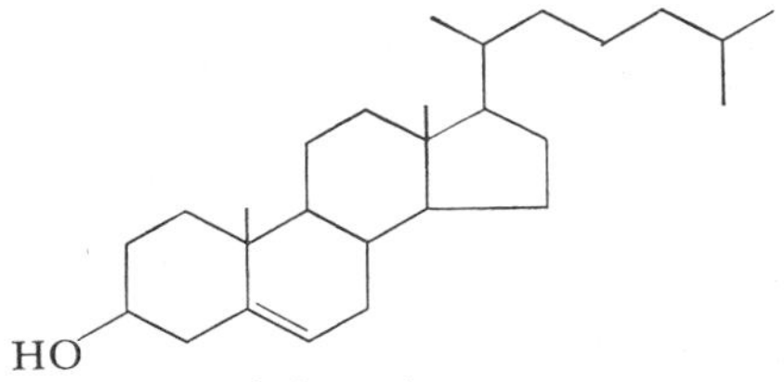

HO

cholesterol

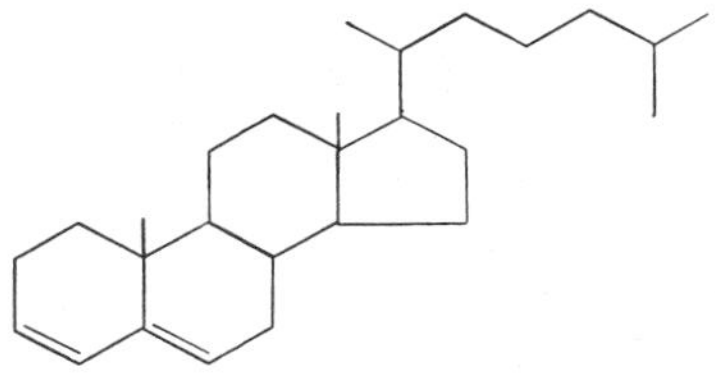

compound from chromatograph Y

To which part of the chromatograph should be turn his attention? What might he do to fix things up?

19. Predict the relative molar responses (sensitivities) in a flame-ionization detector for the following compounds: decane, decanone, diethyladipate, and 2,2,5-trimethylheptane.
20. It is planned to do a gas chromatographic analysis of an aqueous butanoic acid solution of a concentration which turns out to be just below the limit of detection for a flame-ionization detector. This problem can be solved if one uses a different detector and a special derivative. How?
21. A friend is analyzing trace gases using a gas chromatograph with a thermal-conductivity detector and helium carrier gas. A series of carbon dioxide standards is prepared by mixing small amounts of carbon dioxide with carefully purified helium. When these standards are injected into the column in order to calibrate the detector, the detector-output peak profiles shown on page 391 are obtained. Explain what might be wrong.

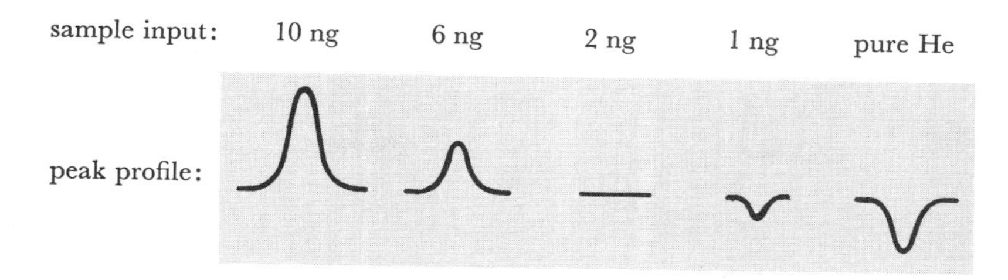


22. (a) What mobile-phase pH and what type of ion exchanger would you choose in order to attempt the ion-exchange chromatographic fractionation of a mixture of carboxylic acids?
    (b) Which acid would be eluted first, one with a $pK_a$ of 4 or one with a $pK_a$ of 5? Explain your answer.
    (c) Which acid would be eluted first, propanoic or octadecanoic? Explain your answer. (These acids have approximately equal $pK_a$ values.)
23. Should the ion-exchange chromatographic separation of a mixture of amines be possible? If so, under what conditions? Predict the order of elution of the following compounds: ammonia, *n*-propylamine, methylethylamine, trimethylamine, tetramethylammonium chloride. Refer to Appendix 2.
24. How much trouble would it cause the water-softening industry if people *liked* $Ca^{2+}$ and $Mg^{2+}$ in their water and wanted $Na^+$ and $K^+$ removed? Explain your answer.
25. Generally speaking, should the affinity of an ion-exchange resin for various species increase or decrease as one moves vertically downward in the periodic table of elements? Explain your answer.
26. A small amount of radioactive silver-108 ion is adsorbed in an ion-exchange column. It is necessary that the silver ion be eluted very rapidly in an absolute minimal volume of eluent. The form of the resin after the elution is of no importance. Suggest a solution to this problem.
27. A cation-exchange column is saturated with very strongly held $Fe^{3+}$. It is desired to recover the $Fe^{3+}$ and to convert the resin to the $H^+$ form. A friend suggests washing the column with 10 $F$ sulfuric acid, but you choose 6 $F$ hydrochloric acid instead. Why?
28. The normal retention time of trace amounts of $Na^+$ on a certain ion-exchange chromatographic column is 2.4 hr. A molecule of mobile phase (which is unretained by the resin) passes through the column in 23 min. It has been determined experimentally that a mobile-phase $Na^+$ concentration of 0.080 $M$ is sufficient to keep all the exchange sites in the $Na^+$ form. While the column is being washed with a 0.080 $M$ $Na^+$ solution, a trace of radioactive sodium ion is added to the eluent at the column inlet. When does the radioactive sodium ion first appear in the eluate? Explain your answer.
29. Should it be possible to separate amino acids by use of an *anion* exchange column? If so, explain how you would proceed; state, in particular, what eluents you would use.
30. The compounds listed below occur naturally, but do not happen to be included in Figure 14–12. On the basis of their structures and acidities,

speculate on their probable positions in the amino-acid elution sequence.

(a) tryptophan

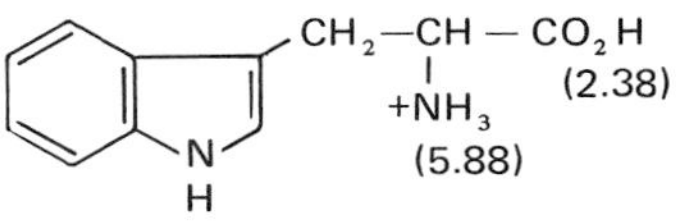

(indole ring, N–H) $-CH_2-CH-CO_2H$ (2.38); $+NH_3$ (5.88)

(b) hydroxyproline

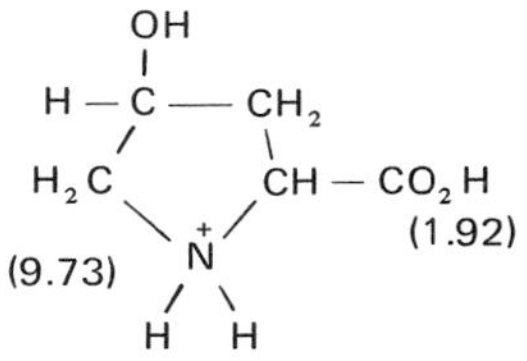

OH

H – C — $CH_2$

$H_2C$ $\qquad$ $CH-CO_2H$ (1.92)

(9.73) $\overset{+}{N}$

H H

(c) β-alanine

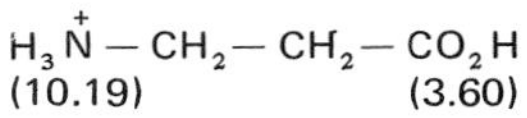

$H_3\overset{+}{N}-CH_2-CH_2-CO_2H$

(10.19) $\qquad$ (3.60)

(d) taurine

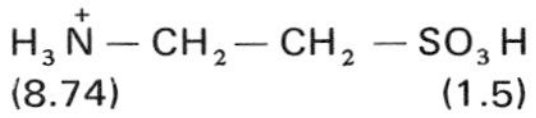

$H_3\overset{+}{N}-CH_2-CH_2-SO_3H$

(8.74) $\qquad$ (1.5)

(e) ornithine

$CH_2-\overset{+}{N}H_3$ (10.76)

$(CH_2)_2$

$H_3\overset{+}{N}-CH-CO_2H$

(8.69) $\qquad$ (1.71)

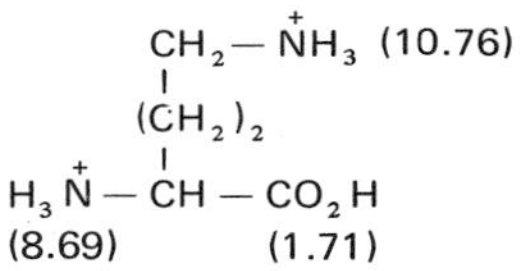

31. A molecular-exclusion chromatographic column is made from an old 50-ml buret. Having no other information, what statement can you immediately make about elution volumes for this column?
32. Imagine that you can make a porous material with any range of pore sizes that you wish. In order to obtain maximum resolution of a particular pair of compounds, do you create a material with a wide range of pore sizes or a very narrow range of pore sizes? Why?
33. Gel filtration is sometimes used in large-scale pharmaceutical preparation work. When large-scale procedures are frequently repeated, it often pays to adopt some highly specialized conditions. Can you conceive of any situation, for example, where it might be advantageous to construct a column in which materials with large and small pores are mixed? Discuss the characteristics of such a column with respect to solutes of various sizes.
34. Explain in your own words why particles with solid, inert cores having thin shells of chromatographic media around them offer advantages to the chromatographer? What possible disadvantages can you imagine?

## SUGGESTIONS FOR ADDITIONAL READING


1. K. H. Altgelt and L. Segal, eds.: *Gel Permeation Chromatography.* Dekker, New York, 1971.
2. D. D. Bly: Gel permeation chromatography. Science *168*:527, 1970.
3. P. R. Brown: *High Pressure Liquid Chromatography.* Academic Press, New York, 1973.
4. H. Determann: Principles of gel chromatography. *In* J. C. Giddings and R. A. Keller, eds.: *Advances in Chromatography.* Volume 8, Dekker, New York, 1969, pp. 3–45.
5. E. Heftmann, ed.: *Chromatography.* Second edition, Reinhold Publishing Corporation, New York, 1967.
6. F. Helfferich: *Ion Exchange.* McGraw-Hill Book Company, New York, 1962.
7. C. Horvath: Columns in gas chromatography. *In* L. S. Ettre and A. Zlatkis, eds.: *The Practice of Gas Chromatography.* Wiley-Interscience, New York, 1967.
8. J. C. Kirchner: *Thin-Layer Chromatography.* Volume XII of *Technique of Organic Chemistry*, E. S. Perry and A. Weissberger, eds. Wiley-Interscience, New York, 1967.
9. J. J. Kirkland, ed.: *Modern Practice of Liquid Chromatography.* Wiley-Interscience, New York, 1971.
10. I. M. Kolthoff and P. J. Elving, eds.: *Treatise on Analytical Chemistry.* Part I, Volumes 2 and 3, Wiley-Interscience, New York, 1959.
11. A. B. Littlewood: *Gas Chromatography.* Second edition, Academic Press, New York, 1970.
12. H. Purnell: *Gas Chromatography.* John Wiley and Sons, New York, 1962, Chapters 3 and 10.
13. K. Randerrath: *Thin-Layer Chromatography.* Academic Press, New York, 1966.
14. L. R. Snyder: *Principles of Adsorption Chromatography.* Dekker, New York, 1968.
15. E. Stahl, ed.: *Thin-Layer Chromatography, A Laboratory Handbook.* Second edition, Springer, New York, 1969.

# 15 INTRODUCTION TO SPECTROCHEMISTRY

Some of the earliest means used for the characterization of objects and substances were based upon the observation of color. Even today, the description of any object, from automobiles to computers, usually includes a statement regarding color. Color arises from the absorption and emission of light by matter, each form of matter displaying its own absorption and emission properties and, therefore, its own color. From this, it is a simple extension to realize that chemical species can be characterized by means of their absorption or emission behavior. In fact, this behavior is so important that an entire subdiscipline of chemistry called **spectrochemical analysis** has been developed around it. In this chapter, we will examine the fundamental principles of spectrochemistry; in the following chapters, we will consider a number of specific techniques which employ these principles for purposes of qualitative and quantitative chemical analysis.

## WHAT IS SPECTROCHEMICAL ANALYSIS?

The term "spectrochemical" derives from two other words: spectrum and chemical. In spectrochemical analysis, we employ the spectrum of electromagnetic radiation to determine chemical species and to characterize their interactions with electromagnetic radiation. As shown in Figure 15–1, a spectrum is a plot of some measurable property of the radiation, $f(\nu)$, as a function of the frequency of the radiation, $\nu$. From a spectrum, two important pieces of information can be obtained. First, from the shape of the spectrum, a chemical species can often be identified qualitatively. Second, from the magnitude of $f(\nu)$ at chosen frequencies, the amount of a chemical species present can be determined quantitatively.

These statements become more meaningful when it is recalled that, for a photon (quantum) of electromagnetic radiation, frequency is related to energy through the Planck equation,

$$E = h\nu \tag{15–1}$$

where $E$ is the energy of the photon, $\nu$ is its frequency, and $h$ is Planck's constant ($6.624 \times 10^{-27}$ erg sec). Therefore, a photon of electromagnetic radiation has a definite energy and can cause transitions between the quantized energy states in atoms, molecules, and other chemical species. To cause such a transition, the energy of the photon must be equal to the difference between the energy states involved in the transition. Thus, we can examine chemical species by using electromagnetic radiation as a probe, the frequency of the radiation being related to the energy change associated with the observed transition.

Because energy states differ among chemical species, it is to be expected that the energy changes involved in the transitions will also differ. This implies that a spectrum will be a highly individual property of each substance, and that observation of the spectrum can be employed to advantage for identification of the substance. In effect, the

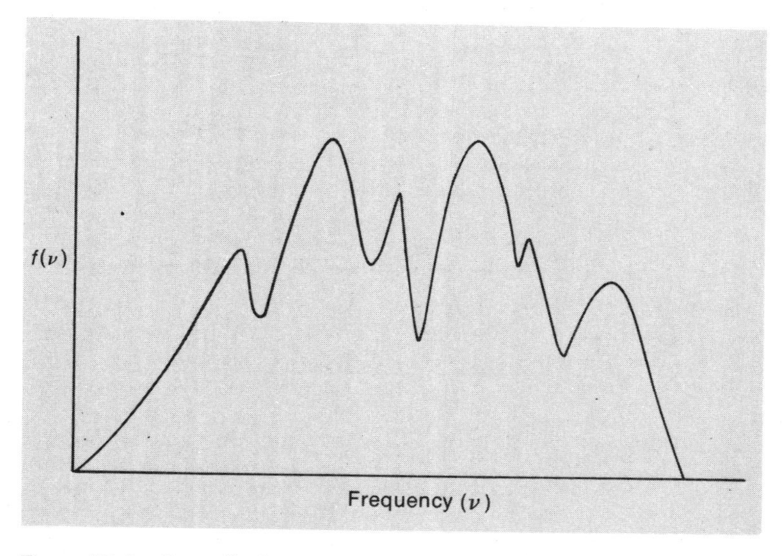


*Figure 15–1.* Generalized spectrum consisting of a plot of some function of frequency, $f(\nu)$, versus $\nu$, the frequency of electromagnetic radiation. In spectrochemical analysis, $f(\nu)$ is usually related to the power of the radiation at each frequency.

spectrum is a map of the transitions which occur between energy states of the chemical species. Furthermore, the number of times each transition occurs during a fixed interval is related to the total number of chemical species which can undergo that transition. Therefore, if the measured parameter, $f(\nu)$, in a spectrum can be related to the total number of transitions, the spectrum can be used to determine the concentration of species present.

There is no restriction on the frequency of electromagnetic radiation employed in spectrochemistry. In fact, spectrochemical analysis can utilize frequencies ranging from those of audio waves (10 to 10,000 Hz)* to those of gamma rays ($10^{22}$ Hz). Yet, over this enormous range of frequencies, the principles of spectrochemical analysis remain unchanged, the only difference being the magnitude of the energy changes which are probed. These energy changes are often associated with particular types of transitions – rotational, vibrational, electronic, nuclear, and so on. Figure 15–2 shows the range of frequencies commonly used in spectrochemistry, along with the kind of transition probed in each frequency region.

Although we can associate definite types of transitions with each frequency region delineated in Figure 15–2, another reason for establishing these frequency regions is that the instrumentation required to make meaningful measurements in each region is different. For example, the optical instrumentation useful for the measurement of electronic transitions in the ultraviolet and visible spectral regions cannot be used at all for observation of nuclear transitions in the $\gamma$-ray region or of molecular rotations in the microwave region. For this reason, spectrochemical analysis has grown exceedingly diverse and now includes a number of individual branches, each of which deals with a specific spectral region. Because the experimental measurements and tools are different for each spectral region, it is customary to consider each region individually, so that the study of spectrochemistry often becomes an examination of seemingly unrelated

*1 Hz (hertz) ≡ 1 cycle per second.

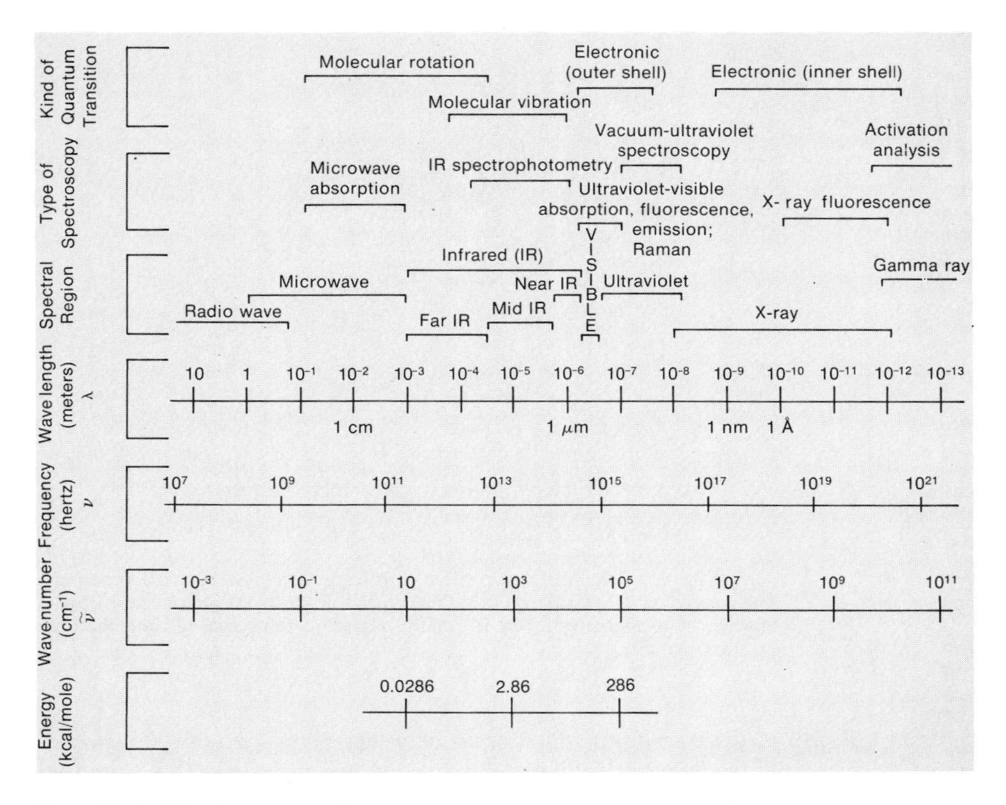


*Figure 15–2.* The electromagnetic spectrum.

techniques. In our discussion, however, we will retain the thought that all spectrochemical techniques have many common underlying principles that can be applied equally to all areas.

## A SPECTROCHEMICAL VIEW OF ELECTROMAGNETIC RADIATION

Because the interaction of radiation with chemical species is the foundation upon which spectrochemical analysis is based, several important characteristics of electromagnetic radiation must be understood before we proceed further. To assist in this discussion, it will be helpful to visualize an electromagnetic wave.

Figure 15–3 portrays a sinusoidally oscillating electromagnetic wave, traveling through space in an arbitrary direction $x$. As its name implies, the wave consists of oscillating electric ($E$) and magnetic ($M$) fields, which are orthogonal to each other and which travel at a constant velocity ($c$), equal to approximately $3 \times 10^{10}$ cm per sec in vacuum. Let us assume that the wave has a constant frequency of oscillation ($\nu$) and that it can be observed at some instant in time. Because the wave travels at a constant velocity ($c$), the spacing between its maxima – that is, its **wavelength** – will be a constant which is characteristic of the wave. This wavelength will simply be the distance the wave travels during one period ($1/\nu$) of its oscillation. Therefore, the wavelength ($\lambda$) can be calculated from the following well-known formula:

$$\text{distance} = \text{velocity} \times \text{time}$$

$$\lambda = c \times \frac{1}{\nu} = \frac{c}{\nu} \tag{15–2}$$

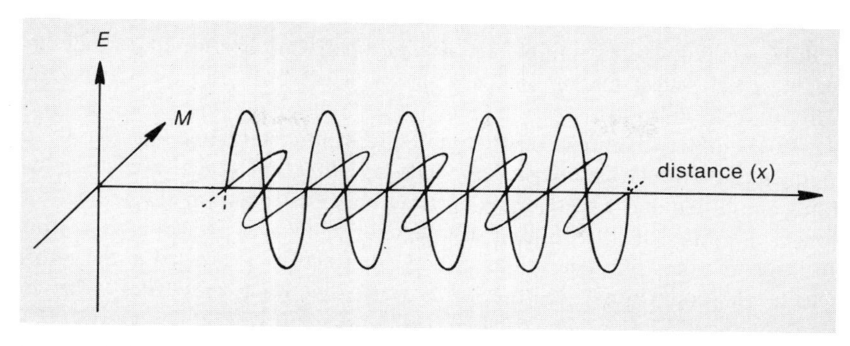


*Figure 15–3.* Portrayal of an electromagnetic wave, showing the electric ($E$) and magnetic ($M$) vectors; the wave is represented at some instant in time and is traveling through space in direction $x$. In this representation, the wave shown is plane polarized.

Because of this relationship to frequency ($\nu$), wavelength ($\lambda$) is often used as the horizontal axis for a spectrum such as that illustrated in Figure 15–1. Wavelength is expressed in terms of meters (m), centimeters (cm), micrometers ($\mu$m), nanometers (nm), or angstrom units (Å), depending on the spectral region. The relationship among these units is given in Table 15–1. Note that wavelength is inversely proportional to frequency and, therefore, to energy as well. Because it is sometimes desirable to refer to the energy of a spectrochemical transition, a quantity proportional to energy is often used instead of wavelength or frequency. This quantity, called a **wavenumber**, has units of $cm^{-1}$ (reciprocal centimeters), is given the symbol $\tilde{\nu}$, and is defined by the expression

$$\tilde{\nu} = \frac{1}{\lambda} = \frac{\nu}{c} \tag{15-3}$$

where $\lambda$ is the wavelength in cm. Although $\tilde{\nu}$ is *proportional* to frequency, it is in fact *not* a frequency and should never be so called. Properly formulated, frequency is expressed in units of $sec^{-1}$ or Hz (hertz), whereas $\tilde{\nu}$ is expressed in units of reciprocal length. These three quantities – wavelength, frequency, and wavenumber – can be used interchangeably when spectra are displayed. Generally, the mode of display is chosen according to convenience, although conversions between units can be easily made by applying the formulas presented above.

Several other characteristics of electromagnetic waves are important to spectrochemical analysis. One such characteristic, monochromaticity, refers to the spectral purity of the wave. For an idealized wave such as that depicted in Figure 15–3, only a single frequency exists. Such a wave is said to be **monochromatic**, which literally means "single colored." Actually, few truly monochromatic waves are ever employed in spectrochemistry. More often, the radiation used contains a range of frequencies spread over a certain spectral interval. To describe the breadth of this frequency range, the term **bandwidth** or **spectral bandwidth** is used. Although bandwidth properly refers only to a frequency range, it is also used commonly to denote a wavelength interval. As we shall

**Table 15–1**

| | |
|---|---|
| 1 cm | $= 10^{-2}$ m |
| 1 $\mu$m | $= 10^{-6}$ m |
| 1 nm | $= 10^{-9}$ m |
| 1 Å | $= 10^{-10}$ m |

find later, the bandwidth of electromagnetic radiation is often of considerable importance in spectrochemical measurements and can affect both the qualitative and quantitative validity of an analysis.

In all spectrochemical measurements, it is important to determine the amplitude and frequency of the electromagnetic radiation. Unfortunately, the accurate measurement of both amplitude and frequency is possible only for radiation at microwave frequencies or lower, because of the limited frequency-response of available detectors. In spectral regions of higher frequency, the variable that can be measured is the **radiant power ($P$)**, which is proportional to the *square* of the wave amplitude. Radiant power has spectrochemical importance because it is the amount of energy transmitted in the form of electromagnetic radiation per unit time. If $E$ is the energy of a photon, the radiant power can be expressed by the relation

$$P = E\Phi = h\nu\Phi \qquad (15\text{–}4)$$

where $\Phi$ is the photon flux – that is, the number of photons (quanta) per unit time.

The radiant power of a beam of electromagnetic radiation is often referred to as its *intensity*. Actually, **intensity** is properly defined as radiant power from a point source per unit solid angle, typically with units of watts per steradian. Although one can correctly speak of the intensity of a source of radiation, it is incorrect to describe radiation striking a sample in terms of its intensity, especially if the radiation is **collimated** – that is, if it appears to originate from an infinitely distant source. Unfortunately, the term *intensity* is widely used in a qualitative sense or even as an equivalent to *radiant power*. To avoid these ambiguities, we will use the term *intensity* only in its correct sense.

## TYPES OF INTERACTIONS OF RADIATION WITH MATTER

Let us consider how radiation can interact with chemical species to provide information about the species. We will specifically exclude several types of interactions from our discussion. These interactions, termed reflection, refraction, and diffraction, are used extensively in spectrochemical instrumentation and are better treated in a text on that topic. Here, we will concern ourselves with the ways in which radiation interacts more directly with the chemical sample itself. Three distinct examples of this sort of interaction can be cited – absorption, luminescence, and emission. Our discussions of these interactions will be intentionally general and will apply equally well to any spectral region of electromagnetic radiation (see Figure 15–2) and to samples of every form (atoms, ions, or molecules) in any phase (solid, liquid, or gas).

### Absorption

If a beam of electromagnetic radiation is sent into a chemical sample, it is possible for the sample to absorb a portion of the radiation. This phenomenon is depicted in Figure 15–4A, which shows a beam of radiation having a radiant power $P_0$ being directed into a sample. Each specific frequency, $\nu_1$, $\nu_2$, and so on, which comprises the beam of radiation will, of course, have its own energy, $h\nu_1$, $h\nu_2$, and so on. If the chemical sample contains a species whose energy states differ by any of these exact energies, the sample will absorb radiation at those frequencies. This behavior is illustrated in Figure 15–4B, in which a chemical species having energy levels **G** and **E** is portrayed. If the species (atom, molecule, or ion) exists in the lower (ground) energy state **G** before its encounter with

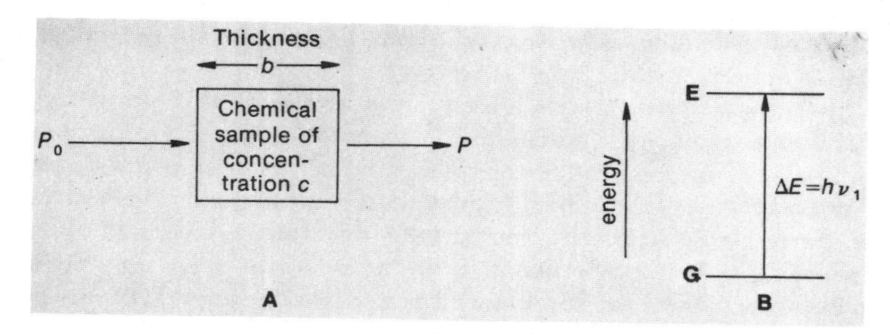


*Figure 15–4.* *A*, Diagram showing the relationship between the incident ($P_0$) and transmitted ($P$) radiant power for an absorbing substance at a concentration $c$ in a spectrophotometer cell with a sample path length $b$.

*B*, Diagram illustrating the change in energy ($\Delta E$) of a species during absorption of a photon of energy $h\nu_1$. In the act of absorption, the species is promoted from the ground state (G) to an excited state (E).

the beam of radiation, it can be **excited** to the upper state **E**. In this excitation, the species must absorb a quantity of energy $h\nu_1$, which is exactly equal to the difference in energy between states **G** and **E**.

Energy required to excite the species in the upper energy state **E** is drawn from the beam of radiation, so that the total radiant power of the beam is diminished at the frequency of absorption, $\nu_1$. Therefore, after encountering a number of absorbing species in the sample, the beam will exit from the sample with a reduced radiant power $P$. It should be recognized that *only* those frequencies capable of being absorbed by the sample will be attenuated in a purely absorbing sample; all other frequencies will pass through the sample with no power loss. This result suggests the possibility that components in the sample can be identified from the absorption spectrum, that is, from the frequency components of the radiation which are absorbed. Understandably, the diminution in the radiant power of the beam at each frequency should be related to the number of absorbing chemical species present in the sample. This information, which provides the basis for quantitative analysis, will be treated in more detail later in this chapter.

### Luminescence

When a quantity of radiant energy (a photon) is absorbed from a beam of radiation by a chemical species, the species is promoted to an *excited state*, **E**, as shown in Figure 15–4B. However, the excited-state species has a limited lifetime, and would prefer to rid itself of this extra energy and return to the ground state **G**. To attain this state of tranquility, an excited chemical species must dispose of a quantity of energy equal to the difference in energy between the excited state and the ground state. This energy can be released in several ways: it can be transferred to other species, it can be converted into other forms of energy (such as thermal or electrical energy), or it can be emitted in the form of electromagnetic radiation. When the energy gained by a chemical species during absorption is emitted in the form of radiation, the process is called **luminescence**. This process is portrayed in Figure 15–5.

In Figure 15–5A, a beam of electromagnetic radiation of power $P_0$ is sent into a chemical sample, just as in the previously considered case of absorption. If any components of the sample have suitable energy levels, a portion of the incoming radiation will be absorbed so that the transmitted beam will have a slightly reduced power $P$. Thus,

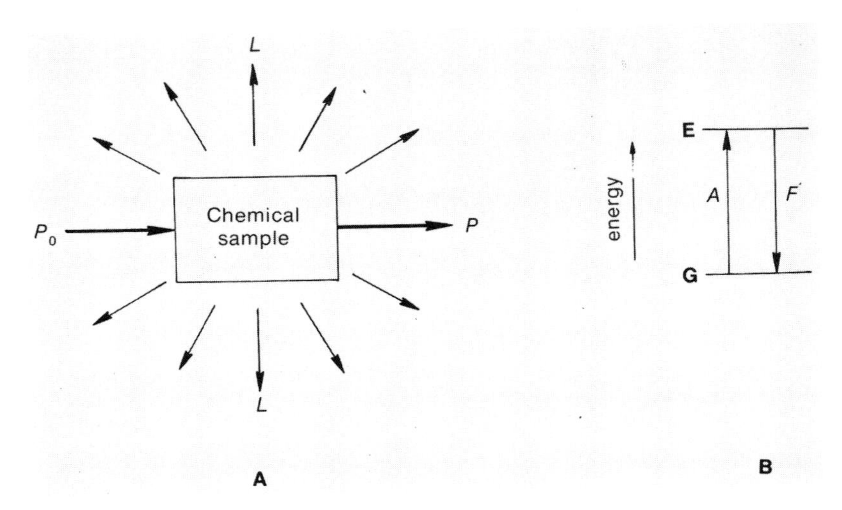


*Figure 15–5.* *A*, Diagram showing isotropic luminescence produced by absorption of radiation from incident beam of power $P_0$. Radiant power ($L$) of luminescence is some fraction of the radiant power absorbed ($P_0 - P$).

*B*, Changes in energy of a chemical species during absorption ($A$) and resonance fluorescence ($F$). Resonance fluorescence is a special luminescence process.

the difference in power ($P_0 - P$) between the incident ($P_0$) and transmitted ($P$) beams is used to excite chemical species present in the sample. This process is portrayed as the absorption step labeled $A$ in Figure 15–5B. The excited species now present will spontaneously undergo deactivation, one possible way being by emission of radiation. If the energy is emitted immediately, the radiated photon will be equal in energy and frequency to the radiation which was initially absorbed. This so-called **resonance fluorescence** is one type of luminescence and is labeled $F$ in Figure 15–5B.

Although the incoming radiation ($P_0$) and the transmitted radiation ($P$) are directional, the luminescent radiation ($L$) has an equal probability of traveling in any direction, as indicated in Figure 15–5A, and is thus said to be **isotropic**.

Like absorption, the phenomenon of luminescence can be used for both qualitative and quantitative analysis. Clearly, the radiant power of luminescence will depend both on the concentration of the luminescing chemical species and on the frequency of the incident radiation.

## Emission

Species in a chemical sample can, of course, be excited by means other than the absorption of radiant energy. Thermal, chemical, electrical, and other forms of energy can all be used to excite atoms, molecules, and ions to higher energy states. If this excitation results in the liberation of electromagnetic radiation from the sample, the process is termed **emission**. This process is portrayed in Figure 15–6A, in which energy in a nonradiant form is introduced into a sample. Provided that the energy is of sufficient magnitude, a number of species in the sample can be excited to a higher energy state **E**, as shown in Figure 15–6B. Following excitation, the species are in a situation similar to that which exists during the processes of absorption and luminescence, where their excited-state energy can be lost by either radiational or nonradiational means. If, as in the case of luminescence, the energy is released through radiation, it will be liberated in all

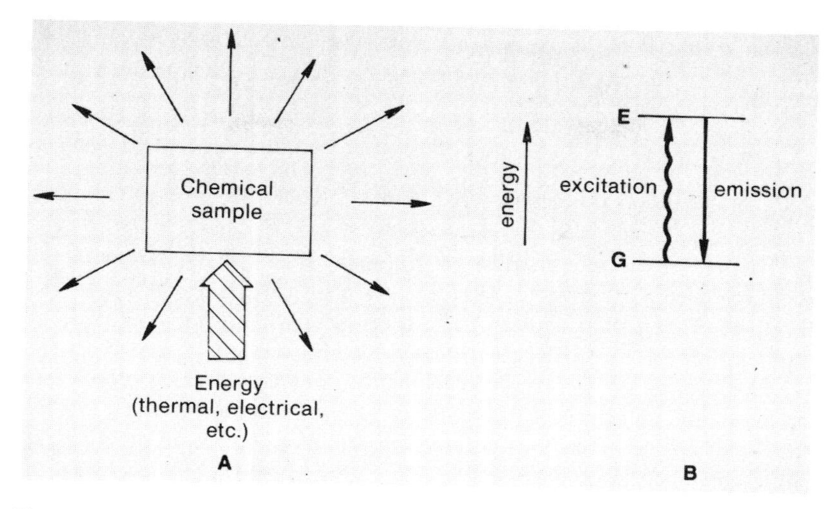


*Figure 15–6.* *A*, Diagram illustrating isotropic emission from a chemical species excited through the input of thermal, electrical, or other energy.

*B*, Energy changes that occur during excitation of and emission by a chemical species; the wavy line used for excitation denotes a radiationless process.

directions. This emission process is very important in spectrochemical analysis, not only for the examination of chemical samples but also in the generation of radiation used in the analysis. Because other forms of energy can be conveniently converted to radiant energy by this means, the process of emission is essential to the operation of sources of radiation, as will be discussed later. As one would surmise, luminescence is a special kind of emission process.

## SPECTROCHEMICAL MEASUREMENTS

Early applications of spectral measurements to chemical analysis were generally limited to visual observations of color and "brightness" of chemical samples. Because of the need to rely on subjective estimations, these measurements were semiquantitative at best and could only be used as crude guides in qualitative identification. Central to such difficulties were the inherent limitations in resolution and range of man's senses.

One of the most serious limitations was the necessity to visually detect the radiation (light) which was employed. Not only is the response of the human eye limited to a very narrow spectral region – the visible region shown in Figure 15–2 – but the eye is relatively insensitive to small changes in light level. Therefore, quantitative results are difficult to obtain. In addition, the eye responds differently to various frequencies of visible radiation, so that a comparison of light levels at two different frequencies (colors) is a rather uncertain endeavor. As we are all aware, eyes are subject to fatigue, so that the visual examination of a large number of samples is undesirable and would probably result in increasingly greater error as fatigue set in. Finally, the eye (as well as the observer) has a rather limited response time, probably on the order of tenths of a second. Response time will often limit the number of samples that can be examined over a period of time. Furthermore, response time can affect the precision and accuracy of an analysis.

Scientists involved in the study of spectrochemical analysis have sought to remove the limitations discussed above through the use of instruments. Instruments have caused a revolution in spectrochemical analysis, converting it from a mistrusted, seldom-used art to a science that is the most widely used of all analytical techniques. In its most general form, an instrument for use in spectrochemical analysis can be visualized as shown in Figure 15–7.

### A Generalized Spectrochemical Instrument

In the generalized instrument of Figure 15–7, an input device is used to convert chemical information in a sample into information in the form of electromagnetic radiation. Therefore, this device is a **transducer** which *encodes* the chemical information into another form, namely, electromagnetic information.

This concept of a transducer is highly general and can be applied to other kinds of chemical and physical instrumentation as well. A loudspeaker is an electrical input–acoustical output transducer, whereas a glass electrode and flame-ionization detector are both chemical input–electrical output transducers. These examples illustrate that a transducer changes the *form* of the information but, ideally, not its content.

To decode or extract the chemical information encoded in the radiation, a second transducer, commonly called the detector, is required. In the spectrochemical instrument, the function of a detector is to provide a measurable electrical signal which is proportional to some property, usually radiant power, of the radiation incident upon the detector. In effect, the detector converts the information present in the electromagnetic radiation to another form, generally electrical, which is more amenable to signal-processing techniques.

The electrical signal from the detector, which contains the encoded information about the sample, must next be rendered interpretable to a human observer. For this operation, a readout system is employed; the final output of the readout device can take many forms, from a simple meter deflection to columns of numbers or graphical displays from a computer. However, its function remains the same – to convert the information present in the incoming electrical signal to a form which is meaningful.

A final but indispensable component of our generalized spectrochemical instrument is the control system. This device is responsible for coordinating all the events and operations that take place in the instrument, from selection and introduction of the sample to interpretation of the readout signal. In most conventional instruments, this control function is mainly performed by a human operator. To a varying extent, the operator will be assisted by partially automated control systems, although he will still be primarily responsible for monitoring and decision-making. In future spectrochemical

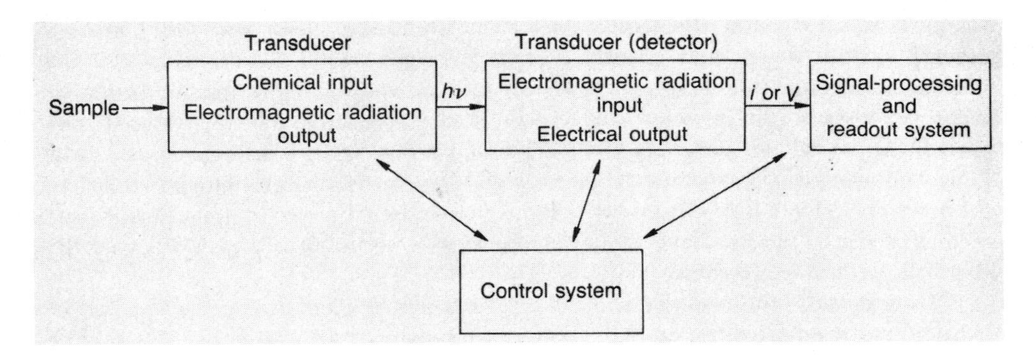


*Figure 15–7.* Generalized spectrochemical instrument; arrows indicate the flow of information through the instrument, and $i$ or $V$ represents an electrical current or voltage, respectively.

instruments, it can be expected that the control function will be automated to an ever greater extent, with on-line digital computer systems furnishing the necessary manipulatory and decision-making power.

**An Absorption Spectrophotometer.** As an example of a specific type of instrument that can be employed in spectrochemical analysis, let us consider the device shown in Figure 15–8. Components within the dashed box work together as a chemical input–electromagnetic radiation output transducer, as called for in Figure 15–7. For the instrument in Figure 15–8, a broad-band source of radiation is used. This broad-band source, which can be any of several available types (such as a tungsten or deuterium lamp), emits continuous radiation over a wide spectral range. To select the frequencies of radiation to be sent through the sample, a device called an optical monochromator is employed. This monochromator permits one to choose a narrow range of spectral frequencies and to vary the frequencies which are passed. From the monochromator, selected radiation is sent to the sample, where the radiant power ($P_0$) of the radiation is altered according to the chemical nature of the sample. That portion of the radiation transmitted by the sample ($P$) is converted by the photodetector into a proportional electrical signal, generally in the form of a voltage or current. Finally, the electrical signal is displayed on a suitable meter or recorder, possibly calibrated in terms of the concentration of the sample. In the present situation, it is assumed that all control operations are carried out manually.

Because the instrumentation requirements for various spectral regions can be very different, there is often little similarity in the design and appearance of instruments used for different methods of spectrochemical analysis, even though all instruments fit the generalized block diagram of Figure 15–7. Instruments employed for one spectral region or another vary widely in size, cost, complexity, and even principle of operation.

## FUNDAMENTAL LAWS OF SPECTROCHEMISTRY

We have briefly discussed the nature of three processes commonly employed in spectrochemical analysis: absorption, luminescence, and emission. In this section, we will derive and discuss some of the quantitative relationships governing these processes. These relationships, which will be used extensively in subsequent chapters, are fundamental to all spectrochemical techniques and supply the foundation upon which much of quantitative spectrochemistry is based.

### Quantitative Laws of Absorption

Quantitative methods of analysis based upon the absorption of radiant energy by matter require the measurement of radiant power and a quantitative understanding of the

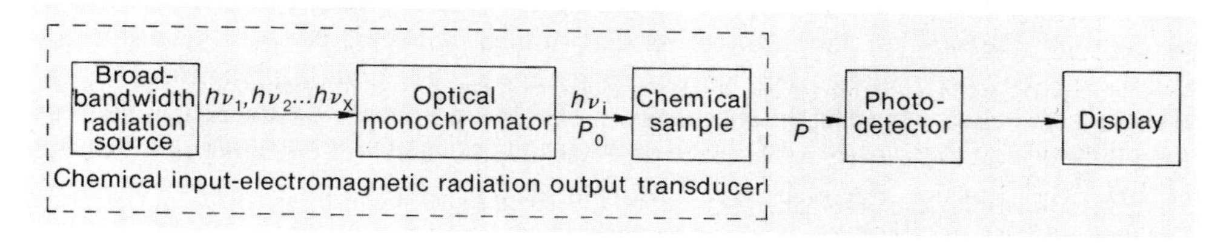


*Figure 15–8.* Block diagram of a versatile spectrochemical instrument for the measurement of absorption in the ultraviolet-visible spectral region.

laws which govern the extent of absorption. The power of an electromagnetic wave is quantized, and the wave can be viewed as a beam of photons, with the number of photons per unit time being proportional to radiant power.*

Adopting this view and considering the transmission of electromagnetic radiation through a container of absorbing species, we can state simply that the extent of absorption depends on the number of encounters between photons and species capable of absorbing them. As the photons pass through the medium, the rate at which they are absorbed depends on the number of photon-absorber collisions, which in turn depends on the power of the electromagnetic radiation and on the concentration of the sample species. For example, if we wish to double the rate of photon absorption, we can double the number of photon-absorber collisions by doubling the number of photons or by doubling the number of absorbers. In either case, collisions will occur twice as often, and the rate of absorption will go up by a factor of two. If the population densities of both photons and absorbers were doubled, they would bump into each other four times as often, and the rate of absorption would increase by a factor of four.

Relationships between radiant power, concentration, and rate of absorption are embodied in the **Lambert-Beer law**, which is often referred to simply as **Beer's law** and which can be expressed mathematically as

$$\log \frac{P_0}{P} = abc \qquad (15\text{–}5)$$

where $P_0$ is the power of the radiation incident upon the sample and $P$ is the power of the radiation transmitted by the sample. If the parameter, $b$, commonly known as the **sample path length**, is expressed in centimeters and the concentration factor, $c$, in grams of absorbing substance per liter of solution, the constant $a$ is designated as the **absorptivity**† and has units of liter $g^{-1}$ $cm^{-1}$.

Several features of equation (15–5) have experimental significance. First, it can be seen that the concentration can be determined from a measurement of relative beam power with and without the absorber in the beam. Absolute measurements of beam power are not necessary, a fact which considerably simplifies absorption spectrophotometric procedures. Second, it is clear that the path length of the radiation through the sample medium must be accurately known. This condition is usually met if the sample is placed in a **spectrophotometer cell** which is as transparent as possible to the radiation being employed. The composition, size, and form of the cell depend upon the nature and concentration of the sample and on the spectral region being utilized for the measurements.

Frequently, it is desirable to specify $c$ in terms of molar concentration, with $b$ remaining in units of centimeters. In the latter situation, Beer's law is rewritten as

$$\log \frac{P_0}{P} = \varepsilon bc \qquad (15\text{–}6)$$

where $\varepsilon$ (in units of liter $mole^{-1}$ $cm^{-1}$) is called the **molar absorptivity**.

---

*To speak of an electromagnetic wave as a beam of photons is to oversimplify the situation somewhat. For example, an understanding of scattering phenomena or of deviations from Beer's law caused by changes in refractive index (discussed later) requires that the electromagnetic radiation be treated as a wave.

†Absorptivity ($a$) and absorbance ($A$) are terms recommended originally by the Joint Committee on Nomenclature in Applied Spectroscopy, established by the Society for Applied Spectroscopy and the American Society for Testing Materials (see H. K. Hughes: Anal. Chem., *24*:1349, 1952). In the older literature, absorptivity was frequently referred to as absorbancy index, specific extinction, or extinction coefficient; and absorbance was often called optical density, absorbancy, or extinction.

The quantity log $(P_0/P)$ is defined as **absorbance*** and is given the symbol $A$. Because $A$ is directly proportional to the concentration of the absorbing species, some instruments for absorption measurements are calibrated to read directly in absorbance units. The ratio $(P/P_0)$ is called the **transmittance**, $T$, whereas $100(T)$ is the **per cent transmittance**. It is not uncommon to find commercial instruments calibrated directly in units of transmittance or per cent transmittance.

**Deviations from Beer's Law.** The applicability of Beer's law can be tested for any particular system if one measures the absorbance for each of a series of samples of known concentration of the absorbing species. A plot of the experimental data in terms of absorbance ($A$) versus concentration ($c$) will yield a straight line passing through the origin if Beer's law is obeyed. More often than not, a plot of the data over a wide range of concentration of an absorbing substance will give a graph such as that shown in Figure 15–9, indicating that Beer's law is applicable only up to concentration $c_1$. Nevertheless, if a suitable calibration curve is prepared from a series of samples containing known concentrations of the absorbing species, it is still possible to determine the concentration of the absorbing substance in an unknown.

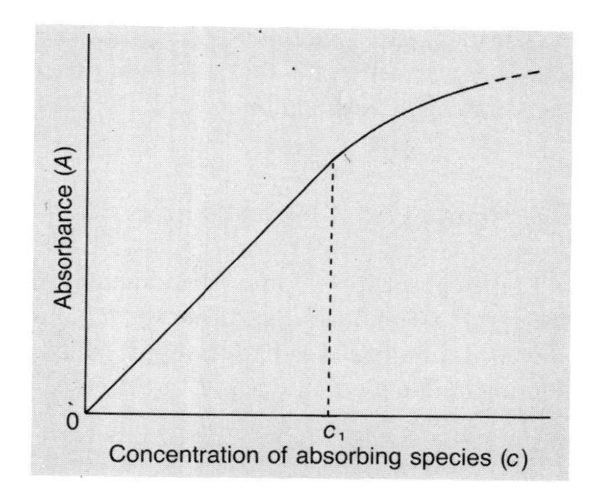


*Figure 15–9.* Relationship between absorbance ($A$) and concentration ($c$) of an absorbing substance. Such a graph is often called a *Beer's law plot* or, more generally, a *working curve*.

Deviations from Beer's law are of three types – real, chemical, and instrumental. Real deviations arise from changes in the refractive index of a medium which occur because of variations in the concentrations of its components. A rigorous derivation of Beer's law assumes a constant refractive index for the absorbing medium; any error in this assumption produces a consequent uncertainty in the experimental results. Generally, errors caused by changes in refractive index are minimal, so that *real* deviations from Beer's law are neglected in most absorption spectrochemical analyses.

Chemical deviations from Beer's law are caused by shifts in the position of a chemical or physical equilibrium involving the absorbing species. Consider, for example, the following reaction between an absorbing species A and another species B to form species C and D:

$$A + B \rightleftharpoons C + D$$

*See second footnote on page 404.

Obviously, any change affecting the position of this equilibrium will alter the concentration of A. In fact, a certain change in the original concentration of A itself *might* not result in a proportional change in the final concentration of A available to absorb radiation, because the position of equilibrium depends on other species besides A alone. To avoid this problem, conditions should be established so that any equilibrium involving A is relatively unaffected by changes in the concentration of A. For example, the concentration of either C or D could be made sufficiently great to keep the position of equilibrium shifted well toward the formation of A. Specific examples of this behavior will be encountered in later chapters.

Use of non-monochromatic radiation is the most common instrumental cause of deviations from Beer's law. Because Beer's law is rigorously applicable only for absorption of radiation of a single frequency, some error will nearly always exist when real instrument components are used. In most spectral regions, it is difficult or impossible to obtain truly monochromatic radiation. Specific examples of this sort of error will be discussed when we consider absorption of radiation in particular spectral regions.

Some experimental procedures used in absorption spectrophotometry involve calculations based upon Beer's law, and others do not. As already implied, it is frequently possible to perform highly accurate quantitative determinations even when the chemical system departs from this fundamental law of absorption. For a successful analysis, the chief requirements are that the radiation-absorbing properties of a chemical system be measurable and reproducible.

## Quantitative Law of Luminescence

Like absorption, the phenomenon of luminescence is useful for quantitative spectrochemical analysis. However, in luminescence spectrometry, it is the absolute power of luminesced radiation that is measured, rather than a power ratio. The expression relating luminescence power ($L$) to sample concentration ($c$) can be shown to be*

$$L = kP_0abc \tag{15-7}$$

where $k$ is a proportionality constant, $P_0$ is the radiant power of the exciting radiation incident on the sample (see Figure 15–5A), $a$ is the absorptivity of the sample for the exciting radiation, and $b$ is the path length of the exciting radiation through the luminescing sample.

According to equation (15–7), the detected luminescence signal ($L$) can be increased for any given sample concentration ($c$) by an increase in either the incident radiant power of the exciting source ($P_0$) or the radiation path through the sample ($b$). Both of these latter variables are experimentally adjustable, a feature which often makes luminescence measurements more sensitive than those employing absorption.

Strictly speaking, equation (15–7) is valid only at low sample concentrations and fails at higher values of $c$. This behavior leads to typical luminescence *working curves* such as that shown in Figure 15–10. As with absorption measurements, concentrations above $c_1$ in Figure 15–10 can still be determined, albeit less conveniently, merely by construction of the working curve from measurements on standard solutions of the luminescing species.

---

*A complete derivation of equation (15–7), showing important assumptions, can be found in D. G. Peters, J. M. Hayes, and G. M. Hieftje: *Chemical Separations and Measurements*. W. B. Saunders Company, Philadelphia, 1974, p. 618.

*Figure 15–10.* Relationship between radiant power of luminescence ($L$) and the concentration ($c$) of the luminescing species. Note that $c_1$ is the concentration beyond which non-linearity becomes significant.

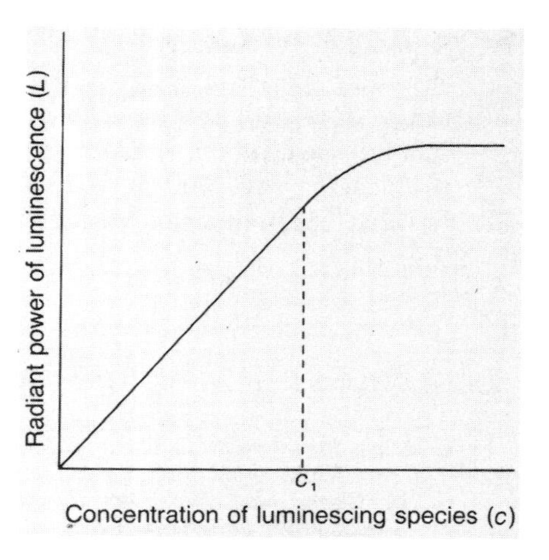


## QUESTIONS AND PROBLEMS

1. Define or explain each of the following terms: spectrum, radiant power, intensity, wavelength, wavenumber, monochromatic, spectral bandwidth, photon, absorption, luminescence, emission, ground state, excited state, excitation, isotropic, transducer, source, readout system, absorptivity, molar absorptivity, absorbance, transmittance, per cent transmittance, Beer's law.
2. Calculate the frequency, $\nu$, in reciprocal seconds (hertz) corresponding to each of the following wavelengths of electromagnetic radiation: (a) 222 nm, (b) 530 nm, (c) 17 Å, (d) 0.030 cm, (e) $1.30 \times 10^{-7}$ cm, (f) 6.1 $\mu$m.
3. Calculate the wavenumber, $\tilde{\nu}$, in $cm^{-1}$ for each of the wavelengths listed in problem 2. To what spectral region does each of these wavenumber values correspond?
4. Calculate the wavelength, in centimeters and in nanometers, corresponding to each of the following frequencies of electromagnetic radiation: (a) $1.97 \times 10^9$ hertz, (b) $4.86 \times 10^{15}$ hertz, (c) $7.32 \times 10^{19}$ hertz.
5. Calculate the wavenumber, in $cm^{-1}$, for each of the frequencies in problem 4 and give the spectral region in which each is found.
6. Calculate the energy in ergs of a photon of each of the following wavelengths: (a) 803 nm, (b) 3.68 $\mu$m, (c) 9.95 Å, (d) 11.5 cm.
7. Calculate the energy in ergs of a photon of wavelength 2615 Å. What is the total energy of a mole of such photons? If 1 mole of photons of wavelength 100 nm has a total energy of 286 kcal, what is the energy in kilocalories of 1 mole of photons of wavelength 2615 Å? Calculate the number of ergs in 1 kcal.
8. Calculate the absorbance which corresponds to each of the following values of per cent transmittance: (a) 36.8 per cent, (b) 22.0 per cent, (c) 82.3 per cent, (d) 100.0 per cent, (e) 4.20 per cent.
9. Calculate the per cent transmittance which corresponds to each of the following absorbance values: (a) 0.800, (b) 0.215, (c) 0.585, (d) 1.823, (e) 0.057.

10. Suppose that the per cent transmittance of a sample containing an absorbing species is observed to be 24.7 per cent in a spectrophotometer cell with a path length, $b$, of 5.000 cm. What will be the per cent transmittance of the same sample in a cell with each of the following path lengths: (a) 1.000 cm, (b) 10.00 cm, (c) 1.000 mm?
11. In what spectral region would the radiation from a "black light" be found?

## SUGGESTIONS FOR ADDITIONAL READING


1. E. J. Bair: *Introduction to Chemical Instrumentation*. McGraw-Hill Book Company, New York, 1962.
2. G. L. Clark, ed.: *The Encyclopedia of Spectroscopy*. Reinhold Publishing Corporation, New York, 1960.
3. G. W. Ewing: *Instrumental Methods of Chemical Analysis*. Third edition, McGraw-Hill Book Company, New York, 1969.
4. I. M. Kolthoff and P. J. Elving, eds.: *Treatise on Analytical Chemistry*. Part I, Volumes 5 and 6, Wiley-Interscience, New York, 1959.
5. G. H. Morrison, ed.: *Trace Analysis: Physical Methods*. Wiley-Interscience, New York, 1965.
6. R. A. Sawyer: *Experimental Spectroscopy*. Dover, New York, 1963.
7. D. A. Skoog and D. M. West: *Principles of Instrumental Analysis*. Holt, Rinehart and Winston, New York, 1971.
8. H. A. Strobel: *Chemical Instrumentation*. Addison-Wesley Publishing Company, Reading, Massachusetts, 1973.
9. F. J. Welcher, ed.: *Standard Methods of Chemical Analysis*. Sixth edition, Volume III-A, Van Nostrand, Princeton, New Jersey, 1966, pp. 3-282.
10. H. H. Willard, L. L. Merritt, and J. A. Dean: *Instrumental Methods of Analysis*. Fifth edition, Van Nostrand, Princeton, New Jersey, 1974.
11. J. D. Winefordner, ed.: *Spectrochemical Methods of Analysis*. Wiley-Interscience, New York, 1971.

# 16 INSTRUMENTATION AND MOLECULAR ANALYSIS IN THE ULTRAVIOLET AND VISIBLE REGIONS

In this chapter we will examine methods of spectrochemical analysis which rely upon the absorption, emission, or luminescence of visible and ultraviolet radiation. Although it might seem awkward to discuss together two apparently different spectral regions, the distinction between ultraviolet and visible radiation, unlike that between most other regions of the spectrum, is really quite artificial, being based primarily on the spectral response of the human eye.

Interestingly, the human eye responds only to a narrow band of wavelengths of electromagnetic radiation. This band, called the visible region of the spectrum (Figure 15–2, page 396), extends from approximately 400 to 700 nm. Not by coincidence, this range of wavelengths is the region of greatest intensity of solar radiation reaching the earth. Therefore, the traditional distinction between the ultraviolet and visible spectral region is based on a physiological rather than a chemical or physical response. In spectrochemical analysis, however, we are concerned with chemical and physical phenomena, so that the physiological distinction is unimportant. From a spectrochemical standpoint, the nature of the transitions arising from the absorption, emission, or luminescence of ultraviolet and visible radiation is the same. In addition, the instrumentation used in these spectral regions is similar, so that it is simpler to treat the regions together.

When ultraviolet or visible radiation interacts with atoms, molecules, or ions, transitions can occur between energy levels associated with valence or outer-shell electrons. Because these are the electrons involved in chemical bonding, spectroscopic observations made with ultraviolet or visible radiation can often be correlated with structural characteristics of a molecule or with association among atoms. Clearly, ultraviolet or visible spectra will differ greatly depending upon whether atoms or molecules are being observed and upon whether the chemical sample exists in the gaseous, liquid, or solid state.

Because of these variations, we will discuss atomic and molecular spectroscopy separately in this book. In the present chapter, after a consideration of the basic instrumentation required for ultraviolet-visible spectrometry, the qualitative and quantitative aspects of molecular spectroscopy based on absorption and luminescence of ultraviolet-visible radiation are examined. In the next chapter, we will investigate methods of atomic or elemental analysis which utilize ultraviolet-visible radiation, including atomic emission and absorption flame spectrometry. In addition, there is a section on the uses of electrical discharges as emission sources for elemental analysis.

## INSTRUMENTATION FOR ULTRAVIOLET-VISIBLE SPECTROMETRY

Referring back to Figure 15–7 on page 402, let us consider what instrumentation is required for spectrochemistry in the ultraviolet-visible region. To begin, the chemical input–electromagnetic radiation output transducer, which converts chemical information about the sample into the form of electromagnetic radiation, will vary in kind and complexity, depending on whether we desire to measure emission, absorption, or luminescence and on whether we wish to perform an atomic or molecular analysis. Accordingly, we will defer discussion of the configuration as well as certain components of this transducer to later sections which deal with specific techniques. However, one component of this transducer that is common to all ultraviolet-visible spectrochemical techniques is a **frequency selector.** A frequency selector separates or disperses electromagnetic radiation into relatively narrow bands of wavelengths or frequencies which can then be examined individually or simultaneously to determine the encoded information about the sample.

To determine the radiant power of the beam at each isolated band of frequencies, an electromagnetic radiation input–electrical output transducer **(detector)** is employed, as discussed in Chapter 15, page 402. Detectors, like frequency selectors, for the ultraviolet-visible region are of several types, any of which can be utilized for a specific kind of analysis. Let us examine these frequency selectors and detectors in some detail.

### Frequency Selectors

Three broad categories of frequency selectors are commonly used in the ultraviolet-visible spectral region – monochromators, polychromators, and filters. Each has advantages and disadvantages which must be carefully weighed for any particular application. Important differences among these frequency selectors are cost, ability to separate adjacent spectral intervals, and amount of transmitted radiation of the chosen frequency.

**Filters.** Generally, filters are the least expensive frequency selectors. Two kinds of filters exist – absorption filters and interference filters. Absorption filters function by absorbing some part of the spectrum of incident radiation, so that the transmitted radiation is deficient in that portion of the spectrum. Often, individual absorption filters are placed in series so that only narrow spectral bands are transmitted. This is illustrated in Figure 16–1, which shows the fraction of radiation transmitted by a pair of absorption filters as a function of the wavelength of radiation. Combining the two filters results in one having a peaked transmittance curve. In Figure 16–1, the **spectral bandwidth** ($\delta\lambda$) of the filter is defined as the range of transmitted wavelengths measured at half the peak height. Absorption filters typically have spectral bandwidths ranging from 30 to 50 nm, so their resolving power is not great. The wavelength at the peak of the transmittance curve for the combination filter ($\lambda_0$) is termed the **central wavelength** or **nominal wavelength** of the filter. Another parameter depicted in Figure 16–1 is the **transmission factor** ($\tau$), which is defined as the ratio of the output radiant power ($P$) to the input power ($P_0$) at the central wavelength:

$$\tau = \frac{P}{P_0} \tag{16–1}$$

Transmission factors for absorption filters are usually small, on the order of 0.05 to 0.2.

Interference filters, as the name implies, operate on the principle of interference of waves of electromagnetic radiation. Although somewhat more expensive than absorption

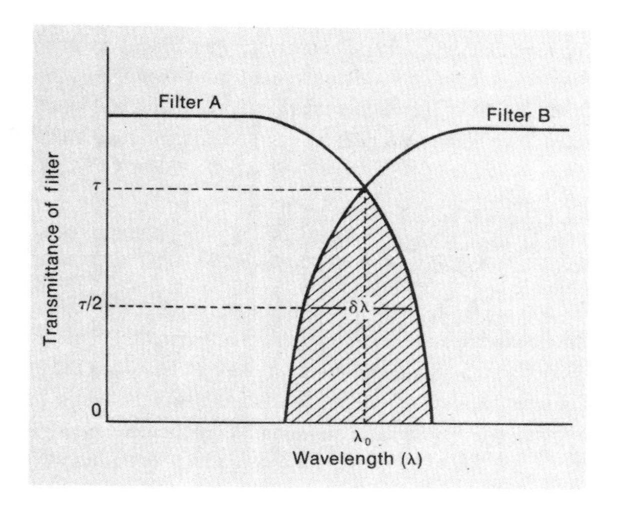


*Figure 16–1.* Transmittance curve for a combination filter (shaded region). Filters A and B, whose transmittance curves are shown, combine to produce a filter with a peaked transmittance curve; $\tau$ is the transmission factor for the combination filter, $\lambda_0$ is the nominal (central) wavelength of the combination filter, and $\delta\lambda$ is the spectral bandwidth of the combination filter.

filters, interference filters often have considerably narrower spectral bandwidths and greater transmission factors. It is not uncommon to find interference filters with spectral bandwidths between 5 and 20 nm, and transmission factors greater than 0.6.

Whenever absorption or interference filters are used, it is necessary to employ a different filter for each spectral interval to be examined. Therefore, filters are most often used in applications in which only a single spectral interval is needed, such as in quantitative spectrometry.

**Monochromators.** A monochromator (literally, a device which produces a single color), like a filter, isolates only one spectral interval at a time. However, unlike a filter, a monochromator enables the spectral interval to be set anywhere in the ultraviolet-visible region. With a monochromator, the spectral bandwidth can be as small as 0.01 nm; the transmission factor for a monochromator is a function of spectral bandwidth but is generally less than that of a filter system.

Basically, a monochromator consists of a dispersing device, focusing optics, and a pair of slits. Figure 16–2 reveals the arrangement of these components for a typical prism monochromator. Radiation arriving at the entrance slit is sent to the prism through a lens. Within the prism, the radiation is separated by refraction into its components, ultraviolet radiation being refracted (bent) most and visible red light being refracted least. Each refracted component of the radiation is then focused at the **focal plane**, where the dispersed spectrum appears. A movable exit slit placed at the focal plane can then be positioned to allow selection of any wavelength or frequency in the dispersed spectrum. In practice, the dispersed spectrum is rarely scanned by moving the exit slit. Instead, the prism is rotated so that the desired portion of the dispersed spectrum falls on the exit slit. With suitably calibrated mechanical linkages, the position of the prism can be related to the central wavelength impinging upon the exit slit; the numerical value of this wavelength can then be displayed for observation.

Prism monochromators of the kind shown in Figure 16–2 produce a spectral dispersion which is nonlinear with respect to location on the focal plane. Therefore, considerable care is necessary to ensure that the desired central wavelength actually falls on the exit slit. In addition, high-quality prisms must be free of bubbles and defects and must be as transparent as possible. To overcome these problems, a diffraction grating is often substituted for the prism as the dispersing element in a monochromator. With a grating monochromator, radiation is dispersed linearly in wavelength. Moreover, the

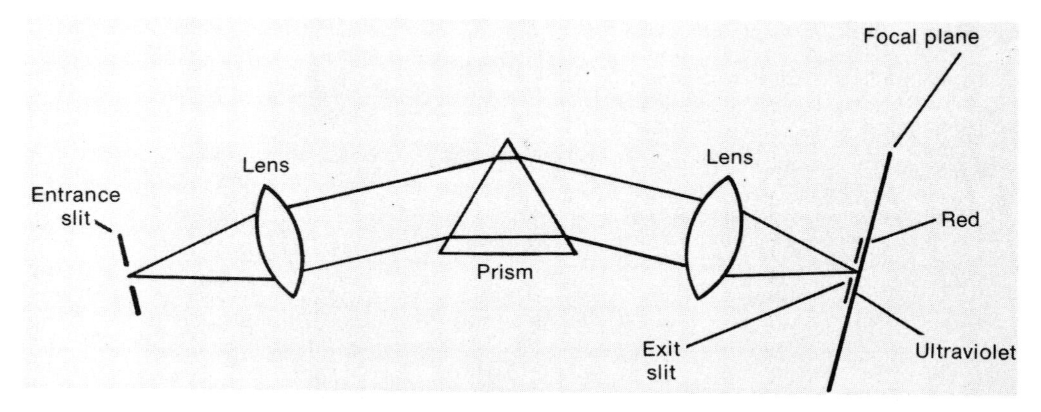


*Figure 16–2.* Schematic diagram of a simple prism monochromator. (See text for discussion.)

efficiency of a grating monochromator can be higher and its cost lower than for a corresponding prism instrument.

A grating monochromator typical of those in common use today is depicted in Figure 16–3. Notice that a grating monochromator frequently uses only reflecting surfaces in its optical system, whereas a prism monochromator (Figure 16–2) utilizes transmitting optics. Although this distinction between prism and grating systems is not always found, it is common. When a choice is possible, reflecting optics are often preferred because of their freedom from undesirable optical effects such as chromatic aberration.

**Polychromators.** Whereas filters and monochromators serve to isolate specific, narrow spectral regions for measurement, a polychromator allows the simultaneous observation of many or all spectral elements in a beam of radiation. Physically, a polychromator (literally, a device producing many colors) resembles a monochromator in that it has an entrance slit, focusing optics, and a dispersing device. In fact, one type of

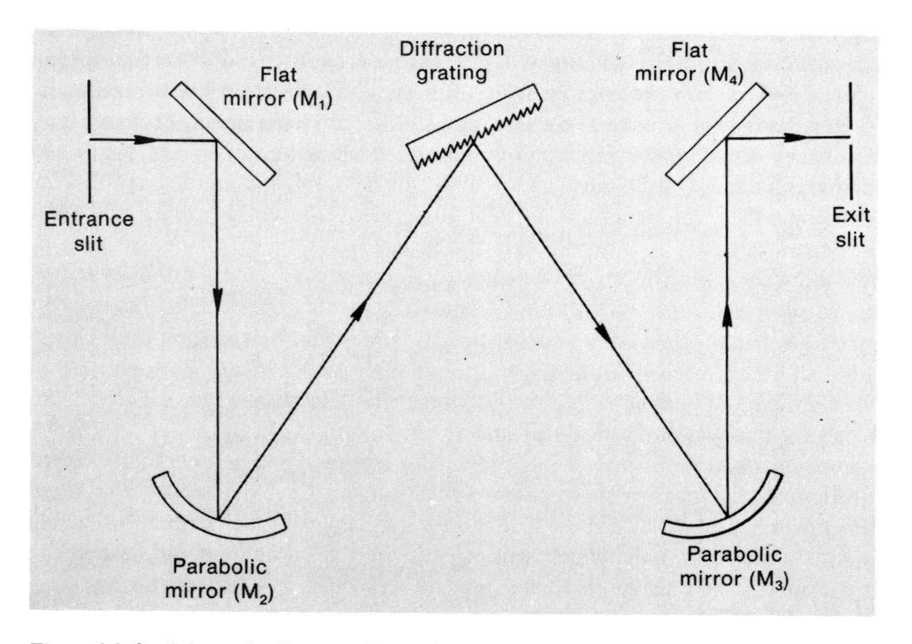


*Figure 16–3.* Schematic diagram of a grating monochromator. (See text for discussion.)

polychromator, called a **spectrometer**, is little more than a monochromator with several exit slits. In a spectrometer, a radiation detector is placed at every exit slit so that each isolated spectral interval can be individually and simultaneously observed.

Another often-encountered form of polychromator is a **spectrograph**. A spectrograph has no exit slit but, rather, employs a continuous radiation detector such as a photographic plate or film placed along the focal plane. An example of a spectrograph can be seen in Figure 16–4. In contrast to a spectrometer, the spectrograph of Figure 16–4 produces a continuous record of *all* components of a spectrum, so that a great deal of information is gathered over a given time interval. In effect, a spectrograph resembles a spectrometer having an infinite number of exit slits and detectors. However, a spectrograph requires a specific type of detector such as a photographic emulsion which is capable of recording a continuous spectrum.

### Detectors for Ultraviolet-Visible Radiation

Radiation detectors used in the ultraviolet-visible spectral region fall into two categories, those which provide spatial resolution and those which do not. The human eye is an example of the first kind of detector; the "electric eye" on an automatic camera is an example of the second. When either kind of detector is used with one of the frequency selectors described in the preceding section, this distinction translates, respectively, into an ability or lack of ability to detect more than one spectral interval at a time. For instance, a photographic film provides spatial resolution so that, when employed as the detector for a polychromator, it enables many spectral intervals over a broad range of wavelengths to be recorded simultaneously. Other detectors such as photomultipliers do not offer spatial resolution, so that they are better utilized with a filter, with a monochromator, or as one of the several individual detectors in a spectrometer. For

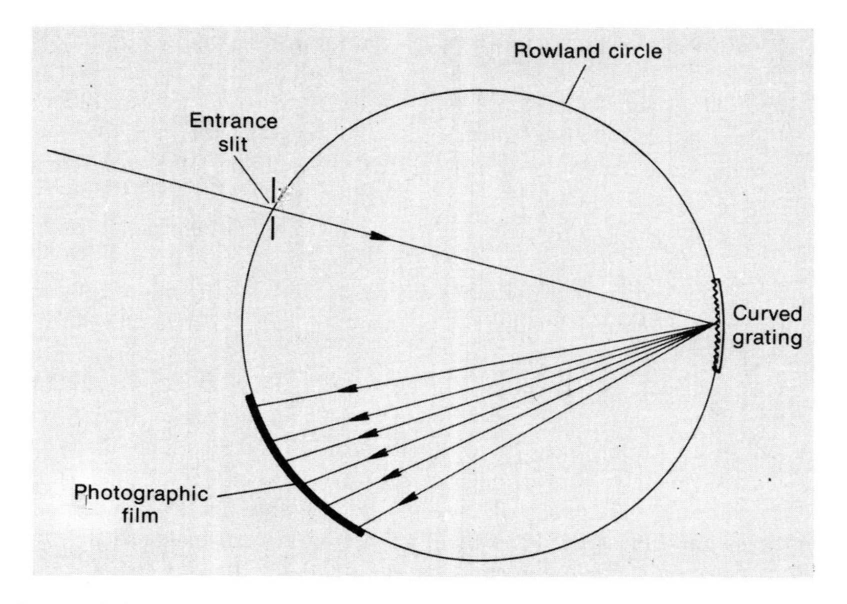


*Figure 16–4.* Diagram of a curved-grating (Rowland circle) spectrograph. Note that the curved grating, photographic film, and entrance slit all lie on the same circle. The radius of curvature of the grating is equal to the diameter of the circle.

convenience in further discussion, we will refer to these detectors as either multi-wavelength or single-wavelength devices, even though their spectral-discrimination ability arises from their combination with a suitable frequency selector.

**Single-Wavelength Detectors.** A list of some single-wavelength detectors often employed in the ultraviolet-visible spectral region is presented in Table 16–1. These detectors differ in their sensitivity, linearity, spectral response, response time, and price. Ideally, each detector should be extremely sensitive and should respond linearly to the radiant power incident upon it. Furthermore, the detector should respond equally to all spectral frequencies and should respond rapidly to a change in the level of incident radiation. The degree to which each detector approaches these ideals is summarized in Table 16–1.

**Table 16–1. Single-Wavelength Detectors for Ultraviolet and Visible Radiation**

| Detector | Usable Wavelength Range (nm) | Response Time (seconds) | Sensitivity | Price (approximate) |
|---|---|---|---|---|
| Human eye | 400–700 | 0.1 | Moderate | |
| Photomultiplier tube | 220–650 (depends on cathode) | $10^{-8}$ | Very high | $100 |
| Avalanche photodiode | 450–1200 | $10^{-10}$ | Moderate | $200 |
| Photoresistor (CdS) | 400–800 | 0.1 | Low | $1–10 |
| Silicon photodiode | 350–1200 | $10^{-8}$ | High | $10–200 |
| Vacuum photodiode | 220–650 (depends on cathode) | $10^{-9}$ | High | $50 |

**Multi-Wavelength Detectors.** In the ultraviolet-visible region, the photographic emulsion is the most common multi-wavelength detector. Because the emulsion is usually supported on glass or plastic, the actual detector takes the form of a photographic plate or film, respectively. Although other multi-wavelength detectors are now appearing, photographic detection is still dominant. An example of a spectrum (that of sodium) recorded on photographic film is displayed in Figure 16–5.

As mentioned earlier, the use of a photographic emulsion offers the advantages of being able both to detect a *range* of wavelengths simultaneously and to resolve the spectral intervals within the range extremely well. Furthermore, a photographic emulsion directly *integrates* the radiation incident upon it, so that low radiation levels can be detected if sufficiently long exposure times are employed. Moreover, a spectrum recorded on an emulsion can be stored for extended periods of time if suitable precautions are taken. These advantages over single-wavelength detectors would seem to make the emulsion an attractive choice for the observation and recording of any spectrum.

Unfortunately, photographic detection has several undesirable features. First, because the exposing, developing, fixing, and drying of a photographic emulsion is a lengthy process, the time needed to obtain a spectrum is quite long, usually from several minutes to a few hours. Second, even after the photographic processing is completed, spectral information present in the emulsion must be converted to a proportional electrical signal, often by means of a tedious operation requiring use of a **microphotometer**. Third, photographic detection is notoriously nonlinear with respect to integrated radiant power.

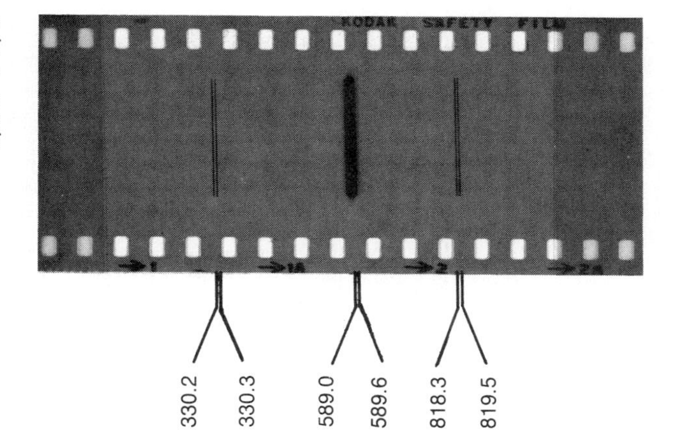


*Figure 16–5.* A photographically recorded emission spectrum of sodium. Notice the relative simplicity of this "line" spectrum, which shows three doublets; the numbers are the wavelengths of the lines in nanometers (nm).

Therefore, it is necessary to calibrate each individual emulsion for its response to radiation if quantitative results are to be achieved. Because of these disadvantages and inconveniences, photographic emulsions are not frequently used in modern spectrochemical instruments except in those applications where direct integration and recording of continuous spectra are essential.

In this section we have briefly considered specific examples of two building blocks common to all instruments for ultraviolet-visible spectrometry – the frequency selector and the radiation detector. As we investigate various analytical methods which employ ultraviolet-visible radiation, it will be of interest to observe the ways in which these components are used. It should become obvious that the spectrochemical techniques discussed are remarkably similar, differing primarily in their methodology and application.

## MOLECULAR ANALYSIS BY ULTRAVIOLET-VISIBLE ABSORPTION SPECTROPHOTOMETRY

Characterization of chemical species by means of their ultraviolet or visible absorption spectra is common in all areas of research and development, from the investigation of the quantum properties of unstable molecules in the upper atmosphere to the determination of the number of cobalt atoms present in a molecule of vitamin $B_{12}$. Ultraviolet-visible spectrophotometry is useful for both qualitative and quantitative analysis, although we will focus primarily upon the latter.

For the majority of quantitative determinations, ultraviolet-visible spectrophotometry is performed with liquid samples. However, the technique is equally applicable to gaseous or solid samples. As we shall see, the state of the sample plays an important role in governing the nature of the transitions observed in ultraviolet-visible absorption spectra.

### Energy Transitions in Molecules

We can discuss the process of absorption in terms of a transition between two energy states of a chemical species. Figure 16–6 shows three electronic energy levels of a molecule: the ground state (G) and two excited singlet states ($S_1$ and $S_2$). In turn, each electronic state has associated with it a number of vibrational levels labeled $v_0$, $v_1$, $v_2$, and so on. Because the energy difference between electronic levels is considerably greater

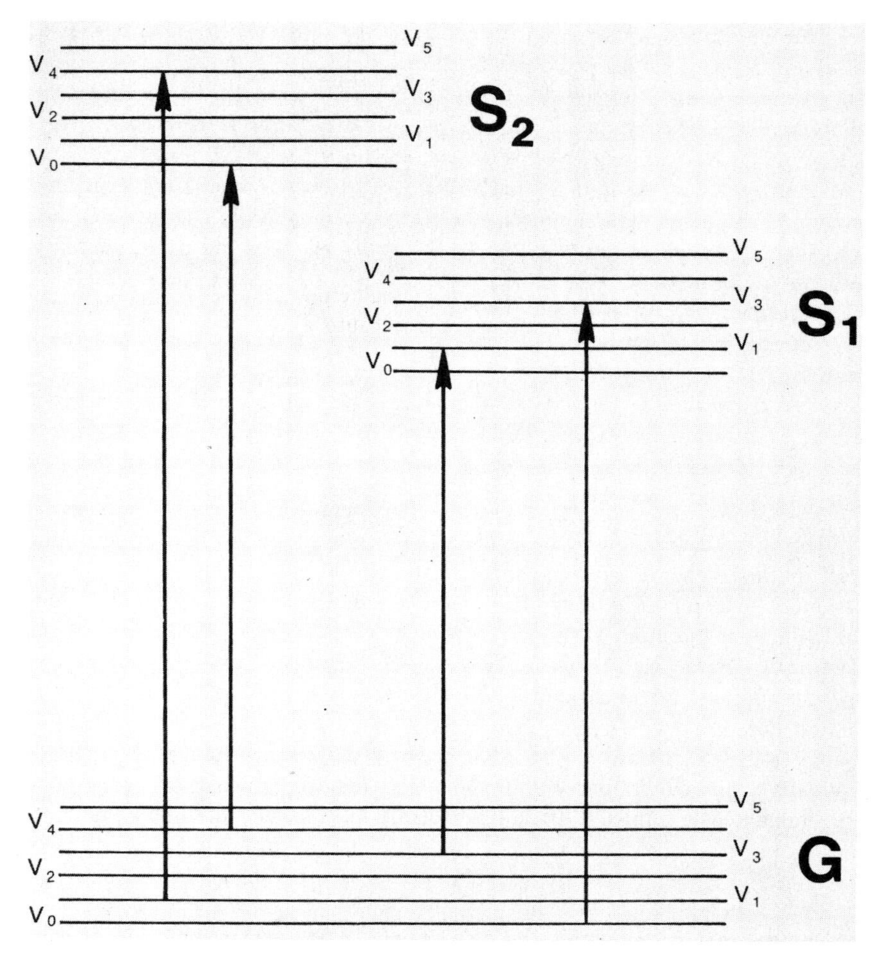


*Figure 16–6.* Electronic energy-level diagram for a hypothetical molecule, showing the energy changes (arrows) that occur upon absorption of electromagnetic radiation; **G** represents the ground electronic state, $\mathbf{S_1}$ and $\mathbf{S_2}$ are excited electronic states, and $v_0$, $v_1$, $v_2$, and so on denote the various vibrational levels for each electronic state.

than the thermal energy ($kT$) at room temperature, the Boltzmann distribution law indicates that most molecules will reside in the ground electronic state. However, the difference in energy between vibrational levels is much smaller. Therefore, at or near room temperature, most molecules are found in the ground electronic state but might be in excited vibrational states.

Absorption of ultraviolet or visible radiation usually involves raising a molecule from one of several vibrational levels in the ground electronic state to a vibrational level of an excited electronic state. Several transitions of this kind are indicated in Figure 16–6. Each of these transitions corresponds to a definite energy change and, therefore, can be caused only by absorption of a photon having exactly that energy. Thus, we would expect the ultraviolet or visible spectrum of a molecule to exhibit a set of sharp absorption lines, one for each of the discrete transitions. This sort of spectrum is called a **line spectrum.** However, for each vibrational level, there are many rotational states, making possible a seemingly infinite number of transitions. In practice, these transitions are seldom resolved, because the energy differences between rotational levels are small. Consequently, an ultraviolet or visible absorption spectrum usually appears as a series of smooth, broad peaks, each peak representing the average of a number of individual transitions. Depending upon the resolving power of the instrument being used and upon the state of the sample, the individual spectral bands which are observed can represent either discrete transitions between specific vibrational levels in different electronic states or merely the gross features of the electronic transitions themselves.

If the molecular sample is in the gaseous state, it is often possible to observe individual transitions between different vibrational levels in two electronic states. This gives rise to so-called vibrational fine-structure in electronic absorption spectra. In a liquid sample, however, solvent-solute and solute-solute interactions obscure this fine-structure, so that only the gross features of an electronic transition can be seen. An example of such behavior is shown in Figure 16–7, in which the ultraviolet absorption spectra of acetaldehyde in the gaseous state and in an aqueous solution are displayed. As for liquids, the ultraviolet and visible absorption spectra of solids usually consist of smooth peaks representing electronic transitions within the molecules of the solid. In the solid lattice, vibrational interactions among neighboring molecules make the observation of vibrational fine-structure impossible.

### Capabilities of Spectrophotometry

Ultraviolet and visible molecular spectrophotometry offers several significant strengths as an analytical technique. First, its sensitivity is excellent. It is frequently

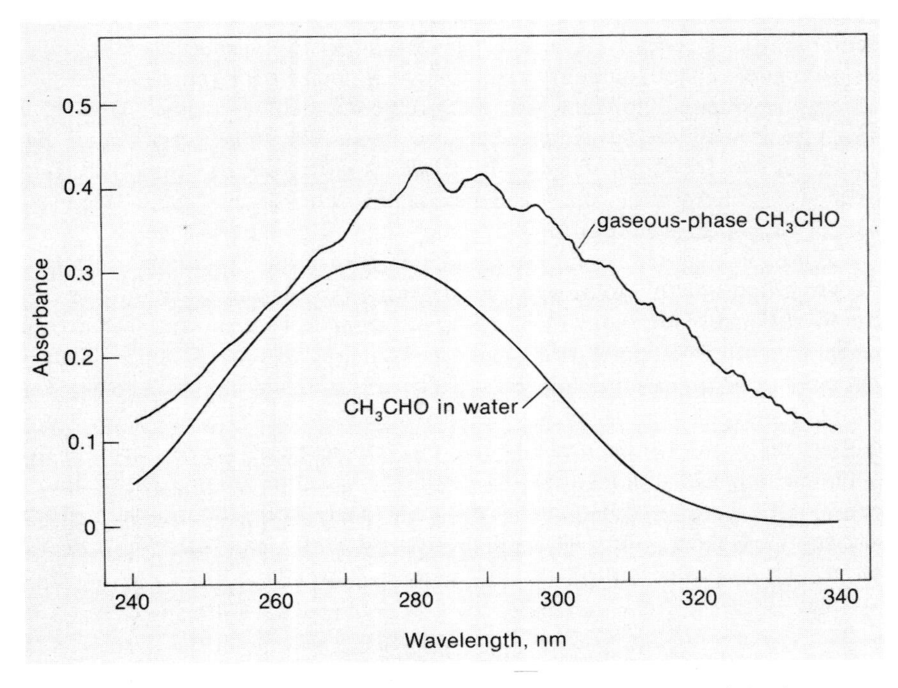


*Figure 16–7.* Ultraviolet absorption spectra for acetaldehyde ($CH_3CHO$) in the gaseous state and in an aqueous solution. Vibrational fine-structure, in the form of a number of readily resolved, small humps on top of the electronic absorption band, is clearly visible in the gas-phase spectrum. In an aqueous solution, solute-solvent and solute-solute interactions obscure this fine-structure, so that only a smooth absorption band is seen. The spectrum of acetaldehyde in the gaseous state has been displaced upward 0.1 absorbance unit to keep the two curves from overlapping. Notice that the wavelength of the absorption maximum for acetaldehyde is influenced by its environment.

possible to quantitatively determine less than $10^{-7}$ to $10^{-6}$ $M$ concentrations of a molecular species. This virtue is important in two areas of application. In one area, namely **trace analysis**, ultraviolet-visible spectrophotometry permits the quantitative measurement of minor sample constituents which are present at levels near 1 part per million by weight. In another area, termed **microanalysis**, major components can be quantitatively determined in a sample of extremely small size. Compared with traditional titrimetric and gravimetric methods, spectrophotometry is rapid and convenient. With most modern spectrophotometric instrumentation, it is possible to analyze five to ten samples per minute. Moreover, it is relatively easy to automate spectrophotometric procedures, from the first step of sample introduction to the final calculation of the concentrations of several constituents in a sample. Finally, spectrophotometric methods yield, in addition to analytical data, fundamental information about the structure of molecules and the nature of chemical bonding. In fact, such studies have been responsible for much of our present knowledge of the quantum properties of matter.

### Spectrophotometric Measurements

Let us now examine the basic instrumentation and methodology of ultraviolet-visible spectrophotometry. Because we are dealing with absorption, quantitative work is based upon Beer's law. Also, instrumentation requirements are defined by the considerations presented in Chapter 15 (see Figures 15–7 and 15–8). A typical ultraviolet-visible spectrophotometer is shown in Figure 16–8.

Most probably, the source of radiation in the spectrophotometer will be a tungsten, deuterium (hydrogen), or quartz-halogen lamp. Because each of these sources generates its maximum radiation intensity in a different region of the ultraviolet-visible spectrum, it is not uncommon to find two lamps incorporated into the more sophisticated spectrophotometers, each lamp being used in its optimum spectral region. Each source emits a broad spectrum of radiation, so that a portion of its output spectrum must be isolated for illumination of the chemical sample. For this purpose, a frequency selector such as a monochromator or filter is employed.

Radiation transmitted through the sample cell falls upon a suitable detector, such as a photomultiplier or photodiode, which in turn converts the radiation to a proportional electrical signal for readout. For protection from excessive and extraneous radiation, the detector is often equipped with a shutter. Whenever a sample is being introduced into or removed from the spectrophotometer, the shutter should be closed to avoid damage to the photodetector.

Let us now consider the measurements necessary to perform a quantitative determination with the instrument of Figure 16–8. From Beer's law, we know that the concentration of the desired chemical species (**analyte**) is proportional to its absorbance at a specific frequency or wavelength; in turn, the absorbance is the logarithm of the ratio of incident radiant power to transmitted radiant power:

$$A = abc = \log \frac{P_0}{P} \qquad (16\text{–}2)$$

To evaluate concentration, we must determine the radiant power incident on the sample ($P_0$) and the radiant power transmitted by the sample ($P$). In the simplest case, we can measure $P_0$ by first removing the sample and its container from the sample compartment of the spectrophotometer. When the sample is reinserted into the spectrophotometer, $P$ can be determined.

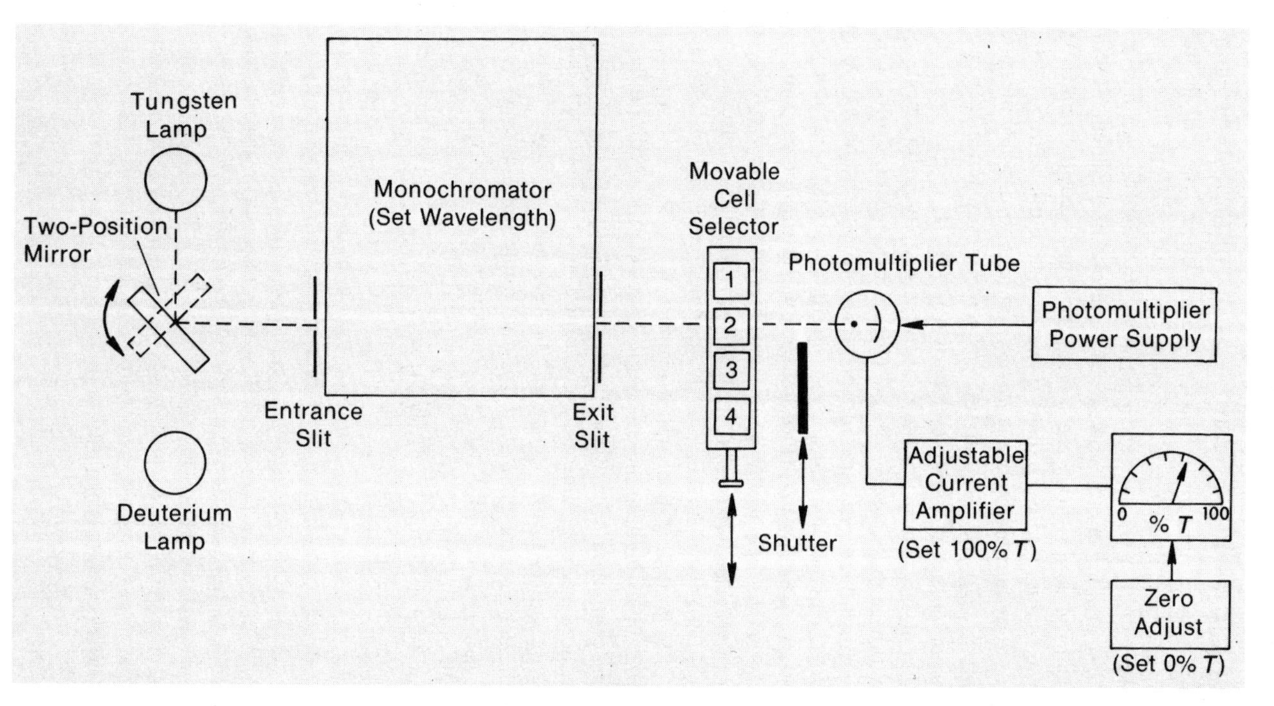


*Figure 16–8.* Schematic diagram of a generalized single-beam ultraviolet-visible spectrophotometer having two sources and a movable cell selector. Selection of the desired source is accomplished with the aid of a two-position mirror. To set the readout meter to 0 per cent transmittance, the shutter is pushed in front of the photomultiplier tube (to block the light path) and the zero adjust control is manipulated. To set the readout meter to 100 per cent transmittance, the shutter is pulled open, a spectrophotometer cell containing a blank solution is positioned in the light path, and the adjustable current amplifier is manipulated. To obtain the per cent transmittance for an unknown solution, it is only necessary to position the cell containing it in the light path, to open the shutter, and to read the meter.

In practice, the simple procedure outlined above can produce considerable error. To understand the difficulty, consider a typical sample cell for ultraviolet-visible spectrophotometry as shown in Figure 16–9. A properly designed cell will have accurately fixed dimensions, will be made of a suitable transparent material (such as quartz or Vycor glass), and will have parallel, optically clear windows on opposite sides. Ideally, such a cell should have no effect upon the measurements taken.

However, as portrayed in Figure 16–10, there can be an *apparent* absorption of radiation even when the spectrophotometer cell contains none of the analyte species. This apparent absorption is caused by reflection of incident radiation from the cell walls or by scattering of incident radiation from dust, fingerprints, or other foreign material present on the cell walls or suspended in the solvent; absorption of the incident radiation by another component in the sample or by the solvent itself produces the same effect. These radiation losses decrease the effective radiant power incident upon the analyte from the original value ($P_0$) to a lower value $(P_0)_{effective}$. For correct absorbance measurements, it is essential to employ $(P_0)_{effective}$ for the 100 per cent transmittance (zero absorbance) point. This can be accomplished if a sample blank is utilized.

**A sample blank** is merely a solution identical to the sample in all respects except that it contains none of the analyte species. When placed in a spectrophotometer cell, the blank should scatter, absorb, refract, and reflect radiation in the same way as the sample solution. Thus, the radiant power transmitted by the blank should be equal to that effectively incident upon the analyte, $(P_0)_{effective}$. Of course, this procedure assumes a perfect match of the cells in which the sample and blank solutions are held, a criterion that can never be exactly met but can be approached through the purchase of so-called matched cells. The difficulty of having exactly matched cells can be the factor which limits the accuracy of high-quality spectrophotometers.

Therefore, correction for undesirable radiation losses can be accomplished if the 100 per cent transmittance point is set with a blank solution in the cell compartment. To simplify this procedure, many commercial spectrophotometers have specially designed cell compartments in which several cells can be held and rapidly inserted into the radiation path. With such an arrangement, a cell containing the blank can be placed into position to set $(P_0)_{effective}$ (the 100 per cent transmittance level). Next, one or more cells containing solutions to be analyzed can be sequentially inserted to obtain respective values for $P$ (or per cent transmittance). With care, this method is capable of yielding spectrophotometric measurements precise to ±1 per cent.

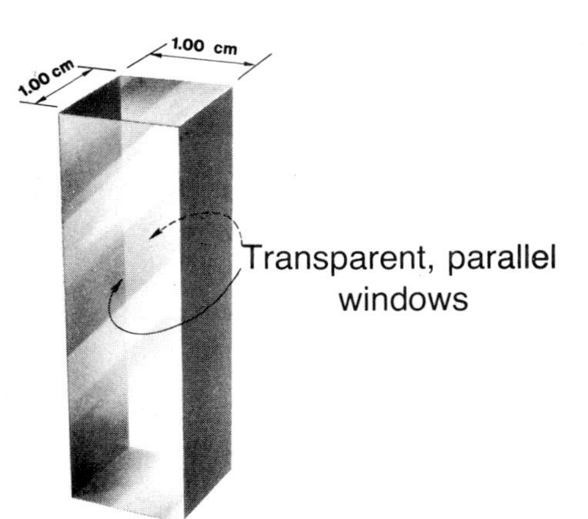


*Figure 16–9.* Sketch of a typical 1.00-cm sample cell used in ultraviolet-visible spectrophotometry. For simplicity, the cell has been drawn with infinitely thin walls.

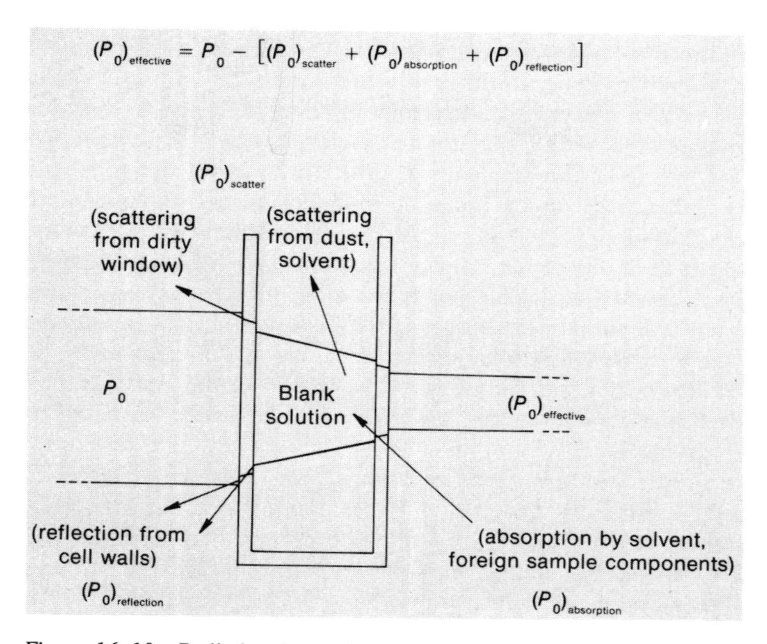


*Figure 16–10.* Radiation losses in a spectrophotometer cell in which no analyte is present. The width of the radiation beam is a schematic indication of the radiant power. (See text for discussion.)

**Spectrophotometric Errors**

Several sources of error exist in ultraviolet-visible spectrophotometry. Among the sources of error are those arising from non-obeyance of Beer's law (see Chapter 15, page 405). Others involve the characteristics of ultraviolet-visible spectrophotometric instrumentation. Instrumental errors include variations in source power, detector response, electrical noise, and cell positioning; and there is always the human error associated with the reading of absorbance or transmittance scales. Although these errors can often be minimized or eliminated through the use of calibration procedures, the cumulative error from all sources is usually between 0.2 and 1 per cent. Let us first consider the scale-reading error.

Many existing ultraviolet-visible spectrophotometers present their readout in terms of transmittance on a meter, chart recorder, or similar system. These readout devices all generate a common source of error, that of reading the transmittance exactly. For a properly designed meter, the reading error will be constant and will probably equal the width of the needle on the meter face. For a recorder, the constant readout error will probably be the width of the pen trace. If we assume that the transmittance reading obtained from each of these devices has a constant uncertainty, how much error will this uncertainty cause in the calculated concentration of the species we are seeking to determine?

To calculate the error in concentration, we must examine the effect that an uncertainty in transmittance has on the value of the concentration. To do this, let us write Beer's law in the form

$$T = \frac{P}{P_0} = 10^{-abc} \qquad (16\text{–}3)$$

which shows that the relationship between transmittance and concentration is an exponential one. From the plot of this relationship in Figure 16–11, we can determine the effect that a constant error in transmittance will have on the calculated value of $c$.

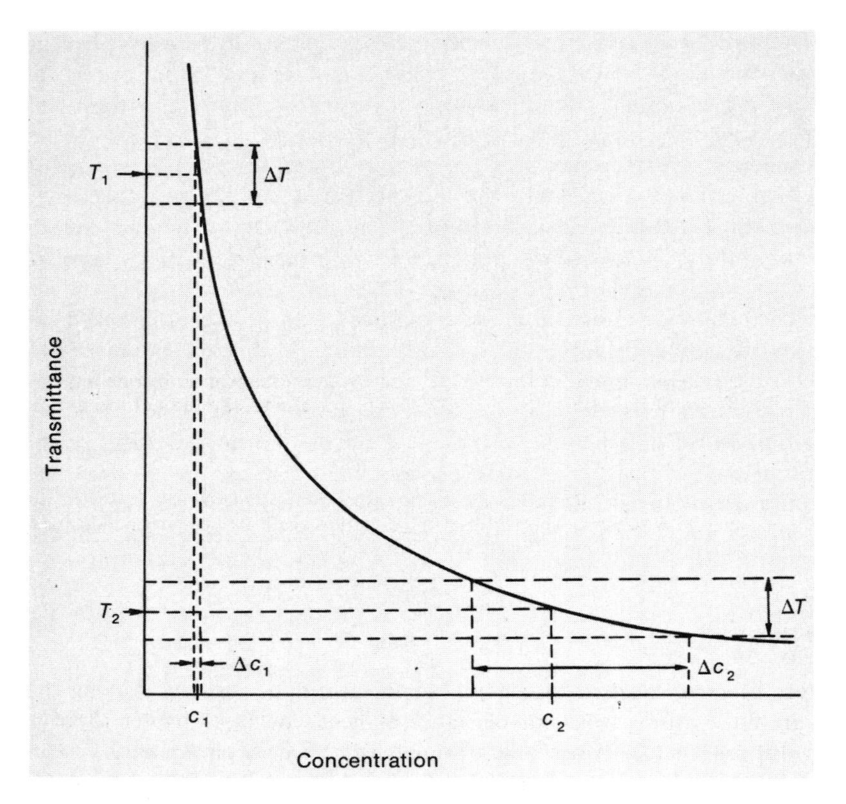


*Figure 16–11.* Plot of exponential form of Beer's law showing the relative error in concentration ($\Delta c/c$) produced by a constant error is transmittance ($\Delta T$) for different transmittance values. (See text for discussion.)

Consider first a spectrophotometric measurement made on a solution with a low concentration ($c_1$) of the desired species. According to Figure 16–11, this solution will have a high transmittance value ($T_1$). Assuming that the uncertainty of reading a meter, recorder, or other device corresponds to a constant absolute error in $T$, we can determine from Figure 16–11 the resulting error in $c$. From the extrapolated lines in Figure 16–11, it is seen that the absolute error in concentration ($\Delta c_1$) is small when a solution has a high transmittance. However, because the concentration ($c_1$) itself is low, the relative error in concentration ($\Delta c_1/c_1$) is quite large.

Let us compare the preceding situation with that encountered when a high concentration ($c_2$) of the analyte is present. This solution will have a relatively low transmittance value ($T_2$). If a constant error in $T$ is again assumed, there is a large absolute error ($\Delta c_2$) in the concentration; and, even though the concentration itself is large, the relative error ($\Delta c_2/c_2$) is also large.

These results suggest that, somewhere between the extremes of samples having high and low transmittances, we might expect a transmittance value for which the relative error in concentration ($\Delta c/c$) is minimal. In fact, the relative error is minimal at 36.8 per cent transmittance or an absorbance of 0.434.* A plot showing the relative error in concentration as a function of per cent transmittance for an absolute reading error of 1 per cent in the transmittance is shown in Figure 16–12.

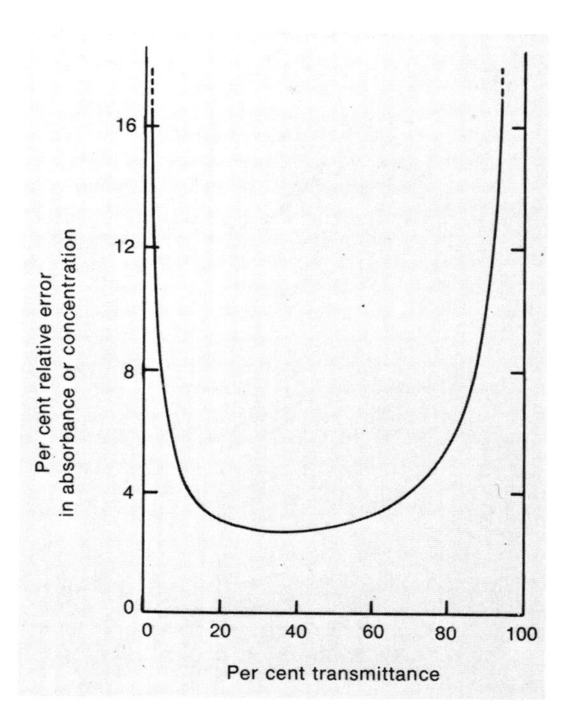


*Figure 16–12.* Per cent relative error in absorbance or concentration, as a function of per cent transmittance, due to a 1 per cent absolute error in the observed per cent transmittance. (See text for discussion.)

*It must be emphasized that the foregoing considerations apply only to measurements made with spectrophotometers having a constant reading error in transmittance.* Although many present-day instruments fall into this category, the trend is toward spectrophotometers which incorporate scale-expansion capability or digital readout, in which readout error is not the accuracy-limiting factor. In fact, spectrophotometric measurements limited by readout error are the *least* accurate and can always be improved if one simply substitutes a more sensitive readout device. In this case, electrical noise can become the limiting source of error. For those instruments in which random noise is the accuracy-limiting factor, the optimum results are obtained near 11 per cent transmittance (or an absorbance of 0.96).[a–c]

Other errors in ultraviolet-visible spectrophotometric methods arise from drift in the output power of the source or variation in the wavelength response of the detector. These errors can be minimized if a spectrophotometer having double-beam optics is employed. The optical system and operation of a typical double-beam spectrophotometer are illustrated in Figure 16–13. In the optical system of Figure 16–13A, the incident radiation ($P_0$) is alternately sent to two cells, one containing the sample solution and the

---

*For a detailed quantitative discussion of the effects of scale-reading errors, refer to D. G. Peters, J. M. Hayes, and G. M. Hieftje: *Chemical Separations and Measurements*. W. B. Saunders Company, Philadelphia, 1974, pp. 638-639.

[a]J. D. Ingle, Jr., and S. R. Crouch: Anal. Chem., *44*:1375, 1972.
[b]H. K. Hughes: Appl. Opt., *2*:937, 1963.
[c]J. D. Ingle, Jr.: Anal. Chem., *45*:861, 1973.

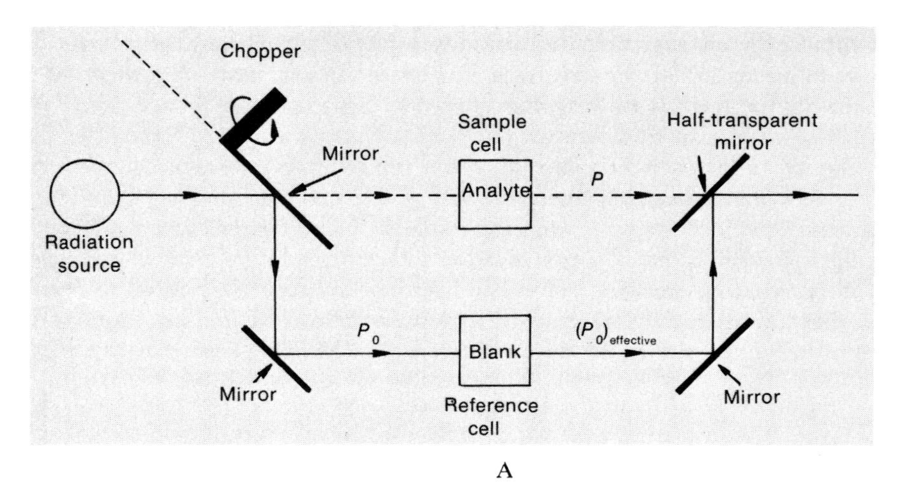


A

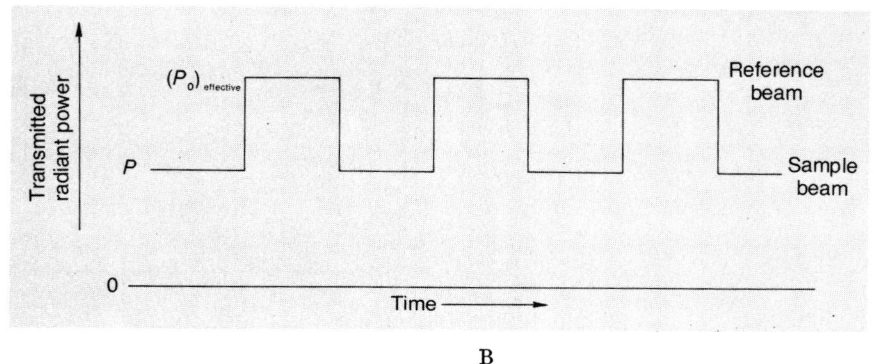


B

*Figure 16–13.* Schematic diagram of the optical system and operation of a double-beam spectrophotometer.

A. Simplified diagram of sample chamber, with beam-switching chopper (rotating mirror). When the mirror is in the position shown, radiation is directed through the reference cell; when the rotating mirror reaches the position shown by the dashed line, radiation passes through the sample cell.

B. Detector signal for a double-beam spectrophotometer. (See text for discussion.)

other a reference solution. This reference solution serves as a blank and should be identical to the sample solution except that it contains none of the analyte species. Because the species being determined absorbs a larger fraction of the incident radiation than does the blank, the radiant power transmitted by the sample cell ($P$) is smaller than that transmitted by the reference cell $(P_0)_{effective}$. Therefore, as shown in Figure 16–13B, the periodic variation from sample to blank will produce an output from the detector having the appearance of a square wave and of amplitude proportional to $[(P_0)_{effective} - P]$.

As an example of the effectiveness of a double-beam spectrophotometer in correcting for instrument errors, consider a change in source power ($P_0$). If $P_0$ should increase, the radiation transmitted by the sample, and therefore the signal from the detector, would appear to increase. However, this change will be largely compensated by a proportional increase in the radiant power transmitted by the reference cell, so that the difference $[(P_0)_{effective} - P]$ is less susceptible to a variation in the source power. Analogous considerations apply to any variation of detector response. For this reason, double-beam spectrophotometers are used almost always when it is important to scan a large region of the spectrum.

### Deviations from Beer's Law

Deviations from the fundamental law of absorbance are especially prominent in ultraviolet-visible spectrophotometry and, therefore, deserve consideration here. Of the instrumental deviations, the most common is that arising from the use of non-monochromatic radiation. The derivation of Beer's law assumes that monochromatic radiation is employed. If this requirement is not satisfied, two situations can arise. In the first and most common case, the molar absorptivity of the analyte changes over the range of wavelengths of radiation incident upon the sample. Of course, the radiant powers at each of these wavelengths will be linearly additive. However, to satisfy Beer's law, the *logarithms* of the radiant powers must add, so that a deviation from Beer's law results. This problem is difficult to avoid in spectrophotometric measurements, and necessitates a compromise between the use of a wide spectral bandpass to obtain high levels of radiant power and a narrow bandpass to have adherence to Beer's law.

In the second situation involving non-monochromatic radiation, the range of wavelengths incident upon the sample is so much wider than the absorption band of the analyte that some wavelengths pass nearly unattenuated through the sample. This can be considered as the extreme of the preceding case. Here, the molar absorptivity of the analyte changes drastically (and, in fact, approaches zero) over part of the band of radiation employed. In this case, the non-absorbed portion of the incoming radiation generates a constant background level which is independent of the analyte concentration. Thus, the absorbance of the sample solution would increase less rapidly than expected, and a plot of absorbance versus concentration would look like that shown in Figure 15–9. Curvature seen in Figure 15–9 is negligible except at high values of concentration or absorbance. This is because the non-absorbed background radiation is sufficiently small that its contribution to the transmitted power is not significant until the absorbance of the sample solution is large. Such behavior is termed a *negative deviation* from Beer's law because the plot of absorbance versus concentration bends toward the concentration axis. Note that, although a curve such as Figure 15–9 does not obey Beer's law, it can still be used for analytical purposes because it accurately represents the change in observed absorbance with sample concentration.

As suggested in Chapter 15 (page 405), chemical deviations from Beer's law are generally encountered when the absorbing substance undergoes a change in its degree of dissociation, hydration, or complexation upon dilution or concentration. For example, the extent of dissociation of a weak acid changes with concentration, so that Beer's law will not be applicable to a solution of a weak acid in which the radiation is absorbed by the anion of the weak acid. Dilution or concentration of a potassium chromate solution causes a shift in the position of equilibrium between chromate ion and dichromate ion:

$$2\,CrO_4^{2-} + 2\,H^+ \rightleftharpoons Cr_2O_7^{2-} + H_2O$$

Therefore, chromate solutions fail to follow Beer's law unless the pH is sufficiently high to keep essentially all the anions in the chromate form. If this is not done, the absorbance of chromate at 372 nm will show a negative Beer's law deviation. By contrast, the dichromate absorbance at 348 nm will deviate in a positive direction, as illustrated in Figure 16–14.

### Choice of Wavelength

In order to minimize errors and achieve the highest possible analytical sensitivity, it is essential that the optimum wavelength be selected for an ultraviolet-visible spectrophotometric determination. There are two important considerations in choosing this

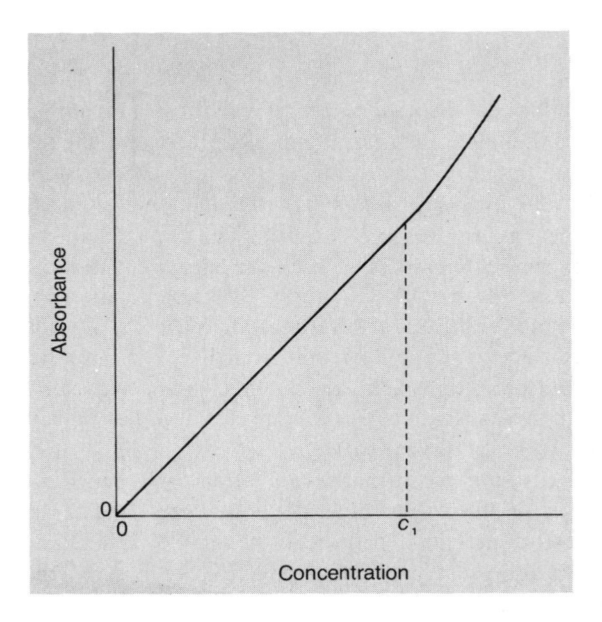


*Figure 16–14.* Plot showing positive deviation from Beer's law; $c_1$ is the maximum concentration for which Beer's law is valid. (See text for discussion.)

wavelength: the absorbance of the analyte itself and the absorbance of interfering components. If foreign components present in the sample absorb the same band of frequencies or wavelengths as the desired species, it is best to choose another absorption band of the analyte for spectrophotometric measurements. This usually presents no problem, since most molecular species that absorb in the ultraviolet-visible region have several absorption bands suitable for analytical use. Whatever band is finally chosen should be selected on the basis of a compromise between sensitivity and concentration range of the desired species, this selection being made to yield transmittance readings in the range producing the smallest spectrophotometric error.

The shape of molecular absorption bands in the ultraviolet-visible region varies greatly and must be considered in the final selection of the particular wavelength used for a spectrophotometric analysis. Suppose, for example, that the desired species has the hypothetical absorption spectrum shown in Figure 16–15. This absorption spectrum consists of an intense, sharp peak and a less intense, broad band. Obviously, selection of the wavelength corresponding to the maximum of the sharp peak will provide greater

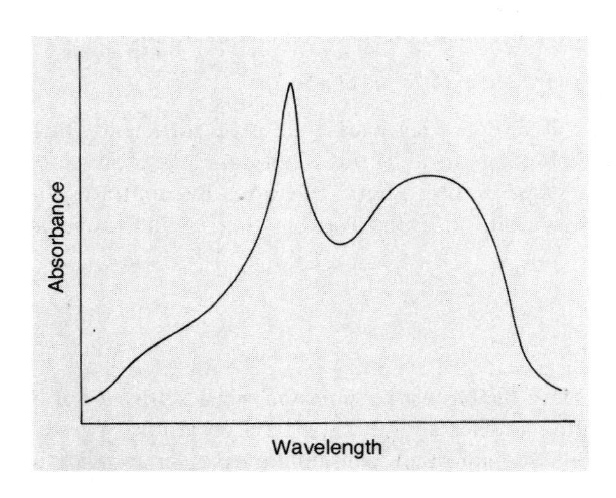


*Figure 16–15.* Hypothetical absorption spectrum of an analyte. To minimize the spectrophotometric error, absorbance measurements should be made at wavelength corresponding to the flat portion of the broad band. (See text for discussion.)

sensitivity. However, a small variation in the position of the frequency selector will cause a shift in the wavelengths of radiation incident upon the sample cell and, consequently, a large change in the absorbance reading. Because a slight wavelength shift is not uncommon in absorption spectrophotometers, it is better to choose a wavelength for analysis which lies in the central portion of the broad band rather than one either at the top of a sharp peak or on the sloping side of a peak or band.

### Analysis of Multicomponent Samples

It is often possible to determine in the same sample the concentrations of several species having overlapping absorption spectra. The procedure relies upon the fact that, if two or more species contribute to the absorbance at a fixed wavelength, the total absorbance equals the sum of the individual absorbances. Thus, if two components $x$ and $y$ contribute to the absorbance at wavelength $\lambda_1$, we have

$$(A_{\text{total}})_1 = (A_{\text{x}})_1 + (A_{\text{y}})_1 \qquad (16\text{-}4)$$

Applying Beer's law to each of the two components and choosing a spectrophotometer cell with a sample path length ($b$) of 1.00 cm, we can rewrite the preceding relation as

$$(A_{\text{total}})_1 = a_{\text{x}_1}c_{\text{x}} + a_{\text{y}_1}c_{\text{y}} \quad (\text{all at } \lambda_1) \qquad (16\text{-}5)$$

A similar equation can be formulated for the total absorbance at another wavelength $\lambda_2$:

$$(A_{\text{total}})_2 = a_{\text{x}_2}c_{\text{x}} + a_{\text{y}_2}c_{\text{y}} \quad (\text{all at } \lambda_2) \qquad (16\text{-}6)$$

Numerical values of the absorptivities $a_{\text{x}_1}$ and $a_{\text{x}_2}$ can be determined at $\lambda_1$ and $\lambda_2$, respectively, by measurements upon a standard solution of $x$ – that is, upon a solution in which $c_{\text{x}}$ is known and $c_{\text{y}}$ is zero. Similarly, the values of $a_{\text{y}_1}$ and $a_{\text{y}_2}$ at the same two wavelengths can be determined with a standard solution of $y$.

When we measure the absorbance of an unknown mixture of two species at two different wavelengths, we have two equations containing only $c_{\text{x}}$ and $c_{\text{y}}$ as unknowns. These equations can be solved simultaneously for the composition of the unknown sample. Preferably, one of the wavelengths should be chosen such that $x$ absorbs much more than $y$ and the other wavelength such that $y$ absorbs to a much greater extent than $x$.

In principle, this procedure can be extended to samples containing several ($n$) species; it is only necessary to make $n$ measurements of the absorbance of the sample solution at $n$ different wavelengths. These data provide $n$ equations containing $n$ unknowns, each unknown being the concentration of one species in the mixture. However, this procedure is not very satisfactory for samples containing more than three components because accuracy declines drastically.

### Selected Determinations

Because there are many examples of useful analyses based upon the absorption of ultraviolet or visible radiation, those procedures discussed here have been chosen arbitrarily and are meant only to be illustrative. We shall consider the determination of organic compounds and species of current ecological interest. All the methods are of practical value in modern analytical laboratories, and each procedure possesses one or more features that are applicable to other determinations as well.

In addition, ultraviolet-visible spectrophotometry is valuable for the characterization of chemical equilibria in solution, because measurements of absorbance can be made without altering or disturbing a system under examination. This approach is often utilized to determine the formulas of complex ions in solution and to evaluate the formation constants for these species. Although we will not develop here the details of the theory and experimental procedures used in such studies, several problems at the end of this chapter are concerned with this subject.

**Spectrophotometric Determination of Organic and Biological Compounds.** Perhaps the most common application of ultraviolet or visible absorption spectrophotometry is the determination of organic or biological compounds. The high absorptivities and characteristic absorption spectra of many of these species often make spectrophotometry the technique of choice for both qualitative and quantitative purposes. Furthermore, the presence of specific functional groups can sometimes be ascertained from the absorption spectrum of a compound.

In the ultraviolet-visible region, absorption bands of organic and biological compounds are fairly broad. For example, an aldehyde group having a maximum absorptivity at 210 nm will exhibit appreciable absorption at wavelengths as long as 230 nm and as short as 200 nm. In addition, the position of a certain band will change depending on the structure of the molecule in which the group is found. Thus, the benzene absorption at 255 nm is shifted to 262 nm in toluene and to 265 nm in chlorobenzene.*

Quantitative determinations by measurement of ultraviolet or visible absorption are not limited to unknown constituents which themselves undergo suitable electronic transitions. For example, no absorptions are found for alcohol (–OH) groups throughout the entire wavelength range from 200 to 1000 nm. However, many alcohols react with phenyl isocyanate to form phenyl alkyl carbamates,

$$C_6H_5{-}N{=}C{=}O + C_2H_5OH \longrightarrow C_6H_5{-}N(H){-}C({=}O){-}OC_2H_5$$

phenyl isocyanate — ethyl-N-phenylcarbamate

which can be determined quantitatively by measurement of the absorbance at 280 nm.

A common method for the determination of amino acids involves their reaction with ninhydrin to produce an intensely blue-colored compound whose absorbance can be measured at 575 nm. This method is employed in several commercial amino-acid analyzers, in which proteins under investigation are hydrolyzed into their component amino acids which are, in turn, separated and measured spectrophotometrically.

**Water-Pollution Analysis – Measurement of Phosphate.** One of the most sensitive techniques available for the determination of phosphate in water is the so-called *molybdenum blue* method. In this procedure, phosphate reacts with ammonium molybdate to form ammonium phosphomolybdate, $(NH_4)_3P(Mo_3O_{10})_4$. In turn, the latter compound can be reduced with a wide variety of reagents, including hydroquinone, tin(II), and iron(II), to yield an intensely colored substance called molybdenum blue, the

*For a more detailed treatment of the relationship between organic structure and ultraviolet-visible absorption spectra, consult H. H. Jaffe and M. Orchin: *Theory and Applications of Ultraviolet Spectroscopy*. John Wiley and Sons, New York, 1962; and A. I. Scott: *Interpretation of the Ultraviolet Spectra of Natural Products*. Pergamon Press, Oxford, 1964.

concentration of which can be measured spectrophotometrically to provide a determination of phosphate. Molybdenum blue is a complex polymer which consists of a mixture of molybdenum(VI) and molybdenum(V), but it is not a stoichiometrically well-defined compound. It is interesting to note that compounds containing the same element in two different oxidation states display characteristically intense colors and correspondingly high absorptivities. Both of the two main reactions in this procedure, namely, the formation of ammonium phosphomolybdate and the subsequent reduction to molybdenum blue, can yield incorrect products. Consequently, these reactions must be carried out under carefully prescribed conditions with respect to the acidity of the solution, the amount of ammonium molybdate reagent used, and the temperature and time of the reaction.

**Air-Pollution Analysis – Measurement of Sulfur Dioxide.** Sulfur dioxide is probably the most important of all air pollutants because of its ubiquitous nature and its adverse effect on the upper respiratory tract. One of the standard methods for the detection and measurement of sulfur dioxide is the spectrophotometric procedure devised by West and Gaecke.[a] In this procedure, one collects the sulfur dioxide by bubbling the air sample through a 0.1 $F$ sodium tetrachloromercurate(II) solution to form the stable, nonvolatile product disulfitomercurate(II):

$$HgCl_4^{2-} + 2\,SO_2 + 2\,H_2O \rightarrow Hg(SO_3)_2^{2-} + 4\,Cl^- + 4\,H^+$$

The stabilized product is then treated with the acid-bleached dye *p*-rosaniline [(4-amino-3-methylphenyl)bis(4-aminophenyl)methanol] in the presence of formaldehyde to yield a bright red-violet substance whose absorbance is measured at 569 nm. This procedure is suitable for the accurate determination of sulfur dioxide in air at concentrations as low as 0.005 ppm by volume; the only known interference is nitrogen dioxide ($NO_2$), which, if present at concentrations above 2 ppm, must be removed prior to the analysis.

## MOLECULAR ANALYSIS BY LUMINESCENCE SPECTROMETRY

Although luminescence spectrometry is not used as much as absorption spectrophotometry, the former provides the basis for some of the most sensitive molecular analytical techniques. **Luminescence** is the emission of radiation from a species after that species has absorbed radiation. As we shall see, absorption is a necessary, but not sufficient, prerequisite for luminescence. To understand luminescence and its importance to spectrochemical analysis, let us consider the molecular events leading to a luminescent transition by referring to the energy-level diagram of Figure 16–16.

This diagram resembles that shown earlier for absorption, except that we will consider not only the singlet states, $\mathbf{S_1}$ and $\mathbf{S_2}$, but also the associated triplet states, $\mathbf{T_1}$ and $\mathbf{T_2}$. An electronic triplet state, it should be recalled, has two electrons whose spins are unpaired. By contrast, in a singlet state all the electrons in a molecule are paired; the ground state of a molecule is therefore almost always a singlet state. Because two unpaired electrons have a slightly lower energy than two paired electrons, a triplet state is somewhat lower in energy than the corresponding singlet state, as illustrated in Figure 16–16. To simplify matters, let us examine only one absorption process (labeled $A$), which will be assumed to be a transition from the ground electronic state (**G**) to an excited vibrational level of an excited singlet state, say $\mathbf{S_2}$.

---

[a] P. W. West and G. C. Gaecke: Anal. Chem., *28*:1816, 1956.

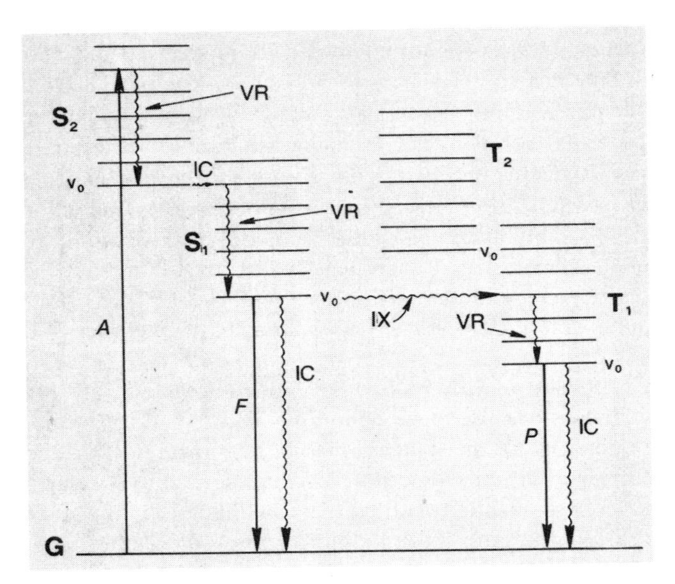


*Figure 16–16.* Electronic energy-level diagram for a molecule with ground state (G) and excited singlet (S) and triplet (T) states. Radiationless transitions between states are represented by wavy arrows; *A* is absorption, *F* is fluorescence, *P* is phosphorescence, VR is vibrational relaxation, IX is intersystem crossing, and IC is internal conversion. (See text for discussion.)

Following this transition, the molecule can rid itself of the absorbed energy by several alternative pathways. The particular pathway used will be governed by the kinetics of various competing processes, some of which are indicated in Figure 16–16. For example, the molecule residing in an excited vibrational level of electronic state $\mathbf{S}_2$ can lose energy by radiating a photon equal in energy to the difference between its existing state and the ground state. However, in solution this radiative loss has a much smaller rate constant ($k \simeq 10^8$ sec$^{-1}$) than the competing process of **vibrational relaxation** (labeled **VR** in Figure 16–16). Vibrational relaxation involves the transfer of vibrational energy to neighboring molecules and, in solution, occurs very rapidly ($k \sim 10^{13}$ sec$^{-1}$).

In solution, an excited molecule will rapidly relax to the lowest vibrational level of the electronic state in which it resides, in the present case $\mathbf{S}_2$. At this point, the kinetically favored event is **internal conversion** (labeled IC in Figure 16–16), which shifts the molecule from the lowest vibrational level of $\mathbf{S}_2$ to an excited vibrational level of the lower singlet state $\mathbf{S}_1$. This occurrence is made possible by the relatively small energy difference between the excited singlet states of most molecules and the high degree of coupling that exists between their vibrational levels. Following internal conversion, the molecule is rapidly deactivated by vibrational relaxation to the lowest vibrational level of state $\mathbf{S}_1$. Because vibrational relaxation and internal conversion have such high rates, *most excited molecules in solution will decay to the lowest vibrational level of the lowest excited singlet state before any analytically useful radiation is emitted.*

Once a molecule reaches the lowest vibrational level of the lowest excited singlet state, it can return to the ground state in several ways, among which is the emission of radiation. Internal conversion can still occur, of course. However, it is less probable since for most molecules the energy separation between the first excited singlet state and the ground state is greater than that between excited states. The radiative loss of energy from a singlet to the ground state is termed **fluorescence** and is labeled *F* in Figure 16–16. Because several other processes compete with fluorescence, it is necessary to employ a figure of merit, termed the **fluorescence quantum efficiency**, to indicate the fraction of excited molecules that fluoresce.

Fluorescence is obviously an analytically useful process and its basic characteristics can be deduced from the preceding discussion. Because fluorescence almost always occurs after some loss of vibrational or electronic energy, the wavelength of the fluoresced radiation will be longer (that is, its frequency will be lower) than that of the absorbed radiation. For this reason, the fluorescence spectrum of a molecule is shifted to longer wavelengths from the absorption spectrum. The lifetime of the fluorescence can also be of importance and will be equal to that of the excited singlet state (between $10^{-9}$ and $10^{-7}$ second).

As an alternative to internal conversion and fluorescence, a molecule in state $\mathbf{S_1}$ can undergo **intersystem crossing** (IV in Figure 16–16), which involves an electron spin-flip within the molecule to place it in an excited vibrational level of triplet state $\mathbf{T_1}$. Following this, vibrational relaxation will be very rapid, dropping the molecule to the lowest vibrational level of $\mathbf{T_1}$. Because intersystem crossing is a spin-forbidden process (that is, because it is relatively improbable from quantum mechanical considerations), it can compete with fluorescence and internal conversion only for certain molecules. In these cases, the rate of intersystem crossing approximates that of the competing processes ($k$ between $10^6$ and $10^9$ $sec^{-1}$).

Any transition from triplet state $\mathbf{T_1}$ to the ground state **G** is also a spin-forbidden process. For this reason, the lifetime of the triplet state is relatively long, ranging from $10^{-6}$ to 10 seconds, depending on the specific molecule. The particular lifetime will be governed by the dominant mode of energy loss; if radiative, this process is termed **phosphorescence** (*P* in Figure 16–16) but, if nonradiative, it again involves internal conversion and vibrational relaxation. Phosphorescence is therefore a luminescence process in which a molecule undergoes a transition from a triplet state to the ground state. Because of its spin-forbidden character, phosphorescence has a much longer lifetime than fluorescence; the lifetime is equal to the triplet-state lifetime of between $10^{-6}$ and 10 seconds.

As mentioned earlier, triplet state $\mathbf{T_1}$ is lower in energy than singlet state $\mathbf{S_1}$ and is therefore closer to the ground state. For this reason, internal conversion to the ground state is much more efficient than phosphorescence for most molecules, and at room temperature is the dominant pathway for the loss of triplet-state energy. To utilize phosphorescence for analytical work, therefore, it is customary to cool the sample, often to liquid nitrogen temperatures (77°K), in an effort to increase the **phosphorescence quantum efficiency**. The latter is defined as the fraction of excited molecules which phosphoresce.

From the preceding discussion, it should be clear that the two luminescence phenomena – fluorescence and phosphorescence – are competitive and, furthermore, must compete with other modes of energy loss by an excited molecule. Whichever process dominates is dependent upon both the type of molecule being examined and its environment. In some cases, the environment can be controlled to increase the yield of the desired event; the yield of either event can be determined from its quantum efficiency.

## Luminescence Instrumentation

A block diagram for a generalized luminescence spectrometer is shown in Figure 16–17. Because luminescence is an isotropic property – that is, because there is emission in all directions – it is possible to detect the emitted radiation at any desired direction

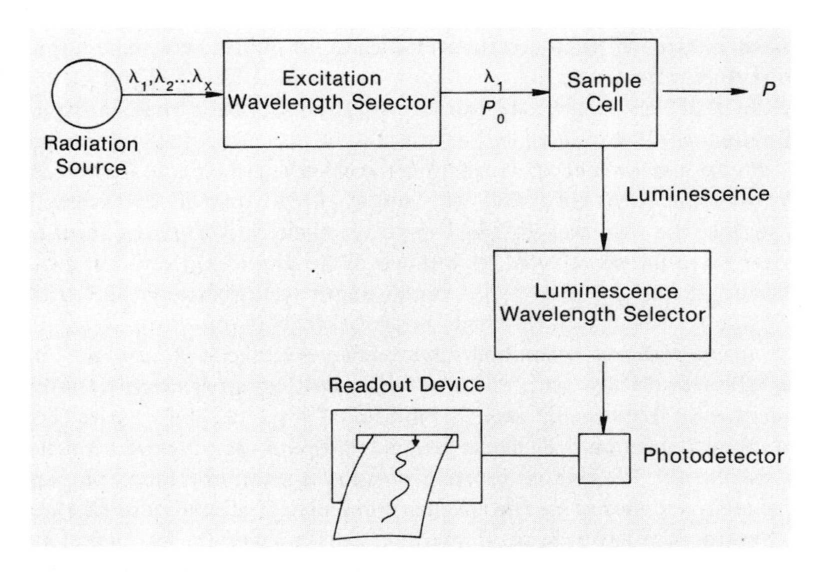


*Figure 16–17.* Generalized block diagram of a luminescence spectrometer. (See text for discussion.)

from the sample. In order to minimize the likelihood of interference from radiation used to excite the sample molecules, the luminescence is often observed at right angles to the exciting radiation. Although this 90° orientation is not found in all luminescence spectrometers, it is by far the most common configuration. Because the luminescence power is proportional to the incident source power (see Chapter 15, page 406), the exciting radiation is usually provided by an intense source such as a xenon arc lamp. A wavelength selector such as a filter or monochromator is used to obtain a narrow band of wavelengths from the source suitable for excitation of the desired molecular species in the sample. When excited, these molecules can either fluoresce or phosphoresce, and the spectrum of the luminescence radiation is determined with the aid of another wavelength selector and an appropriate photodetector and readout system.

There are two classes of luminescence spectrometers: those which employ filters and those which employ monochromators for spectral isolation. Obviously, instruments which utilize monochromators are more versatile and provide greater selectivity; because a narrower wavelength band is isolated for both excitation and observation of the luminescence spectrum, there is less chance of interference by other sample components. Also, with the monochromator systems it is possible to obtain either a complete luminescence or a complete excitation spectrum. One obtains an **excitation spectrum** by scanning the desired wavelength region with the excitation monochromator while the emission monochromator is held at a fixed wavelength. Thus, an excitation spectrum is similar, but not identical, to a spectrophotometric absorption spectrum. An excitation spectrum indicates what wavelengths can be used to *pump* the specific transition being observed with the emission monochromator. Compared with monochromator-based instruments, the filter devices are simpler to operate and less expensive. In addition, a filter fluorimeter or phosphorimeter provides higher sensitivity for quantitative determinations. Both types of luminescence spectrometers are widely used.

### Advantages of Fluorimetry and Phosphorimetry

**Fluorimetry** and **phosphorimetry** are terms used to denote luminescence spectrometry based on fluorescence or phosphorescence, respectively. Both techniques have several advantages over absorption spectrophotometric methods. First, fluorimetry and phosphorimetry offer greater selectivity and freedom from spectral interferences. This is because there are far more chemical species which absorb ultraviolet or visible radiation than those which fluoresce or phosphoresce. Moreover, in fluorimetry and phosphorimetry, the absorption wavelength as well as the emission wavelength can be varied, so that it is possible to further reduce spectral interferences by judicious choice of *both* excitation (absorption) and luminescence wavelengths.

Second, fluorimetry and phosphorimetry are generally more sensitive than absorption methods. This is because in the luminescence techniques a direct measurement of the power of the emitted radiation can be made. By contrast, it is necessary in absorption methods to determine the *difference* between two large radiation levels, the incident power $P_0$ and the transmitted power $P$. Because it is always easier to measure a small signal against no background than it is to measure the difference between two large signals, fluorimetry and phosphorimetry provide greater sensitivity than absorption spectrophotometry. This added sensitivity gives the luminescence methods still another advantage compared with absorption spectrophotometry. Whereas Beer's law plots are often linear over a 10-to-100-fold range of concentration, it is not uncommon to find the relationship between luminescence power and concentration being linear over three or four orders of magnitude in concentration. Although this extended linear range is not necessary to quantitative analysis, it is often quite useful practically and requires fewer points on a calibration curve.

One additional advantage is often enjoyed in phosphorimetry – that of **time-resolution**. Because phosphorescence has a relatively long lifetime, it is possible to discriminate among different molecular species on the basis of their luminescence-time behavior. This provides an added dimension useful in both qualitative identification and quantitative determination of particular species by minimizing interferences from fluorescence and phosphorescence of other sample constituents as well as fluorescence from the analyte itself. In addition, scattering from the sample is eliminated through the use of time-resolution. To obtain time-resolution in phosphorimetry, two common mechanical configurations are employed. In one arrangement, a slotted can rotates around the sample cell at varying speeds. Thus, the sample is illuminated by the primary source with the can in one position; the phosphorescent radiation can reach the detector only after the can has rotated through an angle of 90°. Changing the rotation rate of the can then varies the time delay between absorption by the sample and measurement of its phosphorescence. The second common configuration employs two rotating slotted disks, located on opposite sides of the sample cell. If the slots are displaced from each other, the time delay between excitation of the sample and measurement of its luminescence can be altered by adjustment of the rotation rate of the disks.

Despite the obvious advantages of time-resolution, phosphorimetry is not as widely utilized as fluorimetry. This fact is due to the added complexity of phosphorimetric instrumentation, the smaller number of species that phosphoresce, and the inconvenience of having to cool the samples to obtain adequate phosphorescence quantum efficiencies. Until recently, it was necessary in phosphorimetry to employ special solvent mixtures which would form clear, rigid glasses upon cooling.[a] One such solvent frequently utilized is EPA, a 5:5:2 mixture of diethyl *e*ther, iso*p*entane, and ethyl *a*lcohol. Recently, the use of frozen "snows" of aqueous solutions has been investigated with some success.[b]

[a] J. D. Winefordner and P. A. St. John: Anal. Chem., *35*:2212, 1963.

[b] R. J. Lukasiewicz, J. J. Mousa, and J. D. Winefordner: Anal. Chem., *44*:1339, 1972.

### Errors in Fluorimetry and Phosphorimetry

As discussed earlier, the power of fluorescent or phosphorescent radiation emitted by a sample is a direct function of the quantum efficiency. Thus, the quantum efficiency for the desired luminescence process must be constant and reproducible if a successful fluorimetric or phosphorimetric method of analysis is to be developed. When the quantum efficiency is reduced appreciably, the luminescence is said to be **quenched**. Unfortunately, many extraneous substances can affect the quantum efficiency and can quench luminescence. In particular, heavy atoms or paramagnetic species strongly affect the rate of intersystem crossing which, in turn, alters the quantum efficiency for fluorescence or phosphorescence, thereby causing an error in the analysis. In phosphorimetry, of course, it is highly desirable to increase intersystem crossing, whereas in fluorimetry it is not. Therefore, to avoid quenching in most fluorimetric procedures, heavy atoms or paramagnetic species must be excluded from the sample solution. Oxygen, being paramagnetic, is a particularly serious offender and is usually removed by the bubbling of nitrogen through solutions to be analyzed.

Unlike absorption methods, fluorimetry and phosphorimetry involve the direct measurement of a radiation signal. Although it is possible to compensate for drifts in such parameters as source power and cell positioning in absorption spectrophotometry, such compensation is not as convenient for the luminescence techniques. For example, any drift or variation with wavelength of the primary source power reflects itself in a corresponding drift in the luminescence power. In most high-quality luminescence spectrometers, this problem is overcome either through adequate stabilization of the source power or through the monitoring of source power and the application of appropriate corrections.

Any factors which affect absorption will also influence fluorescence and phosphorescence since absorption is necessary before either of these processes can occur. Therefore, all sources of error previously discussed for absorption spectrophotometry apply equally to luminescence methods of analysis. In addition, a phenomenon called the **inner-filter effect**, which is peculiar to luminescence processes, must be considered. The inner-filter effect can be understood with the aid of Figure 16–18. In Figure 16–18, assume that the common 90° viewing configuration is used and that the exciting radiation enters the cell from the left. Because the sample solution will absorb the exciting radiation as it passes through the cell, the power of the radiation will be less near the right side of the cell ($P_R$) than at the left side ($P_L$). Since the luminescence power is directly proportional to the power incident on the sample, the luminescence power will be less on the right side of the cell ($L_R$) than on the left ($L_L$). Furthermore, the luminescence from

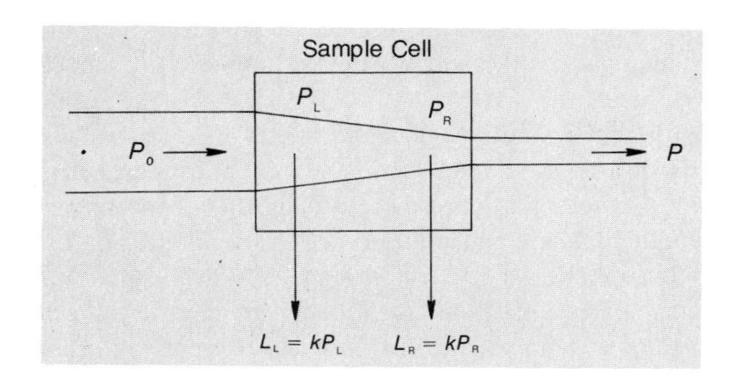


*Figure 16–18.* Pictorial representation of the *inner-filter effect* caused by attenuation of the radiation beam passing through the sample; $P_L$ and $P_R$ are the powers of the transmitted radiation at the left and right sides of the cell, respectively, and $L_L$ and $L_R$ are the powers of luminescence from the left and right sides of the cell, respectively. The power of the radiation incident upon the sample cell is $P_0$, and the power of the radiation transmitted through the cell is $P$. (See text for discussion.)

the right side of the cell will not be directly proportional to concentration, because the radiation exciting the molecules near that side of the cell does not remain constant but rather varies with concentration. The inner-filter effect generally limits the maximum concentration of the sample species which can be determined by means of fluorimetry or phosphorimetry. In addition, it can cause the shape of a plot of luminescence power versus concentration to be extremely nonlinear or even ambiguous, as shown in Figure 16–19.

### Luminescence Determinations

Most luminescence determinations are performed on samples of clinical, biological, or forensic interest, although inorganic luminescence analysis is not uncommon. Many drugs possess rather high quantum efficiencies for luminescence, so that their determination by means of fluorimetry or phosphorimetry is both sensitive and practicable. Quinine, for example, can be detected at levels below one part per billion by weight and is often used as a calibration standard for fluorescence analysis. Many inorganic ions can be determined by means of fluorimetry or phosphorimetry if complexed with a luminescing organic ligand. Such ligands can be quite selective in their affinities for certain metals or nonmetals, so that luminescence methods for the determination of inorganic substances can be both specific and extremely sensitive.

**Fluorimetric Drug Analysis.**[a,b] Quinine is just one of a large family of drugs which can be sensitively determined by means of fluorimetry or phosphorimetry. Like quinine, some of these physiologically active agents fluoresce directly; others can form

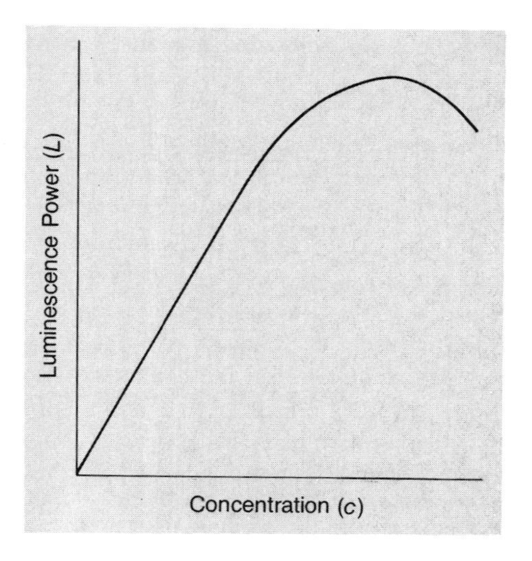


*Figure 16–19.* Dependence of luminescence power ($L$) on the concentration ($c$) of the luminescing species. The anomalous curvature of the plot is caused by the inner-filter effect. (See text for discussion.)

[a] J. Axelrod, R. O. Brady, B. Witkop, and E. V. Evarts: Ann. N.Y. Acad. Sci., *66*:435, 1957.
[b] G. K. Aghajanian and H. L. Bing: Clin. Pharmacol. Therap., *5*:611, 1964.

luminescent complexes with other organic substances. As an example, consider the fluorimetric determination of lysergic acid diethylamide (LSD), whose structure is

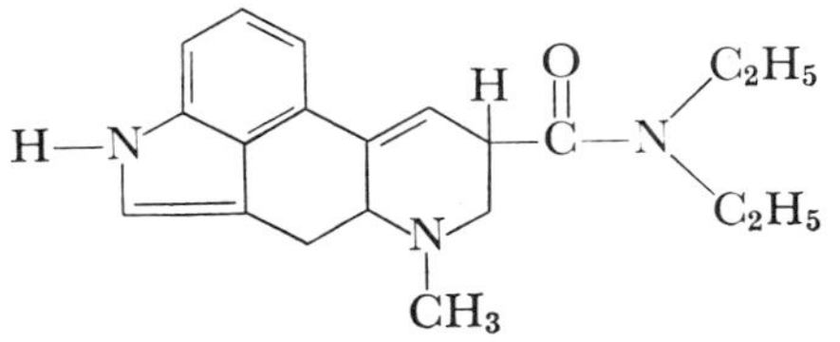

The increasingly common unauthorized use of LSD as a hallucinogenic agent has made it mandatory for law enforcement agencies to devise methods for the detection and determination of LSD in the presence of other drugs. LSD is one of the most active of all hallucinogens; as little as 50 $\mu$g taken orally is sufficient to produce hallucinatory effects. Thus, the method of analysis for LSD must be extremely sensitive. Because LSD is known to be highly fluorescent, a luminescence method of analysis meets this criterion. In the established fluorimetric procedure, a sample of blood plasma or urine (approximately 5 ml) is made slightly alkaline and extracted with a 98:2 volume-to-volume mixture of *n*-heptane and isoamyl alcohol. After extraction of LSD into the organic phase, the LSD is extracted back into an aqueous solution of 0.004 *F* hydrochloric acid. The organic solvent is then separated, and the fluorescence of the acid extract is measured directly, an excitation wavelength of 335 nm and a fluorescence wavelength of 435 nm being used.

This procedure is extremely selective, with little interference from other hallucinogens or from metabolites of LSD. With care, this technique can be used to measure plasma concentrations of LSD as low as one nanogram per milliliter, more than sufficiently sensitive to detect a *single* oral dose of LSD.

**Inorganic Fluorimetric Analysis.** Although most inorganic ions do not fluoresce directly, many of these species do form chelate complexes with organic molecules, some of which are highly fluorescent. Elements which have been analyzed in this way are Al, Au, B, Be, Ca, Cd, Cu, Eu, Ga, Gd, Ge, Hf, Hg, Mg, Nb, Pd, Rh, Ru, S, Sb, Se, Si, Sm, Sn, Ta, Tb, Th, Te, W, Zn, and Zr. Although a large number of organic molecules form fluorescent chelate complexes with these elements, three of the more common reagents are 8-hydroxyquinoline, 2,2′-dihydroxyazobenzene, and dibenzoylmethane. Unfortunately, some of these chelating agents are rather nonspecific and form complexes with many inorganic ions, so that interferences are common. Often, a prior separation procedure such as ion-exchange chromatography must be employed. However, the sensitivity of a fluorimetric determination often adequately compensates for the lack of specificity.

An example of an inorganic fluorimetric determination is the routine clinical method for the measurement of magnesium ion in serum and urine after complexation with 8-hydroxyquinoline. This procedure entails addition of the sample of serum or urine to a buffered solution (pH 6.5) of 8-hydroxyquinoline. The fluorescence of the resulting chelate complex is measured at 510 nm with an excitation wavelength of 380 nm. The only interfering ion is calcium, which can be removed by precipitation with potassium oxalate. Results compare very favorably with those obtained by means of atomic absorption and flame emission spectrometry.

## QUESTIONS AND PROBLEMS

1. Define or explain each of the following terms: transmission factor of a filter, analyte, sample blank, luminescence, resonance fluorescence,

internal conversion, intersystem crossing, vibrational relaxation, triplet state, and inner-filter effect.

2. What is the distinction between *trace analysis* and *microanalysis*?
3. Some simple instruments for absorption spectrophotometry employ test tubes as sample cells. Comment on this practice. Will a double-beam optical system compensate for errors introduced by use of test-tube cells? Why or why not?
4. Define the fluorescence quantum efficiency ($\varphi_F$) by means of an expression that incorporates rate constants for the processes depicted in Figure 16–16. Do the same for the phosphorescence quantum efficiency ($\varphi_P$). For assistance, consult the references listed at the end of the chapter.
5. Explain how the time-resolution capability of a rotating-can phosphorimeter can eliminate interference from scattering and fluorescence by the sample.
6. For some molecules, the absorption and fluorescence spectra are nearly mirror images of each other. Using the energy-level diagram of Figure 16–16, explain why this is so.
7. Explain why the luminescence spectrum of a molecule will be similar but not identical to the absorption spectrum of the molecule.
8. List the differences between fluorescence and phosphorescence.
9. Spectrophotometry is a valuable tool for the evaluation of equilibrium constants. For example, the equilibrium constant for the reaction

$$AuBr_4^- + 2\,Au + 2\,Br^- \rightleftharpoons 3\,AuBr_2^-$$

can be established by preparation of a mixture of $AuBr_4^-$ and $AuBr_2^-$ in contact with a piece of pure gold metal, followed by spectrophotometric measurement of the concentration of $AuBr_4^-$ from its absorption maximum at 382 nm. In one experiment, a solution *initially* containing a *total* of $6.41 \times 10^{-4}$ milliequivalent per milliliter of dissolved gold (present as both $AuBr_4^-$ and $AuBr_2^-$) in 0.400 $F$ hydrobromic acid was allowed to equilibrate in the presence of pure gold metal. The absorbance of the resulting solution was found to be 0.445 in a 1-cm cell at 382 nm. In separate experiments, the absorbance of an $8.54 \times 10^{-5}$ $M$ $AuBr_4^-$ solution in 0.400 $F$ hydrobromic acid was determined to be 0.410 at 382 nm, and $AuBr_2^-$ was observed to exhibit no absorption at this wavelength.

(a) Calculate the molar absorptivity of $AuBr_4^-$ at 382 nm.

(b) Calculate the equilibrium concentrations of $AuBr_4^-$ and $AuBr_2^-$.

(c) Evaluate the equilibrium constant for the reaction, neglecting activity effects.

(d) Predict how ionic strength will affect the magnitude of the equilibrium constant based on concentrations alone.

10. The titanium(IV)-peroxide complex has an absorption maximum at 415 nm, whereas the analogous vanadium(V)-peroxide complex exhibits its absorption maximum near 455 nm. When a 50-ml aliquot of $1.06 \times 10^{-3}$ $M$ titanium(IV) was treated with excess hydrogen peroxide and the final volume adjusted to exactly 100 ml, the absorbance of the resulting solution (containing 1 $F$ sulfuric acid) was 0.435 at 415 nm and 0.246 at 455 nm when measured in a 1-cm cell. When a 25-ml aliquot of

$6.28 \times 10^{-3}$ *M* vanadium(V) was similarly treated and diluted to 100 ml, the absorbance readings were 0.251 at 415 nm and 0.377 at 455 nm in a 1-cm cell. A 20-ml aliquot of an unknown mixture of titanium(IV) and vanadium(V) was treated as the previous standard solutions were, including the dilution to 100 ml, and the final absorbance readings were 0.645 at 415 nm and 0.555 at 455 nm. What were the titanium(IV) and vanadium(V) concentrations in the original aliquot of the unknown solution?

11. A 5.00-ml aliquot of a standard iron(III) solution, containing 47.0 mg of iron per liter, was treated with hydroquinone and orthophenanthroline to form the iron(II)-orthophenanthroline complex, and finally diluted to exactly 100 ml. The absorbance of the resulting solution was measured in a 1-cm spectrophotometer cell and found to be 0.467 at 510 nm. Calculate the per cent transmittance of the solution and calculate the molar absorptivity of the iron(II)-orthophenanthroline complex.
12. A mixture of dichromate and permanganate ions in 1 *F* sulfuric acid was analyzed spectrophotometrically at 440 and 545 nm as a means for the simultaneous determination of these two species, and the observed values of the absorbances were 0.385 and 0.653, respectively, at each wavelength for a 1-cm cell. Independently, the absorbance in a 1-cm cell of an $8.33 \times 10^{-4}$ *M* solution of dichromate in 1 *F* sulfuric acid was found to be 0.308 at 440 nm and only 0.009 at 545 nm. Similarly, a $3.77 \times 10^{-4}$ *M* solution of permanganate, placed in a 1-cm cell, exhibited an absorbance of 0.035 at 440 nm and 0.886 at 545 nm. Calculate the molar absorptivity of dichromate at 440 nm, the molar absorptivity of permanganate at 545 nm, and the concentrations of dichromate and permanganate in the unknown mixture.
13. Dissociation constants for acid-base indicators can be evaluated by means of spectrophotometry. The acid dissociation constant for methyl red was determined as follows. A known quantity of the indicator was added to each of a series of buffer solutions of various pH values, and the absorbance of each solution was measured at 531 nm, at which wavelength only the red acid form of the indicator absorbs radiation. The experimental data were as follows:

| pH of Buffer | Absorbance |
|---|---|
| 2.30 | 1.375 |
| 3.00 | 1.364 |
| 4.00 | 1.274 |
| 4.40 | 1.148 |
| 5.00 | 0.766 |
| 5.70 | 0.279 |
| 6.30 | 0.081 |
| 7.00 | 0.017 |
| 8.00 | 0.002 |

Calculate the acid dissociation constant for methyl red indicator.

14. The *method of continuous variations* or *Job's method* is a procedure commonly used to determine the number of ligands coordinated to a metal ion. In this technique, the sum of the moles of metal ion and ligand is kept *constant*, as is the solution volume, but the individual quantities of metal and ligand are varied. A plot is constructed of the

absorbance due to the metal-ligand complex as a function of the mole fraction of the metal ion. Such a plot consists of two straight-line portions, intersecting at a point which corresponds to the mole fraction of metal ion in an unknown complex. Suppose that the method of continuous variations was employed to establish the identity of the complex formed between iron(II) and 2,2′-bipyridine

2,2′-bipyridine

and that the following data were obtained:

| Mole Fraction of Iron(II) | Absorbance at 522 nm |
|---|---|
| 0.08 | 0.231 |
| 0.12 | 0.346 |
| 0.17 | 0.491 |
| 0.22 | 0.632 |
| 0.28 | 0.691 |
| 0.36 | 0.615 |
| 0.45 | 0.531 |
| 0.56 | 0.422 |
| 0.65 | 0.334 |
| 0.75 | 0.241 |
| 0.83 | 0.162 |
| 0.91 | 0.087 |

(a) Construct a plot of absorbance versus the mole fraction of iron(II).
(b) Determine the formula of the complex formed between iron(II) and 2,2′-bipyridine by extrapolating the two straight-line portions to their point of intersection and interpreting the result.
(c) If the sum of the concentrations of iron(II) and 2,2′-bipyridine (in all their forms in solution) in this experiment remained constant at $2.74 \times 10^{-4}$ $M$ and all absorbances were measured in a 1-cm cell, calculate the molar absorptivity of the complex.

15. A 0.150 $F$ solution of sodium picrate in a 1 $F$ sodium hydroxide medium was observed to have an absorbance of 0.419, due only to the absorption by picrate anion. In the same spectrophotometer cell and at the same wavelength as in the previous measurement, a 0.300 $F$ solution of picric acid was found to have an absorbance of 0.581. Calculate the dissociation constant for picric acid.
16. Suppose that you desire to determine spectrophotometrically the acid dissociation constant for an acid-base indicator. A series of measurements is performed in which the *total* concentration of indicator is 0.000500 $F$. In addition, all the spectrophotometric measurements are obtained with a cell of 1-cm path length and at the same wavelength. Other components, in addition to the indicator, are introduced as listed in the following table of data, but none of these exhibits any measurable absorption. Calculate the dissociation constant for the indicator.

| Solution Number | Other Component | Absorbance |
|---|---|---|
| 1 | 0.100 *F* HCl | 0.085 |
| 2 | pH 5.00 buffer | 0.351 |
| 3 | 0.100 *F* NaOH | 0.788 |

17. A solution known to contain both ferrocyanide and ferricyanide ions was examined spectrophotometrically at a wavelength of 420 nm, where only ferricyanide absorbs. A portion of the solution was placed into a 1-cm cell and found to have a transmittance of 0.118. The molar absorptivity of ferricyanide at 420 nm is 505 liter $mole^{-1}$ $cm^{-1}$. A platinum indicator electrode was inserted into the solution of ferricyanide and ferrocyanide, and its potential was observed to be +0.337 v versus the normal hydrogen electrode. Calculate the concentrations of ferricyanide and ferrocyanide in the original solution.
18. A two-color acid-base indicator has an acid form which absorbs visible radiation at 410 nm with a molar absorptivity of 347 liter $mole^{-1}$ $cm^{-1}$. The base form of the indicator has an absorption band with a maximum at 640 nm and a molar absorptivity of 100 liter $mole^{-1}$ $cm^{-1}$. In addition, the acid form does not absorb significantly at 640 nm, and the base form exhibits no measurable absorption at 410 nm. A small quantity of the indicator was added to an aqueous solution, and absorbance values were observed to be 0.118 at 410 nm and 0.267 at 640 nm for a 1-cm spectrophotometer cell. Assuming that the indicator has a $pK_a$ value of 3.90, calculate the pH of the aqueous solution.
19. In the so-called *mole-ratio method* for determining the identities of colored metal ion complexes, a series of solutions is prepared in which the concentration of the metal ion is kept constant while the concentration of the ligand is varied. Then a plot is constructed of the absorbance at a suitable wavelength versus the *ratio* of total moles of ligand to total moles of metal cation. From the point of intersection of the straight-line portions of the resulting curve, the formula of a metal-ion complex can be found. In a study of the chloro complexes of cobalt(II) formed in acetone, D. A. Fine (J. Am. Chem. Soc., *84*:1139, 1962) prepared solutions containing a constant cobalt(II) concentration of $3.75 \times 10^{-4}$ *M* and various concentrations of lithium chloride. When absorbance readings were taken in a 5-cm cell at 640 nm, the following data were obtained:

| Chloride Concentration, $M \times 10^4$ | Absorbance at 640 nm |
|---|---|
| 1.99 | 0.118 |
| 3.83 | 0.225 |
| 4.95 | 0.292 |
| 6.08 | 0.352 |
| 6.83 | 0.405 |
| 7.50 | 0.423 |
| 8.30 | 0.422 |
| 9.08 | 0.403 |
| 9.84 | 0.386 |
| 10.50 | 0.369 |
| 11.44 | 0.361 |
| 12.45 | 0.372 |
| 13.36 | 0.382 |
| 14.29 | 0.393 |

(a) Construct a plot of absorbance versus the mole ratio of chloride to cobalt(II). What can you deduce regarding the identities of any chloro complexes of cobalt(II)?

(b) Extrapolate the straight-line portions of the curve to their points of intersection. Assuming that each point of intersection corresponds to the absorbance of a pure complex ion, calculate the molar absorptivity of each complex ion identified above. What do you think about the validity of this approach?

(c) Using experimental data from the plot of absorbance versus mole ratio, along with the results obtained in parts (a) and (b), what can you determine about the value(s) for the formation constant(s) for any of the chloro complexes of cobalt(II)?

20. In the fluorimetric determination of the excretion of therapeutic doses of penicillin, a sample of urine is first extracted with chloroform. A 10.00-ml portion of the penicillin-containing chloroform extract is then added to 5.00 ml of a benzene solution of 10 mg of 2-methoxy-6-chloro-9-($\beta$-aminoethyl)-aminoacridine,

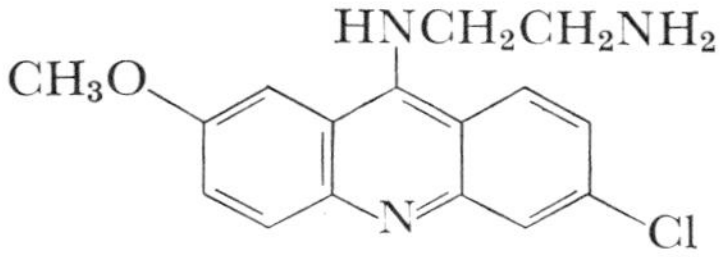

$HNCH_2CH_2NH_2$

$CH_3O$ Cl

along with 2.00 ml of acetone and 5.00 ml of a 1 volume per cent solution of glacial acetic acid in benzene. During a waiting period of 1 hour, the aminoacridine forms a condensation product with penicillin. Next, the condensation product is isolated by means of a series of extractions. Finally, an acidified aqueous solution of the condensation product is placed in a cell, the solution is irradiated with ultraviolet light at 365 nm, and the yellow luminescence at 540 nm is measured with a detector, the signal appearing in the form of a current reading on a galvanometer. In the analysis of 10.00 ml of urine according to this procedure, the galvanometer readout device registered 28.78 microamperes. A sample blank (containing no penicillin) gave a reading of 9.13 microamperes, whereas two different standard solutions containing 0.625 and 1.500 $\mu$g of penicillin per 10.00 ml gave galvanometer readings of 18.89 and 32.54 microamperes, respectively, when treated according to the above procedure. Calculate the penicillin content of the urine sample in micrograms per milliliter.

## SUGGESTIONS FOR ADDITIONAL READING

*General References*


1. R. E. Dodd: *Chemical Spectroscopy*. Elsevier, Amsterdam, 1962.
2. I. M. Kolthoff and P. J. Elving, eds.: *Treatise on Analytical Chemistry,* Part I, Volume 5, Wiley-Interscience, New York, 1959, pp. 2707-3078.
3. D. A. Skoog and D. M. West: *Principles of Instrumental Analysis*. Holt, Rinehart and Winston, New York, 1971, Chapters 2, 3, 4, and 8.
4. F. J. Welcher, ed.: *Standard Methods of Chemical Analysis*. Sixth edition, Volume III-A, Van Nostrand, Princeton, New Jersey, 1966, pp. 3-37, 78-104.
5. H. H. Willard, L. L. Merritt, and J. A. Dean: *Instrumental Methods of Analysis*. Fifth edition, Van Nostrand, Princeton, New Jersey, 1974, Chapters 3 and 4.
6. J. D. Winefordner, ed.: *Spectrochemical Methods of Analysis*. Wiley-Interscience, New York, 1971, Chapters 5-8.

*Ultraviolet-Visible Spectrophotometry*

1. R. P. Bauman: *Absorption Spectroscopy*. John Wiley and Sons, New York, 1962.
2. C. R. Hare: Visible and ultraviolet spectroscopy. *In* T. H. Gouw, ed.: *Guide to Modern Methods of Instrumental Analysis*. Wiley-Interscience, New York, 1972, Chapter 5.
3. E. B. Sandell: *Colorimetric Determination of Traces of Metals*. Third edition, Wiley-Interscience, New York, 1959.

*Fluorescence and Phosphorescence*

1. G. G. Guilbault, ed.: *Fluorescence; Theory, Instrumentation, and Practice*. Dekker, New York, 1967.
2. D. M. Hercules, ed.: *Fluorescence and Phosphorescence Analysis*. Wiley-Interscience, New York, 1966.
3. S. Udenfriend: *Fluorescence Assay in Biology and Medicine*. Academic Press, New York, Volume 1, 1962; Volume 2, 1969.
4. J. D. Winefordner, S. G. Schulman, and T. C. O'Haver: *Luminescence Spectrometry in Analytical Chemistry*. Wiley-Interscience, New York, 1972.

# 17 ATOMIC ELEMENTAL ANALYSIS

In this chapter, we will consider techniques for elemental analysis which utilize ultraviolet and visible radiation. Among these are flame spectrometric methods based on atomic emission and absorption, DC arc spectrometry, and high-voltage spark spectrometry. Because flame spectrometric methods are most widely used, we will deal with them in greatest detail.

## ELEMENTAL ANALYSIS

In order to perform an elemental analysis by means of a spectrochemical technique, one must be able to observe spectral information that is unambiguously characteristic of each element being determined. Such information would obviously not be produced by translational, rotational, or vibrational motions of molecules, because these motions involve several or all of the atoms in the molecule. Similarly, the ultraviolet (valence) electronic spectra of most molecules do not directly reflect elemental composition, because the electronic energy levels between which transitions occur are combinations of the atomic energy levels of several elements.

There appear to be two ways to perform a successful elemental analysis through spectrochemical means: (1) by observing transitions between atomic levels which are not involved in bonding and which are characteristic of a specific element or (2) by separating the atoms comprising a molecule, isolating them in the gas phase, and observing the electronic transitions characteristic of the free atoms.

In the first category, energy states not affected by bonding include those of inner-shell electrons and of the atomic nucleus. Transitions between the innermost electronic levels involve energy differences corresponding to x-ray frequencies and are highly characteristic of the atom undergoing the transition. A technique which utilizes these transitions for elemental analysis is **x-ray fluorescence spectrometry**.*

Nuclear transitions are also free from the effects of chemical bonding and can be utilized for elemental analysis. Naturally radioactive elements undergo distinctive nuclear transitions resulting in the emission of alpha particles, beta particles, and gamma rays. The energy spectra of these radiations can be used to identify and determine the elements. This procedure provides the basis for radiotracer experiments in biology, biochemistry, and medicine. For nonradioactive elements, a technique called **neutron activation analysis**† can be employed to yield similar results. In neutron activation analysis, the sample to be analyzed is bombarded with slow (thermal) neutrons, some of which are captured by the atomic nuclei to produce radioactive isotopes. Because these

*X-ray fluorescence spectrometry is an important tool for the routine elemental analysis of many industrial samples. For a detailed treatment, refer to the book by L. S. Birks: *X-Ray Spectrochemical Analysis*. Second edition, John Wiley and Sons, New York, 1969.

†For a detailed treatment of neutron activation analysis, consult the monograph by P. Kruger: *Principles of Activation Analysis*. Wiley-Interscience, New York, 1971.

artificially activated nuclei emit characteristic radiation just as the naturally radioactive elements do, they can then be determined. Neutron activation analysis is an extremely sensitive technique and can detect as little as $10^{-15}$ g of an element. However, its use requires the availability of a source of slow neutrons, so that it is somewhat restricted in its application.

In the second category of spectrochemical techniques for elemental analysis, atoms are separated from each other and are caused to produce characteristic absorption, emission, or fluorescence spectra independently. This separation of atoms is surprisingly easy and can be achieved in a variety of ways. Among the spectrochemical techniques employing this principle are mass spectrometry and those methods to be discussed in this chapter.

### Atomic Spectrochemical Analysis

Among the techniques which employ ultraviolet-visible radiation for elemental analysis, the methods used to produce free atoms vary considerably. However, in all cases, the atom-producing medium must be energetic – so energetic, in fact, that the liberated atoms are often appreciably excited. Therefore, the device producing the atoms serves not only as an **atom reservoir** or **atom cell** but often as an **excitation source** as well. Generally, electrical atomization devices such as the arc or spark are more energetic than a thermal atomization system such as a flame, so that the number of excited atoms is larger with the former. For this reason, the arc and spark are nearly always used as excitation sources from which atomic emission is observed. In contrast, flames can be employed as atom reservoirs for emission, absorption, and fluorescence spectrometry.

Regardless of whether atomic emission, absorption, or fluorescence is used for an analysis, the observed spectral features are similar. Because the atoms are essentially isolated from each other, their spectra consist of narrow lines, generally less than 0.1 Å in width. Furthermore, the spectra are relatively simple, especially for the lighter elements. Consider, for example, the emission spectrum for mercury displayed in Figure 17–1. This spectrum of mercury atoms excited in an electrical discharge shows a number of spectral lines, well separated from each other and located at specific wavelengths, the most prominent of which have been marked. These emission lines are highly characteristic of mercury and can be used for its qualitative detection in a sample. Even if other elements are present in the sample, the identification of mercury is still possible, because each element produces a characteristic spectrum of narrow lines which does not obscure the line spectra of other elements. Atomic spectrochemical analysis is therefore a powerful tool for qualitative elemental analysis. Furthermore, the power of each spectral line can be related to the concentration of a specific component of the sample. Generally, quantitative spectrochemical analysis is performed empirically, with calibrated standards and fixed experimental conditions.

## FLAME SPECTROMETRIC ANALYSIS

Let us begin our discussion of atomic spectrometry with the flame spectrometric methods, at present the most widely used of all techniques for elemental analysis. In these methods, a chemical flame is employed as an atom cell, and the emission, absorption, or fluorescence properties of the liberated atoms are examined spectrometrically. Because the procedures used to measure emission, absorption, and fluorescence differ somewhat, the field of flame spectrometry is commonly divided into

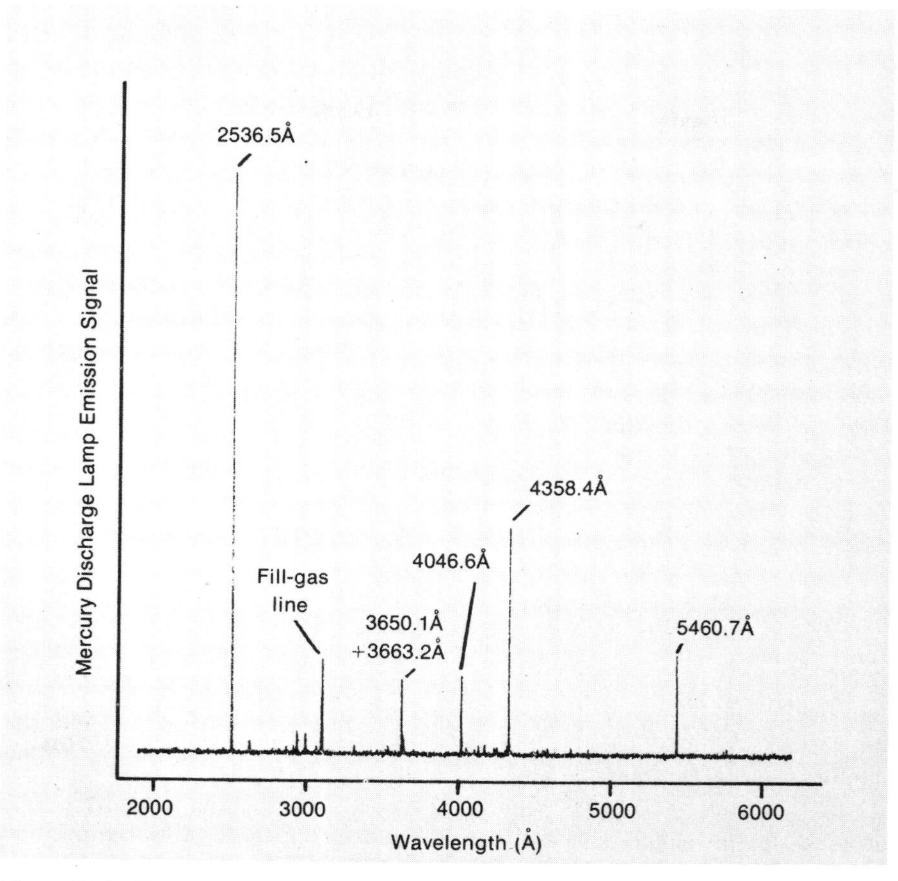


*Figure 17–1.* Emission spectrum of mercury atoms excited in a low-power electrical discharge. Prominent, characteristic spectral lines are indicated. A fill-gas emission line is produced by excitation of an inert gas present in the mercury discharge lamp.

three areas – flame emission spectrometry, atomic absorption spectrometry, and atomic fluorescence flame spectrometry. Because atomic fluorescence flame spectrometry is not at present a widely employed technique for routine elemental analysis, we will focus further discussion only on the first two methods.

We have seen in Chapter 15 (pages 398 to 401) that emission and absorption are distinctly different. Yet, in flame emission and atomic absorption spectrometry, the form of the chemical sample is the same – namely, *free atoms in a flame*. In order for these techniques to be employed successfully, the generation of free atoms in a flame must be efficient, reproducible, and predictable. However, the process of converting a chemical sample into free atoms in a flame is exceedingly complex and depends upon a number of factors that must be carefully controlled.

### Formation of Atoms in a Flame

As shown in Figure 17–2, a typical system for the production of free atoms in flame spectrometry consists of a **spray chamber** and a **burner**. A sample solution is aspirated into the spray chamber with the aid of a **nebulizer** which is controlled by the flow of an

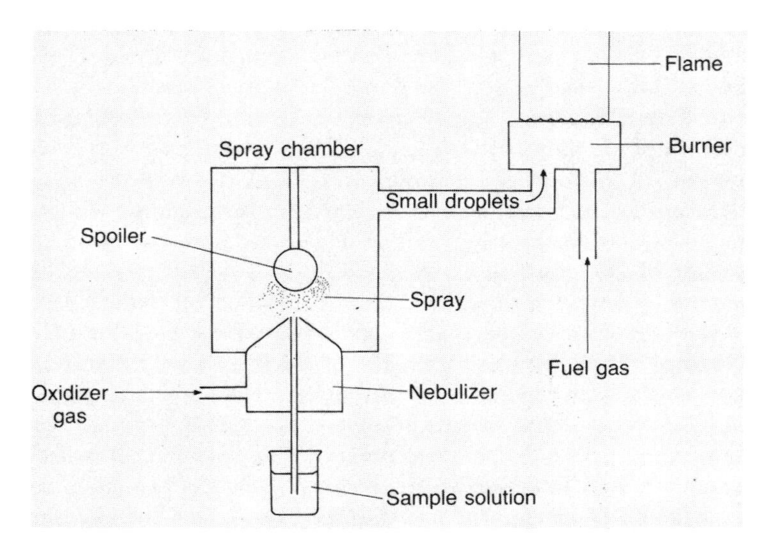


*Figure 17–2.* Premix burner-nebulizer system used in flame spectrometry.

oxidizer gas such as air. The cloud of droplets so produced strikes an obstruction in the spray chamber termed a **spoiler**, which traps the larger droplets or breaks them into smaller droplets. Small droplets are carried by the oxidizer gas from the spray chamber into the burner, where the droplets and oxidizer gas are mixed with a fuel gas. Next, the droplets pass through holes in the top of the burner along with the oxidizer-fuel gas mixture which, when ignited, forms the flame.

In the flame, a complex sequence of events must occur to convert the species dissolved in the sample droplets into free atoms. First, the droplets must evaporate (desolvate) to form small particles of the dried solute. Then the solute particles must vaporize to yield free atoms. In the process of vaporization, the solid particles of solute can be transformed into not only free atoms but molecules and molecular fragments as well; however, the latter species cannot contribute to atomic emission or absorption. Furthermore, some of the free atoms are ionized in the flame and cannot undergo the desired electronic transitions. Clearly, if analytical results of high sensitivity and reproducibility are to be obtained, all factors affecting atom production should be optimized and held as constant as possible. What are some of the experimental variables influencing the liberation of atoms in a flame?

**Nebulization.** The efficiency of nebulization (spraying) controls the fraction of the sample solution converted into tiny droplets which can ultimately reach the flame. In turn, nebulization is governed by such parameters as the viscosity and surface tension of the sample solution, the flow rate of nebulizer gas, and the design of the nebulizer itself. For example, the use of organic solvents which lower the viscosity of the sample solution frequently enhances the nebulization process. On the other hand, sample solutions containing high solute concentrations have relatively high viscosities which cause a decrease in the amount of sample nebulized and a consequent reduction in the number of free atoms in the flame. For accurate results to be obtained in a flame spectrometric analysis, the nature and overall composition of standard (calibrating) solutions must match those of the unknown sample as closely as possible.

**Desolvation.** In the flame, desolvation of the nebulized droplets critically influences the number of free atoms. Desolvation is affected by a number of experimental parameters, including the flame temperature, the nature of the solvent, and the residence time of the droplets in the flame. Thus, it is desirable to employ a combination of oxidizer and fuel gases that burns at a high temperature, and to use a slow rate of gas flow so that the droplets spend as much time in the flame and undergo as much desolvation as possible. The use of organic solvents often enhances desolvation as well, since many of these solvents evaporate more rapidly than water and, additionally, tend to cool the flame less as they evaporate.

**Vaporization.** Vaporization of the solid particles of solute which remain after desolvation is the final step leading to the formation of free atoms. Vaporization is controlled by many of the same variables as desolvation. An increase in flame temperature or the residence time of particles in the flame will increase the fraction of particles that completely vaporize. However, vaporization is strongly affected by the nature of the vaporizing particle itself. For example, a particle of aluminum oxide vaporizes much more slowly than a particle of sodium chloride of similar size. Therefore, if we compare the behavior of a cloud of sodium chloride particles and a cloud of aluminum oxide particles of the same size, the fraction of sodium chloride particles which vaporize after a specified time will be greater than the fraction of aluminum oxide particles at any given flame temperature. This effect becomes very important in the flame spectrometric analysis of complex solutions. Certain concomitants in a solution can significantly alter the process of vaporization and atom production of an element under observation to generate so-called vaporization interferences.

One well-documented case of vaporization interference is the effect of phosphate on the flame spectrometric determination of calcium. It has been found that a calcium solution containing phosphate will generate a smaller flame spectrometric signal than will a solution of identical calcium concentration containing no phosphate. This behavior can be attributed to the formation of a calcium pyrophosphate species in the flame. Because calcium pyrophosphate volatilizes more slowly than most other calcium salts (such as the chloride), it will liberate fewer calcium atoms and will lead to a depression of the observed signal.

Many such vaporization interferences can be reduced or eliminated if the atoms are observed at elevated positions in the flame. To reach these upper flame regions, solute particles must travel farther, providing them additional time in which to volatilize. A well-designed nebulizer system also helps to minimize vaporization interferences by producing small droplets and, consequently, small solute particles which vaporize more readily.

Finally, the vaporization interference can be minimized or eliminated through the use of substances called "releasing agents" which enhance the release of calcium atoms from the slowly vaporizing phosphate-containing particle. For example, large amounts of lanthanum ion, added to a solution containing calcium and phosphate, preferentially combine with phosphate, leaving the calcium atoms free to vaporize. Another releasing agent is EDTA which, when added to the sample solution, complexes calcium to prevent the formation of the calcium-phosphate species. When the solution is sprayed into a flame, the calcium-EDTA complex is readily destroyed, thereby releasing the calcium atoms. A third kind of releasing agent merely provides a matrix or substrate which decomposes or vaporizes rapidly. For example, a particle composed primarily of glucose tends to break up readily in the flame because of its organic nature. If a large amount of glucose is added to a solution containing phosphate and an alkaline earth metal, the desolvated particle will consist primarily of glucose with calcium and phosphate spread throughout the particle. When the particle breaks up, the calcium-phosphate species, if such exist, will then be extremely small and readily vaporized.

Because of the high incidence of vaporization interferences in flame spectrometry, the practicing analyst should consult appropriate reference books before attempting to determine specific elements.

**Ionization.** Any ionization which occurs in the flame reduces the total number of atoms available for observation by means of flame spectrometry. Ionization is usually important only in high-temperature flames (above 2000°K) and for elements having relatively low ionization energies such as the alkali and alkaline earth metals.

Ionization interferences arise because of the effect of one element on the degree of ionization of another. For example, rubidium partially ionizes in the flame according to the equilibrium

$$Rb \rightleftharpoons Rb^+ + e$$

so that a fixed fraction of the rubidium will be in the neutral atomic state and will generate a proportional flame spectrometric signal. However, if potassium atoms are added to the same flame, they will also ionize

$$K \rightleftharpoons K^+ + e$$

and increase the electron concentration in the flame. In turn, this increased electron concentration shifts the rubidium ionization equilibrium back toward the formation of rubidium atoms, so that the rubidium atomic signal will increase.

Clearly, a reliable rubidium determination could not be performed on a sample containing an unknown amount of potassium. To avoid this problem, samples can be doped with an excess of another alkali metal, such as lithium, whose concentration is not sought. This added element ionizes in the flame, just as do rubidium and potassium, to greatly increase the electron concentration and to shift the ionization equilibria of all other elements essentially completely to the free atomic state. This approach has the doubly beneficial effect of increasing the number of free atoms and removing the interference.

From the foregoing, we see that atom formation in a flame is a complex process and that the use of a flame for analytical purposes requires control of a number of variables, many of which are interdependent. In the selection of optimum conditions for a flame spectrometric analysis, the choice of the flame and burner is particularly critical.

### Flames Used in Atomic Spectrometry

There are several desirable characteristics of an analytical flame which derive from the considerations of the preceding section. Of course, the flame should be stable, safe, and inexpensive to maintain; it should also have a relatively high temperature and a slow rise-velocity, both of which enhance the efficiency of desolvation and vaporization and lead to a larger emission or absorption signal. In addition, a flame should generate a low spectral background and a reducing atmosphere. In the flame, many metals tend to form stable oxides. Because these oxides are refractory and not readily dissociated at ordinary flame temperatures, it is necessary to reduce the oxides to promote the formation of free atoms. This reduction can be enhanced in almost any flame if the flow rate of fuel gas is greater than that needed for stoichiometric combustion. Under these conditions, the flame is said to be **fuel rich**. Fuel-rich flames produced by hydrocarbon fuels such as acetylene provide excellent reducing atmospheres because of the presence of many carbon-containing radical species. Temperatures of several common flame-gas mixtures employed in flame spectrometry are given in Table 17–1.

Table 17–1. Common Flame-Gas Mixtures Employed in Flame Spectrometry*

| Fuel Gas | Oxidizer Gas | Measured Temperature (°C)† |
|---|---|---|
| Acetylene | Air | 2125–2400 |
| Acetylene | Nitrous oxide | 2600–2800 |
| Acetylene | Oxygen | 3060–3135 |
| Hydrogen | Oxygen | 2550–2700 |
| Hydrogen | Air | 2000–2050 |

* From the compilation by R. N. Kniseley. *In* J. A. Dean and T. C. Rains, eds.: *Flame Emission and Atomic Absorption Spectrometry.* Vol. 1, Dekker, New York, 1969, p. 191. (Courtesy of Marcel Dekker, Inc.)

† Exact temperature depends on the flow rates of the fuel and oxidizer gases and on the design of the burner.

### Sample Preparation

Because flame spectrometric methods usually rely on a burner-nebulizer system for the production of free atoms, samples to be analyzed must be in the form of a solution. Although this requirement presents no problem for many samples, others are dissolved only with great difficulty. Digestion, fusion, or combustion might be required before a suitable sample solution can be obtained. For example, clinical samples are usually digested first with hot nitric acid, perchloric acid, or a mixture of the two acids to destroy the organic matter; then, after addition of reagents to prevent interferences, the solution can be diluted to the desired concentration. Other samples, particularly silicates, are extremely hard to dissolve and must be treated with hydrofluoric acid or be fused with an appropriate flux to render them soluble. A large number of detailed procedures have been developed for the treatment of hard-to-dissolve samples. These procedures can be found in references cited at the end of this chapter.

## FLAME SPECTROMETRIC TECHNIQUES

Although the characteristics of atom formation in a flame and the existence of interferences are common to all flame spectrometric methods, each of the methods has its own particular capabilities and limitations. In the following discussion, we will consider each of the methods independently, drawing comparisons where appropriate.

### Flame Emission Spectrometry

Flame emission spectrometry is instrumentally the simplest of the flame spectrometric techniques. Recalling the generalized spectrochemical instrument of Chapter 15 (page 402), we see that the chemical input-electromagnetic radiation output transducer in flame emission spectrometry is the flame itself. This fact becomes clear when the generalized instrument of Figure 15–7 is compared with the schematic diagram of a typical flame emission spectrometer shown in Figure 17–3. In the latter instrument, the atoms liberated in the flame are excited to emit their characteristic radiation. This radiation, which can be used for both quantitative and qualitative analysis, is focused by a

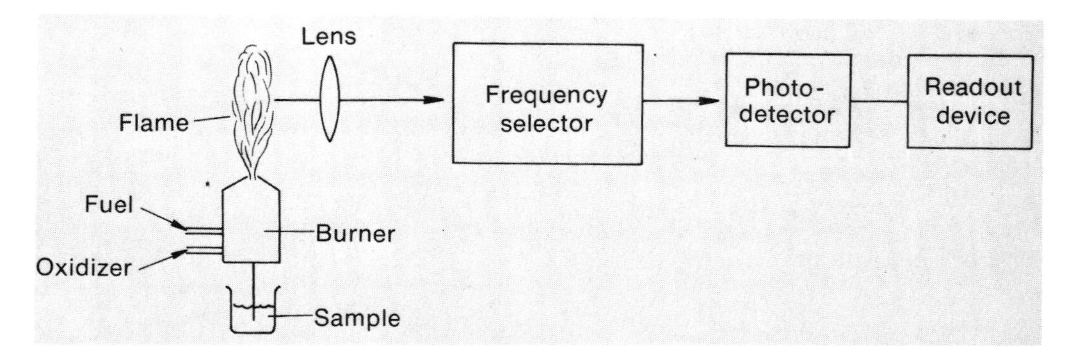


*Figure 17–3.* Generalized, schematic diagram of a simple flame emission spectrometer.

simple lens onto a frequency selector such as a filter or monochromator. The frequency-selected radiation is then detected and converted to an electrical signal by a suitable photodetector such as a photomultiplier tube. The resulting electrical signal, which is proportional to the power of the emitted radiation, is displayed on a readout device.

**Flame Emission Working Curves.** Quantitative analysis by means of flame emission spectrometry is usually performed with the aid of a series of working or calibration curves, one for each element to be determined. An example of such a working curve for potassium is shown in Figure 17–4. As predicted by the considerations of Chapter 15, the relationship between the total power of emitted radiation (or the readout value) and solution concentration is linear up to relatively high concentrations – in this case approximately 85 parts per million by weight of potassium ion. However, above this concentration, the curve bends toward the concentration axis because of the phenomenon known as **self-absorption.** Self-absorption is the absorption of emitted radiation by cooler atoms near the edge of the flame; in other words, atoms near the center of the flame, being hotter, emit radiation which can be absorbed by atoms of the same element residing at the edge of the flame. This effect is most significant when the atom concentration of the flame is high. In fact, it can be shown theoretically that, whereas the relationship between emitted power and concentration should be linear at low concentrations, the power of the emitted radiation will increase only in proportion to the square root of the concentration at higher concentration values. Just as in the case of deviations from Beer's law, the onset of self-absorption at high concentrations does not preclude quantitative analysis by means of flame emission spectrometry. It is only necessary to construct a working curve such as that shown in Figure 17–4.

At low concentrations of potassium ion, the curve in Figure 17–4 again bends, but away from the concentration axis, in this case because of ionization. For a low total concentration of potassium in the flame, the number of electrons added to the flame by ionization of potassium will be small. This low electron concentration causes the ionization equilibrium

$$K \rightleftharpoons K^+ + e$$

to be shifted toward the right. Therefore, at low potassium concentrations, a smaller fraction of the total number of potassium species residing in the flame exists as free atoms. This causes the atomic emission to be lower than that expected, producing a bend in the working curve at low potassium ion concentrations and reducing the sensitivity for the determination of potassium. A similar situation prevails for any other alkali or alkaline earth metal that is to be determined. Like ionization interferences, this problem

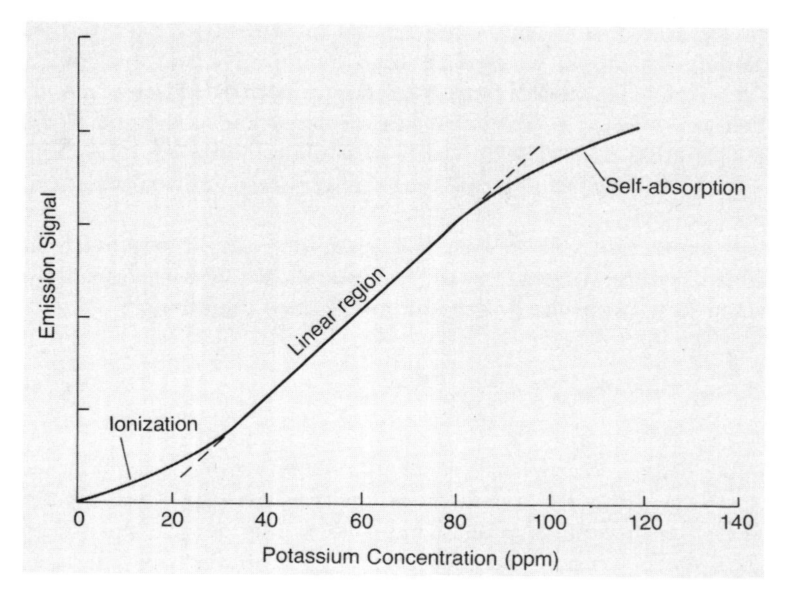


*Figure 17–4.* Working curve for determination of potassium by means of flame emission spectrometry.

can be overcome by addition of an excess of another easily ionizable element whose determination is not desired.

**Flame Emission Instruments.** Some of the high-temperature flames often used in flame emission spectrometry are listed in Table 17–1, along with their maximum temperatures. With these extremely hot flames, excellent sensitivities can be obtained for the determination of a great number of elements. Minimum detectable concentrations of these elements, together with the wavelength of the most sensitive emission line for each, are included in Figure 17–9 (to be discussed in detail later). Most elements can be readily detected at concentration levels less than one part per million (ppm). Furthermore, using flame emission spectrometry, it is possible with care to perform quantitative analyses at levels of precision between 1 and 5 per cent. In the simplest instruments (called **flame photometers**) for flame emission spectrometry, which are designed for routine, quantitative determinations involving no spectral interferences, the frequency selector is merely a filter. The lower cost, convenience, and large radiation throughput of a filter all combine to provide high sensitivity and operational simplicity. It is necessary only to insert the sample solution in the atomizing system and to observe the emission signal on a meter or strip-chart recorder; calibration is possible so that the readout can be directly in concentration units.

To perform qualitative analyses by means of flame emission spectrometry, a monochromator must be used as the frequency selector. In operation, the monochromator is set to scan across the wavelength region of interest, and the emission lines characteristic of each element appear as peaks against the flame background. By observing the wavelengths at which emission peaks do occur, one can identify the elements present from compiled listings or tables such as Figure 17–9. A flame emission spectrum for a sample containing sodium, magnesium, and calcium is shown in Figure 17–5. Although the concentration of each element in solution was identical, the lines identified with each element differ in intensity and wavelength. The simplicity of such a line spectrum usually makes qualitative analysis by means of flame emission spectrometry relatively straightforward, although complications can arise from the flame background.

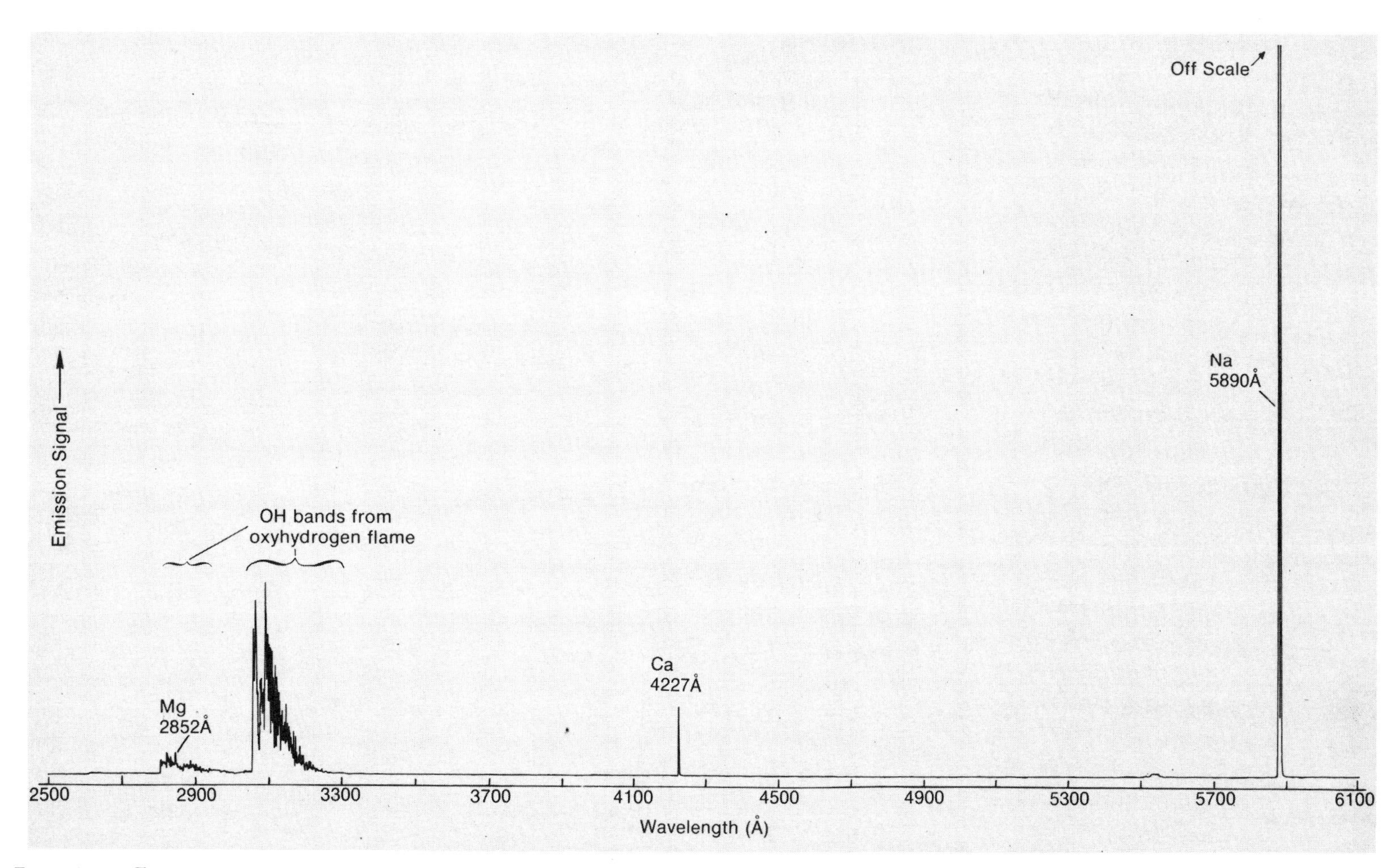


*Figure 17–5.* Flame emission spectrum of a solution which contains 10 ppm each of the elements sodium, magnesium, and calcium. Prominent, analytically useful spectral lines are indicated. Notice that the emission line at 2852 Å for magnesium is nearly buried by the –OH emission bands from the oxygen-hydrogen (oxyhydrogen) flame.

Flame background spectra differ in complexity, in intensity, and in the wavelengths of maximum emission, and are strongly affected by the fuel-oxidizer gas-flow ratio. For elements present in the sample at low concentrations, qualitative or quantitative analysis can be seriously hindered by the background spectrum of the flame. For this reason, it is good practice to record a background emission spectrum for a blank solution and to compare this background spectrum with that of the actual sample solution. If care is taken, this procedure enables very accurate and sensitive qualitative and quantitative flame emission analyses to be performed.

Another common complication which can arise in flame emission spectrometry is a spectral interference between elements in a sample. A number of elements possess emission lines lying only small fractions of an angstrom from each other, so that more than one emission line might be measured simultaneously. To minimize spectral interferences, nonroutine flame emission spectrometry should usually be performed with a high-quality monochromator, preferably one capable of resolving wavelength intervals as close together as 0.1 Å. This resolution should be compared to that required for ultraviolet-visible molecular spectrophotometry, in which bandwidths of 100 Å are often adequate. Of course, in some applications such as the determination of sodium and potassium in clinical samples discussed below, few other elements exist in appreciable concentration in the sample. This permits the use of a low-background flame and a low-resolution spectral dispersion system; even an interference filter provides adequate spectral isolation in some cases, as mentioned earlier.

**Applications of Flame Emission Spectrometry.** Flame emission spectrometry is widely used to determine the concentrations of sodium, potassium, calcium, and magnesium in clinical samples. The convenience, accuracy, sensitivity, and speed of this technique suit it well to routine analyses. To perform the analysis, the sample is first dissolved with nitric or perchloric acid if considerable protein is present (as in blood serum). A releasing agent (lanthanum) and an ionization suppressant (lithium) are then added, and the solution is accurately diluted to an appropriate volume with high-purity deionized water. Many body fluids contain a significant quantity of phosphate, so that the use of a releasing agent is essential. Finally, the prepared sample solution is analyzed with a flame emission spectrometer, such as a flame photometer having separate channels (detectors) or replaceable filters for each element to be determined.

Flame emission spectrometry is also valuable for the determination of metal ions in waste water and for the measurement of water hardness. In these applications, a qualitative analysis is often desired, in which case a preliminary spectrum is obtained with a monochromator-equipped flame emission spectrometer. Elements present are ascertained from the resulting spectrum, and suitable emission lines are selected for quantitative analysis. After the preparation of standards and the construction of working curves, the unknown sample is introduced into the flame emission spectrometer and a quantitative determination is performed.

### Atomic Absorption Spectrometry

In principle, atomic absorption spectrometry is similar to molecular absorption spectrophotometry. In atomic absorption spectrometry, the absorption spectra of isolated atoms are used to obtain information on the kind and number of atoms present in a chemical sample. This information, in turn, indicates the elemental composition of the sample. The instrumentation used in atomic absorption spectrometry provides a clear indication of the characteristics of this powerful technique.

**Instrumentation for Atomic Absorption Spectrometry**. As shown in Figure 17–6, the instrumentation for atomic absorption spectrometry resembles that used in ultraviolet-visible spectrophotometry. In an atomic absorption spectrometer, the flame is illuminated with a primary source of radiation, a portion of which is absorbed by ground-state atoms. Radiation transmitted by the flame passes through a monochromator and on to a photodetector-readout system which displays the final signal in terms of either transmittance or absorbance. Although similar in appearance to an ultraviolet-visible spectrophotometer, an atomic absorption spectrometer differs with respect to both the sample cell and the source of primary radiation.

Sample Cell. In atomic absorption spectrometry, the sample cell is simply the flame itself. Because atomic absorption spectrometry relies upon Beer's law, the sensitivity of the technique depends on the path length of primary radiation through the flame. For this reason, **slot burners** which provide a long path length have been developed for atomic absorption spectrometry. For such a burner, the exit orifice is a narrow slot between 5 and 10 centimeters in length, and the premixed fuel and oxidizer gases leave the bowl of the burner through this slot, thereby defining the geometrical shape of the resulting flame. As shown in Figure 17–8, the source of primary radiation is directed down the long axis of the flame.

Primary Source. Unlike the broad absorption bands found in molecular spectrophotometry, the absorption spectra for atomic species consist of extremely narrow lines, usually on the order of 0.01 Å wide. For Beer's law to be valid, the bandwidth of the radiation to be absorbed by the atoms must be narrower than the absorption line for the absorbing species. This means that either the line width of the primary radiation source or the bandpass of the frequency selector (monochromator) must be less than 0.01 Å. However, because all but the most expensive monochromators have bandpasses of 0.1 Å or more, it is the primary source which must provide radiation of a sufficiently narrow bandwidth. The source of narrow-band radiation most commonly used in atomic absorption spectrometry is the **hollow-cathode lamp**. In the hollow-cathode lamp, a low-power electrical discharge is sustained between an inert electrode (anode) and a second electrode (cathode) made from the element to be determined. Atoms from the cathode are thereby excited, to produce a very pure line spectrum of the desired element in addition to the line spectrum of the inert fill-gas (argon or neon) present in the hollow-cathode lamp. Because of inter-element influences, a hollow-cathode lamp is rarely made to emit the spectra of more than four elements; the most stable and intense lamps are made for a single element.

Because each spectral line from a hollow-cathode lamp has an extremely narrow bandwidth, spectral interferences in atomic absorption spectrometry are less common than in flame emission spectrometry. In flame emission spectrometry, frequency selection must be performed entirely by the monochromator; because the monochromator passes a relatively broad range of wavelengths, it is not unlikely that the monochromator will pass,

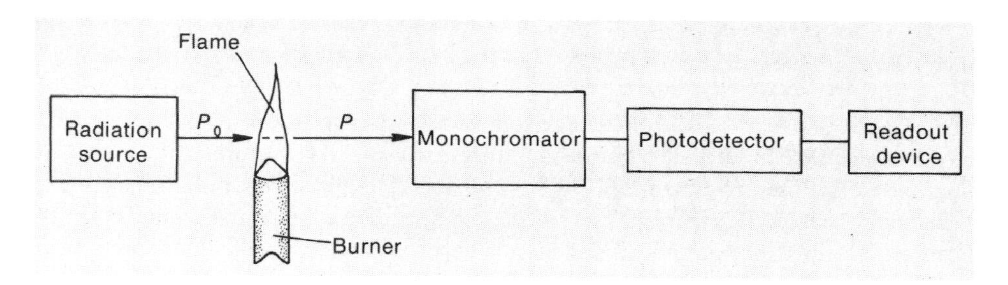


*Figure 17–6.* Simplified schematic diagram of an atomic absorption spectrometer.

in addition to the line of the desired element, a line of an interfering element or an emission line or band from the flame itself. However, in atomic absorption spectrometry, the line spectrum of the hollow-cathode lamp governs the effective bandwidth so that fewer spectral interferences result.

Monochromator. In general, the wavelength selector used in atomic absorption spectrometry can be simpler than that used in flame emission spectrometry. Because the range of wavelengths detected in atomic absorption spectrometry is determined primarily by the hollow-cathode source and not by the monochromator, the monochromator serves primarily to minimize the detected background radiation from the flame and to remove extraneous lines emitted by the hollow-cathode fill-gas. Usually, a monochromator with a bandpass of 0.5 Å is sufficient to prevent significant nonlinearity of Beer's law plots caused by background radiation from the flame.

Another problem in atomic absorption spectrometry is that excited-state atoms in the flame undergo emission. Unless eliminated, this emission signal will cause a concentration-dependent background and a consequent curvature of Beer's law plots. This difficulty is usually solved through the use of a **chopper**, of which there are two kinds: in electronic chopping, the radiation from the hollow-cathode source is alternately turned on and off, whereas in mechanical chopping a rotating shutter is used to achieve the same effect.

The operation of a rotating-shutter chopper is illustrated in Figure 17–7. In Figure 17–7A, the rotating shutter is shown positioned to block the radiation from the hollow-cathode lamp. Under these conditions, the only radiation reaching the photodetector is that emitted by the flame ($P_F$). When the shutter is opened, as shown in Figure 17–7B, the radiation reaching the detector is the *sum* of the emission from the flame ($P_F$) and the radiation from the hollow-cathode lamp transmitted by the flame ($P_C$). As the rotating shutter alternately passes and blocks the radiation from the hollow-cathode source, the photodetector signal will appear as shown in Figure 17–7C. The alternating, square-wave portion of this signal will have an amplitude proportional to the radiation from the hollow-cathode lamp transmitted by the flame.

### Comparison Between Atomic Absorption and Flame Emission Spectrometry

Although flame emission and atomic absorption spectrometry both employ a flame as an atom cell, the two methods differ in several important respects.

**Sensitivity.** For the most part, atomic absorption spectrometry and flame emission spectrometry are not competitive but complementary. Elements best determined by means of flame emission spectrometry are generally not those best determined by atomic absorption spectrometry – and the converse is true. The reason for this difference is rather simple.

Elements having low excitation energies – that is, elements which are easily excited – will emit very efficiently when placed in a high-temperature flame. This emission signal, when measured against the relatively low background of the flame, provides a sensitive method for the detection of that element. By contrast, an element having a high excitation energy will not be efficiently excited in a chemical flame; instead, most of the atoms of this element will reside in the ground electronic state. These ground-state atoms are amenable to determination by means of an absorption technique.

In most cases *flame emission spectrometry is more sensitive for the determination of elements having resonance spectral lines* between 400 and 800 nm, whereas atomic*

---

*A resonance spectral line is one which arises from a transition between the ground state and an excited state.

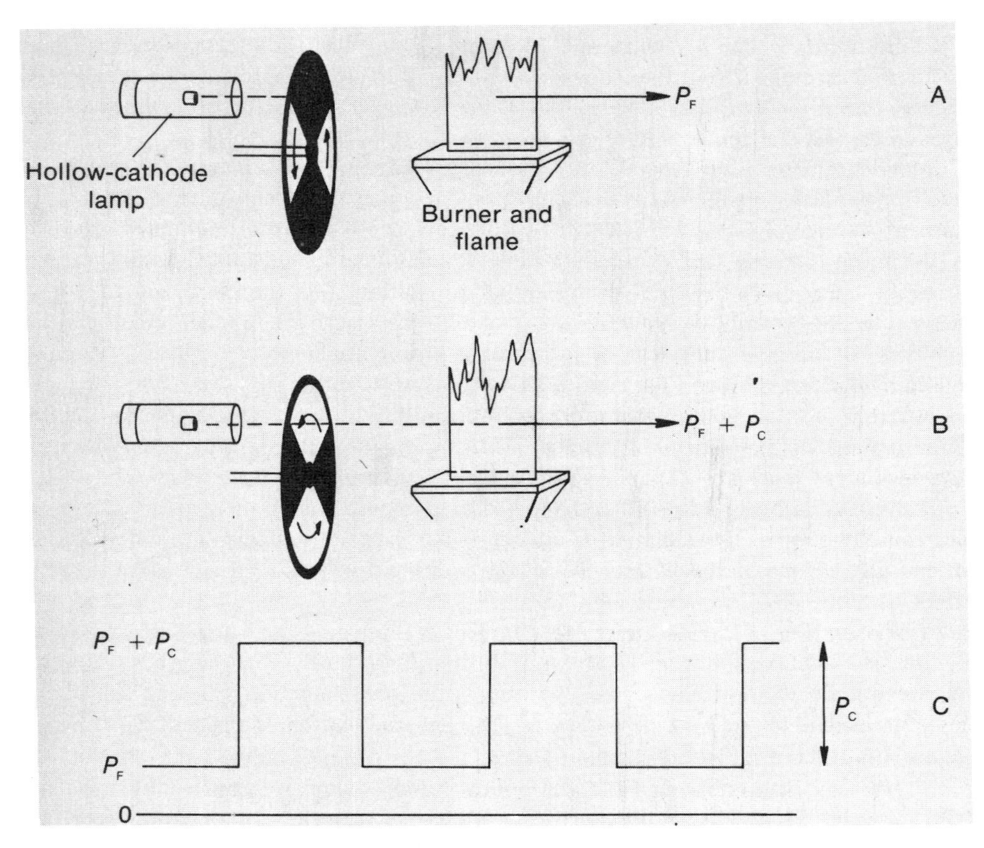


*Figure 17–7.* Elimination of emission background in atomic absorption spectrometry through the use of a rotating chopper; $P_F$ is the radiant power emitted by the flame and $P_C$ is the desired radiant power from the hollow-cathode lamp transmitted by the flame. (See text for discussion.)

*absorption spectrometry is more sensitive for elements whose resonance lines lie between 200 and 300 nm.* Elements whose resonance lines are between 300 and 400 nm (and certain other elements as well) can be determined equally well by means of either technique. Figure 17–9, to be discussed later, shows the elements best determined by each procedure.

It has been argued that atomic absorption spectrometry should be more sensitive than flame emission spectrometry for the determination of all elements. This is because most atoms of any element will reside in the ground electronic state even at flame temperatures, so that a greater number of atoms is capable of absorption than of emission. In fact, for many elements the sensitivity of atomic absorption spectrometry is inferior to that of flame emission spectrometry. This is largely due to the necessity in atomic absorption spectrometry, as in all absorption methods, to measure a small difference between two large signals, a feat which is always more difficult than the measurement of a small signal by itself.

**Interferences.** As mentioned earlier, vaporization interferences should be equally troublesome in both atomic absorption and flame emission spectrometry. Although exaggerated claims have frequently been made that interferences are less in atomic absorption spectrometry, the observed differences can be traced to the types of burners originally used for each technique.

**Qualitative and Quantitative Analysis.** When qualitative analysis is desired, flame emission spectrometry is clearly superior to atomic absorption spectrometry. Qualitative analysis with flame emission spectrometry merely requires that the entire emission spectrum of the flame be scanned, whereas atomic absorption spectrometry necessitates the use of a different hollow-cathode lamp for each element to be detected. Of course, qualitative analysis by means of atomic absorption spectrometry is possible if a continuous primary source is employed, but then a monochromator having an extremely narrow bandpass must be used if adequate sensitivity is to be obtained.

**Effect of Changes in Flame Temperature.** In a typical flame, a large fraction of the atoms of most elements will be in the ground electronic state, although the exact fraction will depend on the flame temperature. If the flame temperature should vary, the resulting change in the number of excited atoms will have a relatively greater effect on the excited-state population, because of its initially smaller number, than on the ground-state population. Therefore, it would seem that flame emission spectrometry would be more strongly affected by changes in flame temperature than would atomic absorption spectrometry. However, the dominant effect of such a change is an alteration in the extent of atom formation, so that the two techniques are affected nearly equally. As discussed earlier, atom formation in the flame depends upon the desolvation of droplets, the vaporization of the resulting solute particles, and the establishment of favorable conditions for equilibria involving the atoms. The overall temperature-dependence of all these processes will, in most cases, outweigh that of the excited-state and ground-state populations.

**Instrumental Requirements.** Instrument systems for atomic absorption spectrometry are generally more expensive than those for flame emission spectrometry. A typical atomic absorption spectrometer costs between \$3000 and \$15,000, whereas simple special-purpose (clinical) instruments for flame emission spectrometry are often priced at less than \$2000. Of course, the more sophisticated atomic absorption systems employ choppers, complex signal-processing equipment, instrumental correction for curvature of Beer's law plots, and direct, digital readout of concentration. With a high-quality atomic absorption spectrometer, precision levels of 0.1 per cent can often be obtained in quantitative analysis. Figure 17–8 shows a schematic diagram of a modern atomic absorption spectrometer.

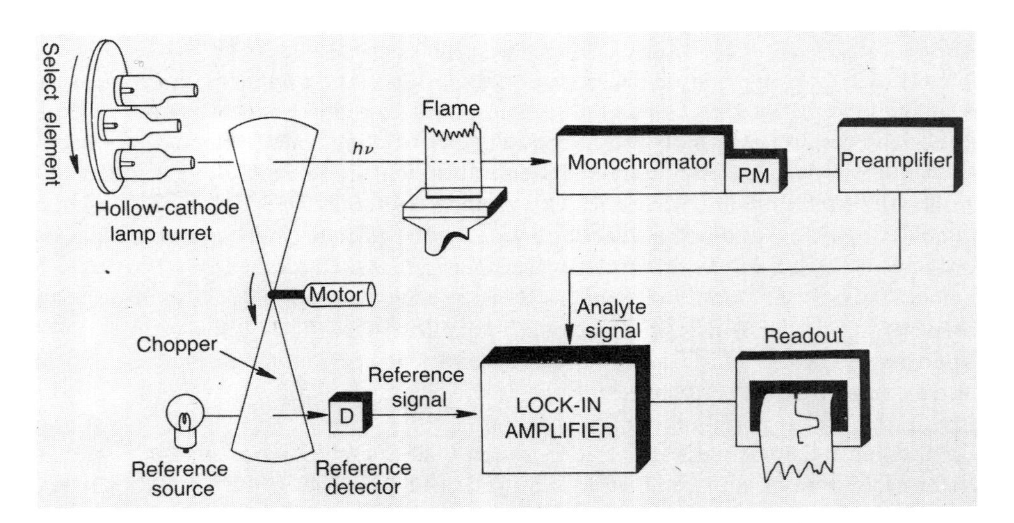


*Figure 17–8.* Modern atomic absorption spectrometer, with a multiple hollow-cathode lamp turret which can be rotated to select the desired lamp and a sophisticated signal-processing system; PM is a photomultiplier tube.

From the foregoing discussion, it should now be clear that atomic absorption and flame emission spectrometry are indeed complementary rather than competitive. In recognition of this, most modern atomic absorption spectrometers incorporate a provision for analysis by means of flame emission spectrometry. Well established procedures are available for the analysis of most elements in the periodic table by either one technique or the other. The periodic table shown in Figure 17–9 indicates the elements determinable by means of flame emission or atomic absorption spectrometry, which of the two methods is better, the optimum wavelengths for use with each technique, and the minimum concentrations detectable by each method. It is possible to detect most elements at the parts-per-million to parts-per-billion concentration level with either flame emission or atomic absorption spectrometry. This high degree of sensitivity, the excellent precision attainable, and the convenience of using solution samples have made the combination of these two techniques the most popular current method for elemental analysis.

## FLAMELESS ATOMIZATION

In the foregoing discussion, only flames were considered as atomization devices. However, flames are imperfect atom cells for several reasons. They are somewhat dangerous to operate, and necessitate the storage and handling of potentially hazardous fuel and oxidizer gases, a requirement which makes then undesirable in industrial, teaching, and clinical laboratories. Flames are also relatively expensive to employ because large volumes of fuel and oxidizer gases are consumed. In addition, spectroscopic techniques based on the use of flames require relatively large amounts of sample solution, are often troubled by interferences, and are generally restricted to samples of relatively low concentration. Certainly, it would be desirable if a cheaper, safer, and more efficient atomizer could be found. A number of atomizers having these characteristics have been developed and are being used more and more widely.

The new flameless atomizers take several forms. In general, they consist of a rod, a loop, a boat, or a trough, made of conductive carbon or metal upon which the sample is placed. The conductive sample support is electrically heated by passage of a large current that vaporizes and partially atomizes the sample. These devices are often extremely efficient atomizers; they utilize extremely small samples; and their use reduces the amount of sample preparation necessary for an analysis. Although flameless atomization techniques can be used for atomic fluorescence, until now most applications have been to atomic absorption because of the ready availability of atomic absorption spectrometers.

A typical flameless atomization system, illustrated in Figure 17–10, consists of a small carbon tube supported between two graphite electrodes. In operation, 1 to 50 $\mu$l of a liquid sample is placed within the tube and the entire assembly is heated by passage of a low current through it. At this current, the temperature of the tube is just sufficient to remove the solvent from the sample. Then, a higher current is passed through the assembly to ash the sample if it contains a volatile matrix. Finally, the sample itself is vaporized by passage of a current of several hundred amperes through the tube. At this current, the tube can reach temperatures greater than 3000°K, sufficient for the vaporization and atomization of most elements.

In commercial flameless atomization systems, all the operations described above are performed automatically according to a programmed sequence. If the sample is observed by means of atomic absorption spectrometry, the detected signal appears as shown in Figure 17–11. Note the "false absorption" which appears during the sample ashing and drying periods. This "false absorption" is caused by scattering of radiation from the

Sample key (Cr): Atomic Absorption Wavelength 357.9 nm; Atomic Absorption Detection Limit 0.002 μg/ml; ·Atomic Emission Wavelength 425.4 nm; Atomic Emission Detection Limit 0.004 μg/ml

Elements best analyzed by atomic absorption shown in outline

Elements best analyzed by flame emission shaded gray

Elements suitable for analysis by either method shown in black

Elements unsuitable for analysis by either absorption or emission in shaded boxes

Criteria for preference based on significant difference in detection limit

| Element | Atomic Absorption Wavelength (nm) | Atomic Absorption Detection Limit (μg/ml) | Atomic Emission Wavelength (nm) | Atomic Emission Detection Limit (μg/ml) |
|---|---|---|---|---|
| H | | | | |
| He | | | | |
| Li | 670.8 | 0.001 | 670.8 | 0.00002 |
| Be | 234.9 | 0.002 | 234.9 | 0.1 |
| B | 249.7 | 2.5 | 518.0' | 0.05 |
| C | | | | |
| N | | | | |
| O | | | | |
| F | | | | |
| Ne | | | | |
| Na | 589.0 | 0.0008 | 589.0 | 0.00005 |
| Mg | 285.2 | 0.0003 | 285.2 | 0.005 |
| Al | 396.2 | 0.04 | 396.2 | 0.005 |
| Si | 251.6 | 0.1 | 251.6 | 3.0 |
| P | NA | NA | 526.0' | 0.05 |
| S | NA | NA | 392.0 | 2.0 |
| Cl | | | | |
| Ar | | | | |
| K | 766.5 | 0.003 | 766.5 | 0.0005 |
| Ca | 422.7 | 0.0005 | 422.7 | 0.0001 |
| Sc | 391.2 | 0.1 | 402.4 | 0.08 |
| Ti | 364.3 | 0.1 | 399.8 | 0.2 |
| V | 318.4 | 0.02 | 437.9 | 0.01 |
| Cr | 357.9 | 0.002 | 425.4 | 0.004 |
| Mn | 279.5 | 0.001 | 403.1 | 0.005 |
| Fe | 248.3 | 0.004 | 372.0 | 0.02 |
| Co | 240.7 | 0.002 | 345.4 | 0.03 |
| Ni | 232.0 | 0.005 | 352.5 | 0.02 |
| Cu | 324.7 | 0.004 | 324.7 | 0.01 |
| Zn | 213.9 | 0.001 | 481.1 | 8.0 |
| Ga | 287.4 | 0.05 | 417.2 | 0.01 |
| Ge | 265.2 | 0.2 | 265.2 | 0.4 |
| As | 193.7 | 0.1 | 193.7 | 12.0 |
| Se | 196.0 | 0.1 | NA | NA |
| Br | | | | |
| Kr | | | | |
| Rb | 780.0 | 0.005 | 780.0 | 0.001 |
| Sr | 460.7 | 0.004 | 460.7 | 0.0002 |
| Y | 410.2 | 0.4 | 597.2' | 0.03 |
| Zr | 360.1 | 4.0 | 360.1 | 5.0 |
| Nb | 405.9 | 1.0 | 405.9 | 1.2 |
| Mo | 313.3 | 0.03 | 390.3 | 0.1 |
| Tc | | | | |
| Ru | 349.9 | 0.06 | 372.8 | 0.3 |
| Rh | 343.5 | 0.02 | 343.5 | 0.03 |
| Pd | 247.6 | 0.01 | 363.5 | 0.05 |
| Ag | 328.1 | 0.0005 | 328.1 | 0.008 |
| Cd | 228.8 | 0.0006 | 326.1 | 0.8 |
| In | 303.9 | 0.03 | 451.1 | 0.003 |
| Sn | 224.6 | 0.03 | 284.0 | 0.1 |
| Sb | 217.6 | 0.03 | 252.8 | 0.6 |
| Te | 214.3 | 0.05 | 238.3 | 200.0 |
| I | | | | |
| Xe | | | | |
| Cs | 852.1 | 0.10 | 455.5 | 0.05 |
| Ba | 553.5 | 0.02 | 553.5 | 0.002 |
| La | 550.1 | 5.0 | 441.8' | 0.01 |
| Hf | 307.3 | 14.0 | 531.2 | 19.0 |
| Ta | 271.5 | 3.0 | 474.0 | 4.0 |
| W | 400.9 | 3.0 | 400.9 | 0.6 |
| Re | 346.0 | 0.6 | 346.0 | 0.1 |
| Os | 305.9 | 0.3 | 442.0 | 1.7 |
| Ir | 284.9 | 1.0 | 550.0' | 0.4 |
| Pt | 265.9 | 0.05 | 265.9 | 3.8 |
| Au | 242.8 | 0.02 | 267.6 | 0.5 |
| Hg | 253.7 | 0.2 | 253.7 | 10.0 |
| Tl | 276.8 | 0.02 | 535.1 | 0.02 |
| Pb | 217.0 | 0.01 | 405.8 | 0.1 |
| Bi | 223.1 | 0.05 | 223.1 | 2.0 |
| Po | | | | |
| At | | | | |
| Rn | | | | |
| Fr | | | | |
| Ra | | | | |
| Ac | | | | |
| Ce | NA | NA | 494.0 | 98.0 |
| Pr | 495.1 | 4.0 | 495.1 | 0.07 |
| Nd | 463.4 | 0.6 | 492.5 | 0.7 |
| Pm | | | | |
| Sm | 429.7 | 2.0 | 476.0 | 0.2 |
| Eu | 459.4 | 0.1 | 466.2 | 0.001 |
| Gd | 368.4 | 4.0 | 622.0 | 0.07 |
| Tb | 432.7 | 2.0 | 534.0' | 0.03 |
| Dy | 421.2 | 0.2 | 404.6 | 0.05 |
| Ho | 410.4 | 0.1 | 410.4 | 0.1 |
| Er | 400.8 | 0.1 | 400.8 | 0.07 |
| Tm | 371.8 | 0.4 | 371.8 | 0.08 |
| Yb | 398.8 | 0.02 | 398.8' | 0.006 |
| Lu | 331.2 | 3.0 | 451.9 | 1.3 |
| Th | NA | NA | 492.0 | 11.0 |
| Pa | | | | |
| U | 358.5 | 23.0 | 544.8 | 5.5 |
| Np | | | | |
| Pu | | | | |
| Am | | | | |
| Cm | | | | |
| Bk | | | | |
| Cf | | | | |
| Es | | | | |
| Fm | | | | |
| Md | | | | |
| No | | | | |
| Lw | | | | |

·The elements Na, K, and Rb were determined in an air-acetylene flame. All others in a nitrous oxide-acetylene flame. Detection limits in many cases can be improved by the addition of an easily ionized substance such as sodium or potassium.

'Band Emission

*Figure 17–9.* Periodic table of elements detectable by means of flame emission and atomic absorption spectrometry. Courtesy of Instrumentation Laboratory, Inc., Lexington, Massachusetts.

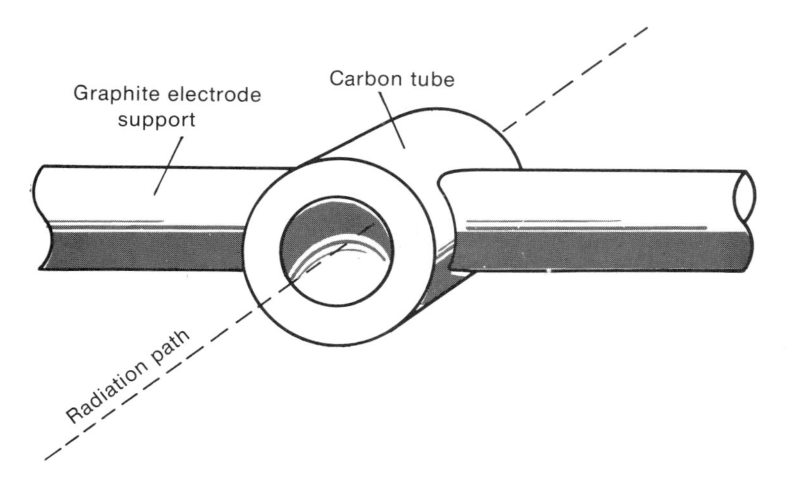


*Figure 17–10.* Carbon-tube type of flameless atom reservoir. Courtesy of Varian Techtron, Walnut Creek, California.

hollow-cathode source by the smoke and ash produced during ashing. If the ashing and drying periods are not separated in time from atomization of the sample, an erroneous absorption reading will result. This signal overlap is one of the present problems plaguing flameless atomizers. To minimize this difficulty, special instrumental arrangements for "background subtraction" must be employed. A discussion of this instrumentation is, unfortunately, beyond the scope of this text.

As atom cells, flameless atomization systems have several significant advantages over chemical flames. Perhaps the greatest virtue is that of absolute sensitivity. Although the techniques of flame spectrometry are capable of providing high sensitivity for very low concentrations, they require at least two milliliters of sample solution for a reliable result;

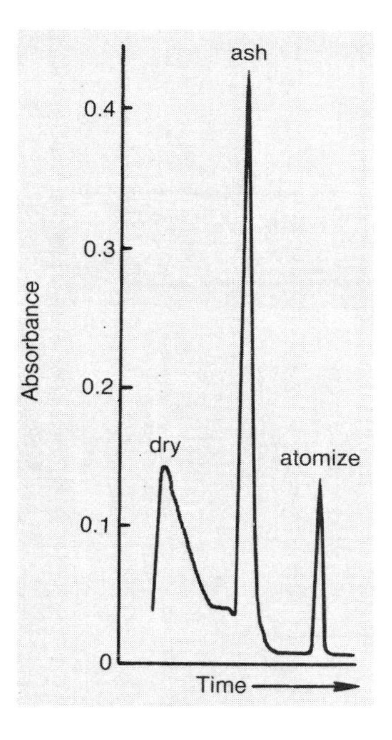


*Figure 17–11.* Atomic absorption signal produced by atomization of vanadium in fuel oil (diluted 1:10 with xylene) using the carbon-tube atomizer shown in Figure 17–10. Total sample volume, 5 μl; concentration of vanadium in sample, 1.4 ppm. (See text for discussion.)

at a detection limit of one part per billion, this volume of solution corresponds to $2 \times 10^{-9}$ gram of a particular element. By contrast, flameless atomization systems are often able to detect as little as $10^{-14}$ gram of an element. Absolute detection limits by weight obtained through the use of the carbon-tube atomizer in atomic absorption spectrometry are shown in Table 17–2; these can be compared with the detection limits for the flame methods given in Figure 17–9.

At present, the most serious problem encountered in flameless atomization involves interelement interferences. Because the regions of atom observation in the flameless systems are not in direct contact with the heated atomizer surface, they do not reach the high temperatures found in most flames, so that atom recombination often occurs and interferences are common. The design of the carbon-tube atomizer shown in Figure 17–10 reduces interferences somewhat by partially confining the vaporized sample within the heated orifice of the tube. Atomization is thereby enhanced and atom recombination reduced. Furthermore, the atomized sample can provide a larger signal by residing within the region of observation for a longer time. A small hydrogen flame surrounding the atomizer has been found to further minimize vaporization interferences by providing a reducing environment. However, even with these precautions, some elements are not amenable to flameless atomization techniques. For example, boron cannot be determined with the carbon-tube atomizer because boron carbide, an extremely refractory compound, is formed.

Other problems encountered with present flameless atomizers are sample carryover and limited useful lifetime. Carbon-tube atomizers are often porous and tend to absorb a portion of each sample, carrying the portion over to the next determination. This difficulty can be largely circumvented by the use of pyrolytic carbon, which is less permeable than other forms of carbon, and by the placement of a drop of an inert, pure organic liquid such as xylene inside the tube before each determination. Apparently,

**Table 17–2. Absolute and Concentration Detection Limits Attainable with a Flameless Atomizer***

| Element | Absolute Grams | Concentration, ng/ml in 5 μl Sample | Element | Absolute Grams | Concentration, ng/ml in 5 μl Sample |
|---|---|---|---|---|---|
| Ag | $2 \times 10^{-13}$ | 0.04 | Li | $5 \times 10^{-12}$ | 1.0 |
| Al | $3 \times 10^{-11}$ | 6.0 | Mg | $6 \times 10^{-14}$ | 0.012 |
| As | $1 \times 10^{-10}$ | 20 | Mn | $5 \times 10^{-13}$ | 0.1 |
| Au | $1 \times 10^{-11}$ | 2.0 | Mo | $4 \times 10^{-11}$ | 8.0 |
| Be | $9 \times 10^{-13}$ | 0.18 | Na | $1 \times 10^{-13}$ | 0.02 |
| Bi | $7 \times 10^{-12}$ | 1.4 | Ni | $1 \times 10^{-11}$ | 2.0 |
| Ca | $3 \times 10^{-13}$ | 0.06 | Pb | $5 \times 10^{-12}$ | 1.0 |
| Cd | $1 \times 10^{-13}$ | 0.02 | Pd | $2 \times 10^{-10}$ | 40 |
| Co | $6 \times 10^{-12}$ | 1.2 | Pt | $2 \times 10^{-10}$ | 40 |
| Cr | $5 \times 10^{-12}$ | 1.0 | Rb | $6 \times 10^{-12}$ | 1.2 |
| Cs | $2 \times 10^{-11}$ | 4.0 | Sb | $3 \times 10^{-11}$ | 6.0 |
| Cu | $7 \times 10^{-12}$ | 1.4 | Se | $1 \times 10^{-10}$ | 20 |
| Eu | $1 \times 10^{-10}$ | 20 | Sn | $6 \times 10^{-11}$ | 12 |
| Fe | $3 \times 10^{-12}$ | 0.6 | Sr | $5 \times 10^{-12}$ | 1.0 |
| Ga | $2 \times 10^{-11}$ | 0.4 | Tl | $3 \times 10^{-12}$ | 0.6 |
| Hg | $1 \times 10^{-10}$ | 20 | V | $1 \times 10^{-10}$ | 20 |
| K | $9 \times 10^{-13}$ | 0.18 | Zn | $8 \times 10^{-14}$ | 0.016 |

* Courtesy of Varian Techtron, Walnut Creek, California.

xylene forms a coating on the carbon surface and prevents penetration of the sample solution.

Although longer-lived carbon-tube atomizers are now being introduced, most are usable for only a hundred determinations or less. This limited lifetime necessitates the frequent replacement of the tube and, unless the replacement tube is identical to the original, a readjustment of the atomizing-current program. This readjustment, when required, is often critical because of its influence on interelement interferences. A recent innovation to prolong the carbon-tube lifetime involves passing methane gas through the heated tube. This procedure results in the pyrolysis of the methane and the deposition of a fresh carbon surface on the tube.

## ELECTRICAL-DISCHARGE SPECTROMETRY

Topics to be discussed in this section usually come under the heading "optical emission spectroscopy." Although the techniques do utilize emitted radiation in the optical (ultraviolet-visible) region of the spectrum, the term is hardly descriptive. At present, a large number of different electrical discharges are employed to excite atomic emission. Among these are the DC arc, AC arc, high-voltage spark, radio-frequency and microwave plasmas, plasma jet, and plasma torch. Because the most frequently used are the DC arc, the high-voltage spark, and the radio-frequency plasma, we will confine our discussion to these three sources. Each of these discharges excites a sample in a different way and by a slightly different mechanism. In addition, the instrumentation differs considerably so that it will be most convenient to examine each discharge separately.

### DC Arc

A DC arc source consists of a high-current (5 to 30 amperes), low-voltage (10 to 25 volts) electrical discharge supported between two electrodes. Because of the high temperature of the discharge (2000 to 4000°K), a sample placed within the discharge is partially vaporized into free atoms which are excited and emit characteristic spectra.

Excitation of a sample by a DC arc discharge is partly thermal and partly electrical in origin. Unfortunately, it is difficult to control the variables – primarily arc current and arc resistance – which govern the temperature of the discharge and, thus, the excitation of the sample. For example, refractory samples vaporize slowly in the arc, thereby changing the character and resistance of the discharge. In addition, the high temperature of the discharge gradually erodes the electrodes between which the discharge occurs, also causing the arc resistance to change. Consequently, there is a lack of reproducibility from sample to sample, an instability of the discharge during the analysis of a single sample, and a strong dependence on the sample matrix. These characteristics limit the attractiveness of the DC arc as an excitation source for quantitative analysis. However, the extreme sensitivity and the ease of handling certain samples make the DC arc an attractive source for some applications.

**DC Arc Spectra.** Radiation emitted by a DC arc discharge has several distinguishable components, including atomic spectra, molecular spectra, and background emission from the electrode material itself. Because the DC arc operates at an extremely high temperature, the electrodes are heated to incandescence and emit intense and spectrally continuous background radiation, resembling that from a blackbody.

Molecules and radicals are quite efficiently excited in a DC arc discharge and produce their distinctive band spectra, often with superimposed vibrational fine-structure. These band spectra often cover broad spectral regions, and can obscure much of the atomic radiation emitted by the discharge. One notable example is the presence of the so-called cyanogen bands, emitted by CN radicals formed by reaction of carbon electrodes with atmospheric nitrogen.

Atomic line spectra from a DC arc discharge are considerably different in appearance from those obtained in flame emission spectrometry, primarily because of the difference in temperature between the two sources. In a flame, atoms are usually raised only to lower excited electronic levels so that the emission spectra are quite simple. By contrast, the thermal and electrical energy available in the DC arc is sufficient to cause population of extremely high energy levels; thus, a considerably greater number of transitions is possible and a more complex emission spectrum is obtained, especially for the heavier elements. For example, the emission spectrum of iron contains more than 4000 lines. In the case of uranium, the number of emission lines is so great that the spectrum appears to be nearly continuous and obscures all but the most intense emission lines of other elements present.

To further complicate the situation, many emission lines of ions are observed from the DC arc source. The high temperature of the DC arc is capable of ionizing a number of elements, notably the alkali metals and alkaline earths, having relatively low ionization energies. For these elements, the observed emission spectra exhibit not only atomic emission lines but ion emission lines as well.

Because of the high excitation efficiency and complex emission spectra obtained with the DC arc source, two facts are evident. First, the DC arc is an excellent source for qualitative analysis. Second, to allow qualitative analysis to be performed efficiently and reliably, a high-quality spectral dispersion device must be employed to separate the desired atomic emission lines from the large number of other spectral features present.

**Dispersion and Detection of Spectra.** To permit observation of a large number of emission lines from atoms and to allow correction for any spectral background, a high-dispersion polychromator is usually used in DC arc spectrometry. Compared with monochromators commonly employed in flame spectrometry, the DC arc polychromator is physically quite large in order to provide greater spatial dispersion of the spectrum and easier separation of desired lines from background and band radiation.

Photographic and photoelectric detection systems are both employed in DC arc spectrometry, the photographic system being most common for qualitative analysis and the photoelectric detector system for quantitative applications. As discussed in Chapter 16 (page 414), photographic detection has an advantage in that it permits examination of a broad spectral range at one time. This makes it possible to detect a large number of elements simultaneously and, furthermore, to obtain readings of the background emission adjacent to the spectral lines being observed. Also, a photographic emulsion automatically integrates the emission from the arc, thereby averaging out any arc instability. An example of a photographically recorded spectrum for sodium was presented in Figure 16–5 (page 415).

When photoelectric detection is employed, it is usually necessary to electronically integrate the emission signal for a period of time to minimize the effect of arc instability. In photoelectric detection systems, a number of photomultipliers are placed along the focal curve or plane of a polychromator. For quantitative work, a background reading must be obtained at a wavelength immediately adjacent to each spectral line being observed; thus, two photodetectors are needed for each spectral line of interest. This, of course, greatly increases the cost of a photoelectric detection system. Such a **direct-reading spectrometer** can cost up to $100,000. Its use is only justified for applications in which certain elements must be determined frequently and rapidly. One such application is in the metals industry, where analyses of steel and aluminum alloys, for example, must be performed routinely and rapidly to maintain proper process control.

**Qualitative Analysis Using a DC Arc.** DC arc spectrometry is one of the most sensitive of all techniques for elemental analysis. Concentrations between parts per million and parts per billion can be detected *simultaneously* for as many as 70 elements. To perform a qualitative analysis using DC arc excitation and photographic readout, the wavelength scale on the photographic emulsion must be calibrated. To do this, one can photograph in two separate experiments the emission spectrum of a known element (such as iron) on the top half of the emulsion, and the emission spectrum of the sample on the bottom half of the emulsion. As mentioned earlier, iron has a very rich arc spectrum, consisting of a large number of lines whose wavelengths are accurately known. Therefore, the iron spectrum serves as a calibration scale from which the wavelengths and identities of the spectral lines of a chemical sample can be determined.

A sometimes inconvenient feature of the DC arc discharge which can be turned to advantage in qualitative analysis is fractional distillation of the sample. Because of the extremely high temperatures found in the DC arc discharge, solid samples are often melted and fractionally vaporized. More volatile species such as alkali metals, alkaline earths, and other light elements vaporize first, with heavier or more refractory elements vaporizing later. By successive exposure of several spectra above and below each other on the same photographic emulsion, the emission spectra of these lighter elements can be detected before the complex spectra of heavier elements can interfere. An extension of this method is found in the "carrier distillation" technique, in which a low-boiling element such as gallium is deliberately added to the sample. When it vaporizes, the low-boiling element carries with it a number of other elements, thereby separating them from more refractory species.

Because of the high temperature gradient found in a DC arc discharge, the incidence of self-absorption is high. Atoms in the core of the discharge, being at a higher temperature than those toward the outside of the discharge, emit radiation which is absorbed by atoms at the outer edge of the discharge. Furthermore, spectral line broadening is extreme at the high temperatures and electric fields found in the central portion of the DC arc, so that radiation emitted by atoms within this region has a broader spectral bandwidth than the absorption line width of the atoms surrounding the arc. Under these conditions, the central portion of the band of radiation emitted by the atoms within the arc is absorbed by those atoms surrounding the arc. This extreme example of self-absorption, termed **self-reversal**, can complicate qualitative analysis if its existence is not recognized. A self-reversed line, such as that shown in Figure 17–12, can appear to be two separate lines on either side of the true emission line. (Compare this self-reversed line with the sodium doublet line at 589 nm in Figure 16–5.)

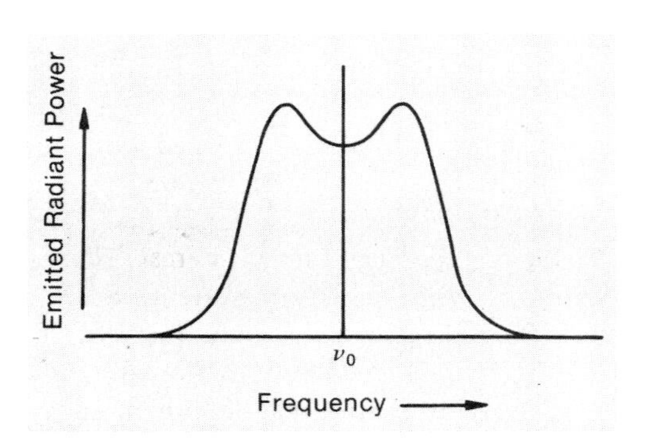


*Figure 17–12.* Profile of a self-reversed spectral line; the center of the line is at frequency $\nu_0$. (See text for discussion.)

**Quantitative Analysis by DC Arc Spectrometry.** Quantitative analysis with a DC arc source rarely achieves precision levels of better than 5 to 10 per cent because of arc instability and matrix interferences. Because of arc instability, there is nonuniform atomization of the sample, so that it is difficult to obtain a precise quantitative estimate of its composition. The intensity of atomic emission is dependent on the sample matrix, because the sample matrix strongly affects the resistance and mode of excitation of the DC arc. Fortunately, matrix effects can be minimized if a **matrix buffer** is employed. When such a buffer, usually a low-boiling substance such as lithium chloride, is added in large quantity to a sample, the vaporization and excitation behavior is dependent primarily on the nature of the buffer and not on the sample matrix.

Another way to minimize the effects of arc instability and the sample matrix is through the use of an **internal standard**, an element whose vaporization and excitation characteristics very closely match those of the element to be determined. The internal standard is added in constant concentration to all samples and standards. Any instabilities or variations in the arc will affect the intensities of emission lines for the sample and internal standard equally. Analytical working curves can then be plotted as the *ratio* of the line intensities of the internal standard and sample versus the concentration ratio of the two. Use of an internal standard provides improvements in precision up to an order of magnitude.

**Sample Preparation.** Solids, liquids, and gases can all be analyzed with the aid of the DC arc source. Special electrodes, standards, and procedures have been developed for many different types of samples. We will discuss only a few representative examples.

Conductive Metal Samples. These are the most convenient samples to analyze by means of DC arc spectrometry. Conductive metals are usually cast or machined into appropriate electrodes which can be used directly in the DC arc discharge. At times, both electrodes are formed from the conductive sample; at other times, the sample is employed as a cathode and another material, often spectroscopically pure graphite, as the anode (counter electrode). In all procedures, extreme care must be exercised to prevent contamination of samples, especially because of the high sensitivity of DC arc spectrometry.

Powders and Nonconductive Solids. Although many samples are nonconductive, they can be pulverized and mixed with a conductive material such as high-purity graphite powder. The mixture of sample and conductive material must be as homogeneous as possible and must be packed carefully into a specially formed sample electrode. Several sample electrodes used for powders or nonconductive samples are shown in Figure 17–13. These electrodes are usually made of high-purity graphite; in addition, some of the electrodes are shaped to reduce conduction of heat from the sample cup so that vaporization will be improved. Typical counter electrodes are also depicted in Figure 17–13. Counter electrodes are often pointed, in an attempt to confine the arc discharge to a small area of the sample electrode.

Liquid Samples. It is neither convenient nor safe to introduce liquid samples directly into a DC arc discharge. Therefore, liquid samples must be handled with the aid of special techniques or electrodes. Three common procedures involve the use of the flat-topped electrode, the dipping-wheel electrode, and the porous-cup electrode.

In the first procedure, a cylindrical, flat-topped electrode formed from low-porosity graphite is employed. The electrode is preheated in a specially designed holder to a temperature just sufficient to volatilize the solvent from the liquid sample, which is introduced from a dropper onto the electrode surface. As each drop evaporates, a new portion of sample solution is added to the electrode surface until an appreciable crust of dried sample material is built up. An arc is struck between two such electrodes prepared for each sample to provide extremely high sensitivities for the elements present in the

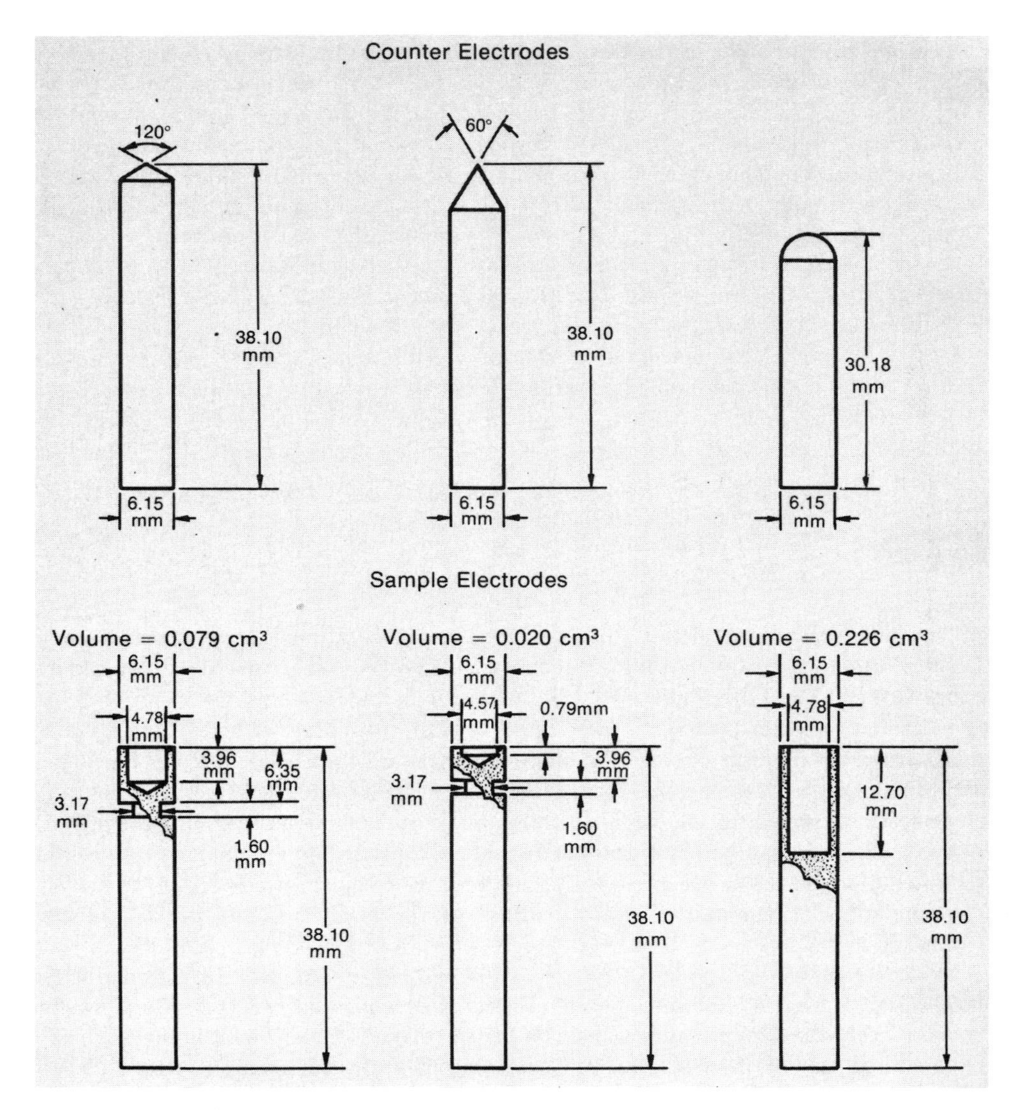


*Figure 17–13.* Examples of sample-containing and counter electrodes used for the arc or spark discharge analysis of powdered samples. Dimensions are in millimeters (mm).

original solution. This procedure is quite time-consuming and tedious, but provides extremely high sensitivity and respectable levels of precision.

The dipping-wheel electrode is most often used for viscous liquid samples, such as lubricating oils. As shown in Figure 17–14, the dipping-wheel electrode system consists of a slowly rotating disk and a counter electrode. The disk rotates through a pool of the liquid to be analyzed, bringing it into position directly beneath the counter electrode. An arc struck between the disk and the counter electrode then excites the elements in solution which are carried up by the disk from the bath. With viscous liquid samples, such as oils, the dipping-wheel electrode offers both excellent sensitivity and precision for trace metal analysis.

The porous-cup electrode consists of a tubular electrode having a porous graphite bottom. A liquid sample placed in the porous cup slowly seeps through the bottom of the electrode, and an arc is struck between it and the counter electrode beneath to excite the

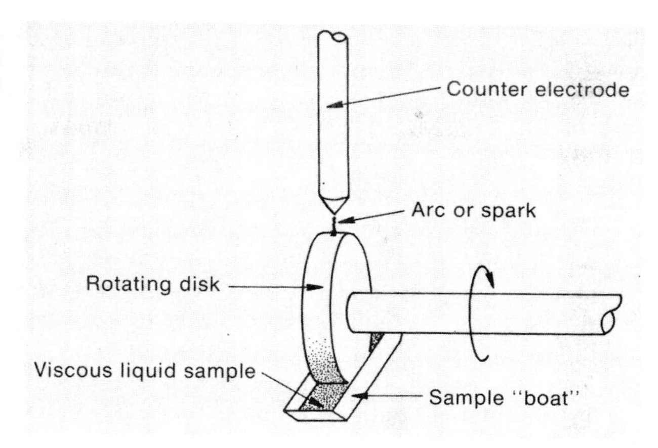


*Figure 17-14.* Dipping-wheel electrode arrangement for the analysis of viscous liquids in an electrical discharge.

elements present in the sample. Although this procedure is less sensitive and precise than either of the two preceding methods, the minimal handling of the sample makes this an attractive technique for solution analysis.

Analysis of Gases. Analysis of gases in a DC arc is a relatively simple process and merely involves directing the gases into an arc struck between two inert electrodes. To increase sensitivity and reduce the occurrence of interferences from dust particles and other foreign material, the arc is often surrounded by a chamber. One such chamber is the Stallwood jet, into which the gases to be analyzed are introduced tangentially. The gases swirl about the arc, so that elements present in the gases enter the arc periodically and are efficiently excited. Electrodes (usually high-purity graphite) are introduced into the Stallwood jet through appropriately placed holes in the top and bottom of the chamber. A Stallwood jet enclosure can also be used for the analysis of solid samples when it is desired to employ a controlled gas environment around the arc to eliminate cyanogen bands and to minimize self-absorption.

### High-Voltage Spark Spectrometry

The emission characteristics of a high-voltage spark differ markedly from those of the DC arc. Unlike the continuous DC arc, the spark appears intermittently and utilizes extremely high voltages and currents. A high-voltage spark is an oscillatory discharge, often having positive and negative peak currents exceeding 1000 amperes. At the onset of the discharge, the sample electrode is usually negative, so that the sample will be efficiently vaporized by positive-ion bombardment. During subsequent oscillations of the spark, the electrodes change polarity so that the sample is recycled between the sample and counter electrodes to provide efficient excitation. This enhances the quantitative analytical utility of the spark.

The extremely high currents found in a single spark discharge cause localized but extremely intense heating of the sample, with temperatures in excess of 10,000°K sometimes being attained. Such heating populates the very high energy electronic levels of atoms in the discharge. Thus, the emission spectra are even more intense and complex than those produced by the DC arc. This increases the difficulty of detecting individual emission lines and renders qualitative analysis more tedious.

Although the duration of a single spark discharge is quite short (usually about 100 microseconds), the period between discharges is relatively long (tens of milliseconds or longer). This delay gives the electrodes a reasonably long time to cool off between spark discharges, resulting in less fractional distillation and sample consumption, but also in lower sensitivity, than when the DC arc is used. Because a relatively long time exists between individual spark discharges, the atoms vaporized from the sample have sufficient time to diffuse or blow away before the appearance of the next discharge. Because there is little accumulation of atomic vapor around the electrode, self-absorption and self-reversal occur less frequently.

The intermittent nature of the high-voltage spark makes it a superior source for quantitative analysis. Because the spark starts anew for each discharge, it is not confined to specific "hot spots" on the electrode, but instead tends to randomly and reliably sample the entire electrode surface. This improved sampling provides better statistics for quantitative analysis. With care, relative precision levels between 1 and 5 per cent can be routinely obtained with the high-voltage spark.

**Instrumentation for High-Voltage Spark Spectrometry.** Dispersion and detection systems employed in high-voltage spark spectrometry are, for the most part, identical to those in DC arc spectrometry. Also, the procedures and precautions used in both techniques are very similar. However, because quantitative analysis is more frequently performed with the high-voltage spark, a direct-reading spectrometer is often used. With such a system, 20 or more elements can be routinely determined at precision levels better than 5 per cent.

Sample handling and electrode preparation are similar in both high-voltage spark and DC arc spectrometry; all techniques discussed earlier can be used for high-voltage spark spectrometry. However, because of the intermittent nature of the high-voltage spark, it is possible to analyze a solution by spraying it directly into the discharge. The high energy of the discharge is capable of evaporating solvent from the sample droplets, vaporizing the dry sample, and exciting the resulting atoms quite efficiently.

### Radio-Frequency Plasma Spectrometry

The radio-frequency plasma is a flame-like electrical discharge which shows increasing promise as an excitation source for qualitative and quantitative elemental analysis. A schematic diagram of a radio-frequency plasma appears in Figure 17–15. A high current, oscillating at a radio frequency between 10 and 50 MHz, generates an intense magnetic field which interacts with the charged species present within the coil. When no sample is present, an inert supporting gas such as argon sustains the plasma. Electrons produced by ionization of the supporting gas move with the radio-frequency field and, in their oscillation, collide with neutral supporting gas atoms and ionize them.

A cooled quartz tube, placed within the radio-frequency coil, is used to contain the plasma. Argon gas, flowing tangentially through the quartz tube, forces the ionized gases partially beyond the coil, so that the plasma is visible above the quartz tube. Elemental emission is ordinarily viewed in this upper region.

This source possesses an extremely high temperature (above 5000°K), so that vaporization and atomization of samples occur readily and the intensity of atomic emission is high. Furthermore, the source is far more stable than either the high-voltage spark or the DC arc, and can provide analytical results precise to ±1 per cent. Liquid samples can be sprayed directly into the plasma, so that sample preparation is minimized. Furthermore, the time required for an analysis is greatly reduced so that it becomes feasible to employ simple, inexpensive, single-channel monochromator systems.

The radio-frequency plasma is simple, safe, and inexpensive to operate and maintain. Its high sensitivity and multi-element capability make it a strong contender to replace some of the less attractive, currently used emission sources.

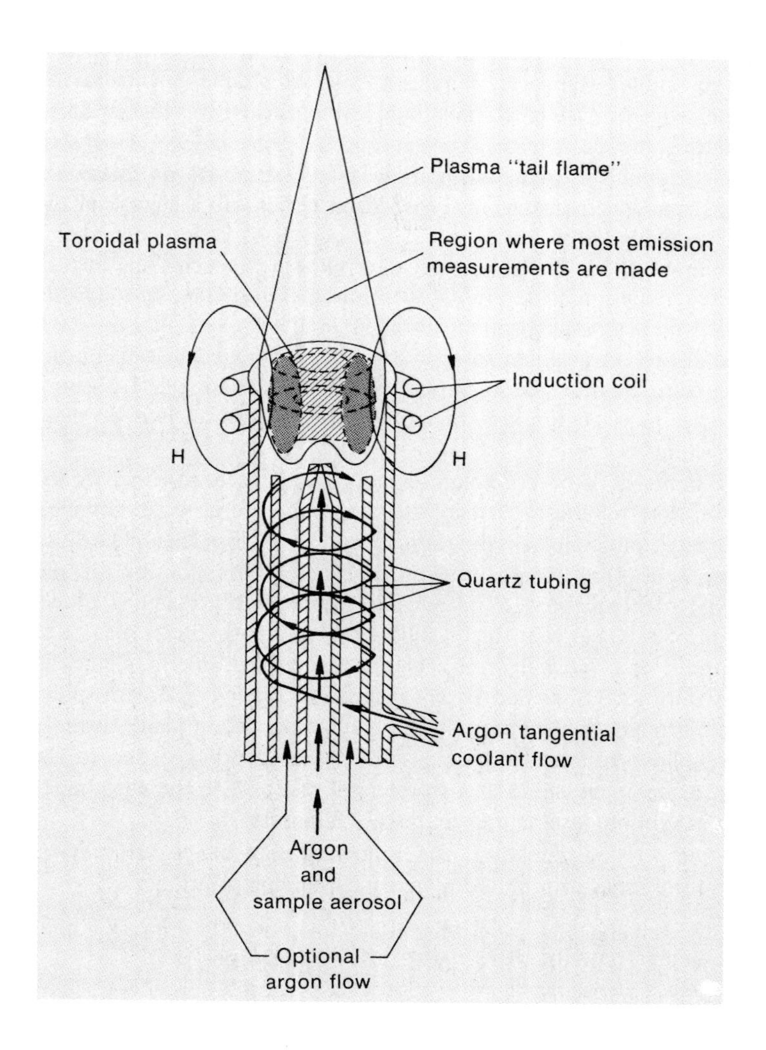


*Figure 17–15.* Schematic diagram of an inductively coupled radio-frequency plasma source. The toroidal plasma is formed within the quartz tubing by a high-frequency magnetic field (*H*). The magnetic field, in turn, is produced by current flowing in the induction coil. Most atomic emission measurements are taken just above the doughnut-shaped plasma core in the "tail flame." (Reproduced, with permission, from an article by V. A. Fassel and R. N. Kniseley: Anal. Chem., *46*:1110A, 1974.)

## QUESTIONS AND PROBLEMS

1. Explain why the simultaneous analysis of several elements by means of atomic absorption spectrometry would be more difficult than by flame emission spectrometry.
2. In Figure 17–5 (page 452), the emission line for magnesium is seen to be nearly obscured by the -OH band background emission of the flame. Will this make the determination of magnesium in the oxygen-hydrogen flame impossible? Why or why not? Which flame spectrometric technique will be affected most and which least by this spectral interference?
3. What effect will a change in the flow rate of fuel gas produce in a practical flame spectrometric analysis? What effect will a change in the flow rate of oxidizer gas have?
4. If the hollow-cathode lamp primarily determines the spectral bandwidth of detected radiation in atomic absorption spectrometry, as stated on page 455, why is a monochromator needed at all?
5. Most metals are in an ionic form in a solution, whereas in a flame the emission from neutral atoms is observed. To go from ions to atoms, these species must acquire one or more electrons. From where do these electrons come?
6. A novel flameless atom reservoir is now being used for the atomic absorption determination of mercury contamination in water supplies. The atomizer consists of a reduction cell and an enclosed sample cell. In operation, a 10.00-ml sample of water is placed in the reduction cell and is diluted to 100 ml; then 25 ml of concentrated sulfuric acid is added, and 10 ml of a 10 per cent tin(II) sulfate solution in 0.25 $F$ sulfuric acid is introduced as a reducing agent. Mercury is reduced to the elemental (atomic) state and is carried to an absorption cell by a stream of air bubbling through the solution in the reduction cell. Finally, using a mercury hollow-cathode lamp as a source, the absorbance due to the mercury atoms is measured at a wavelength of 2537 Å; the absorbance reaches a maximum level in approximately 3 minutes.

   In such a determination, the following absorbance values were obtained for a series of standard mercury-containing solutions:

   | Total Amount of Mercury in Standard ($\mu$g) | Absorbance |
   |---|---|
   | 0.00 | 0.002 |
   | 0.30 | 0.090 |
   | 0.60 | 0.175 |
   | 1.00 | 0.268 |
   | 2.00 | 0.440 |

   Two water samples, when treated according to the above procedure, gave absorbance values of 0.040 and 0.305, respectively. What was the total amount of mercury in each of the samples? What was the concentration (in $\mu$g/ml) of mercury in each of the original water samples?
7. In Figure 17–9 (page 459), the halogens and rare gases are conspicuously absent from the list of elements which can be determined by means of flame emission or atomic absorption spectrometry. Explain

why these elements are not amenable to flame spectrometric analysis. Could these elements be determined by means of DC arc or high-voltage spark spectrometry? Why or why not?

8. Would either the DC arc or the high-voltage spark be suitable as flameless atomization systems for use in atomic absorption spectrometry? Why or why not?
9. Which would be superior – the DC arc or the high-voltage spark – as an electrical discharge source for the spectrometric analysis of a disk of a very inhomogeneous alloy? Explain and justify your answer.
10. In the flame emission spectrometric analysis of clinical samples, lithium is often employed both as an ionization suppressant and as an internal standard for the determination of sodium and potassium. Because the lithium emission is affected by variations in flame composition and temperature in a manner similar to that of sodium and potassium, the effects of these variations can be minimized if the observed sodium or potassium emission is compared with that produced by a constant amount of lithium added to each solution. To construct a working curve for sodium, the ratio of the emission signal for sodium to that for lithium is plotted versus the concentration of sodium; the same approach can be followed for potassium. Construct working curves for the following data:

| Standard Solution | Relative Signals (arbitrary units) | | |
|---|---|---|---|
| | Na | K | Li |
| (a) 0.1 ppm Na, 0.1 ppm K, 1000 ppm Li | 0.11 | 0.15 | 86 |
| (b) 0.5 ppm Na, 0.5 ppm K, 1000 ppm Li | 0.52 | 0.68 | 80 |
| (c) 1.0 ppm Na, 1.0 ppm K, 1000 ppm Li | 1.2 | 1.5 | 91 |
| (d) 5.0 ppm Na, 5.0 ppm K, 1000 ppm Li | 5.9 | 7.7 | 91 |
| (e) 10.0 ppm Na, 10.0 ppm K, 1000 ppm Li | 10.5 | 14 | 81 |

What are the sodium and potassium concentrations in a sample which produces the following emission signals: Na, 1.4; K, 0.73; Li, 95? Now, plot *normal* working curves (that is, graphs of just the emission signal for sodium or potassium versus the concentration of the respective element); determine the sodium and potassium concentrations in the unknown sample from these working curves. How much error would have been incurred if the internal standard technique had *not* been employed? Was the internal-standard approach essential in this determination? Why or why not?

11. To determine the lead content of urine by means of atomic absorption spectrometry, the method of standard addition can be utilized. Exactly 50.00 ml of urine was pipetted into each of two 100-ml separatory funnels. To one funnel was added 300 $\mu$l of a standard solution containing 50.0 mg of lead per liter. Then the pH of each mixture was adjusted to 2.8 by dropwise addition of hydrochloric acid. Next, 500 $\mu$l of a freshly prepared 4 per cent solution of ammonium pyrrolidine dithiocarbamate in methyl-*n*-amyl ketone was added to each funnel, and

the aqueous and nonaqueous phases were thoroughly shaken to extract the lead. Finally, the lead-containing organic phase was examined by means of atomic absorption spectrometry, a lead hollow-cathode lamp with an emission line at 283.3 nm being used. If the absorbance of the extract from the unknown urine sample was 0.325, and if the absorbance of the extract from the unknown spiked with a known amount of lead was 0.670, what was the lead concentration (in mg/liter) of the original urine sample?

12. Atomic absorption spectrometry can be employed to determine traces of wear metals in used lubricating oils. To perform an analysis of a sample of used lubricating oil, 5.000 grams of the thoroughly mixed oil is weighed into a 25.00-ml volumetric flask, the oil is dissolved in 2-methyl-4-pentanone, and additional 2-methyl-4-pentanone is added up to the calibration line on the flask. Then the resulting solution is aspirated into an air-acetylene flame; for the determination of copper and lead, separate hollow-cathode lamps with emission lines at 324.7 and 283.3 nm, respectively, are used. A series of standard solutions, containing known amounts of copper and lead in the correct mixture of *unused* lubricating oil and 2-methyl-4-pentanone, can be used to establish calibration curves. Utilizing the following information, determine the *weight per cent* of copper and lead in a 5.000-g sample of used lubricating oil:

| Solution | Absorbance at 283.3 nm (due to Pb) | Absorbance at 324.7 nm (due to Cu) |
|---|---|---|
| (a) standard: 19.5 μg Pb/ml, 5.25 μg Cu/ml | 0.356 | 0.514 |
| (b) standard: 4.00 μg Pb/ml, 4.00 μg Cu/ml | 0.073 | 0.392 |
| (c) standard: 12.1 μg Pb/ml, 6.27 μg Cu/ml | 0.220 | 0.612 |
| (d) standard: 8.50 μg Pb/ml, 1.05 μg Cu/ml | 0.155 | 0.101 |
| (e) standard: 15.2 μg Pb/ml, 2.40 μg Cu/ml | 0.277 | 0.232 |
| (f) unknown | 0.247 | 0.371 |

## SUGGESTIONS FOR ADDITIONAL READING

*Flame Methods*

1. F. Burriel-Martí and J. Ramírez-Muñoz: *Flame Photometry: A Manual of Methods and Applications*. Elsevier, New York, 1957.
2. J. A. Dean: *Flame Photometry*. McGraw-Hill Book Company, New York, 1960.
3. J. A. Dean and T. C. Rains, eds.: *Flame Emission and Atomic Absorption Spectrometry*. Dekker, New York, Volume I (Theory), 1969; Volume II (Components and Techniques), 1971.
4. R. Herrmann and C. T. J. Alkemade: *Chemical Analysis by Flame Photometry*. Translated by P. T. Gilbert, Jr., Wiley-Interscience, New York, 1963.

5. R. Mavrodineanu, ed.: *Analytical Flame Spectroscopy*. Springer-Verlag, New York, 1970.
6. R. Mavrodineanu and H. Boiteux: *Flame Spectroscopy*. John Wiley and Sons, New York, 1965.
7. W. Slavin: *Atomic Absorption Spectroscopy*. Wiley-Interscience, New York, 1968.
8. J. D. Winefordner, ed.: *Spectrochemical Methods of Analysis*. Part I (Flame Spectrometric Methods), Wiley-Interscience, New York, 1971.

*Electrical Discharge Methods*

1. L. H. Ahrens and S. R. Taylor: *Spectrochemical Analysis*. Second edition, Addison-Wesley Publishing Company, Reading, Massachusetts, 1961.
2. W. R. Brode: *Chemical Spectroscopy*. Second edition, John Wiley and Sons, New York, 1943.
3. P. W. J. M. Boumans: *Theory of Spectrochemical Excitation*. Plenum Publishing Company, New York, 1966.
4. I. M. Kolthoff and P. J. Elving, eds.: *Treatise on Analytical Chemistry*. Part I, Volume 6, Wiley-Interscience, New York, 1959.
5. R. A. Sawyer: *Experimental Spectroscopy*. Third edition, Dover, New York, 1963.

# 1 APPENDIX

## SOLUBILITY PRODUCT CONSTANTS*

## (All values are valid at or near room temperature)

| Substance | Formula | $K_{sp}$ |
|---|---|---|
| Aluminum hydroxide | $Al(OH)_3$ | $2 \times 10^{-32}$ |
| Barium arsenate | $Ba_3(AsO_4)_2$ | $7.7 \times 10^{-51}$ |
| Barium carbonate | $BaCO_3$ | $8.1 \times 10^{-9}$ |
| Barium chromate | $BaCrO_4$ | $2.4 \times 10^{-10}$ |
| Barium fluoride | $BaF_2$ | $1.7 \times 10^{-6}$ |
| Barium iodate | $Ba(IO_3)_2 \cdot 2H_2O$ | $1.5 \times 10^{-9}$ |
| Barium oxalate | $BaC_2O_4 \cdot H_2O$ | $2.3 \times 10^{-8}$ |
| Barium sulfate | $BaSO_4$ | $1.08 \times 10^{-10}$ |
| Beryllium hydroxide | $Be(OH)_2$ | $7 \times 10^{-22}$ |
| [a]Bismuth hydroxide | BiOOH | $4 \times 10^{-10}$ |
| Bismuth iodide | $BiI_3$ | $8.1 \times 10^{-19}$ |
| Bismuth phosphate | $BiPO_4$ | $1.3 \times 10^{-23}$ |
| Bismuth sulfide | $Bi_2S_3$ | $1 \times 10^{-97}$ |
| Cadmium arsenate | $Cd_3(AsO_4)_2$ | $2.2 \times 10^{-33}$ |
| Cadmium hydroxide | $Cd(OH)_2$ | $5.9 \times 10^{-15}$ |
| Cadmium oxalate | $CdC_2O_4 \cdot 3H_2O$ | $1.5 \times 10^{-8}$ |
| Cadmium sulfide | CdS | $7.8 \times 10^{-27}$ |
| Calcium arsenate | $Ca_3(AsO_4)_2$ | $6.8 \times 10^{-19}$ |
| Calcium carbonate | $CaCO_3$ | $8.7 \times 10^{-9}$ |
| Calcium fluoride | $CaF_2$ | $4.0 \times 10^{-11}$ |
| Calcium hydroxide | $Ca(OH)_2$ | $5.5 \times 10^{-6}$ |
| Calcium iodate | $Ca(IO_3)_2 \cdot 6H_2O$ | $6.4 \times 10^{-7}$ |
| Calcium oxalate | $CaC_2O_4 \cdot H_2O$ | $2.6 \times 10^{-9}$ |
| Calcium phosphate | $Ca_3(PO_4)_2$ | $2.0 \times 10^{-29}$ |
| Calcium sulfate | $CaSO_4$ | $1.9 \times 10^{-4}$ |
| Cerium(III) hydroxide | $Ce(OH)_3$ | $2 \times 10^{-20}$ |
| Cerium(III) iodate | $Ce(IO_3)_3$ | $3.2 \times 10^{-10}$ |
| Cerium(III) oxalate | $Ce_2(C_2O_4)_3 \cdot 9H_2O$ | $3 \times 10^{-29}$ |
| Chromium(II) hydroxide | $Cr(OH)_2$ | $1.0 \times 10^{-17}$ |
| Chromium(III) hydroxide | $Cr(OH)_3$ | $6 \times 10^{-31}$ |
| Cobalt(II) hydroxide | $Co(OH)_2$ | $2 \times 10^{-16}$ |
| Cobalt(III) hydroxide | $Co(OH)_3$ | $1 \times 10^{-43}$ |
| Copper(II) arsenate | $Cu_3(AsO_4)_2$ | $7.6 \times 10^{-36}$ |
| Copper(I) bromide | CuBr | $5.2 \times 10^{-9}$ |
| Copper(I) chloride | CuCl | $1.2 \times 10^{-6}$ |
| Copper(II) iodate | $Cu(IO_3)_2$ | $7.4 \times 10^{-8}$ |
| Copper(I) iodide | CuI | $5.1 \times 10^{-12}$ |
| Copper(I) sulfide | $Cu_2S$ | $2 \times 10^{-47}$ |
| Copper(II) sulfide | CuS | $9 \times 10^{-36}$ |

* See footnotes at end of table.

| Substance | Formula | $K_{sp}$ |
|---|---|---|
| Copper(I) thiocyanate | $CuSCN$ | $4.8 \times 10^{-15}$ |
| Gold(III) hydroxide | $Au(OH)_3$ | $5.5 \times 10^{-46}$ |
| Iron(III) arsenate | $FeAsO_4$ | $5.7 \times 10^{-21}$ |
| Iron(II) carbonate | $FeCO_3$ | $3.5 \times 10^{-11}$ |
| Iron(II) hydroxide | $Fe(OH)_2$ | $8 \times 10^{-16}$ |
| Iron(III) hydroxide | $Fe(OH)_3$ | $4 \times 10^{-38}$ |
| Lead arsenate | $Pb_3(AsO_4)_2$ | $4.1 \times 10^{-36}$ |
| Lead bromide | $PbBr_2$ | $3.9 \times 10^{-5}$ |
| Lead carbonate | $PbCO_3$ | $3.3 \times 10^{-14}$ |
| Lead chloride | $PbCl_2$ | $1.6 \times 10^{-5}$ |
| Lead chromate | $PbCrO_4$ | $1.8 \times 10^{-14}$ |
| Lead fluoride | $PbF_2$ | $3.7 \times 10^{-8}$ |
| Lead iodate | $Pb(IO_3)_2$ | $2.6 \times 10^{-13}$ |
| Lead iodide | $PbI_2$ | $7.1 \times 10^{-9}$ |
| Lead oxalate | $PbC_2O_4$ | $4.8 \times 10^{-10}$ |
| [b]Lead oxide | $PbO$ | $1.2 \times 10^{-15}$ |
| Lead sulfate | $PbSO_4$ | $1.6 \times 10^{-8}$ |
| Lead sulfide | $PbS$ | $8 \times 10^{-28}$ |
| Magnesium ammonium phosphate | $MgNH_4PO_4$ | $2.5 \times 10^{-13}$ |
| Magnesium arsenate | $Mg_3(AsO_4)_2$ | $2.1 \times 10^{-20}$ |
| Magnesium carbonate | $MgCO_3 \cdot 3H_2O$ | $1 \times 10^{-5}$ |
| Magnesium fluoride | $MgF_2$ | $6.5 \times 10^{-9}$ |
| Magnesium hydroxide | $Mg(OH)_2$ | $1.2 \times 10^{-11}$ |
| Magnesium oxalate | $MgC_2O_4 \cdot 2H_2O$ | $1 \times 10^{-8}$ |
| Manganese(II) hydroxide | $Mn(OH)_2$ | $1.9 \times 10^{-13}$ |
| [c]Mercury(I) bromide | $Hg_2Br_2$ | $5.8 \times 10^{-23}$ |
| [c]Mercury(I) chloride | $Hg_2Cl_2$ | $1.3 \times 10^{-18}$ |
| [c]Mercury(I) iodide | $Hg_2I_2$ | $4.5 \times 10^{-29}$ |
| [d]Mercury(II) oxide | $HgO$ | $3.0 \times 10^{-26}$ |
| [c]Mercury(I) sulfate | $Hg_2SO_4$ | $7.4 \times 10^{-7}$ |
| Mercury(II) sulfide | $HgS$ | $4 \times 10^{-53}$ |
| [c]Mercury(I) thiocyanate | $Hg_2(SCN)_2$ | $3.0 \times 10^{-20}$ |
| Nickel arsenate | $Ni_3(AsO_4)_2$ | $3.1 \times 10^{-26}$ |
| Nickel carbonate | $NiCO_3$ | $6.6 \times 10^{-9}$ |
| Nickel hydroxide | $Ni(OH)_2$ | $6.5 \times 10^{-18}$ |
| Nickel sulfide | $NiS$ | $3 \times 10^{-19}$ |
| Palladium(II) hydroxide | $Pd(OH)_2$ | $1 \times 10^{-31}$ |
| Platinum(II) hydroxide | $Pt(OH)_2$ | $1 \times 10^{-35}$ |
| Radium sulfate | $RaSO_4$ | $4.3 \times 10^{-11}$ |
| Silver arsenate | $Ag_3AsO_4$ | $1 \times 10^{-22}$ |
| Silver bromate | $AgBrO_3$ | $5.77 \times 10^{-5}$ |
| Silver bromide | $AgBr$ | $5.25 \times 10^{-13}$ |
| Silver carbonate | $Ag_2CO_3$ | $8.1 \times 10^{-12}$ |
| Silver chloride | $AgCl$ | $1.78 \times 10^{-10}$ |
| Silver chromate | $Ag_2CrO_4$ | $2.45 \times 10^{-12}$ |
| Silver cyanide | $Ag[Ag(CN)_2]$ | $5.0 \times 10^{-12}$ |
| Silver iodate | $AgIO_3$ | $3.02 \times 10^{-8}$ |
| Silver iodide | $AgI$ | $8.31 \times 10^{-17}$ |
| Silver oxalate | $Ag_2C_2O_4$ | $3.5 \times 10^{-11}$ |
| [e]Silver oxide | $Ag_2O$ | $2.6 \times 10^{-8}$ |

| Substance | Formula | $K_{sp}$ |
|---|---|---|
| Silver phosphate | $Ag_3PO_4$ | $1.3 \times 10^{-20}$ |
| Silver sulfate | $Ag_2SO_4$ | $1.6 \times 10^{-5}$ |
| Silver sulfide | $Ag_2S$ | $2 \times 10^{-49}$ |
| Silver thiocyanate | $AgSCN$ | $1.00 \times 10^{-12}$ |
| Strontium carbonate | $SrCO_3$ | $1.1 \times 10^{-10}$ |
| Strontium chromate | $SrCrO_4$ | $3.6 \times 10^{-5}$ |
| Strontium fluoride | $SrF_2$ | $2.8 \times 10^{-9}$ |
| Strontium iodate | $Sr(IO_3)_2$ | $3.3 \times 10^{-7}$ |
| Strontium oxalate | $SrC_2O_4 \cdot H_2O$ | $1.6 \times 10^{-7}$ |
| Strontium sulfate | $SrSO_4$ | $3.8 \times 10^{-7}$ |
| Thallium(I) bromate | $TlBrO_3$ | $8.5 \times 10^{-5}$ |
| Thallium(I) bromide | $TlBr$ | $3.4 \times 10^{-6}$ |
| Thallium(I) chloride | $TlCl$ | $1.7 \times 10^{-4}$ |
| Thallium(I) chromate | $Tl_2CrO_4$ | $9.8 \times 10^{-13}$ |
| Thallium(I) iodate | $TlIO_3$ | $3.1 \times 10^{-6}$ |
| Thallium(I) iodide | $TlI$ | $6.5 \times 10^{-8}$ |
| Thallium(I) sulfide | $Tl_2S$ | $5 \times 10^{-21}$ |
| [f]Tin(II) oxide | $SnO$ | $1.4 \times 10^{-28}$ |
| Tin(II) sulfide | $SnS$ | $1 \times 10^{-25}$ |
| Titanium(III) hydroxide | $Ti(OH)_3$ | $1 \times 10^{-40}$ |
| [g]Titanium(IV) hydroxide | $TiO(OH)_2$ | $1 \times 10^{-29}$ |
| Zinc arsenate | $Zn_3(AsO_4)_2$ | $1.3 \times 10^{-28}$ |
| Zinc carbonate | $ZnCO_3$ | $1.4 \times 10^{-11}$ |
| Zinc ferrocyanide | $Zn_2Fe(CN)_6$ | $4.1 \times 10^{-16}$ |
| Zinc hydroxide | $Zn(OH)_2$ | $1.2 \times 10^{-17}$ |
| Zinc oxalate | $ZnC_2O_4 \cdot 2H_2O$ | $2.8 \times 10^{-8}$ |
| Zinc phosphate | $Zn_3(PO_4)_2$ | $9.1 \times 10^{-33}$ |
| Zinc sulfide | $ZnS$ | $1 \times 10^{-21}$ |

Although water appears in the formulas of a number of substances, it is not included in the solubility-product expression.

[a] $BiOOH \rightleftharpoons BiO^+ + OH^-$; $K_{sp} = [BiO^+][OH^-]$

[b] $PbO + H_2O \rightleftharpoons Pb^{2+} + 2\,OH^-$; $K_{sp} = [Pb^{2+}][OH^-]^2$

[c] All mercury(I) compounds contain the dimeric species $Hg_2^{2+}$. Therefore, the solubility reaction and solubility-product expression are represented in general by:

$$(Hg_2)_mX_n \rightleftharpoons m\,Hg_2^{2+} + n\,X^{-2m/n}; \qquad K_{sp} = [Hg_2^{2+}]^m[X^{-2m/n}]^n$$

[d] $HgO + H_2O \rightleftharpoons Hg^{2+} + 2\,OH^-$; $K_{sp} = [Hg^{2+}][OH^-]^2$

[e] $\frac{1}{2}\,Ag_2O + \frac{1}{2}\,H_2O \rightleftharpoons Ag^+ + OH^-$; $K_{sp} = [Ag^+][OH^-]$

[f] $SnO + H_2O \rightleftharpoons Sn^{2+} + 2\,OH^-$; $K_{sp} = [Sn^{2+}][OH^-]^2$

[g] $TiO(OH)_2 \rightleftharpoons TiO^{2+} + 2\,OH^-$; $K_{sp} = [TiO^{2+}][OH^-]^2$

More comprehensive tables of solubility-product data are given in the following references:

1. L. G. Sillén and A. E. Martell: *Stability Constants of Metal-Ion Complexes.* Chem. Soc. (London), Spec. Publ. No. 17 (1964); No. 25 (1971).
2. L. Meites, ed.: *Handbook of Analytical Chemistry.* McGraw-Hill Book Company, New York, 1963.

# APPENDIX

## EQUILIBRIUM CONSTANTS FOR PROTON-TRANSFER REACTIONS OF ACIDS AND THEIR CONJUGATE BASES IN WATER*

| System: Acid | System: Base | Proton-Transfer Equilibrium (Acid ⇌ Conjugate Base + $H^+$) | $pK_a$ for acid | $pK_b$ for conjugate base |
|---|---|---|---|---|
| **Acetic** | | $CH_3COOH \rightleftharpoons CH_3COO^- + H^+$ | 4.76 | 9.24 |
| | **α-Alanine** | $CH_3CH(^+NH_3)COOH \rightleftharpoons CH_3CH(^+NH_3)COO^- + H^+$ | 2.34 | 11.66 |
| **α-Alanine** | | $CH_3CH(^+NH_3)COO^- \rightleftharpoons CH_3CH(NH_2)COO^- + H^+$ | 9.87 | 4.13 |
| | **Ammonia** | $NH_4^+ \rightleftharpoons NH_3 + H^+$ | 9.26 | 4.74 |
| | **Aniline** | $C_6H_5NH_3^+ \rightleftharpoons C_6H_5NH_2 + H^+$ | 4.60 | 9.40 |
| **Arsenic** | | $H_3AsO_4 \rightleftharpoons H_2AsO_4^- + H^+$ | 2.19 | 11.81 |
| | | $H_2AsO_4^- \rightleftharpoons HAsO_4^{2-} + H^+$ | 6.94 | 7.06 |
| | | $HAsO_4^{2-} \rightleftharpoons AsO_4^{3-} + H^+$ | 11.50 | 2.50 |
| [a]**Arsenious** | | $H_3AsO_3 \rightleftharpoons H_2AsO_3^- + H^+$ | 9.29 | 4.71 |

* See footnotes at end of table.

| | | | |
|---|---|---|---|
| **Benzoic** | $C_6H_5COOH \rightleftharpoons C_6H_5COO^- + H^+$ | 4.20 | 9.80 |
| [b]**Boric** | $H_3BO_3 \rightleftharpoons H_2BO_3^- + H^+$ | 9.24 | 4.76 |
| **Bromoacetic** | $BrCH_2COOH \rightleftharpoons BrCH_2COO^- + H^+$ | 2.90 | 11.10 |
| *n*-**Butanoic** | $CH_3CH_2CH_2COOH \rightleftharpoons CH_3CH_2CH_2COO^- + H^+$ | 4.82 | 9.18 |
| **Carbonic** | $H_2CO_3 \rightleftharpoons HCO_3^- + H^+$<br>$HCO_3^- \rightleftharpoons CO_3^{2-} + H^+$ | 6.35<br>10.33 | 7.65<br>3.67 |
| **Chloroacetic** | $ClCH_2COOH \rightleftharpoons ClCH_2COO^- + H^+$ | 2.86 | 11.14 |
| **Chromic** | $H_2CrO_4 \rightleftharpoons HCrO_4^- + H^+$<br>$HCrO_4^- \rightleftharpoons CrO_4^{2-} + H^+$ | −0.98<br>6.50 | 14.98<br>7.50 |
| **Citric** | $HOOCCH_2C(OH)(COOH)CH_2COOH \rightleftharpoons HOOCCH_2C(OH)(COO^-)CH_2COOH + H^+$ | 3.13 | 10.87 |
| | $HOOCCH_2C(OH)(COO^-)CH_2COOH \rightleftharpoons HOOCCH_2C(OH)(COO^-)CH_2COO^- + H^+$ | 4.77 | 9.23 |
| | $HOOCCH_2C(OH)(COO^-)CH_2COO^- \rightleftharpoons {}^-OOCCH_2C(OH)(COO^-)CH_2COO^- + H^+$ | 6.40 | 7.60 |

| System | | Proton-Transfer Equilibrium | $pK_a$ | $pK_b$ |
|---|---|---|---|---|
| Acid | Base | (Acid ⇌ Conjugate Base + $H^+$) | for acid | for conjugate base |
| | **Diethylamine** | $(CH_3CH_2)_2NH_2^+ \rightleftharpoons (CH_3CH_2)_2NH + H^+$ | 10.93 | 3.07 |
| | **Dimethylamine** | $(CH_3)_2NH_2^+ \rightleftharpoons (CH_3)_2NH + H^+$ | 10.77 | 3.23 |
| | **Ethanolamine** | $HOCH_2CH_2NH_3^+ \rightleftharpoons HOCH_2CH_2NH_2 + H^+$ | 9.50 | 4.50 |
| | **Ethylamine** | $CH_3CH_2NH_3^+ \rightleftharpoons CH_3CH_2NH_2 + H^+$ | 10.67 | 3.33 |
| | **Ethylenediamine** | $^+NH_3CH_2CH_2NH_3^+ \rightleftharpoons NH_2CH_2CH_2NH_3^+ + H^+$<br>$NH_2CH_2CH_2NH_3^+ \rightleftharpoons NH_2CH_2CH_2NH_2 + H^+$ | 7.18<br>9.96 | 6.82<br>4.04 |
| [c]**Ethylenediamine-tetraacetic** ($H_4Y$) (see page 125 for more information) | | $H_4Y \rightleftharpoons H_3Y^- + H^+$<br>$H_3Y^- \rightleftharpoons H_2Y^{2-} + H^+$<br>$H_2Y^{2-} \rightleftharpoons HY^{3-} + H^+$<br>$HY^{3-} \rightleftharpoons Y^{4-} + H^+$ | 1.99<br>2.67<br>6.16<br>10.26<br>(in 0.1 *F* KCl, 20°C) | 12.01<br>11.33<br>7.84<br>3.74 |
| **Formic** | | $HCOOH \rightleftharpoons HCOO^- + H^+$ | 3.75 | 10.25 |
| | **Glycine** | $^+NH_3CH_2COOH \rightleftharpoons {}^+NH_3CH_2COO^- + H^+$ | 2.35 | 11.65 |
| **Glycine** | | $^+NH_3CH_2COO^- \rightleftharpoons NH_2CH_2COO^- + H^+$ | 9.60 | 4.40 |
| | **Hydrazine** | $H_2NNH_3^+ \rightleftharpoons H_2NNH_2 + H^+$ | 7.99 | 6.01 |
| **Hydrazoic** | | $HN_3 \rightleftharpoons N_3^- + H^+$ | 4.72 | 9.28 |
| **Hydrocyanic** | | $HCN \rightleftharpoons CN^- + H^+$ | 9.21 | 4.79 |
| **Hydrofluoric** | | $HF \rightleftharpoons F^- + H^+$ | 3.17 | 10.83 |
| **Hydrogen sulfide** | | $H_2S \rightleftharpoons HS^- + H^+$<br>$HS^- \rightleftharpoons S^{2-} + H^+$ | 6.99<br>14.96 | 7.01<br>−0.96 |

| Acid | Base | Equilibrium | pKa | pKb |
|---|---|---|---|---|
| | **Hydroxylamine** | $HONH_3^+ \rightleftharpoons HONH_2 + H^+$ | 5.98 | 8.02 |
| **Hypochlorous** | | $HClO \rightleftharpoons ClO^- + H^+$ | 7.53 | 6.47 |
| **Iodic** | | $HIO_3 \rightleftharpoons IO_3^- + H^+$ | 0.79 | 13.21 |
| **Lactic** | | $CH_3CH(OH)COOH \rightleftharpoons CH_3CH(OH)COO^- + H^+$ | 3.86 | 10.14 |
| **Malonic** | | $HOOCCH_2COOH \rightleftharpoons HOOCCH_2COO^- + H^+$<br>$HOOCCH_2COO^- \rightleftharpoons {}^-OOCCH_2COO^- + H^+$ | 2.86<br>5.70 | 11.14<br>8.30 |
| | **Methylamine** | $CH_3NH_3^+ \rightleftharpoons CH_3NH_2 + H^+$ | 10.72 (in 0.5 $F$ $CH_3NH_3^+NO_3^-$) | 3.28 |
| | **Methylethylamine** | $(CH_3)(CH_3CH_2)NH_2^+ \rightleftharpoons (CH_3)(CH_3CH_2)NH + H^+$ | 10.85 | 3.15 |
| [d]**Nitrilotriacetic** ($H_3Y$) | | $H_3Y \rightleftharpoons H_2Y^- + H^+$<br>$H_2Y^- \rightleftharpoons HY^{2-} + H^+$<br>$HY^{2-} \rightleftharpoons Y^{3-} + H^+$ | 1.89<br>2.49<br>9.73 | 12.11<br>11.51<br>4.27 |
| **Nitrous** | | $HNO_2 \rightleftharpoons NO_2^- + H^+$ | 2.80 (in 1 $F$ $NaClO_4$) | 11.20 |
| **Oxalic** | | $H_2C_2O_4 \rightleftharpoons HC_2O_4^- + H^+$<br>$HC_2O_4^- \rightleftharpoons C_2O_4^{2-} + H^+$ | 1.19<br>4.21 | 12.81<br>9.79 |
| **Phenol** | | $C_6H_5OH \rightleftharpoons C_6H_5O^- + H^+$ | 9.98 | 4.02 |

| System: Acid | System: Base | Proton-Transfer Equilibrium (Acid $\rightleftharpoons$ Conjugate Base + $H^+$) | $pK_a$ for acid | $pK_b$ for conjugate base |
|---|---|---|---|---|
| **Phosphoric** | | $H_3PO_4 \rightleftharpoons H_2PO_4^- + H^+$ | 2.13 | 11.87 |
| | | $H_2PO_4^- \rightleftharpoons HPO_4^{2-} + H^+$ | 7.21 | 6.79 |
| | | $HPO_4^{2-} \rightleftharpoons PO_4^{3-} + H^+$ | 12.32 | 1.68 |
| **Phosphorous** | | $H_3PO_3 \rightleftharpoons H_2PO_3^- + H^+$ | 1.29 (18°C) | 12.71 (18°C) |
| | | $H_2PO_3^- \rightleftharpoons HPO_3^{2-} + H^+$ | 6.70 (18°C) | 7.30 (18°C) |
| ***o*-Phthalic** | | (benzene ring: COOH, COOH) $\rightleftharpoons$ (benzene ring: $COO^-$, COOH) + $H^+$ | 2.95 | 11.05 |
| | | (benzene ring: $COO^-$, COOH) $\rightleftharpoons$ (benzene ring: $COO^-$, $COO^-$) + $H^+$ | 5.41 | 8.59 |
| **Picric** | | (benzene ring: OH, $O_2N$, $NO_2$, $NO_2$) $\rightleftharpoons$ (benzene ring: $O^-$, $O_2N$, $NO_2$, $NO_2$) + $H^+$ | 0.29 | 13.71 |
| | **Piperidine** | (piperidine ring: $\overset{+}{N}$, H, H) $\rightleftharpoons$ (piperidine ring: N–H) + $H^+$ | 11.20 (in 0.5 *F* $KNO_3$) | 2.80 (in 0.5 *F* $KNO_3$) |
| **Propanoic** | | $CH_3CH_2COOH \rightleftharpoons CH_3CH_2COO^- + H^+$ | 4.87 | 9.13 |
| | ***n*-Propylamine** | $CH_3CH_2CH_2NH_3^+ \rightleftharpoons CH_3CH_2CH_2NH_2 + H^+$ | 10.74 (20°C) | 3.26 (20°C) |

| | | | |
|---|---|---|---|
| Pyridine | $C_5H_5NH^+ \rightleftharpoons C_5H_5N + H^+$ | 5.23 | 8.77 |
| Salicylic | $C_6H_4(OH)COOH \rightleftharpoons C_6H_4(OH)COO^- + H^+$ | 2.97 | 11.03 |
| Succinic | $HOOCCH_2CH_2COOH \rightleftharpoons HOOCCH_2CH_2COO^- + H^+$ | 4.21 | 9.79 |
| | $HOOCCH_2CH_2COO^- \rightleftharpoons {}^-OOCCH_2CH_2COO^- + H^+$ | 5.64 | 8.36 |
| Sulfamic | $NH_2SO_3H \rightleftharpoons NH_2SO_3^- + H^+$ | 0.99 | 13.01 |
| Sulfuric | $HSO_4^- \rightleftharpoons SO_4^{2-} + H^+$ | 1.92 | 12.08 |
| Sulfurous | $H_2SO_3 \rightleftharpoons HSO_3^- + H^+$ | 1.76 | 12.24 |
| | $HSO_3^- \rightleftharpoons SO_3^{2-} + H^+$ | 7.21 | 6.79 |
| Tartaric | $HOOCCH(OH)\text{—}CH(OH)COOH \rightleftharpoons HOOCCH(OH)\text{—}CH(OH)COO^- + H^+$ | 3.04 | 10.96 |
| | $HOOCCH(OH)\text{—}CH(OH)COO^- \rightleftharpoons {}^-OOCCH(OH)\text{—}CH(OH)COO^- + H^+$ | 4.37 | 9.63 |
| Triethylamine | $(CH_3CH_2)_3NH^+ \rightleftharpoons (CH_3CH_2)_3N + H^+$ | 10.77 [in 0.4 $F$ $(C_2H_5)_3NH^+NO_3^-$] | 3.23 |
| Trimethylamine | $(CH_3)_3NH^+ \rightleftharpoons (CH_3)_3N + H^+$ | 9.80 | 4.20 |
| Tris(hydroxymethyl)-aminomethane | $(HOCH_2)_3CNH_3^+ \rightleftharpoons (HOCH_2)_3CNH_2 + H^+$ | 8.08 | 5.92 |

Unless stated otherwise, all data pertain to 25°C and zero ionic strength. However, for numerical problems in this book, the constants may be applied to other conditions without correction for differences in temperature or ionic strength.

Each listed $pK_a$ value refers to a proton-transfer equilibrium in which an acid reacts with water (acting as a Brønsted-Lowry base). For example, $pK_a$ for acetic acid is 4.76; this value pertains to the equilibrium

$$CH_3COOH + H_2O \rightleftharpoons CH_3COO^- + H_3O^+$$

Correct proton-transfer equilibria for other acids can be similarly written.

Each listed $pK_b$ value refers to a proton-transfer equilibrium in which a base reacts with water (acting as a Brønsted-Lowry acid). For example, $pK_b$ for acetate is 9.24; this value pertains to the equilibrium

$$CH_3COO^- + H_2O \rightleftharpoons CH_3COOH + OH^-$$

Correct proton-transfer equilibria for other bases can be formulated in the same way.

For acids such as arsenic acid ($H_3AsO_4$) that undergo stepwise proton-transfer reactions, the use of additional subscripts to distinguish among the several $pK_a$ values is necessary; for arsenic acid, we can write $pK_{a1} = 2.19$, $pK_{a2} = 6.94$, and $pK_{a3} = 11.50$. Similar subscripts can be used to designate $pK_b$ values for bases that can accept more than one proton, *e.g.*, carbonate ion.

Notice that the *sum* of $pK_a$ for an acid and $pK_b$ for its conjugate base must be 14.00 for water solutions; see page 60.

[a] Arsenious acid behaves essentially as a monoprotic acid in water. To indicate this behavior, the abbreviated proton-transfer equilibrium is often written

$$HAsO_2 \rightleftharpoons AsO_2^- + H^+$$

[b] Boric acid behaves as a monoprotic acid in water. To convey this behavior, the abbreviated proton-transfer reaction is frequently written

$$HBO_2 \rightleftharpoons BO_2^- + H^+$$

[c] Ethylenediaminetetraacetic acid is a tetraprotic acid with the structural formula

```
HOOC—CH2                          CH2—COO-
        \+                      +/
         HN—CH2—CH2—NH
        /                        \
-OOC—CH2                          CH2—COOH
```

but, for purposes of brevity, it is usually represented as $H_4Y$.

[d] Nitrilotriacetic acid is a triprotic acid with the structural formula

```
      CH2—COOH
   + /
   HN—CH2—COO-
     \
      CH2—COOH
```

but, for the sake of brevity, it is often represented as $H_3Y$.

Comprehensive data concerning acid-base equilibria are given in the following references:

1. L. Meites, ed.: *Handbook of Analytical Chemistry*. McGraw-Hill Book Company, New York, 1963.
2. L. G. Sillén and A. E. Martell: *Stability Constants of Metal-Ion Complexes*. Chem. Soc. (London), Spec. Publ. No. 17 (1964); No. 25 (1971).

# APPENDIX

## LOGARITHMS OF STEPWISE AND OVERALL FORMATION CONSTANTS FOR METAL ION COMPLEXES*

| Ligand | Metal Ion | log $K_1$ | log $K_2$ | log $K_3$ | log $K_4$ | log $K_5$ | log $K_6$ | Conditions |
|---|---|---|---|---|---|---|---|---|
| Acetate, $CH_3COO^-$ | $Ag^+$ | 0.73 | −0.09 | | | | | |
| | $Cd^{2+}$ | 1.30 | 0.98 | 0.14 | −0.42 | | | 3 $F$ $NaClO_4$ |
| | $Cu^{2+}$ | 1.79 | 1.15 | | | | | 3 $F$ $NaClO_4$ |
| | $Hg^{2+}$ | | 8.43 ($\beta_2$) | | | | | |
| | $Pb^{2+}$ | 2.19 | 0.72 | 0.61 | | | | 2 $F$ $NaClO_4$ |
| Acetylacetonate, $CH_3-\overset{\overset{O}{\|}}{C}-CH=\overset{\overset{O^-}{\|}}{C}-CH_3$ | $Al^{3+}$ | 8.6 | 7.9 | 5.8 | | | | 30°C |
| | $Cd^{2+}$ | 3.83 | 2.76 | | | | | 30°C |
| | $Co^{2+}$ | 5.40 | 4.11 | | | | | 30°C |
| | $Cu^{2+}$ | 8.22 | 6.73 | | | | | 30°C |
| | $Fe^{2+}$ | 5.07 | 3.60 | | | | | 30°C |
| | $Fe^{3+}$ | 11.4 | 10.7 | 4.6 | | | | |
| | $Mg^{2+}$ | 3.63 | 2.54 | | | | | 30°C |
| | $Mn^{2+}$ | 4.18 | 3.07 | | | | | 30°C |
| | $Ni^{2+}$ | 6.06 | 4.71 | 2.32 | | | | 20°C |
| | $UO_2^{2+}$ | 7.74 | 6.43 | | | | | 30°C |
| | $Zn^{2+}$ | 4.98 | 3.83 | | | | | 30°C |
| Ammonia, $NH_3$ | $Ag^+$ | 3.37 | 3.84 | | | | | |
| | $Cd^{2+}$ | 2.65 | 2.10 | 1.44 | 0.93 | −0.32 | −1.66 | 2 $F$ $NH_4NO_3$; 30°C |
| | $Co^{2+}$ | 1.99 | 1.51 | 0.93 | 0.64 | 0.06 | −0.74 | 30°C |
| | $Co^{3+}$ | 7.3 | 6.7 | 6.1 | 5.6 | 5.05 | 4.41 | 2 $F$ $NH_4NO_3$; 30°C |

* See footnotes at end of table.

| Ligand | Metal Ion | log $K_1$ | log $K_2$ | log $K_3$ | log $K_4$ | log $K_5$ | log $K_6$ | Conditions |
|---|---|---|---|---|---|---|---|---|
| Ammonia, $NH_3$ (*continued*) | $Cu^+$ | 5.93 | 4.93 | | | | | 2 *F* $NH_4NO_3$; 18°C |
| | $Cu^{2+}$ | 4.31 | 3.67 | 3.04 | 2.30 | −0.46 | | 2 *F* $NH_4NO_3$; 18°C |
| | $Hg^{2+}$ | 8.8 | 8.7 | 1.00 | 0.78 | | | 2 *F* $NH_4NO_3$; 22°C |
| | $Ni^{2+}$ | 2.36 | 1.90 | 1.55 | 1.23 | 0.85 | 0.42 | 1 *F* $NH_4NO_3$ |
| | $Zn^{2+}$ | 2.18 | 2.25 | 2.31 | 1.96 | | | 30°C |
| Bromide, $Br^-$ | $Bi^{3+}$ | 4.30 | 1.25 | 0.32 | | | | 1 *F* $HNO_3$ |
| | $Cd^{2+}$ | 2.23 | 0.77 | −0.17 | 0.10 | | | |
| | $Hg^{2+}$ | 8.94 | 7.94 | 2.27 | 1.75 | | | 0.5 *F* $NaClO_4$ |
| | $Pb^{2+}$ | 1.65 | 0.75 | 0.88 | 0.22 | | | 3 *F* $NaClO_4$ |
| | $Zn^{2+}$ | 0.22 | −0.32 | −0.64 | −0.26 | | | 0.7 *F* $HClO_4$; 20°C |
| Chloride, $Cl^-$ | $Bi^{3+}$ | 2.43 | 2.00 | 1.35 | 0.43 | 0.48 | | |
| | $Cd^{2+}$ | 2.00 | 0.70 | −0.59 | | | | |
| | $Cu^+$ | | 4.94 ($\beta_2$) | | | | | |
| | $Fe^{3+}$ | 1.48 | 0.65 | −1.0 | | | | |
| | $Hg^{2+}$ | 6.74 | 6.48 | 0.95 | 1.05 | | | 0.5 *F* $NaClO_4$ |
| | $Pb^{2+}$ | 1.10 | 1.16 | −0.40 | −1.05 | | | |
| Cyanide, $CN^-$ | $Ag^+$ | | 19.85 ($\beta_2$) | | | | | |
| | $Cd^{2+}$ | 5.48 | 5.14 | 4.56 | 3.58 | | | 3 *F* $NaClO_4$ |
| | $Co^{2+}$ | | | | | | 19.09 ($\beta_6$) | 5 *F* $CaCl_2$ |
| | $Cu^+$ | | 24.0 ($\beta_2$) | 4.59 | 1.70 | | | |
| | $Hg^{2+}$ | 18.00 | 16.70 | 3.83 | 2.98 | | | 0.1 *F* $NaNO_3$; 20°C |
| | $Ni^{2+}$ | | | | 30.3 ($\beta_4$) | | | |
| | $Zn^{2+}$ | | | | 16.72 ($\beta_4$) | | | |
| Ethylenediamine, $H_2NCH_2CH_2NH_2$ | $Ag^+$ | 4.70 | 3.00 | | | | | 0.1 *F* $NaNO_3$; 20°C |
| | $Cd^{2+}$ | 5.63 | 4.59 | 2.07 | | | | 1 *F* $KNO_3$ |
| | $Co^{2+}$ | 5.93 | 4.73 | 3.30 | | | | 1 *F* KCl |
| | $Co^{3+}$ | 18.7 | 16.2 | 13.81 | | | | 1 *F* $NaNO_3$; 30°C |
| | $Cu^+$ | | 10.8 ($\beta_2$) | | | | | |
| | $Cu^{2+}$ | 10.75 | 9.28 | | | | | 1 *F* $KNO_3$ |
| | $Fe^{2+}$ | 4.28 | 3.25 | 1.99 | | | | 1 *F* KCl; 30°C |

| Ligand | Metal ion | | | | | | | Conditions |
|---|---|---|---|---|---|---|---|---|
| | $Hg^{2+}$ | 14.3 | 9.0 | | | | | 0.1 *F* $KNO_3$ |
| | $Ni^{2+}$ | 7.72 | 6.36 | 4.33 | | | | 1 *F* KCl |
| | $Zn^{2+}$ | 5.77 | 5.06 | 3.28 | | | | 20°C |
| Ethylenediaminetetraacetate (EDTA), $(^{-}OOCCH_2)_2NCH_2CH_2N(CH_2COO^{-})_2$ [see Table 6–1, page 128] | | | | | | | | |
| Fluoride, $F^-$ | $Al^{3+}$ | 6.13 | 5.02 | 3.85 | 2.74 | 1.63 | 0.47 | 0.53 *F* $KNO_3$ |
| | $Ce^{3+}$ | 3.99 | | | | | | |
| | $Fe^{3+}$ | 5.17 | 3.92 | 2.91 | | | | 0.5 *F* $NaClO_4$ |
| [a]Hydroxide, $OH^-$ | $Al^{3+}$ | 8.98 | | | 32.43 ($\beta_4$) | | | |
| | $Bi^{3+}$ | 12.42 | | | | | | 3 *F* $NaClO_4$ |
| | $Cd^{2+}$ | 6.38 | | | | | | 1 *F* $KNO_3$ |
| | $Co^{2+}$ | 2.80 | | | | | | |
| | $Cu^{2+}$ | 6.66 | | | | | | |
| | $Fe^{2+}$ | 4.5 | | | | | | 1 *F* $NaClO_4$ |
| | $Fe^{3+}$ | 10.95 | 10.74 | | | | | 3 *F* $NaClO_4$ |
| | $Hg^{2+}$ | 10.77 | | | | | | 3 *F* $NaClO_4$ |
| | $Ni^{2+}$ | 3.08 | | | | | | |
| | $Pb^{2+}$ | 6.9 | | 13.95 ($\beta_3$) | | | | |
| | $Zn^{2+}$ | 4.34 | | 14.23 ($\beta_3$) | 1.26 | | | |
| 8-Hydroxyquinolate (oxinate), [structure: quinoline ring with N and $O^-$] | $Cd^{2+}$ | 9.43 | 7.68 | | | | | 50% dioxane |
| | $Ce^{3+}$ | 9.15 | 7.98 | | | | | 50% dioxane |
| | $Co^{2+}$ | 10.55 | 9.11 | | | | | 50% dioxane |
| | $Cu^{2+}$ | 13.49 | 12.73 | | | | | 50% dioxane |
| | $Fe^{2+}$ | 9.83 | 9.01 | | | | | 70% dioxane |
| | $Fe^{3+}$ | | | 38.00 ($\beta_3$) | | | | 50% dioxane |
| | $Mg^{2+}$ | 6.38 | 5.43 | | | | | 50% dioxane |
| | $Mn^{2+}$ | 8.28 | 7.17 | | | | | 50% dioxane |
| | $Ni^{2+}$ | 11.44 | 9.94 | | | | | 50% dioxane |
| | $Pb^{2+}$ | 10.61 | 8.09 | | | | | 50% dioxane |
| | $UO_2^{2+}$ | 11.25 | 9.64 | | | | | 50% dioxane |
| | $Zn^{2+}$ | 9.96 | 8.90 | | | | | 50% dioxane |

| Ligand | Metal Ion | log $K_1$ | log $K_2$ | log $K_3$ | log $K_4$ | log $K_5$ | log $K_6$ | Conditions |
|---|---|---|---|---|---|---|---|---|
| Iodide, $I^-$ | $Bi^{3+}$ | | | | | | 19.4 ($\beta_6$) | 2 $F$ $NaClO_4$; 20°C |
| | $Cd^{2+}$ | 2.10 | 1.33 | 1.06 | 0.92 | | | |
| | $Cu^+$ | | 8.85 ($\beta_2$) | | | | | |
| | $Hg^{2+}$ | 12.87 | 10.95 | 3.67 | 2.37 | | | 0.5 $F$ $NaClO_4$ |
| | $Pb^{2+}$ | 1.26 | 1.54 | 0.62 | 0.50 | | | 1 $F$ $NaClO_4$ |
| Nitrilotriacetate (NTA), $N(CH_2COO)_3^{3-}$ | $Ba^{2+}$ | 6.41 | | | | | | 20°C |
| | $Ca^{2+}$ | 8.17 | 3.43 | | | | | 20°C |
| | $Cd^{2+}$ | 9.54 | 5.7 | | | | | 0.1 $F$ KCl; 20°C |
| | $Co^{2+}$ | 10.6 | 3.9 | | | | | 0.1 $F$ KCl; 20°C |
| | $Cu^{2+}$ | 12.68 | | | | | | 0.1 $F$ KCl; 20°C |
| | $Fe^{2+}$ | 8.83 | | | | | | 0.1 $F$ KCl; 20°C |
| | $Fe^{3+}$ | 15.87 | 8.45 | | | | | 0.1 $F$ KCl; 20°C |
| | $Mg^{2+}$ | 7.00 | | | | | | 20°C |
| | $Mn^{2+}$ | 7.44 | 3.7 | | | | | 0.1 $F$ KCl; 20°C |
| | $Ni^{2+}$ | 11.26 | 4.7 | | | | | 0.1 $F$ KCl; 20°C |
| | $Pb^{2+}$ | 11.39 | | | | | | 0.1 $F$ $KNO_3$; 20°C |
| | $Sr^{2+}$ | 6.73 | | | | | | 20°C |
| | $Zn^{2+}$ | 10.67 | 3.0 | | | | | 0.1 $F$ KCl; 20°C |
| Oxalate, $C_2O_4^{2-}$ | $Al^{3+}$ | | 13 ($\beta_2$) | 3.3 | | | | 18°C |
| | $Cd^{2+}$ | 4.00 | 1.77 | | | | | |
| | $Co^{2+}$ | 4.79 | 1.91 | | | | | |
| | $Cu^{2+}$ | 6.19 | 4.04 | | | | | |
| | $Fe^{2+}$ | | 4.52 ($\beta_2$) | 0.70 | | | | 0.5 $F$ $NaClO_4$ |
| | $Fe^{3+}$ | 9.4 | 6.80 | 4 | | | | |
| | $Mg^{2+}$ | | 4.38 ($\beta_2$) | | | | | |
| | $Mn^{2+}$ | 3.82 | 1.43 | | | | | |
| | $Ni^{2+}$ | 5.16 | 1.35 | | | | | |
| | $Pb^{2+}$ | | 6.54 ($\beta_2$) | | | | | |
| | $Zn^{2+}$ | 5.00 | 2.36 | | | | | |

| Ligand | Metal ion | | | | | Remarks |
| --- | --- | --- | --- | --- | --- | --- |
| Pyridine, $C_5H_5N$ | $Ag^+$ | 2.00 | 2.11 | | | |
| | $Cd^{2+}$ | | 2.14 ($\beta_2$) | | 2.50 ($\beta_4$) | 0.1 $F$ $KNO_3$ |
| | $Co^{2+}$ | 1.14 | 0.4 | | | 0.5 $F$ $HNO_3$ |
| | $Cu^{2+}$ | 2.52 | 1.86 | 1.31 | 0.85 | 0.5 $F$ $KNO_3$ |
| | $Hg^{2+}$ | 5.1 | 4.9 | | | 0.5 $F$ $HNO_3$ |
| | $Ni^{2+}$ | 1.78 | 1.05 | | | 0.5 $F$ $HNO_3$ |
| | $Zn^{2+}$ | 1.41 | −0.30 | 0.50 | 0.32 | 0.1 $F$ KCl |
| Thiocyanate, $SCN^-$ | $Ag^+$ | | 8.39 ($\beta_2$) | 1.23 | 0.28 | |
| | $Cu^+$ | | 11.00 ($\beta_2$) | −0.10 | −0.42 | |
| | $Fe^{3+}$ | 2.14 | 1.31 | | | 5 $F$ $NaNO_3$ |
| | $Hg^{2+}$ | | 17.26 ($\beta_2$) | 2.71 | 1.72 | 0.5 $F$ $NaClO_4$ |
| Thiosulfate, $S_2O_3^{2-}$ | $Ag^+$ | 8.82 | 4.64 | 0.69 | | 20°C |
| | $Cd^{2+}$ | 3.92 | 2.52 | | | |
| | $Cu^+$ | 10.35 | 1.92 | 1.44 | | 0.8 $F$ $Na_2SO_4$ |
| | $Hg^{2+}$ | | 29.27 ($\beta_2$) | 2.40 | 1.35 | |
| | $Pb^{2+}$ | 2.56 | 2.32 | 1.46 | −0.09 | 3 $F$ $NaClO_4$ |

Unless indicated otherwise, the data pertain strictly to 25°C and zero ionic strength; however, for all numerical problems in this book, these constants can be used without corrections for differences in temperature and ionic strength. Note that in several instances the logarithms of $\beta_n$ values are listed instead of the logarithms of $K_n$ values; for example, an entry such as 8.43 ($\beta_2$) denotes that $\log \beta_2 = 8.43$. Discussions of the formulation and significance of stepwise ($K_n$) and overall ($\beta_n$) formation constants are presented in Chapter 6, pages 121 - 123.

[a] Reactions involving metal cations and hydroxide ions do not fit into the usual classification of complex-formation equilibria but should be thought of as examples of simple Brønsted-Lowry proton-transfer reactions.

Each metal ion species in the above tabulation is represented by the general symbol $M^{n+}$, where $n$ is the charge of the metal cation. However, it should be recalled that a metal ion is hydrated (solvated). For example, although the iron(III) ion is abbreviated as $Fe^{3+}$, the actual species may be more reasonably written as $Fe(H_2O)_6^{3+}$. A usual set of complexation equilibria involves the stepwise substitution of a (hypothetical monodentate) ligand, L, for one of the water molecules, as in the following reactions:

$$Fe(H_2O)_6^{3+} + L \rightleftharpoons Fe(H_2O)_5L^{3+} + H_2O$$

$$Fe(H_2O)_5L^{3+} + L \rightleftharpoons Fe(H_2O)_4L_2^{3+} + H_2O \cdots$$

$$\cdots Fe(H_2O)L_5^{3+} + L \rightleftharpoons FeL_6^{3+} + H_2O$$

In the case of reactions between a metal ion such as $Fe(H_2O)_6^{3+}$ and $OH^-$, however, one should visualize a process in which the hydroxide ion (Brønsted-Lowry base) accepts a proton from one of the water molecules (Brønsted-Lowry acid) coordinated to iron(III), *e.g.*,

$$Fe(H_2O)_6^{3+} + OH^- \rightleftharpoons Fe(H_2O)_5OH^{2+} + H_2O$$

If the water molecules bound to iron(III) are omitted from the preceding equilibrium, we have

$$Fe^{3+} + OH^- \rightleftharpoons FeOH^{2+}$$

Clearly, no actual exchange or substitution of $OH^-$ for $H_2O$ is involved in such an equilibrium.

It may be noted that the so-called acid dissociation constants for several metal cations can be calculated from the data in this table. Thus, the first step in the acid dissociation of $Fe(H_2O)_6^{3+}$ can be formulated as follows:

$$Fe(H_2O)_6^{3+} + H_2O \rightleftharpoons Fe(H_2O)_5OH^{2+} + H_3O^+$$

In abbreviated form, this reaction becomes

$$Fe^{3+} + H_2O \rightleftharpoons FeOH^{2+} + H^+$$

Now, the equilibrium constant for this acid dissociation of iron(III) may be calculated as follows:

$$Fe(H_2O)_6^{3+} + OH^- \rightleftharpoons Fe(H_2O)_5OH^{2+} + H_2O; K_1 = 9.1 \times 10^{10}$$

$$2\ H_2O \rightleftharpoons H_3O^+ + OH^-; K_w = 1.0 \times 10^{-14}$$

$$Fe(H_2O)_6^{3+} + H_2O \rightleftharpoons Fe(H_2O)_5OH^{2+} + H_3O^+; K_{a1} = K_1K_w = 9.1 \times 10^{-4}$$

or

$$Fe^{3+} + OH^- \rightleftharpoons FeOH^{2+}; K_1 = 9.1 \times 10^{10}$$

$$H_2O \rightleftharpoons H^+ + OH^-; K_w = 1.0 \times 10^{-14}$$

$$Fe^{3+} + H_2O \rightleftharpoons FeOH^{2+} + H^+; K_{a1} = K_1K_w = 9.1 \times 10^{-4}$$

Notice that the acid strength of $Fe(H_2O)_6^{3+}$ or $Fe^{3+}$ in water is greater than that of acetic acid, for example, and almost as great as the second dissociation constant for sulfuric acid ($K_{a2} = 1.2 \times 10^{-2}$).

Good sources of information about formation constants for metal ion complexes are:

1. L. G. Sillén and A. E. Martell: *Stability Constants of Metal-Ion Complexes.* Chem. Soc. (London), Spec. Publ. No. 17 (1964); No. 25 (1971).
2. K. B. Yatsimirskii and V. P. Vasil'ev: *Instability Constants of Complex Compounds* (translated from Russian). Consultants Bureau, New York, 1960.
3. L. Meites,ed.: *Handbook of Analytical Chemistry.* McGraw-Hill Book Company, New York, 1963.

# 4 APPENDIX

## STANDARD AND FORMAL POTENTIALS FOR HALF-REACTIONS*

(All values pertain to 25°C and are quoted in volts with respect to the normal hydrogen electrode, taken to have a standard potential of zero.)

| Half-Reaction | $E^0$ |
|---|---|
| *Aluminum* | |
| $Al^{3+} + 3e \rightleftharpoons Al$ | −1.66 |
| $Al(OH)_4^- + 3e \rightleftharpoons Al + 4OH^-$ | −2.35 |
| *Antimony* | |
| $Sb_2O_5 + 6H^+ + 4e \rightleftharpoons 2SbO^+ + 3H_2O$ | +0.581 |
| $Sb + 3H^+ + 3e \rightleftharpoons SbH_3$ | −0.51 |
| *Arsenic* | |
| $H_3AsO_4 + 2H^+ + 2e \rightleftharpoons HAsO_2 + 2H_2O$ | +0.559 |
| $HAsO_2 + 3H^+ + 3e \rightleftharpoons As + 2H_2O$ | +0.248 |
| $As + 3H^+ + 3e \rightleftharpoons AsH_3$ | −0.60 |
| *Barium* | |
| $Ba^{2+} + 2e \rightleftharpoons Ba$ | −2.90 |
| *Beryllium* | |
| $Be^{2+} + 2e \rightleftharpoons Be$ | −1.85 |
| *Bismuth* | |
| $BiCl_4^- + 3e \rightleftharpoons Bi + 4Cl^-$ | +0.16 |
| $BiO^+ + 2H^+ + 3e \rightleftharpoons Bi + H_2O$ | +0.32 |
| *Boron* | |
| $H_2BO_3^- + 5H_2O + 8e \rightleftharpoons BH_4^- + 8OH^-$ | −1.24 |
| $H_2BO_3^- + H_2O + 3e \rightleftharpoons B + 4OH^-$ | −1.79 |
| *Bromine* | |
| $2BrO_3^- + 12H^+ + 10e \rightleftharpoons Br_2 + 6H_2O$ | +1.52 |
| $Br_2(aq) + 2e \rightleftharpoons 2Br^-$ | +1.087[a] |
| $Br_2(l) + 2e \rightleftharpoons 2Br^-$ | +1.065[a] |
| $Br_3^- + 2e \rightleftharpoons 3Br^-$ | +1.05 |
| *Cadmium* | |
| $Cd^{2+} + 2e \rightleftharpoons Cd$ | −0.403 |
| $Cd(CN)_4^{2-} + 2e \rightleftharpoons Cd + 4CN^-$ | −1.09 |
| $Cd(NH_3)_4^{2+} + 2e \rightleftharpoons Cd + 4NH_3$ | −0.61 |
| *Calcium* | |
| $Ca^{2+} + 2e \rightleftharpoons Ca$ | −2.87 |
| *Carbon* | |
| $2CO_2 + 2H^+ + 2e \rightleftharpoons H_2C_2O_4$ | −0.49 |
| *Cerium* | |
| $Ce^{4+} + e \rightleftharpoons Ce^{3+}$ (1 *F* $HClO_4$) | *+1.70* |
| $Ce^{4+} + e \rightleftharpoons Ce^{3+}$ (1 *F* $HNO_3$) | *+1.61* |
| $Ce^{4+} + e \rightleftharpoons Ce^{3+}$ (1 *F* $H_2SO_4$) | *+1.44* |
| *Cesium* | |
| $Cs^+ + e \rightleftharpoons Cs$ | −2.92 |
| *Chlorine* | |
| $Cl_2 + 2e \rightleftharpoons 2Cl^-$ | +1.3595 |
| $2ClO_3^- + 12H^+ + 10e \rightleftharpoons Cl_2 + 6H_2O$ | +1.47 |
| $ClO_3^- + 2H^+ + e \rightleftharpoons ClO_2 + H_2O$ | +1.15 |

* See footnotes at end of table.

| Half-Reaction | $E^0$ |
|---|---|
| $HClO + H^+ + 2e \rightleftharpoons Cl^- + H_2O$ | +1.49 |
| $2 HClO + 2 H^+ + 2e \rightleftharpoons Cl_2 + 2 H_2O$ | +1.63 |
| *Chromium* | |
| $Cr_2O_7^{2-} + 14 H^+ + 6e \rightleftharpoons 2 Cr^{3+} + 7 H_2O$ | +1.33 |
| $Cr^{3+} + e \rightleftharpoons Cr^{2+}$ | −0.41 |
| $Cr^{2+} + 2e \rightleftharpoons Cr$ | −0.91 |
| $CrO_4^{2-} + 4 H_2O + 3e \rightleftharpoons Cr(OH)_3 + 5 OH^-$ | −0.13 |
| *Cobalt* | |
| $Co^{3+} + e \rightleftharpoons Co^{2+}$ | +1.842 |
| $Co(NH_3)_6^{3+} + e \rightleftharpoons Co(NH_3)_6^{2+}$ | +0.1 |
| $Co(OH)_3 + e \rightleftharpoons Co(OH)_2 + OH^-$ | +0.17 |
| $Co^{2+} + 2e \rightleftharpoons Co$ | −0.277 |
| $Co(CN)_6^{3-} + e \rightleftharpoons Co(CN)_6^{4-}$ | −0.84 |
| *Copper* | |
| $Cu^{2+} + 2e \rightleftharpoons Cu$ | +0.337 |
| $Cu^{2+} + e \rightleftharpoons Cu^+$ | +0.153 |
| $Cu^{2+} + I^- + e \rightleftharpoons CuI$ | +0.86 |
| $Cu^{2+} + 2 CN^- + e \rightleftharpoons Cu(CN)_2^-$ | +1.12 |
| $Cu(CN)_2^- + e \rightleftharpoons Cu + 2 CN^-$ | −0.43 |
| $Cu(NH_3)_4^{2+} + e \rightleftharpoons Cu(NH_3)_2^+ + 2 NH_3$ | −0.01 |
| $Cu^{2+} + 2 Cl^- + e \rightleftharpoons CuCl_2^-$ | +0.463 |
| $CuCl_2^- + e \rightleftharpoons Cu + 2 Cl^-$ | +0.177 |
| *Fluorine* | |
| $F_2 + 2e \rightleftharpoons 2 F^-$ | +2.87 |
| *Gold* | |
| $Au^{3+} + 2e \rightleftharpoons Au^+$ | +1.41 |
| $Au^{3+} + 3e \rightleftharpoons Au$ | +1.50 |
| $Au(CN)_2^- + e \rightleftharpoons Au + 2 CN^-$ | −0.60 |
| $AuCl_2^- + e \rightleftharpoons Au + 2 Cl^-$ | +1.15 |
| $AuCl_4^- + 2e \rightleftharpoons AuCl_2^- + 2 Cl^-$ | +0.926 |
| $AuBr_2^- + e \rightleftharpoons Au + 2 Br^-$ | +0.959 |
| $AuBr_4^- + 2e \rightleftharpoons AuBr_2^- + 2 Br^-$ | +0.802 |
| *Hydrogen* | |
| $2 H^+ + 2e \rightleftharpoons H_2$ | 0.0000 |
| $2 H_2O + 2e \rightleftharpoons H_2 + 2 OH^-$ | −0.828 |
| *Iodine* | |
| $I_2(aq) + 2e \rightleftharpoons 2 I^-$ | +0.6197[b] |
| $I_3^- + 2e \rightleftharpoons 3 I^-$ | +0.5355 |
| $I_2(s) + 2e \rightleftharpoons 2 I^-$ | +0.5345[b] |
| $2 IO_3^- + 12 H^+ + 10e \rightleftharpoons I_2 + 6 H_2O$ | +1.20 |
| $2 ICl_2^- + 2e \rightleftharpoons I_2 + 4 Cl^-$ | +1.06 |
| *Iron* | |
| $Fe^{3+} + e \rightleftharpoons Fe^{2+}$ | +0.771 |
| $Fe^{3+} + e \rightleftharpoons Fe^{2+}$ (1 *F* HCl) | *+0.70* |
| $Fe^{3+} + e \rightleftharpoons Fe^{2+}$ (1 *F* $H_2SO_4$) | *+0.68* |
| $Fe^{3+} + e \rightleftharpoons Fe^{2+}$ (0.5 *F* $H_3PO_4$ − 1 *F* $H_2SO_4$) | *+0.61* |
| $Fe(CN)_6^{3-} + e \rightleftharpoons Fe(CN)_6^{4-}$ | +0.36 |
| $Fe(CN)_6^{3-} + e \rightleftharpoons Fe(CN)_6^{4-}$ (1 *F* HCl or $HClO_4$) | *+0.71* |
| $Fe^{2+} + 2e \rightleftharpoons Fe$ | −0.440 |
| *Lead* | |
| $Pb^{2+} + 2e \rightleftharpoons Pb$ | −0.126 |
| $PbSO_4 + 2e \rightleftharpoons Pb + SO_4^{2-}$ | −0.3563 |
| $PbO_2 + SO_4^{2-} + 4 H^+ + 2e \rightleftharpoons PbSO_4 + 2 H_2O$ | +1.685 |
| $PbO_2 + 4 H^+ + 2e \rightleftharpoons Pb^{2+} + 2 H_2O$ | +1.455 |
| *Lithium* | |
| $Li^+ + e \rightleftharpoons Li$ | −3.045 |
| *Magnesium* | |
| $Mg^{2+} + 2e \rightleftharpoons Mg$ | −2.37 |
| $Mg(OH)_2 + 2e \rightleftharpoons Mg + 2 OH^-$ | −2.69 |

| Half-Reaction | $E^0$ |
|---|---|
| *Manganese* | |
| $Mn^{2+} + 2e \rightleftharpoons Mn$ | −1.18 |
| $MnO_4^- + 4H^+ + 3e \rightleftharpoons MnO_2 + 2H_2O$ | +1.695 |
| $MnO_4^- + 8H^+ + 5e \rightleftharpoons Mn^{2+} + 4H_2O$ | +1.51 |
| $MnO_2 + 4H^+ + 2e \rightleftharpoons Mn^{2+} + 2H_2O$ | +1.23 |
| $MnO_4^- + e \rightleftharpoons MnO_4^{2-}$ | +0.564 |
| $Mn^{3+} + e \rightleftharpoons Mn^{2+}$ (8 $F$ $H_2SO_4$) | *+1.51* |
| *Mercury* | |
| $2Hg^{2+} + 2e \rightleftharpoons Hg_2^{2+}$ | +0.920 |
| $Hg^{2+} + 2e \rightleftharpoons Hg$ | +0.854 |
| $Hg_2^{2+} + 2e \rightleftharpoons 2Hg$ | +0.789 |
| $Hg_2SO_4 + 2e \rightleftharpoons 2Hg + SO_4^{2-}$ | +0.6151 |
| $HgCl_4^{2-} + 2e \rightleftharpoons Hg + 4Cl^-$ | +0.48 |
| $Hg_2Cl_2 + 2e \rightleftharpoons 2Hg + 2Cl^-$ (0.1 $F$ KCl) | *+0.334* |
| $Hg_2Cl_2 + 2e \rightleftharpoons 2Hg + 2Cl^-$ (1 $F$ KCl) | *+0.280* |
| $Hg_2Cl_2 + 2K^+ + 2e \rightleftharpoons 2Hg + 2KCl(s)$ (saturated calomel electrode) | *+0.2415* |
| *Molybdenum* | |
| $Mo^{6+} + e \rightleftharpoons Mo^{5+}$ (2 $F$ HCl) | *+0.53* |
| $Mo^{4+} + e \rightleftharpoons Mo^{3+}$ (4 $F$ $H_2SO_4$) | *+0.1* |
| *Neptunium* | |
| $Np^{4+} + e \rightleftharpoons Np^{3+}$ | +0.147 |
| $NpO_2^+ + 4H^+ + e \rightleftharpoons Np^{4+} + 2H_2O$ | +0.75 |
| $NpO_2^{2+} + e \rightleftharpoons NpO_2^+$ | +1.15 |
| *Nickel* | |
| $Ni^{2+} + 2e \rightleftharpoons Ni$ | −0.24 |
| $NiO_2 + 4H^+ + 2e \rightleftharpoons Ni^{2+} + 2H_2O$ | +1.68 |
| *Nitrogen* | |
| $NO_2 + H^+ + e \rightleftharpoons HNO_2$ | +1.07 |
| $NO_2 + 2H^+ + 2e \rightleftharpoons NO + H_2O$ | +1.03 |
| $HNO_2 + H^+ + e \rightleftharpoons NO + H_2O$ | +1.00 |
| $NO_3^- + 4H^+ + 3e \rightleftharpoons NO + 2H_2O$ | +0.96 |
| $NO_3^- + 3H^+ + 2e \rightleftharpoons HNO_2 + H_2O$ | +0.94 |
| $NO_3^- + 2H^+ + e \rightleftharpoons NO_2 + H_2O$ | +0.80 |
| $N_2 + 5H^+ + 4e \rightleftharpoons N_2H_5^+$ | −0.23 |
| *Osmium* | |
| $OsO_4 + 8H^+ + 8e \rightleftharpoons Os + 4H_2O$ | +0.85 |
| $OsCl_6^{2-} + e \rightleftharpoons OsCl_6^{3-}$ | +0.85 |
| $OsCl_6^{3-} + e \rightleftharpoons Os^{2+} + 6Cl^-$ | +0.4 |
| $Os^{2+} + 2e \rightleftharpoons Os$ | +0.85 |
| *Oxygen* | |
| $O_3 + 2H^+ + 2e \rightleftharpoons O_2 + H_2O$ | +2.07 |
| $H_2O_2 + 2H^+ + 2e \rightleftharpoons 2H_2O$ | +1.77 |
| $O_2 + 4H^+ + 4e \rightleftharpoons 2H_2O$ | +1.229 |
| $H_2O_2 + 2e \rightleftharpoons 2OH^-$ | +0.88 |
| $O_2 + 2H^+ + 2e \rightleftharpoons H_2O_2$ | +0.682 |
| *Palladium* | |
| $Pd^{2+} + 2e \rightleftharpoons Pd$ | +0.987 |
| $PdCl_6^{2-} + 2e \rightleftharpoons PdCl_4^{2-} + 2Cl^-$ | +1.288 |
| $PdCl_4^{2-} + 2e \rightleftharpoons Pd + 4Cl^-$ | +0.623 |
| *Phosphorus* | |
| $H_3PO_4 + 2H^+ + 2e \rightleftharpoons H_3PO_3 + H_2O$ | −0.276 |
| $H_3PO_3 + 2H^+ + 2e \rightleftharpoons H_3PO_2 + H_2O$ | −0.50 |
| *Platinum* | |
| $PtCl_6^{2-} + 2e \rightleftharpoons PtCl_4^{2-} + 2Cl^-$ | +0.68 |
| $PtBr_6^{2-} + 2e \rightleftharpoons PtBr_4^{2-} + 2Br^-$ | +0.59 |
| $Pt(OH)_2 + 2H^+ + 2e \rightleftharpoons Pt + 2H_2O$ | +0.98 |
| *Plutonium* | |
| $PuO_2^+ + 4H^+ + e \rightleftharpoons Pu^{4+} + 2H_2O$ | +1.15 |

| Half-Reaction | $E^0$ |
|---|---|
| $PuO_2^{2+} + 4\,H^+ + 2\,e \rightleftharpoons Pu^{4+} + 2\,H_2O$ | +1.067 |
| $Pu^{4+} + e \rightleftharpoons Pu^{3+}$ | +0.97 |
| $PuO_2^{2+} + e \rightleftharpoons PuO_2^+$ | +0.93 |
| *Potassium* | |
| $K^+ + e \rightleftharpoons K$ | −2.925 |
| *Radium* | |
| $Ra^{2+} + 2\,e \rightleftharpoons Ra$ | −2.92 |
| *Rubidium* | |
| $Rb^+ + e \rightleftharpoons Rb$ | −2.925 |
| *Selenium* | |
| $SeO_4^{2-} + 4\,H^+ + 2\,e \rightleftharpoons H_2SeO_3 + H_2O$ | +1.15 |
| $H_2SeO_3 + 4\,H^+ + 4\,e \rightleftharpoons Se + 3\,H_2O$ | +0.740 |
| $Se + 2\,H^+ + 2\,e \rightleftharpoons H_2Se$ | −0.40 |
| *Silver* | |
| $Ag^+ + e \rightleftharpoons Ag$ | +0.7995 |
| $Ag^{2+} + e \rightleftharpoons Ag^+$ (4 *F* $HNO_3$) | +*1.927* |
| $AgCl + e \rightleftharpoons Ag + Cl^-$ | +0.2222 |
| $AgBr + e \rightleftharpoons Ag + Br^-$ | +0.073 |
| $AgI + e \rightleftharpoons Ag + I^-$ | −0.151 |
| $Ag_2O + H_2O + 2\,e \rightleftharpoons 2\,Ag + 2\,OH^-$ | +0.342 |
| $Ag_2S + 2\,e \rightleftharpoons 2\,Ag + S^{2-}$ | −0.71 |
| *Sodium* | |
| $Na^+ + e \rightleftharpoons Na$ | −2.714 |
| *Strontium* | |
| $Sr^{2+} + 2\,e \rightleftharpoons Sr$ | −2.89 |
| *Sulfur* | |
| $S + 2\,H^+ + 2\,e \rightleftharpoons H_2S$ | +0.141 |
| $S_4O_6^{2-} + 2\,e \rightleftharpoons 2\,S_2O_3^{2-}$ | +0.08 |
| $SO_4^{2-} + 4\,H^+ + 2\,e \rightleftharpoons H_2SO_3 + H_2O$ | +0.17 |
| $S_2O_8^{2-} + 2\,e \rightleftharpoons 2\,SO_4^{2-}$ | +2.01 |
| $SO_4^{2-} + H_2O + 2\,e \rightleftharpoons SO_3^{2-} + 2\,OH^-$ | −0.93 |
| $2\,H_2SO_3 + 2\,H^+ + 4\,e \rightleftharpoons S_2O_3^{2-} + 3\,H_2O$ | +0.40 |
| $2\,SO_3^{2-} + 3\,H_2O + 4\,e \rightleftharpoons S_2O_3^{2-} + 6\,OH^-$ | −0.58 |
| $SO_3^{2-} + 3\,H_2O + 4\,e \rightleftharpoons S + 6\,OH^-$ | −0.66 |
| *Thallium* | |
| $Tl^{3+} + 2\,e \rightleftharpoons Tl^+$ | +1.25 |
| $Tl^+ + e \rightleftharpoons Tl$ | −0.3363 |
| *Tin* | |
| $Sn^{2+} + 2\,e \rightleftharpoons Sn$ | −0.136 |
| $Sn^{4+} + 2\,e \rightleftharpoons Sn^{2+}$ | +0.154 |
| $SnCl_6^{2-} + 2\,e \rightleftharpoons SnCl_4^{2-} + 2\,Cl^-$ (1 *F* HCl) | +*0.14* |
| $Sn(OH)_6^{2-} + 2\,e \rightleftharpoons HSnO_2^- + H_2O + 3\,OH^-$ | −0.93 |
| $HSnO_2^- + H_2O + 2\,e \rightleftharpoons Sn + 3\,OH^-$ | −0.91 |
| *Titanium* | |
| $Ti^{2+} + 2\,e \rightleftharpoons Ti$ | −1.63 |
| $Ti^{3+} + e \rightleftharpoons Ti^{2+}$ | −0.37 |
| $TiO^{2+} + 2\,H^+ + e \rightleftharpoons Ti^{3+} + H_2O$ | +0.10 |
| $Ti^{4+} + e \rightleftharpoons Ti^{3+}$ (5 *F* $H_3PO_4$) | −*0.15* |
| *Tungsten* | |
| $W^{6+} + e \rightleftharpoons W^{5+}$ (12 *F* HCl) | +*0.26* |
| $W^{5+} + e \rightleftharpoons W^{4+}$ (12 *F* HCl) | −*0.3* |
| $W(CN)_8^{3-} + e \rightleftharpoons W(CN)_8^{4-}$ | +0.48 |
| $2\,WO_3(s) + 2\,H^+ + 2\,e \rightleftharpoons W_2O_5(s) + H_2O$ | −0.03 |
| $W_2O_5(s) + 2\,H^+ + 2\,e \rightleftharpoons 2\,WO_2(s) + H_2O$ | −0.043 |
| *Uranium* | |
| $U^{4+} + e \rightleftharpoons U^{3+}$ | −0.61 |
| $UO_2^{2+} + e \rightleftharpoons UO_2^+$ | +0.05 |
| $UO_2^{2+} + 4\,H^+ + 2\,e \rightleftharpoons U^{4+} + 2\,H_2O$ | +0.334 |
| $UO_2^+ + 4\,H^+ + e \rightleftharpoons U^{4+} + 2\,H_2O$ | +0.62 |

| Half-Reaction | $E^0$ |
|---|---|
| *Vanadium* | |
| $VO_2^+ + 2\,H^+ + e \rightleftharpoons VO^{2+} + H_2O$ | +1.000 |
| $VO^{2+} + 2\,H^+ + e \rightleftharpoons V^{3+} + H_2O$ | +0.361 |
| $V^{3+} + e \rightleftharpoons V^{2+}$ | −0.255 |
| $V^{2+} + 2\,e \rightleftharpoons V$ | −1.18 |
| *Zinc* | |
| $Zn^{2+} + 2\,e \rightleftharpoons Zn$ | −0.763 |
| $Zn(NH_3)_4^{2+} + 2\,e \rightleftharpoons Zn + 4\,NH_3$ | −1.04 |
| $Zn(CN)_4^{2-} + 2\,e \rightleftharpoons Zn + 4\,CN^-$ | −1.26 |
| $Zn(OH)_4^{2-} + 2\,e \rightleftharpoons Zn + 4\,OH^-$ | −1.22 |

The *standard potential* for a redox couple is defined on page 189 as the potential (sign and magnitude) of an electrode consisting of that redox couple under standard-state conditions measured in a galvanic cell against the normal hydrogen electrode at 25°C.

*Formal potentials*, properly designated by the symbol $E^{0\prime}$ and defined on page 193 are italicized in the above table. The solution condition to which each formal potential pertains is written in parentheses following the half-reaction.

[a] The half-reaction and standard potential

$$Br_2(aq) + 2\,e \rightleftharpoons 2\,Br^-; \qquad E^0 = +1.087\ v$$

pertain to the system in which the activity of dissolved molecular bromine, $Br_2$, as well as the activity of the bromide ion, is unity in water. Actually, this is an impossible situation because the solubility of $Br_2$ in water is only about 0.21 *M* at 25°C.

On the other hand, the half-reaction and standard potential

$$Br_2(l) + 2\,e \rightleftharpoons 2\,Br^-; \qquad E^0 = +1.065\ v$$

apply to an electrode system in which excess *liquid* bromine is in equilibrium with an aqueous solution containing bromide ion at unit activity. It follows that an aqueous solution in equilibrium with liquid bromine will be saturated with respect to molecular bromine at a concentration (activity) of 0.21 *M*.

Thus, these two standard potentials are different because the former refers to the (hypothetical) situation in which the concentration (activity) of $Br_2(aq)$ is taken to be unity, whereas the latter refers to the physically real situation for which the concentration (activity) of $Br_2(aq)$ is only 0.21 *M*.

[b] The reason for the difference between the two entries

$$I_2(aq) + 2\,e \rightleftharpoons 2\,I^-; \qquad E^0 = +0.6197\ v$$

and

$$I_2(s) + 2\,e \rightleftharpoons 2\,I^-; \qquad E^0 = +0.5345\ v$$

is essentially the same as that stated in the preceding footnote. The first half-reaction requires (hypothetically) an aqueous molecular iodine concentration or activity of unity, whereas the second half-reaction specifies that excess *solid* iodine be in equilibrium with an aqueous, iodide solution of unit activity. Since the solubility of molecular iodine in water at 25°C is approximately 0.00133 *M*, it is impossible to ever have an aqueous solution containing 1 *M* molecular iodine. Thus, the standard potential for the second half-reaction will be considerably less oxidizing or less positive than the value for the first half-reaction.

Among the best sources of standard and formal potentials are the following references:

1. W. M. Latimer: *Oxidation Potentials.* Second edition, Prentice-Hall, Englewood Cliffs, New Jersey, 1952.
2. J. J. Lingane: *Electroanalytical Chemistry.* Second edition, Wiley-Interscience, New York, 1958, pp. 639–651.
3. L. Meites, ed.: *Handbook of Analytical Chemistry.* McGraw-Hill Book Company, New York, 1963.

# 5 APPENDIX

## TABLE OF LOGARITHMS

| | 0 | 1 | 2 | 3 | 4 | 5 | 6 | 7 | 8 | 9 |
|---|---|---|---|---|---|---|---|---|---|---|
| 1.0 | 0.0000 | 0.0043 | 0.0086 | 0.0128 | 0.0170 | 0.0212 | 0.0253 | 0.0294 | 0.0334 | 0.0374 |
| 1.1 | 0.0414 | 0.0453 | 0.0492 | 0.0531 | 0.0569 | 0.0607 | 0.0645 | 0.0682 | 0.0719 | 0.0755 |
| 1.2 | 0.0792 | 0.0828 | 0.0864 | 0.0899 | 0.0934 | 0.0969 | 0.1004 | 0.1038 | 0.1072 | 0.1106 |
| 1.3 | 0.1139 | 0.1173 | 0.1206 | 0.1239 | 0.1271 | 0.1303 | 0.1335 | 0.1367 | 0.1399 | 0.1430 |
| 1.4 | 0.1461 | 0.1492 | 0.1523 | 0.1553 | 0.1584 | 0.1614 | 0.1644 | 0.1673 | 0.1703 | 0.1732 |
| 1.5 | 0.1761 | 0.1790 | 0.1818 | 0.1847 | 0.1875 | 0.1903 | 0.1931 | 0.1959 | 0.1987 | 0.2014 |
| 1.6 | 0.2041 | 0.2068 | 0.2095 | 0.2122 | 0.2148 | 0.2175 | 0.2201 | 0.2227 | 0.2253 | 0.2279 |
| 1.7 | 0.2304 | 0.2330 | 0.2355 | 0.2380 | 0.2405 | 0.2430 | 0.2455 | 0.2480 | 0.2504 | 0.2529 |
| 1.8 | 0.2553 | 0.2577 | 0.2601 | 0.2625 | 0.2648 | 0.2672 | 0.2695 | 0.2718 | 0.2742 | 0.2765 |
| 1.9 | 0.2788 | 0.2810 | 0.2833 | 0.2856 | 0.2878 | 0.2900 | 0.2923 | 0.2945 | 0.2967 | 0.2989 |
| 2.0 | 0.3010 | 0.3032 | 0.3054 | 0.3075 | 0.3096 | 0.3118 | 0.3139 | 0.3160 | 0.3181 | 0.3201 |
| 2.1 | 0.3222 | 0.3243 | 0.3263 | 0.3284 | 0.3304 | 0.3324 | 0.3345 | 0.3365 | 0.3385 | 0.3404 |
| 2.2 | 0.3424 | 0.3444 | 0.3464 | 0.3483 | 0.3502 | 0.3522 | 0.3541 | 0.3560 | 0.3579 | 0.3598 |
| 2.3 | 0.3617 | 0.3636 | 0.3655 | 0.3674 | 0.3692 | 0.3711 | 0.3729 | 0.3747 | 0.3766 | 0.3784 |
| 2.4 | 0.3802 | 0.3820 | 0.3838 | 0.3856 | 0.3874 | 0.3892 | 0.3909 | 0.3927 | 0.3945 | 0.3962 |
| 2.5 | 0.3979 | 0.3997 | 0.4014 | 0.4031 | 0.4048 | 0.4065 | 0.4082 | 0.4099 | 0.4116 | 0.4133 |
| 2.6 | 0.4150 | 0.4166 | 0.4183 | 0.4200 | 0.4216 | 0.4232 | 0.4249 | 0.4265 | 0.4281 | 0.4298 |
| 2.7 | 0.4314 | 0.4330 | 0.4346 | 0.4362 | 0.4378 | 0.4393 | 0.4409 | 0.4425 | 0.4440 | 0.4456 |
| 2.8 | 0.4472 | 0.4487 | 0.4502 | 0.4518 | 0.4533 | 0.4548 | 0.4564 | 0.4579 | 0.4594 | 0.4609 |
| 2.9 | 0.4624 | 0.4639 | 0.4654 | 0.4669 | 0.4683 | 0.4698 | 0.4713 | 0.4728 | 0.4742 | 0.4757 |
| 3.0 | 0.4771 | 0.4786 | 0.4800 | 0.4814 | 0.4829 | 0.4843 | 0.4857 | 0.4871 | 0.4886 | 0.4900 |
| 3.1 | 0.4914 | 0.4928 | 0.4942 | 0.4955 | 0.4969 | 0.4983 | 0.4997 | 0.5011 | 0.5024 | 0.5038 |
| 3.2 | 0.5051 | 0.5065 | 0.5079 | 0.5092 | 0.5105 | 0.5119 | 0.5132 | 0.5145 | 0.5159 | 0.5172 |
| 3.3 | 0.5185 | 0.5198 | 0.5211 | 0.5224 | 0.5237 | 0.5250 | 0.5263 | 0.5276 | 0.5289 | 0.5302 |
| 3.4 | 0.5315 | 0.5328 | 0.5340 | 0.5353 | 0.5366 | 0.5378 | 0.5391 | 0.5403 | 0.5416 | 0.5428 |
| 3.5 | 0.5441 | 0.5453 | 0.5465 | 0.5478 | 0.5490 | 0.5502 | 0.5514 | 0.5527 | 0.5539 | 0.5551 |
| 3.6 | 0.5563 | 0.5575 | 0.5587 | 0.5599 | 0.5611 | 0.5623 | 0.5635 | 0.5647 | 0.5658 | 0.5670 |
| 3.7 | 0.5682 | 0.5694 | 0.5705 | 0.5717 | 0.5729 | 0.5740 | 0.5752 | 0.5763 | 0.5775 | 0.5786 |
| 3.8 | 0.5798 | 0.5809 | 0.5821 | 0.5832 | 0.5843 | 0.5855 | 0.5866 | 0.5877 | 0.5888 | 0.5899 |
| 3.9 | 0.5911 | 0.5922 | 0.5933 | 0.5944 | 0.5955 | 0.5966 | 0.5977 | 0.5988 | 0.5999 | 0.6010 |
| 4.0 | 0.6021 | 0.6031 | 0.6042 | 0.6053 | 0.6064 | 0.6075 | 0.6085 | 0.6096 | 0.6107 | 0.6117 |
| 4.1 | 0.6128 | 0.6138 | 0.6149 | 0.6160 | 0.6170 | 0.6180 | 0.6191 | 0.6201 | 0.6212 | 0.6222 |
| 4.2 | 0.6232 | 0.6243 | 0.6253 | 0.6263 | 0.6274 | 0.6284 | 0.6294 | 0.6304 | 0.6314 | 0.6325 |
| 4.3 | 0.6335 | 0.6345 | 0.6355 | 0.6365 | 0.6375 | 0.6385 | 0.6395 | 0.6405 | 0.6415 | 0.6425 |
| 4.4 | 0.6435 | 0.6444 | 0.6454 | 0.6464 | 0.6474 | 0.6484 | 0.6493 | 0.6503 | 0.6513 | 0.6522 |
| 4.5 | 0.6532 | 0.6542 | 0.6551 | 0.6561 | 0.6571 | 0.6580 | 0.6590 | 0.6599 | 0.6609 | 0.6618 |
| 4.6 | 0.6628 | 0.6637 | 0.6646 | 0.6656 | 0.6665 | 0.6675 | 0.6684 | 0.6693 | 0.6702 | 0.6712 |
| 4.7 | 0.6721 | 0.6730 | 0.6739 | 0.6749 | 0.6758 | 0.6767 | 0.6776 | 0.6785 | 0.6794 | 0.6803 |
| 4.8 | 0.6812 | 0.6821 | 0.6830 | 0.6839 | 0.6848 | 0.6857 | 0.6866 | 0.6875 | 0.6884 | 0.6893 |
| 4.9 | 0.6902 | 0.6911 | 0.6920 | 0.6928 | 0.6937 | 0.6946 | 0.6955 | 0.6964 | 0.6972 | 0.6981 |
| 5.0 | 0.6990 | 0.6998 | 0.7007 | 0.7016 | 0.7024 | 0.7033 | 0.7042 | 0.7050 | 0.7059 | 0.7067 |
| 5.1 | 0.7076 | 0.7084 | 0.7093 | 0.7101 | 0.7110 | 0.7118 | 0.7126 | 0.7135 | 0.7143 | 0.7152 |
| 5.2 | 0.7160 | 0.7168 | 0.7177 | 0.7185 | 0.7193 | 0.7202 | 0.7210 | 0.7218 | 0.7226 | 0.7235 |
| 5.3 | 0.7243 | 0.7251 | 0.7259 | 0.7267 | 0.7275 | 0.7284 | 0.7292 | 0.7300 | 0.7308 | 0.7316 |
| 5.4 | 0.7324 | 0.7332 | 0.7340 | 0.7348 | 0.7356 | 0.7364 | 0.7372 | 0.7380 | 0.7388 | 0.7396 |
| 5.5 | 0.7404 | 0.7412 | 0.7419 | 0.7427 | 0.7435 | 0.7443 | 0.7451 | 0.7459 | 0.7466 | 0.7474 |

| | 0 | 1 | 2 | 3 | 4 | 5 | 6 | 7 | 8 | 9 |
|---|---|---|---|---|---|---|---|---|---|---|
| 5.6 | 0.7482 | 0.7490 | 0.7497 | 0.7505 | 0.7513 | 0.7520 | 0.7528 | 0.7536 | 0.7543 | 0.7551 |
| 5.7 | 0.7559 | 0.7566 | 0.7574 | 0.7582 | 0.7589 | 0.7597 | 0.7604 | 0.7612 | 0.7619 | 0.7627 |
| 5.8 | 0.7634 | 0.7642 | 0.7649 | 0.7657 | 0.7664 | 0.7672 | 0.7679 | 0.7686 | 0.7694 | 0.7701 |
| 5.9 | 0.7709 | 0.7716 | 0.7723 | 0.7731 | 0.7738 | 0.7745 | 0.7752 | 0.7760 | 0.7767 | 0.7774 |
| 6.0 | 0.7782 | 0.7789 | 0.7796 | 0.7803 | 0.7810 | 0.7818 | 0.7825 | 0.7832 | 0.7839 | 0.7846 |
| 6.1 | 0.7853 | 0.7860 | 0.7868 | 0.7875 | 0.7882 | 0.7889 | 0.7896 | 0.7903 | 0.7910 | 0.7917 |
| 6.2 | 0.7924 | 0.7931 | 0.7938 | 0.7945 | 0.7952 | 0.7959 | 0.7966 | 0.7973 | 0.7980 | 0.7987 |
| 6.3 | 0.7993 | 0.8000 | 0.8007 | 0.8014 | 0.8021 | 0.8028 | 0.8035 | 0.8041 | 0.8048 | 0.8055 |
| 6.4 | 0.8062 | 0.8069 | 0.8075 | 0.8082 | 0.8089 | 0.8096 | 0.8102 | 0.8109 | 0.8116 | 0.8122 |
| 6.5 | 0.8129 | 0.8136 | 0.8142 | 0.8149 | 0.8156 | 0.8162 | 0.8169 | 0.8176 | 0.8182 | 0.8189 |
| 6.6 | 0.8195 | 0.8202 | 0.8209 | 0.8215 | 0.8222 | 0.8228 | 0.8235 | 0.8241 | 0.8248 | 0.8254 |
| 6.7 | 0.8261 | 0.8267 | 0.8274 | 0.8280 | 0.8287 | 0.8293 | 0.8299 | 0.8306 | 0.8312 | 0.8319 |
| 6.8 | 0.8325 | 0.8331 | 0.8338 | 0.8344 | 0.8351 | 0.8357 | 0.8363 | 0.8370 | 0.8376 | 0.8382 |
| 6.9 | 0.8388 | 0.8395 | 0.8401 | 0.8407 | 0.8414 | 0.8420 | 0.8426 | 0.8432 | 0.8439 | 0.8445 |
| 7.0 | 0.8451 | 0.8457 | 0.8463 | 0.8470 | 0.8476 | 0.8482 | 0.8488 | 0.8494 | 0.8500 | 0.8506 |
| 7.1 | 0.8513 | 0.8519 | 0.8525 | 0.8531 | 0.8537 | 0.8543 | 0.8549 | 0.8555 | 0.8561 | 0.8567 |
| 7.2 | 0.8573 | 0.8579 | 0.8585 | 0.8591 | 0.8597 | 0.8603 | 0.8609 | 0.8615 | 0.8621 | 0.8627 |
| 7.3 | 0.8633 | 0.8639 | 0.8645 | 0.8651 | 0.8657 | 0.8663 | 0.8669 | 0.8675 | 0.8681 | 0.8686 |
| 7.4 | 0.8692 | 0.8698 | 0.8704 | 0.8710 | 0.8716 | 0.8722 | 0.8727 | 0.8733 | 0.8739 | 0.8745 |
| 7.5 | 0.8751 | 0.8756 | 0.8762 | 0.8768 | 0.8774 | 0.8779 | 0.8785 | 0.8791 | 0.8797 | 0.8802 |
| 7.6 | 0.8808 | 0.8814 | 0.8820 | 0.8825 | 0.8831 | 0.8837 | 0.8842 | 0.8848 | 0.8854 | 0.8859 |
| 7.7 | 0.8865 | 0.8871 | 0.8876 | 0.8882 | 0.8887 | 0.8893 | 0.8899 | 0.8904 | 0.8910 | 0.8915 |
| 7.8 | 0.8921 | 0.8927 | 0.8932 | 0.8938 | 0.8943 | 0.8949 | 0.8954 | 0.8960 | 0.8965 | 0.8971 |
| 7.9 | 0.8976 | 0.8982 | 0.8987 | 0.8993 | 0.8998 | 0.9004 | 0.9009 | 0.9015 | 0.9020 | 0.9025 |
| 8.0 | 0.9031 | 0.9036 | 0.9042 | 0.9047 | 0.9053 | 0.9058 | 0.9063 | 0.9069 | 0.9074 | 0.9079 |
| 8.1 | 0.9085 | 0.9090 | 0.9096 | 0.9101 | 0.9106 | 0.9112 | 0.9117 | 0.9122 | 0.9128 | 0.9133 |
| 8.2 | 0.9138 | 0.9143 | 0.9149 | 0.9154 | 0.9159 | 0.9165 | 0.9170 | 0.9175 | 0.9180 | 0.9186 |
| 8.3 | 0.9191 | 0.9196 | 0.9201 | 0.9206 | 0.9212 | 0.9217 | 0.9222 | 0.9227 | 0.9232 | 0.9238 |
| 8.4 | 0.9243 | 0.9248 | 0.9253 | 0.9258 | 0.9263 | 0.9269 | 0.9274 | 0.9279 | 0.9284 | 0.9289 |
| 8.5 | 0.9294 | 0.9299 | 0.9304 | 0.9309 | 0.9315 | 0.9320 | 0.9325 | 0.9330 | 0.9335 | 0.9340 |
| 8.6 | 0.9345 | 0.9350 | 0.9355 | 0.9360 | 0.9365 | 0.9370 | 0.9375 | 0.9380 | 0.9385 | 0.9390 |
| 8.7 | 0.9395 | 0.9400 | 0.9405 | 0.9410 | 0.9415 | 0.9420 | 0.9425 | 0.9430 | 0.9435 | 0.9440 |
| 8.8 | 0.9445 | 0.9450 | 0.9455 | 0.9460 | 0.9465 | 0.9469 | 0.9474 | 0.9479 | 0.9484 | 0.9489 |
| 8.9 | 0.9494 | 0.9499 | 0.9504 | 0.9509 | 0.9513 | 0.9518 | 0.9523 | 0.9528 | 0.9533 | 0.9538 |
| 9.0 | 0.9542 | 0.9547 | 0.9552 | 0.9557 | 0.9562 | 0.9566 | 0.9571 | 0.9576 | 0.9581 | 0.9586 |
| 9.1 | 0.9590 | 0.9595 | 0.9600 | 0.9605 | 0.9609 | 0.9614 | 0.9619 | 0.9624 | 0.9628 | 0.9633 |
| 9.2 | 0.9638 | 0.9643 | 0.9647 | 0.9652 | 0.9657 | 0.9661 | 0.9666 | 0.9671 | 0.9675 | 0.9680 |
| 9.3 | 0.9685 | 0.9689 | 0.9694 | 0.9699 | 0.9703 | 0.9708 | 0.9713 | 0.9717 | 0.9722 | 0.9727 |
| 9.4 | 0.9731 | 0.9736 | 0.9741 | 0.9745 | 0.9750 | 0.9754 | 0.9759 | 0.9763 | 0.9768 | 0.9773 |
| 9.5 | 0.9777 | 0.9782 | 0.9786 | 0.9791 | 0.9795 | 0.9800 | 0.9805 | 0.9809 | 0.9814 | 0.9818 |
| 9.6 | 0.9823 | 0.9827 | 0.9832 | 0.9836 | 0.9841 | 0.9845 | 0.9850 | 0.9854 | 0.9859 | 0.9863 |
| 9.7 | 0.9868 | 0.9872 | 0.9877 | 0.9881 | 0.9886 | 0.9890 | 0.9894 | 0.9899 | 0.9903 | 0.9908 |
| 9.8 | 0.9912 | 0.9917 | 0.9921 | 0.9926 | 0.9930 | 0.9934 | 0.9939 | 0.9943 | 0.9948 | 0.9952 |
| 9.9 | 0.9956 | 0.9961 | 0.9965 | 0.9969 | 0.9974 | 0.9978 | 0.9983 | 0.9987 | 0.9991 | 0.9996 |

# 6 APPENDIX

## FORMULA WEIGHTS OF COMMON COMPOUNDS

| Compound | Formula weight |
|---|---|
| $AgBr$ | 187.78 |
| $AgBrO_3$ | 235.78 |
| $AgCl$ | 143.32 |
| $Ag_2CrO_4$ | 331.73 |
| $AgI$ | 234.77 |
| $AgNO_3$ | 169.87 |
| $Ag_3PO_4$ | 418.58 |
| $AgSCN$ | 165.95 |
| $Ag_2SO_4$ | 311.80 |
| $AlBr_3$ | 266.71 |
| $Al_2O_3$ | 101.96 |
| $Al(OH)_3$ | 78.00 |
| $Al_2(SO_4)_3$ | 342.15 |
| $As_2O_3$ | 197.84 |
| $As_2O_5$ | 229.84 |
| $As_2S_3$ | 246.04 |
| $BaCl_2$ | 208.25 |
| $BaCl_2 \cdot 2\ H_2O$ | 244.28 |
| $BaCO_3$ | 197.35 |
| $BaC_2O_4$ | 225.36 |
| $BaF_2$ | 175.34 |
| $BaI_2$ | 391.15 |
| $Ba(IO_3)_2$ | 487.15 |
| $BaO$ | 153.34 |
| $Ba(OH)_2$ | 171.36 |
| $BaSO_4$ | 233.40 |
| $Bi_2O_3$ | 465.96 |
| $Bi_2S_3$ | 514.15 |
| $CaCl_2 \cdot 2\ H_2O$ | 147.02 |
| $CaCO_3$ | 100.09 |
| $CaF_2$ | 78.08 |
| $Ca(NO_3)_2$ | 164.09 |
| $CaO$ | 56.08 |
| $Ca(OH)_2$ | 74.10 |
| $Ca_3(PO_4)_2$ | 310.18 |
| $CaSO_4$ | 136.14 |
| $Ce(HSO_4)_4$ | 528.41 |
| $CeO_2$ | 172.12 |
| $Ce(NH_4)_4(SO_4)_4 \cdot 2\ H_2O$ | 632.57 |
| $CO_2$ | 44.01 |
| $Cr_2O_3$ | 151.99 |
| $CuO$ | 79.54 |
| $Cu_2O$ | 143.08 |
| $CuS$ | 95.60 |
| $Cu_2S$ | 159.14 |
| $CuSO_4 \cdot 5\ H_2O$ | 249.68 |
| $FeCl_3$ | 162.21 |
| $Fe(NO_3)_3 \cdot 6\ H_2O$ | 349.95 |
| $FeO$ | 71.85 |
| $Fe_2O_3$ | 159.69 |
| $Fe_3O_4$ | 231.54 |
| $Fe(OH)_3$ | 106.87 |
| $FeS_2$ | 119.98 |
| $FeSO_4 \cdot 7\ H_2O$ | 278.05 |
| $Fe_2(SO_4)_3$ | 399.87 |
| $Fe(NH_4)_2(SO_4)_2 \cdot 6\ H_2O$ | 392.14 |
| $HBr$ | 80.92 |
| $HCHO_2$ (formic acid) | 46.03 |
| $HC_2H_3O_2$ (acetic acid) | 60.05 |
| $HC_7H_5O_2$ (benzoic acid) | 122.13 |
| $HCl$ | 36.46 |
| $HClO_4$ | 100.46 |
| $H_2C_2O_4 \cdot 2\ H_2O$ (oxalic acid) | 126.06 |
| $HI$ | 127.91 |
| $HNO_3$ | 63.01 |
| $H_2O$ | 18.02 |
| $H_2O_2$ | 34.01 |
| $H_3PO_4$ | 98.00 |
| $H_2S$ | 34.08 |
| $H_2SO_3$ | 82.08 |
| $H_2SO_4$ | 98.08 |
| $Hg_2Br_2$ | 561.00 |
| $HgCl_2$ | 271.50 |
| $Hg_2Cl_2$ | 472.09 |
| $Hg_2I_2$ | 654.99 |
| $HgO$ | 216.59 |
| $KBr$ | 119.01 |
| $KBrO_3$ | 167.01 |
| $KCl$ | 74.56 |
| $KClO_3$ | 122.55 |
| $KClO_4$ | 138.55 |
| $KCN$ | 65.12 |
| $KCNS$ | 97.18 |
| $K_2CO_3$ | 138.21 |
| $K_2CrO_4$ | 194.20 |
| $K_2Cr_2O_7$ | 294.19 |
| $K_3Fe(CN)_6$ | 329.26 |
| $K_4Fe(CN)_6 \cdot 3\ H_2O$ | 422.41 |
| $KHC_4H_4O_6$ (tartrate) | 188.18 |
| $KHC_8H_4O_4$ (phthalate) | 204.23 |
| $KHCO_3$ | 100.12 |
| $KHSO_4$ | 136.17 |
| $KI$ | 166.01 |
| $KIO_3$ | 214.00 |
| $KMnO_4$ | 158.04 |
| $KNO_2$ | 85.11 |
| $KNO_3$ | 101.11 |
| $K_2O$ | 94.20 |
| $KOH$ | 56.11 |
| $K_3PO_4$ | 212.28 |
| $K_2PtCl_6$ | 486.01 |
| $K_2SO_4$ | 174.27 |
| $LiCl$ | 42.39 |
| $Li_2CO_3$ | 73.89 |

| | | | |
|---|---|---|---|
| $Li_2O$ | 29.88 | $Na_3PO_4$ | 163.94 |
| LiOH | 23.95 | $Na_3PO_4 \cdot 12\ H_2O$ | 380.12 |
| $MgCO_3$ | 84.32 | $Na_2S$ | 78.04 |
| $MgNH_4AsO_4$ | 181.27 | $Na_2SO_3$ | 126.04 |
| $MgNH_4PO_4$ | 137.32 | $Na_2SO_4 \cdot 10\ H_2O$ | 322.19 |
| MgO | 40.31 | $Na_2S_2O_3$ | 158.11 |
| $Mg(OH)_2$ | 58.33 | $Na_2S_2O_3 \cdot 5\ H_2O$ | 248.18 |
| $Mg_2P_2O_7$ | 222.57 | $P_2O_5$ | 141.94 |
| $MgSO_4$ | 120.37 | $PbCl_2$ | 278.10 |
| $MnO_2$ | 86.94 | $PbCrO_4$ | 323.18 |
| $Mn_2O_3$ | 157.87 | $PbI_2$ | 461.00 |
| $NH_3$ | 17.03 | $Pb(IO_3)_2$ | 557.00 |
| $NH_4Cl$ | 53.49 | $Pb(NO_3)_2$ | 331.20 |
| $NH_4NO_3$ | 80.04 | PbO | 223.19 |
| $NH_4OH$ | 35.05 | $PbO_2$ | 239.19 |
| $(NH_4)_2SO_4$ | 132.14 | $PbSO_4$ | 303.25 |
| NO | 30.01 | $SO_2$ | 64.06 |
| $NO_2$ | 46.01 | $SO_3$ | 80.06 |
| $N_2O_3$ | 76.01 | $Sb_2O_3$ | 291.50 |
| $Na_2B_4O_7 \cdot 10\ H_2O$ | 381.37 | $Sb_2O_5$ | 323.50 |
| NaBr | 102.90 | $Sb_2S_3$ | 339.69 |
| $NaBrO_3$ | 150.90 | $SiCl_4$ | 169.90 |
| $NaCHO_2$ (formate) | 68.01 | $SiF_4$ | 104.08 |
| $NaC_2H_3O_2$ (acetate) | 82.03 | $SiO_2$ | 60.08 |
| NaCl | 58.44 | $SnCl_2$ | 189.60 |
| NaCN | 49.01 | $SnCl_4$ | 260.50 |
| $Na_2CO_3$ | 105.99 | $SnO_2$ | 150.69 |
| $Na_2C_2O_4$ | 134.00 | $SrCO_3$ | 147.63 |
| $NaHCO_3$ | 84.00 | SrO | 103.62 |
| $Na_2HPO_4$ | 141.96 | $SrSO_4$ | 183.68 |
| NaHS | 56.06 | $TiO_2$ | 79.90 |
| $NaH_2PO_4$ | 119.97 | $UO_3$ | 286.03 |
| NaI | 149.89 | $U_3O_8$ | 842.09 |
| $NaNO_2$ | 69.00 | $WO_3$ | 231.85 |
| $NaNO_3$ | 84.99 | ZnO | 81.37 |
| $Na_2O$ | 61.98 | $Zn_2P_2O_7$ | 304.68 |
| $Na_2O_2$ | 77.98 | $ZnSO_4$ | 161.43 |
| NaOH | 40.00 | $ZnSO_4 \cdot 7\ H_2O$ | 287.54 |

# 7 APPENDIX

## ANSWERS TO NUMERICAL PROBLEMS

(Except in those problems in the text having stated numerical values for the pertinent equilibrium constants, the equilibrium-constant data compiled in the various preceding appendices have been used to determine the answers to problems given below.)

### CHAPTER 2

2. (a) 9.6 parts per thousand
   (b) 0.083 part per thousand
   (c) 0.99 part per million
   (d) 1.76 parts per million
   (e) 2.19 parts per thousand
   (f) 0.17 part per thousand
4. (a) 60.10 per cent
   (b) 0.26 per cent
   (c) 0.43 per cent
5. -0.56 per cent; -0.92 per cent
6. (a) 21.19 per cent
   (b) $2.9 \times 10^{-5}$ per cent
   (c) 55.86 per cent
   (d) 54.00 per cent
7. 0.0548
8. 1.83 hr
9. 8 mg
10. (a) 0.5 mg
    (b) 100
    (c) 49.8 per cent
11. (a) 7671 hr; 7536 hr
    (b) Yes, only 0.13 per cent chance
    (c) 2.27 per cent
12. ±4.1 ppm
13. (a) ±0.58 per cent
    (b) ±0.73 per cent
    (c) ±0.82 per cent
14. No
15. $c = 3.922A - 0.0973$
16. (a) output = $0.7201 + 9.916c$
    (b) 0.467 ppm
    (c) $0.024_6$; $\pm 0.068_4$

## CHAPTER 3

1. (a) 0.02570 $F$
   (b) 0.1192 $F$
   (c) 0.007597 $F$
   (d) 0.3122 $F$
   (e) 0.08541 $F$
2. (a) 0.2574 g
   (b) 17.72 g
   (c) 0.5531 g
   (d) 12.97 g
   (e) 0.6394 g
3. (a) 0.2000 $N$
   (b) 0.3000 $N$
   (c) 0.1000 $N$
   (d) 0.3000 $N$
   (e) 0.1000 $N$
7. 4.25
8. 0.0247 $M$
9. 7.65 g
10. $3.16 \times 10^{-7}$ $M$
11. 0.388 g
12. (a) +78.12 kcal
    (b) +31.1 kcal
    (c) −16.73 kcal
    (d) −24.14 kcal
    (e) −17.05 kcal
    (f) +59.9 kcal
    (g) +12.1 kcal
13. (a) $K = 1.05 \times 10^{489}$
    (b) $K = 6.62 \times 10^{5}$
    (c) $K = 1.15 \times 10^{-39}$
    (d) $K = 4.46 \times 10^{18}$
    (e) $K = 9.12 \times 10^{-17}$
    (f) $K = 1.82 \times 10^{230}$
    (g) $K = 4.46 \times 10^{22}$
15. 22.1 kcal
17. $[PCl_3] = [Cl_2] = 1.89$ $M$;
    $[PCl_5] = 0.11$ $M$
18. (a) 0.018
    (b) 0.033
    (c) 0.26
    (d) 1.50
19. (a) 0.161
    (b) 0.426
    (c) 0.474
20. (a) 0.630
    (b) 0.708
    (c) 0.603
    (d) 0.884
    (e) 0.424

## CHAPTER 4

1. (a) $4.26 \times 10^{-7}$ *M*
   (b) $1.48 \times 10^{-5}$ *M*
   (c) $6.03 \times 10^{-12}$ *M*
   (d) $2.88 \times 10^{-9}$ *M*
   (e) $7.41 \times 10^{-4}$ *M*
2. (a) $1.91 \times 10^{-11}$ *M*
   (b) $7.25 \times 10^{-5}$ *M*
   (c) $2.95 \times 10^{-12}$ *M*
   (d) $1.55 \times 10^{-2}$ *M*
   (e) $5.13 \times 10^{-8}$ *M*
3. (a) 1.86
   (b) 4.68
   (c) 10.13
   (d) -0.38
   (e) 6.38
5. (a) 0.40
   (b) 12.67
   (c) 2.47
   (d) 4.67
   (e) 1.19
   (f) 9.24
   (g) 9.44
   (h) 6.11
   (i) 7.17
   (j) 1.91
   (k) 2.34
   (l) 10.57
6. $[H^+] = 1.62 \times 10^{-7}$ *M*; pH = 6.79
7. 1.00 to 27.6
8. 1.00 to 7.25
11. $[CO_3^{2-}] = 0.11$ *M*;
    $[HCO_3^-] = 0.47$ *M*
12. 1.00 to 1.70; pH = 5.17
13. pH = 1.24;
    $[HPO_4^{2-}] = 6.2 \times 10^{-8}$ *M*
14. $[NH_4^+] = 0.283$ *M*;
    $[NH_3] = 7.78 \times 10^{-7}$ *M*;
    pH = 3.70
15. $[H^+] = 1.89 \times 10^{-3}$ *M*;
    $[C_2H_5COO^-] = 1.89 \times 10^{-3}$ *M*;
    $[C_2H_5COOH] = 0.265$ *M*
16. 5.06
17. $[H^+] = 2.74 \times 10^{-2}$ *M*;
    $[NO_2^-] = 2.74 \times 10^{-2}$ *M*;
    $[HNO_2] = 0.473$ *M*
18. $[F^-] = 2.10 \times 10^{-2}$ *M*;
    $[HF] = 0.649$ *M*;
    pH = 1.68
19. $[NH_3] = 0.0260$ *F*;

$[NH_4^+] = 6.9 \times 10^{-4}\ M$;
$[H^+] = 1.45 \times 10^{-11}\ M$;
$[OH^-] = 6.9 \times 10^{-4}\ M$

20. (b) $[HC_2O_4^-] = 0.00834\ M$;
    $[H_2C_2O_4] = 0.000639\ M$;
    $[C_2O_4^{2-}] = 0.00967\ M$
    (c) $[H_2C_2O_4] = 3.58 \times 10^{-6}\ M$;
    $[HC_2O_4^-] = 2.31 \times 10^{-3}\ M$;
    $[C_2O_4^{2-}] = 1.42 \times 10^{-3}\ M$

## CHAPTER 5

2. 180.0 g
3. 0.4021 *F*
5. Points on titration curve correspond to pH 2.54, 4.35, 6.15, 8.14, 11.12, and 12.22
6. Points on titration curve correspond to pH 11.33, 9.86, 9.44, 9.08, 8.66, 5.08, 1.64, and 1.24
7. $1.84 \times 10^{-5}\ F$
8. (a) −0.99 per cent
   (b) 8.43
   (c) 0.0836
9. (a) 336
   (b) $1.25 \times 10^{-5}$
   (c) 8.76
11. Points on titration curve correspond to pH 4.00, 4.11, 4.32, 5.74, 6.96, 7.00, 9.00, and 9.25
12. 6.005 g
13. 0.3449 *F*
14. 1.158 *F* acid; 1.242 *F* base
15. 118.7
16. 4.804 per cent
17. 0.1835 *F*; 0.06805 *F*
18. 35.52 per cent
19. 0.2932 per cent
20. 28.00 ml
21. $[CO_3^{2-}] = 0.048\ M$;
    $[HCO_3^-] = 0.024\ M$
24. 19.99 per cent ethylamine;
    24.34 per cent diethylamine;
    55.67 per cent triethylamine
25. 5.173 per cent sulfanilamide;
    5.085 per cent sulfathiazole

## CHAPTER 6

4. $Zn(NH_3)^{2+}$, 0.24; $Zn(NH_3)_2{}^{2+}$, 0.18; $Zn(NH_3)_3{}^{2+}$, 0.15; $Zn(NH_3)_4{}^{2+}$, 0.05; $[Zn(NH_3)^{2+}] = 3.87 \times 10^{-4}\ M$; $[Zn(NH_3)_2{}^{2+}] = 2.90 \times 10^{-4}\ M$; $[Zn(NH_3)_3{}^{2+}] = 2.42 \times 10^{-4}\ M$; $[Zn(NH_3)_4{}^{2+}] = 8.10 \times 10^{-5}\ M$

7. $[Cu(NH_3)^{2+}] = 7.65 \times 10^{-3}$ *M*; $[Cu(NH_3)_2{}^{2+}] = 0.0895$ *M*; $[Cu(NH_3)_3{}^{2+}] = 0.246$ *M*; $[Cu(NH_3)_4{}^{2+}] = 0.123$ *M*
8. 0.589 *F*
9. $[Zn(NH_3)_2{}^{2+}] = 7.85 \times 10^{-7}$ *M*; $[Zn(NH_3)_3{}^{2+}] = 2.56 \times 10^{-5}$ *M*
10. $[Ag^+] = 1.19 \times 10^{-9}$ *M*; $[Ag(NH_3)^+] = 1.10 \times 10^{-6}$ *M*; $[Ag(NH_3)_2{}^+] = 0.00300$ *M*
11. $[Ag^+] = 3.65 \times 10^{-19}$ *M*; $[CN^-] = 0.0423$ *M*; $[Ag(CN)_2{}^-] = 0.0462$ *M*
12. $[Ag^+] = 2.00 \times 10^{-10}$ *M*; $[CN^-] = 1.33 \times 10^{-6}$ *M*; $[Ag(CN)_2{}^-] = 0.0250$ *M*
13. $[H_4Y] = 4.49 \times 10^{-20}$ *M*; $[H_3Y^-] = 1.42 \times 10^{-12}$ *M*; $[H_2Y^{2-}] = 9.74 \times 10^{-6}$ *M*; $[HY^{3-}] = 0.0213$ *M*; $[Y^{4-}] = 3.69 \times 10^{-3}$ *M*
14. $3.92 \times 10^{-8}$ *M*
15. $7.55 \times 10^{-19}$ *M*
16. $5.62 \times 10^{10}$
17. $1.49 \times 10^{-12}$
20. 0.008832 *F*
21. 0.0844 *M*
22. 28.9 mg
23. 111.1 ppm $CaCO_3$; 35.76 ppm $Ca^{2+}$; 5.30 ppm $Mg^{2+}$
24. 31.45 per cent
25. 7.886 per cent

## CHAPTER 7

3. (a) $1.78 \times 10^{-4}$
   (b) $2.52 \times 10^{-13}$
   (c) $1.42 \times 10^{-16}$
   (d) $1.24 \times 10^{-11}$
4. (a) $8.8 \times 10^{-14}$
   (b) $7.5 \times 10^{-3}$
   (c) $1.3 \times 10^{-5}$
   (d) $2.6 \times 10^{-3}$
   (e) $6.2 \times 10^{-4}$
6. $4.4 \times 10^{-12}$ *M*
7. $3.5 \times 10^{-5}$ *M*
8. $[Ag^+] = 1.39 \times 10^{-13}$ *M*
9. 11.5 *F*
10. $1.08 \times 10^{-16}$
11. $7.29 \times 10^{-4}$
12. 0.096 *F*
13. $S = 1.82 \times 10^{-4}$ *F*; $[SO_4{}^{2-}] = 6.05 \times 10^{-7}$ *M*; $[HSO_4{}^-] = 1.82 \times 10^{-4}$ *M*
14. $[Pb^{2+}] = 1.8 \times 10^{-4}$ *M*; $[Sr^{2+}] = 4.3 \times 10^{-3}$ *M*; $[HSO_4{}^-] = 4.4 \times 10^{-3}$ *M*; $[SO_4{}^{2-}] = 8.8 \times 10^{-5}$ *M*
15. $1.0 \times 10^{-8}$ *M*
16. $5.3 \times 10^{-3}$ *M*; $6.5 \times 10^{-3}$ *M*; $1.73 \times 10^{-4}$ *M*
17. $1.4 \times 10^{-9}$
18. (a) AgCl
    (b) $[Cl^-] = 5.05 \times 10^{-6}$ *M*
    (c) 50 per cent
19. $1.31 \times 10^{-6}$ *M*
20. (a) $3.2 \times 10^{-10}$
    (b) $2.6 \times 10^{-6}$ *M*

21. 11.98 per cent
22. 16.18 per cent
23. 31.65 per cent
24. 46.80 per cent
25. 25.74 per cent
26. 0.3270 g $K_2CO_3$; 0.6730 g $KHCO_3$
27. 31.20 per cent
28. 1.38 per cent $Na_2O$; 2.67 per cent $K_2O$
29. 77.8 per cent Ag; 0.0197 g $PbCl_2$
30. 91.00 per cent
31. 26.96 weight per cent
32. 22.70 per cent
33. 66.02 per cent NaOH; 30.22 per cent NaCl; 3.76 per cent $H_2O$
34. $2.241 \times 10^{-5}$ *M*

## CHAPTER 8

3. (a) +0.925 v
   (b) −0.950 v
   (c) +0.495 v
   (d) +0.325 v
   (e) −1.342 v
4. +0.336 v, $K = 2.0 \times 10^{12}$, copper electrode is anode;
   −0.64 v, $K = 2 \times 10^{-13}$, platinum electrode is cathode;
   +0.414 v, $K = 1.59 \times 10^{13}$, left-hand platinum electrode is anode;
   −1.361 v, $K = 8.0 \times 10^{-47}$, left-hand platinum electrode is cathode;
   +0.487 v, $K = 6.3 \times 10^{13}$, zinc electrode is anode;
   −0.109 v, $K = 1.4 \times 10^{-9}$, platinum electrode is cathode
5. (a) 0.328
   (b) $3.23 \times 10^{4}$
   (c) $4.95 \times 10^{18}$
   (d) 0.122
   (e) $6.75 \times 10^{71}$
6. $1.9 \times 10^{-8}$
7. $1.29 \times 10^{-3}$ *M*
8. 708
9. $1.0 \times 10^{-12}$
10. $1.48 \times 10^{-19}$
11. (b) $E^0 = -0.73$ v; $K = 1.8 \times 10^{-25}$
    (c) −0.73 v
    (d) copper electrode
    (e) $4.6 \times 10^{-16}$ *M*
12. $3.08 \times 10^{-12}$
13. $[Co^{2+}] = 2.5 \times 10^{-3}$ *M*; $[Tl^{+}] = 0.500$ *M*
14. $1.20 \times 10^{-7}$ *M*
15. (a) $[Ce^{3+}] = [Fe^{3+}] = 0.00833$ *M*; $[Fe^{2+}] = 0.0042$ *M*; $[Ce^{4+}] = 2.1 \times 10^{-15}$ *M*
    (b) $[Ce^{3+}] = [Fe^{3+}] = 0.0081$ *M*; $[Fe^{2+}] = 2.2 \times 10^{-15}$ *M*; $[Ce^{4+}] = 0.0038$ *M*
16. 0.271 v
17. $2.6 \times 10^{-3}$ *M*
18. $[V^{3+}] = 4.54 \times 10^{-5}$ *M*; $[V^{2+}] = 0.0750$ *M*; $[Cd^{2+}] = 0.0375$ *M*

19. 0.0191 $M$
20. 0.490 $M$
21. 7.07
22. (a) +0.46 v
    (b) $1.30 \times 10^{-14}$ $M$
    (c) 40.00 ml
    (d) +1.16 v
    (e) −1.1 per cent
23. (b) $1.76 \times 10^{-7}$ $M$
    (c) 50.0057 ml
    (d) +0.011 per cent
25. (a) $1.74 \times 10^{15}$
    (b) +0.550 v
    (c) $3.43 \times 10^{-10}$ $M$
    (d) 5.00 ml
    (e) +0.722 v

## CHAPTER 9

1. 0.02186 $F$
2. 0.05676 $F$
3. 85.24 per cent
4. (a) 17.86 per cent
   (b) 8.359 per cent FeO; 16.25 per cent $Fe_2O_3$
5. 37.93 mg
6. 59.73 per cent
7. 3.332 g
8. 0.5842 g
9. $2.81 \times 10^{-6}$
11. $[Fe^{3+}] = 0.0385$ $M$; $[VO_2^{+}] = 0.0142$ $M$
12. 0.02642 $F$
13. 0.08130 $F$
14. 118.3 ml
15. 0.7607 weight per cent
16. 0.1383 $F$
17. 0.1175 $F$
18. 0.05668 per cent
19. 2
21. 14.59 mg glycerol; 32.72 mg ethylene glycol
22. 39.26 per cent
23. 22.44 mg; 4.51 fluid ounces
24. 3.64 per cent
25. 5.334 per cent

## CHAPTER 10

1. (a) ±7.5 per cent
   (b) ±0.129 mv
2. (a) $9.064 \times 10^{-6}$ $M$

(b) +1.5 per cent
3. 46.67 per cent; 46.87 per cent; −0.43 per cent
5. 0.000337 *M*
6. $1.0 \times 10^{13}$
7. 2.25
8. $7.11 \times 10^{-5}$ *M*
9. (a) theoretical volume = 31.09 ml; experimental volume = 30.93 ml
   (b) −0.1939
   (c) $6.79 \times 10^{-6}$ *M*
   (d) $4.183 \times 10^{-3}$ *M*
   (e) $1.31 \times 10^{-18}$
10. 4.05
11. $1.69 \times 10^{-9}$
12. (a) $7.75 \times 10^{-4}$ *M*
    (b) $7.17 \times 10^{-4}$ *M* to $8.40 \times 10^{-4}$ *M*

## CHAPTER 11

1. (a) +0.350 v
   (b) +0.120 to +0.299 v; +0.475 v
2. −0.053 v; 0.0986; 154 mg
3. 0.1360
4. 0.413 per cent
5. $[Ni^{2+}]$ = 0.06232 *M*; $[Co^{2+}]$ = 0.3022 *M*
6. 57.97 mg
7. 4.15 ml per gallon
8. 89.46 per cent
9. 117.6 mg
10. 22.94 mg $NH_4Cl$; 5.394 per cent
11. −2
12. $OsO_4 + 8\ H^+ + 5\ e \rightleftharpoons Os^{3+} + 4\ H_2O$
13. $1.77 \times 10^{-4}$ *M*
14. 0.42 per cent
15. $1.15 \times 10^{-3}$ *M*

## CHAPTER 12

5. (a) *a*
   (b) *b*
7. 0.976
8. 94.7 per cent
9. (a) $9.90 \times 10^{-3}$
   (b) $7.51 \times 10^{-4}$
   (c) 99.010 per cent; 99.925 per cent
10. 3.06 extractions required; therefore, 4 extractions must be performed
11. (a) 0.632
12. 14.905
13. 5 hr
16. (a) 86.99

(b) 750
17. (a) a dimer
(b) 76.1
(c) $\Delta H = -14.12$ kcal; the energy of a hydrogen bond in the formic acid dimer is 7.06 kcal/mole

## CHAPTER 13

2. **R** = 0.0958; $t_M$ = 25 min; $t_S$ = 236 min
3. 108 sec
4. 36
5. 0.4167; 0.4167
6. (a) 69.93 cm/sec
(b) 0.0946
(c) 6.61 cm/sec
8. a factor of 3; not generally applicable
11. 29.33 min
12. for component A: 6.5 ml, 13 min, 0.231; for component B: 9 ml, 18 min, 0.167
13. 0.167 cm/sec
16. 22,500
17. 1.265 sec; 12.65 sec; 126.5 sec
18. 0.089 mm
21. 1267.5 days
22. $\sqrt{2}$
26. 21.8 ml/min; 2625 theoretical plates
27. (a) 11.0, 25.0, 40.0 cm/sec
(b) 1313, 1287, 1077 theoretical plates; 1.526, 1.555, 1.872 mm
(c) $A = 0.56$, $B = 7.2$, $C = 0.0283$ for $v$ in cm/sec and $H$ in mm
(d) $8.8 \leqslant v \leqslant 28.8$ cm/sec
(e) 15.95 cm/sec
(f) 37.7 cm/sec, a time-saving of a factor of 2.36
28. $\mathcal{R} = 0.25$; 67,500 plates required for $\mathcal{R} = 1.0$
29. 47,000 plates; $L = 4.70$ m

## CHAPTER 14

1. (a) 0.47
(b) 6.71 cm from origin
2. B
6. I, "no" $H_2O$; II, 0.70 monolayer; III, 1.39 monolayers; IV, 2.32 monolayers; V, 3.48 monolayers
9. 2.84 atm
10. 45.28 min
11. 335.7 ml
12. 6779
13. 259°C
14. no; at 210°C, $t'_R$ = 52.81 min (>45 min)
15. 2448
16. yes; its retention index is 2648 (two additional $-CH_2-$ units)

17. 1535
18. get a glass-lined injector
19. 10, 9, 8, ~10.5
28. 2.4 hr after its introduction
30. (a) before arginine
    (b) before threonine
    (c) just after phenylalanine
    (d) before aspartic acid
    (e) before lysine
31. The largest elution volume should be less than 50 ml

## CHAPTER 15

2. (a) $1.35 \times 10^{15}$ Hz
   (b) $5.66 \times 10^{14}$ Hz
   (c) $1.76 \times 10^{17}$ Hz
   (d) $1.00 \times 10^{12}$ Hz
   (e) $2.31 \times 10^{17}$ Hz
   (f) $4.92 \times 10^{13}$ Hz
3. (a) $4.50 \times 10^4$ $cm^{-1}$; ultraviolet
   (b) $1.89 \times 10^4$ $cm^{-1}$; visible
   (c) $5.86 \times 10^6$ $cm^{-1}$; x-ray
   (d) 33.3 $cm^{-1}$; far infrared
   (e) $7.70 \times 10^6$ $cm^{-1}$; x-ray
   (f) $1.64 \times 10^3$ $cm^{-1}$; middle infrared
4. (a) 15.2 cm; $1.52 \times 10^8$ nm
   (b) $6.17 \times 10^{-6}$ cm; 61.7 nm
   (c) $4.10 \times 10^{-10}$ cm; $4.10 \times 10^{-3}$ nm
5. (a) 0.0656 $cm^{-1}$; microwave
   (b) $1.62 \times 10^5$ $cm^{-1}$; ultraviolet
   (c) $2.44 \times 10^9$ $cm^{-1}$; x-ray
6. (a) $2.48 \times 10^{-12}$ erg
   (b) $5.40 \times 10^{-13}$ erg
   (c) $2.00 \times 10^{-9}$ erg
   (d) $1.73 \times 10^{-17}$ erg
7. $7.602 \times 10^{-12}$ erg; $4.579 \times 10^{12}$ ergs; 109.4 kcal; $4.186 \times 10^{10}$ ergs/kcal
8. (a) 0.434
   (b) 0.658
   (c) 0.0846
   (d) 0
   (e) 1.377
9. (a) 15.8
   (b) 61.0
   (c) 26.0
   (d) 1.50
   (e) 87.7
10. (a) 75.6
    (b) 6.10
    (c) 97.2

## CHAPTER 16

9. (a) 4800 liter $mole^{-1}$ $cm^{-1}$
   (b) $9.27 \times 10^{-5}$ *M*; $3.63 \times 10^{-4}$ *M*
   (c) $3.23 \times 10^{-6}$
10. [Ti(IV)] = $2.74 \times 10^{-3}$ *M*; [V(V)] = $6.38 \times 10^{-3}$ *M*
11. 34.1 per cent transmittance; $1.11 \times 10^4$ liter $mole^{-1}$ $cm^{-1}$
12. For $Cr_2O_7^{2-}$, $\epsilon$ = 370 liter $mole^{-1}$ $cm^{-1}$; for $MnO_4^-$, $\epsilon$ = 2350 liter $mole^{-1}$ $cm^{-1}$; $[Cr_2O_7^{2-}]$ = $9.73 \times 10^{-4}$ *M*; $[MnO_4^-]$ = $2.73 \times 10^{-4}$ *M*
13. $K_a = 7.98 \times 10^{-6}$
14. (b) $Fe(C_{10}H_8N_2)_3^{2+}$
    (c) $1.05 \times 10^4$ liter $mole^{-1}$ $cm^{-1}$
15. $K_a = 0.470$
16. $K_a = 6.08 \times 10^{-6}$
17. $[Fe(CN)_6^{3-}]$ = $1.84 \times 10^{-3}$ *M*; $[Fe(CN)_6^{4-}]$ = $4.51 \times 10^{-3}$ *M*
18. 4.80
19. (a) $CoCl_2$ and $CoCl_3^-$ are present
    (b) $\epsilon$ = 235 liter $mole^{-1}$ $cm^{-1}$ for $CoCl_2$; $\epsilon$ = 191 liter $mole^{-1}$ $cm^{-1}$ for $CoCl_3^-$
    (c) $\beta_2 = 2.67 \times 10^{10}$
20. 0.126 μg/ml

## CHAPTER 17

6. 0.12 and 1.16 μg, respectively; 0.012 and 0.116 μg/ml, respectively
10. $Na^+$, 1.13 ppm; $K^+$, 0.44 ppm; ~5 per cent error; no, because of the relatively low precision of the readings, which were only to two significant figures
11. 0.283 mg/liter
12. 0.0019 per cent copper; 0.0068 per cent lead

# INDEX

Page numbers in *italics* refer to illustrations; those followed by (t) refer to tables.

# 1. PERIODIC CHART OF THE ELEMENTS

| 1 H 1.0079 |
|---|

| 1A | 2A | 3B | 4B | 5B | 6B | 7B | 8B | 8B | 8B | 1B | 2B | 3A | 4A | 5A | 6A | 7A | |
|---|---|---|---|---|---|---|---|---|---|---|---|---|---|---|---|---|---|
| | | | | | | | | | | | | | | | | | 2 He 4.00260 |
| 3 Li 6.941 | 4 Be 9.01218 | | | | | | | | | | | 5 B 10.81 | 6 C 12.011 | 7 N 14.0067 | 8 O 15.9994 | 9 F 18.99840 | 10 Ne 20.179 |
| 11 Na 22.98977 | 12 Mg 24.305 | | | | | | | | | | | 13 Al 26.98154 | 14 Si 28.086 | 15 P 30.97376 | 16 S 32.06 | 17 Cl 35.453 | 18 Ar 39.948 |
| 19 K 39.098 | 20 Ca 40.08 | 21 Sc 44.9559 | 22 Ti 47.90 | 23 V 50.9414 | 24 Cr 51.996 | 25 Mn 54.9380 | 26 Fe 55.847 | 27 Co 58.9332 | 28 Ni 58.71 | 29 Cu 63.546 | 30 Zn 65.38 | 31 Ga 69.72 | 32 Ge 72.59 | 33 As 74.9216 | 34 Se 78.96 | 35 Br 79.904 | 36 Kr 83.80 |
| 37 Rb 85.4678 | 38 Sr 87.62 | 39 Y 88.9059 | 40 Zr 91.22 | 41 Nb 92.9064 | 42 Mo 95.94 | 43 Tc 98.9062 | 44 Ru 101.07 | 45 Rh 102.9055 | 46 Pd 106.4 | 47 Ag 107.868 | 48 Cd 112.40 | 49 In 114.82 | 50 Sn 118.69 | 51 Sb 121.75 | 52 Te 127.60 | 58 I 126.9045 | 54 Xe 131.30 |
| 55 Cs 132.9054 | 56 Ba 137.34 | 57-71 see La series | 72 Hf 178.49 | 73 Ta 180.9479 | 74 W 183.85 | 75 Re 186.2 | 76 Os 190.2 | 77 Ir 192.22 | 78 Pt 195.09 | 79 Au 196.9665 | 80 Hg 200.59 | 81 Tl 204.37 | 82 Pb 207.2 | 83 Bi 208.9804 | 84 Po | 85 At | 86 Rn |
| 87 Fr | 88 Ra 226.0254 | 89-103 see Ac series | 104 | 105 Ha | | | | | | | | | | | | | |

| | | | | | | | | | | | | | | | |
|---|---|---|---|---|---|---|---|---|---|---|---|---|---|---|---|
| La Series | 57 La 138.9055 | 58 Ce 140.12 | 59 Pr 140.9077 | 60 Nd 144.24 | 61 Pm | 62 Sm 150.4 | 63 Eu 151.96 | 64 Gd 157.25 | 65 Tb 158.9254 | 66 Dy 162.50 | 67 Ho 164.9304 | 68 Er 167.26 | 69 Tm 168.9342 | 70 Yb 173.04 | 71 Lu 174.97 |
| Ac Series | 89 Ac | 90 Th 232.0381 | 91 Pa 231.0359 | 92 U 238.029 | 93 Np 237.0482 | 94 Pu | 95 Am | 96 Cm | 97 Bk | 98 Cf | 99 Es | 100 Fm | 101 Md | 102 No | 103 Lr |

For B, C, H, Pb, Sm, and S, the precision of the atomic weight is limited by variations in the isotopic abundances in terrestrial samples. For B, Li, and U, variations in atomic weights in processed material may occur because of commercial isotopic separations. Values are reliable to ±1 in the last digit, or ±3 if the last digit is smaller.